T0224820

Magnetic Components

Peter Zacharias

Magnetic Components

Basics and applications

 Springer

Peter Zacharias
Electrical Power Supply Systems
University of Kassel
Kassel, Germany

ISBN 978-3-658-37205-7   ISBN 978-3-658-37206-4   (eBook)
https://doi.org/10.1007/978-3-658-37206-4

This book is a translation of the original German edition „Magnetische Bauelemente" by Zacharias, Peter, published by Springer Fachmedien Wiesbaden GmbH in 2020. The translation was done with the help of artificial intelligence (machine translation by the service DeepL.com). A subsequent human revision was done primarily in terms of content, so that the book will read stylistically differently from a conventional translation. Springer Nature works continuously to further the development of tools for the production of books and on the related technologies to support the authors.

This Springer imprint is published by the registered company Springer Fachmedien Wiesbaden GmbH, part of Springer Nature.
The registered company address is: Abraham-Lincoln-Str. 46, 65189 Wiesbaden, Germany

Preface

Electrical engineering as a discipline and economic sector has developed rapidly over the last 200 years. Most of the machines used in industry today are electrical machines which use the force effects of magnetic fields. Magnetic or inductive components are also used to convert electrical energy and signals. Even many controllable components of electrical circuits and systems use controllable electromagnetic components.

Since the beginning of the 1980s, power electronics has been experiencing a rapid development due to constantly improving semiconductor switches. With the highly developed silicon switches and the switches based on wide-bandgap semiconductor technology, almost ideal switches are available for power electronics. The general expectation is the continuity of miniaturization of electrical and electronic structures by means of less and less lossy components. Inductive and capacitive components should become smaller and smaller with increasing frequency – this is the expectation. A more detailed analysis shows, however, that this reduction seems to be limited, in particular by the effects of the electromagnetic field. The change of electric and magnetic fields to high frequencies weakens the utilization of the material used, increases component losses, and leads to the induction of undesirable voltages, to name just a few factors. At the same time, there have been several developments in the field of magnetic components, both for hard magnetic and soft magnetic materials. Due to the scientific focus on the development of semiconductor switching elements, the development of magnetic and inductive components has been somewhat marginalized within the last decades, although there is plenty of know-how from previous decades one could learn from.

From my own experience in the development of power electronic converters, I concluded that the magnetic or inductive components of power electronics are increasingly representing a shortage in the development of electronic assemblies. In many cases, the excellent properties of semiconductor components can only be fully exploited when they are supplemented by adapted passive components. At the same time, the knowledge of the diverse design possibilities of these components might fall into oblivion. Corresponding university lectures have been increasingly replaced by other lectures in the past, driven by technical development. Therefore, the lecture "Magnetic Components" was established at the University of Kassel 10 years ago. The present book was initially intended as a script. One or other former listeners of the lecture will recognize the charts from the blackboard. The variety of theoretical and methodological approaches to analysis and design but also the examples of the application soon exceeded the scope of a lecture script. This history of origins helps to determine the structure of the book. The first nine chapters describe basic analytical relationships and methods with which magnetic components can be analyzed and designed. Chapter 10 contains examples of the development practice of electrical and electronic converters. No claim of completeness can be made here. A selection was made from the available material of the everyday scientific life. Analytical mathematical description methods were preferred when dealing with the questions. Finite element method calculations can only be found as results for illustration purposes. The book is to be used as an explanation of what can be achieved with certain approaches and with which methods the results of FEM calculations can be interpreted or examined for plausibility. FEM-calculations with their numerical problems have been covered in the literature by other authors.

On the one hand, the book is intended to support the study of electrical engineering, and on the other hand, it is intended to draw interest and curiosity for the diversity in the world of electromagnetic components. The developer of components or systems shall be provided with systematic and methodical approaches which give access to the conditioning of most different questions in the development of electromagnetic components. Even though there are many similar problems in moving or rotating electrical machines, these are not dealt with in this book. The book deals exclusively with static electrical machines and storages, such as transformers and chokes. The materials, aspects, and methods are partly very different. I hope that the aesthetics inherent in electromagnetic fields will attract more students to study the design possibilities of electromagnetic components and to participate in the development of technology.

Kassel, Germany   Peter Zacharias
September 2021

Symbol Directory

Coded Notation

Variable type	Notation	Examples
Scalar	Letters without accent	$U,\ i,\ Q,\ t,\ \varphi,\ V$
Vector	Letter with a superimposed arrow	$\vec{A},\vec{B},\vec{E},\vec{H},\vec{J},\vec{r},\vec{s}$
Parallel vectors	$\parallel$	$\vec{x} \parallel \vec{y}$
Vertical vectors	$\perp$	$\vec{x} \perp \vec{y}$
Pointer	Underlined letter	$\underline{i},\underline{I},\underline{u},\underline{U},\underline{\varphi},\underline{\psi},\underline{v}$
Instantaneous value	Small letters or marking	$i,\vec{H}(t),u(t),\varphi,\varphi(t)$
Peak value	Letter with circumflex	$\hat{i},\hat{B},\hat{H},\hat{u},\hat{\varphi},\hat{\psi}$
Effective value, RMS value	Capital letters, RMS index	$I,\ U,\ E_{RMS},\ H_{RMS}$
Real part		$\mathrm{Re}\left(\underline{y}\right)$
Imaginary part		$\mathrm{Im}\left(\underline{y}\right)$
Element in a set		$x \in M$

Important abbreviations

Abbreviation	Meaning
EMF	Electromotive force
MMK	Magnetomotive force
EMC	Electromagnetic compatibility

Operations with Variables

Operation	Operator	Example
Addition /	+	$c = a + b$
Subtraction	−	$c = a - b$
Multiplication	.	$c = a \cdot b, c = ab$
Division	÷,	$a \div b, c = \frac{a}{b}$
Serial resistance	Addition	$R_1 + R_2$
Resistance parallel connection	"Double slash"	$R_1 // R_2$
Scalar product	.	$\vec{B} = \mu \cdot \vec{H}, (\vec{B} \parallel \vec{H})$
Vector product	×	$\vec{F} = q \cdot \vec{v} \times \vec{B}, (\vec{F} \perp \vec{v}, \vec{F} \perp \vec{B})$
Dyadic product	$\otimes$	
Sum of element 1 to n	$\sum\limits_{i=1}^{n}$	$y = \sum\limits_{i=1}^{n} x_i$
Product from element 1 to n	$\prod\limits_{i=1}^{n}$	$y = \prod\limits_{i=1}^{n} x_i$
Indefinite integral over f(x)	$\int$	$y(x) = \int f(x) \cdot dx$
Determined integral over f(x)	$\int\limits_{a}^{b}$	$Y = \int\limits_{a}^{b} f(x) \cdot dx$
Closed loop integral over f(x)	$\oint$	$Y = \oint f(x) \cdot dx$
Area integral over f(x,y,z)	$\int\limits_{A}$	$Y = \int\limits_{A} f(x, y, z) \cdot dA$
Volume integral over f(x,y,z)	$\int\limits_{V}$	$Y = \int\limits_{A} f(x, y, z) \cdot dV$
Differential of f(x)	df	$df = f' \cdot dx, f' = \frac{df}{dx}$
Partial differentiation	∂f	$\frac{\partial f}{\partial x}$
Total differential	Δf	$\Delta f = \frac{\partial f}{\partial x} + \frac{\partial f}{\partial y} + \frac{\partial f}{\partial z}$

Variables Used

Due to the limited alphabet and the complexity of the matter, it was not possible to avoid duplicate use of letters altogether. The current meaning can be seen in the context.

Latin Letters

Letter	Meaning	Unit of measurement
a, b, c, d, e, g,	Measure	1 m
A	Area	1 m^2
A$_W$	Winding cross-section	1 m^2
A$_{Fe}$	Magnetization cross-section	1 m^2
B	Induction	1 T = 1 Vs/m^2
B$_{sat}$	Saturation induction	1 T = 1 Vs/m^2
B$_r$	Remanence induction	1 T = 1 Vs/m^2
C	Capacity	1 F = 1 As/V
c$_0$	Speed of light in vacuum	29 792,458 m/s
D	Displacement flux density	1 As/m^2
E	Electric field strength	1 V/m
f	Frequency	1 Hz = 1 s-1
f	Function	–
G	Electrical conductance	1 S/m
H	Magnetic field strength	1 A/m
H$_C$	Coercive field strength	1 A/m
I, i	Electricity	1 A
J, j	Current density	1 A/m^2
j	Imaginary parameter	–
k	Factor, coupling factor	–
L	Inductance	1 H = 1 Vs/A
M	Mutual inductance	1 H = 1 Vs/A
N	Number of turns	–
P, p	Power, active power	1 W = 1 VA
Q, q	Reactive power	1 var
Q, q	Charge	As
R	Electrical resistance	1 Ohm = 1 V/A
R$_m$	Magnetic resistance	1 A/Vs
s	Path	1 m
t	Time	1 s
T	Period duration	1 s
U, u	Electric voltage	1 V
V, v	Magnetic voltages	1 A
V	Volume	1 m^3
W	Work, energy	1 Ws = 1 J
x, y, z	Coordinate, path	1 m
X	Reactance	1 Ohm = 1 V/A

(continued)

Letter	Meaning	Unit of measurement
Y	Reactive conductance	1 S = 1 A/V
Z	Apparent resistance, impedance	1 Ohm = 1 V/A
EMF	Electromotive force	1 V
MMF	Magnetomotive force	1 A

Greek Letters

Letter	Meaning	Unit of measurement
α, β, γ	Angle	π rad = 180°
α, β	Exponents of the Steinmetz formula	–
δ	Air gap	1 m
Δ	Change	–
ε	Dielectric constant, permittivity	1 As/(Vm)
ε_r	Relative dielectric constant	–
ε_0	Electric space constant	8.854187 pF/m
η	Efficiency	1 %
ϑ	Temperature	1 K, 1 °C
Θ	Magnetic force	1 A
λ	Wavelength	1 m
$\lambda \| , \lambda_\perp$	Magnetostriction parallel and perpendicular to the magnetic field	–
Λ	Magnetic conductance, permeance	1 Vs/A
μ	Permeability	Vs/(Am)=H/m
μr	Relative permeability	–
μ_0	Permeability in vacuum	$4\pi \cdot 100$ nH/m
ν	Run variable, index	–
ξ	General variable	–
π	Circular number	3,14159
ρ	Specific resistance	1 Ωm
σ	Conductivity	1 S/m
φ	Phase angle	1 °
φ, Φ	Magnetic flux	1 Vs
ψ, Ψ	Flux linkage	1Vs
ω	Electrical angular frequency	rad/s
Ω	Mechanical angular frequency	rad/s

Circuit Symbols

Designation	Circuit symbol	
Switch		
Resistance		
Capacitor		
Inductor general, without core		
Inductor with core		
Controllable inductor		
Transformer		
Ideal transformer		
Voltage source, magnetic force	Common	Specific magnetic
Current source, flow source		
Operational amplifier		

(continued)

Designation	Circuit symbol
Gyrator	
Diode	
Bipolar transistor, npn	
IGBT	
Field-effect transistor (n-channel)	
Thyristor	

Contents

The Magnetostatic Field

1

Magnetism was first discovered in the ancient world, when people noticed that lodestones, naturally magnetized pieces of the mineral magnetite, could attract iron [1]. The word magnet comes from the Greek term μαγνῆτις λίθος magnētis lithos [2], "the Magnesian stone [3], lodestone." In ancient Greece, Aristotle attributed the first of what could be called a scientific discussion of magnetism to the philosopher Thales of Miletus, who lived from about 625 BC to about 545 BC [4]. The ancient Indian medical text Sushruta Samhita describes using magnetite to remove arrows embedded in a person's body [5].

In ancient China, the earliest literary reference to magnetism lies in a 4th-century BC book named Guiguzi [6]. The 2nd-century BC annals, Lüshi Chunqiu, also notes: "The lodestone makes iron approach; some (force) is attracting it" [7]. The earliest mention of the attraction of a needle is in a 1st-century work Lunheng (Balanced Inquiries): "A lodestone attracts a needle" [8]. The 11th-century Chinese scientist Shen Kuo was the first person to write—in the Dream Pool Essays—of the magnetic needle compass and that it improved the accuracy of navigation by employing the astronomical concept of true north. By the 12th century, the Chinese were known to use the lodestone compass for navigation. They sculpted a directional spoon from lodestone in such a way that the handle of the spoon always pointed south.

Alexander Neckam, by 1187, was the first in Europe to describe the compass and its use for navigation. In 1269, Peter Peregrinus de Maricourt wrote the Epistola de magnete, the first extant treatise describing the properties of magnets. In 1282, the properties of magnets and the dry compasses were discussed by Al-Ashraf, a Yemeni physicist, astronomer, and geographer [9].

Leonardo Garzoni's only extant work, the Due trattati sopra la natura, e le qualità della calamita, is the first known example of a modern treatment of magnetic phenomena. Written in years near 1580 and never published, the treatise had a wide diffusion. In particular, Garzoni is referred to as an expert in magnetism by Niccolò Cabeo, whose

© Springer Fachmedien Wiesbaden GmbH, part of Springer Nature 2022
P. Zacharias, *Magnetic Components*,
https://doi.org/10.1007/978-3-658-37206-4_1

Philosophia Magnetica (1629) is just a re-adjustment of Garzoni's work. Garzoni's treatise was known also to Giovanni Battista Della Porta and William Gilbert.

In 1600, William Gilbert published his De Magnete, Magneticisque Corporibus, et de Magno Magnete Tellure (On the Magnet and Magnetic Bodies, and on the Great Magnet the Earth). In this work he describes many of his experiments with his model earth called the terrella. From his experiments, he concluded that the Earth was itself magnetic and that this was the reason compasses pointed north (previously, some believed that it was the pole star (Polaris) or a large magnetic island on the north pole that attracted the compass).

An understanding of the relationship between electricity and magnetism began in 1819 with work by Hans Christian Ørsted, a professor at the University of Copenhagen, who discovered by the accidental twitching of a compass needle near a wire that an electric current could create a magnetic field. This landmark experiment is known as Ørsted's Experiment. Several other experiments followed, with André-Marie Ampère, who in 1820 discovered that the magnetic field circulating in a closed-path was related to the current flowing through a surface enclosed by the path; Carl Friedrich Gauss; Jean-Baptiste Biot and Félix Savart, both of whom in 1820 came up with the Biot–Savart law giving an equation for the magnetic field from a current-carrying wire; Michael Faraday, who in 1831 found that a time-varying magnetic flux through a loop of wire induced a voltage, and others finding further links between magnetism and electricity. James Clerk Maxwell synthesized and expanded these insights into Maxwell's equations, unifying electricity, magnetism, and optics into the field of electromagnetism. In 1905, Albert Einstein used these laws in motivating his theory of special relativity [10], requiring that the laws held true in all inertial reference frames.

Electromagnetism has continued to develop into the 21st century, being incorporated into the more fundamental theories of gauge theory, quantum electrodynamics, electroweak theory, and finally the standard model.

The magnetic field appeared for a long time as a mysterious force, inexplicably occurring between certain bodies but not between others. The first known mention of magnetism, which was observed through its force effects, goes back to Thales of Miletus (624–546 BC), known through the Thales circle. It seems that the magnetic compass has been used in Asia for a relatively long time as an additional aid for orientation in shipping. In China, this application has been known since the so-called time of the warring empires, between 475 BC and 221 BC. Petrus Peregrinus de Maricourt reported experiments with magnets, which he used to discover and describe the bipolarity and indestructibility of magnetic dipoles in 1269 [11]. Around 1600 William Gilbert (1544–1603), a court physician to Queen Elizabeth, discovered the analogy of the earth's magnetic field to magnetic stones [1]. He explained the compass by assuming that the earth itself acts like a large magnet.

Hans Christian Oersted (1777–1851) discovered in 1820 that an electric current through a wire deflects compass needles perpendicular to the wire [2]. André Marie Ampère, shortly after learning of Oersted's discovery, showed that current-carrying wires exert forces on each other. In the same year, Jean-Baptiste Biot (1774–1862) and Felix Savart (1791–1841) showed that forces exerted by a current-carrying wire on a magnet decrease by 1/r. As early as 1821 Michael Faraday (1791–1867) made a current-carrying wire rotate continuously around a magnet. He built the first electric motor. In 1831 he reported on induction. He showed that alternating currents in one circuit also "induce" currents in an adjacent circuit. A little later in 1834, he discovered the transformer effect and self-induction. In the same year, Heinrich Friedrich Emil Lenz (1804–1865) formulated his law for determining the direction of induced currents (Lenz's rule).

In 1938 Faraday described the analogy of induced electricity in insulators and induced magnetism in magnetic materials. In 1845/46, Faraday reported that the plane of the polarity of light is rotated when it passes through glass along field lines of an electromagnet runs (Faraday rotation) and published the conjecture that light is of electromagnetic origin. Wilhelm Eduard Weber (1804–1891) suggested in 1847 that diamagnetism is due to the induction of molecular currents according to the Faraday principle. William Thomson, Lord Kelvin (1824–1907) introduced the terms magnetic permeability and susceptibility in 1850. In 1864, James Clerk Maxwell (1831–1879) completed his treatise on the joint description of electricity and magnetism, thus providing the first closed theory of electromagnetism.

This means that the study of magnetism as a "mysterious" type of force effects began with permanent magnetic phenomena. If you study permanent magnets, you will quickly discover that there are attractive and repulsive force effects that cause different polarity. These poles always occur in pairs. If you split a permanent magnet, the two parts again have two poles. This pair-wise occurrence of the poles - in contrast to single electric charges with different polarity - was the reason why the magnetic field is free of sources. Even in the vicinity of an electric current, experimental forces can be determined on a magnetic needle, iron filings or a second current. Therefore, this "new" effect is interpreted by a field coupled with a moving charge. This is called the magnetic field. This chapter is concerned with the calculation of magnetic fields, their force effects and phenomena of their temporal change. The name "magnet" probably comes from the place Magnesia in Asia Minor, in whose area ferrous ores were found in ancient times, where such, at that time mysterious, force effects were discovered. First of all, field sizes are introduced in the following, which allow us to describe the magnetic field $\vec{H}$ completely. Chap. 1 is devoted to static descriptions, while Chap. 2 is devoted to the effects of magnetic fields that change over time.

1.1 Magnetic Field Quantities and Characteristics of Magnetic Circuits

The magnetic field with its field strength H is a force field that starts at the north pole of a magnet and whose field lines close over the south pole (Fig. 1.1a). It is characteristic that these lines are always closed, that is, they have neither beginning nor end. Figure 1.1a shows a permanent bar magnet (permanent magnet) with its H field lines. If you divide the

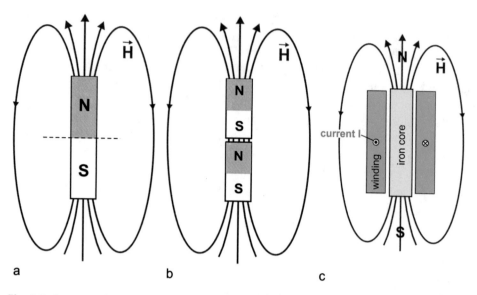

Fig. 1.1 Structure of the magnetic field using the example of a bar magnet. (**a**) Basic structure, (**b**) when the bar magnet is divided and (**c**) when an electromagnet

bar magnet (Fig. 1.1b), you get two submagnets, each of which again has a north pole and a south pole. If you provide an iron core with a winding through which a current I flows, you can observe that a magnetic field similar to that of a bar magnet builds up around the winding. When the current is switched off, the magnetic field disappears again. If you leave the core out, you will notice a field whose force effect is not as strong as with the core.

If a freely rotating small magnetic needle is brought into the vicinity of a straight conductor through which current flows, a torque on this magnetic needle is detected. This torque or force could be used to measure the strength of the magnetic field. By determining the magnitude and direction of the setting force on the magnetic needle, the field lines of the magnetic field strength H can be determined (Fig. 1.1).

The magnetic field H with its force effects is described by directional lines, the distance between which is a measure of the strength of the magnetic field. The direction of a line corresponds to the setting direction of a magnetic needle, while the density of the lines is proportional to the magnitude of the field strength $\vec{H}$.

If the magnetic needle is positioned at various points and the magnetic needle is moved in the direction of its adjustment with a constant adjustment force (Fig. 1.2), a long, straight conductor through which current flows (the return conductor is very far away) will form a circle in a plane through which the conductor passes vertically. If the experiment is repeated at other distances from the conductor, concentric circles are obtained for other setting forces (Fig. 1.2a).

One draws the density of the lines proportional to the magnitude of the mechanical force. These circles belong in the spatial arrangement to coaxial cylinders (Fig. 1.3a). With

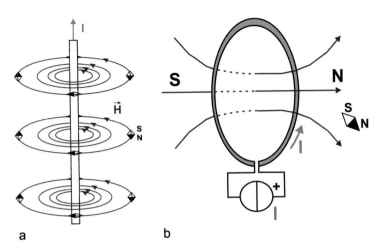

Fig. 1.2 Conductors through which a current flows are connected according to the "right screw" rule and interactions with rotating magnetic needles. (**a**) straight conductor, (**b**) conductor loop

other arrangements through which current flows (conductor ring, coil), the Fig. 1.3b, c. The figures in Fig. 1.4 show the alignment of iron filings in the field of conductor arrangements as shown in Fig. 1.3. The closed lines can still be seen around the individual conductors; inside, all subfields overlap to form an almost homogeneous field. Outside the coil, the forces are so small that they are not sufficient to align the iron filings. In Fig. 1.4d, e other materials have been brought into the field. It can be seen that iron, nickel and cobalt (ferromagnetics) try to concentrate the magnetic field on themselves (Fig. 1.4e), while all other substances do not show such recognizable effects (Fig. 1.4d).

Characteristic are the always (independent of material transitions) closed lines, which serve to represent the magnetic field. The field images give the impression of a vortex around the current. The vortex soul is the current. One also speaks of a vortex field).

The entire self-contained phenomenon that swirls around the current is called the magnetic flux Φ. The space that is filled by the magnetic field and that encloses the current is called the magnetic circuit. The magnetic flux is not a vectorial but a scalar quantity, therefore it has an orientation as counting direction and is therefore for a magnetic circuit the adequate quantity to the electric current in the electric circuit.

The physical (counting) direction of the magnetic flux Φ is defined in the direction in which the pole of the magnetic needle points to the north pole in the earth's field (Fig. 1.1). This sense of direction, together with that of the associated current, results in a right-hand screw (Figs. 1.1b and 1.2). If you think of the tip of a screw with a right-hand thread pointing in the direction of the current, its forward rotation indicates the physical sense of direction of the flux belonging to this current. In contrast to the electrostatic field, the magnetic field is source-free! This means that there are *no* "magnetic charges" (no *magnetic monopoles*) as with the electrostatic field.

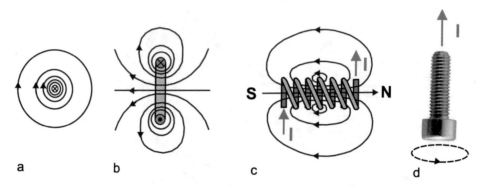

Fig. 1.3 Positive direction of the field lines according to a screw with right-hand thread. (**a**) in the case of a single conductor flowing into the image plane, (**b**) in the case of a conductor loop, (**c**) in the case of a short cylindrical coil, (**d**) screw with right-hand thread for comparison (The direction of the field lines also indicates the direction of the field size H. The density of the field lines is proportional to the field size)

For quantitative information on the size of the magnetic field, a unit must be defined. The International System of Units is based on four basic units (meter, kilogram, second, ampere). Therefore, the units of all magnetic field quantities are also based on these quantities. The unit of magnetic flux can be derived from the law of induction, which will be discussed later (Chap. 2). The unit of flux Φ is the volt-second (1 Vs), which is also called 1 Weber (1 Wb).

$$[\Phi] = 1 \ \text{Vs} = 1 \ \text{Wb}$$

The Weber (Wb) unit is primarily used in Anglo-American literature.

As described, the magnetic flux is an integral quantity over a certain area. The density of the field lines and thus the density of the flux lines is not the same everywhere. To describe the flux Φ passing through an area A, the relationship between the scalar quantity Φ and the vectorial quantities 'flux density' $\vec{B}$ and 'area element' $\vec{dA}$ is

$$\Phi = \int_A \vec{B} \cdot \vec{dA} = \int_A B \cdot \cos{(\beta)} dA$$

This relationship is shown graphically with the angular relationship for the scalar product Fig. 1.5. In a vacuum, there is a simple relationship between the magnetic field strength $\boldsymbol{H}$ and the magnetic flux density $\vec{B}$, also called induction, which is

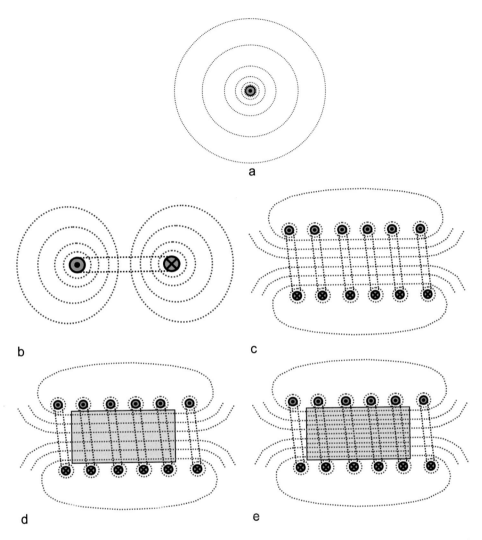

Fig. 1.4 Static field images of (**a**) (vertical) (**a**) current filament, (**b**) of a current loop, (**c**) magnetic field of a coil in the air; (**d**) of a coil with brass inside, (**e**) of a coil with iron to concentrate the lines of force (magnetic field lines) according to [3]

$$\vec{B} = \mu_0 \cdot \vec{H}$$

The value μ_0 designates the permeability of the vacuum with

Fig. 1.5 Relationship between
the field values induction B,
magnetic flux Φ and area A

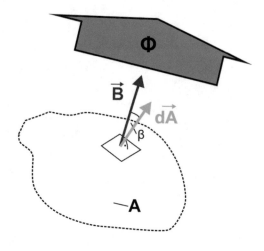

$$\mu_0 = 4\pi \cdot 10^{-7} \cdot \frac{Vs}{Am} = 1.2566 \cdot 10^{-6} \cdot \frac{Wb}{Am}$$

To describe other substances relative to the conditions in a vacuum, the dimensionless
value of relative permeability μ_r was introduced. Thus the relation B(H) changes into

$$\vec{B} = \mu_0 \cdot \mu_r \cdot \vec{H}$$

The relative permeability μ_r is to be set practically equal to 1 for gases, while for the solid
matter it can also deviate from it significantly upwards or slightly downwards (see Chap. 3).
For simple calculations, one assumes linear conditions and thus constant values for μ_r. This
applies strictly to paramagnetic materials $(\mu_r > 1)$ and diamagnetic materials $(\mu_r < 1)$. In
both cases, the difference to 1 is only small. For ferromagnetic or ferrimagnetic materials
$(\mu_r >> 1)$, linearity is only approximately valid in certain operating points. These material
properties are dealt with separately in Chap. 3.

At this point, it should only be pointed out that many materials do not behave
isotropically with regard to the propagation of the magnetic field. This means that the
permeability μ_r depends on the direction in space. In this case, it can no longer be a scalar
quantity but has the properties of a tensor.

$$\vec{B} = \mu_0 \cdot \|\mu_r\| \cdot \vec{H}$$

If this equation is written in the Cartesian coordinates *(x; y; z)*, one obtains

$$\vec{B} = \mu_0 \cdot \|\mu_r\| \cdot \vec{H} = \mu_0 \cdot \begin{bmatrix} \mu_{rx} & \mu_{rxy} & \mu_{rxz} \\ \mu_{ryx} & \mu_{ry} & \mu_{ryz} \\ \mu_{rzx} & \mu_{rzy} & \mu_{rz} \end{bmatrix} \cdot \begin{bmatrix} H_x \\ H_y \\ H_z \end{bmatrix}$$

If the main axes for the permeability coincide with the coordinates *(x;y;z)*, one obtains

$$\vec{B} = \mu_0 \cdot \|\mu_r\| \cdot \vec{H} = \mu_0 \cdot \begin{bmatrix} \mu_{rx} & 0 & 0 \\ 0 & \mu_{ry} & 0 \\ 0 & 0 & \mu_{rz} \end{bmatrix} \cdot \begin{bmatrix} H_x \\ H_y \\ H_z \end{bmatrix}$$

In the following, isotropy and linearity are assumed. When measuring the magnetic flux with a flux meter, the measuring arrangement is designed so that the deflection of the indicating instrument is proportional to the flux passing through a coil area. The measurement uses the effect of the law of induction, which is explained and analysed in the next chapter. In the "classical measuring technique", a so-called ballistic galvanometer was used to measure the magnetic flux Φ. This is an aperiodically damped galvanometer without elastic restoring forces. When the measuring coil is moved towards the measuring position from a sufficiently distant initial position, this arrangement has an integrating function. The measurement result is then theoretically proportional to the measured integrated induced electric charge

$$M_{end} = Q = \int_{t1}^{t2} \frac{u_{ind}}{R} dt = \int_{t1}^{t2} k \frac{d\phi}{R \cdot dt} \cdot v \cdot dt = \int_{S1}^{S2} \frac{k}{R} \frac{d\phi}{ds} \cdot \frac{ds}{dt} dt \sim \phi$$

In modern measuring systems, all measuring functions including integration are carried out digitally. However, the principle remains the same.

In the magnetic field, such a loop is used to measure the vertically continuous partial flux Φ (Fig. 1.6). If the measuring loop is moved along the flux lines and the cross-section of the loop is changed so that the deflection of the instrument remains constant, the track of the loop forms a flux tube with constant flux. If you divide the entire flux Φ into flux tubes with the same Φ, you get a selected flux field. The river tubes result in lines in one plane as shown in Fig. 1.6.

If you now divide the partial flux $\Delta\Phi$ of a tube by the partial area vertically penetrated by it $\Delta A_\perp$, you obtain (at the transition to differentially narrow flux tubes) the differential form of the definition equation for the flux density or induction B:

$$B = \frac{d\Phi}{dA_\perp}$$

Fig. 1.6 Magnetic "flux tube": cross-sections with the same magnetic fluxes, arranged in rows according to the field lines [after 3]

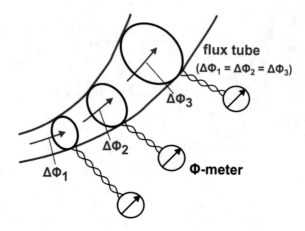

The flux density $\vec{B}$ is (like the current density $\vec{J}$, the displacement current density $\vec{D}$, etc.) a vector. That means vectors have $\vec{B}$ both magnitude and a direction of action. The magnetic flux on $\Phi = \int_A \vec{B} \cdot \vec{dA}$ the other hand is a scalar with a magnitude and a sign.

The physical unit/dimension of flux density is the tesla:

$$[B] = \left[\frac{\Phi}{A}\right] = \frac{Vs}{m^2} = 1 \text{ T} = 1 \text{ Tesla} = 10^4 G = 10^4 \text{Gauss}.$$

The former subunit 1 Gauss $= 1G = 10^{-4}$ T is no longer included in the table of legal units (now: *Système International d'Unités* (SI)), but is f. e. still used in English literature and astrophysics. The unit Tesla was named after Nikola Tesla, the pioneer of electrical engineering, who in the nineteenth century produced groundbreaking inventions in the field of electrical engineering and especially magnetism. The magnetic field of the earth, for comparison, has a flux density of about 30 µT at the equator and a value twice as high at the poles.

It can be summed up:

- The flux density lines are always closed in themselves. This applies regardless of whether the magnetic field is formed in a vacuum or any material, or whether the 'magnetic circuit' is composed of several materials.

From this fundamental property of the flux, which is completely analogous to the property of the electric current, the same statement of the freedom of source results as it was formulated for the electric current as Kirchhoff's first theorem:

- The total flux flowing into a volume part must be equal to the flux flowing out of this volume part (more precisely: the flux lines are without beginning or end, infinite. With special arrangements they can run into infinity in both directions).

Fig. 1.7 Induction lines $\vec{B}$ running through a volume element with the surface A_O

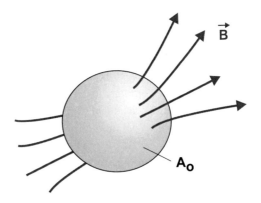

The word "flowing" is not to be taken literally here because nothing flows. The magnetic flux is the characterization of a spatial state created by a magnetic field, the flux lines are only its model representation and allow easier access to human cognition. A fundamental theoretical difference is also that the magnetic field strength is a directly time-dependent quantity, whereas flux and flux density depend on the prehistory. The dependencies B (H) are, even in the case of paramagnetic materials, only unambiguous if the start of integration is defined.

For a volume which is penetrated by flux density lines (Fig. 1.7), the expression of the source-free magnetic field is

$$\sum_{v} \Delta \Phi_v = \sum_{v} B_v \cdot \Delta A_{\perp} = 0$$

The transition to differential quantities yields Stokes'theorem

$$\int_{Ao} \vec{B} \cdot \vec{dA} = 0$$

In other words:

• The envelope integral of the flux density (induction) is zero.
Another expression of the same facts is:

• The sums of the incoming fluxes Φ_v and outgoing magnetic fluxes Φ_μ to a volume element are equal.

$$\sum_{v=1}^{n} \Phi_v = \sum_{v=1}^{m} \Phi_\mu$$

The equations that express the physical property of electric current and magnetic flux, the *source freedom* or *continuity,* in terms of formulas, represent important basic relationships for field calculations. It is therefore important not only to know them as mathematical equations but above all to grasp *them in terms of content,* as emphasized by the mnemonics here.

Since a charge is also surrounded by an electric field, the theorem applies:

- A moving electrical charge is always associated with an *electromagnetic field.*

If the charge moves at a constant speed, the magnitude of the field sizes does not change.

We are constantly surrounded by the magnetic field of the earth. At the beginning of the nineteenth century, research into the earth's magnetic field received strong impulses, for example, the Magnetic Association was founded in Göttingen. Carl Friedrich Gauss succeeded in establishing a comprehensive theory of geomagnetism. Based on the potential field theory, he was able to prove in 1839 that the main part of the earth's magnetic field originates from the earth's interior.

1.2   The Circular Integral in the Magnetic Field - the Magnetomotive Force (MMF) - the Magnetic Force

In addition to the unity between the magnetic field vortex and the electric current and the associated freedom from the source of the magnetic field, there is another law of nature for an arbitrary circulation, that is, a closed path in the magnetic field.

If one assumes an orbit in a homogeneous $\vec{B}$-field (i.e. independent of the $\vec{B}$ location), which is composed of path segments along or perpendicular to the $\vec{B}$-lines (Fig. 1.8), one can create a vivid representation of the problem 'closed line integral'. For a clear description of the procedure the following definitions are made:

$$\left|\overline{AB}\right| = a, \left|\overline{BC}\right| = b, \left|\overline{CD}\right| = c, \left|\overline{DA}\right| = d.$$

Along the upper section AB of the circuit $\vec{s_1}$, the line integral due to parallelism of and $\vec{B}\,\vec{s_1}$

$$\int_{A}^{B} \vec{B} \cdot d\vec{s} = B \cdot a$$

From C to D it is the same size, but negative, there and $\vec{B}$ antiparalleld $\vec{s}$:

$$\int\limits_{C}^{D} \vec{B} \cdot d\vec{s} = -B \cdot c = -B \cdot a$$

The side edges 'b' and 'd' do not contribute to the amount of $\int \vec{B} \cdot d\vec{s}$, da and $\vec{B} d\vec{s}$ are perpendicular to each other. Therefore, the closed line integral is

$$\oint \vec{B} \cdot d\vec{s} = B \cdot a + B \cdot b \cdot \cos\frac{\pi}{2} + B \cdot c \cdot \cos\pi + B \cdot d \cdot \cos\frac{\pi}{2} = B \cdot a + 0 - B \cdot a + 0$$

$$= 0$$

The same results for the triangular path in Fig. 1.9a Here is

$$\oint \vec{B} \cdot d\vec{s} = \int\limits_{A}^{B} B \cdot \cos\beta \cdot ds + \int\limits_{B}^{C} B \cdot \cos\pi \cdot ds + \int\limits_{C}^{A} B \cdot \cos\alpha \cdot ds = 0$$

You can divide any revolution (e.g. Fig. 1.9b) into small and smallest rectangular and triangular circulations. Then the following also applies to this circuit for homogeneous material with linear properties

$$\oint \vec{B} \cdot d\vec{s} = \mu_0 \cdot \oint \vec{B} \cdot d\vec{s}$$

In a homogeneous field of a vectorial quantity, the circular integral of this vector along any path is always zero. In inhomogeneous fields, this is not so generally true. In an electrostatic field, for example, the closed line integral of the *electric* field strength is zero even in inhomogeneous fields. Among other things, this is a characteristic of a potential field.

Experimentally determined experience now shows that even in an inhomogeneous magnetic field the circulation integral of induction B is independent of the path equal to zero, − but only if the circulation path does not include any current. Figure 1.10 shows for the conditions of a magnetic vortex field around a conductor different integration paths with different results for the closed lineintegral.

The mathematical formulation of the facts in Fig. 1.10 provides in detail

$$\oint\limits_{S1} \vec{B} \cdot d\vec{s} = 0$$

Fig. 1.8 Rectangular
circulation path s_1 in a
homogeneous magnetic B-field
with sections parallel,
antiparallel and perpendicular to
the field lines

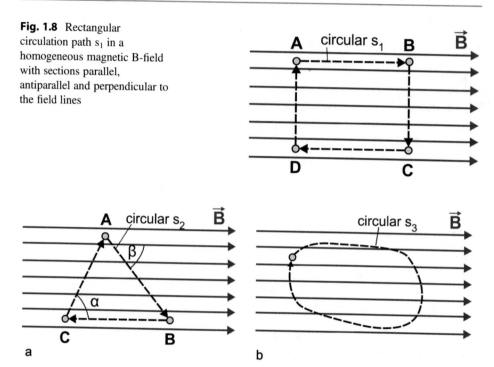

Fig. 1.9 Circulation paths in the magnetic field. (**a**) Triangular circulation path s_2 in a homogeneous magnetic B-field and (**b**) any circulation path s_3

$$\oint_{S2} \vec{B} \cdot d\vec{s} = \mu \cdot I \text{ respectively } \oint_{S2} \vec{H} \cdot d\vec{s} = I$$

$$\oint_{S3} \vec{B} \cdot d\vec{s} = 0$$

In general, it can be said that

- The closed line integral in any path around a current is equal to the circular integral along a field line around that current.
- It is completely irrelevant to the value of the closed line integral at which point the current flows through the surface clamped by the closed path.
- If several currents flow through this area, the same consideration as for one current results: The closed line integral is equal to the signed sum of the circulation integrals around the individual currents.

Applications of these rules are shown in Fig. 1.11. In Fig. 1.11a, path s1 comprises only one current, while s_2 comprises two currents. A current is counted positively if its direction

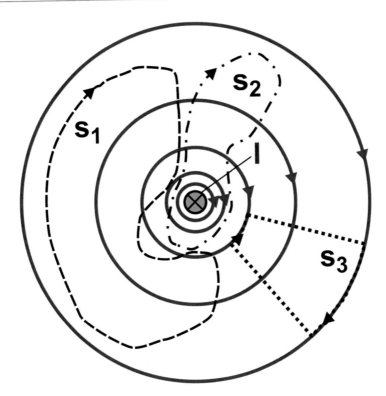

Fig. 1.10 Further circulation paths for closed line integrals in the inhomogeneous magnetic field in homogeneous material, s_1, s_3: Circulations in the inhomogeneous field outside a current or s_2: around a current across material boundaries according to [3]

is pointing in the direction of the thumb of the right hand, while the fingers of the right fist are pointing in the direction of the field lines. Figure 1.11b shows this principle once again separately for 3 currents.

The general conclusion is that

$$\oint_s \vec{B} \cdot d\vec{s} = \mu \cdot \sum_{v=1}^{n} I_v = \mu \cdot \oint_s \vec{H} \cdot d\vec{s}$$

This is where the so-called magnetomotive force (MMF) is derived from:

$$\Theta = \sum I_v$$

The MMF is defined for a circulation path in the magnetic field as the sum of the signed currents covered by that closed path. Since the currents flow through the area surrounded

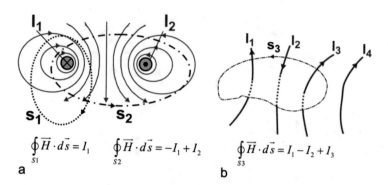

$$\oint_{S1} \vec{H}\cdot d\vec{s} = I_1 \qquad \oint_{S2} \vec{H}\cdot d\vec{s} = -I_1 + I_2 \qquad \oint_{S3} \vec{H}\cdot d\vec{s} = I_1 - I_2 + I_3$$

a   b

Fig. 1.11 The closed line integral around several currents. (**a**) 2 currents with flux lines and 2 paths, (**b**) several currents from one path

by the path, this current sum is also called the magnetic force. If the path N includes equal currents I, then the magnetic force is

$$\Theta = N \cdot I$$

The unit of magnetomotive force (MMF) is the ampere [A].

The law of magnetic force is also known as Ampère's law. It was discovered by André-Marie Ampère (* 20. January 1775 in Lyon, France; † 10. June 1836 in Marseille). The term magnetomotive force (MMF) to characterize the drive for the magnetic flux (analogous to EMF as a drive source for a current) is explained by the following arrangements (Fig. 1.12). In this figure, a copper strip bent into a cylinder is used to generate the magnetic flux. A large current flows through this copper strip with only one turn.

In the following Fig. 1.12 this is replaced by a coil of 10 turns with the same outer shape. If you now send the current I or $0.1 \cdot I$ through each of the 10 turns so that the product $I \cdot N$ remains the same, you can see that the magnetic field is the same in both figures. The total flux Φ, including, for example, the partial flux $\Delta \Phi$ in the thickly drawn flux tube, is the same in both cases. So it seems that the flux depends not only on the strength of the current but also on the product $I \cdot N$. The MMK $\Theta = I \cdot N$ is therefore decisive for the flux.

The closed line integral of magnetic induction (flux density) in a vacuum is proportional to the MMK. The closed line (cyclic) integral (Ampére's law) of the magnetic field strength is equal to the MMK, namely

$$\text{MMC} : \Theta = \sum_{v=1}^{n} I_v = \oint_{S} \vec{H} \cdot d\vec{s}$$

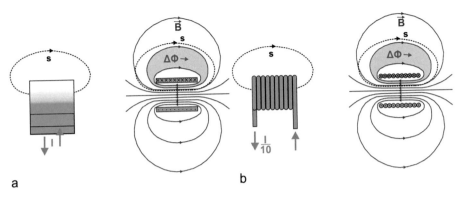

Fig. 1.12 To form the closed line integral with several turns. (a) N = 1, Θ = I, (b) N = 10, Θ = I

Whether the included currents belong to a quasi-stationary current flow field due to the conductivity of a material or to a field of displacement currents is irrelevant. This is expressed by Maxwell's first equation (here in the integral form), which is a generalisation of the correlation found by *Hans Christian Oersted (Ørsted)* (* *14. August 1777 in Rudkøbing; † 9. March 1851 in Copenhagen). In* general, the Ampére's theorem applies to galvanic and dielectric currents everywhere in space:

$$\Theta = \sum I_v = \oint_S \vec{H} \cdot d\vec{s} = \int_A \left(\sigma \vec{E} + \varepsilon \frac{d\vec{E}}{dt} \right) \cdot d\vec{A}$$

The path s includes area A. Both the conductivity σ and the permittivity ε are tensors in the general case of directional dependence of these material values. The now obsolete cgs unit of magnetic field strength H (abbreviated with the unit symbol Oe) was named after Oersted. Since 1970, it is no longer considered an official unit but is still used by many Anglo-American manufacturers and authors. The official unit of magnetic field strength is ampere/meter [A/m].

The following applies

$$1 \text{ Oe} = \frac{1000}{4\pi} \cdot \frac{A}{m} \approx 79.577 \frac{A}{m}$$

By multiplication with the magnetic vacuum permeability μ_0, a magnetic flux density of $100 \mu T$ in the SI unit system or 1 G in the Gaussian CGS unit system is obtained in a vacuum:

$$\mu_0 \cdot 1 \ \text{Oe} = 1 \ \text{G}$$
$$1 \ \text{G} = 4\pi \cdot 10^{-7} \frac{\text{Vs}}{\text{Am}} \cdot \frac{1000 \ \text{A}}{4\pi \ \text{m}} = 10^{-4} \text{T}$$

(The earth's magnetic field has a magnetic flux density of approximately between $0.3\text{G} = 30\mu\text{T}$ (equator) and $0.6\text{G} = 60\mu\text{T}$ (poles).)

These interrelationships are presented here since many diagrams and values of magnetic materials can still be found in the cgs system even in current catalogues from the British and American regions.

1.3   The Magnetic Circuit and its Elements

In this section, a clear calculation method is introduced in the analogy between the electric field and the magnetic field, which allows us to calculate magnetic arrangements with predominantly high permeability in the magnetic flux ranges similar to electric circuits with the help of Kirchhoff's rules.

A speciality of the magnetic field is that there are no materials (neither gases nor vacuum) in nature that are non-conductive for the magnetic field. Therefore, the calculation methods presented here always remain approximations. In a first approximation, the propagation of the magnetic field out of the arrangement is not considered here. In a later chapter, methods are then shown which improve an approximation to reality without having to resort to methods of field calculation based on FEM (*finite element method*) or systems of difference equations. It is not about competition or replacement of field calculation programs, but primarily about aids for a simple dimensioning of magnetic arrangements and also aids in the interpretation of calculation results.

Figure 1.13 shows the basic shape of a magnetic circuit, which has obvious similarities to the basic electrical circuit. The MMK $\Theta = \text{I} \cdot \text{N}$ drives a magnetic flux $\Phi 1$ through a magnetization cross-section A1. This is followed by sections with cross-sections A2, A3, A4, A5 and A6.

Since it is an unbranched magnetic circuit, Kirchhoff's node rule provides the following information when applied to any connection

$$\Phi_1 = \Phi_2 = \Phi_3 = \Phi_4 = \Phi_5 = \Phi_6 = \Phi$$

It is therefore obvious to assign a resistance to the driving force to the individual sections - similar to the electrical circuit. The magnetic flux Φ is applied as a quantity analogous to the electric current. Furthermore, a quantity of a "magnetic voltage drop" V is introduced, analogous to the electrical voltage drop. The magnetic circuit shown in Fig. 1.13 above can then be broken down into its individual components (Fig. 1.14).

The magnetic voltage drop can be directly derived from the field quantities

Fig. 1.13 Example of an unbranched magnetic circuit with concentrated guidance of the flux Φ, driven by a magetic force Θ and passing through different cross-sections A_x in a core with an air gap

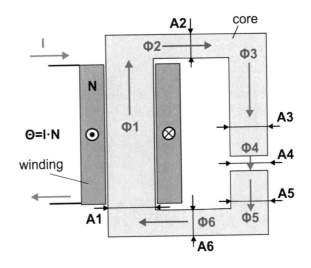

Fig. 1.14 Representation of an elementary magnetic resistance (reluctance) R_m to derive/define its design equation

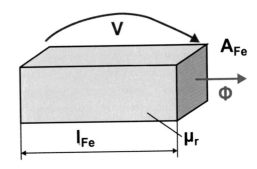

$$V = \int\limits_a^b \vec{H} \cdot d\vec{s} \approx \int\limits_a^b H \cdot ds = \frac{\Phi}{\mu_r \cdot \mu_0 \cdot A_{Fe}} l_{Fe}$$

For the connection between V and Φ on an element, a proportional connection can be assumed, which is called magnetic resistance (reluctance) in analogy to the electric current field:

$$R_m = \frac{l_{Fe}}{\mu_0 \cdot \mu_r \cdot A_{Fe}}$$

After introducing this element, Kirchhoff's two rules can be formulated as follows for the static state of a magnetic circuit:

- For a node with several magnetic fluxes, the following applies

$$\sum_{v=1}^{n} \Phi_v = 0$$

- For a closed mesh circulation in a magnetic circuit, a balance is valid between the sum of all original magnetic forces (MMF) Θ_i and the sum of all magnetic forces drops V_j:

$$\sum_{i=1}^{n} \Theta_i = \sum_{j=1}^{m} V_j$$

Initially, the effects of the law of induction on mutual induction are not considered here. An extension of the equivalent circuit diagram will therefore have to take place at the transition to the stationary and unsteady alternating magnetic field. With the agreements made above, traceability of magnetic to electrical circuits is possible (Fig. 1.15).

The abstract shape of this magnetic circuit is then shown in the following figure (Fig. 1.16). The magnetic resistance of the air gap is shown separately as a leakage channel. Here the magnetic field lines emerge from the material, which is usually ferro- or ferrimagnetic. The longer the air gap in relation to the side length of the magnetization cross-section, the greater the deviation of the actual value from the value calculated according to the above formula. Since for air applies: $\mu_{r.\ Luft} \approx 1 \neq 0$ the actual effective magnetic resistance of the air gap is always less than the calculated one. In addition to the air gaps, magnetic field lines also emerge from the surface of the core and form a magnetic shunt to the air gap. This must be taken into account when dimensioning and is the subject of the following chapters.

The magnetic flux Φ results from the above equivalent circuit using Kirchhoff's adapted rules for

$$\Phi = \frac{\Theta}{R_{mges}} = \frac{I \cdot N}{R_{m1} + R_{m2} + R_{m3} + R_{m4} + R_{m5} + R_{m\delta}}$$

The magnetic flux is proportional to the exciting current I since the MMF is $\theta = I.N$. It is covered by the winding N times (here completely). This is expressed with the term flux linkage Ψ:

$$\Psi = \Phi \cdot N$$

The flux linkage Ψ characterizes the coil structure with regard to the coupling (interlinking) of individual windings with different flux proportions. Figure 1.17 shall illustrate this. Here, a bar magnet is located in a single-layer winding, so that each winding is coupled with a different flux. The flux tubes are obviously not evenly coupled to the coil.

For the flux linkage must then be written:

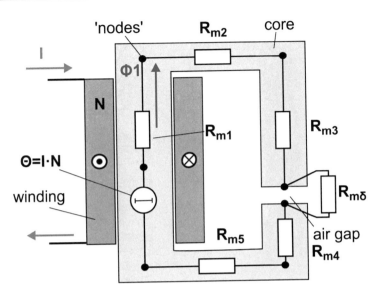

Fig. 1.15 Transition from the geometric arrangement to symbolic elements of the magnetic circuit

Fig. 1.16 Magnetic equivalent circuit diagram of an unbranched magnetic circuit according to Fig. 1.15

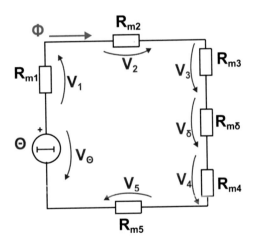

$$\Psi = \sum_{i=1}^{N} \Phi_i$$

To characterize a magnetic coupling of a winding or several conductor loops in relation to the actual current used, the term inductance L and permeance Λ (magnetic conductance) is used in addition to reluctance (magnetic resistance).

Fig. 1.17 Coupling of the flux of a bar magnet with the windings of a single-layer winding

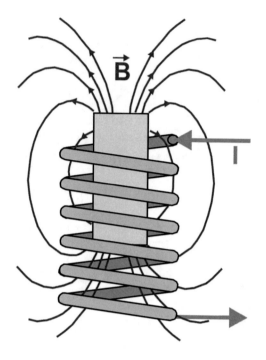

$$L = \frac{\Psi}{I} = \frac{\sum_{i=1}^{N} \Phi_i}{I} \approx \frac{N \cdot \Phi}{I} = \frac{N^2}{R_{mges}} = N^2 \cdot \Lambda_{ges}$$

With the relationship between reluctance R_m and permeance Λ

$$\Lambda = \frac{1}{R_m} = A_L,$$

one obtains with the value AL a characteristic quantity for magnetic cores, which will play a role repeatedly later.

1.4 Energy and Forces in the Magnetic Field

The Lorentz force

The magnetic field was defined by its force effects. The question, therefore, arises how large the energy content of the electric field is and how force effects can be determined. The energy density of the magnetic field results from the basic field quantities magnetic field strength H and induction B analogous to the electrostatic field to

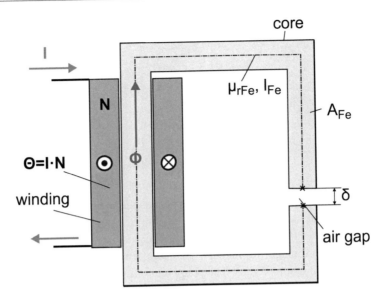

Fig. 1.18 Unbranched magnetic circuit with air gap δ

$$\frac{\Delta W}{\Delta V} = \frac{B \cdot H}{2} = \frac{\mu_0 \cdot \mu_r \cdot H^2}{2} = \frac{B^2}{2 \cdot \mu_0 \cdot \mu_r}$$

If you look at the simple magnetic circuit of the previous section, you will see that the flux $\Phi = \Phi_1 = \Phi_2 = \Phi_3 = \Phi_4 = \Phi_5$ is the same everywhere, but the flux density/induction is different in each section of the circuit (Fig. 1.18).

If you want to find out the total stored energy of the circuit, you can use the above formula and write it down:

$$W_{m.ges} = \Delta W_{Fe} + \Delta W_\delta = \frac{\left(\frac{\Phi}{A_{Fe}}\right)^2}{2 \cdot \mu_0 \cdot \mu_{rFe}} A_{Fe} \cdot l_{Fe} + \frac{\left(\frac{\Phi}{A_{Fe}}\right)^2}{2 \cdot \mu_0} A_{Fe} \cdot \delta = \frac{\Phi^2}{2 \cdot \mu_0 \cdot A_{Fe}} \left(\frac{l_{Fe}}{\mu_{rFe}} + \delta\right)$$

Since the magnetic flux is the same everywhere, this equation means that if the permeability is sufficiently high, the major part of the total energy is stored in the air gap. The flux is driven by an original magnetic voltage and can be calculated as

$$\Phi = \frac{\Theta}{R_{mFe} + R_{m\delta}} = \frac{\mu_0 \cdot I \cdot N \cdot A_{Fe}}{\frac{l_{Fe}}{\mu_{rFe}} + \delta},$$

resulting in

$$W_{m.ges} = \frac{\mu_0 \cdot A_{Fe} \cdot I^2 \cdot N^2}{2\left(\frac{l_{Fe}}{\mu_{rFe}} + \delta\right)}$$

Now, if the air gap is increased, the total energy content is changed. This process is associated with a force effect F on the interfaces A_{Fe}:

$$F = \frac{dW_{m.ges}}{d\delta} = -\frac{\mu_0 \cdot A_{Fe} \cdot I^2 \cdot N^2}{2\left(\frac{l_{Fe}}{\mu_{rFe}} + \delta\right)^2}$$

The force effect is therefore approximately dependent on the square of the distance. It has a negative sign and therefore acts against the change of the air gap. The two end faces of the air gap are pulled together. The maximum force occurs at $\delta = 0$. The greater the permeability of the ferromagnetic material, the more sensitive is the magnitude of the maximum force to minimum air gaps. The pole point of the function $F(\delta)$ is then located more closely below the surface of the opposing magnetic poles. Consequently, $F(\delta = 0)$ becomes larger.

For a rough estimate of the magnetic forces, the formula

$$F \approx \frac{d}{d\delta}\left(\frac{B^2}{2\mu_0} \cdot A_{Fe} \cdot \delta\right) = \frac{B^2}{2\mu_0} \cdot A_{Fe}$$

is used. However, the conditions of a constant induction are difficult to achieve under linear conditions. The sign of the force is positive and thus actually describes a force in the direction of the change in position. This contradicts experience. Nevertheless, the formula is often used.

The formula indicates that with a permanent flux in ferromagnetic materials (e.g. remanence) the forces acting on the interfaces can become very large. Magnetic forces at interfaces are ultimately the basis for the construction of electrical machines. On the other hand, moving parts of soft magnetic materials can also "stick" due to excessive remanence, for example, after the attraction of relays. This can be countered by inserting a non-magnetic foil.

1.5   The Biot-Savart Law

The Biot-Savart's law establishes a relationship between the magnetic field strength H and the current density J and allows the spatial calculation of magnetic field strength distributions based on the knowledge of the spatial current distributions. Here the law is treated as a relationship between the magnetic flux density B and the electric current

Fig. 1.19 Specifications for the
definition of Biot-Savárt's law

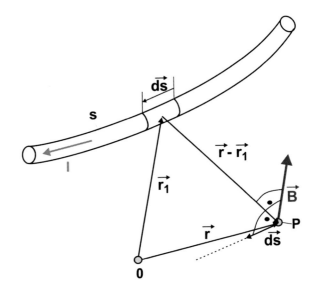

density J. The law assumes that the relationship between B and H is linear and that the permeability does not depend on the previous history (Fig. 1.19).

This law was named after the two French mathematicians Jean-Baptiste Biot (* 21 April 1774 in Paris; † 3 February 1862 in Paris) and Félix Savart (* 30 June 1791 in Charleville-Mézières, Ardennes; † 16 March 1841 in Paris). Together with Ampère's law, it represents one of the fundamental laws of magnetostatics. It allows a general differential description of the relationships.

A conductor of the length $d\vec{s}$ at the location $\vec{r_1}$ through which a current I flows (Fig. 1.19) generates the $\vec{r}$ magnetic flux density $d\vec{B}(r)$ at the location P with the location vector:

$$d\vec{B}(r) = \frac{\mu_0}{4\pi} \cdot I \cdot \left(d\vec{s} \times \frac{\left(\vec{r} - \vec{r}_1\right)}{\left|\vec{r} - \vec{r}_1\right|^3} \right).$$

The total magnetic flux density at point P is obtained by summing up all existing pieces of the path s.

$$B(r) = \int_S d\vec{B} = \frac{\mu_0 \cdot I}{4\pi} \cdot \int_S \left(d\vec{s} \times \frac{\left(\vec{r} - \vec{r}_1\right)}{\left|\vec{r} - \vec{r}_1\right|^3} \right)$$

The resulting path integral can generally be transformed into a volume integral using Stokes' theorem, using the following relations

$$I \cdot d\vec{s} = \left(\vec{J} \times \vec{A} \right) \cdot d\vec{s} = \vec{v} \cdot dq = \vec{v} \cdot \rho \cdot dV = \vec{J} \cdot dV$$

Here J is the electric current density, A is the area penetrated by J, v is the velocity of the charge carriers, q is the charge quantity, ρ is the charge carrier density and dV is a volume element of the conductor under consideration. Thus one finally obtains the integral form of Biot-Savárt's law:

$$B(r) = \frac{\mu_0}{4\pi} \int\limits_V \vec{J}\left(\vec{r_1}\right) \times \frac{\left(\vec{r} - \vec{r}_1\right)}{\left|\vec{r} - \vec{r}_1\right|^3} \cdot dV$$

In this way, the effective induction at a certain point in space can be calculated, which is caused by a corresponding current distribution. If the current distribution changes, the induction at a certain point in space will also change. This also has consequences for current measurement methods when only field strength/flux density is measured at a point near the conductor. If the current distribution changes - such as due to the skin effect - the measurement signal also changes for the same total current. Although the corresponding error can be reduced by selecting a more suitable location for the sensor, it cannot be eliminated in principle. Biot-Savárt's law is applied in numerical calculations of electromagnetic fields.

1.6   The Lorentz Force

The force effect on charges in the electromagnetic field is described by the Lorentz force (named after the Dutch mathematician and physicist Hendrik Antoon Lorentz (* 18. July 1853 in Arnhem; † 4. February 1928 in Haarlem). The force effect comprises the Coulomb force and the Lorentz force in the narrower sense, which is the one we are talking about here. The Lorentz force in the narrower sense is therefore the magnetic component of the general Lorentz force. Figure 1.20 is used for explanation. When an electric charge q moves $\vec{v}$ through a homogeneous magnetic field $\vec{B}$ at a speed, a force $\vec{F}$ is created. The vectors form a right-hand system, which is described by the vector product:

$$\vec{F_L} = q \cdot \vec{v} \times \vec{B} = \overline{\left(q \cdot v \cdot B \cdot \sin\left(\sphericalangle\left(\vec{v} ; \vec{B} \right) \right) \right)}$$

That is, $\vec{F}_L$ is perpendicular to the $\vec{v}$ plane determined by $\vec{B}$.

A free-flying charge changes its path perpendicular to the direction of flight from $\vec{v}$. (Fig. 1.21). This means for a charge carrier in a vacuum that it moves on a circular path. The

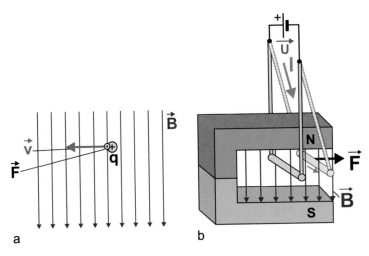

Fig. 1.20 Relationships in the formation of the Lorentz force (the force vector is on the image plane) (**a**) on a positive charge carrier and (**b**) on a conductor through which a current flows (the direction of the current is the technical direction of the current)

centrifugal force on the charge carrier is $\vec{F}_R$ compensated $\vec{F}_L$ by the Lorentz force, resulting in a circular path. The balance of forces then means

$$\vec{F}_L = \vec{F}_R$$
$$q \cdot v \cdot B \cdot \sin\left(\angle \vec{v}, \vec{B}\right) = m \cdot \omega^2 \cdot r$$

At the same time, the following applies to the speed and radius of the circular path
$$v = \omega \cdot r \text{ and } r = \frac{v}{\omega} \Delta W_{rot} = \frac{m}{2} r^2 \omega^2$$
The trajectory frequency can thus be easily calculated.

$$\omega = \frac{q \cdot B}{m}$$

These relationships are used in the cyclotron as particle accelerators to achieve high speeds. The function of the cyclotron is briefly described below because of its clear combination of elementary relationships (Fig. 1.22). In the homogeneous field of an electromagnet, there are 2 dish-shaped electrodes connected to an AC voltage source. Near the centre, there is a source (Q) that emits charged particles. Electrons are assumed to be particles. When they come into the space between the electrodes, they are accelerated according to the potential difference and take an energy

Fig. 1.21 Formation of a circular path during the uniform movement of a charge carrier in a homogeneous field

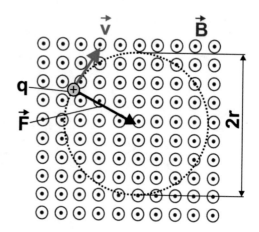

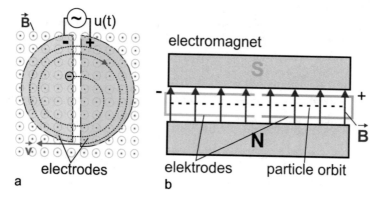

Fig. 1.22 Functionality of a cyclotron (**a**) Top view, (**b**) Side view

$$\Delta W_{kin} = e \cdot \Delta U = \frac{m_e}{2}\left(v_{i+1}^2 - v_i^2\right) = \frac{m_e}{2}\left(v_i + \Delta v + v_i\right)\left(v_i + \Delta v - v_i\right) = m_e\left(v_i + \frac{\Delta v}{2}\right)\Delta v$$

$$\Delta v \approx \frac{e \cdot \Delta U}{m_e \cdot v_i}$$

$$\Delta W_{rot} = \frac{m_e}{2}r^2\omega^2$$

$$v_{i+1} = \sqrt{\frac{2(i+1)e \cdot \Delta U}{m_e}}$$

No energy is absorbed during the circular movement within an electrode. The magnetic field only holds the circular orbit. If a DC voltage were applied to the electrodes, braking would start the next time the electrode enters the gap between the electrodes. If, however, a polarity change has occurred, further acceleration can take place according to the above-mentioned pattern. With increasing speed, the increase in speed becomes smaller and smaller with an ever-larger path radius

$\Delta v \approx \frac{e \cdot \Delta U}{m_e \cdot v_i}$ at $v_{i+1} = \sqrt{\frac{2(i+1)e \cdot \Delta U}{m_e}}$.

an ever-increasing orbit radius results at the i-th cycle with the orbit radius

$$r = \frac{v}{\omega} = \frac{v_0 + \sum\limits_{i=1}^{n} \Delta v}{\omega}$$

To achieve high speeds, the polarity reversal of electrodes A and B must be synchronized with the movement. The highest speed is reached at the edge of an electrode and the particle with high speed can continue to be used.

The Lorentz force in the strict sense is the basis for all electromagnetic motors. If it is possible to translate the force effect into a continuous movement, a motor has been created. In a very elementary way, unipolar motors (homopolar motors) function by simply mechanically using the Lorentz force.

A consistent application of electromagnetic force effects in a mechanical construction leads to very simple motor designs that also function without reversing the polarity of the magnetic field (unipolar motor, homopolar motor). The best known homopolar motor is certainly Barlow's wheel [4, 5] (Fig. 1.23). Here, too, the moving part itself is passed through by current. The current flow is made possible by sliding contacts or a conductive liquid. The formation of the magnetic field between the poles of a permanent magnet allows the Lorentz force to build up at the part of the wheel through which the current flows. This causes the wheel to move.

Figure 1.24 shows a homopolar motor based on the same principle as it can be built with a simple battery. The correlations of the Lorentz force immediately explain its function. There is no reversal of polarity. The current is only conducted from the positive to the negative pole via sliding contacts on the positive and negative pole via brackets. A strong permanent magnet is located in the lower part, around which a corresponding magnetic field is built up. This generates the Lorentz force on the moving charge carriers and thus on the conductor. This creates a continuous movement, even without reversing the polarity as in the classic DC machines with a commutator. The direction of action of the Lorentz force generates a constant torque so that the conductor clip moves around the battery. However, the amazingly simple design has the disadvantage of a very low power density, which is why the principle is essentially only used for illustrative purposes.

1.7 The Vector Potential

The magnetic field is not a potential field. Nevertheless, it is possible to use similar procedures under certain conditions. The vector potential A(r) is a mathematical tool in electrodynamics that was introduced to simplify the handling of magnetic induction or flux density and to adapt it to the potential field.

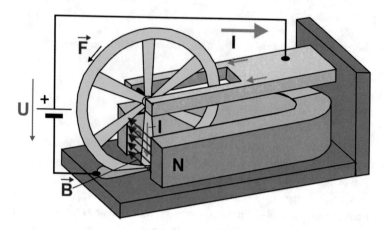

Fig. 1.23 Barlow's wheel as a unipolar motor using the Lorentz force as drive

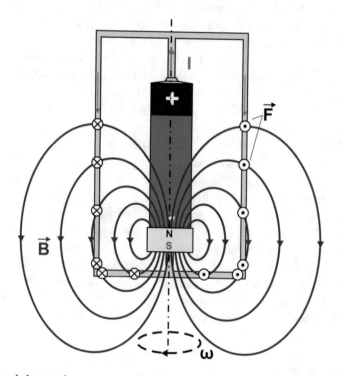

Fig. 1.24 Simple homopolar motor using the Lorentz force as drive

Mathematically, the vector potential (in contrast to the scalar potential) is a vector field with rotation according to the following definition.

$$\vec{B}\left(\vec{r}\right) = rot\left(\vec{A}\left(\vec{r}\right)\right) = \nabla \times \vec{A}\left(\vec{r}\right) = \det \begin{bmatrix} e_x & \dfrac{\partial}{\partial x} & A_x \\ e_y & \dfrac{\partial}{\partial y} & A_y \\ e_z & \dfrac{\partial}{\partial z} & A_z \end{bmatrix} = \begin{bmatrix} \dfrac{\partial A_z}{\partial y} - \dfrac{\partial A_y}{\partial z} \\ \dfrac{\partial A_x}{\partial z} - \dfrac{\partial A_z}{\partial x} \\ \dfrac{\partial A_y}{\partial x} - \dfrac{\partial A_x}{\partial y} \end{bmatrix}$$

The components e_x, e_y and e_z are unit vectors in the x, y and z directions.

The vector potential itself represents a vector field, which in ∇ turn leads to a vector field via the vector product with the Nabla operator. The analogy of the connections between the electric field, which is connected to the Nabla operator via the scalar product ($\vec{E}\left(\vec{r}\right) = \nabla \cdot \varphi\left(\vec{r}\right)$), and the magnetic field leads to the name vector potential. Vector potentials can be used, among other things, to decouple the Maxwell equations used to describe the electromagnetic field and thus make them easier to solve.

References

1. Gilbert, W. et al.: On the magnet, magnetic bodies also, and on the great magnet the earth, Übersetzer: Silvanus Phillips Thompson, Chiswick Press, London, 1900
2. Dibner, B.: Oersted and the discovery of electromagnetism. Burndy Library, Norwalk CT 1961
3. Lunze, K.: Einführung in die Elektrotechnik. VEB Verlag Technik Berlin, 1985
4. Barlow, B.: curious electro-magnetic Experiment. The Philosophical Magazine and Journal, Ausgabe 59, 1822
5. Demtröder, W.: Experimentalphysik. Bd.2: Elektrizität und Optik. Springer, Berlin 2004, ISBN 3-540-20210-2
6. Guiguzi, China's First Treatise on Rhetoric: A Critical Translation and Commentary. Trans. Hui Wu. Carbondale, Ill.: Southern Illinois University Press, 2016.
7. Sellmann, James D. 2002. Timing and Rulership in Master Lü's Spring and Autumn Annals (Lüshi chunqiu). Albany: State University of New York Press.
8. Lunheng (The Balanced Inqiries). Classical Chinese text by Wang Chong (80 AD)
9. Schmidl, Petra G. (1996–97). "Two Early Arabic Sources On The Magnetic Compass". Journal of Arabic and Islamic Studies. 1: 81–132
10. Einstein, A.: On the Electrodynamics of Moving Bodies. Annalen der Physik 17 (1905), pp. 891 - 921
11. Sturlese, L.; Thomson, R. B.: Petrus Peregrinus de Maricourt: Opera. Epistula de Magnete. Nova compositio astrolabii particularis. Scuola Normale Superiore, Pisa 1995, ISBN 88-7642-049-5

The Magnetodynamic Field

2

2.1 The Law of Induction

Contrary to the static relationships of the magnetic field in the previous chapter, magnetodynamics is considered as the electrical feedback of magnetic fields on conductors and magnets. The origin of electric tensions by temporal changes of magnetic field (induction phenomena) was discovered in 1831 by Michael Faraday (* 22 September 1791 at Newington, Surrey; † 25 August 1867 at Hampton Court Green, Middlesex). The induction effects occur in two forms: time-varying flux and movement of a conductor in a magnetic field that is constant in time. Both processes can occur simultaneously [1–3]. They are first treated separately and then combined in one equation. In Chap. 1, it was found that the integral of the magnetic flux density $\vec{B}$ over a closed enveloping area A_O is zero:

$$\Phi = \int_{AO} \vec{B} \cdot d\vec{A} = 0$$

That is to say: *The magnetic field is source-free.* That means: there are no magnetic charges.

For comparison: In the electric field, the following applies

$$Q = \int_{AO} \vec{D} \cdot d\vec{A},$$

where Q is the charge enclosed by enveloping area A_O. Sources of the electric field are electric charges.

© Springer Fachmedien Wiesbaden GmbH, part of Springer Nature 2022
P. Zacharias, *Magnetic Components*,
https://doi.org/10.1007/978-3-658-37206-4_2

The equation $\Phi = \int_{AO} \vec{B} \cdot d\vec{A} = 0$ has formal similarity with the relationship $I = \int_{AO} \vec{J} \cdot d\vec{A} = 0$ that applies in the stationary flux field. Despite this analogy, the magnetic flux describes a spatial state and not a transport problem of actually flowing media. The chosen and retained term 'flux' should help to give people an idea of the state that is not in itself 'comprehensible'.

The law of **electromagnetic induction**, briefly induction law, describes the relationship between electric and magnetic fields. It states that when the magnetic flux through a surface changes, a rotational voltage is created at the edge of this surface. In frequently used formulations, the law of induction is described by representing the edge line of the surface as an interrupted conductor loop at whose open ends the voltage can be measured.

The analytical description, which is useful for understanding, is divided - like the general formulation possibilities of Maxwell's equations - into two possible forms of representation:

- The integral form or also *global form* of the law of induction: Here the global properties of a spatially extended field area are described (via an integration path).
- The differential form or *local form* of the law of induction: Here the properties of individual local field points are described in the form of densities. The volumes of the global form tend towards zero, and the field strengths occurring are treated as differential quantities.

Both forms of representation describe the same facts. Depending on the specific application and the problem, it may be useful to use one or the other form. Both forms of representation are described below.

Law of Induction in Differential Form

Basic idea of electromagnetic induction: If the surface density B of the magnetic flux through a surface changes, it is thereby surrounded by a circuital vector electric field E, which would induce a current counteracting the flux change when the circuit is closed. Fig. 2.1 shows the interaction of the field quantities in two different situations.

The following $\frac{d\vec{B}}{dt} > 0$; $\vec{E} > 0$ is assumed. The law of induction in differential form then has the mathematical form:

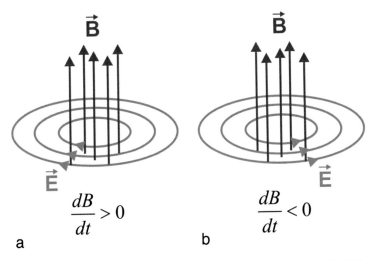

Fig. 2.1 During the temporal change of the magnetic flux density **B**, an electric field **E** is generated - depending on the direction of the change - that surrounds the magnetic field as a 'circuital vector field'. (**a**) Change of the flux density > 0, (**b**) change of the flux density < 0

$$rot\,\vec{E} = \det \begin{bmatrix} e_x & \dfrac{\partial}{dx} & E_x \\[2ex] e_y & \dfrac{\partial}{dy} & E_y \\[2ex] e_z & \dfrac{\partial}{dz} & E_z \end{bmatrix} = \begin{bmatrix} \dfrac{\partial E_z}{\partial y} - \dfrac{\partial E_y}{\partial z} \\[2ex] \dfrac{\partial E_x}{\partial z} - \dfrac{\partial E_z}{\partial x} \\[2ex] \dfrac{\partial E_y}{\partial x} - \dfrac{\partial E_x}{\partial y} \end{bmatrix} = -\dfrac{d\vec{B}}{dt}$$

This is equivalent to the presence of circuital electric vector fields around a changing magnetic field. The presence of electric circuital vector field or a magnetic flux density that changes over time is the essential characteristic of electromagnetic induction. In electric fields without induction (e.g. in the field of stationary charges), there are no closed field lines of electric field strength E. The law of induction in differential form is mainly used for theoretical derivations and in the numerical calculation of magnetic fields.

With regard to this EMF, Kirchhoff's mesh theorem.

'The signed sum of the original voltages is equal to the signed sum of the voltage drops for one mesh circulation'.

In general, recent literature no longer emphasizes the difference between EMF and voltage drops, but also assigns voltage sources to voltage drops according to the conventions shown in Fig. 2.2.

a) $\varphi_a - \varphi_b = U_{ab} = E \cdot \Delta s$ b) $EMF_{ba} = -U_{EMF\,ab} = U_{ab}$

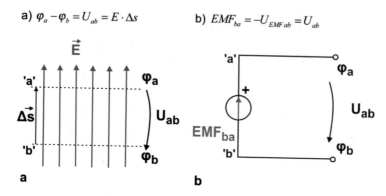

Fig. 2.2 Relationship between electrical source voltage (electromotive force/EMF)/electrical field strength and voltage drop/potential. (**a**) In the electric field and (**b**) in a symbolic circuit

Law of Induction in the Integral Form

Stokes' theorem provides the relationship between differential and integral form of the law of induction. Global vortex and source strengths are thereby transformed into local vortex or source densities of individual points in space. The theorem of Stokes delivers here

$$\int_A -\frac{d}{dt}\vec{B} \cdot d\vec{A} = \iint_A rot\,\vec{E} \cdot d\vec{A} = \oint_S \vec{E} \cdot d\vec{s}$$

If the conductor loop in Fig. 2.3 covers a certain area A, an EMF E_{ind} is induced (Fig. 2.2).

The voltage drop between the two points 'b' and 'a' (start and end points of a contour) along the drawn path is the sum of all products $-\vec{E} \cdot d\vec{s}$. A sum of infinitesimally small (differential) products is expressed symbolically by the integral sign $\int$ - a Latin 's'. As Fig. 2.3 shows, $U_{ab} = EMF_{ab}$ also applies at the same time.

$$U_{ab} = -\int_b^a \vec{E} \cdot d\vec{s} = -\oint_S \vec{E} \cdot d\vec{s} = \int_A \frac{d}{dt}\vec{B} \cdot d\vec{A} = -EMF_{ab} = -U_{ba}$$

To determine the signs, a path direction $\vec{s}$ is chosen when forming the closed loop integral, so that it points in the direction of the fingers of the right fist, while $d\vec{A}$ and $\vec{B}$ in the direction of the thumb. In this form, one speaks of Faraday's law of induction. If one applies the differentiation rules to the area integral, the product rule leads to a part caused by the induction in a resting conductor loop and a part caused by the movement of the conductor loop in space.

Fig. 2.3 Induced electromotive force (EMF) or voltage drop at a conductor loop in a variable magnetic field (points a and b are on top of each other in the analytical description of the cyclic integral)

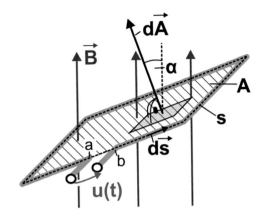

$$U_{ab} = U_{ind} = -\int_{a}^{b} \vec{E} \cdot d\vec{s} = \int_{A} \frac{d}{dt}\left(\vec{B} \cdot d\vec{A}\right) = \int_{A}\left(\frac{d\vec{B}}{dt} \cdot d\vec{A} + \vec{B} \cdot \frac{d\vec{A}}{dt}\right) = U_{ind.\,Re\,st}$$

$$+ U_{ind.Movement} = EMF_{ba}$$

Figure 2.4 is intended to illustrate these relationships. Figure 2.4a shows the assumption that was initially made: A static conductor loop with the number of turns N is in a variable magnetic field so that a voltage is induced. In Fig. 2.4b, the conductor loop rotates so that the scalar product

$$\vec{B} \cdot d\vec{A} = B \cdot dA \cdot \cos\left(\alpha(t)\right)$$

changes as α changes. This component is used in all moving electrical machines for acceleration or deceleration. It is due to the Lorentz force on moving charge carriers.

Basically, this formal approach can be continued. Induction B can also change if the permeability μ changes. This can be achieved by various external influences, as shown in Chap. 3, among others. If one assumes a linearized relationship in an operating point

$$B(H) = k_0 + k_1 \cdot H,$$

a general representation is obtained with the magnetic field strength along the path a....b

a) $u(t) = \dfrac{d}{dt}\left(B \cdot A \cdot \cos\left(\dfrac{\pi}{2}\right)\right)$

$u(t) = A \cdot \dfrac{d}{dt}B(t)$

b) $u(t) = \dfrac{d}{dt}\left(B \cdot A \cdot \cos\left(\dfrac{\pi}{2}\right)\right)$

$u(t) = B \cdot \dfrac{d}{dt}A(t) = B \cdot b \cdot v$

c) $u(t) = \dfrac{d}{dt}(B \cdot A \cdot \cos(\omega t))$

$u(t) = -B \cdot A \cdot \omega \cdot \sin(\omega t)$

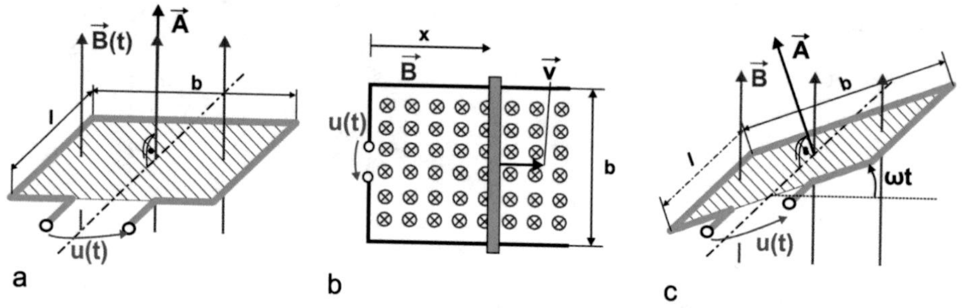

Fig. 2.4 Induction of an electrical voltage by (**a**) a variable induction vector $\vec{B}$, or (**b**) a variable area vector $\vec{A}$ parallel to the induction or (**c**) with a rotating area vector

$$U_{ind} = -\int_a^b \vec{E} \cdot d\vec{s} = \int_A \frac{d}{dt}\left(\vec{B} \cdot d\vec{A}\right)$$

$$= \int_A \left(\left(k_1 \frac{d\vec{H}}{dt} + \vec{H} \cdot \frac{dk_1}{dt}\right) \cdot d\vec{A} + \left(k_0 + k_1 \cdot \vec{H}\right) \cdot \frac{d\vec{A}}{dt}\right).$$

The permeability changes for ferromagnetic and ferrimagnetic materials as a differential quantity, or an increase in the B(H) characteristic curve, if the operating point H is changed (compare Chap. 3). In any case, such materials exhibit a pronounced saturation. This means, e.g. in the case of ferrites assumed to be isotropic, that the ratio of B and H of a 'main field' changes when a 'control field' is perpendicular to it. Assuming that one of the two fields already controls the material in one direction up to near saturation, the amount of the induction vector is limited. However, a vectorial superposition with another field hardly increases the magnitude of the B-vector. This is equivalent to a changed permeability. Figure 2.5 shows the operating principle at 2 crossed, almost homogeneous magnetic fields. It should be noted that this principle only works if the relationship of the B (H) characteristic is non-linear. In the linear region of the superposition, no interaction takes place, as is to be expected in orthogonal linear systems.

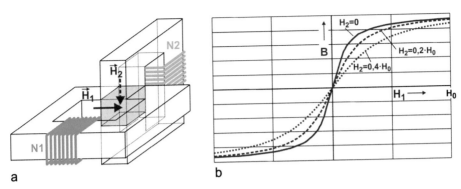

Fig. 2.5 Change of the effective permeability by superposition with an orthogonal second magnetic field. (**a**) Principle arrangement for field superposition, (**b**) mutual influence of orthogonally superimposed magnetic fields [4]

2.2 Self-Induction

In voltage induction by magnetic fields, a distinction is made between self induction and mutual induction. In self induction, a voltage is induced by the build-up of a magnetic field through a flowing variable current, which is opposite to the cause (Lenz's rule). For technical documentation, the current circuit symbols from IEC 60617 are used. The inductive component L is called inductance (see Chap. 1), which describes the ratio of flux linkage Ψ and current I

$$L = \frac{\Psi}{I} = \frac{N \cdot \Phi}{I}$$

L reflects the magnetic storage properties of the electrical component. If not noted separately, a linear relationship $B = \mu H$ is assumed below. For an assumed magnetic component with a constant magnetization cross section A and the length l as well as the number of windings N, the following applies

$$L = \frac{\Psi}{I} = \frac{N \cdot \Phi}{I} = \frac{N^2 \cdot \Phi}{\Theta} = \frac{N^2 \cdot B \cdot A}{H \cdot l} = \frac{N^2 \cdot \mu \cdot H \cdot A}{H \cdot l} = \frac{N^2 \cdot \mu \cdot A}{l} = \frac{N^2}{R_m} = \Lambda \cdot N^2 = A_L \cdot N^2$$

The inductance therefore corresponds to the reciprocal value of the magnetic resistance (Λ permeance, A_L value) multiplied by N^2. If one now transforms and differentiates this equation, one obtains the relationship between voltage drop and flowing current as

$$\Psi = N \cdot \Phi = N \cdot B \cdot A = L \cdot I$$
$$\frac{d\Psi}{dt} = N \cdot \frac{d\Phi}{dt} = N \cdot \frac{dB}{dt} \cdot A = L \cdot \frac{dI}{dt}$$
$$u_{ind} = L \cdot \frac{dI}{dt}$$

At the same time, the following applies to the energy stored in the magnetic field

$$W_{magn} = \frac{L}{2} I^2$$

This relationship can be used to determine the inductance/magnetic resistance of areas of a magnetic field. Even if a conductor is considered, it is fulfilled by a magnetic field at sufficiently low frequencies. Thus, one can also assign an inductance to a straight conductor.

Assume a cylindrical conductor with radius R_1 and homogeneous current density (Fig. 2.6)

$$J = \frac{I}{A} = \frac{I}{\pi \cdot R_1^2}.$$

A concentric path with radius 'r' inside the conductor comprises the partial current

$$I(r) = \frac{I}{\pi \cdot R_1^2} \pi \cdot r^2 = I \cdot \frac{r^2}{R_1^2}$$

This results in the magnetic field strength in the conductor to

$$H(r) = \frac{I(r)}{2\pi \cdot r} = \frac{I \cdot r}{2\pi \cdot R_1^2}.$$

The energy stored in the magnetic field is

$$W_i = \frac{\mu_r \cdot \mu_0}{2} \int_0^{R1} H^2(r) \cdot l \cdot 2\pi r dr = \frac{\mu_r \cdot \mu_0 \cdot l}{4\pi \cdot R_1^4} I^2 \int_0^{R1} r^3 dr = \frac{\mu_r \cdot \mu_0 \cdot l}{16\pi} I^2 = \frac{L_i}{2} I^2$$
$$L_i = \frac{\mu_r \cdot \mu_0 \cdot l}{8\pi}$$

The internal inductance of a conductor (in the case of direct current or at low frequencies) is therefore

Fig. 2.6 Determination of the internal inductance of a conductor

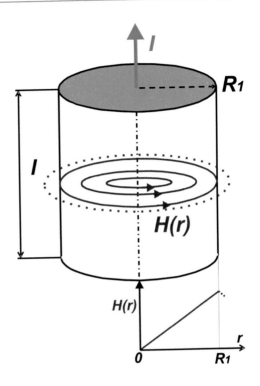

$$L_i = \frac{\Psi}{I} = \frac{\Phi}{I} = \mu_r \cdot \frac{\mu_0}{8\pi} \cdot l = \frac{\mu_r}{2} \cdot l \cdot 10^{-7} \frac{H}{m}.$$

This means: *internal* inductance L_{int} of a single cylindrical straight conductor is independent of its radius and depends only on its length. In general, this value is neglected. However, it does play a role for high permeabilities, such as ferromagnetic materials, and also for very high frequencies. For high currents (i.e. with large cross sections and few turns), the internal inductance also plays a not negligible role since the inductive voltage drop may be equal to or greater than the ohmic one. At higher frequencies of the current, the current is increasingly concentrated in the envelope of the cylinder shown (skin effect). At sufficiently high frequencies, the internal inductance of a conductor is then negligible.

The inductance of a component is composed of the internal inductance of the conductor and the inductance of the magnetic field around the conductor, so that the following applies

$$L = L_{int} + L_{ext}$$

However, the external inductance L_{ext} can only be determined if the geometry of the conductor for the returning current is known. In the following, the inductance L of an arrangement is assumed, taking into account the above-mentioned findings. For a model-like description of the u-i-t behaviour (Fig. 2.7), the counting direction is defined with the

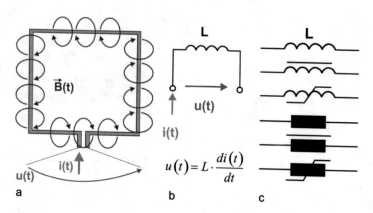

Fig. 2.7 Self induction by the flux of a time varying current in a conductor loop with the number of turns N. (**a**) Basic structure, (**b**) symbolic representation, (**c**) symbols used in the book (general inductance, inductance with core, inductance with non-linear core)

load reference arrow system. Various symbols are also in use, a selection of which is shown in Fig. 2.7.

With inductance, a voltage drop is generated, which corresponds to the differentiated current waveform except for a factor. This means, for example, that if the current with a switch ($\frac{di}{dt} \rightarrow -\infty$) is interrupted, very high voltages ($u_{ind} \rightarrow -\infty$) are generated by the reduction of the magnetic field. This effect is used in spark plugs for combustion engines to ignite the mixture of air and fuel by voltage flashovers between two electrodes. The voltage load on the switch itself must be limited by suitable measures.

An inductance L represents an ideal, loss-free component. With superconducting inductors, the resistance R is zero, but considerable energy is required to maintain low temperatures. However, real inductors also have a winding resistance R, which is given by the winding material (Fig. 2.8).

According to the counting agreements, the following applies to the voltage drop across a component during self induction

$$u(t) = u_L(t) + u_R(t) = L \cdot \frac{di}{dt} + R \cdot i$$

When a real inductance is connected to a voltage source (Fig. 2.9), the initial value i $(t = 0) = 0$, since the current cannot change abruptly. Consequently, the voltage drop is also $u_R = 0$. For t = 0, the following must therefore apply

$$U_0 = L \cdot \frac{di}{dt}\bigg|_{t=0} \quad \text{or rather} \quad \frac{di}{dt}\bigg|_{t=0} = \frac{U_0}{L}.$$

The differential equation for this unbranched circuit is

Fig. 2.8 Consideration of the
winding resistance of the coil

$$u(t) = u_L(t) + u_R(t) = L \cdot \frac{di}{dt} + R \cdot i$$

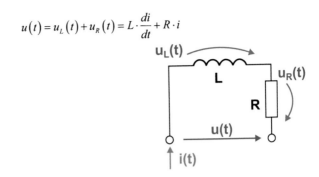

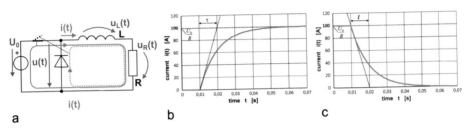

a   b   c

Fig. 2.9 Switching on and off a current through an inductor L = 1 mH, R = 0.1 Ω, U_0 = 10 V. (**a**)
Wiring diagram, (**b**) inrush current, (**c**) breaking current

$$U_0 = R \cdot i_{on} + L \cdot \frac{di_{on}}{dt} \text{ with } i_{on}(0) = 0$$

The general solution results from the sum of the homogeneous and the particulate solution
to $i(t) = i_{hom}(t) + i_{part}(t)$. With the initial condition, the homogeneous solution is
parameterized. The solution for $t \to \infty$ is used to determine the particulate solution. One
obtains

$$i_{on}(t) = -\frac{U_0}{R} \cdot e^{-\frac{L}{R}t} + \frac{U_0}{R} = \frac{U_0}{R} \cdot \left(1 - e^{-\frac{L}{R}t}\right)$$

When the switch is opened again, the current continues to flow through the inductance L. In
order for the circuit to close, a diode is required as a "free wheel", for example, which is
ideally assumed here ($U_F = 0$). The corresponding differential equation is then

$$0 = R \cdot i_{off} + L \cdot \frac{di_{off}}{dt} \text{ with } i_{off}(0) = \frac{U_0}{R}$$

The solution of this differential equation is

$$i_{on}(t) = \frac{U_0}{R} \cdot e^{-\frac{R}{L}t} + 0 = \frac{U_0}{R} \cdot e^{-\frac{R}{L}t} = \frac{U_0}{R} \cdot e^{-\frac{t}{\tau}} \text{ with } \tau = \frac{L}{R}$$

The quantity τ is the time constant of the self-discharge of the inductive storage. If you calculate time periods $\Delta t << \tau$, you can neglect the 'volatile part', i.e. the homogeneous solution. The approximate solution of a curve with switched elements is then made up of straight sections. The larger the time constant, the closer the component with its properties is to an ideal inductance.

If the same consideration is applied to the Laplace-transformed differential equation of a sinusoidal voltage $u_0(t)$, one obtains in complex notation with the complex operator 'p'

$$\underline{U}_0 = R \cdot \underline{I} + p \cdot L \cdot \underline{I}$$

The stationary harmonic solution for the current is then with $p = j\omega$

$$\underline{I} = \frac{U_0}{R + p \cdot L} \text{ or rather } \underline{I}(j\omega) = \frac{U_0}{R + j\omega \cdot L} = \frac{U_0}{R} \cdot \frac{1}{1 + j\omega \cdot \frac{L}{R}} = \frac{U_0}{R} \cdot \frac{1}{1 + j\omega \cdot \tau}$$

The mapping of voltage to current is expressed by the frequency response of RMS value and phase angle (Fig. 2.10)

$$I(\omega) = \frac{U_0}{R} \cdot \frac{1}{\sqrt{1 + (\omega \cdot \tau)^2}} \text{ or rather } \varphi = \arctan(-\omega \cdot \tau)$$

The impedance of an ideal inductance is $j\omega L$ and is called reactance. The real component of the series connection R_s is called resistance or series resistance. The complex impedance is called impedance $\underline{Z} = R_s + j\omega L$. These quantities are used to characterize the quality or the quality factor of a real inductance:

$$Q = \frac{|\text{Im}\{\underline{Z}\}|}{\text{Re}\{\underline{Z}\}} = \frac{\omega L}{R_s}$$

With a simple series connection of R_s and L, the quality Q would increase with the frequency ω. This is not the case for various reasons. With higher frequency, R_s becomes larger and larger, e.g. because of the skin effect. Moreover, additional eddy current losses occur in the winding and core, which act like a resistance parallel to L. It is difficult to produce inductors with $Q > 100$. Figure 2.11 shows the course of the quality factor Q_L of an inductance of 100 µH over frequency compared to the quality of a polypropylene capacitor of 1 µF. It becomes clear that it is much easier to achieve high qualities with capacitors.

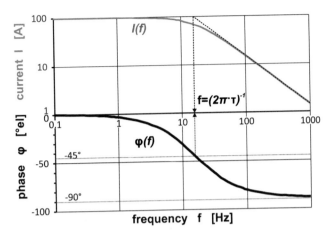

Fig. 2.10 Frequency response of effective current and phase for the example in Fig. 2.9

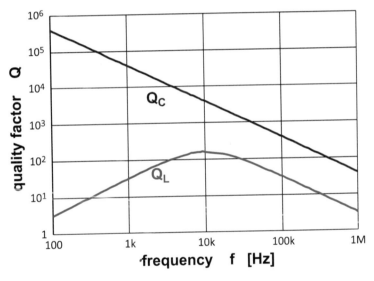

Fig. 2.11 Quality factor curve of a typical inductance $L = 250\ \mu H$ ($R_s = 50\ m\Omega$, $R_p = 5\ k\Omega$) compared to a typical polypropylene film capacitor $C = 1\ \mu F$ ($R_s = 4\ m\Omega$, $R_p = 10^{11}\ \Omega$)

For comparison: $Q = 0.5$ marks the critical damped case. Electrical oscillating circuits achieve typical values of $Q = 100$, while mechanical oscillators (oscillating crystals) achieve quality values of 10^5–10^6. In the case of oscillators, the oscillating circuits used for this purpose must therefore be de-attenuated.

With gyrators, which can form the dual impedances to a given impedance, low-loss inductors can be simulated in principle with low-loss capacitors. However, this is

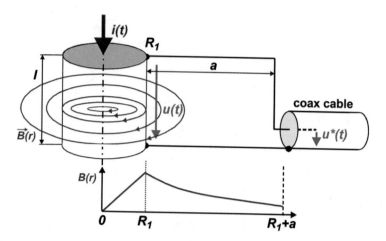

Fig. 2.12 Measurement of an AC voltage on a straight section of conductor

particularly important in communications engineering and at very high frequencies. The internal consumption of the gyrator as an electronic circuit far exceeds the avoided losses in the inductance. In power engineering, this possibility is therefore not important.

A very common task where the effect of self-induction can be observed is to measure the current by measuring a voltage drop. Figure 2.12 shows an example of such a situation. At a conductor section with radius R_1, the length l and the resistivity ρ, a voltage drop $u(t)$ is caused by the current $i(t)$. This voltage drop is composed of the ohmic voltage drop, the voltage drop at the internal inductance of the conductor and the induced voltage in the conductor loop with the area $A = a - l$, so that one can write

$$u^*(t) = u(t) + \frac{d}{dt}\left(\int_A \vec{B} \cdot \overrightarrow{dA}\right) = i \cdot \rho \cdot \frac{l}{\pi R_1^2} + L_i \frac{di}{dt} + \frac{d}{dt}\int_{R_1}^{R_1+a} \frac{\mu_0 \cdot i}{2\pi \cdot r} \cdot l \cdot dr$$

Thus, the measured voltage is composed of an ohmic and an inductive voltage drop. In the example shown, there is an increase of the effect of the internal inductance by the external magnetic field.

$$u^*(t) = i \cdot R + \left(\frac{\mu_0 \cdot l}{8\pi} + \frac{\mu_0 \cdot l}{2\pi} \ln\left(\frac{R_1 + a}{R_1}\right)\right) \cdot \frac{di}{dt} = i \cdot R + L_{eff} \cdot \frac{di}{dt}$$

Among other things because the internal inductance is frequency-dependent due to the skin effect, external frequency compensation of the measurement error can only be achieved within limits. For high current pulses, short shunts with large width and small thickness provide a path in this direction if the feedback effects on the measuring circuit are

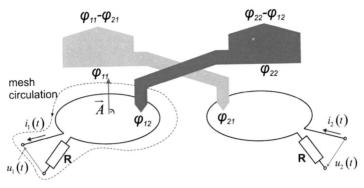

mesh equation 1, general: $u(t) = i(t) \cdot R + \dfrac{d}{dt} \int_A \vec{B} \cdot d\vec{A} = i(t) \cdot R + u_{ind}(t)$

mesh equation 1, special: $u_1(t) = \dfrac{d}{dt}(\varphi_{11} - \varphi_{12}) + i_1(t) \cdot R_1 = \dfrac{d\varphi_{11}}{dt} - \dfrac{d\varphi_{12}}{dt} + i_1(t) \cdot R_1$

Fig. 2.13 Coupling of two coils via their magnetic fluxes (the coupled flux is opposite to the main flux)

acceptable. The example in Fig. 2.12 leads thematically to the next section, which deals with the coupling between two circuits.

2.3   Mutual Induction

With mutual-induction, a conductor loop/coil is in the alternating magnetic field of another coil (Fig. 2.13). In the coil L_1, a voltage is then generated by self-induction, to which the voltage induced by the other magnetic field is added. This process is symmetrical, so that the statement also applies to the second coil L_2. Thus, following applies to the voltages of the conductor loops N_1 and N_2

$$u_1 = N_1 \cdot \frac{d}{dt}(\varphi_{11} + \varphi_{12}) = u_{11} + u_{12}$$

$$u_2 = N_2 \cdot \frac{d}{dt}(\varphi_{22} + \varphi_{21}) = u_{22} + u_{21}$$

The double indices should have the following meaning:

1. Digit: Place,
2. Digit: Origin.

This means, for example with φ_{12}: flux in N_1 caused by N_2.

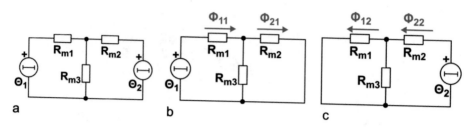

Fig. 2.14 Magnetic equivalent circuit for coupled coils. (**a**) Overall circuit and application of the superposition principle to the flux components caused by Θ_1 (**b**) and Θ_1 (**c**)

One can also express the relationships between the flux proportions as proportional factors/flux division factors.

This allows to write $k_{12} = \frac{\varphi_{12}}{\varphi_{22}}$ or $k_{21} = \frac{\varphi_{21}}{\varphi_{11}}$, so that a transformation of the equation system for positive coupling is possible.

$$u_1 = N_1 \cdot \frac{d}{dt}(\varphi_{11} + k_{12} \cdot \varphi_{22}) = \frac{d}{dt}(\psi_{11} + N_1 \cdot k_{12} \cdot \varphi_{22}) = \frac{d}{dt}\left(\psi_{11} + \frac{N_1}{N_2} \cdot k_{12} \cdot \psi_{22}\right)$$

$$u_2 = N_2 \cdot \frac{d}{dt}(\varphi_{22} + k_{21} \cdot \varphi_{11}) = \frac{d}{dt}(\psi_{22} + N_2 \cdot k_{21} \cdot \varphi_{11}) = \frac{d}{dt}\left(\psi_{22} + \frac{N_2}{N_1} \cdot k_{21} \cdot \psi_{11}\right)$$

Furthermore, taking into account self-induction

$$u_1 = \frac{d}{dt}\left(L_1 \cdot i_1 + \frac{N_1}{N_2} \cdot k_{12} \cdot L_2 \cdot i_2\right) = L_1 \cdot \frac{d}{dt}i_1 + \frac{N_1}{N_2} \cdot k_{12} \cdot L_2 \cdot \frac{d}{dt}i_2$$

$$u_2 = \frac{d}{dt}\left(L_2 \cdot i_2 + \frac{N_2}{N_1} \cdot k_{21} \cdot L_1 \cdot i_1\right) = L_2 \cdot \frac{d}{dt}i_2 + \frac{N_2}{N_1} \cdot k_{21} \cdot L_1 \cdot \frac{d}{dt}i_1$$

The flux division factors k_{12} and k_{21} correspond to the flux division factors when applying the superposition principle according to Fig. 2.14.

This figure also allows to easily determine the self-inductances L_1 and L_2. One thus obtains the relationships

$$k_{12} = \frac{R_{m3}}{R_{m1} + R_{m3}}, k_{21} = \frac{R_{m3}}{R_{m2} + R_{m3}}, L_1 = \frac{N_1^2}{R_{m1} + \frac{R_{m2} \cdot R_{m3}}{R_{m2} + R_{m3}}} \text{ and } L_2 = \frac{N_2^2}{R_{m2} + \frac{R_{m1} \cdot R_{m3}}{R_{m1} + R_{m3}}}.$$

If these are inserted into the respective second summands of the system of equations, the following results

$$\frac{N_1}{N_2}k_{12}\cdot L_2 = \frac{R_{m3}}{R_{m1}+R_{m3}}\frac{\dfrac{N_2 N_1}{R_{m1}\cdot R_{m3}}}{R_{m2}+\dfrac{R_{m1}\cdot R_{m3}}{R_{m1}+R_{m3}}}$$

$$= \frac{R_{m3}}{R_{m1}R_{m2}+R_{m2}R_{m3}+R_{m1}R_{m3}}N_1 N_2 = M_{12}$$

$$\frac{N_2}{N_1}k_{21}\cdot L_1 = \frac{R_{m3}}{R_{m2}+R_{m3}}\frac{\dfrac{N_1 N_2}{R_{m2}\cdot R_{m3}}}{R_{m1}+\dfrac{R_{m2}\cdot R_{m3}}{R_{m2}+R_{m3}}}$$

$$= \frac{R_{m3}}{R_{m1}R_{m2}+R_{m2}R_{m3}+R_{m1}R_{m3}}N_1 N_2 = M_{21}=M_{12}=M$$

This gives the well-known system of equations for voltages on two magnetically coupled coils to

$$u_1 = L_1\frac{di_1}{dt}\pm M\frac{di_2}{dt}$$

$$u_2 = L_2\frac{di_2}{dt}\pm M\frac{di_1}{dt}\ \text{with}\ M = k_{12}\cdot\frac{N_1}{N_2}\cdot L_2 = k_{21}\cdot\frac{N_2}{N_1}\cdot L_1 = \sqrt{k_{12}\cdot k_{21}\cdot L_1\cdot L_2}$$

In both equations, ± signs are inserted because the windings can be designed in the same direction or in opposite directions. All linear coupled magnetic arrangements such as transformers, and also transformer-coupled electrical interference from magnetic fields of foreign devices, function according to this principle. The mutual inductance has no equivalent in electric field arrangements. Even if one takes into account that there are piezoelectric transformers, electrical oscillations are first converted into mechanical oscillations there, and then converted back into electrical oscillations in a coupled mechanical oscillator. A problem of these oscillators is on the one hand the efficiency and on the other hand the fact that good insulators are usually bad heat conductors, so that the resulting power dissipation from the piezoelectric transformers is difficult to transport.

2.4 Transformers

In transformers, a bidirectional conversion of electrical energy with very low losses is possible. The maximum efficiency of energy transformation tends to increase as the components become larger. Transformers are treated in several following chapters under different aspects, so that here in this section only the basic connections and terms for linear transformers are presented. To simplify the presentation, a loss-free transformer is used. Starting point of the considerations should be Fig. 2.15.

Both windings N_1 and N_2 are located on a common core, so that there is a magnetic flux component (main flux) that passes through both windings. Since the surrounding air has permeability different from zero, there are also flux portions that only close via one

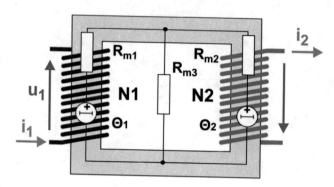

Fig. 2.15 Arrangement of 2 windings on a ferromagnetic core with leakage flux over the air path between the windings (counting arrows were chosen so that a simple chain connection of the four poles is possible)

winding, not coupled with others. These flux components are called leakage flux. The magnetic equivalent circuit diagram shown in Fig. 2.15 is essentially the same as that shown in Fig. 2.14a.

In complex notation, one writes as in Sect. 2.3.

$$u_1 = j\omega \cdot L_1 \cdot i_1 \pm j\omega \cdot M \cdot i_2$$

$$u_2 = j\omega \cdot L_2 \cdot i_2 \pm j\omega \cdot M \cdot i_1 \text{ with } M = k_{12} \cdot \frac{N_1}{N_2} \cdot L_2 = k_{21} \cdot \frac{N_2}{N_1} \cdot L_1 = \sqrt{k_{12} \cdot k_{21} \cdot L_1 \cdot L_2}$$

$$\begin{bmatrix} u_1 \\ u_2 \end{bmatrix} = \begin{bmatrix} j\omega L_1 & \pm j\omega M \\ \pm j\omega M & j\omega L_2 \end{bmatrix} \cdot \begin{bmatrix} i_1 \\ i_2 \end{bmatrix} = j\omega \cdot \begin{bmatrix} L_1 & \pm M \\ \pm M & \pm L_2 \end{bmatrix} \cdot \begin{bmatrix} i_1 \\ i_2 \end{bmatrix}$$

$$\begin{bmatrix} u_1 \\ u_2 \end{bmatrix} = j\omega \cdot \begin{bmatrix} \dfrac{N_1^2}{R_{m1} + R_{m2}//R_{m3}} & \pm \dfrac{R_{m3}}{R_{m1}R_{m2} + R_{m2}R_{m3} + R_{m1}R_{m3}} N_1 N_2 \\[4mm] \pm \dfrac{R_{m3}}{R_{m1}R_{m2} + R_{m2}R_{m3} + R_{m1}R_{m3}} N_1 N_2 \dfrac{N_2^2}{R_{m1tot}} & \dfrac{N_2^2}{R_{m2} + R_{m1}//R_{m3}} \end{bmatrix} \cdot \begin{bmatrix} i_1 \\ i_2 \end{bmatrix}$$

The transformation of the impedance form into the cascaded 2-port matrix form provides for positive sign of coupling

$$
\begin{bmatrix} u_1 \\ u_2 \end{bmatrix} = \begin{bmatrix} j\omega L_1 & j\omega M \\ j\omega M & j\omega L_2 \end{bmatrix} \cdot \begin{bmatrix} i_1 \\ i_2 \end{bmatrix} = \underline{Z} \cdot \underline{I} = j\omega \cdot \begin{bmatrix} L_1 & M \\ M & L_2 \end{bmatrix} \cdot \begin{bmatrix} i_1 \\ i_2 \end{bmatrix}
$$

$$
\begin{bmatrix} u_1 \\ i_1 \end{bmatrix} = \begin{bmatrix} A_{11} & A_{12} \\ A_{21} & A_{22} \end{bmatrix} \cdot \begin{bmatrix} u_2 \\ i_2 \end{bmatrix} = \underline{A} \cdot \begin{bmatrix} u_2 \\ i_2 \end{bmatrix}
$$

$$
\begin{bmatrix} u_1 \\ i_1 \end{bmatrix} = \frac{1}{j\omega M} \begin{bmatrix} j\omega L_1 & -\omega^2 (L_1 L_2 - M)^2 \\ 1 & j\omega L_2 \end{bmatrix} \cdot \begin{bmatrix} u_2 \\ i_2 \end{bmatrix} = \begin{bmatrix} \dfrac{L_1}{M} & \dfrac{j\omega (L_1 L_2 - M)^2}{M} \\ 1 & \dfrac{L_2}{M} \end{bmatrix} \cdot \begin{bmatrix} u_2 \\ i_2 \end{bmatrix}
$$

For the characterization in general, the quantities coupling factor 'k' (cf. Sect. 2.3) and the inductive leakage factor 'σ' are used with following definitions

$$
k = \frac{M}{\sqrt{L_1 \cdot L_2}}
$$

$$
\sigma = 1 - k^2 = 1 - \frac{M^2}{L_1 \cdot L_2}
$$

This makes it easier to write the system of equation to

$$
\begin{bmatrix} u_1 \\ i_1 \end{bmatrix} = \begin{bmatrix} \dfrac{L_1}{M} & \dfrac{j\omega M \cdot \sigma}{k^2} \\ \dfrac{1}{j\omega M} & \dfrac{L_2}{M} \end{bmatrix} \cdot \begin{bmatrix} u_2 \\ i_2 \end{bmatrix} = \begin{bmatrix} \dfrac{L_1}{M} & \dfrac{j\omega M \cdot (1 - k^2)}{k^2} \\ \dfrac{1}{j\omega M} & \dfrac{L_2}{M} \end{bmatrix} \cdot \begin{bmatrix} u_2 \\ i_2 \end{bmatrix}
$$

The formulas derived in Sect. 2.3 for the example Fig. 2.16 are also applicable here. The inductance values L_1, L_2 and M are in accordance with these definitions

$$
L_1 = \frac{N_1^2}{R_{m1} + \frac{R_{m2} \cdot R_{m3}}{R_{m2} + R_{m3}}}, L_2 = \frac{N_2^2}{R_{m2} + \frac{R_{m1} \cdot R_{m3}}{R_{m1} + R_{m3}}} \text{ and } M = \frac{R_{m3} \cdot N_1 \cdot N_2}{R_{m1} R_{m2} + R_{m2} R_{m3} + R_{m1} R_{m3}}. \text{ Some selected cases are}
$$

interesting here.

a) Secondary open circuit $i_2 = 0$

The ratios then applicable can be used to determine the transformation ratio for the voltages 'ü' and the input impedance 'Z_1'

Fig. 2.16 Magnetic equivalent circuit diagram of the winding arrangement according to Fig. 2.15

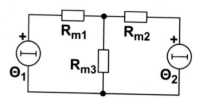

$$u_1|_{i_2=0} = \begin{aligned}\frac{L_1}{M} \cdot u_2|_{i_2=0} = \frac{L_1}{k\sqrt{L_1 \cdot L_2}} \cdot u_2|_{i_2=0} = \frac{1}{k}\sqrt{\frac{L_1}{L_2}} \cdot u_2|_{i_2=0} \rightarrow \gamma \\ = \frac{u_1}{u_2}\Big|_{i_2=0} = \frac{1}{k}\sqrt{\frac{L_1}{L_2}} \end{aligned}$$

$$i_1|_{i_2=0} = \frac{u_2}{j\omega M} = \frac{1}{j\omega M}\cdot\frac{M}{L_1}\cdot u_1|_{i_2=0} = \frac{1}{j\omega L_1}\cdot u_1|_{i_2=0} \rightarrow Z_1|_{i_2=0} = \frac{u_1}{i_1}\Big|_{i_2=0} = j\omega L_1$$

b) Secondary short circuit $u_2 = 0$

$$u_1\Big|_{u_2=0} = \frac{j\omega M\cdot\sigma}{k^2}\cdot i_2|_{u_2=0} \dot{i}_1\Big|_{u_2=0} = \frac{L_2}{M}\cdot i_2|_{u_2=0} \rightarrow \frac{i_2}{i_1}\Big|_{u_2=0} = \frac{M}{L_2} = k\sqrt{\frac{L_1}{L_2}} = \frac{1}{\gamma}Z_1|_{u_2=0} \qquad = \frac{u_1}{i_1}$$

By using these relationships, the system of equations of the chain equations for the transformer can be represented as follows

$$\begin{bmatrix} u_1 \\ i_1 \end{bmatrix} = \frac{1}{k}\cdot\begin{bmatrix} \gamma & j\omega\sigma\cdot\sqrt{L_1 L_2} \\ \frac{1}{j\omega\cdot\sqrt{L_1 L_2}} & \frac{1}{\gamma} \end{bmatrix}\cdot\begin{bmatrix} u_2 \\ i_2 \end{bmatrix}$$

If you translate this system of equations into an electrical circuit, you get Fig. 2.17. It is also one of the elementary electrical equivalent circuit diagrams (Π equivalent circuit diagram, corresponds to a delta circuit). The drawn inductances L_i are also the representatives of the inductances measured from the primary or secondary side. These can be expressed by the magnetic resistances R_{mv} or the permeances (magnetic conductance values) Λ_v.

$$L_{v1} = N_1^2 \cdot \Lambda_v = \frac{N_1^2}{R_{mv}} \text{ or } L_{v1} = N_1^2 \cdot \Lambda_v = \frac{N_1^2}{R_{mv}}.$$

Using the model example 2.16, one then obtains the following relationships for the parameters used (related to the primary side)

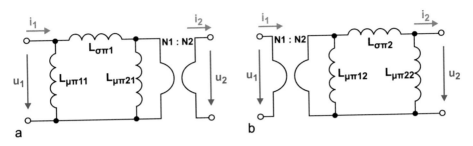

Fig. 2.17 Electrical circuit for mapping the electrical behaviour of a loss-free transformer with a basic magnetic structure according to Fig. 2.16 with an ideal transformer inserted. (**a**) Inductance values related to the primary side, (**b**) related to secondary side

$$L_{\mu\pi11} = \frac{N_1^2}{R_{m1}}, L_{\mu\pi21} = \frac{N_1^2}{R_{m2}}, L_{\sigma\pi1} = \frac{N_1^2}{R_{m3}}, k = \frac{R_{m3}}{\sqrt{(R_{m1}+R_{m3})(R_{m2}+R_{m3})}}$$

This allows the quantities for the two-port equations to be determined with reference to Fig. 2.17 as

$$L_{11} = L_{\mu\pi11}//\left(L_{\sigma\pi1}+L_{\mu\pi21}\right)$$
$$L_{21} = L_{\mu\pi21}//\left(L_{\sigma\pi1}+L_{\mu\pi11}\right)$$
$$k = \frac{R_{m3}}{\sqrt{(R_{m1}+R_{m3})(R_{m2}+R_{m3})}} = \frac{1}{\sqrt{\left(\frac{R_{m1}}{R_{m3}}+1\right)\left(\frac{R_{m2}}{R_{m3}}+1\right)}}$$

$$\sigma = 1 - k^2 = 1 - \frac{1}{\left(\frac{R_{m1}}{R_{m3}}+1\right)\left(\frac{R_{m2}}{R_{m3}}+1\right)}$$

$$\gamma = \frac{1}{k}\sqrt{\frac{L_{11}}{L_{11}}} = \frac{1}{k}\frac{L_{\mu\pi11}\left(L_{\sigma\pi1}+L_{\mu\pi21}\right)}{L_{\mu\pi21}\left(L_{\sigma\pi1}+L_{\mu\pi11}\right)}$$

With a Δ–Y conversion, a T equivalent circuit diagram can be generated from this Π equivalent circuit diagram, which behaves electrically identical. However, whereas the Π equivalent circuit diagram in the example in Fig. 2.18a shows a physical equivalent as a component, this no longer applies after conversion to a T equivalent circuit diagram in Fig. 2.18b. A different magnetic circuit fits the T equivalent circuit. This problem is discussed in detail in Chap. 5.

The substitute parameters can be transformed into each other by a simple star-delta transformation, whose transformation equations are listed in Fig. 2.18.

The Π equivalent circuit diagram is particularly suitable for 'loosely' coupled windings. This refers to windings that are arranged on two different sections of the magnetic circuit as

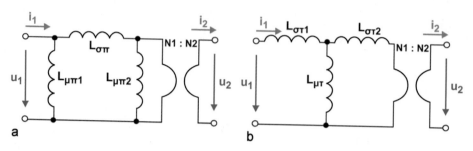

Fig. 2.18 Equivalent electrical equivalent circuit diagrams for a loss-free transformer. (**a**) Π equivalent circuit diagram, (**b**) T equivalent circuit diagram

shown in Fig. 2.15. For closely coupled windings, these are arranged one above the other on a core branch. The leakage flux then arises because the flux of all windings of a winding is not completely interlinked. Although this effect also occurs with loose coupling, it is lower in the order of magnitude of the leakage flux by a factor of 10...20. In the T equivalent circuit diagram of closely coupled windings, it is then also possible to make a formal assignment of the magnetic energy stored in the stray field to the primary or secondary winding. Chapter 5 is devoted to these considerations in detail. With more strongly branched magnetic fields, the two common forms of equivalent circuit diagrams have no physical significance.

For transformers with ferromagnetic core ($\mu_r >> 1$), the coupling factor is only slightly different from 1. In addition, the relation $L_{\mu\pi1} \approx L_{\mu\pi2} >> L_{\sigma\pi}$ or $L_{\mu\pi} >> L_{\sigma\tau1} \approx L_{\sigma\tau2}$. While the leakage inductances $L_{\sigma x}$ can be regarded as constant, this is not the case for $L_{\mu x}$. All core-related inductances are strongly dependent on the flux density.

The above example refers to a loss-free transformer. If the coupling factor k is selected here identically to 1, the scatter-free transformer is obtained. For the coupling factor $k = 1$ (or scattering factor $\sigma = 0$), the chain form of the two port equations is for positive coupling

$$\begin{bmatrix} u_1 \\ i_1 \end{bmatrix} = \frac{1}{k} \cdot \begin{bmatrix} \gamma & j\omega\sigma \cdot \sqrt{L_1 L_2} \\ \frac{1}{j\omega \cdot \sqrt{L_1 L_2}} & \frac{1}{\gamma} \end{bmatrix} \cdot \begin{bmatrix} u_2 \\ i_2 \end{bmatrix} = \begin{bmatrix} \gamma & 0 \\ \frac{1}{j\omega \cdot \sqrt{L_1 L_2}} & \frac{1}{\gamma} \end{bmatrix} \cdot \begin{bmatrix} u_2 \\ i_2 \end{bmatrix}$$

The electrical equivalent circuit diagram for this case is shown in Fig. 2.19a.

If we now assume an ideal magnetic conductor for the magnetic core ($\mu_r \rightarrow \infty$, i.e. L_1, $L_2 \rightarrow \infty$), the chain equation is further simplified to

$$\begin{bmatrix} u_1 \\ i_1 \end{bmatrix} = \frac{1}{k} \cdot \begin{bmatrix} \gamma & j\omega\sigma \cdot \sqrt{L_1 L_2} \\ \frac{1}{j\omega \cdot \sqrt{L_1 L_2}} & \frac{1}{\gamma} \end{bmatrix} \cdot \begin{bmatrix} u_2 \\ i_2 \end{bmatrix} = \begin{bmatrix} \gamma & 0 \\ 0 & \frac{1}{\gamma} \end{bmatrix} \cdot \begin{bmatrix} u_2 \\ i_2 \end{bmatrix}$$

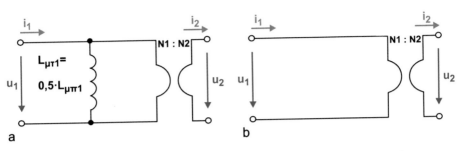

Fig. 2.19 Idealized leakage free transformer. (**a**) With the transformation ratio $\gamma = \sqrt{\frac{L_1}{L_2}} = \frac{N_1}{N_2}$ and (**b**) ideal transformer with $L_1 \to \infty$ or $L_2 \to \infty$

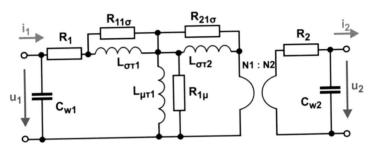

Fig. 2.20 Broadband equivalent circuit diagram for a transformer

The corresponding equivalent electrical circuit is shown in Fig. 2.19b. In order to identify this element from the theory of electrical networks in the equivalent circuits, the symbol shown here is derived from an inductor and reminds one of a gyrator. The different stages of idealization are used to simplify problems.

At the ideal transformer, one can easily summarize the (idealized) basic properties of this component

- It transforms voltages with the transmission ratio $\gamma = \frac{u_1}{u_2} = \sqrt{\frac{L_1}{L_2}} = \frac{N_1}{N_2}$.
- Currents with $\frac{i_1}{i_2} = \frac{1}{\gamma}$ are also transformed.
- It is itself (ideally) loss-free and does not store energy.
- A secondary impedance $\underline{Z}_2$ is measured on the primary side as $\underline{Z}_1 = \gamma^2 \cdot \underline{Z}_2$
- Primary and secondary sides are electrically insulated from each other.

One or more of the idealized basic properties are the reason for the many different applications of inductive components. In reality, the transformer is a very complex structure in transmission technology. If one considers large ranges of voltage u, current i and frequency f, an electrical equivalent circuit diagram as shown in Fig. 2.20 must actually be used (for the T equivalent circuit).

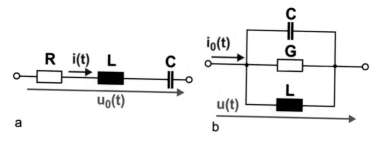

Fig. 2.21 Basic circuits of resonant circuits. (**a**) Design of a 'pure' series resonant circuit and **b** of a 'pure' parallel resonant circuit (**b**)

The analytical effort starts with the parameterization of the model and continues with all calculations. Chapters 3–9 are intended to provide methodological assistance in dealing with such problems. Figure 2.20 shows an approach to calculate not only the magnetic properties but also the ohmic winding losses (R_1, R_2), the eddy current losses in the windings ($R_{11\sigma}$, $R_{21\sigma}$) and the core losses, which can also be considered frequency-dependent for larger frequency ranges. The turns of a winding also form one or more capacitances, the effect of which is also approximately taken into account in Fig. 2.20. This results in a very complex behaviour of a real transformer.

2.5   Oscillating Circuits with Concentrated Elements

If inductive storage is combined with capacitive storage, oscillating circuits are obtained. Their stationary and transient behaviour is shown below. In principle, a distinction can be made between series resonance and parallel resonance (Fig. 2.21). These circuit are dual arrangements, which can be treated accordingly. The mesh equation of the series resonance inverter is similar to the structure of the node equation of the parallel resonant circuit.
    Written as a differential equation, one obtains for the two cases

Series Resonant Circuit

$$u_0 = R \cdot i + L \cdot \frac{di}{dt} + \frac{1}{C} \int i \cdot dt$$

Parallel Resonant Circuit

$$i_0 = G \cdot i + C \cdot \frac{du}{dt} + \frac{1}{L} \int u \cdot dt$$

Both equations lead to a linear integro-differential equation, which can be transformed by differentiation into an ordinary differential equation and solved with the exponential approach. For the series resonant circuit, the solution is derived below. The solution for the parallel resonant circuit is analogous. By differentiation, one obtains the usual differential equation

$$\frac{du_0}{dt} = R \cdot \frac{di}{dt} + L \cdot \frac{d^2i}{dt^2} + \frac{1}{C}i = \frac{1}{C}i + R \cdot \frac{di}{dt} + L \cdot \frac{d^2i}{dt^2}$$

The general solution $i_a(t)$ of this equation consists of the linear combination of homogeneous solution $i_h(t)$ and particulate solution $i_p(t)$.

$$i_a(t) = i_h(t) + i_p(t)$$

The homogeneous solution is determined for the case that the "forcing function" disappears with the exponential approach

$$0 = \frac{1}{C}i_h + R \cdot \frac{di_h}{dt} + L \cdot \frac{d^2i_h}{dt^2} \rightarrow i_h(t) = A \cdot e^{\lambda_1 \cdot t} + B \cdot e^{\lambda_2 \cdot t}$$

From the coefficients of the differential equation, the characteristic equation is derived as a quadratic equation

$$\frac{1}{C} + R \cdot \lambda_x^1 + L \cdot \lambda_x^2 = 0 \text{ with the normal form } \frac{1}{L \cdot C} + \frac{R}{L} \cdot \lambda_x + \lambda_x^2 = 0.$$

This provides the two roots λ_1 and λ_2.

$$\lambda_{1,2} = -\frac{R}{2L} \pm \sqrt{\frac{R^2}{4L^2} - \frac{1}{LC}}$$

The solutions of the characteristic equation contain terms, which are more widely used in practical applications and whose individual naming is worthwhile.

Damping Constant δ
This provides the enveloping function of the decaying oscillation

$$\delta = \frac{R}{2L} = \frac{1}{\tau_D}$$

Resonant Frequency ω_0

With this frequency, a special state is reached in the oscillating circuit. At maximum, the electrical and magnetic storages achieve equal values of the stored energy.

$$\omega_0 = \sqrt{\frac{1}{LC}}$$

Natural Oscillation Frequency ω_e

By this frequency, current and voltage of the system free oscillate.

$$\omega_e = \sqrt{\frac{1}{LC} - \frac{R^2}{4L^2}} = \sqrt{\omega_0^2 - \delta^2}$$

One can write for the homogeneous (volatile) solution with undefined coefficients and obtain

$$i_h(t) = A \cdot e^{\left(-\frac{R}{2L} + \sqrt{\frac{R^2}{4L^2} - \frac{1}{LC}}\right) \cdot t} + B \cdot e^{\left(-\frac{R}{2L} - \sqrt{\frac{R^2}{4L^2} - \frac{1}{LC}}\right) \cdot t} = e^{-\frac{R}{2L} t}$$

$$\cdot \left(A \cdot e^{\left(\sqrt{\frac{R^2}{4L^2} - \frac{1}{LC}}\right) \cdot t} + B \cdot e^{-\left(\sqrt{\frac{R^2}{4L^2} - \frac{1}{LC}}\right) \cdot t} \right)$$

The free constants A and B are determined from the initial conditions. If, for example, $U_c(0) = U_0$ and $i_0(0)$, a linear system of equations is obtained to determine A and B to

$$i_h(0) = 0 = A + B$$

$$u_c = U_0 = -L \frac{di_h(t)}{dt}\bigg|_{t=0} = \left(+\frac{R}{2} - \sqrt{\frac{R^2}{4L} - \frac{L}{C}} \right) \cdot A + \left(\frac{R}{2} + \sqrt{\frac{R^2}{4L} - \frac{L}{C}} \right) \cdot B$$

Releasing, inserting and combining require some work. The result has the form of a damped oscillation. Current $i_h(t)$ and voltage $u_{Ch}(t)$ characterize the state of the accumulators in the system. The homogeneous solution for the equations of state with the simplifying notations is

$$i_h(t) = -\frac{U_0}{\omega_e \cdot L} \cdot e^{-\delta t} \cdot \sin(\omega_e t)$$

$$u_{Ch}(t) = U_0 \cdot e^{-\delta t} \cdot \left(\cos\left(\omega_e t\right) + \frac{\delta}{\omega_e} \sin\left(\omega_e t\right) \right) = U_0 \cdot e^{-\delta t} \cdot \cos\left(\omega_e t - \arctan\left(\frac{\delta}{\omega_e}\right) \right)$$

Depending on the solution of the characteristic equation, 3 different cases must be distinguished

1. Underdamped, oscillation ($\omega_0^2 - \delta^2 > 0$)
   The state variables oscillate with a natural oscillation frequency $\omega_e \leq \omega_0$. This corresponds to the case discussed above.
2. Critical damped case ($\omega_0^2 - \delta^2 = 0$)
   Here the natural oscillation frequency $\omega_e = 0$. The final state is reached here the fastest. From the characteristic equation, the condition for this behaviour also results. For the resistance R then applies

$$R = 2 \cdot \sqrt{\frac{L}{C}} = 2 \cdot Z_W$$

Here Z_W is the characteristic impedance of the oscillating circuit.
One obtains a double solution for the root λ from the characteristic equation, so that one has to construct another solution. This solution is then

$$i_h(t) = -\frac{U_0}{L} \cdot t \cdot e^{-\delta t}$$
$$u_{Ch}(t) = U_0 \cdot (1 + \delta t)e^{-\delta t}$$

3. Overdamped case ($\omega_0^2 - \delta^2 < 0$)
   This also means that there is no more oscillation, i.e. there is no change in the sign of the state variables. However, the approach to the asymptotic limit value is slower than in the critical damped case. The solution for the current here is

$$i_h(t) = -\frac{U_0}{\sqrt{\delta^2 - \omega_0^2} \cdot L} \cdot e^{-\delta t} \cdot \sinh\left(\sqrt{\delta^2 - \omega_0^2} \cdot t \right).$$

The 3 cases are shown in comparison in Fig. 2.22.
For the representation of processes in oscillating circuits with switched excitation, the so-called phase plane is also frequently used. In this case, current or voltage and the corresponding first derivative are plotted on a diagram. In a series resonant circuit, this corresponds to the voltage of the capacitor and its current. This representation used to be quite common in power electronics, where resonant circuits were used to a much greater extent because of the available switches. In resonant circuit inverters, the commutation reactive power was provided by thyristors from the resonant circuits. The phase plane is a

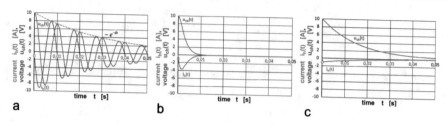

Fig. 2.22 Curves of current and capacitor voltage of a series resonant circuit (**a**) in the underdamped case $R = 0.1 \cdot Z_W$, (**b**) in the critical damped case $R = 2 \cdot Z_W$ and (**c**) in the overdamped case $R = 10 \cdot Z_W$ ($U_{C0} = 10$ V, $Z_W = 1$ Ohm, $f_0 = 100$ Hz)

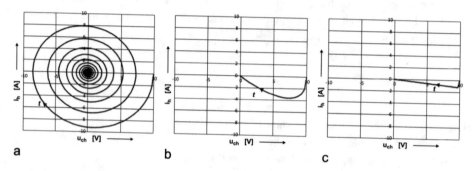

Fig. 2.23 Current and voltage at the capacitor for (**a**) oscillation case, (**b**) critical damped case and (**c**) creep case for the same conditions as in Fig. 2.22 as representation in the phase plane

useful means of tracking changes in the state of the resonant circuit. The same processes as in Fig. 2.22 are therefore shown again in Fig. 2.23 in the phase plane.

Among the non-homogenious solutions, the one with sinusoidal (harmonic) excitation is of particular importance. The non-homogenious solution for this is simply found in complex notation. If the excitation sinusoidal voltage with a certain frequency is given, the non-homogenious solution for the current is found in complex notation

$$\underline{I} = \frac{\underline{U}_0}{R + j\omega L + \frac{1}{j\omega C}}$$

The conditions at the oscillating circuit are generally described by its impedance, i.e. by its complex resistance $\underline{Z}$. By definition is

$$\underline{Z} = R + j\omega L + \frac{1}{j\omega C} \text{ with } \omega = 2\pi f.$$

The impedance is displayed either as a frequency response of magnitude and phase angle Fig. 2.24 or as a locus curve $\underline{Z}(\omega)$ (Fig. 2.25a). The locus curve $\underline{Z}(\omega)$ is a straight line parallel to the imaginary axis. The locus curve of the complex coupling coefficient $\underline{Y}(\omega)$

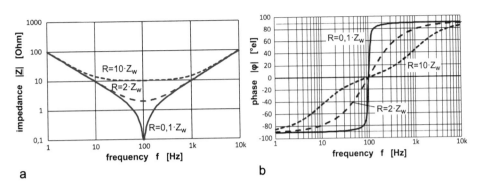

Fig. 2.24 Impedance $\underline{Z}(\omega)$ of a series resonant circuit with different damping. (**a**) Amplitude diagram, (**b**) phase diagram

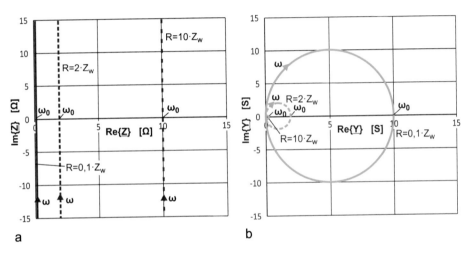

Fig. 2.25 Frequency locus curves of a series resonant circuit. (**a**) Impedance $\underline{Z}(\omega)$, (**b**) Admittance $\underline{Y}(\omega)$

therefore becomes a circle through the coordinate origin (Fig. 2.25b). All illustrations were created for a series resonant circuit. Parallel resonant circuit and series resonant circuit are dual circuits. The appearance of impedance and admittance locus curves is therefore reversed.

As already mentioned, the natural oscillation frequency changes with damping, in contrast to the resonance frequency. As can be seen from the instructions given above, the following applies

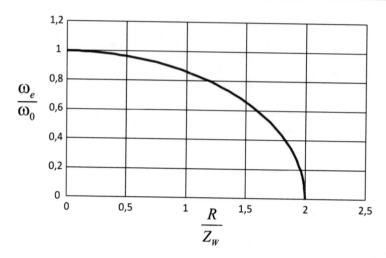

Fig. 2.26 Relative change of the natural oscillation frequency compared to the resonance frequency
with variable damping

$$\omega_e = \sqrt{\frac{1}{LC} - \frac{R^2}{4L^2}} = \frac{1}{\sqrt{LC}} \sqrt{1 - \frac{R^2 C}{4L}}$$

$$\frac{\omega_e}{\omega_0} = \frac{f_e}{f_0} = \sqrt{1 - \frac{R^2 C}{4L}} = \sqrt{1 - \left(\frac{R}{2 \cdot Z_W}\right)^2}$$

Figure 2.26 shows this relationship in a standardized form. For R < <Zw, the dependence
is small, but becomes stronger with higher damping. The forms of representation used here
are all used in the book and are explained in the simple basic references in the first chapters
of this book.

2.6   Energy Propagation Between Lines and in the Space

Poynting Theorem

In common parlance, it is assumed that the transport of electrical energy takes place in the
electrical lines. However, if this were to be the case, e.g. in analogy to water (for energy: in
oil pipelines), the transport of electrons, which actually takes place, would be assigned an
energy transport in the case of direct current. Mostly copper is used as conductor material.
When single conductors are laid, current densities of approximately 10 A/mm^2 are
achieved. From this, the electrical field strength $\vec{E}$ can be calculated with the conductivity
of copper of $\kappa = 56$ MS/m

$$\vec{E} = \frac{1}{\kappa}\vec{J} = \frac{V \cdot m}{56 \cdot 10^6 A} \cdot 10 \frac{A}{mm^2} = 0.1786 \frac{V}{m}$$

The speed of charge carriers in solids results from electric field strength $\vec{E}$ and mobility b. The charge carriers in metals are electrons with a mobility of about 50 $cm^2 V^{-1} s^{-1}$, from which the speed $\vec{v}$ of the electrons can be estimated.

$$\vec{v} = b \cdot \vec{E} = 50 \frac{cm^2}{V \cdot s} \cdot 0.1786 \frac{V}{m} = 0.893 \frac{mm}{s} = 3.148 \frac{m}{h}$$

The transport movement of electrons in solid materials is therefore very slow. For 100 km, electrons need 1323 days under these conditions. This would not be a really convincing argument for electric energy transportation, if one thinks of the comparison with crude oil as energy carrier. The crude oil is constantly being pumped up as long as there is still some available. But how does the energy transport in alternating current and in free space/ vacuum explain itself? With alternating current on the grid, with this conception no transport takes place, because the electrons oscillate at about 9 μm. In a vacuum capacitor, there are no charge carriers and still energy transfer takes place. The Poynting theorem provides a uniform explanation.

The Poynting theorem (after John Henry Poynting (* 9 September 1852 in Monton; † 30 March 1914 in Birmingham)) describes the energy balance in electrodynamics and represents a special form of the law of conservation of energy. The basic statement is that an electromagnetic field can do work while it becomes "weaker". The energy transport takes place directed in a vector field with a power density $\vec{S}(t)$. This vector is also called the Poynting vector

$$\vec{S} = \vec{E} \times \vec{H}$$

It points in the direction of the energy transport. It is propagating at the speed of light. The different situations for energy transport in the electric and magnetic circuit are shown in Fig. 2.27.

The conclusion from these considerations is that the electrical energy is transported in the electromagnetic field between the electrical or magnetic conductors. The electrons are charge carriers, but not energy carriers. Energy carrier (in the sense of transporter) is the electromagnetic field that propagates in free space at the speed of light. In the case of conducted energy transport, the conductors serve to ensure low losses in the generation of the electromagnetic fields with correspondingly high power densities due to high conductivities. The actual energy transport takes place between the conductors.

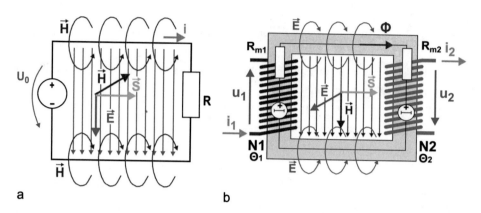

a   b

Fig. 2.27 Energy transport in the electromagnetic field. (**a**) In a basic electric circuit and (**b**) in a magnetic circuit (transformer)

Homogeneous Isotropic 2-Wire Cable

In the simplest case, wires and cables are used to transmit power and messages. For an idealized case, the properties of such cables are derived below. In borderline cases, cables and wires should work without loss. The energy (also as signal carrier) is transmitted in the electromagnetic field. Because both the storage properties of the electric field and the magnetic field are used, both capacitive and inductive properties can be expected in cables as components of electrical circuits. Fig. 2.28 shows a section with its representatives for the individual field components and the equivalent electrical circuit.

The analytical presentation can be greatly simplified by making the following definitions, which are close to most conditions.

- The transmission medium has a uniform (isotropic, homogeneous) structure.
- The cable coatings L', R', C' and G' are therefore constant over the entire cable.
- The conduction is considered in the steady state (homogeneous solution of the differential equation) and is excited with a sinusoidal quantity (a reference to the volatile, homogeneous solution of the differential equation follows this illustration).
- The displacement current density contributes practically nothing to the current in the direction of the line.

These are

$$R' = \frac{dR}{dx}, L' = \frac{dL}{dx}, C' = \frac{dC}{dx}, G' = \frac{dG}{dx}$$

The use of mesh and node equations provides the following approach for the element shown in Fig. 2.28

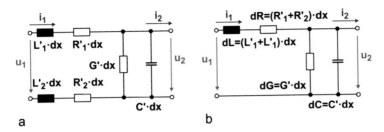

Fig. 2.28 Structure of a cable element with resistance and inductance per unit legth as well as capacitance coating and conductance per unit length for each cable element 'dx'. (**a**) Conductive layers are shown separately, (**b**) combined for the calculation

$$u(x_1) = R' \cdot i(x_1) + L' \cdot \frac{d}{dt} i(x_1) + u(x_1 + dx)$$

$$= R' \times i(x_1) + L' \times \frac{d}{dt} i(x_1) + u(x_1) + \frac{\partial u(x_1)}{\partial x} dx$$

$$i(x_1) + \frac{\partial i(x_1)}{\partial x} dx = i(x_1) + (G' + j\omega C) \cdot \left(u(x_1) + \frac{\partial u(x_1)}{\partial x} dx \right)$$

This approach leads by differentiation according to x as well as mutual insertion on the line equations

$$\frac{\partial u}{\partial x} + R' \cdot i + L' \cdot \frac{\partial}{\partial t} i = 0$$

$$\frac{\partial i}{\partial x} + G' \cdot u + C' \cdot \frac{\partial}{\partial t} u = 0$$

The conversion and separation of the two variables to u or i provides, for example, the telegraph equations for the time-dependent voltage distribution

$$\frac{\partial^2}{\partial t^2} u + \left(\frac{R'}{L'} + \frac{G'}{C'} \right) \frac{\partial}{\partial t} u + \frac{R'G'}{L'C'} u = \frac{1}{L'C'} \cdot \frac{\partial^2}{\partial x^2} u$$

$$\frac{\partial^2}{\partial t^2} i + \left(\frac{R'}{L'} + \frac{G'}{C'} \right) \frac{\partial}{\partial t} i + \frac{R'G'}{L'C'} i = \frac{1}{L'C'} \cdot \frac{\partial^2}{\partial x^2} i$$

With the cable parameters per unit length, characteristic parameters can be summarized appropriately.

Propagation coefficient
$$\gamma = \sqrt{(R' + j\omega L')(G' + j\omega C')} = \alpha + j\beta$$

Characteristic impedance $\underline{Z}_w = \sqrt{\frac{R' + j\omega L'}{G' + j\omega C'}}$

The sizes α and β are interpreted as

Damping coefficient

$$\alpha = \sqrt{\tfrac{1}{2}\left[R'G' - \omega^2 L'C' + \sqrt{(R'^2 + \omega^2 L'^2)(G'^2 + \omega^2 C')}\right]}$$

and

Phase coefficient

$$\beta = \sqrt{\tfrac{1}{2}\left[\omega^2 L'C' - R'G' + \sqrt{(R'^2 + \omega^2 L'^2)(G'^2 + \omega^2 C')}\right]}$$

If a signal with an angular frequency ω is coupled into a line, this signal moves in the direction of the line at a propagation speed or

phase velocity

$$v = \tfrac{\omega}{\beta}.$$

In a lossless line, $R' = 0$ and $G' = 0$, so that for a lossless homogeneous line, the propagation velocity can be specified as

$$v = \frac{\omega}{\beta} = \frac{\omega}{\sqrt{\tfrac{1}{2}\left[\omega^2 L'C' - R'G' + \sqrt{(R'^2 + \omega^2 L'^2)(G'^2 + \omega^2 C')}\right]}} = \frac{1}{\sqrt{L'C'}}$$

The corresponding wavelength λ added to the frequency f is then

$$v = \lambda \cdot f = \frac{\omega}{\beta} \rightarrow \lambda = \frac{2\pi}{\beta}$$

The propagation velocity is a constant variable under these conditions. A coaxial cable RG58 ($Z_w = 50$ Ohm) has a capacitance of approximately $C' = 100$ pF/m. This results in an inductance coating of $L' = 250$ nH/m, resulting in a signal propagation speed of $v = 2 \cdot 10^8$ m/s. That is 2/3 of the speed of light. A signal needs 5 ns to cover 1-m cable distance.

In the energy supply, 50 Hz/60 Hz are standard for energy transmission. For a 380-kV overhead line, the coverings $L' = 800$ nH/m and $C' = 14$ pF/m are to be expected. This results in a characteristic impedance $Z_w = 239$ Ohm, and with a propagation speed of $v = 298.8 \cdot 10^6$ m/s, it almost reaches the speed of light. This results in wavelengths of $\lambda = 5976$ (4980) km for 50 (60) Hz. Cable lengths of $l < \lambda/4$ are called 'electrically short'. With these, the wave-like propagation need not be taken into account. With the above data, the vast majority of transmission lines can therefore be regarded as electrically short.

A signal that almost always consists of a mixture of different frequencies can be transported without distortion. If the end of the line is terminated with a resistance equal to the real characteristic impedance Z_w, it can be received there undistorted. If we now consider the characteristic impedance $\underline{Z}_w$ as a complex quantity, then

$$\underline{Z}_w = \sqrt{\frac{R' + j\omega L'}{G' + j\omega C'}} = \sqrt{\frac{L'}{C'}} \cdot \sqrt{\frac{\frac{R'}{j\omega L'} + 1}{\frac{G'}{j\omega C'} + 1}}$$

The condition that $\underline{Z}_w$ becomes real is obviously also fulfilled if the following applies

$$\frac{R'}{L'} = \frac{G'}{C'}$$

This is the general condition for a distortion-free line. If this condition is not fulfilled, individual frequency components have different propagation speeds and the signal changes its shape as it passes through the line. Each signal can be imagined as a Fourier series of waves in the superposition of individual waves of different frequencies. If the above condition is violated, the individual waves each propagate at a certain (different) phase velocity. The resulting group velocity changes according to the definition

$$v_g = v_p - \lambda \cdot \frac{\partial v_p}{\partial \lambda}$$

and leads to a deformation of the signal, which propagates on the line as a wave. At $\frac{\partial v_p}{\partial \lambda} = 0$, the group velocity is identical to the phase velocity and the shape of the envelope is preserved. If this is not the case, dispersion is present (frequency-dependent propagation speeds). In this case, the envelope of the wave packet pulls apart as it passes through the line. Test leads are prepared accordingly. Broadband probes for oscilloscopes are therefore extremely high-tech products. Another important requirement to be met during signal transmission is the termination of the line with the appropriate characteristic impedance to avoid reflections. When reflections occur, only part of the energy is absorbed (i.e. passed on or dissipated) at the downstream unit. Another part is reflected at the connection point and moves back in the line.

Assume that a voltage source $U_0(t)$, which performs a jump $0 \rightarrow U_{max}$ for $t = 0$, is connected via a resistor R_1 to the input of a line with the characteristic impedance Z_w and $R_1 = Z_w$. A voltage with a maximum magnitude of $0.5 \cdot U_{0max}$ is then generated at the line input according to the voltage divider rule. A front of the voltage pulse then runs through the line to the output of the line. The line acts there momentarily like a source with the internal resistance Z_w, so that there a jump of the following type

$$u_2(\Delta t) := \left(0 \rightarrow U_{0\,max} \cdot \left(0,\ 5 + r_2 \frac{R_2}{Z_w + R_2} \right) \right)$$

can be observed. r_2 is the reflection coefficient at the line output with

$$r_2 = \frac{R_2 - Z_w}{R_2 + Z_w}$$

The reflection coefficient r_2 can be greater, less or equal to zero. In the case of adjustment $R_2 = Z_w$, $r_2 = 0$ and no reflection takes place. For a returning wave, the reflection coefficient r_1 acts at the input accordingly. If we now consider the stationary state for $t \to \infty$ for a loss-free line terminated with a resistor R_2, we obtain the relationship

$$u_1(t \to \infty) = u_2(t \to \infty) = U_{0\,max} \cdot \frac{R_2}{R_1 + R_2}$$

A transition process must therefore take place until the stationary state is reached. This consists of the sequence of individual partial waves, which in turn act on the signal at the output and input of the line. Methodically, this process can be easily recorded with a pulse timetable (lattice diagram [5]) even without a simulation. Figure 2.29 shows the conditions for a signal pulse on a line which is excited with a source resistance $R_1 = Z_w$, terminated with a characteristic impedance Z_w and the cases $R_2 > Z_w$ and $R_2 < Z_w$. At the input, the voltage $U_0/2$ appears first due to the voltage division Z_w to R_1. This voltage pulse runs through the line and reaches the output after a time $\tau = \frac{l}{v}$. If $R_2 = Z_w$, the pulse is not reflected and the system remains in this state. If $R_2 > Z_w$, a wave is reflected at R_2 with r_2 and with the *same sign* ($\Delta u_2 = 0.25 \cdot U_0$), so that the value $u_2(t \geq \tau) = 0.75 \cdot U_0$ is obtained by superposition. The returning wave hits an input terminated with $R_1 = Z_w$ and adds to the input signal. This means that the input and output signal are stationary equal. For the case $R_2 < Z_w$, a returning wave with opposite sign is generated by reflection. Here, too, this wave meets an adapted input, is superimposed on the input signal and thus establishes the stationary equality between input signal and output signal.

In the example Fig. 2.29, the compensation process is finished after a reflection, because the returning wave is not reflected at the input. Therefore, the adjustment of the generator is as important for a clean signal transmission as that of the load side. If a generator with $R_1 = Z_w/3$ is connected, a reflection factor of $r_1 = -0.5$ is obtained at the line input (Fig. 2.30).

The returning wave at $R_2 = 3 \cdot Z_w$ is then reflected with the factor $r_2 = 0.5$ and returns to the input. The reflection on both sides results in a transient response similar to that of an oscillating circuit.

The curves for input and output voltage of lines can also be constructed without simulation according to a "pulse timetable". Such a schedule, which includes both place and time, is shown in Fig. 2.31. It can easily be converted into a calculation table. The following definitions are used

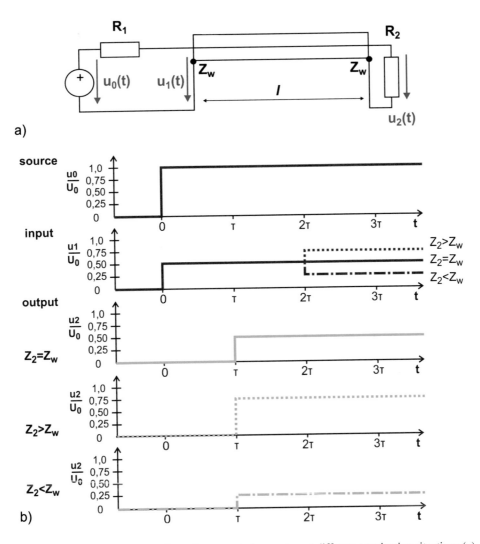

Fig. 2.29 Pulses at different points of a transmission system at different termination situations (**a**) model structure, (**b**) pulse diagrams with drawn superpositions of the reflected components for $R_2 = Z_w$, $R_2 = 3 \cdot Z_w$ and $R_2 = Z_w/3$

Source voltage at $t = 0 - \Delta$	$U_{0 - \Delta}$	Input voltage	u_1
Source voltage at $t = 0 + \Delta$	$U_{0 + \Delta}$	Output voltage	u_2
Internal resistance of source	R_1	Cable length	l
Characteristic impedance of the line	Z_w	Phase velocity	v
Load Resistance	R_2	Line runtime	$\tau = l/v$

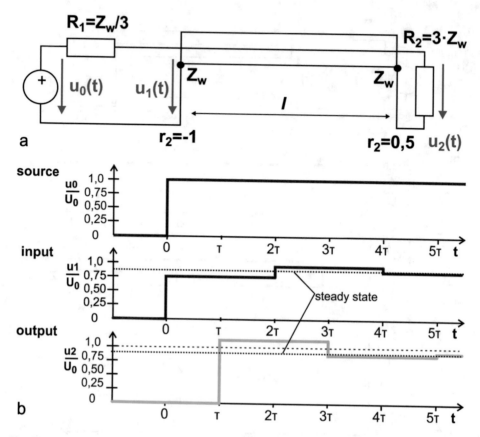

Fig. 2.30 Operation of a loss-free line with mismatch on both sides. (a) Circuit, (b) pulse diagram

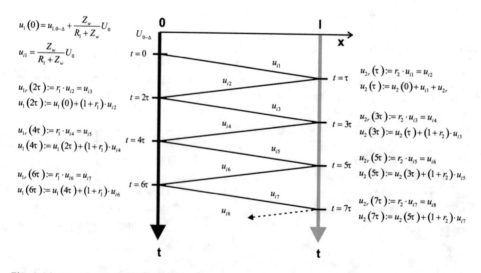

Fig. 2.31 "Lattice diagram" for a line to calculate the input and output signals of a line after a voltage jump at the input

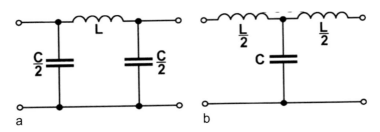

Fig. 2.32 Common simple line equivalent circuits. (**a**) as π equivalent circuit or (**b**) as τ replacement circuit, where $L = L' \cdot l$ and $C = C' \cdot l$

Returning voltage wave	u_{1r}, u_{2r}
Input side reflection coefficient	$r_1 = \frac{R_1 - Z_w}{R_1 + Z_w}$
Output side reflection coefficient	$r_2 = \frac{R_2 - Z_w}{R_2 + Z_w}$
Stationary end value $u_1(t \to \infty) = u_1(t \to \infty) = u\infty$	$u_\infty = \frac{R_2}{R_1 + R_2} \cdot U_{0+\Delta}$

The fading processes on lines after switching operations can be observed universally on lines for signals and the transmission of electrical energy. The oscillations shown in the examples can be very disturbing in applications. Therefore, the matching conditions must be determined and adhered to in order to avoid interference.

In power engineering, the so-called natural power P_N of the double line also has to do with wave propagation. The natural power is called

$$P_N = \frac{U_0^2}{Z_w}.$$

It is the power with which a line can be loaded in power engineering, while at the same time itself causing neither inductive nor capacitive power. If the cable is loaded with an effective resistance $Z_2 = Z_w$, this power can be passed directly to the connected load. If the line is terminated with a characteristic impedance $Z_2 = Z_w$, this is referred to as the 'case of matching'. The 'incoming' energy is completely converted into heat. Between the case of total reflection and that of matching, there is also partial reflection. Any point where a change in the characteristic impedance occurs in a line acts as a disturbance of the energy transfer, which leads to a partial reflection of the incoming energy. In simple experimental line models, the differential segmentation is simulated by π- or τ-equivalent circuit diagrams with the same characteristic impedance as a corresponding piece of a double line (Fig. 2.32). For circular frequencies $f \gg \frac{1}{2\pi\sqrt{LC}}$, this circuit behaves very similarly to the lines, so that effects such as the Ferranti effect, which describes voltage peaks in unloaded lines, can also be well reproduced.

In practical application in power electronics, line reflections can be problematic, e.g. when supplying motors from pulse inverters via long supply lines. The motors are supplied with basic frequencies of up to several 100...1000 Hz from a switched inverter, depending on the achievable rotational frequencies. With pure drive inverters, current smoothing is usually carried out by the machine itself. Nevertheless, increased lifetime problems of the insulations are observed due to the increased speeds of the voltage change (du/dt > 1 kV/μs). For a short time, these lead to increased displacement currents, which cause a faster aging of the insulation materials. A countermeasure here is, for example, a du/dt filter. If long supply lines are also used, voltage peaks occur in addition due to line reflections, since the lines cannot be connected with their characteristic impedance. These lead to a further amplification of the problem.

With the input and output variables of the quadripole shown in Fig. 2.28, the following can be formulated in complex notation from the mesh and node point theorems for the two-port equations

$$u_1(x_1) = (R' + j\omega L') \cdot i_1(x_1) + u_2(x_1 + dx)$$
$$i_2(x_1 + dx) = i_1(x_1) + (G' + j\omega C) \cdot u_2(x_1 + dx)$$

This system of equations represents the Laplace-transformed approach to the telegraph eq. A boundary value problem is obtained from an initial boundary value problem. With this solution, the transmission equations of the line quadripole can be formulated.

The cascaded form of equations of the 2-port then read in matrix notation

$$\begin{bmatrix} \underline{U}_1 \\ \underline{I}_1 \end{bmatrix} = \begin{bmatrix} \cosh\left(\underline{\gamma}l\right) & \underline{Z}_w \cdot \sinh\left(\underline{\gamma}l\right) \\ \dfrac{1}{\underline{Z}_w} \cdot \sinh\left(\underline{\gamma}l\right) & \cosh\left(\underline{\gamma}l\right) \end{bmatrix}$$

If the line is now terminated with an impedance $\underline{Z}_2$ to connect a source with internal resistance $\underline{Z}_1$, the voltage ratios in the x-direction are

$$\frac{\underline{U}_x}{\underline{U}_0} = \frac{\cosh\left(\underline{\gamma}(l-x)\right) + \frac{\underline{Z}_w}{\underline{Z}_1} \cdot \sinh\left(\underline{\gamma}(l-x)\right)}{\left(1 + \frac{\underline{Z}_1}{\underline{Z}_2}\right) \cdot \cosh\left(\underline{\gamma}l\right) + \left(\frac{\underline{Z}_1}{\underline{Z}_w} + \frac{\underline{Z}_w}{\underline{Z}_2}\right) \cdot \sinh\left(\underline{\gamma}l\right)}$$

For the beginning and end of the line, this results in the special cases

$$\frac{U_1}{U_0} = \frac{Z_2 \cosh\left(\underline{\gamma}l\right) + \frac{Z_w}{Z_1} Z_2 \cdot \sinh\left(\underline{\gamma}l\right)}{\left(Z_2 + Z_1\right) \cdot \cosh\left(\underline{\gamma}l\right) + \left(Z_2 \frac{Z_1}{Z_w} + Z_w\right) \cdot \sinh\left(\underline{\gamma}l\right)} \quad \text{and}$$

$$\frac{U_2}{U_0} = \frac{Z_2}{\left(Z_1 + Z_2\right) \cdot \cosh\left(\underline{\gamma}l\right) + \left(Z_2 \frac{Z_1}{Z_w} + Z_w\right) \cdot \sinh\left(\underline{\gamma}l\right)}$$

as well as the complex voltage ratio

$$\frac{U_2}{U_1} = \frac{Z_2}{Z_2 \cdot \cosh\left(\underline{\gamma}l\right) + \frac{Z_w}{Z_1} \cdot Z_2 \cdot \sinh\left(\underline{\gamma}l\right)}$$

In the undamped case

$$\underline{\gamma} = \alpha + j\beta = 0 + j\sqrt{\omega^2 L'C'} = j\omega\sqrt{L'C'}$$

From this follows

$$\frac{U_2}{U_1} = \frac{Z_2}{Z_2 \cdot \cos\left(\omega\sqrt{L'C'}l\right) + \frac{Z_w}{Z_1} \cdot \sin\left(\omega\sqrt{L'C'}l\right)}$$

The relative output voltage for an unloaded line, i.e. for $Z_2 \rightarrow \infty$, is then

$$\lim_{Z_2 \rightarrow \infty} \left(\frac{U_2}{U_1}\right) = \lim_{Z_2 \rightarrow \infty} \left(\frac{Z_2}{Z_2 \cdot \cos\left(\omega\sqrt{L'C'}l\right) + \frac{Z_w}{Z_1} \cdot \sin\left(\omega\sqrt{L'C'}l\right)}\right) = \frac{1}{\cos\left(\omega\sqrt{L'C'}l\right)}$$

Figure 2.33a shows a corresponding graphic representation of this relationship. Clear resonance phenomena can be seen in the form of voltage peaks. If one represents the input impedance $\underline{Z}_e$ as a function of the load impedance $\underline{Z}_2$, one obtains

$$\underline{Z}_e = \frac{U_1}{I_1} = \frac{Z_2 + Z_w \cdot \tanh\left(\underline{\gamma}l\right)}{1 + \frac{Z_2}{Z_w} \cdot \tanh\left(\underline{\gamma}l\right)}$$

Thus, the limit value $\underline{Z}_e$ for an non-loaded output of the line is

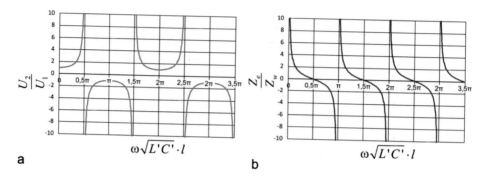

Fig. 2.33 Frequency dependence of an unloaded line. (**a**) Voltage transmission factor, (**b**) input impedance

$$\lim_{Z_2 \to \infty} \underline{Z}_e = \lim_{Z_2 \to \infty} \left(\frac{\underline{U}_1}{\underline{I}_1} \right) = \lim_{Z_2 \to \infty} \left(\frac{\underline{Z}_2 + \underline{Z}_w \cdot \tanh\left(\underline{\gamma} l\right)}{1 + \frac{\underline{Z}_2}{\underline{Z}_w} \cdot \tanh\left(\underline{\gamma} l\right)} \right) = \frac{\underline{Z}_w}{\tanh\left(j\omega\sqrt{L'C'} \cdot l\right)}$$

$$= \frac{\underline{Z}_w}{\tan\left(\omega\sqrt{L'C'} \cdot l\right)}$$

The graphic representation of this relationship is shown in Fig. 2.33b. These relationships can be used to design devices that are to be used as transformers or impedance converters, using the wave propagation time within the device.

From the diagrams shown in Fig. 2.33, special frequencies and modes of operation of line sections can be derived. For example, at $\omega\sqrt{L'C'} \cdot l$, i.e. at $l = \frac{\lambda}{4}$ the input impedance is zero, while the output voltage is infinite. This is the behaviour of a series resonant circuit. Other applications can be derived analogously (Lecher line). These cases are summarized in Table 2.1.

In the low-frequency range (l < λ/4), a cable has mainly inductive properties. The network impedance is not constant, but changes constantly due to the constant switching operations and changing load cases in the network. For this reason, the frequency response is also not constant. Since this also affects low-voltage distribution technology, typical equivalent circuits are used for comparative calculations (Fig. 2.34).

3- and More Wires

In 3-wire and multi-wire systems, coupling via counter induction is added to self-induction (Fig. 2.35). Such a situation exists for any device with a 1-phase connection + protective conductor (Lx - N - PE). In this case, energy is only transmitted through the cable pair Lx - N. The PE protective conductor has mainly protective and safety functions. The transmitted interference currents/interference voltages are limited in order to guarantee system safety in future expansion stages. A diagram of a cable section is shown in Fig. 2.35. In a pure

Table 2.1 Effect of line sections in electrical circuits with lengths around λ/4

Designation	Condition	Scheme
Series resonant circuit	$Z_2 \to \infty, l = \frac{\lambda}{4}$	0　　　　　　　$\frac{\lambda}{4}$
Parallel resonant circuit	$Z_2 = 0, l = \frac{\lambda}{4}$	0　　　　　　　$\frac{\lambda}{4}$
Capacitor	$Z_2 \to \infty, l < \frac{\lambda}{4}$	0　　　　　　　$\frac{\lambda}{4}$
Inductance	$Z_2 \to 0, l < \frac{\lambda}{4}$	0　　　　　　　$\frac{\lambda}{4}$
Capacitor	$Z_2 \to 0, l > \frac{\lambda}{4}$	0　　　　　　　$\frac{\lambda}{4}$
Inductance	$Z_2 \to 0, l > \frac{\lambda}{4}$	0　　　　　　　$\frac{\lambda}{4}$

single-phase system, there is only one channel in which the outward and return currents are the same. In a three-wire system, the signed sum of all 3 phase currents is zero.

If one additionally assumes that each pair of conductors is also coupled by a mutual inductance, a differential element of a 3-wire system is obtained in analogy to Fig. 2.28b.

Fig. 2.34 50-Hz network model for a typical connection point of the low voltage network (4-wire system) according to DIN EN 50006/VDE 08383

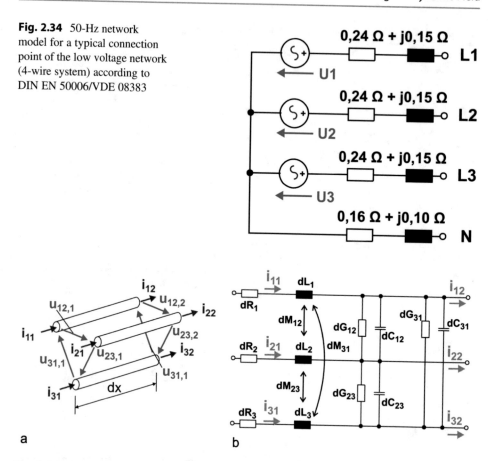

Fig. 2.35 Three-conductor system for power and signal transmission. (**a**) Schematic layout, (**b**) equivalent circuit diagram of a differential part of the three-phase line, taking into account the mutual induction

Instead of 2 equations, such a system is now described by 6 linear independent equations. As a result, the complexity increases significantly. It becomes somewhat simpler if the third conductor is designed as a "ground plane" with a disappearing resistance and inductance coating ($R' = 0$, $L' = 0$), as is usual with strip conductors on a ceramic or plastic carrier as opposed to a copper surface on the back.

If the case described above is maintained, energy is transferred between the conductors L_1 and N, while PE is only conducted in parallel. Signals in L_1 and N opposite PE then move in other wave channels. By designing the inductive and capacitive line layouts, it is possible to achieve that phase velocities and attenuations are different in the different channels. This is the subject of filter technology for ensuring the electromagnetic compatibility of devices. Signals in the "main channel" L-N are referred to as push-pull signals (*differential mode*), while signals against PE are referred to as "*common mode*"signals. By

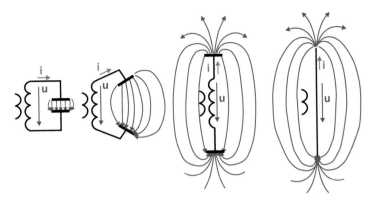

Fig. 2.36 Transition from resonant circuit to dipole antenna for the generation of electromagnetic waves in space with schematic representation of the E-field. (**a**) Resonant circuit, (**b, c**) resonant circuit with electric free field, (**d**) dipole antenna

cleverly combining self-induction and mutual induction, both types of signals can be influenced almost separately. Such cases are discussed in Chap. 10.

Propagation of Electromagnetic Waves in Free Space

Based on the previous cases of this section, the case of an inductively excited resonant circuit illustrated in Fig. 2.36 will be presented. Energy is coupled into the inductance of the parallel resonant circuit from an AC voltage source. If the electrodes of the capacitor are now pulled further and further apart, a dipole is finally obtained as a structure that combines the inductive and capacitive properties of a resonant circuit.

The space surrounding the dipole antenna is now part of the oscillating circuit. The electromagnetic field as the energetic state of the space cannot be built up at will. The speed of propagation is linked to the magnetic and dielectric properties of the vacuum or the space permeated by the field. In this case, the velocity of propagation is

$$c = \frac{1}{\sqrt{\varepsilon \cdot \mu}} = \frac{1}{\sqrt{\varepsilon_r \cdot \mu_r}} \cdot \frac{1}{\sqrt{\varepsilon_0 \cdot \mu_0}} = \frac{1}{\sqrt{\varepsilon_r \cdot \mu_r}} \cdot c_0 = \frac{1}{\sqrt{\varepsilon_r \cdot \mu_r}} \cdot 299792458 \, \frac{m}{s}$$

Here c_0 is the speed of light in a vacuum. Similar to lines, a characteristic impedance Z_w can be defined for the propagation of the electromagnetic field in space. It amounts to

$$Z_w = \sqrt{\frac{\mu}{\varepsilon}} = \sqrt{\frac{\mu_r}{\varepsilon_r}} \cdot \sqrt{\frac{\mu_0}{\varepsilon_0}} = \sqrt{\frac{\mu_r}{\varepsilon_r}} \cdot Z_{w0} = \sqrt{\frac{\mu_r}{\varepsilon_r}} \cdot 376{,}73\Omega.$$

Depending on the shape of the emitting resonant circuit/antenna, very different field shapes can be generated. By coupling several resonant circuits, the field can also be further shaped and directed, so that a distinct directional characteristic of the far field can be achieved by simple means, especially at very high frequencies.

2.7 The Maxwell Equations

This section summarizes the relationships between the electrical and magnetic quantities that ultimately form the electromagnetic field. The equations of the Scottish physicist James Clerk Maxwell (* 13 June 1831 in Edinburgh; † 5 November 1879 in Cambridge) describe the classical phenomena of electromagnetism.

The equations worked out in the period 1861–1864 describe how electric and magnetic fields are related to each other and to electric charges and electric current under given boundary conditions. Together with the general Lorentz force, they explain practically all phenomena of classical electrodynamics. These equations in their differential and integral form and their semantics are the basis of this book and are therefore presented in this subchapter with respect to their definitions and relationships [6–8].

The electromagnetic field is composed of two coupled fields, each with a different character.

- The electric field is a source field, so that the field lines have a beginning and an end. The sources are positive and negative charges, which can occur individually.
- The magnetic field is a circuital vector field. In circuital vector fields, the field lines are closed curves. There is no beginning and no end. There are no "magnetic charges" (magnetic monopoles) but only magnetic dipoles. Circuital vector fields without explicit dipoles occur, for example, in induction processes.

The following physical quantities are used (Table 2.2).

The material equations for the description of the relationships of the electric and magnetic field quantities to each other are

(a) Current density ↔ electric field strength

$$\vec{J} = \sigma \cdot \vec{E}$$

(b) Current ↔ current density

$$I = \int_A \vec{J} \cdot d\vec{A}$$

(c) Displacement flux density ↔ electric field strength

$$\vec{D} = \varepsilon \cdot \vec{E}$$

(d) Charge ↔ displacement flux density

$$Q = \int_A \vec{D} \cdot d\vec{A}$$

Table 2.2 Basic parameters of Maxwell's equations

Physical quantity	Formula symbol	Unit of measurement	Type
Electric field strength	E	V/m	Vector
Voltage	U	V	Scalar
Current density	J	A/m^2	Vector
Displacement current density	D	A·s/m^2	Vector
Charge	Q	A·s	Scalar
Current	I	A	Scalar
Magnetic field strength	H	A/m	Vector
Magnetic potential drop	V	A	Scalar
Magnetic flux density	B	V·s/m^2	Vector
Magnetic flux	Φ	V·s	Scalar
Conductivity	σ	S/m	Scalar
Permeability	μ	V·s/(A·m)	Scalar
Permittivity	ε	A·s/(V·m)	Scalar
Charge density	ρ	A·s/m^3	Scalar
Speed	v	m/s	Vector

Table 2.3 Maxwell's equations in integral and differential form

Integral form of presentation	Differential form of representation
Ampére's law	
$\oint_s \vec{H} \cdot d\vec{s} = \int_A \left(\vec{J} - \rho \cdot \vec{v} \right) \cdot d\vec{A} + \frac{d}{dt}\int_A \vec{D} \cdot d\vec{A}$	$rot\,\vec{H} = \vec{\nabla} \times \vec{H} = \left(\vec{J} - \rho \cdot \vec{v} \right) + \frac{d}{dt}\vec{D}$
Law of Induction	
$\oint_s \vec{E} \cdot d\vec{s} = -\frac{d}{dt}\int_A \vec{B} \cdot d\vec{A}$	$rot\,\vec{E} = \vec{\nabla} \times \vec{E} = -\frac{d}{dt}\vec{B}$
Electric field as source field	
$\int_{AO} \vec{D} \cdot d\vec{A} = \int_V \rho \cdot dV = Q$	$div\,\vec{D} = \vec{\nabla} \cdot \vec{D} = \rho$
Magnetic field as a source-free field	
$\int_{AO} \vec{B} \cdot d\vec{A} = 0$	$div\,\vec{B} = \vec{\nabla} \cdot \vec{B} = 0$

(e) Magnetic flux density $\leftrightarrow$ magnetic field strength

$$\vec{B} = \mu \cdot \vec{H}$$

(f) Magnetic flux $\leftrightarrow$ magnetic flux density

$$\Phi = \int_A \vec{B} \cdot d\vec{A}$$

The Maxwell equations are in their integral and differential form (Table 2.3).

From these basic equations, all relationships between the electrical and magnetic quantities in electrical engineering can be derived. The integral form appears to be more descriptive and is therefore the preferred form in this book.

References

1. Küpfmüller, K.; Kohn, G.: Theoretische Elektrotechnik und Elektronik. 14 Auflage. Springer, 1993, ISBN 3-540-56500-0.
2. Meschede, D.: Gerthsen Physik. Springer-Verlag 25. Auflage 2015, ISBN 978-3-662-45976-8
3. Philippow, E: Taschenbuch Elektrotechnik, Band 1, Verlag Technik, Berlin 1981
4. Wanlass, S., D.; Wanlass, C. L.; Wanlass, L. K.: "The Paraformer; A new passive power conversion device", IEEE Wescon Technical Papers, 1968.
5. Kothari, D. P.; Nagrath, I. J.: Power Systems Engineering. 2nd Ed., McGraw-Hill 2008, p. 653 ff.
6. Schwab, A. J.: Begriffswelt der Feldtheorie, Elektromagnetische Felder, Maxwell'sche Gleichungen. 6 Auflage. Springer, 2002, ISBN 3-540-42018-5.
7. Einstein, A.: Zur Elektrodynamik bewegter Körper, Annalen der Physik und Chemie 17, 30. Juni 1905, Seiten 891-921
8. Simonyi, K.: Theoretische Elektrotechnik, 9. Auflage, VEB Deutscher Verlag der Wissenschaften, Berlin, 1989

Magnetic Properties of Materials

3

Electrical and magnetic phenomena are interrelated. It is therefore clear that the magnetic material properties play an important role in many areas of electrical engineering. In the following explanations, the formal (macroscopic) description of magnetic properties will be explained first. Then the explanation of these properties is based on atomistic processes.

3.1 Macroscopic Description of Magnetic Properties

Magnetic field strength H and magnetic flux density or induction B are used to describe magnetic phenomena. H is the cause and is attributed to the movement of charge carriers. B can be interpreted, for example, by forces on moving charge carriers. In the SI unit system, the magnetic field strength is given in A/m; the unit of induction is tesla ($1 \text{ T} = 1 \text{Vs/m}^2$). For the relationship between the vectorial quantities $\vec{B}$ and $\vec{H}$ applies in a vacuum

$$\vec{B} = \mu_0 \cdot \vec{H}$$

with the magnetic field constant (induction constant, vacuum permeability)

$$\mu_0 = 4\pi \cdot 10^{-7} \frac{\text{Vs}}{\text{Am}} = 0.4 \cdot \pi \frac{\mu H}{m}.$$

When describing the relationship between B and H in the presence of matter, the influence of a magnetically isotropic substance is first considered. Isoptrop means that the material properties are independent of the spatial orientation of the field. Furthermore, it is assumed that there is a linear relationship between B and H. Under these restrictive conditions, the following applies

© Springer Fachmedien Wiesbaden GmbH, part of Springer Nature 2022
P. Zacharias, *Magnetic Components*,
https://doi.org/10.1007/978-3-658-37206-4_3

$$\vec{B} = \mu \cdot \vec{H} = \mu_r \cdot \mu_0 \cdot \vec{H}$$

where μ_r is the relative permeability of the substance in the magnetic field. If you only want to consider the influence of the material, you can also divide this relationship into a part that is assigned to the vacuum and a part that is only caused by the material. In order to be able to introduce a distinction here, the term polarization J is introduced here, which indicates the influence of the material.

The induction is then composed of the vacuum induction and the polarization:

$$\vec{B} = \mu_0 \cdot \vec{H} + \vec{J} \text{ respectively } \vec{B} = \mu_0 \cdot \left(\vec{H} + \vec{M} \right)$$

Where J is the magnetic polarization. M is called magnetization. Consequently, the following can be formulated for the polarization J

$$\vec{J} = (\mu_r - 1) \cdot \mu_0 \cdot \vec{H} = \chi_m \cdot \mu_0 \cdot \vec{H} = \mu_0 \cdot \vec{M}$$

These relationships are summarised in Table 3.1.

According to the values of the permeability number or the susceptibility χ_m, the following cases are distinguished (see also the comparison with the electric field quantities in Table 3.1):

1.	Diamagnetism:	$\chi_m < 0$	$\mu_r < 1$
2.	Paramagnetism/Antiferromagnetism	$\chi_m \approx (>) \, 0$	$\mu_r \approx (>) \, 1$
3.	Ferromagnetism/Ferrimagnetism	$\chi_m \gg 1$	$\mu_r \gg 1$

For technical applications, the phenomena of ferro- and ferrimagnetism listed under 3. are particularly important. They exhibit a pronounced non-linearity (saturation). Diamagnetism rarely plays a technical role, as χ_m is only very slightly different from zero. The antiferromagnetic properties are included here in the sense of a complete description of the magnetic order states. In the case of a magnetically anisotropic material (e.g. a single crystal) it is to be expected that the vectors of the magnetic field and induction do not point in the same direction. The direction-dependent permeability is then described using a tensor $\| \mu \|$:

$$\vec{B} = \| \mu \| \cdot \vec{H} = \begin{bmatrix} \mu_{xx} & \mu_{xy} & \mu_{xz} \\ \mu_{yx} & \mu_{yy} & \mu_{yz} \\ \mu_{zx} & \mu_{zy} & \mu_{zz} \end{bmatrix} \cdot \vec{H}$$

In it, $\| \mu \|$ is a matrix with 9 elements. The number of independent coefficients of the permeability matrix required to describe the magnetic properties of a single crystal depends

Table 3.1 Basic quantities and basic relationships of the magnetic field

Physical quantity	Formula-sign	Unit	Basic relationship
Magnetic field strength (common)	H	A/m	$\vec{H} = \frac{B}{\mu}$
Magnetization (material related)	M	A/m	$\vec{M} = \frac{B}{\mu_0} - \vec{H}$
Magnetic induction(common)	B	$T = Vs/m^2$	$\vec{B} = \mu \cdot \vec{H} = \mu_0 \cdot \left(\vec{H} + \vec{M}\right)$
Magnetic polarization (material related)	J	$T = Vs/m^2$	$\vec{J} = \mu_0 \cdot (\mu_r - 1) \cdot \vec{H} = \mu_0 \cdot \chi_m \cdot \vec{H}$
Induction constant	μ_0	Vs/Am	$\mu_0 = 4\pi \cdot 10^{-7} \frac{Vs}{Am}$
Relative permeability (common)	μ_r	–	$\mu_r = \frac{B}{\mu_0 H}$
Magnetic susceptibility (material related)	χ_m	–	$\chi_m = (\mu_r - 1) = \frac{M}{H} = \frac{J}{\mu_0 \cdot H}$

on the symmetry properties of the crystal. For a crystal of the triclinic system, 6 independent coefficients are required. For crystals with cubic elementary cells, it is sufficient to specify a single coefficient of permeability for all 3 elements of the main diagonal.

As an illustration, 2 examples are given for all 3 groups of substances in Table 3.2.

Diamagnetic materials are especially noble gases, halogens as well as boron, silicon, germanium and selenium. The metals copper, silver and gold are also diamagnetic. The highest value of the amount of susceptibility of diamagnetic substances at room temperature is found in bismuth. Metals in the superconducting state are identified by $\chi_m = -1$, that is, there is no magnetic induction inside a superconductor.

Paramagnetic are among others the alkali metals, most transition metals and rare earth metals. Of the gaseous elements, only oxygen is paramagnetic. The numerical values of the magnetic susceptibility of dia- and paramagnetic substances are very small, that is, the values of the permeability number are very close to one. Accordingly, dia- and paramagnetic substances are also frequently described as non-magnetic.

Only the elements iron (Fe), cobalt (Co) and nickel (Ni) are *ferromagnetic* at room temperature. At low temperatures, some metals from the rare earth group (e.g. gadolinium (Gd)) also exhibit ferromagnetic behaviour. Ferromagnetism can also occur in alloys that are free of iron, cobalt and nickel.

Ferro- and ferrimagnetic materials are used in electrical engineering, among other things as the core material of coils. This allows the self-inductance L of a coil to be significantly increased. In the case of an elongated cylindrical coil (l >> d), the inside of which is filled with a material with the permeability number μ_r, the following applies approximately

Table 3.2 Examples of different material types with their susceptibilities

Diamagnetic materials

Material	Composition	Susceptibility X
Alumina	Al_2O_3	$-1.81 \cdot 10^{-5}$
Copper	Cu	$-0.96 \cdot 10^{-5}$
Gold	Au	$-3.44 \cdot 10^{-5}$
Bismuth	Bi	$-16.8 \cdot 10^{-5}$
Water	H_2O	$-0.9 \cdot 10^{-5}$
Silver	Ag	$-2.38 \cdot 10^{-5}$
Zinc	Zn	$-1.56 \cdot 10^{-5}$
Paramagnetic materials		
Aluminium	Al	$2.07 \cdot 10^{-5}$
Chrome	Cr	$3.13 \cdot 10^{-5}$
Manganese sulphate	$MnSO_4$	$3.70 \cdot 10^{-5}$
Molybdenum	Mo	$1.19 \cdot 10^{-5}$
Sodium	Na	$8.48 \cdot 10^{-5}$
Titanium	Ti	$1.81 \cdot 10^{-5}$
Zircon	Zr	$1.09 \cdot 10^{-5}$
Ferromagnetic materials		
Iron	Fe	250...680
Grey cast iron	GGG-35.3... -70	500...2500
Nickel	Ni	280...2500
Cobalt	Co	80...200
Mu-Metal	Alloys of Ni, Fe, Cu, Cr, Mo	80,000...100,000
Ferrites	–	4...20,000
Amorphous and nanocrystalline alloys	–	20,000...800,000

$$L = \mu_r \cdot \mu_0 \cdot \frac{N^2 \cdot A_{Fe}}{l} = \mu_r \cdot \mu_0 \cdot \frac{N^2 \cdot \pi \cdot d^2}{4 \cdot l}$$

(N: number of turns, A_{Fe}: area, l: length, d: diameter). The permeability number μ_r of the coil core thus increases the self-inductance of a coil. For a loss-free (ideal) coil, the inductive voltage drop

$$u_L(t) = L \cdot \frac{di_L}{dt} \text{ or in complex notation } \underline{u}_L = j\omega L \cdot \underline{i}_L.$$

When a sinusoidal current is injected in an inductor, a voltage drop occurs that precedes the current by 90° ($\pi/2$). Since every real inductance is also lossy, in reality, the phase shift is <90°. If one assumes that the losses are caused by a resistor R_s which is in series with the inductance Ls, then one measures the voltage

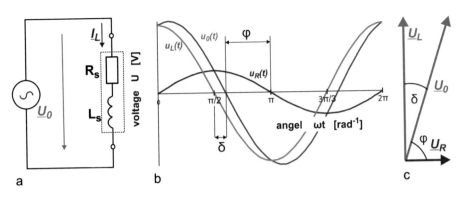

Fig. 3.1 To define the permeability value $\mu_r{}^*$ and the loss factor tanδ in text. (**a**) Coil with connected AC voltage source, (**b**) Voltage curves on the model components, (**c**) Phasor diagram

$$\underline{u}_L = (R_s + j\omega L) \cdot \underline{i}_L$$

This means for the impedance $\underline{Z}_L$ of a real coil:

$$\underline{Z}_L = \frac{\underline{u}_L}{\underline{i}_L} = R_s + j\omega L_s$$

This situation is illustrated in Fig. 3.1. If the design equation for an inductance L is taken as a basis and the impedance is interpreted as a consequence of the inductance, a complex expression for the inductance is obtained. This leads to the definition of a complex permeability

$$\underline{\mu}_r = (\mu_s{}' - j\mu_s{}'')$$

$$\underline{L}^* = \frac{\underline{Z}}{j\omega} = \frac{R_s + j\omega L_s}{j\omega} = \frac{R_s}{j\omega} + L_s = \underline{\mu}\frac{N^2}{l}A = (\mu_s{}' - j\mu_s{}'')\mu_0\frac{N^2}{l}A$$

However, the physical cause of the losses is only at very low frequencies characterized by series resistance. In general, the losses are caused by transformer interaction with the core and the winding (e.g. as eddy currents). These are better represented by a parallel connection of an inductor L_p and a parallel resistor in a model. Both representations can be converted into each other for an arbitrary, but fixed, frequency by equating the impedances of series and parallel connection $\underline{Z}_s = \underline{Z}_p$ and then doing the same for the real and imaginary parts of both equivalent circuits. With

$$R_s + j\omega L_s = \frac{j\omega R_p L_p}{R_p + j\omega L_p}$$

this means for a conversion to the equivalent series connection

$$R_s = \frac{\omega^2 R_p L_p^2}{R_p^2 + j\omega L_p^2} \text{ and } L_s = \frac{R_r^2 L_p}{R_p^2 + j\omega L_p^2}.$$

In general, a series connection is assumed for characterization. The losses cause a phase shift φ between current and voltage. For this applies

$$\tan(\varphi) = \frac{\text{Im}(\underline{Z})}{\text{Re}(\underline{Z})} = \frac{\omega L_s}{R_s} \text{ Or } \varphi = \arctan\left(\frac{\omega L_s}{R_s}\right).$$

Naturally, φ is usually close to 90°. A clearer illustration of the losses can be obtained by the so-called loss angle δ with $\delta + \varphi = 90°$ (see Fig. 3.1b). The loss angle is then

$$\delta = \arctan\left(\frac{R_s}{\omega L_s}\right) \text{ respectively } \tan(\delta) = \frac{R_s}{\omega L_s}.$$

With the above-derived definition for the complex permeability μ_s, one can also write

$$\tan\delta = \frac{\mu_s''}{\mu_s'}$$

In a real coil, there is also a voltage component u_R, which is in phase with the current i_L. This voltage component is attributed to the ohmic resistance of the coil winding and losses in the core material. This is an interpretation and not the representation of a physical fact. Basically, there are practically always losses that can be assigned to a serial resistor as well as those that can be assigned more to a parallel loss resistor. Chap. 7 provides further explanations. The material characteristics are usually presented with the assumption of a series connection. Measuring devices usually allow switching between series or parallel equivalent circuits.

Figure 3.2 shows the frequency-dependent course of the real and imaginary parts of the complex permeability for 2 ferrites. In general, the application range of ferrite is limited to a frequency range for which $\mu_s'/\mu_s'' > 20$. The definitions here correspond to a substitute series connection of inductance and loss resistance. One can convert these dependencies into a parallel circuit which better meets the physical conditions, as shown in Chap. 7.

In the Phasor diagram (Fig. 3.1c) the total voltage $\underline{U}_0$ is given by

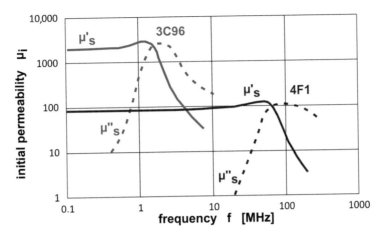

Fig. 3.2 Frequency dependence of real and imaginary part of the permeability for a MnZn ferrite (3C96) and a NiZn ferrite (4F1) [1]

$$\underline{U}_0 = \underline{U}_L + \underline{U}_R$$

For many applications - especially in heavy current engineering - it must be taken into account that in ferro- and ferrimagnetic materials there is a non-linear relationship between the induction and the magnetic field strength. Furthermore, it must be assumed that induction is not only dependent on the instantaneous value of the field strength, but also on the field strength prevailing at an earlier point in time. Therefore, the relationship between B and H is not unique for each time point, but depends on previous history.

The behaviour of ferro- and ferrimagnetic materials can be described by a magnetisation characteristic curve or a hysteresis loop as shown in Fig. 3.3. In this loop, the part marked 1 is the initial magnetization curve that is generated when the magnetic field strength is applied to and increased in a completely demagnetized material (e.g. after an annealing process).

Part 2 of the hysteresis loop describes the course B(H) when the magnetic field strength is brought from a positive value (above H = 0) to a negative value. If the magnetic field strength is first negative and then positive, this results in the branch of the hysteresis loop marked 3. It is assumed here that the material is exposed to high peak values of the magnetic field (full modulation).

A hysteresis loop as shown in Fig. 3.3 has four intersections (H_C, -H_C, B_r and -B_r) with the H and B axes, which are used for characterization. The *coercivity* H_C is the field strength required to make a pre-existing induction disappear. The *remanence induction* B_r is the induction that remains after the complete disappearance of the external magnetic field. Furthermore, the *saturation induction* B_s can also be specified; it is obtained by induction values measured at high field strengths to H = 0.

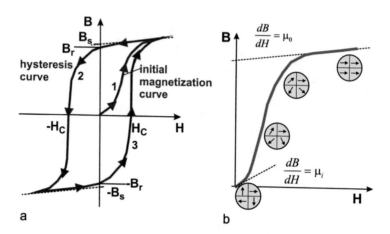

Fig. 3.3 Magnetization curve in ferro- and ferrimagnetics. (**a**) Hysteresis curve with H_C - coercive field strength, B_s - saturation induction, B_s – remanent field strength; (**b**) Initial magnetization curve with schematized behavior of the Weiss domains under the influence of the magnetic field [after 2]

According to the coercive field strength H_C, a rough classification of ferro- and ferrimagnetic materials can be made. Materials with a low coercive field strength are called *soft magnetic*, while a high coercive field strength is characteristic of *hard magnetic* material. In the case of ferromagnetic or ferrimagnetic materials, which are intended for information storage, a hysteresis loop that is as rectangular as possible is desirable in addition to sufficient coercive field strength. Here the remanence induction is almost equal to the saturation induction.

The ambiguous hysteresis curve can also be observed with "infinitely slow" remagnetization. It is due to work that must be introduced into the crystal structure in order to change its position according to the spontaneous magnetization of the grains (so-called Weiss domains). This results in elastic and nonelastic processes. The latter imply an energy input that must be used for a remagnetization process. The loading-unloading-loading cycle of a magnetic storage device is therefore carried out with an efficiency of <100%. At higher frequencies, further loss mechanisms are added, such as induced eddy currents. These can be represented as parallel resistance in an equivalent electrical circuit (Fig. 3.4). Although the B(H) characteristic is generally treated like a U(I) characteristic, significant differences to this are neglected. With the U(I) characteristic curve, both axes are in the same "world" or dimension. U and I are time-dependent variables. In the B (H) curve, H is a quantity whose value depends on time and can be changed instantly. This is not the case with magnetic flux density. The equation of determination

$$B = \frac{1}{N \cdot A_{Fe}} \cdot \int u_L dt$$

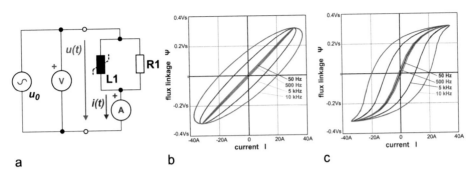

Fig. 3.4 Voltage-time integral, interpreted as flux linkage at a linear inductor 10 mH//1 kOhm and a non-linear inductor with a <u>unique</u> magnetization characteristic of the same amplitude inductance parallel to 1 kOhm. (The voltage U was increased proportionally to the frequency) **(a)** Simulation circuit, **(b)** Combination of two linear elements L_1 and R_1, **(c)** Combination of a non-linear inductance L_1 and a linear resistor R_1

states that the induction depends on the integral of the applied voltage over time and thus also depends on the history. The losses occur instantaneously and can therefore be assigned to any point in time. The stored energy content also results from prehistory. Consequently, if a diagram B(H) is drawn, it corresponds to the so-called phase plane when representing first and second order differential equations. In this plane, the parallel connection of a linear inductor and a resistor also has an "ambiguous" course, as Fig. 3.4. Actually, it is a parameter representation of the induction B=B(H(t)). The time t could also be entered as a parameter of the B(H) curve. The abscissa then follows a time-dependent quantity while the ordinate follows an integral over time. The apparent ambiguity is an expression of the fact that there is a phase shift $\neq 0$ und $\neq (2 \cdot v + 1) \cdot \pi$ between the quantities. Figure 3.4 also shows how the dependency B(H) transforms at different frequencies and how a non-linear relationship for the inductive component $B(H_{L1})$ is the basis for a hysteresis characteristic curve even with linear resistance. The area enclosed by the B(H) characteristic $\oint BdH = W/V$ has the unit of measurement of an energy per volume and generally represents (i.e. without regard to the cause of loss) the energy input for a remagnetization cycle.

A special circuit must be used to simulate a process with two superimposed variables u (t) and a direct current (Fig. 3.5). Even, in reality, it is not possible to connect a voltage source in parallel to a two-pole and superimpose a current on it. The state of the inductance is then determined either by the voltage source or the current source. In complex measuring systems, the problem can be solved by control engineering. With inductors, the problem can be solved technically simply by adding another winding for (DC) premagnetisation. The two circuits are then electrically isolated from each other. A circuit arrangement was developed and tested to simulate such processes (Fig. 3.5). It consists of a linear inductor L_0 which controls the current i_L flowing through an inductor in parallel with a loss resistance. L_1 represents a pure material influence J(H). With the voltage drop resulting from this, the

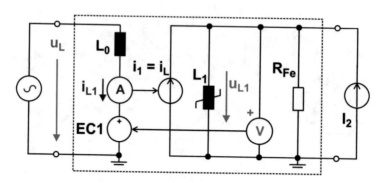

Fig. 3.5 A modelling approach for a core with separate bias winding for a magnetising current I_0

voltage drop of the "inner" nonlinear component L_1 is mapped via a voltage-controlled voltage source.

With the current source I_2, the current caused by the voltage source u_L can be superimposed. In this way, one can easily simulate the interactions realistically. (Numerically problematic is sometimes the interconnection of 2 current sources with one inductor.) Further explanations are given in Chap. 6.

In Fig. 3.6 this simulation method was applied to the data of a standard electrical sheet at 50 Hz. The loss resistance was assumed to be constant and scaled to the losses at 1.6 T and $I_2 = 0$. It can be clearly seen that the characteristic curve changes in shape with increasing modulation. Since $I_2 = 0$, the characteristic curve remains symmetrical.

Premagnetisation with smooth direct current leads to an 'operating point' on the characteristic curve around which magnetisation takes place (Fig. 3.7). In Fig. 3.7a a sinusoidal voltage source with different frequencies and voltages was used for remagnetization. In Fig. 3.7b a square-wave voltage with the same effective voltage was used. The remagnetization trajectory obviously changes. Since it cannot generally be assumed that the loss resistance is constant or independent of induction, the bias current is another factor influencing the losses. The causes of losses are very complex and are based on material properties down to the atomic level. The resulting magnetization curves sometimes have very different shapes, of which four types are shown in Fig. 3.8. They are obtained by determining the ambiguous dependence B(H) using a suitable measuring method with du/dt → 0.

Figure 3.8a shows a form often referred to as the "normal form" and used in explanations. The values B_r/B_s are in the range 0.7...0.9. H_C has values of several 100 A/m. The forms 3.8a, b have basically the same basic form. In the curve form known as the F form, the coercive field strength H_C has very low values. This form is preferred for HF transformers and, depending on the application, also for switching chokes and magnetic amplifiers. In the Z-shape shown in Fig. 3.8c, the ratio $B_r/B_s \approx 1$. This is a frequently desired behaviour in non-linear magnet applications. The differential permeability is extremely high here, which accelerates the switching process. The low H_C values, which

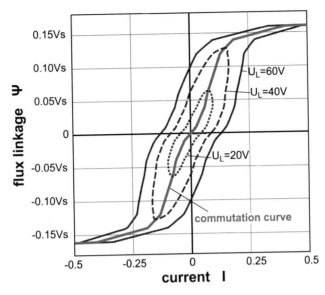

Fig. 3.6 Calculated B(H)-trajectory with variable gain for a standard transformer lamination and development of the commutation curve

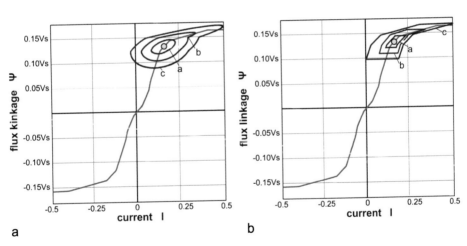

Fig. 3.7 Calculated curves for partial remagnetization at an average DC current of 0.15A. (**a**) for sinusoidal voltage $U_{RMSa} = 14.14$ V, $U_{RMSb} = 28.28$ V, $U_{RMSc} = 42.42$ V and (**b**) for rectangular voltages of equal rms values

are also characteristic of the Z-shape, make it easy to "reset" the magnetization state to the initial value. With hard magnets, which are characterised by very high H_C values, a high remanence or a rectangular curve shape is aimed for. However, this is not referred to as a Z-shape.

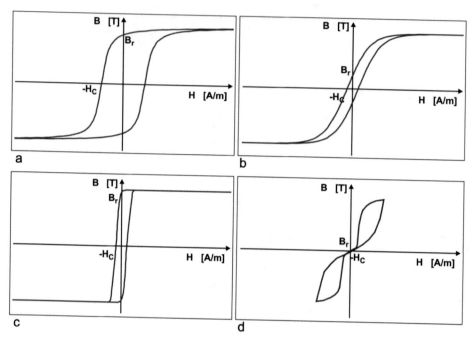

Fig. 3.8 Four typical hysteresis loops of ferro- or ferrimagnetic materials (selection). (**a**) "normal" hysteresis loop, (**b**) flat (F) loop with extended linear component, (**c**) rectangular (Z) loop, (**d**) Perminvar loop

If the magnetization curve has a significantly lower initial permeability over a larger field strength range, then the B(H) curve tends to behave as shown in Fig. 3.8d. The hysteresis curve constricts near the coordinate origin.

Obviously many soft magnetic materials are also materials with highly non-linear properties. For many applications, the aim is to achieve the most linear dependence possible over a wide range of applications $\Psi(I)$, in other words, the most constant inductance possible. To achieve this, the core is equipped with an air gap. The effect of this is shown in Fig. 3.9. Figure 3.9a shows the series connection of a source I·N with a non-linear magnetic resistance R_{mFe} formed by the core and magnetic resistance of the air gap $R_{m\delta}$. The magnetic flux is constant in this unbranched magnetic circuit. The balance of the magnetic voltage drops then results

$$I \cdot N = V_{Fe} + V_{\delta} = \Phi \cdot R_{mFe} + \Phi \cdot R_{m\delta}.$$

For a given flux, the magnetic voltage drops are to be added for this series connection. This is shown in Fig. 3.9b.

The green characteristic curve is obtained by adding the magnetic voltage drops at the core and the air gap point by point. The gradient in the coordinate origin becomes

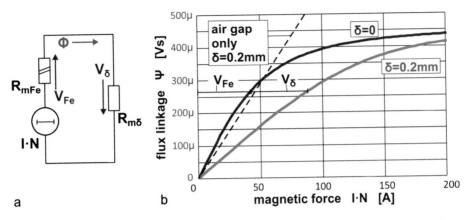

Fig. 3.9 Shearing of a characteristic curve by inserting an air gap. (**a**) magnetic equivalent circuit diagram, (**b**) resulting $\Psi(I)$ characteristic curve

significantly smaller by this measure. Considering that $\Psi = N \cdot \Phi$ is, the slope in the origin is

$$\frac{\Delta\Psi}{\Delta V} = \frac{N \cdot \Delta\Phi}{N \cdot \Delta I} = \frac{1}{R_{mges}} = \Lambda_{ges}$$

Λ_{ges} is the permeance or the magnetic conductance of the entire arrangement. The inductance of a winding with N turns is

$$L = \frac{N^2}{R_{mges}} = N^2 \cdot \Lambda_{ges}.$$

If the green characteristic curve is differentiated, the normalized differential inductance is obtained (see Chap. 7)

$$L^*(I) = \frac{L(I)}{N^2}.$$

The curves of L^* for Fig. 3.9 belonging to different air gaps are shown in Fig. 3.10. This procedure is called 'shearing' of the characteristic curve. A linearization occurs. This means that the deviation from the linearity remains below a set maximum value over a larger range of the modulation. The inductance value is always smaller than the maximum value set by the material. Also, the change in differential inductance is not as strong in the vicinity of saturation. The differential or small-signal inductance at $I = 0$ is defined in normalized form as a so-called AL value:

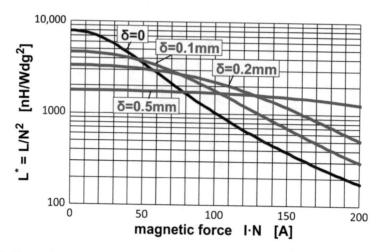

Fig. 3.10 Change in differential inductance L* (A_L value) with the insertion of air gaps into a core

$$A_L = L^*(I = 0) = \frac{L(I = 0)}{N^2} = \frac{\Delta\Psi}{N^2 \cdot \Delta I}$$

This value is given by the manufacturers for the characterization of cores. The unit of measurement here is implicitly nH/Wdg^2 (the unit of measurement is often not shown separately). The inductance of a coil with a specific core is therefore obtained by multiplying the A_L value by the square of the number of turns. For A_L values below the maximum at $\delta = 0$, these values can be tolerated relatively narrowly, as they depend essentially on the mechanical machining tolerance.

Permanent magnets have extremely high H_C values with high remanence values. The magnetization state here is 'frozen' to a certain extent. This picture actually applies, because at high temperatures this state of order disappears and the flux of the magnets goes to zero. This can be dangerous for some applications. For example, synchronous machines lose the ability to build up torque. Figure 3.11 shows the stylized magnetization characteristics of a NdFeB magnet. Both the polarization of the material J(H) and the corresponding B(H) characteristic are shown. Due to the high coercive field strength H_C, it is clearly visible here that the B(H) characteristic curve becomes proportional $\mu_0 \cdot H$ at high field strength values. For soft magnetic materials, the coercivity is <100 A/m. This corresponds to a flux density increase $<125\mu T$ for the material-free space and is therefore hardly noticeable. For a certain range for H shown in Fig. 3.11a, the permanent magnet can also be regarded as a magnetic flux source with internal magnetic resistance. Figure 3.11b shows such a two-pole. Thus, permanent magnetic sections can also be inserted into (linear) magnetic equivalent circuits.

Figure 3.12 shows the demagnetization branch of the characteristic curve for several materials. The product B-H has the unit Ws/m^3, that is, it corresponds to an energy density. The maximum energy density corresponds to the remagnetization energy. A permanent

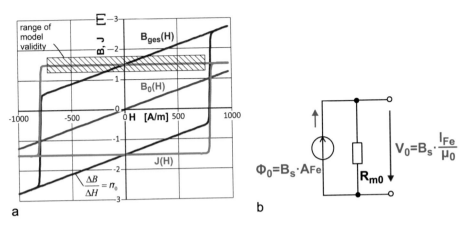

Fig. 3.11 The B(H) characteristic of a permanent magnet together with the polarization J(H) and the characteristic of the 'inner' magnetic vacuum resistance (**a**) and model of an ideal flux source Φ_0 as two-pole (**b**) (A_{Fe}: magnetization cross-section, l_{Fe}: magnetization length)

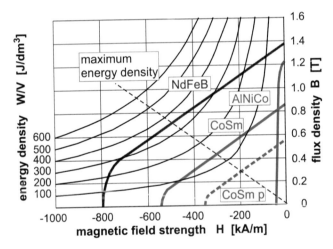

Fig. 3.12 Demagnetization characteristic curve for different materials in comparison and corresponding energy density

magnet is expected to have the highest possible "resistance" to external fields and thus high re-magnetization energy. The so-called quality product $W = (B \cdot H)_{max}$ is used as a measure for this. Therefore, the lines for $B \cdot H =$ const were drawn in the diagram.

In other words, this energy is stored in the permanent magnet - even without current flowing. The energy densities stored in the permanent magnet thus receive values ≤ 600 Ws/dm^3. For comparison: In an electric field with a dielectric of barium titanate ($BaTiO_3$), a

Table 3.3 Energy density in different media [2, 3, 9, 14]

Medium	Max. energy density [Ws/dm^3]	Condition	Spec. Resistance
Nd$_2$Fe$_{14}$B	500	T = 20 °C, T$_C$ = 310–370 °C	1.6 Ω·mm^2/m
CoSm	200	T = 20 °C, T$_C$ = 700–800 °C	50 and ca. 100 × 10^{-6} Ω·cm
Cunife (20Fe20Ni60Cu)	12	T = 20 °C, T$_C$ = 410 °C	1.8·10^{-7} Ωm
Alnico8 (34Fe7Al15Ni35Co4Cu5Ti)	36	T = 20 °C, T$_C$ = 860 °C	1.8·10^{-7} Ωm
Barium titanate (BaTiO$_3$)	4.43	T = 20 °C, E = 100 kV/cm ε$_r$ = 1000	–
Electrolytic capacitors	20–200	T = 20 °C	–
Double-layer capacitors	200–5000	T = 20 °C	–
Lead-acid battery	60…600,000	T = 20 °C	–
Li-ion battery	300…1,500,000	T = 20 °C	–
Diesel	36,000,000	–	–

ferroelectric with a high permittivity number $\varepsilon_r \approx 1000...5000$ and breakdown field strengths of approximately 10 kV/mm, approximately *443 Ws/dm^3* [3, 4] can be achieved.

In Table 3.3 further energy densities are given for illustration and classification. However, this energy is "frozen" in the material of the permanent magnet. The state can only be changed again with considerable energy expenditure. If one estimates the energy required for a remagnetization cycle with

$$\Delta W = V \cdot \oint H dB \approx 4 \cdot V \cdot B_r H_C = V \cdot 4800 \frac{Ws}{dm^3}$$

it is clear that permanent magnets have stored a state of energy, which cannot be used in the sense of a reversible energy storage device with high efficiency like a capacitor or an electrochemical battery. NdFeB magnets are currently the strongest magnets. They are used, for example, to build very compact synchronous machines. High coercive field strength is very important in such and similar applications because when the machines are braked, a high current with a magnetic field opposite to the field causing it occurs. If the values of H$_C$ are too low, this can lead to demagnetisation of the excitation magnets. Then no more torque can be built up in the motor and neither drive nor recuperation is possible.

Because of such aspects, the coercivity is used as a criterion for the classification into hard magnetic and soft magnetic materials (Fig. 3.13). The terms hard magnetic and soft magnetic originate from the early days of research and use of magnetic materials. For

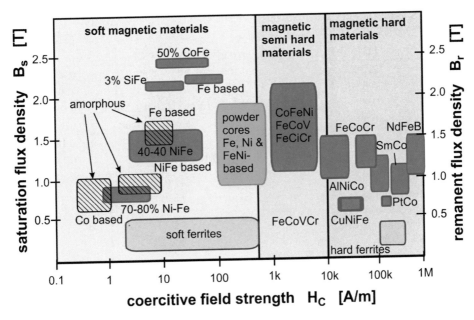

Fig. 3.13 Classification and examples of magnetic materials (based on [2], further parameters in table 3.4)

metals, the rule was that low H_C and B_r values were associated with mechanically soft materials. If necessary, a reduction of these values could also be achieved with a Fe-alloy, for example, by soft annealing. Although in the meantime a large number of materials (alloys, ceramics) have been added for which this rule does not apply, the designation is retained for classification purposes. In general, it can be stated that magnetizability is an ordered state of matter which can be massively changed, disturbed or destroyed by temperature influence. Therefore, all applications are subject to possible application temperatures (Table 3.4).

3.2 Atomistic Models of Magnetism

In the previous section, many relationships were discussed on the basis of the macroscopic properties of the B(H) curves. The magnetic polarization of matter under the influence of a magnetic field is due to the formation and orientation of magnetic dipoles at the atomic level. If the dipoles interact with each other in a suitable way, the orientation of the dipole moments can coincide in certain areas. Such magnetic order states exhibit particularly pronounced magnetic phenomena.

A magnetic dipole can be defined by a circular current $I = e_0 \cdot v$ formed by an electron on a circular path, which encloses an area A (Fig. 3.14). In quantum mechanical physics, the orbit of a charged particle first mechanically generates an angular momentum $\vec{L}$. For this

Table 3.4 Selection of soft and hard magnetic materials with orienting parameters

Material	Bs [T]	Br [T]	H_C [A/m]	T_C [°C]	ρ [$\times 10^{-6}$ Ωm]
Vacodur 50	2.2–2.4	1.7	~100	950	0.4
VC6030F	0.82	–	<10	365	1.3
$Nd_2Fe_{14}B$	1.5	1.5	$5\ldots20\cdot10^5$	310–370	1.6
CoSm	0.9	0.9	$(2..8)\cdot10^5$	700–800	0.5–1
CuNiFe	0.54	0.54	$44\cdot10^3$	410	0.18
AlNiCo	0.76	0.76	$(8\ldots20)\cdot10^4$	860	0.18

Fig. 3.14 Formation of magnetic dipoles and moments at the atomic level using Bohr's atomic model

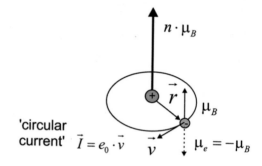

purpose, it is assumed that the mass point executes a plane circular motion around the origin. Then the angular momentum vector is perpendicular to the circular plane, that is, in the direction of the axis of the circular motion. It is formed from the radius $\vec{r}$ and the momentum of the particle $\vec{p}$ via the vector product $\vec{L} = \vec{r} \times \vec{p}$. For the absolute value is then

$$L = r \cdot m \cdot v = m \cdot r^2 \cdot \omega$$

Where ω is the angular velocity, m is the mass and r is the distance r of this mass from the axis of rotation. The angular momentum indicates the 'momentum' of the rotation. The angular momentum vector $\vec{L}$ points in the direction in which a right-hand screw would move ahead if the direction of rotation were the same. The right-hand rule applies. The magnetic moment is derived from this. In quantum mechanical terms, the angular momentum of a charged point particle with mass m and charge q generates the magnetic moment

$$\vec{\mu} = \mu \cdot \frac{\vec{L}}{\hbar},$$

where the so-called magnetron of the particle is $\mu = \frac{q}{2\,m} \cdot \hbar$. If we now use the elementary charge of an electron e_0 for q and its mass m_e for m, where $\hbar$ the reduced Planck's quantum of action is, we obtain Bohr's magnetron

$$\mu_B = \frac{e_0}{2\,m_e} \cdot \hbar = 9.274 \cdot 10^{-24}\,\frac{Ws}{T}$$

It should be noted that, due to the negative charge of the electron, its magnetic moment is always $\vec{\mu}$ directed $\vec{L}$ opposite to its orbital angular momentum. A magnetic (dipole) moment has its lowest energy in the magnetic field when it is opposed to the field, that is, orbital angular momentum and spin are aligned parallel to the field direction.

The orbital torques then result as multiples of Bohr's magnetron from the orbital angular momentum

$$\mu = \frac{e_0 \cdot \omega}{2\pi} \cdot \pi \cdot r^2 = \frac{n \cdot e_0 \cdot h}{4\pi \cdot m_e}$$

The first fraction in the equation can be interpreted as an atomic circular current. Trajectory moments are integer multiples of Bohr's magnetron μ_B. In addition, there is the magnetic dipole moment of the electron (intrinsic angular momentum, spin). It is then $\mu_e = -\mu_B$.

If a medium contains N magnetic dipoles with the same orientation per unit volume, the magnetization is the sum of the individual effects. If the dipoles have different orientations, a vectorial addition of the dipole moments in the volume element must be carried out. In the atomic range, three types of magnetic dipole moments can be distinguished:

1. the magnetic dipole moment of the electrons associated with the intrinsic angular momentum (spin) The magnitude of this physical quantity is called Bohr's magnetron μ_B
2. the magnetic dipole moment of the electrons associated with the angular momentum L.
3. the magnetic moment of nucleons. As a result of the large mass of the nuclear particles - compared to the electrons - and their slow motions, their magnetic moment is relatively small ($\mu_K \sim \mu_B\,/2000$) and therefore generally negligible.

The arrangement of the electron spins in the electron shell of an atom follows Hund's rule. This rule states that when the levels s, p and d etc. are filled up, first only electrons from a uniform spinning device are incorporated. Only then do the electrons follow with the other spinning electrons.

Under suitable conditions, the interaction of the resulting moments of neighbouring atoms (or ions) can lead to order states in the crystalline near-order. The formation of the magnetic order states takes place within spatially limited areas, so-called domains or Weiß's domains (Fig. 3.15a). The structure and arrangement of the domains follows the principle of minimizing the stored (potential) energy. A bar magnet in comparison has a

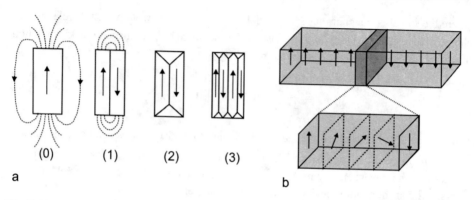

Fig. 3.15 Magnetic order states in the crystalline near region. (**a**) Formation of domains as Weiss domains by subdivision, which are arranged in such a way that the energy content of the arrangement is minimized, (**b**) domains with 180°-Bloch wall, in which the polarization direction changes at a short distance [after 2]

strong external field. For a permanent magnetic material, this means that it has been brought to this state by an external, magnetizing field and that 'internal' friction of the grains of the structure prevents it from assuming a lower-energy state. When one heats such a magnet, one stimulates thermal lattice vibrations. As a result, the magnetic order state is increasingly cancelled. The magnetic field closes over the remaining order states in the material, resulting in a much lower energy state of this new order. The shorter the transition from the completely disordered state to the ordered state with "frozen" magnetic order states, the smaller the crystallites formed. This plays a role, for example, in the sintering of ferrites, in which the ferrites lose 40–50% of their volume. The longer the heat treatment is allowed to continue, the larger crystallites are formed. The dimensions reach several millimetres if you wait long enough. Since the aim is to achieve small grain sizes in the interest of low losses, the management of the thermal process is of particular importance.

For metals, cooling rates of 10^6 K/s can be achieved by casting the alloy as a thin strip onto a cooled roll. At such a speed, crystallization processes are undermined and the metal is obtained as an amorphous material without any internal close order. Subsequent heat treatment can be used to achieve grain formations up to the desired size. Under the right conditions, domains of the same magnetic polarization/orientation are formed in these grains.

The domains with a uniform internal orientation of the magnetic moment are adjacent to each other with so-called Bloch walls, in which the direction of orientation changes over a short distance.

The maximum value of the resulting magnetic moment is to be expected with a half-filled shell. Diamagnetic materials do not exhibit a resulting magnetic moment. Paramagnetic substances, however, contain atoms (or molecules) whose orbital or spin moments are not compensated. The existence of such permanent atomic dipoles can - under suitable conditions - lead to the formation of order states with ferro-, ferri- and antiferromagnetic behaviour.

The atoms (or molecules) *of diamagnetic substances* have no permanent magnetic moment. If such substances are exposed to a magnetic field, magnetic moments are induced which are opposite to the external magnetic field. Atoms with a large extension of the electron shell allow a high diamagnetic susceptibility to be expected. The diamagnetic susceptibility is, in the first approximation, independent of temperature. The magnetic flux density field lines are "pushed" out of the material in diamagnetic materials (Table 3.5).

In *paramagnetic substances,* the single atoms (or molecules) have a resulting magnetic moment. In the absence of a magnetic field, such substances are not magnetized, since the orientations of the dipoles are statistically distributed as a result of thermal movement. By applying a magnetic field, a partial orientation of the magnetic dipoles in the field direction is achieved. The vectorial addition of the dipole moments yields a magnetization of the medium which coincides with the field direction. For some paramagnetic substances, the paramagnetic susceptibility is inversely proportional to the absolute temperature T. This relationship is called Curie law. Paramagnetic properties with such a temperature dependence include substances containing ions of transition elements. This behaviour is particularly pronounced in the case of rare earth salts.

Some substances show very pronounced states of order. As can be seen in Table 3.5, three special forms of magnetic order states can be distinguished.

Ferromagnetic behaviour is characterised by the fact that the atomic moments within a Weiss domain are rectified. In *ferrimagnetic* materials, there is a partial compensation of the magnetic moments. The *antiferromagnetic* order state is characterized by a complete compensation of the magnetic moments.

Ferromagnetic substances are characterized by a spontaneous magnetization, that is, by parallel orientation of the resulting atomic moments within the Weiss domains. The occurrence of ferromagnetism is bound to certain conditions. The atomic distance in relation to the extent of the configuration of the 3d electrons plays a special role. The interaction energy is defined in such a way that the parallel position of the magnetic moments of neighbouring atoms is promoted in the case of positive energy values. With negative interaction energy, an antiparallel position of the magnetic moments can be expected. If the amount of interaction energy is very small, paramagnetic behaviour or a ferromagnetic material with a low Curie temperature (gadolinium, Gd) results.

Formally, the ferro- and antiferromagnetic order states can be understood as boundary cases of ferrimagnetic behaviour. However, it should be noted that the ferromagnetic order state is bound to the existence of conduction electrons. Ferri- and antiferromagnetic behaviour, on the other hand, can also occur in insulators.

Ferrites are ceramic (polycrystalline) materials. During their production, suitable metal oxides in powder form are pressed into moulds and fired. The composition and the sintering process are decisive for the properties of the end product. An example of cubic ferrite material is manganese ferrite with the composition Mn_8^{2+} [Fe_6^{3+}] O_{32} [5]. As a result of the indirect exchange interaction, the manganese ions located on tetrahedron sites have a spin

Table 3.5 Overview of types of magnetism for different materials

Type of Material	Conditions	Magnetic Moment	Relative permeability	Field lines in material
Diamagnetism (intert gases, ion crystals, Bi, Si, Ge, Cu, Ag)	**Atoms with closed shells**		$\mu_r < 1$	
Paramagnetism (alkaline metals, rear earth elementss, O_2, Al, Sn, Pt)	**Atoms with not closed shells**		$\mu_r > 1$	
Ferro Magnetism (Fe, Co, Ni, Gd, alloys)	**Atoms with not closed shells**		$\mu_r >> 1$	
Ferri Magnetism (metal oxides, ferrites, cobalt iron oxides)	**Atoms with not closed shells**		$\mu_r >> 1$	
Anti Ferro Magnetism (MnF$_2$, orthoferrites, hematite)	**Atoms with not closed shells**		$\mu_r > 1$	

Table 3.6 Comparison of MnZn and NiZn ferrites

Ferrite	B_{SAT} [mT]	μ_i -	T_C [°C]	ρ [Ωm]	$f_{typ.}$ [MHz]
MnZn	350–500	500–2.10⁴	100–300	0.1–10	…10
NiZn	350–400	15–2000	100–350	10^3–10^8	…1000

orientation which is antiparallel to the spin orientation of the iron ions located on octahedron sites. The magnetic moments for Mn^{2+} and Fe^{3+} can be used to specify the resulting moment per unit cell.

The magnetic moment per unit cell can be increased by replacing some of the manganese ions with non-magnetic ions (e.g. Zn ions). This produces MnZn ferrites, which are widely used in power electronics. The mineral magnetite ($FeO + Fe_2O_3$) also belongs to the ferrite group.

Nickel ferrite has a relatively low magnetic moment. The magnetic moment can also be considerably increased here if some of the nickel ions are replaced by zinc ions. The zinc ions occupy tetrahedral positions and displace some of the iron ions. This procedure results in the second important ferrite group of NiZn ferrites. Table 3.6 shows the properties of the most commonly used MnZn and NiZn ferrites. Within the typical limits given, a wide range of materials with different combinations of properties are offered depending on the application requirements. The specific resistance of the ferrite types mentioned lies in the range of about 10^{-1} Ωm to 10^5 Ωm at 20 °C.

3.3   Temperature Influences on the Magnetic Properties

As stated above, the magnetic properties of substances are order states. When the temperature is increased, these states are more and more superimposed by heat-induced movements. This leads to a temperature dependence for many parameters. Ferrites are generally more sensitive to temperature changes.

All magnetic order states are destroyed when a certain material-specific temperature is exceeded. This limit temperature is called *Curie temperature* T_C for ferro- and ferrimagnetic substances. In the case of antiferromagnetic substances, it is called the Neel temperature T_N. Above this temperature, the substances mentioned exhibit paramagnetic behavior (Fig. 3.16).

This means that the ferromagnetic properties disappear at temperatures $>T_C$. If permanent magnets are heated to temperatures above the Curie temperature, the internal order states disappear permanently and the material is demagnetized. Afterward, even when cooling down, practically no external field can be detected. Even in soft magnetic states, such a strong increase in temperature is generally associated with permanent changes in properties, since internal phase transitions occur on the crystalline structure. For example, preferential directions disappear in grains that are produced during rolling. On the other

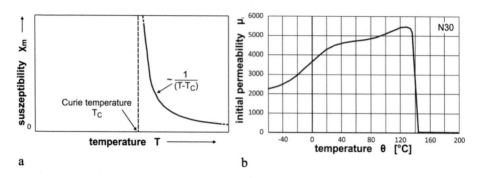

Fig. 3.16 Curie temperature of ferromagnetic materials. (**a**) General/principal, (**b**) Temperature dependence of the initial permeability of ferrite N30 [6]

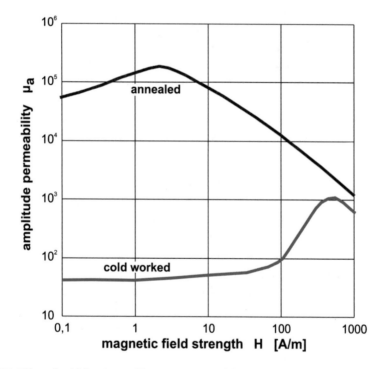

Fig. 3.17 Effect of cold forming and heat treatment on Mu-Metal [7]

hand, the desired outstanding soft magnetic properties of Mu-Metal after mechanical deformation can only be regained by subjecting the material to heat treatment (Fig. 3.17).

At only 1.5 A/m, the coercivity of Mumetall is one order of magnitude below the typical values of electrical steel. Together with a very high maximum permeability, the metal is excellently suited for shielding against magnetic interference fields. In addition to a roll-hardened condition, Mumetall is also supplied soft-annealed or soft deep-drawable, which

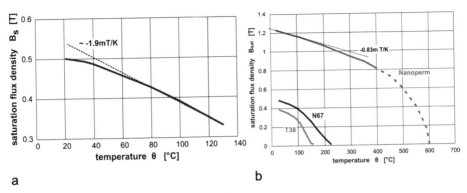

Fig. 3.18 Dependence of saturation induction on temperature. (**a**) for N27 and (**b**) in comparison of ferrites T38, N67 with the metallic material Nanoperm [6, 8]

also allows more complex part geometries. As advantageous as NiFe alloys are for small and medium magnetic fields, this class of materials is less suitable for power-transmitting applications, for example in motors, generators or actuators, due to the limited saturation polarization. In any case, it should be noted that renewed cold forming after final annealing can lead to a loss of important target properties. Heat treatment must therefore be carried out at the end of a forming process.

Soft magnetic materials generally react very sensitively to mechanical processing and thermal treatment. This makes the production of laminated cores, for example, complicated. In order to achieve certain core shapes, the sheets or strips must be cut or cut out mechanically or thermally (laser, EDM (electrodischarge machining)). At the same time, metallic contact on the cut surfaces must be avoided so that the layered materials remain insulated from each other. This is another reason why cores have additional losses at such interfaces.

An increase in temperature always causes a reduction of the saturation induction in ferromagnetic materials (Fig. 3.18). This circumstance must be taken into account when dimensioning components. The specification in the material data sheet usually refers to 20 ° C, while the operating temperature is considerably higher. The maximum usable temperature is limited upwards by the Curie temperature T_C. For most of these materials, this is in the range 130...250 °C. Above this temperature, the materials no longer have ferromagnetic properties. It, therefore, represents a barrier to use. The materials are therefore used up to temperatures at which the loss minimum of the materials occurs (see Fig. 3.21). It must also be taken into account that the insulating materials used have the thermal class corresponding to the application temperature. The usable induction "swing" of metallic materials is usually significantly higher than that of ferrites. These materials are therefore suitable for use with DC chokes with alternating current. Figure 3.18b shows 2 ferrites in comparison to a nanocrystalline material. However, ferrite has a density of 4.9 g/cm^3 compared to about 7.9 g/cm^3 of ferromagnetic alloys. This offers room for optimization.

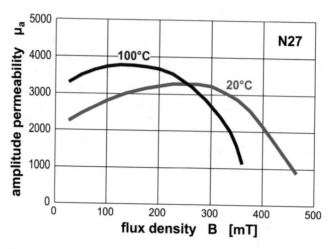

Fig. 3.19 Amplitude permeability as a function of induction for N27 at different temperatures [6]

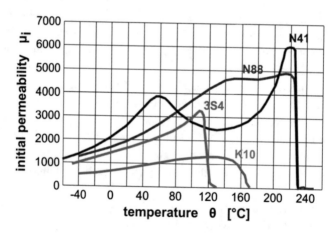

Fig. 3.20 Initial permeability of various ferrites as a function of temperature [1, 6]

Not only the saturation inductance changes with temperature but also the shape and rise of the B(H) curve. This is expressed by the changed amplitude permeability curves in Fig. 3.19. Besides the non-linearity of the B(H) curve, the strong temperature dependence of all material parameters is the reason for the fact that in component design efforts are made to insert linear (air) sections in such a way that at least a partial linearization of the behavior occurs.

No clear trend can be seen in the influence of differential permeability dB/dH, especially for ferrites, as Fig. 3.20 shows. The depicted dependence of initial permeability on temperature has a maximum value shortly before the Curie temperature for many materials. The permeability values at room temperature and maximum use temperature differ quickly by more than 3...4 times for ferrites. Apart from the relatively large tolerances during

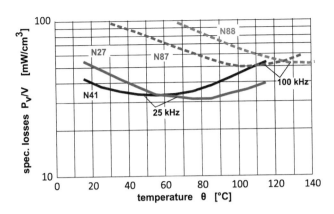

Fig. 3.21 Temperature dependence of the specific magnetization losses for different ferrites [6]

manufacture, this fact must be taken into account when designing the reproducibility of components. Core geometry and winding design play an important role here.

The occurrence of losses during the remagnetization of cores has very different causes, which will be discussed in the following Sect. 3.4. Figure 3.21 shows the specific losses, that is, the power loss per volume as a function of temperature. In the case of ferrites, it can be observed that both positive and negative temperature coefficients occur. It is generally advisable to dimension the components for a maximum temperature that corresponds to the minimum loss. In modern power ferrites, this temperature is in the range of 80 to 140 °C. It is problematic to place the operating point in a range with a positive temperature coefficient since an increase in temperature both increases the specific losses and reduces the saturation induction. This results in a a positive feed back during loss generation. Due to the relatively low thermal conductivity (see Table 3.7), this can lead to problems with longer, poorly cooled magnetic sections. Note that because of the logarithmic representation, the temperature coefficient in Fig. 3.21 can be underestimated.

Table 3.7 Comparison of properties of MnZn and NiZn ferrites (data: Ferroxcube [1])

Parameters	MnZn ferrite	NiZn ferrite	Unit
Modulus of elasticity	90...150	80...150	$\cdot 10^3$ N/mm^2
Compressive strength	200...600	200...700	N/mm^2
Breaking strength	20...65	30...60	N/mm^2
Vickers hardness	600...700	800...900	N/mm^2
Specific resistance	0.1...10	$10^3,,,10^8$	Ωm
Linear thermal expansion	10...12	7...8	$\cdot 10^{-6}$ K^{-1}
Specific heat	700...800	700...800	Ws/(kg·K)
Thermal conductivity	3.5...5	3.5...5	W/(m·K)

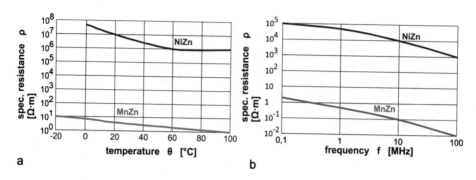

Fig. 3.22 Dependence of the specific resistance of MnZn and NiZn ferrites (data: Ferroxcube [1]). (**a**) Temperature dependence, (**b**) Frequency dependence

An essential distinguishing feature of MnZn and NiZn ferrites is the resistivity of the materials. MnZn is to be classified as a "poor conductor", while NiZn is to be regarded as an insulator. Here too, strong temperature dependencies are evident. As Fig. 3.22 shows, the resistivity of these ferrites in direct current varies greatly with temperature. In MnZn ferrites, it decreases by one order of magnitude in the range 0 °C–100 °C, and in NiZn ferrites by almost 2 orders of magnitude. The electromagnetic field results in various loss mechanisms in these materials, so that lower resistances result at higher frequencies, as shown in Fig. 3.22.

Due to the skin and proximity effect, the measurable resistivity is frequency-dependent, as shown in Fig. 3.22b. Nevertheless, for NiZn ferrites a distance of approximately 4–5 orders of magnitude remains above the level of MnZn ferrites. This predestines them for HF applications and pulse applications with extreme speeds of remagnetization, which result in correspondingly high field strengths in the material.

For these applications, the dielectric constant or permittivity is also of interest, since it can possibly mean a capacitive shunt to the component. The corresponding ratios are shown in Fig. 3.23 for orientation. The permittivity of all ferrites is generally in the order of 10...100 for NiZn materials. This applies in particular to the grain boundaries of the ferrite grains. However, due to their structure, MnZn ferrites also have electric dipoles that are oriented by an electric field and lead to an internal electric polarization. The permittivity numbers are correspondingly very high, as Fig. 3.23 shows. They are a further reason for the increase in the HF conductivity of the material with increasing frequency.

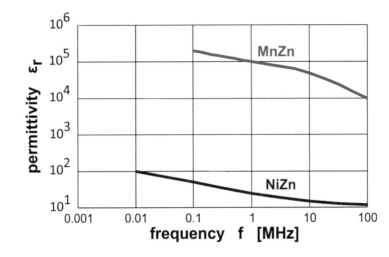

Fig. 3.23 Typical dielectric constant of MnZn and NiZn ferrites as a function of frequency (data: [1])

3.4   Loss Mechanisms in Magnetic Components

With magnetic components, various loss mechanisms occur, which are classified here first of all according to their place of origin as follows.

- Winding losses
- Core losses
- Losses in the environment (eddy current losses)

The winding losses comprise all shares of losses in the winding. First of all, the ohmic line losses are decisive. These increase with increasing frequency due to the skin effect. This causes electromagnetic interactions of individual "streamlines" in the conductor so that the current only flows on the surface of the conductor and not the entire cross-section of a conductor is available. The amount of these losses is determined by the conductor material, wire cross-section, temperature and degree of stranding. Depending on the design, the winding structures are permeated by magnetic fields. These come from adjacent conductors or stray fields. This results in eddy currents through the conductor materials around the magnetic field lines, with corresponding losses. This effect is called the proximity effect. The windings always have a certain winding capacity, through which current flows depending on the frequency. As a result, the dielectric losses that occur must be taken into account in a power loss balance, especially at high frequencies.

The losses in the vicinity of the component are mainly caused by stray magnetic fields of the component. Stray fields (unintended fields) are alternating fields that initially cause induced voltages. These result in losses if the material used in the construction or as shielding is conductive. In high-frequency and/or high-voltage applications, dielectric losses in the environment also play a role. In this section, the main focus is on core-related losses and their causes. The core losses are influenced by very different aspects: Core material, core shape, air gap, temperature, the position of the winding, the distance of the winding to the core, field strength, induction, premagnetisation and signal shape. The material-related losses will be considered in particular in the following. Several effects (hysteresis losses, eddy current losses, magnetic disaccomodation losses and dielectric losses) add up to one result:

$$P_{vcore} = P_{vhys} + P_{vedd} + P_{vdisac} + P_{vdiel}$$

Obviously, for ferromagnetic materials, the closed loop integral gives $\oint HdB$ a value other than zero. It should be noted that this is the characteristic obtained in the case of remagnetization with di/dt $\to 0$. For a volume V, one obtains *hysteresis losses that* increase with frequency to

$$P_{vHys} = \oint HdB \cdot f \approx k_1 \cdot \widehat{B}^2 \cdot f.$$

In a first approximation, the eddy current losses are proportional to the square of the induced voltage. This in turn is proportional to the maximum induction and the frequency that occurs:

$$P_{vWirb} \approx k_2 \cdot \widehat{B}^2 \cdot f^2$$

The magnetic disaccomodation losses (also called residual or viscosity losses) are part of the iron losses in addition to the hysteresis and eddy current losses. They record the time lag of the induction behind a preceding field change (delays in the rotation and diffusion processes). In the case of relatively slow induction changes, processes similar to a highly viscous liquid are observed as a result of non-elastic micro-deformations. For high frequencies, they are negligible compared to the above-mentioned losses (hysteresis and eddy current losses).

A magnetic, thermal or mechanical disturbance suddenly changes the permeability of a ferro- or ferrimagnetic material. The slow decrease in permeability over time following the disturbance is called disaccommodation. The dislocation coefficient is the relative decrease in the initial permeability μ_i of soft magnetic material at constant temperature after demagnetization, divided by the logarithm of the ratio of the two measuring times t_1 and t_2 after demagnetization:

$$D = \frac{\mu_i(t_1) - \mu_i(t_2)}{\mu_i(t_1) \cdot \log_{10}\left(\frac{t_1}{t_2}\right)}$$

In the case of fast remagnetizations, the material can no longer follow here, so that this influence becomes insignificant in the HF range. However, it plays a role as a "creeping process" in magnetic measurements of slow processes, as investigations have also shown on ferrites, which can be used far into the MHz range.

Dielectric losses within the core can only be important at correspondingly high permittivity values and/or frequencies. For MnZn they are possible as influencing variables of lower order. Here an approximate proportionality similar to eddy current losses is present. Since the current density in the electric alternating field can be determined with

$$\underline{J} = (\sigma + j\omega\varepsilon_r\varepsilon_0)\underline{E} \text{ where } \underline{E} \sim j\omega A_{Fe} \cdot B.$$

If one assumes the dielectric losses to be proportional to the square of the dielectric current component, the result is approximately a loss component at very high frequencies which is $\sim f^4$. The loss factor tan δ for core materials as a function of frequency summarizes these losses and provides an indication. As a single characteristic value, it is often only given for a certain frequency and is therefore not very meaningful. In the higher frequency range, the sum introduced above can be written as

$$P_{vKern} \approx k_1 \cdot \widehat{B}^2 \cdot f + k_2 \cdot \widehat{B}^2 \cdot f^2 + k_3 + k_4 \cdot \widehat{B}^2 \cdot f^4$$

With the so-called Steinmetz formula (Charles Proteus Steinmetz, *9. April 1865, Breslau; †26. Oktober 1923, Schenectady) or a modification of it, one can transfer an empirically obtained data sequence for the core losses into an analytical formula. This procedure should not be confused with a physical relationship. An equation is used which has properties as similar as possible to those of the data to be represented. This formula is then adapted to the data according to the Gaussian method of minimizing the quadratic description error using free parameters. This gives an analytical calculation formula that can be differentiated, for example. The procedure is used and described in Chapters 4 and 9. The Steinmetz formula, which has been modified several times in the meantime, reads

$$\frac{P_{vKern}}{V_{Kern}} = k_e \cdot \left(\frac{\widehat{B}}{1T}\right)^\alpha \cdot \left(\frac{f}{1Hz}\right)^\beta$$

By normalizing to the units of measurement, one obtains so-called unit-related size equations. This must be taken into account when using other units. The operating conditions must be set in such a way that the resulting losses do not lead to permissible temperature values being exceeded. The treatment of these relationships in practice can be found in Chap. 4.

3.5   Energy and Forces in the Magnetic Field

The energy stored in the magnetic field is calculated as energy density related to the volume with the B(H) characteristic to

$$\frac{\Delta W_{magn}}{\Delta V} = \int\limits_{\Delta B} H \cdot dB = \int\limits_{B1}^{B2} H \cdot dB$$

For the case $B = \mu \cdot H$, this formula leads to the relationship

$\frac{\Delta W_{magn}}{\Delta V} = \frac{B_2^2 - B_1^2}{2\mu} = \frac{B_2^2 - B_1^2}{2 \cdot \mu_r \cdot \mu_0}$ or for $B_1 = 0$ and $B_2 = B$ the following applies $\frac{\Delta W_{magn}}{\Delta V} = \frac{B^2}{2 \cdot \mu_r \cdot \mu_0}$.

For a given flux density B, the stored energy is therefore inversely proportional to relative permeability μ_r. It follows that in a magnetic circuit with a constant cross-section and an air gap, the stored energy density in the air gap is much higher than that in the ferromagnetic material. At the induction of 1 T, an energy density in an air of 399 Ws/dm^{-3} is thus obtained. Assuming such an arrangement of a magnetic circuit with an air gap, the force effect on an interface can be roughly calculated as follows

$$F = \frac{dW}{d\delta} = \frac{d}{d\delta}\left(\frac{B^2}{2\mu} \cdot V\right) = \frac{d}{d\delta}\left(\frac{B^2}{2\mu} \cdot A_{Fe} \cdot \delta\right) \approx \frac{1}{2\mu_0} B^2 \cdot A_{Fe},$$

if B and H would not change when the field-filled volume element is enlarged. The force would point in the direction of the air gap change. This is not consistent with the observation that the interfaces of a magnetic circuit are drawn towards each other. This means that the assumptions made for the example do not apply to the case of the observation. Therefore some analytical additions are made in the following. If one has a magnetic circuit with $\Theta = I \cdot N$ with a constant cross-section A_{Fe} with high permeable material (μ_{rFe}) with an air gap of the length δ, one can calculate the magnetic flux to

$$\Phi = \frac{I \cdot N}{\frac{l_{Fe}}{\mu_{rFe} \cdot \mu_0 \cdot A_{Fe}} + \frac{\delta}{\mu_0 \cdot A_{Fe}}} = \frac{\mu_0 \cdot A_{Fe} \cdot I \cdot N}{\frac{l_{Fe}}{\mu_{rFe}} + \delta}$$

For the linear case, the flux density is thus

$$B = \frac{\Phi}{A_{Fe}} = \frac{\mu_0 \cdot I \cdot N}{\frac{l_{Fe}}{\mu_{rFe}} + \delta}$$

The energy content of the total magnetic field (material + air gap) is assuming linearity and $B = const.$ Along the whole path $l_{Fe} + \delta$

$$W = W_{Fe} + W_\delta = A_{Fe} \cdot B^2 \left(\frac{l_{Fe}}{2\mu_{rFe}\mu_0} + \frac{\delta}{2\mu_0} \right)^{-1} = \frac{\mu_0 \cdot A_{Fe} \cdot I^2 \cdot N^2}{2} \cdot \left(\frac{l_{Fe}}{\mu_{rFe}} + \delta \right)^{-1}$$

The force acting on the interfaces of the air gap is then:

$$F = \frac{dW}{d\delta} = \frac{\mu_0 \cdot A_{Fe} \cdot I^2 \cdot N^2}{2} \cdot \frac{dW}{d\delta} \left(\frac{l_{Fe}}{\mu_{rFe}} + \delta \right)^{-1} = -\frac{\mu_0 \cdot A_{Fe} \cdot I^2 \cdot N^2}{2} \cdot \left(\frac{l_{Fe}}{\mu_{rFe}} + \delta \right)^{-2}$$

If one assumes here a highly permeable material, so that $l_{Fe}/\mu_r << \delta$ applies, the formula for the force calculation follows

$$|F| \approx \frac{\mu_0 \cdot A_{Fe} \cdot I^2 \cdot N^2}{2 \cdot \delta^2}$$

This term is proportional $\sim \delta^{-2}$ if the air gap is large enough. With an area of $A_{Fe} = 10 \text{ cm}^2$ and $N = 1000$, $I = 1A$ and $\delta = 1$ mm you get a force effect of 628 N. At $\delta = 0$ a maximum force of

$$|F_{max}| = \frac{dW}{d\delta}\bigg|_{\delta=0} = \frac{\mu_0 \cdot \mu_{rFe}^2 \cdot A_{Fe} \cdot I^2 \cdot N^2}{2 \cdot l_{Fe}^2}$$

achieved. Due to the high permeability of the ferromagnetic material, however, this is a purely theoretical value, since even the smallest gaps can have the effect of reducing the force. For the above example, with the additional assumptions $\mu_r = 4000$, $l_{Fe} = 400$ mm, the calculated maximum force is 62.8kN. By assuming a negligible saturation, however, one obtains a maximum induction of

$$B_{max:theor} = \frac{\Phi}{A_{Fe}} = \frac{\mu_0 \cdot I \cdot N}{\frac{l_{Fe}}{\mu_{rFe}}} = 3000 \frac{4\pi \cdot 10^{-7} \frac{Vs}{Am} \cdot 1 \ A \cdot 1000}{0.3 \ m} = 12.57 \ T$$

This value is apparently significantly higher than known saturation flux densities so that the calculation has certainly left the range assumed to be linear. With an assumed saturation flux density of 1.5 T, a maximum force of

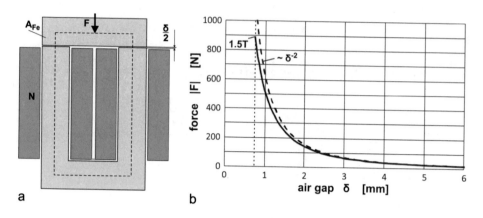

Fig. 3.24 Force effect of a U-core with I-yoke and air gap. (**a**) Schematic structure, (**b**) calculated dependence of the force on the air gap δ compared to the approximate formula

$$F_{max} \approx \frac{1}{2\mu_0} B_{sat}^2 \cdot A_{Fe} = 895 \text{ N}$$

could be achieved theoretically. Figure 3.24 shows the basic structure of such an electromagnet, consisting of a U-core and a yoke so that the air gap length δ is divided into 2 air gaps. It can be seen that the force decreases rapidly with increasing air gap length. For air gaps <0.74 mm one leaves the validity range of the assumption of linear relationships. This example shows that considerable forces can be released. Shell cores, for example, are also available with an air gap, which is produced by grinding off the center leg. If this is used without further processing, one must expect that the core will be excited to vibrate by pulsating currents caused by deformations due to the forces described. The oscillation frequency and thus the frequency of the radiated sound waves corresponds to twice the current frequency due to the quadratic dependence of the forces on the induction. This can result in an unpleasant whistling of the superstructure. Even if you don't hear it, you should keep in mind that humans are also influenced by inaudible sound frequencies and many animals directly perceive a much wider frequency range than humans. Inserting soft plastic panes or silicone glue to dampen this sound radiation helps here.

The above procedure can also be applied differently. For a magnetic component, the following applies to the stored energy

$$W_{magn} = \frac{1}{2} \int_V B \cdot H \cdot dV = \frac{L}{2} I^2$$

A force acting in the x-direction then results from the energy change to

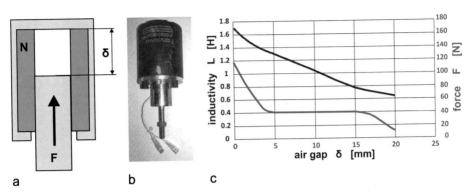

Fig. 3.25 Principle of a linear solenoid. (**a**) Principle structure, (**b**) Designed as a linear solenoid (source: Zacharias 2018), (**c**) Travel dependence of inductance and force at 1.2A [9]

$$F = \frac{d}{dx} W_{magn} = \frac{d}{dx}\left(\frac{L}{2}I^2\right) = \frac{I^2}{2}\cdot\frac{dL}{dx}.$$

This shows a further way to determine magnetic forces. The arrangement shown in Fig. 3.24 can also be treated in this way. This approach is obvious for the modeling of linear solenoids, as shown in Fig. 3.25, for example. The inductance of a coil with the core is greater than that without a core. A moving core outside a coil is pulled into the coil when a current is switched on. Figure 3.25b,c show the design of a linear solenoid and the measured dependence of inductance and force on the path of the armature.

It is difficult to determine the actual inductance for the shown direct current magnet with a solid core. The usual inductance measuring instruments work based on an impedance measurement with alternating voltage. Eddy currents are induced in a solid ferromagnetic core, which counteract the cause and thus falsify the measurement result. In such a case, the DC inductance can also be carried out by force measurement except for a constant.

Force effects on conductors and between conductors are interesting because they practically always affect within magnetic components. Coaxial arrangements are the easiest to calculate (Fig. 3.26). At sufficiently high frequencies, the electromagnetic field only forms between the inner and outer conductors. The energy content of the magnetic field or the inductance can again serve as a starting point for estimating/calculating the effects of forces. Between two coaxial conductors per length energy of

$$W_{magn} \approx \frac{L}{2}\cdot I^2 \cdot l = \frac{\mu}{2\pi}\ln\left(\frac{r_a}{r_i}\right)\cdot I^2 \cdot l$$

For a given internal resistance r_i, the total force F acting on the sheath electrode is equal to

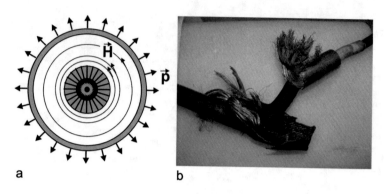

Fig. 3.26 Forces in a coaxial line. (**a**) schematic structure and compressive forces of a coaxial line, (**b**) coaxial line destroyed by magnetic forces ($A_{Cu} = 50$ mm^2) after an 80 kA pulse. (source: Siaenen 2011)

$$F = \frac{dW_{magn}}{dr_a} = \frac{\mu}{2\pi} \cdot \frac{1}{r_a} \cdot I^2 \cdot l$$

A coaxial cable with PE insulation must therefore withstand this electromagnetic pressure, which can be calculated as

$$p = \frac{F}{A_{zyl}} = \frac{\mu_0}{2\pi} \cdot \frac{1}{r_a} \cdot \frac{I^2 \cdot l}{2\pi r_a \cdot l} = \mu_0 \cdot \frac{I^2}{4\pi^2 r_a^2} .$$

This pressure increases quadratically with the current and can reach high values at pulsed currents. At 100kA and $r_a = 1$ cm, this results in a value of

$$p = \mu_0 \cdot \frac{I^2}{4\pi^2 r_a^2} = 4\pi \cdot 10^{-7} \frac{\text{Vs}}{\text{Am}} \cdot \frac{(100 \text{ kA})^2}{4\pi^2 \cdot 1 \cdot 10^{-4} \text{ m}^2} = 3.18 \cdot 10^{+6} Pa = 31.4 \text{ bar}$$

Figure 3.26b shows a coaxial cable, $A_{CU} = 50$ mm^2, which exploded at a fatigue point during a high current pulse [10].

However, forces also act on the surface of a current-carrying single conductor. Assume a cylindrical conductor through which the current I flows. The magnetic field strength is generally

$$H(r) = \frac{I}{2\pi r} \text{ or the energy density } \frac{dW_{magn}}{dV} = \frac{\mu_0 H^2(r)}{2} = \mu_0 \frac{I^2}{8\pi^2 r^2}$$

If we consider the energy stored in a volume element $dV = 1 \cdot 2\pi \cdot r_0 \cdot dr$, we can derive the following specific integral for the total energy, from which we can calculate the force effect and pressure on the conductor surface by differentiation according to the lower limit.

$$W_{magn} = \int_{r_0}^{\infty} \frac{\mu_0 H^2(r)}{2} l \cdot 2\pi r \cdot dr = \int_{r_0}^{\infty} \mu_0 \frac{I^2}{4\pi r} l \cdot dr = \mu_0 \frac{I^2}{4\pi} l \int_{r_0}^{\infty} \frac{dr}{r}$$

$$\frac{dW_{magn}}{dr_0} = F = -\mu_0 \frac{I^2}{4\pi r} l$$

$$p(r_0) = \frac{F}{2\pi r_0 l} = -\mu_0 \frac{I^2}{8\pi^2 r_0^2}$$

By this pressure, the surface of the conductor is stressed resp. compressed. By square dependence of current, here very high values of forces can be achieved. A thin, current-carrying tube can be compressed in this way by a current impulse. These forces are important for liquid and gaseous conductors, such as those used in arc welding. This supports the droplet separation from the welding electrode at high currents. A constriction tends to be reinforced by locally increasing magnetic forces so that the surface forces are more easily overcome. This effect is known as the pinch effect. These forces also act on the arc itself. The arc tends to expand due to thermal effects. Forces in the opposite direction include magnetic forces. The pinch effect caused by the plasma's own magnetic field also occurs here at sufficiently large currents and destabilizes the plasma at very high currents, such as those occurring in nuclear fusion.

The forces on 2 parallel conductors correspond to a fundamental study by J. M. Ampére. Figure 3.27 shows the basic arrangement of two parallel cylindrical conductors with antiparallel currents. If the two conductors are straight, thin, parallel and of infinite length, the simple formula for the length-related magnitude F of the acting forces is

$$\frac{F}{l} = \frac{\mu_0}{2\pi} \cdot \frac{I_1 \cdot I_2}{r}$$

For 2 parallel conductors, [11] gives the formula for calculating the inductance

$$L = \frac{\mu_0}{\pi} l \left(\frac{1}{4} + \ln \left(\frac{r}{\sqrt{r_1 \cdot r_2}} \right) \right) ..$$

From this, the following can be deduced for the acting force

$$\frac{F}{l} = \frac{1}{l} \cdot \frac{d}{dr} \frac{L}{2} I^2 = \frac{\mu_0}{2\pi} \frac{I^2}{r},$$

which corresponds to the above formula. By differentiating the formula for calculating the inductance for 2 conductor rails according to parameter a, the formula for the forces acting on the closely spaced conductor rails can be found

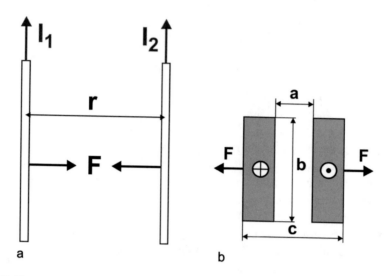

Fig. 3.27 Forces on 2 parallel conductors: (**a**) 2 thin conductors with currents in the same direction and (**b**) 2 busbars with a rectangular cross-section and currents of the same size and opposite direction

$$\frac{F}{l} \approx -\frac{\mu_0}{\pi} \cdot \frac{I^2}{a}.$$

Here too, considerable forces must be expected due to the quadratic current dependence. Continuous currents in the order of 100kA can be found, for example, in the fused-salt electrolysis of aluminium or electroplating. At a distance of a = 10 cm, forces of 40,000 N/ m are therefore acting here, which must be controlled mechanically.

Based on these considerations, the forces on windings in magnetic structures are to be estimated. If several windings of one layer are considered, they are next to each other and have the same orientation of the current. Furthermore, all currents are the same size but at different distances from each other. According to Fig. 3.27 2 windings are drawn towards each other. A free air coil tends to shorten. Due to the quadratic dependence of the forces on the current, the position of windings must be secured mechanically, for example, by taping, for larger currents. With alternating current, the maximum force is reached twice per period. This also applies to the forces in the core. Power transformers for 50 Hz therefore 'hum' at 100 Hz.

If a *single-layer cylindrical coil is* assumed, the windings contract axially, while the magnetic field presses the windings radially outwards (Fig. 3.28). The pressure to be expected is roughly estimated using the formula for long cylindrical coils

Fig. 3.28 Forces acting on the winding (N = 1) of a cylindrical coil

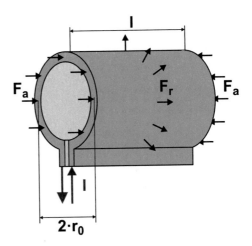

$$p_r \approx \frac{F_r}{A_{zyl}} = \frac{1}{A_{zyl}} \cdot \frac{d}{dr_0}\left(N^2 \frac{\mu_0}{2} \cdot \frac{\pi \cdot r_0^2}{l}\right)I^2 = \frac{\mu_0}{2l^2}N^2 I^2,$$

while the axial forces at the end faces of the coil, related to the length, resulting in

$$\frac{F_a}{2\pi r_0} \approx \frac{1}{2\pi r_0} \cdot \frac{d}{dl_{zyl}}\left(N^2 \frac{\mu_0}{2} \cdot \frac{\pi \cdot r_0^2}{l}\right)I^2 = -\frac{\mu_0 \cdot r_0}{4l^2}N^2 I^2$$

The principal force effects on a single-layer cylindrical coil without core are shown in the axial and radial directions in Fig. 3.28.

In transformers, windings are coupled together. For the following estimations, 2 coupled windings shall be considered. In the case of a short circuit of the secondary winding, the flux through this winding is opposite to the primary flux and of the same magnitude. In this case, the maximum current occurs. The magnetic field then essentially consists of the stray field between the windings. The inside of N_2 in Fig. 3.5. 6a is practically field-free. The magnetic energy of the field between the windings is

$$W_{magn} = \frac{L}{2}I^2 \approx \frac{\mu_0 \cdot \pi (r_2^2 - r_1^2)}{2l} \cdot N^2 \cdot I^2$$

On the surface of the inner winding, the total force F_1 or the surface-related pressure p_1 then acts:

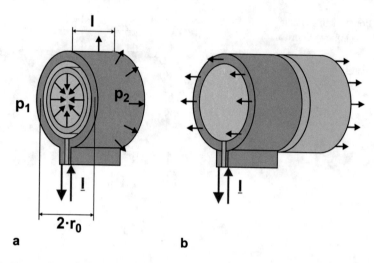

Fig. 3.29 Force effects between a coil with voltage impressing and a short-circuited, magnetically coupled coil. (**a**) Coaxial arrangement of the coils, (**b**) disc coils

$$F_1 = \frac{dW_{magn}}{dr_1} = \frac{L}{2}I^2 \approx -\frac{\mu_0 \cdot \pi \cdot r_1}{l} \cdot N^2 \cdot I^2$$

$$p_1 = \frac{F_1}{2\pi r_1 \cdot l} \approx -\frac{\mu_0 \cdot \pi}{2l^2} \cdot N^2 \cdot I^2 = p_2$$

The consequence of this force effect shall be illustrated with a simple example of a local. One of the most common rated outputs of local network transformers is 630 kVA. This corresponds to a nominal current of 909 A in the 400 V network. A referred short-circuit voltage $u_k = 4\%$ means that the short-circuit current is a maximum of 22.725 kA. The 400 V winding is located inside as $N_2 = 19$ with a winding width $l = 450$ mm and a coil radius $r_1 = 125$ mm. This combination of values results in a total force of $F_1 = 204.4$ kN or pressure on the winding of $p_1 = 578{,}455$ Pa $= 5.78$ bar. The windings must be mechanically designed for these force effects so that they are not destroyed in the event of a fault before the fuses respond. For mechanical reasons, high power transformers in the multi-MW range have u_k values in the range 10...25%. The inductive current limitation limits the force effects in the event of a fault.

Figure 3.29b shows the force effects on two adjacent disc windings. A short circuit generates a field opposite to the cause so that the two windings repel each other. If the secondary winding is arranged between the two halves of the primary winding in the sense of a symmetrical winding structure, axial forces act in the purely symmetrical case. In the case of asymmetries in the structure, additional forces act in a preferred direction which can also be used as acceleration forces [12].

The force effect at the short-circuit location is described briefly below. A short circuit on a double line is assumed for this purpose (Fig. 3.30). The double line has an inductance

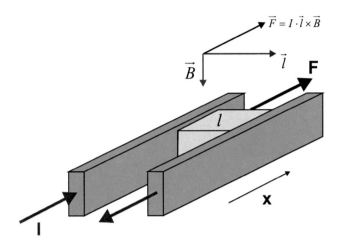

Fig. 3.30 Force effect on a sliding line short circuit in a rail gun

coating (inductance per length) of L'. The energy content of this magnetic field is $W_{magn} = \frac{L'}{2} \cdot l \cdot I^2$ so that the force in the direction of x $F_l = \frac{dW_{magn}}{dl} = \frac{L'}{2} \cdot I^2$ results. If the short-circuit is performed via movable contacts, this short-circuit between two lines can be accelerated. Relatively large currents are required for this. The currents or the rates of current rise are limited by the inductance of the power length reached in each case when the short circuit is moved. By cascade connection of the energy storages for acceleration, very high values (>7 km/s) can still be achieved here, which makes such an arrangement even interesting for the output of microsatellites (first cosmic velocity 7.91 km/s) [12, 13].

3.6 Magnetostriction

The above-mentioned force effects of this chapter are magnetic force effects between solid bodies. Magnetostriction is an effect by which the geometric dimensions of a body are changed under the influence of a magnetic field. *Magnetostrictive* or, conversely, the *magnetoelastic* effect is the interaction between the mechanical quantities stress and strain and the magnetic quantities flux density and magnetic field strength in ferromagnetic bodies. If a ferromagnetic material is brought into a magnetic field, it reacts magnetostrictively with a change in length and, if volume constancy is assumed, also with a change in cross-section. Magnetostriction is associated with strong changes in direction of the molecular magnets. It was first discovered in 1842 in iron. The iron lengthens in the direction of magnetisation and shortens perpendicularly to it. The deformation S of a body proportional to a magnetic field strength H describes the magnetostrictive effect:

Table 3.8 Selection of magnetostriction values of materials [6]

Material	Magnetostriction $\Lambda_\parallel$ [10^{-6}]	Curie temperature T_C [°C]
Fe	−14	770
Ni	−50	358
Co	−93	1120
Metglas 2605SC	60	370
Vitrovac 7	25–40	
Vitrovac 6	0.01–1.0	
Vitroperm	0.1–1	

- Material above Curie temperature T_C: Magnetic moments are lost in thermal movements.
- Cooling below T_C provides spontaneous magnetostriction to a length l_0, but the direction of magnetization varies from domain to domain.
- An applied magnetic field finally aligns magnetic domains according to their orientation and when saturated, completely parallel to the field. Then the saturation magnetostriction ls is reached.

The relative change in length that occurs is a measure of the magnetostrictive effect in a material (see Table 3.8). A distinction is made as to whether this is a change in the direction of the field or transverse to it. The relative change during magnetization is described in the direction of this magnetization by a material characteristic value

$$\lambda_\parallel = \frac{l_s - l_0}{l_0}$$

Since the reversal of the magnetic moments maintains the volume, approximate the change perpendicular to the field $\lambda_\perp = -0.5 \cdot \lambda_\parallel$. For inductive components, magnetic materials with the lowest possible magnetostriction are desirable. On the one hand, the magnetic properties are changed by pressure or tension (e.g. by clamping, bonding or encapsulating cores). The sound emission during magnetization can also be considerable so that components with certain core materials literally whistle (Table 3.8).

Magnetostriction can also be used specifically for the generation of ultrasonic waves (electromagnetic ultrasonic transducers / EMUS transducers). As for many other physical processes, the magnetostriction effect is characterized by the reversal, the change of magnetic properties by external mechanical forces. While materials with positive magnetostriction increase their permeability by a tensile load, materials with negative magnetostriction decrease their permeability by such a tensile load.

Magnetostriction can have positive or negative values; it is independent of the sign of the magnetization. A change in length therefore occurs at twice the frequency of the

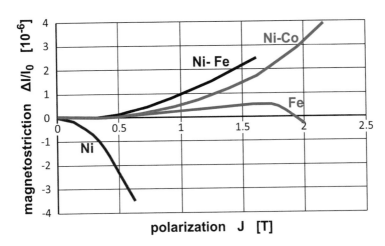

Fig. 3.31 Change in length of ferromagnetic materials with the imprinted magnetic polarization. (according to [2])

exciting (bipolar) magnetic field. Figure 3.31 shows a selection of metals and alloys in comparison.

For iron, both positive and negative values of magnetostriction occur. In the (110)-directions a change of sign of magnetostriction is found near the saturation polarization. In technical (polycrystalline) iron, the course of magnetostriction follows approximately that of the (110) orientations.

Nickel has a consistently negative value of magnetostriction in all crystal orientations. The saturation magnetostriction of polycrystalline nickel is $(\Delta l/l)_S = 35 \cdot 10^{-6}$. High positive values of magnetostriction $((\Delta l/l)_{sat} = ...100.10^{-6})$ are observed in cobalt-iron alloys (with approx. 30% Fe). In nickel-iron alloys, a compensation of positive and negative values of magnetostriction can be achieved. The zero-crossing of magnetostriction occurs in an alloy containing approximately 83% nickel.

For iron-nickel alloys with 36% nickel, the Curie temperature T_C is close to room temperature. In this temperature range, there is therefore a relatively strong decrease in magnetization with increasing temperature. The resulting decrease in length counteracts the thermal expansion, resulting in an exceptionally low coefficient of thermal expansion $(\alpha \sim 10^{-6}\ K^{-1})$. Such a material is called invar alloy. Magnetostriction is caused by magnetic forces that penetrate the crystal structure and lead to deformation [1].

References

1. Ferroxcube: Soft Ferrites and Accessories. Handbook 2013
2. Ivers-Tiffée, E., von Münch, W.: Werkstoffe der Elektrotechnik. Springer-Verlag 2007, ISBN 978-3-8351-0052-7

3. Clarke R., Phase Transition Studies of Pure and Flux-Grown Barium Titanate Crystals, J. Appl. Cryst., 9, pp 335–338, 1976]

4. Claudia Neusel and Gerold A. Schneider: Size-dependence of the dielectric breakdown strength from nano- to millimeter scale in Journal of the Mechanics and Physics of Solids, Vol. 63, February 2014, pages 201–213.

5. Michalowsky, L.: Weichmagnetische Ferrite: zum Aufbau von Präzisions-Hochfrequenzbauelementen für Kommunikationstechnik, Automobilindustrie und Industrieautomatisierung. Expert-Verlag 2006

6. EPCOS: Ferrites and Accessories. Databook 2013

7. VACUUMSCHMELZE: Weichmagnetische Werkstoffe und Halbzeuge, PHT 001, 2002)

8. Magnetec: Werkstoffvergleich NANOPERM – Ferrit

9. Kuhnke/Kendrion: Magnet Hauptkatalog 2019

10. Siaenen, Th.: Zur Präzision der Steuerbarkeit elektromagnetischer Hochleistungslinearmotoren am Beispiel eines symmetrischen Taylor-Experiments. Dissertation. Universität Kassel, 2012

11. Philippow, E.: Taschenbuch Elektrotechnik, Bd. 1, ISBN: 3341002014

12. Siaenen, Th.; Schneider, M.; Zacharias, P.; Löffler, M. J.: Actively Controlling the Muzzle Velocity of a Railgun. IEEE Transactrions on Plasma Science, Vol. 41, No. 5, May 2013

13. Roch, M.; Hundertmark, S.; Löffler, Zacharias, P.: The Modular Augmented Staged Electromagnetic Launcher Operated in the Energy Storage Mode. IEEE Transactrions on Plasma Science, Vol. 43, No. 5, May 2015

14. Hahn, L.; Muhnke, I.: Werkstoffkunde für die Elektrotechnik und Elektronik. Verlag Technik Berlin 1986

Optimization of Soft Magnetic Components

4

4.1 Basic Requirements for Soft Magnetic High Performance Materials

For linear applications such as transformers, alternating current and direct current chokes are generally required:

- Flat hysteresis loop
- High/low permeability, adjustable if possible and close to an F-characteristic
- Linear loop within saturation limits,
- High unipolar induction stroke,
- Low remagnetization losses
- Low dependence of the remagnetization losses on the premagnetization level

For non-linear applications such as switching chokes, transducers, etc. are general requirements:

- Rectangular hysteresis loop (Z-characteristic):
- High remanence ratio B_r/B_{sat}
- Low hysteresis losses
- High flux density stroke ΔB_{sat}

Various metallic and ceramic materials have been developed for this purpose. Figure 4.1 shows a comparison of such materials. Metallic materials generally have a relatively high conductivity ($>10^6$ S/m). This results in relatively high eddy current losses in these materials. These losses can be reduced by limiting the possible expansion range of the eddy currents: For example, thin metal foils insulated against each other or powder grains. Although ferrites are considered insulators, they have a residual conductivity that can be

© Springer Fachmedien Wiesbaden GmbH, part of Springer Nature 2022
P. Zacharias, *Magnetic Components*,
https://doi.org/10.1007/978-3-658-37206-4_4

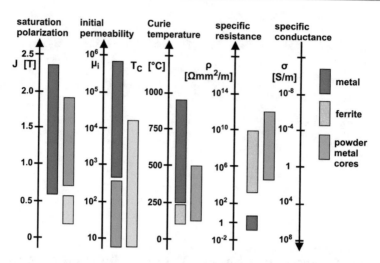

Fig. 4.1 Typical differences between metals, ferrites and powder composites. (according to [1])

disturbing in some applications. Power ferrites for higher frequencies in particular have a relatively high conductivity. (MnZn-Ferrite <10 S/m, NiZn-Ferrite $<10^{-3}$ S/m).

Powder magnetic materials have a very low conductivity as composite materials due to the mutual insulation of the magnetic grains by plastic binders. However, the conductivity of the metal powder particles isolated from each other is still high. The reduction of eddy current losses is mainly due to the particle size. At the same time, the paramagnetic distances of the plastic barriers add up to "air gaps", which limit the relative overall permeability of the materials to values between 5 and 100. Advantages of these materials in addition to the low eddy current losses are

- High saturation induction ($B_{sat} = 1...2$ T)
- Low intrinsic stray field due to the distributed air gap.

For transformer applications (transformer) high permeability, high permeability of the cores is of particular interest in order to keep the magnetizing current small. Here, values of $\mu_r = 1000...30,000$ are achieved with ferrites. Metallic materials in crystalline, amorphous and nanocrystalline form achieve significantly higher values here (Fig. 4.2). These materials have been developed since the 1980s (Vacuumschmelze Hanau, Allied Signals, Hitachi). They are characterized by high permeabilities, high saturation induction and comparatively high specific resistance values. The suitability for higher frequencies is achieved by manufacturing thin films in the range of 15...30 µm. The different forms of the B(H) characteristic (F and Z shape) are achieved by heat treatment in a magnetic field. Cores are available on the market as toroidal tape wound cores and for some years now also as cut tape wound cores. Cores with a cobalt content have the lowest re-magnetization losses even at high frequencies. However, they have lower values of saturation induction and are significantly more expensive than cobalt-free materials.

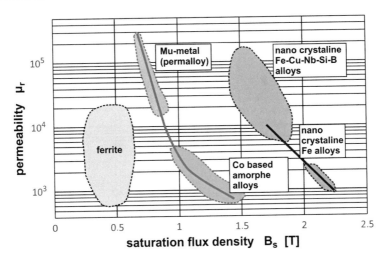

Fig. 4.2 Comparison of soft magnetic materials with flat hysteresis loop. (according to [2])

Figure 4.3 shows that the metallic materials also show their high permeability at medium frequencies up to relatively high magnetic field strengths and can therefore be superior to ferrites for certain applications.

When used as current-compensated chokes, the properties of amorphous and nanocrystalline materials are exploited to a particular advantage over ferrites (Fig. 4.4). In this application, the low-frequency currents or direct currents practically cancel each other out except for small asymmetries. High permeabilities result in a high attenuation effect for HF interference or very small sizes. Even at frequencies well above one MHz, the permeability of metallic materials is still higher than that of ferrites by a factor of 8. With the same attenuation for common-mode interference, components for filtering therefore only have a fraction of the volume if they have the appropriate core materials. The T38 ferrite has one of its typical applications in this area. VP500F is a nanocrystalline material from VAC Hanau with low remanence. VC6025F and VC6030F are amorphous metallic materials with cobalt content.

For higher-frequency applications, the core losses of magnetic components are decisive for the choice of material. For this purpose, the specific losses of various soft magnetic materials are shown in Fig. 4.4 for comparison. It can be seen that the metallic materials come close to the high frequencies properties of ferrites. The question arises for which applications the various magnetic materials can be optimally selected. This requires an analytical description of the material properties.

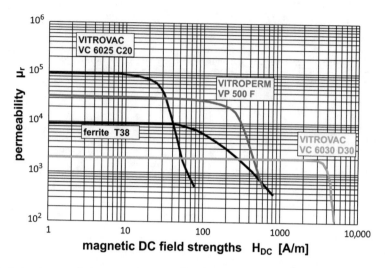

Fig. 4.3 Amorphous and nanocrystalline materials from Vacuumschmelze Hanau in comparison with ferrite T38 with regard to the behavior of the differential permeability under DC bias, measured at f = 10 kHz. (according to data [2], VP 500 F with μr =30,000)

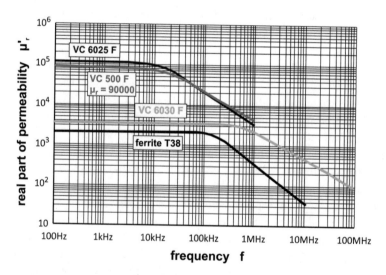

Fig. 4.4 Frequency performance of the permeability of soft magnetic materials for current-compensated chokes. (according to data [2])

4.2 Description of Core Losses During Magnetization

The influence of switching frequency and geometric parameters on the transferable apparent power and transformer losses is to be investigated on a transformer core, which is characterized by its usual characteristics. Figure 4.5a and b show typical dependencies of the losses in a magnetic core. It becomes clear that there is a great variety of materials whose remagnetization losses grow with increasing frequency and increasing induction. Only a decade of frequency is shown. Within this decade, the power loss density of the detected materials covers at least five decades. A quantitative recording of the loss characteristics is therefore necessary for component optimisation.

Many data of the power density p_v of core materials can be represented as $p_v(B_{max}, f)$ by sets of straight lines in a double logarithmic diagram. A function $z(x)$ represented on a double logarithmic scale, which is represented as a straight line, can generally be described analytically with the straight-line equation

$$lg(z) = c + a \cdot lg(x)$$

The non-logarithmic representation describes a relationship of the type

$$lg(y) = c + a \cdot lg(x)$$
$$y = 10^c \cdot x^a$$

If one analyzes Fig. 4.5b, the dependencies $p_v(B_{max})$ with frequency as parameter seem to be parallel shifted straight lines. Two parallel shifted straight lines $lg(p_v) = lg(p_v(B_{max},f))$ with the frequency as a parameter could be described as

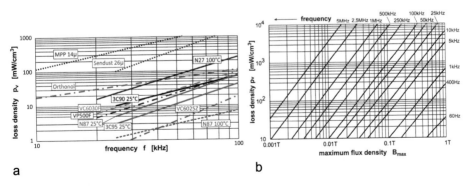

a   b

Fig. 4.5 Frequency- and induction-dependent losses in a magnetic material. (**a**) Trend Power loss densities of soft magnetic materials as a function of frequency at maximum induction of 0.1 T: Metals: VP500F, VC6030F, VC6025Z, Orthonol; **metal powder cores**: MPP14μ, Sendust 26 μ; ferrites: N27, N87, 3C90, 3C95 [3–5], (**b**) Power loss density of Powder Core 26Perm, bound in synthetic material [6]

$$\lg\left(\frac{p_{v1}}{p_{v0}}\right) = c + \alpha \cdot \lg\left(\frac{B_{max}}{B_0}\right) + \beta \cdot \lg\left(\frac{f_1}{f_0}\right)$$

$$\lg\left(\frac{p_{v2}}{p_{v0}}\right) = c + \alpha \cdot \lg\left(\frac{B_{max}}{B_0}\right) + \beta \cdot \lg\left(\frac{f_2}{f_0}\right)$$

or de-logarithmized

$$p_{v1} = p_{v0} \cdot 10^c \cdot \left(\frac{B_{max}}{B_0}\right)^\alpha \cdot \left(\frac{f_1}{f_0}\right)^\beta$$

$$p_{v2} = p_{v0} \cdot 10^c \cdot \left(\frac{B_{max}}{B_0}\right)^\alpha \cdot \left(\frac{f_2}{f_0}\right)^\beta$$

The quantities p_{v0}, B_0 and f_0 are reference quantities that are mainly used to obtain dimensionless expressions as function arguments. If the dependence $\lg(p_v(f))$ can also be represented as a straight line, then this is an approach to analytical data description. It is called the Steinmetz formula. This analytical representation of data is very helpful for optimization calculations. It is "only" a useful form of analytical data representation. Losses arise from different, independent causes and are summarized in the value p_v. The formula therefore only reflects the properties of the measured data in a favourable way and does not have the value of a physically-based equation. The same applies to the many improvements of this approach for pre-magnetized materials and non-sinusoidal magnetization, for example in [2, 7]. For straight-line groups with varying increases in the direction of both the parameters B_{max} and f, as is frequently found with powder magnetic materials (see Fig. 4.2.1b), an approach

$$\lg\left(\frac{p_v}{p_{v0}}\right) = c + \alpha \cdot \lg\left(\frac{B_{max}}{B_0}\right) + \beta \cdot \lg\left(\frac{f}{f_0}\right) + \gamma \cdot \lg\left(\frac{B_{max}}{B_0}\right) \cdot \lg\left(\frac{f}{f_0}\right)$$

suitable for a quantitative description. For curved characteristic curves on a logarithmic scale, a different procedure must be followed. You will find information on this in Chap. 9, and already in the diagram Fig. 4.5b you will notice increasing differences in the slope of $p_v(B)$ with larger frequency differences. It is therefore advisable for optimization calculations to parameterize the approximate formula shown above for a narrowly defined application (target) area. Outside of this area, these formulas are only suitable for trend statements, since large quantitative deviations may occur due to the nature of the dependencies.

A possible procedure for parameter extraction from the measured data is described below (Fig. 4.6) The example shown is an extract from the datasheet of N27 / TDK.

Variant A

We are looking for 3 parameters. Read the 3 values of the power loss density

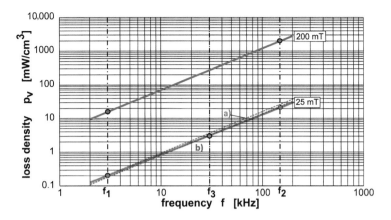

Fig. 4.6 Example for determining the parameters of the Steinmetz formula

(1) (200 mT)

$$\left(\frac{p_{v1}=15\frac{mW}{cm^3}}{1\frac{mW}{cm^3}}\right)=k_e\cdot\left(\frac{B_{max\,1}=0.2\ T}{1\,T}\right)^\alpha\cdot\left(\frac{f_1=3\ kHz}{1\,kHz}\right)^\beta$$

(2) (200 mT)

$$\left(\frac{p_{v2}=2000\frac{mW}{cm^3}}{1\frac{mW}{cm^3}}\right)=k_e\cdot\left(\frac{B_{max\,1}=0.2\ T}{1\,T}\right)^\alpha\cdot\left(\frac{f_2=150\ kHz}{1\ kHz}\right)^\beta$$

(3) (25 mT)

$$\left(\frac{p_{v3}=0.2\frac{mW}{cm^3}}{1\frac{mW}{cm^3}}\right)=k_e\cdot\left(\frac{B_{max\,2}=0.025\,T}{1\,T}\right)^\alpha\cdot\left(\frac{f_1=3\ kHz}{1\ kHz}\right)^\beta$$

If the two Eqs. (1) and (2) and (1) and (3) are divided up, one obtains.

$$\left(\frac{2000}{15}\right)=\left(\frac{150}{3}\right)^\beta\rightarrow\beta=\frac{lg\left(\frac{2000}{15}\right)}{lg\left(\frac{150}{3}\right)}=1.25\,\text{and}$$

$$\left(\frac{15}{0.2}\right)=\left(\frac{200}{25}\right)^\alpha\rightarrow\alpha=\frac{lg\left(\frac{15}{0.2}\right)}{lg\left(\frac{0.2}{0.025}\right)}=2.08$$

The parameter k_{Fe} is determined by inserting it into one of the 3 equations, for example, (2):

$$k_{Fe}=\frac{2000}{0.2^{2.08}\cdot150^{1.25}}\cdot\frac{mW}{cm^3}=107.7\frac{mW}{cm^3}$$

The Steinmetz formula is then

$$p_v(B_{max}, f) = 107.7 \frac{mW}{cm^3} \cdot \left(\frac{B_{max}}{1\ T}\right)^{2.08} \cdot \left(\frac{f}{1\ kHz}\right)^{1.25}$$

The exponents α and β are independent of the standardization variables. The scaling factor k_{Fe} depends on the scaling variables. The orange solid line is the "true" course of $p_v(f, 20\ mT)$. Solution variant a) provides the orange dotted line, which has a description error at $f = 150\ kHz$. It is parallel to the blue line. Good approximations can only be expected in the area of the 3 points that span the solution area. A reduction of the description error can be achieved with the following variant.

Variant B

One selects 2 points (1) and (2) belonging to a parameter B_{max} and also a point (4) between the two frequencies f_1 and f_2 for f_3 (here $f_3 = 30\ kHz$).

(4) (25 mT)

$$\left(\frac{p_{v3} = 3\frac{mW}{cm^3}}{1\frac{mW}{cm^3}}\right) = k_{Fe} \cdot \left(\frac{B_{max2} = 0.025\ T}{1\ T}\right)^{\alpha} \cdot \left(\frac{f_4 = 30\ kHz}{1\ kHz}\right)^{\beta}$$

To determine β, one first proceeds as before and obtains $\beta = 1.25$. If one uses one of the upper points (1) or (2). One receives as an equation for the upper line

$$p_v(f) = 3.81 \frac{mW}{cm^3} \times \left(\frac{f}{1\ kHz}\right)^{1.25} \rightarrow p_v(30\ kHz) = 3.81 \frac{mW}{cm^3} \cdot (30)^{1.25} = 267.5 \frac{mW}{cm^3}$$

This results in $p_v(30\ kHz) = 267.5\ mW/cm^3$, which is almost the same as variant a):

$$\left(\frac{267.5}{3}\right) = \left(\frac{0.2}{0.025}\right)^{\alpha} \rightarrow \alpha = \frac{\lg\left(\frac{267.5}{3}\right)}{\lg\left(\frac{0.2}{0.025}\right)} = 2.16$$

This then results in the formula for a straight line passing through the point (4) on a double logarithmic scale:

$$p_v(B_{max}, f) = 122.7 \frac{mW}{cm^3} \cdot \left(\frac{B_{max}}{1\ T}\right)^{2.16} \cdot \left(\frac{f}{1\ kHz}\right)^{1.25}$$

In this way, the maximum error of the approximation formula becomes smaller than with variant A. It is always true that with this interpolation formula, statements about the further course can only be made in the near periphery since only the properties of the measured data are mapped. The further one moves away from the area of a priori information, the greater the uncertainty and the further the statement assumes the quality of speculation. If the formula were physically based, this would be different.

4.3 Optimization Approach for Transformers

The influence of switching frequency and geometric parameters on the transferable apparent power and transformer losses is to be investigated on a transformer core, which is characterized by its usual characteristics.

As guidelines and basic dependencies are given or known:

- The transformer has a primary and a secondary winding with $N_1 = N_2$. (This is only a simplification, not a limitation). The output of the material is symmetrical and sinusoidal.
- The core losses can be calculated approximately with the Steinmetz formula if the core shape and material are known:

$$P_{vcore} = p_v\left(\widehat{B}, f\right) \cdot V_{Fe} = V_{Fe} \cdot k_{Fe} \cdot \left(\frac{\widehat{B}}{1\ \text{T}}\right)^{\alpha} \cdot \left(\frac{f}{1\ \text{kHz}}\right)^{\beta} \quad \text{with} \quad \widehat{B} = \frac{\sqrt{2} \cdot U_1}{2\ \pi f \cdot N_1 \cdot A_{Fe}}$$

...are in on it:

A_{Fe} Effective magnetization cross-section,
f Inverter frequency,
$\widehat{B}$ max. (bipolar) induction in the effective core cross-section,
k_{Fe} Loss coefficient of the core material,
α, β Exponents for the approximation of the material properties,
U_1 Primary voltage,
V_{Fe} Core volume,
N_1 Primary winding number.

For the core losses one gets like this:

$$P_{vcore} = V_{Fe} \cdot k_{Fe} \cdot \left(\frac{\sqrt{2} \cdot U_1}{2\ \pi f \cdot N_1 \cdot A_{Fe} \cdot 1\ \text{T}}\right)^{\alpha} \cdot \left(\frac{f}{1\ \text{kHz}}\right)^{\beta}$$

The winding losses, however, are approximate:

$$P_{vCu} = 2 \cdot R_{N1} \cdot I_1^2 = 2 \cdot \rho \cdot N_1 \cdot l_m \cdot \frac{I_1^2}{k_f \cdot \frac{A_W}{N_1}} = 2 \cdot \rho \cdot N_1^2 \cdot l_m \cdot \frac{\left(\frac{S}{U_1}\right)^2}{k_f \cdot A_W}$$

Here are:

ρ   Specific resistance of the winding material,
l_m Average coil length,
S   Transformer apparent power,
A_w Cross-sectional area of the winding window,
k_f Fill factor of the space window.

This gives the expression for the total losses of the transformer:

$$P_v = P_{vcore} + P_{vCu} = V_{Fe} \cdot k_{Fe} \cdot \left(\frac{\sqrt{2} \cdot U_1}{2\pi f \cdot N_1 \cdot A_{Fe} \cdot 1\ \text{T}}\right)^\alpha \cdot \left(\frac{f}{1\ \text{kHz}}\right)^\beta + 2 \cdot \rho \cdot N_1^2 \cdot l_m$$
$$\cdot \frac{S^2}{k_f \cdot A_W \cdot U_1^2}$$

Both loss components are dependent on the number of turns N_1, but with opposite tendency, as shown in Fig. 4.7. This results in a pronounced minimum for the core losses.

An optimization task can be formulated in relation to the windings introduced per volt:

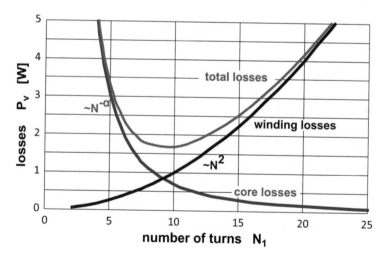

Fig. 4.7 Division of transformer losses into winding losses and core losses for a calculated example

$$P_v = K_1 \cdot N_1^{-\alpha} + K_2 \cdot N_1^2 \Rightarrow \min$$

With

$$K_1 = V_{Fe} \cdot k_e \cdot \left(\frac{\sqrt{2} \cdot U_1}{2\,\pi f \cdot A_{Fe} \cdot 1\ \mathrm{T}} \right)^{\alpha} \cdot \left(\frac{f}{1\ \mathrm{kHz}} \right)^{\beta}$$

$$K_2 = 2 \cdot \rho \cdot l_m \cdot \frac{S^2}{k_f \cdot A_W \cdot U_1^2}$$

By zeroing the first derivative of function $P_v(N_1)$, one finds the necessary condition for a local extremum

$$\frac{dP_v}{dN_1} = -\alpha \cdot K_1 \cdot N_1^{-\alpha-1} + 2 \cdot K_2 \cdot N_1 = 0 \Rightarrow N_{1opt} = \sqrt[\alpha+2]{\frac{\alpha \cdot K_1}{2 \cdot K_2}}.$$

This means that the extreme value is

$$N_{1opt} = \sqrt[(\alpha+2)]{\frac{\alpha \cdot K_1}{2 \cdot K_2}} = \sqrt[(\alpha+2)]{\frac{\alpha \cdot V_{Fe} \cdot k_{Fe} \cdot \left(\frac{\sqrt{2} \cdot U_1}{2\,\pi \cdot A_{Fe} \cdot 1\ \mathrm{T} \cdot 1\ \mathrm{kHz}} \right)^{\alpha} \cdot \left(\frac{f}{1\ \mathrm{kHz}} \right)^{\beta-\alpha}}{4 \cdot \rho \cdot l_m \cdot \frac{S^2}{k_f \cdot A_W \cdot U_1^2}}}$$

In many publications it is assumed for the optimum that core losses and copper losses are equal:

$$P_{vKern} \approx P_{vCu} \quad \text{respectively} \quad K_1 \cdot N_{1opt}^{*\ -\alpha} = K_2 \cdot N_{1opt}^{*\ 2} \rightarrow N_{1opt}^* = \sqrt[2+\alpha]{\frac{K_1}{K_2}}.$$

For the case $\alpha = 2$, both approaches yield the same solution. Assuming a value range $\alpha = 1...3$, the ratio of the "optimal" (i. e. the ratio of the approximation and the "true" optimization) solutions is

$$\frac{N_{1opt}^*}{N_{1opt}} = \frac{\sqrt[2+\alpha]{\frac{K_1}{K_2}}}{\sqrt[\alpha+2]{\frac{\alpha \cdot K_1}{2 \cdot K_2}}} = \sqrt[\alpha+2]{\frac{2}{\alpha}}.$$

Consequently, the two approaches provide very close solutions when considering the restricted value range for $\alpha = 1.7...2.8$ (Fig. 4.8).

If one considers the determined optimum number of turns, then for the dependence of the number of turns this means

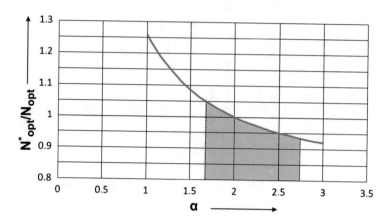

Fig. 4.8 Comparison of the values of the approaches for optimum solutions N_{opt} for minimum losses and N^*_{opt} for ($P_{vCu} = P_{vcore}$)

$$N_{1opt} \sim \left(\frac{f}{1 \text{ kHz}}\right)^{\frac{\beta-\alpha}{\alpha+2}}$$

With the rough approximation $\alpha = 2$ and $\beta = 1$, $N_{1opt} \sim f^{-0.25}$. For power ferrites, the values of the exponents are $\alpha = 1.9...2.85$ and $\beta = 1.2...1.7$. The optimum number of windings therefore falls relatively weakly with increasing frequency.

The optimum of the number of windings is pronounced. Dimensioning to the optimum generally leads to low losses, even if the number of turns cannot be set exactly. Since the increase in total losses is zero at the optimum, even small deviations do not lead to dramatic increases. Optimization is therefore always sensible.

However, the decisive factor for maximum power is that the losses incurred $P_v = P_{vcore} + P_{vCu}$ can be dissipated at a maximum transformer temperature, a certain ambient temperature and a thermal resistance R_{th} specified by the core. As derived in Sect. 8.2, the optimum number of turns determined for the maximum apparent power of the transformer results in

$$S_{max} = \frac{2\pi \cdot A_{Fe} \cdot 1 \text{ T} \cdot 1 \text{ kHz} \cdot \left(\frac{f}{1 \text{ kHz}}\right)^{\frac{\alpha-\beta}{\alpha}}}{(V_{Fe} \cdot k_{Fe})^{\frac{1}{\alpha}} \cdot \sqrt{\frac{2 \cdot \rho \cdot l_m}{k_f \cdot A_W}}} \cdot \left(\frac{\frac{T_{max} - T_{amb}}{R_{th}}}{\left(\frac{2}{\alpha}\right)^{\frac{\alpha}{\alpha+2}} + \left(\frac{\alpha}{2}\right)^{\frac{2}{\alpha+2}}}\right)^{\frac{\alpha+2}{2\alpha}}$$

For the maximum apparent power at a given geometry the following applies

$$S_{max} \sim \left(\frac{f}{1\ \text{kHz}}\right)^{\frac{\alpha-\beta}{\alpha}}$$

This means for an optimized transformer an increase of the transferable apparent power approximately $\sim \sqrt{f}$. However, it is also interesting to note a dependence on geometric variables. If all other variables are combined as constants and only the geometric variables are written down, one obtains

$$S_{max} \sim V_{Fe}^{\frac{-1}{\alpha}} \cdot \sqrt{\frac{k_f \cdot A_{Fe}^2 \cdot A_W}{l_m}}$$

Especially for $\alpha = 2$ and $V_{Fe} = A_{Fe} \cdot l_{Fe}$ one gets the proportionality

$$S_{max} \sim \sqrt{\frac{k_f \cdot A_{Fe} \cdot A_W}{l_{Fe} \cdot l_m}}$$

to estimate the influence of geometric quantities on the transmittable apparent power. The fill factor k_f indicates how much the maximum winding window is available. The theoretical maximum is 1. Since also the winding body etc. have space requirements, realistic values are in the range of $k_f = 0.4...0.7.$, The formula represents, with the exception of one constant, the geometric mean of the electrical and magnetic conductances in the core and can be used as a *figure of merit* (FOM).

The procedure described above implicitly assumes that there are no further restrictions on the target area apart from the warming or the maximum permissible temperature. However, this is not the case for smaller frequencies. Especially at lower frequencies, the limitation of the saturation of the B(H) characteristic is added. In the vicinity of the saturation induction, a change in the flux density leads to significantly greater changes in the current consumption. This means that the form factor of the current and thus the RMS value of the current increases. The simple approach $P_v \rightarrow$min cannot be formulated in the way shown. In the absence of a sufficiently precise analytical description of the power loss density, the problem is solved by introducing a limitation of the optimization space by $B_{max} < B_{sat}$ with the requirement $B_{max} \leq k^* . B_{sat.}$ The factor k^* in the range $0...1$ can be used to influence the extent of the increase in the effective magnetizing current.

In this case, the optimization problem must be formulated differently:

$$P_v = P_{vcore} + P_{vCu} = V_{Fe} \cdot k_{Fe} \cdot \left(\frac{k^* \cdot B_{SAT}}{1\ \text{T}}\right)^{\alpha} \cdot \left(\frac{f}{1\ \text{kHz}}\right)^{\beta} + 2 \cdot \rho \cdot N_1^2 \cdot l_m \cdot \frac{S^2}{k_f \cdot A_W \cdot U_1^2}$$

$$\Rightarrow \text{min}$$

The secondary condition with the effect of a limit is here

$$k^* \cdot B_{SAT} = \frac{\sqrt{2} \cdot U_1}{2\pi f \cdot N_1}.$$

As can be seen from the structure of the determination formula for P_v, this is not a problem for finding a local extreme value. It is rather a matter of fulfilling the equation

$$P_{vzul} = V_{Fe} \cdot k_{Fe} \cdot \left(\frac{k^* \cdot B_{SAT}}{1 \text{ T}}\right)^{\alpha} \cdot \left(\frac{f}{1 \text{ kHz}}\right)^{\beta} + 2 \cdot \rho \cdot N_{1opt}^2 \cdot l_m \cdot \frac{S^2}{k_f \cdot A_W \cdot U_1^2}$$

$$= P_{vzul} = \frac{T_{max} - T_{amb}}{R_{th}}$$

$$P_{vzul} = K_3 \cdot \left(\frac{f}{1 \text{ kHz}}\right)^{\beta} + K_4 \cdot N_{1opt}^2 \cdot S^2 = P_{vzul} = \frac{T_{max} - T_{amb}}{R_{th}}$$

at one edge of the optimization area. For N_{1opt} one then gets

$$N_{1opt} = \sqrt{\frac{T_{max} - T_{amb}}{R_{th} \cdot K_4 \cdot S^2} - \frac{K_3}{K_4 \cdot S^2} \cdot \left(\frac{f}{1 \text{ kHz}}\right)^{\beta}}$$

By increasing the switching frequency, a loss reduction can therefore always be achieved in the area of model validity. Viewed differently, the above-derived relationship for a given maximum permissible power loss of a transformer core provides a dependence for the maximum transferable apparent power with the proportionality

$$S \sim f^{\frac{\alpha - \beta}{\alpha}}.$$

For the approximate values assumed above for $\alpha = 2$ and $\beta = 1$, a core can thus be used to transmit an apparent power increasing with $f^{0.5}$ as the frequency increases. A nine-fold increase in frequency increases the transferable apparent power by a factor of 3. If one also takes into account that the effective wire resistance increases with frequency due to the skin effect and proximity effect, this increase in transmissible power is less than 3 in reality.

The previous calculations assumed a homogeneous filling of the winding space with the winding material. However, this is not the case in reality. The filling factor depends to a considerable extent on the wire thickness used, especially for larger cross-sections. If this is taken into account when calculating the losses, the result is, for example, the course shown in Fig. 4.9. The jumps are explained by the necessary transition to other conductor cross-sections.

As Fig. 4.9 shows, the minimum power dissipation may be significantly influenced by the conductor cross-sections, which are not arbitrarily possible. However, this applies to a much lesser extent to the position of the minimum.

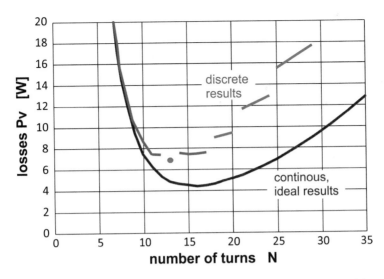

Fig. 4.9 Comparison of simplified calculation of the power loss of a transformer and the calculation taking into account the possible (circular) conductor cross-sections

4.4   Optimization Approach for Chokes with an Air Gap

In the same way as when dealing with the optimization problem of a transformer, one can proceed with loss minimization in inductors. While the transformer is almost always a component with impressed voltage for current transformation, the current is the primary determining variable in an inductor. A distinction is made between two types:

- Smoothing chokes (unipolar or bipolar)
  Here the purpose is to reduce the amplitude of high-frequency current ripples. The frequency of the actual used current is much lower than that of the current ripple. DC/DC converters are a typical example of unipolar current. The flyback converter also belongs to this category. This is an buck-boost converter in which the inductor also performs the function of an insulating transformer. In this case, the current is unipolar, while the voltage is bipolar. Smoothing chokes for feeding current into the mains operate with a basic frequency of 50/60 Hz with bipolar current. With this type of choke, core losses are only caused by the HF component. The current losses are caused jointly by DC and AC components.
- Pure AC chokes are bipolar stressed. The losses in the core are caused by the voltage drop at the same frequency as the main current. Examples here are chokes for AC grid applications. The load here is a sinusoidal current. Applications are current limiters, EMC filters, etc. A further field of application are resonant circuit chokes. In these, the load is suppliedby high-frequency, bipolar alternating currents.

The load quantities are generally non-sinusoidal (as is the case with switching power supplies with rectangular voltage). However, the power loss density is only specified by the manufacturers for sinusoidal quantities. A prediction using a Fourier decomposition is relatively complex and cannot be justified. Because of the non-linear relationships, the superposition principle cannot be applied. The results are not more meaningful than to use the effective value of the load quantity with the basic frequency of a sinusoidal quantity.

The optimization approaches with their basic results are discussed below. For a simplified calculation, we first assume that it is an inductor with a core + air gap δ and is $\mu_r \rightarrow \infty$ the permeability of the core material. Then one can write for the inductance:

$$L = \mu_0 \frac{A_{Fe}}{\delta} N^2$$

from which N can be derived:

$$N^2 = \frac{\delta \cdot L}{\mu_0 \cdot A_{Fe}} \quad \text{with}$$

A_{Fe} Magnetisation cross-section,
δ Air gap,
N   Number of turns.

The core losses for a sinusoidal voltage drop at the choke can be calculated with

$$P_{vcore} = V_{Fe} \cdot k_{Fe} \cdot \left(\frac{\sqrt{2} \cdot U_L}{2 \cdot \pi \cdot f \cdot N \cdot A_{Fe} \cdot 1\ \text{T}} \right)^\alpha \cdot \left(\frac{f}{1\ \text{kHz}} \right)^\beta,$$

The following applies to winding losses:

$$(1.5)\ P_{vCu} = \rho \cdot N^2 \cdot l_m \cdot \frac{I_L^2}{k_f \cdot A_w}$$

...Here are:

ρ Specific resistance of the winding material,
I_L RMS winding current,
l_m Average coil length,
A_w Cross-sectional area of the winding window of a winding,
k_f Space factor of the winding window.

This gives the expression for the total losses of the inductance:

$$P_v = P_{vcore} + P_{vCu} \Rightarrow \text{min}$$

$$P_v = V_{Fe} \cdot k_{Fe} \cdot \left(\frac{\sqrt{2} \cdot U_L}{2\pi \cdot A_{Fe} \cdot N \cdot 1 \text{ T} \cdot 1 \text{ kHz}} \right)^{\alpha} \cdot \left(\frac{f}{1 \text{ kHz}} \right)^{\beta - \alpha} + \frac{\rho \cdot N^2 \cdot l_m \cdot I_L^2}{k_f \cdot A_w} \Rightarrow \text{min}$$

In order to provide this equation with the condition of the desired inductance as well, the number of turns N is replaced. This results in a problem in which the air gap is included as a variable quantity. It should be noted that if the values for δ are too large, the simplified formula becomes very imprecise and consequently the result of the optimization becomes uncertain. The procedure can be used for the area $\delta \leq 0.2 \cdot A_{Fe}^{0.5}$

$$N = \sqrt{\frac{\delta \cdot L}{\mu_0 \cdot A_{Fe}}}$$

$$P_v = V_{Fe} \cdot k_{Fe} \cdot \left(\frac{\sqrt{2} \cdot U_L}{2\pi \cdot A_{Fe} \cdot \sqrt{\frac{\delta \cdot L}{\mu_0 \cdot A_{Fe}}} \cdot 1 \text{ T} \cdot 1 \text{ kHz}} \right)^{\alpha} \cdot \left(\frac{f}{1 \text{ kHz}} \right)^{\beta - \alpha} + \frac{\rho \cdot \frac{\delta \cdot L}{\mu_0 \cdot A_{Fe}} \cdot l_m \cdot I_L^2}{k_f \cdot A_w}$$

$$\Rightarrow \text{min}$$

To achieve the same inductance to L, different numbers of turns must be applied to different air gaps. The above formula illustrates this relationship. This influences the distribution of the losses between the winding and the core. The optimization problem has the following form in relation to the air gap to be inserted δ

$$P_v = \frac{K_5}{\delta^{\frac{\alpha}{2}}} + K_6 \cdot \delta \Rightarrow \text{min}$$

With

$$K_5 = V_{Fe} \cdot k_{Fe} \cdot \left(\frac{\sqrt{2} \cdot U_L}{2\pi \cdot A_{Fe} \cdot \sqrt{\frac{L}{\mu_0 \cdot A_{Fe}}} \cdot 1 \text{ T} \cdot 1 \text{ kHz}} \right)^{\alpha} \cdot \left(\frac{f}{1 \text{ kHz}} \right)^{\beta - \alpha}$$

$$K_6 = \frac{\rho \cdot \frac{L}{\mu_0 \cdot A_{Fe}} \cdot l_m \cdot I_L^2}{k_f \cdot A_w}$$

This means that for the minimum

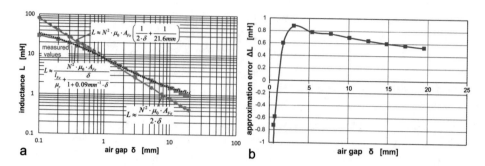

Fig. 4.10 Inductance with an air gap. (**a**) Measured inductance of an AMCC320 core with 108 turns and varied air gap (blue) compared to the approximate formula without boundary field (red) and taking the boundary field into account (green), (**b**) Description error with the formula without taking the boundary field and core inductance into account

$$\frac{dP_v}{d\delta} = \frac{\alpha}{2} \cdot \frac{K_5}{\delta^{\frac{\alpha+2}{2}}} + K_6 = 0 \Rightarrow \delta_{opt} = \left(\frac{\alpha \cdot K_5}{2 \cdot K_6}\right)^{\frac{2}{\alpha+2}}$$

For a larger value range of δ, it is better to write for the function

$$L \approx \left(\mu_0 \frac{A_{Fe}}{\delta} + \frac{1}{R_{m0}}\right) N^2 = \left(\mu_0 A_{Fe} \left(\frac{1}{\delta} + \frac{1}{\delta_0}\right)\right) N^2$$

as an approximation, whose parameter R_{m0} can be easily determined metrologically for a given core shape. One simply measures the inductance of a test winding N* at an air gap $\delta \geq A_{Fe}^{0.5}$ and uses the above approach to determine R_{m0} or δ. Figure 4.10 shows this approximation with an example.

For the number of windings N, one can use R_{m0} for the boundary field if R_{m0} is taken into account:

$$N \approx \sqrt{\frac{L}{\mu_0 \frac{A_{Fe}}{\delta} + \frac{1}{R_{m0}}}}$$

Analogous to the previous procedure, the optimization problem can be formulated as follows:

$$P_v = V_{Fe} \cdot k_{Fe} \cdot \left(\frac{\sqrt{2} \cdot U_L}{2\pi \cdot A_{Fe} \cdot 1T \cdot 1 \text{ kHz}}\right)^\alpha \left(\mu_0 \frac{A_{Fe}}{\delta \cdot L} + \frac{1}{R_{m0} \cdot L}\right)^{\frac{\alpha}{2}} \cdot \left(\frac{f}{1 \text{ kHz}}\right)^{\beta - \alpha} +$$

$$+\frac{\rho \cdot l_m \cdot I_L^2}{k_f \cdot A_w \cdot \left(\mu_0 \frac{A_{Fe}}{\delta \cdot L} + \frac{1}{R_{m0} \cdot L}\right)} \Rightarrow \min$$

There is a loss component, which in this approach depends on the RMS voltage at the choke. The RMS voltage of the fundamental oscillation of the voltage $u_L(t)$ is used as an approximation. A further component depends on the RMS current. The latter is dependent on the current form.

Determination of the Form Factor of the RMS Choke Current

In DC/DC converters, a unipolar current with an alternating component flows in a choke. The course of this current can usually be represented by a polyline. The following applies to discontinuous current range in stationary operation:

$$K_{F1} = \frac{I_{RMS}}{I_d} = \sqrt{\frac{I_d}{3 \cdot \Delta I}\left[\left(1 + \frac{\Delta I}{2I_d}\right)^3 - \left(1 - \frac{\Delta I}{2I_d}\right)^3\right]}$$

The following applies to the non-continous current range

$$K_{F2} = \frac{I_{RMS}}{I_d} = \sqrt{\frac{2 \cdot \Delta I}{3 \cdot I_d}}$$

Figure 4.11 shows the graphical representation of the dependency of the form factor on the ratio $\Delta I/I_d$ at continuous and non-continuous current. It can be seen that the form factor in the non-continuous current mode is increased to a maximum of 1.15 at $\Delta I/I_d = 2$. This means that the winding losses are approximately 30% higher with this form of current than with the direct current. At $\Delta I/I_d = 3$ they are already doubled. The form factor must be taken into account when calculating the current losses and may have a considerable influence on the position of the optimum values of N and δ. The form factor for continous unipolar current is <1.15.

The evaluation of the formulated optimization problem for the optimization of DC chokes is much more complex than for the optimization of transformers. It is therefore carried out using an example. Figure 4.12 shows a graphical representation for a loss of a choke for 20 A, 200 V and 50 kHz with a core AMCC320. For comparison, the two approaches are considered without and with consideration of the boundary field. The example parameters [8] are

$$L = 1 \text{ mH} \qquad U_L = 200 \text{V} \qquad f = 50 \text{ kHz} \qquad I_L = 20 \text{ A}$$

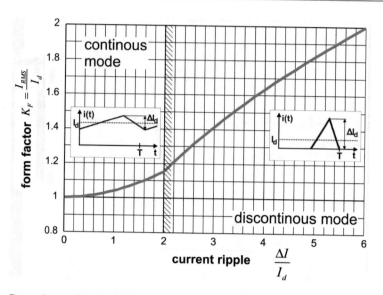

Fig. 4.11 Dependence of the form factor K_F for the choke current on the ratio $\Delta I/I_d$ in continuous and discontinuous operation

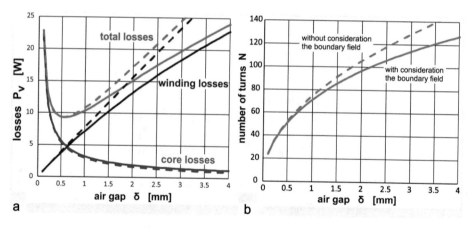

Fig. 4.12 Losses with a 1 mH choke without (dotted line) or with (continuous line) consideration of the boundary field (**a**), the associated number of turns (**b**)

$$A_{Fe} = 902\ \text{mm}^2 \qquad A_{Cu} = 2965\ \text{mm}^2 \qquad m = 2167\ \text{g} \qquad k_f = 0.4$$

$$\rho_{Cu} = 0.0173\ \Omega \cdot \text{mm}^2/\text{m}$$

$$\text{Core losses } P_{vAMCC} = V_e \cdot 6.5 \frac{\text{W}}{\text{kg}} \cdot \left(\frac{\widehat{B}}{1\ \text{T}} \right)^{1.74} \cdot \left(\frac{f}{1\ \text{kHz}} \right)^{1.51}$$

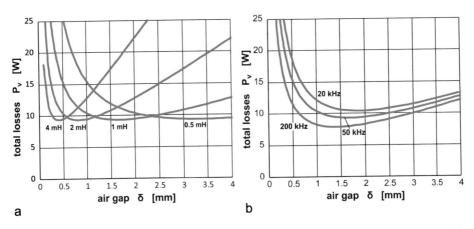

Fig. 4.13 Influence of the specified inductance and frequency on the curves of the dependence of the power dissipation on the air gap (**a**) and the frequency (**b**)

Figure 4.12 shows the progression of the loss shares and the total losses for the setup under these conditions. Here, too, a clear local minimum can be found. The deviations due to the neglect of the boundary field only become noticeable with larger air gaps, so that in the example the optimum can also be found well with the approximation. For larger air gaps δ the approximation yields too many turns.

Characteristic of the optimization task with larger air gaps is an extended range for the loss minimum. In practice, a minimum air gap will be selected in such a way that $P_v = (1.05...1.1)$. P_{vmin} applies, to prevent the stray field of the choke from becoming too large. As Fig. 4.13 shows, the calculated minimum loss remains largely independent of the inductance. However, this only applies if the RMS value of the current is constant. However, it must be taken into account that the form factor of the current becomes larger with a smaller inductance. As a result, the losses increase with the same arithmetic mean value and decreasing inductance. An increase in frequency leads to a reduction in choke losses under otherwise identical conditions.

With DC chokes, the load values for current and voltage are largely decoupled, as can be seen from the previous considerations. This is not the case with chokes for bipolar symmetrical magnetization. The following applies in particular to sinusoidal variables

$$I_L = \frac{U_L}{\omega \cdot L}$$

This results in the following expression for the optimization problem without considering the boundary field

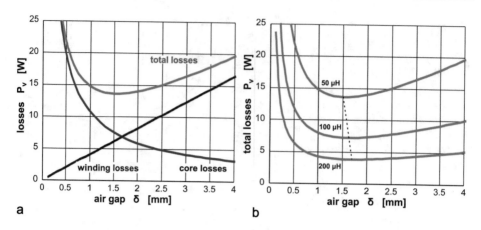

Fig. 4.14 Loss components of a choke. (**a**) AMCC320/500 V/10 kHz with 50 μH, (**b**) Change in total losses with different inductance specifications

$$P_v = V_{Fe} \cdot k_{Fe} \cdot \left(\frac{\sqrt{2} \cdot U_L}{2\pi \cdot A_{Fe} \cdot \sqrt{\frac{\delta \cdot L}{\mu_0 \cdot A_{Fe}}} \cdot 1 \text{ T} \cdot 1 \text{ kHz}} \right)^\alpha \cdot \left(\frac{f}{1 \text{ kHz}} \right)^{\beta - \alpha}$$

$$+ \frac{\rho \cdot \frac{\delta}{\mu_0 \cdot A_{Fe}} \cdot l_m \cdot U_L^2}{4\pi^2 \cdot f^2 \cdot k_f \cdot A_w \cdot L}$$

$$\Rightarrow \text{ min}$$

As can be seen from the results summarized in Fig. 4.14, the results behave differently than with DC chokes. In this case, an essential boundary condition was not considered: The maximum induction must not reach or exceed the saturation induction.

4.5 Special Material Problems in the Design of Inductive Components

For the derivations in Sections 4.2 and 4.3, constant winding resistance and constant specific core losses were assumed. This is practically only approximately the case. Both the specific core losses and the specific resistances are temperature-dependent. This can be taken into account in the calculation by using the values for the maximum operating temperature of the component. Besides, the losses in the windings are frequency-dependent as a result of the two effects described below. This means that with higher frequencies the losses of the winding increase so that this could be taken into account in the calculation with a frequency-dependent resistance of the winding. As a consequence, the losses in a transformer do not decrease continuously with the frequency, as worked out in Sect. 4.2,

Fig. 4.15 Interaction of current in a conductor with its own magnetic field

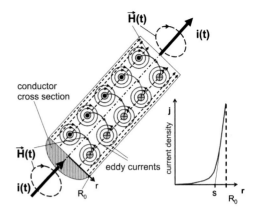

but increase again from a critical frequency. Responsible for this are the skin effect and the proximity effect.

Figure 4.15 will explain the *skin effect* in more detail. The field lines inside a conductor are also swirled around by a magnetic field, just like the entire conductor. A temporal change of this magnetic field leads to a voltage induction in the conductor material, which according to Lenz's rule is opposite to its cause. The result is a weakening of the current inside the conductor or a displacement to the outside, to the surface.

As the current flows at sufficiently high frequencies only in one layer on the conductor surface, this is referred to as the skin effect. The conductive layer has a "thickness" of

$$s = \frac{1}{\sqrt{\pi \cdot f \cdot \sigma \cdot \mu}},$$

σ...conductivity, μ...permeability of the conductor.

Figure 4.16 shows the frequency dependence of this layer thickness for copper and aluminium. If this calculated layer thickness is less than the radius of the conductor, there is an increase in resistance with increasing frequency, since only part of the cross-section is involved in the current flow.

It is therefore advisable to select the wire diameters of individual cores $\leq 2 \cdot s$ and insulate them from each other. If all individual wires maintain the same position in relation to each other, the skin effect continues to work, so that conductors inside a bundle carry less current than conductors outside. Only real stranding (not "twisting") counteracts the skin effect on the entire cross-section.

Another effect to be considered is the proximity effect. It results from the interaction of the conductor material with a "foreign" alternating magnetic field. Adjacent conductors or air gaps in the magnetic core may be surrounded by a very strong stray field, which also penetrates a conductor of a winding (Fig. 4.17). When the field changes over time, these voltages also induce currents in the conductor material (eddy currents), which contribute to additional losses in the conductor. As a first approximation, the larger the volumes

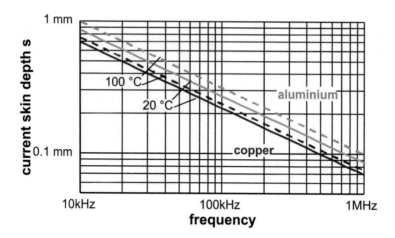

Fig. 4.16 Frequency dependence of the penetration depth of electric current in copper and aluminium at different temperatures

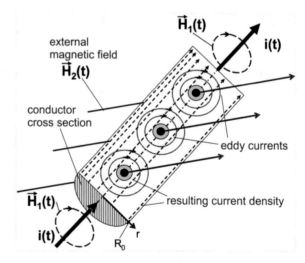

Fig. 4.17 Effect of the proximity effect in a conductor

penetrated by the magnetic field, the greater the additional losses. Fields of eddy currents are directed opposite to their cause. Covered construction or winding parts are thus partially shielded. The interrelationships become very complex as a result. This can be countered by

- in areas of high magnetic leakage field strengths no winding parts are placed and/or
- the winding material is designed as a stranded wire with insulated conductors (HF strand), so that the individual eddy current regions are kept small in their extent. The wire thickness should be selected $<<s = \dfrac{1}{\sqrt{\pi \cdot f \cdot \sigma \cdot \mu}}$.

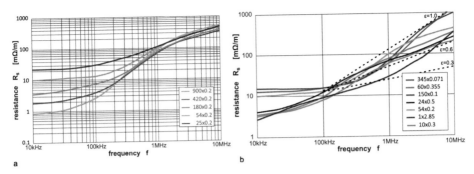

Fig. 4.18 Measured effective serial resistances R_s of straight tensioned conductors. (**a**) consisting of different strands of 0.2 mm enameled copper wire; (**b**) for strands of different cross-sections and diameters, $R_s \approx R_{s*}(1 + (f/f^*))^\varepsilon$ [9]

At high frequencies, the induced electric field strengths in conductor materials are also correspondingly high. This also applies to adjacent conductors. In other words: simple twisting of the conductors does not completely counteract the skin effect, since the eddy currents are also caused by the alternating magnetic field of adjacent conductors. Only stranding in which the individual wires change their position over the length of the strand structure over the cross-section leads to an even utilization of all conductors. At very high frequencies, however, stranding can also lead to conductor materials being brought into the range of influence of alternating magnetic fields which are not active at all in simple bundles. Eddy currents then generate losses there using transformers, which lead to an increase in the real component of the impedance. This fact is shown in Fig. 4.18. Only below a certain frequency can the ohmic component of the impedance be kept low by stranding. Above this frequency, an increase in this part can even be observed. This is due to the internal proximity effect of the strands. Individual wires are "heated" by induced eddy currents from magnetic fields of neighbouring wires. The frequencies must be sufficiently high for this. This effect occurs all the more in multi-layer windings. It can be observed particularly strongly in windings without a ferromagnetic core. The magnetic field strength is then particularly strong in the windings. Figure 4.19 is intended to provide a better understanding. It was designed for induction heating of cylindrical parts with radius 'R' with a penetration depth 's'.

Average Core Losses with Periodically Changing Induction
In pulse width modulated current transformers, inductors are generally used for current smoothing. The current ripple changes depending on the ratio of output voltage: input voltage. As a simple example for the description of the problem and the description of a solution, a 2 level inverter in a half-bridge circuit is used here to feed current into the grid (Fig. 4.20). This can be used, for example, to condition power from photovoltaic generators for feeding into the grid.

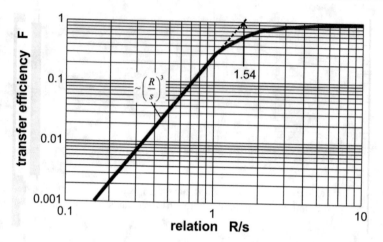

Fig. 4.19 Transfer efficiency for induction heating of cylindrical workpieces with radius R and penetration depth s [10]

Fig. 4.20 Diagram of a 2-level inverter in a half-bridge circuit for feeding electrical energy from a DC voltage source into the grid

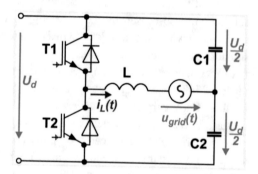

In this case, the current ripple is ideal for

$$\Delta I_L = \frac{U_d\left(1 - \left(\frac{u_{grid}(t)}{U_d} + \frac{1}{2}\right)\right)}{f \cdot L} \cdot \left(\frac{u_{grid}(t) + \frac{1}{2}}{U_d}\right) = \frac{U_d}{f \cdot L} \cdot \left(\frac{1}{4} - \frac{\hat{u}_{grid}^2 \cdot \sin^2(\omega t)}{U_d^2}\right).$$

With $u_{grid} = 0$ the current ripple is maximum. With $u_{grid}(t) = \pm\, 0.5 \cdot U_d$ $\Delta I_L = 0$ (Fig. 4.7) At the same time, the peak-to-peak change in current is a measure of the recorded voltage-time area:

$$\Delta I_L = \frac{1}{L}\int_{t1}^{t2} u_L \cdot dt = \Delta \psi_L = \Delta B \cdot A_{Fe} \cdot N$$

The change in current is therefore proportional to the change in flux. The flux amplitude of a corresponding periodic course can be applied. The average core losses are calculated with the Steinmetz formula to

$$P_{vcore} = V_{Fe} \cdot \frac{1}{T} \int_{t}^{t+T} k_{Fe} \cdot \left(\frac{f}{1 \text{ kHz}}\right)^{\beta} \cdot \left(\frac{\widehat{B}(t)}{1 \text{ T}}\right)^{\alpha} \cdot dt = V_{Fe} \cdot k_{Fe} \cdot \left(\frac{f}{1 \text{ kHz}}\right)^{\beta}$$

$$\cdot \frac{1}{T} \int_{t}^{t+T} \left(\frac{\widehat{B}(t)}{1 \text{ T}}\right)^{\alpha} \cdot dt$$

The periodic course $\widehat{B}(t) \sim \Delta I_L$ or, if the time course of the voltage is known, the course $u(t)$ can be $\widehat{B}(\widehat{u})$. In the above formula, the individual loss components of the remagnetization are summed up/integrated and the average value is calculated. This is - in contrast to the superposition of individual parts from a Fourier transformation, which would also be conceivable here - a mathematically more safe operation. Because of the non-linearity of the loss components, the superposition principle is not applicable. For a fixed waveform, the change in the mean value with changed modulation can simply be recorded via a relative representation in relation to the maximum values of an operating point:

$$\frac{P_{vcore}\left(\widehat{B}(\widehat{u})\right)}{P_{vcore.\max}} = \frac{\frac{1}{T} \int_{t}^{t+T} \left(\frac{\widehat{B}(t)}{1T}\right)^{\alpha} \cdot dt}{\left(\frac{B_{\max}(t)}{1T}\right)^{\alpha}} = \frac{1}{T} \int_{t}^{t+T} \left(\frac{\Delta I_L(t)}{\Delta I_{L.\max}}\right)^{\alpha} \cdot dt$$

For various conditions, Fig. 4.21 shows $\widehat{u}/U_d$ the course of the current change in a standardized form. Thermally, however, the average value of the losses is interesting.

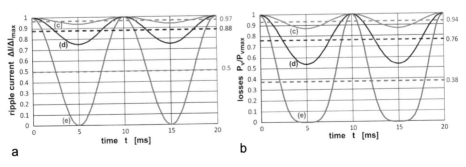

Fig. 4.21 Course of current ripple (**a**) and core power loss (**b**) for different ratios of the amplitude of the mains voltage $\widehat{u}$ and the DC voltage Ud: (**c**) 0.125; (**d**) 0.25; (**e**) 0.5 for grid connected current supply

For the Steinmetz parameter $\alpha = 2.17$, the course of the losses in the core is also given in the normalized form. This means that the density of the core losses changes constantly. This problem can be solved by averaging individual losses. The mean values for the selected modulations are also entered in Fig. 4.21. The procedure described can also be applied to other analytical descriptions of core losses. A secondary condition is a known multiplicative connection of the influences, as in the Steinmetz formula. The factors found in this way must be included in the optimization calculations for the worst case.

4.6   Optimization Approach for Saturable (Switching) Chokes

With chokes whose saturation properties are exploited, the objectives of a design are naturally completely different from those of linear components. The starting point for the formulation of the optimization problem should be a simple application (Fig. 4.22). Shown is the series connection of an IGBT with a non-linear choke to supply a load R with a pulse current of high amplitude and high current steepness. Both thyristors and IGBTs are semiconductor switches whose conductivity during the turn-on process is largely due to the flooding of a junction zone with charge carriers of both polarity. (In the IGBT, this process is co-determined by an intrinsic, field-controlled unipolar component). In solids, the velocity of charge carriers is comparatively low. This fact is responsible for the fact that the voltage drop at the switches is relatively high at very high current slopes. The reason is the relatively low saturation speed of the charge carriers into the junction zone to be flooded.

So at the beginning of an impulse, only a few charge carriers are available. The result is a high voltage drop, which can quickly reach several hundred volts at currents in the kA range. This causes high switching losses which can lead to the destruction of the semiconductors. A saturable choke is used here to relieve the switch (*magnetic assist*)

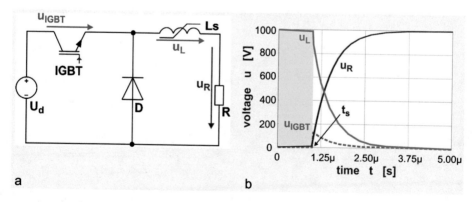

Fig. 4.22 Basic circuit of a buck converter (**a**) to describe the switching performance of a non-linear inductor Ls and (**b**) voltage curves at the main elements

for example. At the beginning of a switching process, practically the entire voltage is first taken over by the choke L_s. The unsaturated choke limits the increase in current to a small value. The flowing current supports the switch-on process, while the voltage drop at the semiconductor switch remains low. When the choke goes into saturation after recording a voltage-time area, the differential inductance changes abruptly by several orders of magnitude and the current rises steeply. The maximum voltage drop and thus the achieved power dissipation reaches much lower values. This type of circuit can be found, for example, when switching on flash lamps, for operating pulse-forming networks of pulsed lasers or for short-circuiters ("crowbar") in high-voltage switchgear. The described principle is also used in magnetic amplifiers and magnetic switches in chains of non-linear LC links for magnetic pulse shortening.

The aim is practically always to minimise costs. Due to the comparatively high material costs of materials suitable for short switching times, this goal boils down to minimizing the core volume. A secondary condition is the setting of a saturation flux concatenation specified by the application. For these applications, toroidal cores made of metallic strip materials with F- or Z-shaped magetisation characteristic curve with low values of the coercivity are used. The saturation occurs with these materials in a small interval ΔB. Toroidal cores are used because with this core shape practically the entire core cross-section has the same flux density so that the core as a whole also goes into saturation. This ensures clean switching operations. Oval cores, for example, have areas that become saturated earlier than others. The inductance in the saturation is increased by this core shape. Ferrites can be used in principle, but usually have a greater coercive field strength and a longer curved area of the B(H) curve in the saturation region. If the core is operated with a magnetic field strength slightly greater than the saturation field strength before the switching process by a bias current, a stroke of the flux linkage of

$$\Delta \Psi = 2 \cdot \Psi_{sat} = \int_0^{ts} u_L(t) \cdot dt = 2 \cdot B_{sat} \cdot A_{Fe} \cdot N \rightarrow A_{Fe} = \frac{\int_0^{ts} u_L(t) \cdot dt}{2 \cdot B_{sat} \cdot N}$$

Here $\Delta \Psi$ is a size specified by the application. At the same time, a saturation inductance L_{sat} should be achieved. If one assumes toroidal cores with inner radius R_i, outer radius R_a and height h, which are provided with a single layer of N windings at a distance a, then one obtains for the expected inductance in the saturated state approximately

$$L_{sat} \approx \frac{N^2 \cdot (2a(h + (R_a - R_i)) + \pi \cdot a^2)}{\pi(R_a + R_i)} \leq L_{target}.$$

A further secondary condition is that the resulting losses can be dissipated from the component. Due to the extremely high form factors for the straining current and voltage curves, the Steinmetz formula is not even remotely applicable. The remagnetization losses of the core depend on the maximum remagnetization speed $\Delta B/dt$ and the pulse duration t_i. The winding losses depend on the I^2t-value of the current pulse with the peak value I_p. Thus, for the total losses at a repetition rate, f_{rep} one can estimate

$$P_v \approx \left(V_{Fe} \cdot p_v \left(\left. \frac{dB}{dt} \right|_{max} ; t_i \right) + R_{Cu} \frac{I_p^2}{2 \cdot t_i} \right) \cdot f_{rep}$$

This is in fact only an estimation. For the core volume, the optimization goal can be formulated as

$$V_{Fe} \approx \pi \cdot h \left(R_a^2 - R_i^2 \right) \Rightarrow \min .$$

At the same time, switching losses are also kept low with this objective. Even if the constraints are all mapped to the optimization target, there is no local minimum in the target area. The optimum solution must therefore be sought at the edges. This task can only be formulated in general terms in a very complicated way. However, it can be solved numerically with relatively little effort. Figure 4.23 shows a graphical representation of an example of the optimization problem. Here a switching choke was optimized for magnetic

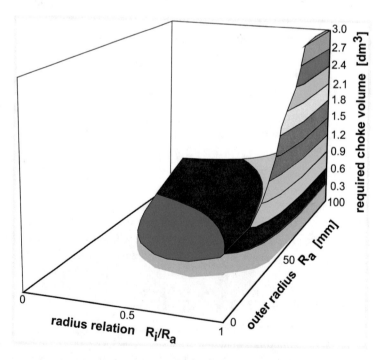

Fig. 4.23 Solution area for volume optimization of a saturable switching choke

pulse shortening in a small excimer laser. It can be seen that the minimum for the volume at the edge is obtained with a radius ratio $R_i/R_a = 0.5$. For toroidal cores with a large inner radius, the volume increases rapidly, since h has to become large to achieve the required voltage-time area.

References

1. Ivers-Tiffée, E., von Münch, W.: Werkstoffe der Elektrotechnik. Springer-Verlag 2007, ISBN 978-3-8351-0052-7
2. VACUUMSCHMELZE: EMC Products based on Nanocrystalline Vitroperm. 2016
3. EPCOS Ferrites and Accessories. Data book 2013
4. VACUUMSCHMELZE: Ringbandkerne 1976
5. FERROXCUBE: Soft Ferrites and Accessories. 2013
6. Micrometals. Alloy Powder Cores.
7. Mühlethaler, J.; Biela, J.; Kolar, J. W., Ecklebe, A.: Improved Core-Loss Calculation for Magnetic Components Employed in Power Electronic Systems; IEEE Transactions on Power Electronics, Vol. 27, No. 2, FEBRUARY 2012
8. Hitachi Metals: POWERLITEC03-15-12 - Copy.doc
9. Fenske, F.: Charakterisierung von Wickelgütern unter Einfluss des Skin- und Proximityeffektes. Universität Kassel, Masterarbeit 2012
10. Conrad, H.; Krampitz, R.: Elektrotechnologie. VEB Verlag Technik Berlin 1983

Transformation of Magnetic and Electrical Circuits

5

Electrical circuits or circuit parts may connected via the windings of a magnetic component and exchange energy with each other. Adequate mathematical models are required for the simulation, which reflect the essential electrical properties of the components. Of interest is therefore the transfer of mixed electrical and magnetic structures into purely electrical networks for the purpose of simulation and dimensioning of components. With the simulation one should be able to "look inside" the magnetic components, to see loads on individual parts. For this purpose, the basics are presented below and extensions of the procedure, as well as possible applications, are shown.

5.1 Comparison of Magnetic and Electrical Circuits

As shown in previous chapters, there are differences in the properties of the magnetic and electric field, but the means of the description of concentrated magnetic and electric circuits are largely similar. On the one hand, they describe two hemispheres of electrical engineering, but on the other hand, they also open up a possibility to connect both – even if this is not obvious at first.

This chapter aims to derive the rules for a transformation from the description of the magnetic circuit to the world of electrical circuits and vice versa by means of a formal but very brief structural analysis of the interrelationships in both forms of description. The approach is very concentrated with the main goal of conveying an understanding of the methodology. Therefore only very simple examples are used, which the reader can easily understand while the losses of windings and cores, represented by according resistances, are neglected. For more comprehensive information and mathematical and thus generally valid proofs, please refer to the literature cited. The starting point should be a simple unbranched magnetic circuit (Fig. 5.1a) consisting of different magnetic sections and a

© Springer Fachmedien Wiesbaden GmbH, part of Springer Nature 2022
P. Zacharias, *Magnetic Components*,
https://doi.org/10.1007/978-3-658-37206-4_5

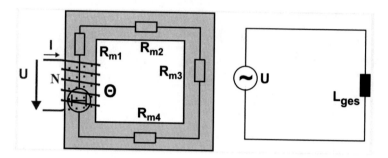

Fig. 5.1 Magnetic and electrical basic "flux circuit" in comparison. (**a**) magnetic circuit, (**b**) electrical unbranched circuit

concentrated winding which can be assigned to a magnetic section and which is connected to an AC voltage source. The equivalent circuit (Fig. 5.1b) consists of an AC voltage source to which an inductor is connected. Ohmic resistances of the windings or as representatives for losses in the core are initially neglected here because they make explanations of the methodology unclear. Insertion of these really important elements will be done later when the rules of transformation are clear.

Looking at the relationships of the magnetic quantities in (Fig. 5.1a), the analysis leads to the following simple equations:

$$\Theta = i \cdot N = \phi \cdot (R_{m1} + R_{m2} + R_{m3} + R_{m4})$$

If this equation is multiplied by the number of windings N, one obtains

$$\Theta \cdot N = i \cdot N^2 = \phi \cdot N \cdot (R_{m1} + R_{m2} + R_{m3} + R_{m4}) = \psi \cdot (R_{m1} + R_{m2} + R_{m3} + R_{m4})$$

A division of this equation by N^2 now yields

$$i \cdot = \frac{\psi}{N^2} \cdot (R_{m1} + R_{m2} + R_{m3} + R_{m4}) = \psi \left(\frac{1}{L1} + \frac{1}{L2} + \frac{1}{L3} + \frac{1}{L4} \right)$$

Thus, the total inductance of the magnetic circuit results from the individual components $L_1 \ldots L_4$.

$$\frac{i}{\psi} = \frac{1}{L_{ges}} = \left(\frac{1}{L1} + \frac{1}{L2} + \frac{1}{L3} + \frac{1}{L4} \right)$$

Consequently, each magnetization section can be assigned a share of the inductance. Figure 5.1b can therefore be generated from a further, more detailed equivalent circuit diagram, as shown in Fig. 5.2 as a graphical representation of the above equation.

Fig. 5.2 Derivation of the electrical equivalent circuit diagram from the elementary quantities of the "iron circuit"

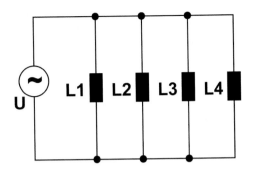

It becomes obvious that the magnetic circuit and the electrical circuit derived directly from it have dual structures to each other (series connection ↔ parallel connection). This basic consideration makes it possible to determine the adequate electrical equivalent circuit diagram of a magnetic arrangement with windings, core areas and leakage sections by graphic inversion of the circuit structure from the magnetic equivalent circuit diagram of a magnetic arrangement with windings, core areas and leakage sections. Such a circuit section is usually integrated into a more complex electrical circuit. Non-linear inductance components can thus be easily characterized by the properties of the material sections of the magnetic circuit. The combination of these elements then takes place in the electrical equivalent circuit diagram. If nonlinearities are taken into account, the superposition principle no longer applies. This means that the result of a calculation is always the result of considering all sources simultaneously. Especially in the case of magnetic sections of material that are pre-magnetized by a direct current, the calculation considers that this direct current is impressed by an ideal current source. In this way, a specific operating point or a corresponding characteristic curve is set in the characteristic field of the inductor. This allows a calculation in a common simulation program for electrical circuits if the $V(\Phi)$ characteristics for the magnetic voltage drop of the magnetized sections are entered.

5.2 Magnetization of a Core Section by a Current-Conducting Winding

The magnetization of a core section through a winding can be described as a quadripole, with the magnetic variables (Φ, V) on one side and the electrical variables (U, I) on the other (Fig. 5.3):

The descriptive equations follow the physical relationships and are derived from magnetic voltage drop $v(t)$ and flux linkage $\psi(t)$

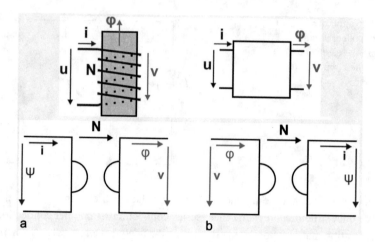

Fig. 5.3 Quadripole conversion between electric and magnetic field quantities in concentrated structures. (**a**) technical structure, (**b**) quadripole diagram

$$v = N \cdot i$$
$$\int u \cdot dt = \psi = N \cdot \phi$$

respectively

$$\begin{bmatrix} \int u \cdot dt \\ v \end{bmatrix} = \begin{bmatrix} \psi \\ v \end{bmatrix} = \begin{bmatrix} 0 & N \\ N & 0 \end{bmatrix} \cdot \begin{bmatrix} i \\ \phi \end{bmatrix}$$

or vice versa

$$\begin{bmatrix} i \\ \phi \end{bmatrix} = \begin{bmatrix} 0 & \dfrac{1}{N} \\ \dfrac{1}{N} & 0 \end{bmatrix} \cdot \begin{bmatrix} \int u \cdot dt \\ v \end{bmatrix} = \begin{bmatrix} 0 & \dfrac{1}{N} \\ \dfrac{1}{N} & 0 \end{bmatrix} \cdot \begin{bmatrix} \psi \\ v \end{bmatrix}$$

The structure of this system of equations largely resembles the transfer equations of a gyrator, a basic element for the representation and analysis of electrical circuits [5.1]. A gyrator transforms the terminal behavior of a network-connected on one side of this quadripole into the dual terminal behavior on the other side. This ability is due to the structure of the descriptive quadripole matrix. These properties are used here analogously to connect different field sizes in an electromagnetic structure. Approaches to this have already been described in [5.1]. In [5.2], the counting arrow directions used here are discussed which differ from those of the original definition in [5.1]. However, they are in line with the physical relationships and the specifications that are usual there.

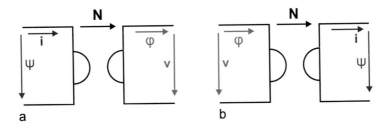

Fig. 5.4 Conversion of an electrical impedance into a (magnetic) reluctance (**a**) and (**b**) vice versa

An "inverting" of the impedance converter Γ^* leads – as the system of equations above with its visualization Fig. 5.4 shows – again to a 'gyrator', but with the reversed direction of action and corresponding matrix for the mapping of the field size pairs. If two such converters are used in a chain circuit (Fig. 5.5), an ideal transformer is obtained.

The ideal transformer maps the input voltage u_1 or the flux linkage ψ_1 to the output voltage u_2 or the flux linkage ψ_2 and the input current i_1 to the output current i_2 in a loss-free and bidirectional manner.

For a chain connection of the four poles, it is advisable to choose the cascaded form instead of the impedance matrix. This requires the following conversion.

$$\begin{bmatrix} \psi \\ v \end{bmatrix} = \begin{bmatrix} 0 & N \\ N & 0 \end{bmatrix} \cdot \begin{bmatrix} i \\ \phi \end{bmatrix} \rightarrow \begin{bmatrix} \psi \\ i \end{bmatrix} = \begin{bmatrix} 0 & N \\ \frac{1}{N} & 0 \end{bmatrix} \cdot \begin{bmatrix} v \\ \phi \end{bmatrix}$$

For the inverse representation, this means.

$$\begin{bmatrix} i \\ \phi \end{bmatrix} = \begin{bmatrix} 0 & \frac{1}{N} \\ \frac{1}{N} & 0 \end{bmatrix} \cdot \begin{bmatrix} \psi \\ v \end{bmatrix} \rightarrow \begin{bmatrix} v \\ \phi \end{bmatrix} = \begin{bmatrix} 0 & N \\ \frac{1}{N} & 0 \end{bmatrix} \cdot \begin{bmatrix} \psi \\ i \end{bmatrix}$$

For the description of the arrangement of Fig. 5.4 can then be written:

$$\begin{bmatrix} \psi_1 \\ i_1 \end{bmatrix} = \begin{bmatrix} 0 & N_1 \\ \frac{1}{N_1} & 0 \end{bmatrix} \cdot \begin{bmatrix} v_1 \\ \phi_1 \end{bmatrix} \quad \text{and} \quad \begin{bmatrix} v_2 \\ \phi_2 \end{bmatrix} = \begin{bmatrix} 0 & N_2 \\ \frac{1}{N_2} & 0 \end{bmatrix} \cdot \begin{bmatrix} \psi_2 \\ i_2 \end{bmatrix}$$

Because $\phi_1 = \phi_2$ and $v_1 = v_2$ must apply:

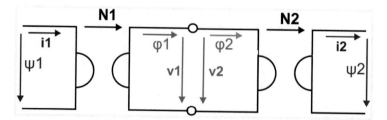

Fig. 5.5 Formation of an ideal transformer by cascaded connection of 2 gyrators

$$\begin{bmatrix} \psi_1 \\ i_1 \end{bmatrix} = \begin{bmatrix} 0 & N_1 \\ \dfrac{1}{N_1} & 0 \end{bmatrix} \cdot \begin{bmatrix} 0 & N_2 \\ \dfrac{1}{N_2} & 0 \end{bmatrix} \cdot \begin{bmatrix} \psi_2 \\ i_2 \end{bmatrix}$$

respectively

$$\begin{bmatrix} u_1 \\ i_1 \end{bmatrix} = \begin{bmatrix} \dfrac{N_1}{N_2} & 0 \\ 0 & \dfrac{N_2}{N_1} \end{bmatrix} \cdot \begin{bmatrix} u_2 \\ i_2 \end{bmatrix}.$$

To distinguish an ideal transformer from symbols usually used in circuit diagrams, the circuit diagram shown in Fig. 5.6 is derived from the above explanations:

The following characteristics of the transformer can also be found in the symbol:

- The transformer insulates electrically.
- Although the ideal transformer transmits energy bi-directionally without loss, it has a specific direction of action for the voltage (or current) transmission. This is illustrated by the directional arrow in the symbol
- The input impedance of the transformer is infinitely large when the output is open circuit, while the input impedance is zero when the secondary short circuit occurs.

However, an ideal transformer according to this definition (Z-matrix) is also able to transfer direct current. It is thus an **abstract idealized** transmission element.

5.3   Duality Principle and Graphic Circuit Inversion

In electrical engineering, two circuits are called dual if there is an identical performance of currents and voltage drops when they are interchanged. For example, when 4 resistors are connected in series, the sum of all 4 voltage drops is the total voltage drop. On the other hand, when 4 resistors are connected in parallel, the sum of all 4 currents results in the total current. In the same way, a dual-component can be found for a basic electrical component if

Fig. 5.6 Circuit symbol with
corresponding counting arrows
for current and voltage for an
ideal transformer

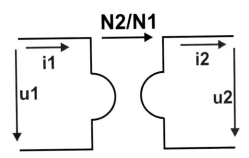

the structure of the relationships between current and voltage at this component is considered, as is the case for capacitance and inductance, for example (Fig. 5.7):

There is a simple formalism for the determination of a circuit dual to a given circuit, which implements the above condition graphically:

- Meshes are replaced by nodes and vice versa
- New nodes are then connected crossing the origin branches
- Components are replaced by their dual components.

Figure 5.8 shows a simple example of an output circuit with 5 meshes. (General note: Each series connection of 2 components also contains a node).

In the concrete execution of the circuit inversion, a node is drawn in each given mesh. The entire outer area of a circuit counts as one mesh and is assigned an account point during the procedure. Then the dual-component Γ_v^* is placed vertically Γ_v on each component so that it extends from one mesh of the given circuit into the adjacent mesh. Then the components are connected Γ_v^* to the previously drawn nodes. The result of this procedure is a new circuit that behaves identically when the observation of currents and voltage drops is reversed. This approach can also be applied to non-linear controlled circuits: backward conducting switches are replaced by backward blocking switches, *"normally-ON"* switches are replaced by *"normally-OFF"* switches, etc. At the end of this chapter is a list of the most common dual switching elements. In some cases, the u-i-t-behavior does not reflect the "natural" behavior of a real component but is forced by the interconnection and control of several components. An example of this is the so-called dual thyristor, which, unlike the thyristor, can only be switched off via its control electrode.

5.4   Coupling of an Electrically-Caused Alternating Magnetic Flux into an Electromagnetic Circuit with Interactions Between Both Types of Fields

If we first look at the arrangement of an electromagnetic circuit with a ferromagnetic core with only one winding, it must be noted that a small part of the magnetic flux generated does not close via the core but via the space outside it. This flux is called leakage flux.

Fig. 5.7 Capacitance C and inductance L as dual components

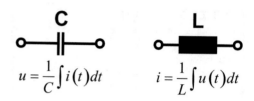

$$u = \frac{1}{C} \int i(t)\,dt \qquad\qquad i = \frac{1}{L} \int u(t)\,dt$$

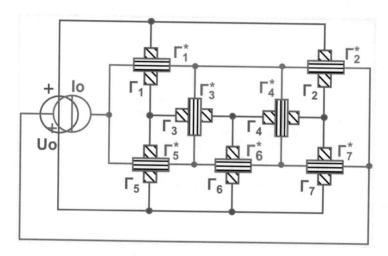

Fig. 5.8 Example of a graphical circuit inversion/detection of a dual circuit for a given circuit

Figure 5.9 shows a schematic drawing of the arrangement as well as a symbolic conversion into a circuit diagram according to the explanations of this chapter.

If the dual circuit is formed according to the rules explained in Sect. 5.3, the following arrangement is obtained in Fig. 5.10. The leakage inductance, that is, the flux component not conducted with the core is represented as a coreless inductance.

In the circuit inversion shown in Fig. 5.10, both two poles and the four poles have been converted in accordance with the rules. The result is a series connection of 2 inductors, which are fed with electrical output variables through a quadripole. After inversion, however, this has magnetic input variables. This is not helpful for electrical network calculations. If, however, one falls back on the effects of the combination of 2 gyrators visualized in Figs. 5.10 and 5.11, this problem can be easily defused, as Fig. 5.11 shows. In this figure, the effect of an air gap (not shown in Fig. 5.9) $L_{\delta 1}$ is also shown.

The operations carried out are

$$\begin{bmatrix} u_1 \\ i_1 \end{bmatrix} = \begin{bmatrix} 0 & N_1 \\ \dfrac{1}{N_1} & 0 \end{bmatrix} \cdot \begin{bmatrix} 0 & 1 \\ 1 & 0 \end{bmatrix} \cdot \begin{bmatrix} u_2 \\ i_2 \end{bmatrix} = \begin{bmatrix} N_1 & 0 \\ 0 & \dfrac{1}{N_1} \end{bmatrix} \cdot \begin{bmatrix} u_2 \\ i_2 \end{bmatrix}$$

The impedance measured from the input side is then

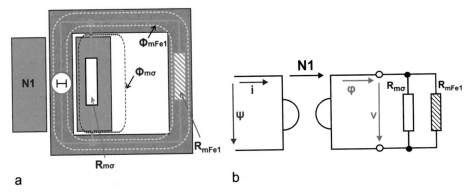

Fig. 5.9 Structure of a choke (**a**) with the representation of the magnetic equivalent circuit diagram and symbolic-analytical representation (**b**) via a gyrator as an intermediary between magnetic and electrical quantities

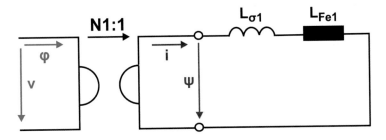

Fig. 5.10   Result of the graphic inversion of Fig. 5.9b

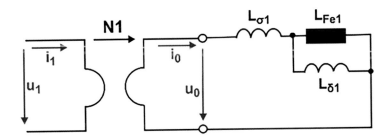

Fig. 5.11   Ferromagnetic core with winding N1, and air gap δ in an electrical equivalent circuit diagram

$$Z = \frac{u_1}{\underline{i}_1} = \frac{N_1^2 \cdot u_2}{\underline{i}_2} = N_1^2 (\underline{X}_{L\sigma1} + \underline{X}_{Fe1}) = N_1^2 \cdot j\omega(L_{\sigma1} + L_{Fe1}) = N_1^2 \cdot j\omega \left(\frac{1}{R_{m\sigma1}} + \frac{1}{R_{mFe1}} \right)$$

The inductances of the equivalent circuit diagram on the right side therefore practically correspond to the magnetic conductance of the arrangement. The magnetic resistance shown in Fig. 5.4.1 as a representative of the leakage field is usually very high due to the differences in the permeabilities of the materials involved. Thus the proportion $L_{\sigma 1}$ is usually very low. It is nevertheless listed here to make the systematic approach clear in detail. Figure 5.4.2 thus represents a model (schematized figure) of a choke with a magnetic leakage flux of the winding.

Now, what happens if you transfer this procedure to an arrangement with 2 windings? The geometrical-design approach is shown in Fig. 5.12a. Each winding has a flux component that does not "flow" over the ferromagnetic core. If one now assigns a leakage flux to each cause of magnetic flux, one obtains, according to Fig. 5.10 in Fig. 5.12b, 2 ideal transformers with the transformation ratios $N_1:1$ or $N_2:1$, in whose output there is a leakage inductance in each case and which together feed a magnetizing inductance. The rearrangement of the elements in Fig. 5.12 leads to the well-known T equivalent circuit diagram of a transformer. In simple 2-winding transformers, it best reflects the physical relationships (Fig. 5.13).

In the following (Fig. 5.14) the procedure is to be extended to a transformer with 2 windings on separate legs of the core. The leakage flux between the windings will be mainly formed here via the air space between the upper and lower yoke area. However, as above, there is also a proportion of magnetic field lines that close in the windings.

The graphic inversion of the magnetic equivalent circuit diagram again provides the electrical equivalent circuit diagram of the arrangement, which is shown in Fig. 5.15.

The following applies to such arrangements $L_{\sigma 0} > > L_{\sigma 1}; L_{\sigma 2}$:

The winding leakage components are usually neglected in the analysis and one obtains Fig. 5.16.

The electrical equivalent circuit diagram is the well-known Π equivalent circuit diagram. It best reflects the physical conditions for strong yoke leakage. 2 separate magnetization inductances are quite useful in this case since the magnetization states of the individual core sections can also be very different – in contrast to the example shown in Fig. 5.4.1. In the case of a secondary short circuit, for example, the leg of the core enclosed by N_2 is practically field-free, while the core section in N_1 carries the entire magnetic flux. In principle, a Δ-Y conversion allows a transfer to the T equivalent circuit diagram. However, the physical reference just described is lost in this case.

If one considers the variation possibilities for a simple magnetically coupled system of 2 windings, one arrives at the following structural picture (Fig. 5.17) of the energy/signal transmission (This is used synonymously here, since computer science uses signals as abstract variables, but in technology, these always have a material carrier. In modern technology these are usually energy states).

The individual units are identified by the following transfer matrices:

Gyrator 1 (winding N1)

$$\|G_1\| = \begin{bmatrix} 0 & N_1 \\ \dfrac{1}{N_1} & 0 \end{bmatrix}$$

Magnetic quadripole MVP

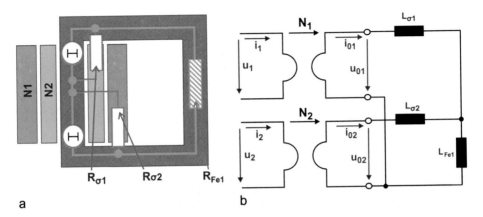

Fig. 5.12 Effect of 2 windings on the leg of a closed ferromagnetic core. (**a**) geometrical arrangement, (**b**) electrical equivalent circuit obtained by graphic circuit inversion of the magnetic equivalent circuit

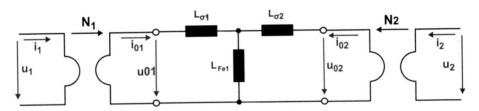

Fig. 5.13   T shaped equivalent circuit diagram of a transformer with 2 windings based on Fig. 5.12

Fig. 5.14 Transformer with strong yoke leakage and with consideration of the winding leakage fields

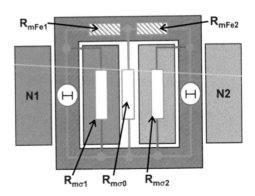

Winding direction K

$$\|MVP\| = \begin{bmatrix} a_{11} & a_{12} \\ a_{21} & a_{22} \end{bmatrix}$$

$$\|K\| = \begin{bmatrix} k & 0 \\ 0 & k \end{bmatrix} \text{ with } k = \pm 1$$

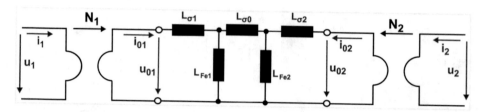

Fig. 5.15 Complete transformer replacement circuit diagram for a 2-winding transformer with strong yoke leakage based on Fig. 5.14

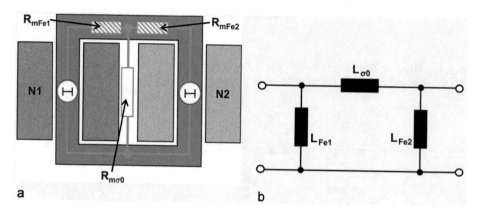

Fig. 5.16 Simplified representation of a transformer with strong yoke leakage. (**a**) magnetic. (**b**) electrical equivalent circuit (without taking into account the transformation to certain numbers of turns)

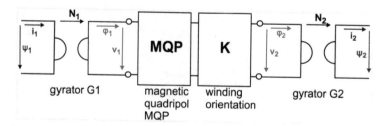

Fig. 5.17 General transmission cascade between two magnetically coupled windings

Gyrator 2

$$\|G_2\| = \begin{bmatrix} 0 & N_2 \\ \dfrac{1}{N_2} & 0 \end{bmatrix}$$

Assuming a linear transfer performance, the following results are obtained for the input and output variables:

$$\begin{bmatrix} \psi_1 \\ i_1 \end{bmatrix} = \|G1\| \cdot \|MVP\| \cdot \|K\| \cdot \|G2\| \cdot \begin{bmatrix} \psi_2 \\ i_2 \end{bmatrix}$$

This means for the first multiplication

$$\begin{bmatrix} \psi_1 \\ i_1 \end{bmatrix} = \begin{bmatrix} \dfrac{N_1 a_{21}}{a_{11}} & \dfrac{N_1 a_{22}}{a_{12}} \\ \dfrac{a_{11}}{N_1} & \dfrac{a_{12}}{N_1} \end{bmatrix} \cdot \|K\| \cdot \|G2\| \cdot \begin{bmatrix} \psi_2 \\ i_2 \end{bmatrix}$$

and finally for the entire expression in the chosen notation for the "integral" sizes

$$\begin{bmatrix} \psi_1 \\ i_1 \end{bmatrix} = \begin{bmatrix} k\dfrac{N_1 a_{22}}{N_2} & kN_1 N_2 a_{21} \\ k\dfrac{a_{21}}{N_1 N_2} & k\dfrac{N_2 a_{11}}{N_1} \end{bmatrix} \cdot \begin{bmatrix} \psi_2 \\ i_2 \end{bmatrix}.$$

By differentiation, the usual notation is obtained as a differential equation system in the cascaded form of the two-port equations.

$$\begin{bmatrix} u_1 \\ \dfrac{di_1}{dt} \end{bmatrix} = \begin{bmatrix} k\dfrac{N_1 a_{22}}{N_2} & kN_1 N_2 a_{21} \\ k\dfrac{a_{21}}{N_1 N_2} & k\dfrac{N_2 a_{11}}{N_1} \end{bmatrix} \cdot \begin{bmatrix} u_2 \\ \dfrac{di_2}{dt} \end{bmatrix}$$

From this, the resistance form commonly used for transformers can be derived by forming:

$$\begin{bmatrix} u_1 \\ u_2 \end{bmatrix} = \begin{bmatrix} \dfrac{a_{22}}{a_{12}} N_1^2 & -k\left(\dfrac{a_{11} a_{22}}{a_{12}} - a_{21}\right) N_1 N_2 \\ \dfrac{N_1 N_2}{k a_{12}} & -\dfrac{a_{11}}{a_{12}} N_2^2 \end{bmatrix} \cdot \begin{bmatrix} \dfrac{di_1}{dt} \\ \dfrac{di_2}{dt} \end{bmatrix}$$

The negative signs in the right column of the Z-Matrix, result from the selected orientation of the counting arrows. Table 5.1 shows some basic configurations of the magnetic circuit with the corresponding analytical interpretations. For the size, k applies $k = \pm 1$, depending on which winding direction was selected. If the magnetic flux lines driven by N_1 are encompassed by the two windings in the same direction of rotation, the positive sign applies, otherwise the negative sign. Thus the above system of equations can also be written as follows:

Table 5.1 Examples for simple magnetic transmission quadrupoles

Designation	Magnetic quadripole equivalent circuit diagram	cascaded shape of the magnetic quadripole matrix	Design example for winding arrangement	electrical transformer equivalent circuit
Ideal transformer		$\begin{bmatrix} 1 & 0 \\ 0 & 1 \end{bmatrix}$	Only abstract Calculation model	
Leakage-free transformer		$\begin{bmatrix} 1 & R_{m1} \\ 0 & 1 \end{bmatrix}$	Only abstract Calculation model	
Transformer With $\mu_r \to \infty$ and distribution		$\begin{bmatrix} 1 & 0 \\ \frac{1}{R_\sigma} & 1 \end{bmatrix}$	Only abstract Calculation model	
Transformer with winding distribution (tight coupling)		$\begin{bmatrix} \dfrac{R_{m1}+R_{\sigma2}}{R_{\sigma2}} & R_{m1} \\ \dfrac{R_{m1}+R_{\sigma1}+R_{\sigma2}}{R_{\sigma1}^2} & \dfrac{R_{m1}+R_{\sigma1}}{R_{\sigma1}} \end{bmatrix}$		
Transformer with strong yoke leakage (loose coupling)		$\left[\begin{array}{cc} \dfrac{R_{\sigma0}+R_{m1}}{R_s} & R_{m1}\left(\dfrac{R_{m1}\left(R_{m2}+R_{\sigma0}\right)}{R_{\sigma0}} + R_{m2} \right) \\ \dfrac{1}{R_{\sigma0}} & \dfrac{R_{\sigma0}+R_{m2}}{R_{\sigma0}} \end{array} \right]$		

$$
\begin{bmatrix} u_1 \\ u_2 \end{bmatrix} =
\begin{bmatrix}
\dfrac{a_{22}}{a_{12}} N_1^2 & \mp \left(\dfrac{a_{11}a_{22}}{a_{12}} - a_{21} \right) N_1 N_2 \\[3mm]
\pm \dfrac{N_1 N_2}{a_{12}} & - \dfrac{a_{11}}{a_{12}} N_2^2
\end{bmatrix}
\cdot
\begin{bmatrix} \dfrac{di_1}{dt} \\[2mm] \dfrac{di_2}{dt} \end{bmatrix}
$$

Table 5.1 is shown with some simple magnetic four poles, to which many calculations can be traced back.

If the parameters of the quadripole for a strong yoke leakage are used in the above equation, neglecting the leakage components in the windings themselves, one obtains

$$
\begin{bmatrix} u_1 \\ u_2 \end{bmatrix} =
\begin{bmatrix}
\dfrac{N_1^2}{R_{m1} + \dfrac{R_{m2}R_\sigma}{R_{m2} + R_\sigma}} & \mp \dfrac{R_\sigma}{R_{m1} + R_\sigma} \dfrac{N_1 N_2}{R_{m2} + \dfrac{R_{m1}R_\sigma}{R_{m1} + R_\sigma}} \\[6mm]
\pm \dfrac{R_\sigma}{R_{m2} + R_\sigma} \dfrac{N_1 N_2}{R_{m1} + \dfrac{R_{m2}R_\sigma}{R_{m2} + R_\sigma}} & - \dfrac{N_2^2}{R_{m2} + \dfrac{R_{m1}R_\sigma}{R_{m1} + R_\sigma}}
\end{bmatrix}
\cdot
\begin{bmatrix} \dfrac{di_1}{dt} \\[2mm] \dfrac{di_2}{dt} \end{bmatrix}
$$

In the following, the individual coefficients of the matrix are to be interpreted. With regard to the structure of the magnetic circuit, it is obvious

$$
\frac{N_1^2}{R_{m1} + \frac{R_{m2}R_\sigma}{R_{m2}+R_\sigma}} = \frac{N_1^2}{R_{m1} + R_{m2}//R_\sigma} = L_1
$$

the self-inductance of the winding of N_1. Analogously, the following applies to the second winding

$$
\frac{N_2^2}{R_{m2} + \frac{R_{m1}R_\sigma}{R_{m1}+R_\sigma}} = \frac{N_2^2}{R_{m2} + R_{m1}//R_\sigma} = L_2
$$

The factor $\frac{R_\sigma}{R_{m1}+R_\sigma} = k_{12}$ $k_{21} = \frac{R_\sigma}{R_{m2}+R_\sigma}$ the coupling factor in the opposite direction. The parameter

$$
\frac{R_\sigma}{R_{m1} + R_\sigma} \cdot \frac{N_1 N_2}{R_{m2} + \frac{R_{m1}R_\sigma}{R_{m1}+R_\sigma}} = M_{12}
$$

is the "mutual inductance" from turn N_2 to N_1, or is therefore

$$
\frac{R_\sigma}{R_{m2} + R_\sigma} \cdot \frac{N_1 N_2}{R_{m1} + \frac{R_{m2}R_\sigma}{R_{m2}+R_\sigma}} = M_{21}
$$

the "mutual inductance" from N_1 to N_2. By replacing the factors with the designations just defined, the equation system

$$\begin{bmatrix} u_1 \\ u_2 \end{bmatrix} = \begin{bmatrix} L_1 & \mp M_{12} \\ \pm M_{12} & -L_2 \end{bmatrix} \cdot \begin{bmatrix} \dfrac{di_1}{dt} \\ \dfrac{di_2}{dt} \end{bmatrix}.$$

One quickly sees that the ratio of the two Z-elements of the minor diagonals of the matrix

$$\left| \frac{z_{12}}{z_{21}} \right| = \frac{\frac{R_\sigma}{R_{m1}+R_\sigma} \frac{N_1 N_2}{R_{m2}+\frac{R_{m1}R_\sigma}{R_{m1}+R_\sigma}}}{\frac{R_\sigma}{R_{m2}+R_\sigma} \frac{N_1 N_2}{R_{m1}+\frac{R_{m2}R_\sigma}{R_{m2}+R_\sigma}}} = \frac{(R_{m2}+R_\sigma)\left(R_{m1}+\frac{R_{m2}R_\sigma}{R_{m2}+R_\sigma}\right)}{(R_{m1}+R_\sigma)\left(R_{m2}+\frac{R_{m1}R_\sigma}{R_{m1}+R_\sigma}\right)} = 1$$

Therefore, a transformer with usually only one mutual inductance M is

$$M = \sqrt{k_{12}k_{21}L_1 L_2} = \sqrt{\frac{R_\sigma}{R_{m1}+R_\sigma} \cdot \frac{R_\sigma}{R_{m2}+R_\sigma} \cdot \frac{N_1^2}{R_{m1}+\frac{R_{m2}R_\sigma}{R_{m2}+R_\sigma}} \cdot \frac{N_2^2}{R_{m2}+\frac{R_{m1}R_\sigma}{R_{m1}+R_\sigma}}}$$

where M in the case under consideration is then calculated as

$$M = \frac{R_\sigma N_1 N_2}{\sqrt{(R_{m1}R_{m2}+R_{m1}R_\sigma+R_{m2}R_\sigma)(R_{m1}R_{m2}+R_{m1}R_\sigma+R_{m2}R_\sigma)}}$$
$$= \frac{R_\sigma N_1 N_2}{R_{m1}R_{m2}+R_{m1}R_\sigma+R_{m2}R_\sigma}.$$

This is not proof of the correctness of the specification for the calculation of M, but only proof that the approach applies to the assumed magnetic structure. If you look at another simple case, which for the sake of simplicity should include a close coupling of the windings – but with consideration of the leakage fields in the winding – you will come to a different result with regard to the previous case. The arrangement as well as the magnetic quadripole matrix in chain form is shown in Fig. 5.18. The leakage channels of the magnetic field can be influenced by the shape of the winding, insertion of para- or ferromagnetic rods, etc.

The transmission equations of the transformer change on this basis to

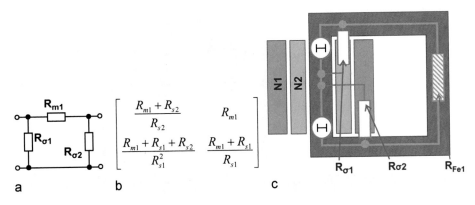

Fig. 5.18 Transformer with winding leakage (close coupling). (**a**) magnetic equivalent circuit diagram, (**b**) magnetic quadripole matrix in cascaded form, (**c**) design approach

$$
\begin{bmatrix} u_1 \\ u_2 \end{bmatrix} =
\begin{bmatrix}
\dfrac{N_1^2}{\dfrac{R_{m1}R_{\sigma1}}{R_{m1}+R_{\sigma1}}} & \mp \dfrac{N_1N_2\left(R_{m1}^2(R_{\sigma1}-R_{\sigma2})+R_{m1}\left(R_{\sigma1}^2-R_{\sigma2}^2\right)+R_{\sigma1}^2R_{\sigma2}\right)}{R_{m1}R_{\sigma1}^2R_{\sigma2}} \\[4ex]
\pm\dfrac{N_1N_2}{R_{m1}} & -\dfrac{N_2^2}{\dfrac{R_{m1}R_{\sigma2}}{R_{m1}+R_{\sigma2}}}
\end{bmatrix}
\cdot
\begin{bmatrix} \dfrac{di_1}{dt} \\ \dfrac{di_2}{dt} \end{bmatrix}
$$

As expected, the elements of the main diagonal of the matrix reflect the self-inductances of both windings. A different picture is found with the mutual inductances:

Only for $R_{s1} = R_{s2}$, that is, with complete magnetic symmetry, the elements $|z_{12}| = |z_{21}|$ and the two mutual inductances M_{12} and M_{21} are the same. Thus, if the magnetic circuits are more complex, a uniform mutual inductance M can not be assumed. It is calculated by taking into account the leakage inductances. As the above derivation shows, in general, the following applies to the elements of the 'Z' matrix $|z_{12}| \neq |z_{21}|$! The structural investigations in this chapter shall help to resolve complex arrangements into simple connections. The following example of a 2-winding transformer shall serve this purpose.

5.5 Modeling Example of a Linear "Leakage Core" Transformer

As you could see, a transformer of the core type has the possibility of both close and loose coupling. One can combine both possibilities to set the required leakage inductance between close coupling (ideally $L_\sigma \to 0$) and loose coupling. Figure 5.19 shows such a design approach. One part N_{11} of the primary winding N_1 is loosely coupled to the secondary winding N_2, while another part N_{12} is closely coupled to it. Using the steps discussed so far, the first step is to quickly construct a simple equivalent magnetic circuit with concentrated elements that takes into account the paths of the leakage magnetic field.

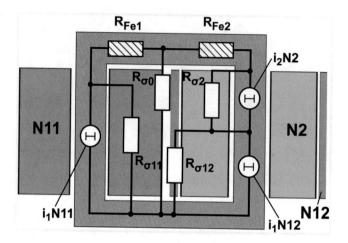

Fig. 5.19 Transformer with close **and** loose coupling of the primary and secondary winding (for simple demonstration, not all conceivable differently magnetized core sections were considered)

The described graphic inversion yields the equivalent electrical circuit in Fig. 5.20.

Here a short investigation of the winding redistribution is to take place. Since the permeability of the core can be assumed to be very high

$$L_{Fe1} \rightarrow \infty \, and \, L_{Fe2} \rightarrow \infty$$

set. Thus, 2 elements are omitted for the network analysis. This leads to a more compact representation without major errors.

Open circuit: $i_2 = 0$, $i_{02} = 0$

$$i_{011} + i_{012} = 0 \rightarrow i_1 = 0$$

This also means

$$u_{011} - u_{012} = 0 \,\, or \, u_{011} = u_{02}$$

The following then applies to the primary voltage

$$u_1 = u_{11} + u_{12} = N_{11}u_{011} + N_{12}u_{12} = (N_{11} + N_{12})u_{011},$$

which, as expected, results in the following for the secondary voltage

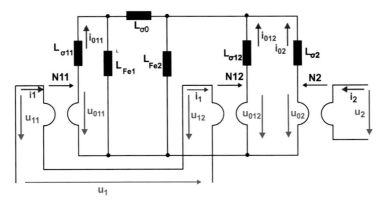

Fig. 5.20 Electrical equivalent circuit of the arrangement in Fig. 5.19

$$u_2 = u_{02}N_2 = u_{011}N_2 = u_1 \frac{N_{11} + N_{12}}{N_2} \ Short \ circuit \ case \ (u_2 = 0, \ u_{20} = 0)$$

The following then applies to the currents entered

$$- i_{02} = i_{011} + i_{012} = i_1N_{11} + i_1N_{12} = i_1(N_{11} + N_{12}),$$

while u_1 is composed of

$$u_1 = u_{11} + u_{12} = u_{011}N_{11} + u_{012}N_{12}.$$

When using sinusoidal voltages, the following can also be written

$$\underline{u_1} = \left(j\omega(L_{\sigma11} + L_{\sigma0})\underline{i_{011}} - j\omega L_{\sigma2}\underline{i_{02}}\right)N_{11} + \left(j\omega L_{\sigma12}\underline{i_{012}} - j\omega L_{\sigma2}\underline{i_{02}}\right)N_{12} \ respectively$$

$$\underline{u_1} = \begin{array}{l} \left(j\omega(L_{\sigma11} + L_{\sigma0})\underline{i_1}N_{11} + j\omega L_{\sigma2}\underline{i_1}(N_{11} + N_{12})\right)N_{11} \\ + \left(j\omega L_{\sigma12}\underline{i_1}N_{12} + j\omega L_{\sigma2}\underline{i_1}(N_{11} + N_{12})\right)N_{12} \end{array}$$

$$\underline{u_1} = j\omega\underline{i_1}\left((L_{\sigma11} + L_{\sigma0})N_{11}^2 + L_{\sigma2}[(N_{11} + N_{12})N_{11} + (N_{11} + N_{12})N_{12}] + L_{\sigma12}N_{12}^2\right)$$

This corresponds to the leakage reactance measured on the primary side with

$$\underline{u_1} = j\omega\underline{i_1}L_{\sigma ges}$$

To achieve a certain flux linkage primarily, a certain number of windings N_1 is required. N_{12} then results in

$$N_{11} = N_1 - N_{12}$$

If this is used for $L_{\sigma tot}$, the change in the total leakage inductance as a function of the winding distribution can be displayed in a standardized form.

$$\frac{L_{\sigma tot}}{N_1^2} = \frac{\left((L_{\sigma 11} + L_{\sigma 0})N_{11}^2 + L_{\sigma 2}[(N_{11} + N_{12})N_{11} + (N_{11} + N_{12})N_{12}] + L_{\sigma 12}N_{12}^2\right)}{N_1^2}$$

$$\frac{L_{\sigma tot}}{N_1^2} = \frac{\left((L_{\sigma 11} + L_{\sigma 0})(N_1 - N_{12})^2 + L_{\sigma 2}N_1^2 + L_{\sigma 12}N_{12}^2\right)}{N_1^2}$$

$$= L_{\sigma 2} + (L_{\sigma 11} + L_{\sigma 0})\left(1 - \frac{N_{12}}{N_1}\right)^2 + L_{\sigma 12}\left(\frac{N_{12}}{N_1}\right)^2$$

To illustrate the effect of the redistribution of the windings, the following quantities for the inductances are assumed as fixed magnetic conductance values of the magnetic circuit:

$$L_{\sigma 11} = L_{\sigma 12} = L_{\sigma 2} = 2\mu H = \frac{1}{50}L_{\sigma 0},$$

You can see the effect of the winding redistribution on a simple graph:

$$\frac{L_{\sigma tot}}{N_1^2} = 2\mu H + 102\mu H\left(1 - \frac{N_{12}}{N_1}\right)^2 + 2\mu H\left(\frac{N_{12}}{N_1}\right)^2$$

The result of this formula calculation is shown in Fig. 5.21.

Depending on the core shape and winding design, very different relations can be set between the two coupling forms. In principle, a negative winding (counter connection) of N_{12} is also possible. Then the leakage inductance increases further. Such measures are common to limit the short-circuit current to small values (e.g. for operating plasma discharges as loads in welding technology or gas discharge lasers or in devices for periodically charging capacitors). Without additional voltage induction, the leakage inductance is then increased. Only in the case of relatively low permeabilities or long lengths of the core sections do these have a decisive influence on the general performance. An approximation $\mu_r \to \infty$ for the ferromagnetic core sections therefore provides a quick overview of the performance of the component in a simplified situation.

Summary of the Procedure

1. A directly magnetized branch is assigned to each winding part.
2. The magnetic circuit is simulated by a magnetic equivalent circuit (MEC) with various concentrated components.

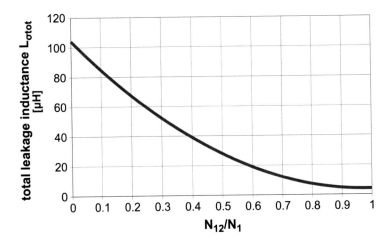

Fig. 5.21 Change in leakage inductance at the transition from loose to fixed coupling ("strong yoke leakage" changes to the approach "weak yoke leakage")

3. The topology for an electrical equivalent circuit (EEC) diagram is formed by graphic circuit inversion.
4. The inductive elements are parameterized or provided with characteristic curves or (see Chap. 7) modeled in their behavior via a mapping process of the information available from experiments.
5. By connecting ideal transformers, the actual number of turns of the windings is implemented in the EEC.
6. The electrical equivalent circuit diagram of the magnetic arrangement is then inserted into an electric network calculation or a simulation system for electrical circuits.

5.6   Modelling Approach for Nonlinear Magnetic Sections

In Sect. 5.2 you could see that the magnetization of a core section by a winding can be described as a quadripole, with the magnetic quantities (Φ, V) on one side and the electrical quantities (U, I) on the other side (Fig. 5.22).

The descriptive equations follow the physical relationships and are derived from magnetic voltage drop v(t) and flux linkage ψ(t)

$$v = N \cdot i$$
$$\int u \cdot dt = \psi = N \cdot \phi$$

respectively

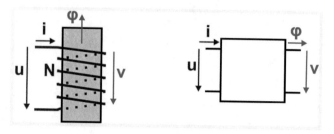

Fig. 5.22 Quadripole conversion between electric and magnetic field quantities in concentrating structures. (a) Physical structure, (b) Diagram

$$\begin{bmatrix} \int u \cdot dt \\ v \end{bmatrix} = \begin{bmatrix} \psi \\ v \end{bmatrix} = \begin{bmatrix} 0 & N \\ N & 0 \end{bmatrix} \cdot \begin{bmatrix} i \\ \phi \end{bmatrix}$$

or in the chain form

$$\begin{bmatrix} \psi \\ i \end{bmatrix} = \begin{bmatrix} 0 & N \\ \frac{1}{N} & 0 \end{bmatrix} \cdot \begin{bmatrix} v \\ \phi \end{bmatrix}$$

These basic relationships and structures can also be applied if the magnetized material shows a nonlinear relationship between φ and v. – Although they are actually only defined for linear networks.

If a linear reluctance $R_{m1}^* = \frac{v}{\phi}$ is connected, one obtains

$$\begin{bmatrix} \psi \\ i \end{bmatrix} = \begin{bmatrix} 0 & N \\ \frac{1}{N} & 0 \end{bmatrix} \cdot \begin{bmatrix} 0 & R_{m1}^* \\ 0 & 1 \end{bmatrix} \cdot \begin{bmatrix} v \\ \phi \end{bmatrix} = \begin{bmatrix} 0 & N \\ 0 & \frac{R_{m1}^*}{N} \end{bmatrix} \cdot \begin{bmatrix} v \\ \phi \end{bmatrix}$$

which is nothing more than

$$\psi = N\phi$$
$$i = \frac{R_{m1}^*}{N}\phi$$

respectively

$$\frac{\psi}{i} = L = \frac{N^2}{R_{m1}^*}$$

If a non-linear reluctance with the dependency v = v(φ) is now connected, one obtains

$$\begin{bmatrix} \psi \\ i \end{bmatrix} = \begin{bmatrix} 0 & N \\ \dfrac{1}{N} & 0 \end{bmatrix} \cdot \begin{bmatrix} v(\phi) \\ \phi \end{bmatrix}$$

This means that a characteristic curve v(φ) is transformed into a characteristic curve i(ψ) with

$$i = \frac{v(\phi)}{N} = \frac{l \cdot H(B \cdot A)}{N}$$

$$\psi = N \cdot B \cdot A = \int u \cdot dt$$

If one knows the characteristic H(B) or v(φ) in the magnetic field (via the connection with the cross-section 'A' and the length 'l' of the section), you also have the necessary information for the description in the electric current field. This applies to reversibly unambiguous characteristic curves B(H) obviously without restriction. In a circuit simulator, the element in question can be defined by entering i(ψ) or L(i), where L here means the differential inductance. If a non-linear reluctance is connected to the gyrator (Fig. 5.23), it also inverts/transforms its φ(v) behavior in the "magnetic world" to a ψ(i) behavior in the world of electrical circuits. The magnetic structure once found would thus also be valid for nonlinear elements after the circuit inversion in the resulting topology.

Figure 5.24 shows some examples of the nanocrystalline material Vitroperm in various qualities compared to a typical ferrite. The achievable saturation induction of these materials is considerably higher than that of ferrites. At the same time, the coercive field strengths H_c are at much lower values. With VITROPERM they are 3 A/m, whereas with the usual MnZn ferrites they are 5...60 A/m in the best case, but can also be much higher.

Hereby, reversibly unambiguous nonlinear characteristics of elements of magnetic circuits can also be integrated almost without problems into electrical simulation systems as nonlinear elements, simply by converting the material characteristics into elementary characteristics of the inductors, which are basically scaled magnetic conductances, according to the described rules. However, this does not yet cover all non-linear elements of interest:

- One group is to be characterized as (current) controllable, non-linear magnetic resistances with which one can influence, for example, the degree of leakage path/coupling between 2 windings for AC applications (Fig. 5.26).
- The other group can be characterized by the distinctly hard magnetic materials with their distinctly ambiguous characteristics

In both cases, the integration of an additional control variable is required to implement the changes in the characteristic curve. For the characteristics with relevant ambiguity, the magnetic materials with so-called Z-characteristics developed for transductor/switching reactor applications are of interest first. Figure 5.25 shows 2 examples from this group with

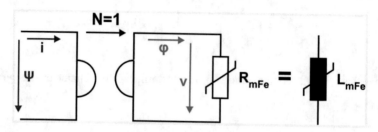

Fig. 5.23 Inversion/transformation of a magnetic section with a reversibly unambigous characteristic

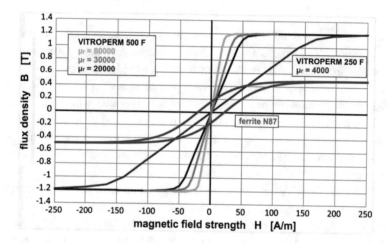

Fig. 5.24 Examples of (due to very low values of $H_c < 3$ A/m almost reversibly clear) characteristics of soft magnetic materials in comparison to ferrite N87 (data basis: Vacuumschmelze Hanau, TDK)

their static magnetization curves. These materials have an almost rectangular magnetization characteristic with extremely low coercive field strength compared to permanent magnetic materials. At the same time, a very high relative permeability is characteristic (see also Table 5.2).

The characteristic curve B(H) of the nonlinear component can be described for many purposes with a good approximation using the tangent hyperbolic function:

$$B(H) \approx B_{SAT} \cdot \tanh\left(\frac{\mu_0 \mu_r (H \pm H_C)}{B_{SAT}}\right) + \mu_0 H$$

Figure 5.26 shows this function in comparison to other, simple variants of an approximation function. The basic properties of magnetization characteristics are different with regard to their slope or curvature. Therefore, the selection of an adapted function type for complete mathematical treatment is useful. The functions shown in Fig. 5.26 are mathematically

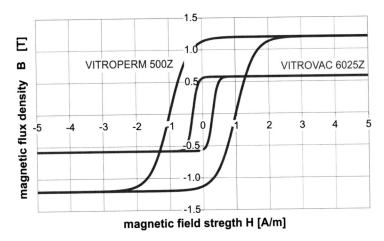

Fig. 5.25 Examples of (due to very low values of Hc almost) reversibly clear characteristics of soft magnetic materials. (Data source: Vacuumschmelze Hanau)

Table 5.2 Comparison of 2 magnetic materials developed for transformers

	VITROPERM 500Z	VITROVAC 6025 Z
Material basis	Nanocrystalline, Fe-based	Amorphous, co-base
Saturation induction (25 °C), B_s	1.2 T	0.58 T
Bipolar induction swing (25 °C) ΔB_{ss}	2.35 T	1.15 T
Bipolar induction swing (90 °C) ΔB_{ss}	2.15 T	1.0 T
Rectangular form factor, B_r/B_s	>94%	>96%
Core losses P_{Fe} (typ.), 50 kHz, $\Delta B = 0.8$ T	100 W/kg	60 W/kg
Coercive field strength H_c	<1 A/m	0.3 A/m
Curie temperature	>600 °C	240 °C
Continuous operating temperature	120 °C	90 °C
Specific resistance	1.2 µΩm	1.35 µΩm
Density	7.35 g/cm^3	7.70 g/cm^3

Source: Vacuumschmelze Hanau GmbH

simple approximations to the behavior of data of characteristic B(H). There is no physical background for the function type here. The maximum value of the increase is reached at +/− H_C and is. If $\mu_r \gg 1$ second summand can be neglected.

$$\frac{dB(H)}{dH} = \frac{1}{a} = \frac{\mu_r \mu_0}{B_{SAT}} \text{ at } H = \pm H_C$$

All functions B(H) strive approximately towards the limit value +/− B_{sat} for large values of H and are differentiable. In reality, this only applies to the polarization P(H) and must be

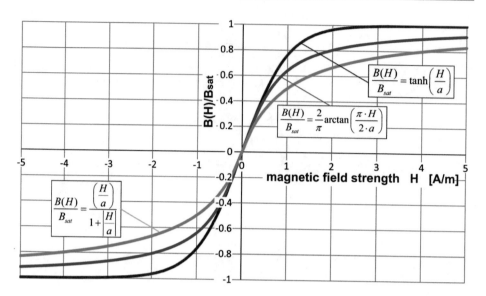

Fig. 5.26 Comparison of different simple analytical functions with the same maximum rise and the same limits at high magnetic field strengths to approximate the real performance of magnetisation characteristics

taken into account by the additive component of the vacuum if necessary, as is the case with permanent magnets (see below).

If one converts the tanh function to the manufacturer's specifications for cores with Z-characteristic, one obtains:

VITROPERM 500Z $\quad B(H) = 1{,}2T \cdot \tanh\left(\frac{1,\,65 \cdot 10^6 \cdot \mu_0 (H \pm 1A/m)}{1,\,2T}\right)$

VITROVAC 6025Z $\quad B(H) = 0{,}58T \cdot \tanh\left(\frac{3 \cdot 10^6 \cdot \mu_0 (H \pm 0,\ 3A/m)}{0,\,58T}\right)$

Due to the low coercive field strength, the "mean characteristic curve" for simulations can be determined here

$$B(H) = B_{SAT} \cdot \tanh\left(\frac{\mu_0 \mu_r H}{B_{SAT}}\right)$$

as an approximation. These cores show a "switching performance" at saturation and are therefore used as cores, for example, in transducers at high frequencies or in pulse applications with high remagnetization rates. For linear applications such as transformers, F cores are used which have a very low remanence induction. The different properties are achieved on the same material with different treatment with magnetic fields during heat treatment.

The coercivity of permanent magnetic materials is 5. . .6 orders of magnitude higher than that of materials for transducer applications, although they also have a rectangular magnetization characteristic. The main problem for simulation purposes and the derivation of the electrical equivalent circuit diagram is here the combination of "2 worlds":

- The magnetic equivalent circuit diagrams and characteristic curves all use the integral form of Maxwell's equations (ψ, i, etc.), while the electrical simulators essentially work with the "differential" parameters (u, di/dt, i, etc.), although they obtain the results by integrating the system of differential equations.
- The common representation of both "worlds" in a simulation system requires a *"gateway"* between both.

Otherwise, the result is a false solution, as the following simple example will show:

If you take a permanent magnet with an infinitely large coercivity H_c, this corresponds to a magnetic flux source. The magnetic equivalent circuit diagram Fig. 5.27 shows this in the 'integral' magnetic world:

A formal transformation of the circuit provides two series-connected inductors L_{pm} and L_{Fe} (Fig. 5.28a). The magnetic flux Φ_0 with its magnetic voltage drop at the two parallel reluctances $R_{mpm}//R_{mFe}$ then responds to a "bias current $I_0 = B_r \cdot A_{Fe}$, where B_r is the remanence induction of the inserted permanent magnetic material with the magnetization cross-section A_{Fe}. The total energy of the arrangement is composedof the sum of the individual energies according to the formula $W_{mag} = 0.5 \cdot I_0^2 \cdot \left(L_{pm} + L_{Fe}\right)$. The point of intersection of the $\Psi(I)$ characteristic curve is shifted on the I axis by the premagnetization. The effect of this measure is shown in Fig. 5.28c) using the course of the $\Psi(I)$ characteristic and the $L*(I)$ characteristic. This method of permanent premagnetization has been used for a long time in the DC range. For example, holding magnets for doors can be compensated with an electric current in their permanent magnetic field so that the holding force is "switched off" as long as the current flows. In the event of an alarm or danger, doors then close with the spring force. When not in use, such a device does not need any current.

Suitable permanent magnetic materials for the operation of chokes in DC converters with unipolar current have only been available for a few years. One solution, for example, is plastic-bound, finely distributed NdFeB magnetic material (powder). This has low remagnetization losses, combined with a high coercivity and sufficient long-term stability.

The magnetization characteristic curve can also be described via a pre-magnetizing direct current. In the equivalent circuit diagram, however, it is completely unclear how this is to be divided up, especially since in reality there is no electrical connection between the main circuit and the "control circuit". This means that the currents of the model – which do not really exist in the circuit – mix with the "real" flowing currents during DC premagnetisation and can lead to nonsensical results. An impedance transformer is required which simultaneously electrically isolates both "worlds" and connects them analytically/logically. This can be achieved by controlled sources, similar to a gyrator. This is presented

Fig. 5.27 Magnetic circuit with permanent magnetic flux through a permanent magnet with very high coercivity

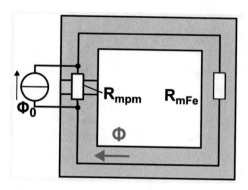

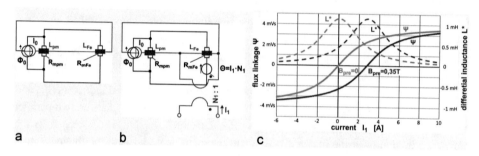

a   b   c

Fig. 5.28 Electrical (transformed) equivalent circuit of a magnetic circuit with permanent premagnetisation. (**a**) premagnetised core without further winding, (**b**) premagnetised core with winding, (**c**) effect of the premagnetisation on the $\Psi(I)$ characteristics and the differential inductance at $A_{Fe} = 3$ cm^2, $l_{Fe} = 16$ cm, $\delta = 1$ mm, $B_{sat.ferrite} = 0.45$ T, $N_1 = 30$

in detail in Chap. 7 for non-linear, DC-premagnetized sections but also those with premagnetization by permanent magnets. Here in the following, only the basic approach is explained. For this purpose, the relationships derived above for the representation of a source-free magnetization characteristic curve are taken as a starting point.

$$i = \frac{v(\phi)}{N} = \frac{l \cdot H(B \cdot A)}{N}$$

$$\psi = N \cdot B \cdot A = \int u \cdot dt$$

Viewed from the electrical circuit, this figure can be generally represented graphically as follows (Fig. 5.29). The component L_0 represents the vacuum part of the space filled by the magnetic field.

In the case of small increments for discretization, that is, (Δt, Δu, Δi)→0 there is initially no logical contradiction. Analog simulation systems of the past have maximum band limits/cut-off frequencies. The same applies to similar extent to digital circuit simulators, with their (variable) step sizes for the calculation. Here, the user can influence,

Fig. 5.29 Basic approach for a transformation unit between non-galvanically coupled, non-linear inductive elements

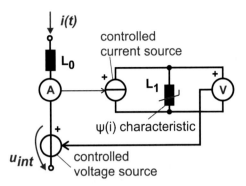

but little a-priori information about the effect of such measures in detail during the simulation.

The simulation system can only react to changes with a certain maximum speed, while the idealized components are assumed to react instantly. This results in a mismatch between the claim of the simulation as an image of reality. This can be illustrated by the fact that the (transforming) approach.

Current measurement → u(i, t, control signal) → controlled voltage drop.

has the following discribed disadvantage when considering the impedance.

For an inductance of any kind applies:

$$\lim_{\omega \to \infty} |X_L| \to \infty,$$

while the structurally given impedances of current measurement in series with a controlled voltage source are always identical to zero.

This problem can be significantly mitigated if only the nonlinear part of the component is externally mapped or the component is composed of a purely linear part L_0, which is located in the electrical simulation circuit, plus a nonlinear part, which is implemented by the transformation shown above. The implementation for a circuit simulator is shown in Fig. 5.30, where an additional control unit (current source) is inserted to map the effect of the additional control winding. Thus, a control circuit can be inserted isolated from the main circuit, so that the actual conditions are taken into account. However, some circuit simulators do not accept current sources in series with an inductor. Then this approach has to be modified.

This measure is also justified by physics. There are practically no magnetic non-conductors. The term polarization denotes the range of materials that behave differently from the vacuum. This separation is also carried out in the proposed and proven model approach and is also common in the description of materials, as will be seen below. After insertion of this part, problems of continuity and differentiability of the characteristic

Fig. 5.30 Extraction of the controlled non-linearity of the magnetic component from the circuit topology of the electrical circuit

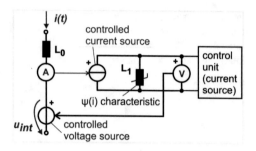

curves and their easy implementation play a major role. If the non-linear, pre-magnetized magnetic section is branched, the individual magnetic branches must be included, if necessary, an outsourced simulation part according to the pattern in Fig. 5.30. Examples of the polarization of two materials and the differential permeability and amplitude permeability derived from it are shown in Fig. 5.31. The strongly non-linear properties of the two materials can be seen.

If one wants to vary material properties and geometric quantities, it is recommended to calculate the quantities to the component properties immediately in the simulation, as shown in Fig. 5.32.

For a calculation example, an arrangement consisting of two toroidal cores is assumed, which are each premagnetized in opposite directions. The windings are applied to the toroidal stripwound cores made of M6X. The windings are calculated so that a power loss density of 150 mWcm^{-3} occurs. Figure 5.33 shows the schematic conversion or the equivalent circuit diagram used for the simulation. The two cores ($60{\times}40{\times}10$) have the same magnetization properties. The inductances of the cores without material L_{0x} result from the permeances of the core geometries without material multiplied by N_1^2. The inductance leakage inductance between N_1 and the cores is neglected here and would be in series with the inductors L_{01} and L_{02}. For AC voltage applications, the control windings are connected in series so that the induced voltages are the opposite. With this arrangement, the effect of the inductance can be controlled by a direct current. For the effect, an available AC voltage source is important, which was assumed here to be 115 V/400 Hz. The controllable inductance, which in this form is also called a *current-controlling magnetic amplifier* or *transductor*, is connected in series with an ohmic load at the voltage source. When the controlling current $\Theta_{0L\#1} = 0$, the inductance is maximum and the load current is minimum. If you now increase the control current or the flooding $\Theta_{0L\#1}$, the two cores are pre-magnetized in different directions and the resulting inductance, which in this case is more than 325 turns around both cores, is reduced. Remarkably, the control effect is achieved completely without semiconductor elements. The magnetic amplifier is one of the oldest control elements in power electronics. The form of a current-controlling magnetic amplifier chosen here as an example is characterized by a relatively low distortion of the sinusoidal shape of the load voltage (Fig. 5.34), which is expressed in low THD$_u$ values (*total harmonic distortion*, the harmonic content of the voltage). A disadvantage is the low

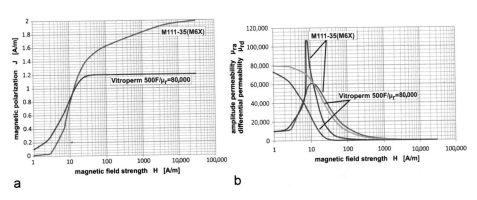

Fig. 5.31 Magnetization properties of a grain-oriented sheet (M6X) and a nanocrystalline material (VP500F). (**a**) magnetic polarization, (**b**) amplitude permeability (blue, violet) and differential permeability (red, green)

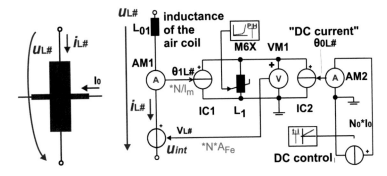

Fig. 5.32 Modeling of a choke pre-magnetized with direct current through an electrical network with controlled sources. In L_1 only the material properties of the grain-oriented sheet M6X are shown as a function of the magnetic field strength H from the magnetic polarization P

current amplification of approximately 1 regarding the nagnetic forces, which can, however, be partly compensated by a correspondingly large number of turns of the controlling DC windings. In the control circuit, the ohmic losses are essential to be applied, while in the load circuit the effective voltage at the load is controlled.

The power amplification (gain) g_p in this example is then about

$$g_P = \frac{P_{load}}{P_{control}} \approx \frac{\frac{(115V)^2}{320\Omega}}{2.8\Omega \cdot (0.5A)^2} \approx 59$$

The efficiency results from the consideration of the component losses P_{Cu} and P_{core} and control losses to

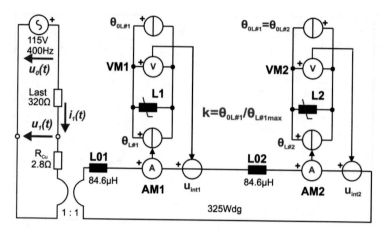

Fig. 5.33 Use of 2 toroidal cores (d_i = 40 mm, d_a = 60 mm, M6X, N_1 = 325) for the construction of a current controlling magnetic amplifier, simulation scheme according to Figs. 5.30, 5.31, and 5.32

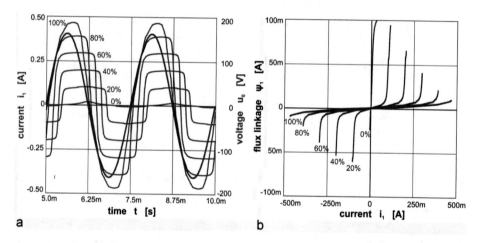

Fig. 5.34 Effect of a current-controlling solenoid amplifier at resistive load in a 115 V/400 Hz network. (**a**) calculated current and voltage curves, (**b**) dynamic $\psi_1(i1)$ characteristic at different modulation k (see Fig. 5.33)

$$\eta_P = \frac{P_{load}}{P_{load} + P_{Cu} + P_{core} + P_{control}}.$$

However, this makes the control unit relatively heavy. If one wants to limit the maximum current or the application area is in a higher frequency range, competitive solutions can be achieved. Of particular interest are the low current distortions that are easy to achieve. In the so-called voltage-controlling magnetic amplifier, which works like a thyristor in mains applications with phase-angle control, one obtains considerably higher amplification values at the price of high values of current distortion. Due to the high weight compared to thyristor applications, such magnetic amplifiers are hardly used anymore. At very high frequencies or currents, the advantage of the current-controlling magnetic amplifier is that only the windings and no additional semiconductor elements limit the current. This results in applications with switch mode power supplies [5.3, 5.4] or in pulse technology (magnetic pulse shortening, etc., Sect. 10.27). In the latter case, the control circuit is used to return the cores to their initial state before each pulse event.

In principle, the voltages at 2 magnetically coupled inductors result via the law of induction to

$$u_1(t) = L_1 \cdot \frac{di_1}{dt} \pm M \cdot \frac{di_2}{dt}$$
$$u_2(t) = \pm M \cdot \frac{di_1}{dt} + L_2 \cdot \frac{di_2}{dt}$$

If this system of equations is projected into an equivalent circuit diagram suitable for a simulation, one obtains the representation shown in Fig. 5.35, consisting of controlled sources. This representation is very similar to the ones explained above but contains no indication of the "nature" of the mutual inductance M. The voltage drop across an inductor serves as an inductive voltage generated by an external magnetic field to control a voltage source as part of an inductor to map the external influence. In the case of branched magnetic circuits with non-linear branches, morphological analysis, as explained in this chapter, must also be carried out to arrive at useful simulation models. The procedure presented here therefore represents a combination of both approaches. A consideration of the nonlinear properties then also enables the mapping of current-dependent coupling factors.

In the following, further aspects of modeling a device with a nonlinear core section are shown with its possibilities and limitations. We will start with a simple toroidal core 80×50×20 of in Vitroperm 500F from VAC. The following Table 5.3 shows the most important company details for the calculations carried out.

The characteristic curve B(H) of the nonlinear component can be described with good approximation using the tangent hyperbolic function:

$$B(H) = B_{SAT} \cdot \tanh\left(\frac{\mu_0 \mu_r H}{B_{SAT}}\right)$$

A corresponding illustration of this approximation is shown in Fig. 5.31. The tanh function only reflects the behavior of the measured data and not the physical context (Fig. 5.36).

Since concrete core parameters can now be used as a basis, this characteristic curve can be converted into the characteristic curve for the non-linear part of the magnetic resistance of the core. The linear component for air/vacuum of the magnetic resistance must be connected in parallel, as shown in Fig. 5.37.

The use of the geometric core values provides the following characteristic curves

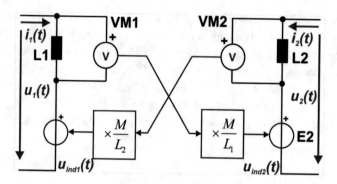

Fig. 5.35 Simulation model for coupled inductors, derived from the transformer equations

Table 5.3 Material data for a toroidal core 50×60×20 from VAC

Outer diameter	d_a	80	mm
Inner diameter	d_i	60	Mm
Height	h	20	Mm
Magnetization cross-section	A_{Fe}	2.28	cm^2
Iron path length	l_{Fe}	20.4	Cm
A_L value@ 10 kHz	$A_{L.10kHz}$	35,000	nH/(Wdg.)^2
A_L valueof an equally sized Paramagnetic core	$A_{L.Luft}$	1.4	nH/(Wdg.)^2
Relative permeability	μ_r	25,000	–

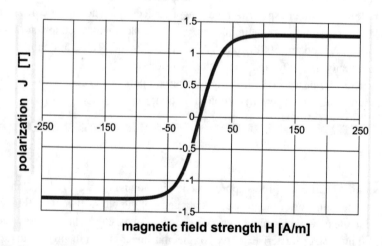

Fig. 5.36 Course of the approximated B(H) characteristic curve for VITROPERM 500F ($\mu_r = 25,000$)

Fig. 5.37 Modeling the
magnetic behavior of the
toroidal core by connecting a
non-linear and a linear
component in parallel

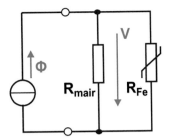

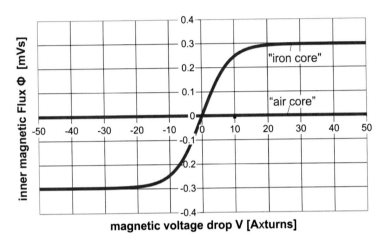

Fig. 5.38 Magnetic characteristic curve components of an iron core VITROPERM 500F 80×60×20

$$\Phi_{Fe}(V) = B_{SAT} \cdot A_{Fe} \cdot \tanh\left(\frac{\mu_0 \mu_r H}{B_{SAT}}\right) = B_{SAT} \cdot A_{Fe} \cdot \tanh\left(\frac{\mu_0 \mu_r \cdot V}{B_{SAT} \cdot l_{Fe}}\right)$$

$$\Phi_{Luft} = A_{Fe} \cdot \mu_0 H = A_{Fe} \cdot \mu_0 \frac{V}{l_{Fe}}$$

Figure 5.38 shows a graphical comparison of the two curves. Since the two magnetic resistors are connected in parallel, the resulting characteristic curve for a given magnetic force must be calculated by adding the resulting magnetic flux of both components. The blue curve therefore only reflects the polarization of the material.

During the transformation of the circuit into the electrical range, a series connection of a non-linear with a linear inductance with a common flux linkage $\Psi(I)$ (see for N = 30 Fig. 5.38) is created from the parallel connection of the magnetic resistors, taking into account the number of windings. By swapping the axes, the characteristic curve $I(\Psi)$ – usually used in network simulators – is obtained (see Fig. 5.39).

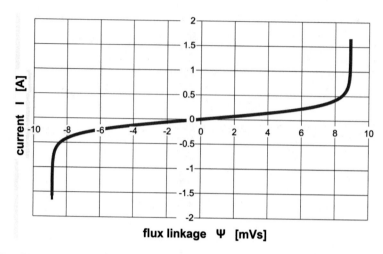

Fig. 5.39 I(Ψ)-characteristic curve of an 80×60×20 toroidal core based on VITROPERM 500F with several turns N = 30, which can be implemented directly in electrical simulators as a characteristic curve

5.7   Consideration of Hard Magnetic Properties

Particularly problematic is the mapping of partial remagnetization processes in hard magnetic materials. Here everything depends on the previous history. (This is why such materials are used as storage media.) Usually, one does not have the information required for a calculation. One would have to be able to calculate a complete model for the material, at least in the direction of magnetisation, to obtain useful results.

At high speeds of remagnetization, a portion of eddy current losses is added to the static magnetization losses according to the hysteresis curve, which can be interpreted as an ohmic portion parallel to the core inductance. Magnetic and electric field as causes of losses are perpendicular to each other. If plausible assumptions can be made as to how the material behaves when the direction of the flux changes, simple aperiodic systems can be studied and mapped well [5.5]. If this information is missing, only a "switching" of the characteristic curves can be represented in the simulation after sufficient remagnetization. Figure 5.40 shows the (static) demagnetization curves of some hard magnetic materials. Only the polarization, that is, the intrinsic part of the material for amplifying the magnetic flux, is taken into account.

Here, too, the characteristic curve B(H) of the nonlinear component can be described with good approximation via the tangent hyperbolic function:

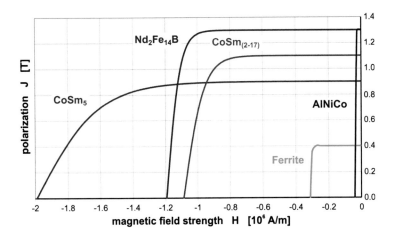

Fig. 5.40 Examples of ambiguous polarization characteristics of hard magnetic materials (demagnetization branch, permanent magnets)

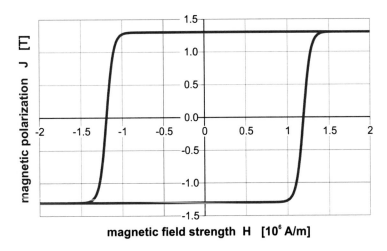

Fig. 5.41 Intrinsic magnetization curve for $Nd_2Fe_{14}B$ magnets (The maximum intrinsic differential relative permeability of the material is about 12)

$$B(H) = B_{SAT} \cdot \tanh\left(\frac{\mu_0 \mu_r (H \pm H_C)}{B_{SAT}}\right)$$

If this is converted to the manufacturer's specifications, the result is, for example, for $Nd_2Fe_{14}B$ (Fig. 5.41):

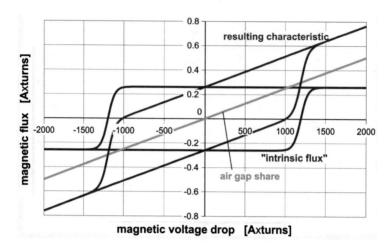

Fig. 5.42 Magnetic characteristic curve for a 2 cm^2 and 1 mm thick disc of Nd$_2$Fe$_{14}$B (blue: internal magnetization state, green: a superimposed portion of the vacuum, red: a characteristic curve of the magnetic two-pole

$$B(H) = 1.3T \cdot \tanh\left(\frac{12 \cdot \mu_0\left(H \pm 1.19 \cdot 10^6 A/m\right)}{1.3T}\right)$$

If we consider a magnetic disk with a cross-section of $A_{Fe} = 2$ cm^2 and a thickness of $l_{Fe} = 1$ mm, we obtain, as in the linear case, the characteristic curve for the magnetic two pole

$$\Phi(V) = 1.3T \cdot A_{Fe} \cdot \tanh\left(\frac{12 \cdot \mu_0\left(\frac{V}{l_{Fe}} \pm 1.19 \cdot 10^6 A/m\right)}{1.3T}\right) + \mu_0 \frac{V}{l_{Fe}} A_{Fe}$$

Summand reflects the influence of the vacuum. A corresponding graphic representation of this relationship is shown in Fig. 5.42. The second axis of the I(Ψ) is swapped to obtain the characteristic curve to be used in a network simulator.

When permanent magnets are installed in a magnetic circuit, a permanent magnetic flux is generated in this circuit. If you look at a magnetic two-pole with an inserted permanent magnet, this causes a shift in the magnetization characteristic B(H) or the characteristic of the differential inductance as a function of the flowing current L(I) (Fig. 5.43). In the basic state (blue line) the characteristic curve is symmetrical to zero. With stronger premagnetization, the magnetization characteristic curve is shifted on the I axis.

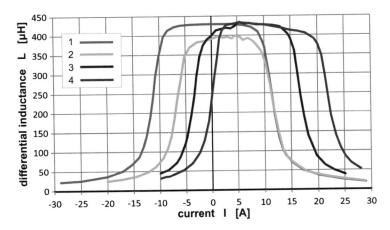

Fig. 5.43 L(I) measurements of four different configurations: 1: with an air gap, 2 with plastic-bonded hard ferrite material and 3 and 4 with different plastic-bonded NdFeB material. For the measurements, the same winding design was used for each of the core halves [5.6]

5.8 Variable Magnetic Sections in Inductive Structures

In the following, a problem area is described where non-linear sections of a magnetic circuit are purposefully brought into a state which results in a controllable inductance. A frequently used design is shown in Fig. 5.44. Here, a magnetic flux meets a core-type winding arrangement from above, which symmetrically pre-magnetizes the core with direct current. In the small-signal range, the voltages induced in the control windings cancel each other out. In the large signal range, the difference of the induced voltages of the "external flux" is measured there $\Phi \sim$.

The same magnetic voltage drop is applied across both pre-magnetized sections. That is, in the case of symmetry, expressed approximately via the tanh function (Fig. 5.45):

$$\Phi_\sim(V) = B_{SAT} \cdot A_{Fe} \cdot \left[\tanh\left(\frac{\mu_0 \mu_r \cdot (V_\sim + I_S N_1)}{B_{SAT} \cdot l_{Fe}} \right) + \tanh\left(\frac{\mu_0 \mu_r \cdot (V_\sim - I_S N_2)}{B_{SAT} \cdot l_{Fe}} \right) \right]$$
$$+ 2\frac{\mu_0 \cdot A_{Fe} \cdot V_\sim}{l_{Fe}}$$

In the graphic inversion, the parallel two pre-magnetized iron branches and the air content are converted into 3 inductances connected in series, which have the following $\Psi(L;I_s)$ characteristic curve based on the number of turns N_1:

$$\Psi(I_\sim; I_S) = B_{SAT} A_{Fe} \cdot \left[\tanh\left(\frac{\mu_0 \mu_r \cdot (I_\sim + I_S N_1)}{B_{SAT} \cdot l_{Fe}} \right) + \tanh\left(\frac{\mu_0 \mu_r \cdot (I_\sim - I_S N_2)}{B_{SAT} \cdot l_{Fe}} \right) \right]$$
$$+ 2\frac{\mu_0 \cdot A_{Fe} \cdot I_\sim}{l_{Fe}}$$

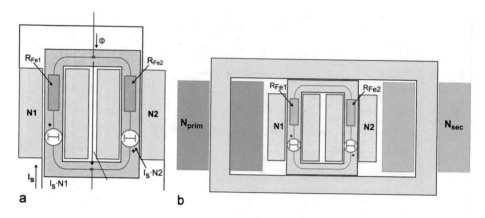

Fig. 5.44 Magnetic resistance pre-magnetizable by the direct current for use as a controllable leakage path in a transformer [5.7]. (**a**) Magnetic equivalent circuit diagram of a flux-controlled core section, (**b**) Position in the leakage path of a transformer between 2 windings

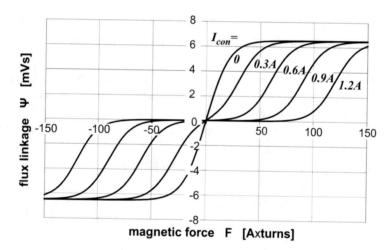

Fig. 5.45 Magnetization of a leakage path controlled by DC bias current according to Fig. 5.6.4 with the parameters (material VAC VP500/μ_r = 25,000, l_{Fe} = 0.5 m (single length), A_{Fe} = 2 × 25 cm², control winding $N_1 = N_2 = 100$)

An analytical explicit conversion of this $\Psi(L_{\sim};I_s)$ characteristic curve into the usual representation of the $L_{\sim}(\Psi;I_s)$ characteristic curve (Fig. 5.46) is not possible; it must then be achieved by a suitable control circuit as described in Chap. 7.

The electrical equivalent circuit diagram of the transformer sketched in Fig. 5.44b would then have the following appearance (Fig. 5.47).

Fig. 5.46 Characteristic of a
controlled inductance according
to Fig. 5.40a. The red dotted
lines show the change in "mean
inductance" at a level of +/−
6 mVs

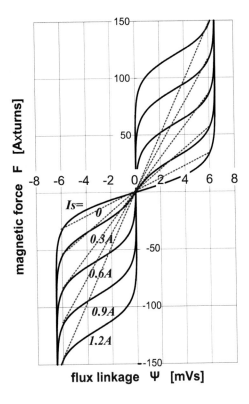

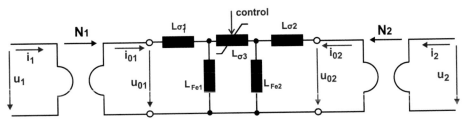

Fig. 5.47 Electrical equivalent circuit diagram of the transformer sketched in Fig. 5.44b. Control characteristic of inductance I ~ (Ψ; $_{Is}$): see Fig. 5.46 without consideration of further air leakage paths

Orthogonal Magnetization for Reluctance Control

In the magnetic field, field components of different causes can be superimposed [5.8]. Depending on the direction, the superposition is vectorial. If conductors are introduced into a material parallel to the main flux (Fig. 5.48), cylindrical-symmetrical magnetic fields are formed in the plane perpendicular to it when current flows through the conductors in the homogeneous material. If these currents are used for premagnetisation, the field lines of the main magnetic field and the control field are perpendicular to each other.

Fig. 5.48 Conductors
embedded in a ferrite core for
orthogonal premagnetisation of
the ferrite material with
dimensioning of the main flux
perpendicular to the image plane

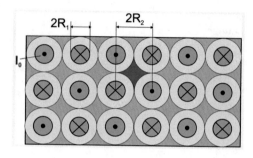

Assuming a maximum value of 100 mW/cm³ for the power loss density, which can be dissipated in the case of core losses without special cooling measures, this results in the following for the permissible current density

$$J_{max} = \sqrt{\frac{\left.\frac{P_v}{V}\right|_{max}}{\rho_{Cu}}} = \sqrt{\frac{100\,\frac{mW}{cm^3}}{0.0173\,\frac{\Omega mm^2}{m}}} = 2.4\,\frac{A}{mm^2}$$

This results for the maximum control current $I_{0.\,max} = \pi \cdot R_1^2 \cdot J_{max}$. Assuming a cylindrical copper conductor with radius R_1, saturation of the surrounding ferromagnetic material must be possible at the maximum permissible current density if a control effect is to be achieved. For the minimum size of R_1, this results in the condition

$$H_{SAT} = \frac{B_{SAT}}{\mu_0\mu_r} = \frac{J_{max} \cdot \pi R_1^2}{2\pi \cdot R_1}$$

This results in a minimum radius for a hole in the ferromagnetic material to

$$R_1 > \frac{2B_{SAT}}{\mu_0\mu_r J_{max}}$$

With $B_{SAT} = 0.4$ T, $\mu_r = 2000$ and $J_{max} = 2.4$ A/mm², a minimum radius of $R_1 > 133$ μm is obtained. The minimum radius increases with saturation induction, while it decreases with permeability and maximum current density. This indicates that miniaturized solutions are possible in thin layers on the surface of a solid where high current densities are possible due to good heat dissipation. If we now assume that parallel holes are drilled into the ferromagnetic material in such a way that the magnetic field lines touch each other at maximum current when saturation occurs (Fig. 5.48), the maximum magnetic conductance is obtained at a bias current $I_0 = 0$ to

$$\Lambda_{max} = \mu_0\mu_r \cdot \frac{4 \cdot R_2^2 - \pi \cdot R_1^2}{l} + \mu_0 \cdot \frac{\pi \cdot R_1^2}{l}$$

$$\Lambda_{min} = \mu_0\mu_r \cdot \frac{4 \cdot R_2^2 - \pi \cdot R_2^2}{l} + \mu_0 \cdot \frac{\pi \cdot R_2^2}{l}$$

$$= \frac{\mu_0}{l}\left(\mu_r\left(4 \cdot R_2^2 - \pi \cdot R_2^2\right) + \pi \cdot R_2^2\right) = \frac{\mu_0}{l}\left(\mu_r(4-\pi)+\pi\right)R_2^2$$

$$\frac{\Lambda_{max}}{\Lambda_{max}} = \frac{\mu_r \cdot 4 \cdot R_2^2 - (\mu_r-1)\cdot\pi\cdot R_1^2}{(\mu_r(4-\pi)+\pi)R_2^2} = \frac{\left(4\mu_r - (\mu_r-1)\cdot\pi\cdot\dfrac{R_1^2}{R_2^2}\right)}{(\mu_r(4-\pi)+\pi)}$$

The radius with the saturated material can be determined from the considerations above to

$$R_2 = \mu_0\mu_r \frac{J_{max}\cdot R_1^2}{2B_{SAT}}$$

$$\frac{\Lambda_{max}}{\Lambda_{max}} = \frac{\left(4\mu_r - (\mu_r-1)\cdot\pi\cdot\dfrac{4B_{SAT}^2}{(\mu_0\mu_r J_{max})^2 R_1^2}\right)}{(\mu_r(4-\pi)+\pi)}$$

Figure 5.49a shows the dependence of the control of the magnetic conductivity on the diameter of a conductor embedded in a homogeneous ferromagnetic material. The calculation is carried out up to the contact of the saturated material zones. For larger material zones, the simple calculation requirements used are violated. However, the chosen approach is helpful for basic considerations. It can be seen that under the assumed conditions the achievable rangeability $\Lambda_{max}/\Lambda_{min} < 4.65$. If for the selected example, it is assumed that at least 90% ($= 4.18$) of the rangeability should be set, $R_1 = 0.368$ mm is sufficient under the selected boundary conditions. The radius R_2 for the zone to be saturated for the boundary condition is then 1.02 mm. The conductors would then have a distance of

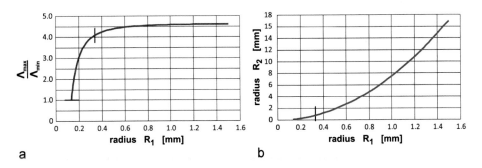

a   b

Fig. 5.49 Control characteristics with orthogonal premagnetisation. (**a**) achievable ratio of maximum and minimum conductance when touching the saturated core regions with radius R_2, (**b**) achievable radius R_2 when the maximum permissible current flows through the cross-section with radius R_1

2.04 mm in the ferromagnetic material (Fig. 5.49b). The total magnetization area is then made up of the sum of the individual areas. If one wants to control a large magnetization area, you must achieve this goal either with a few subareas or with many subareas.

Here it is interesting whether there is an optimal distance to control the magnetic conductance of a section with the lowest possible current flowing through all partial cross sections by series connection. According to the assumptions made above, the maximum control current to be used is $I_{0.\,max} = J_{max} \cdot \pi R_1^2$. Dividing the achievable rangeability by the control current gives as approximation

$$\frac{\Lambda_{max}}{\Lambda_{max} \cdot I_{0.\,max}} = \frac{\left(4\mu_r - (\mu_r - 1) \cdot \pi \cdot \frac{4B_{SAT}^2}{(\mu_0\mu_r J_{max})^2 R_1^2}\right)}{(\mu_r(4 - \pi) + \pi) \cdot J_{max} \cdot \pi R_1^2}$$

Figure 5.50 shows the graphical representation of this relationship for the assumptions made previously. The basic statement is that as many parallel conductors as possible should be embedded tightly in the ferromagnetic material to achieve the highest possible control sensitivity. This is an interesting finding for miniaturized components. Parallel to the surface of electrical sheets, only an insignificant effect can be achieved with this method. The reason is that the electrical steel sheets are insulated from each other and the magnetic field lines of the premagnetisation pass through these non-magnetic insulating layers. Even with high permeability values of the sheets, very high currents are required to achieve saturation effects. The magnetic voltage drops at the insulating layers are responsible for this. If the plates are drilled through to achieve a control effect, the field lines of the premagnetisation lie in the same plane as the main flux. This results in a mixed

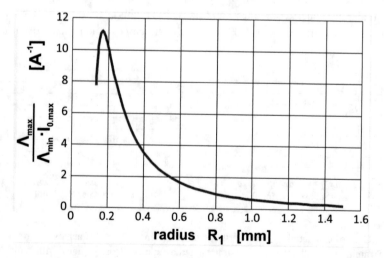

Fig. 5.50 Sensitivity of the conductance control by the bias current $I_{0.\,max}$ to the selected radius R_1 at maximum current density

superposition of the main field and premagnetisation, which ranges from parallel to orthogonal. This allows, for example, to realize concepts of a virtual air gap (see Sect. 10.5).

The shape of the $\Psi(I)$ characteristic curve changes in a different way with orthogonal premagnetisation than with parallel premagnetisation. Let us assume a cylindrical ferrite section $D = 20$ mm and $l = 53$ mm, as in the GUTV64 core, for example. The minimum inner radius of the mounting hole there is 2.5 mm. Assume a ferrite material with $\mu_r = 2000$ and $B_{SAT} = 0.5$ T. Using the tangential hyperbolic function, the $\Phi(\Theta)$ characteristic curve with control parameter I_0 can be estimated, taking into account the location-dependent magnetic field strength H, as

$$\Phi_Z = \int_{R1}^{R2} d\Phi = B = \int_{R1}^{R2} 2\pi r \cdot B_{SAT} \cdot \tanh\left(\frac{\sqrt{H_Z^2 + H_\varphi^2}}{\mu_r \mu_0}\right) \cdot \cos\left(\arctan\left(\frac{H_Z}{H_\varphi}\right)\right) dr$$
$$+ \frac{H_Z \cdot l_{Fe}}{\mu_0 \pi \left(\frac{D^2}{4} - R_1^2\right)}$$

Figure 5.51 shows a graphical representation of the results of the numerical integration. The assumptions were made for this in addition to the approximating tanhyp-function: $D = 20$ mm, $R_1 = 2.5$ mm, $l_{Fe} = 53$ mm and $\mu_r = 2000$. The core section is surrounded by a winding $I_1 \cdot N_1$ along the length l_{Fe}.

If we compare Fig. 5.51 with the control characteristic curve Fig. 5.45, it is noticeable that the typical double-S curves, which make up the characteristic curves in the latter, cannot be observed in Fig. 5.51. Here the magnetization curve is "panned" without creating

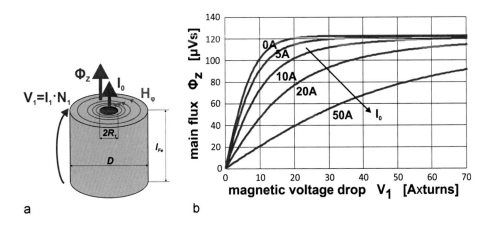

Fig. 5.51 Change of the magnetization characteristic curve with orthogonal premagnetization of a cylindrical core. (**a**) Basic structure, (**b**) calculated courses of the magnetization characteristic curve at different premagnetization currents I_0

additional nonlinearities. The control method should therefore also be suitable for applications in which low current harmonics are to be generated. Due to the vectorial addition of the field strengths, the orthogonal form of the control has a lower impact than the parallel superposition of the fields. A promising approach could be the miniaturization of components with ferromagnetic core and embedded conductors for orthogonal premagnetization.

It can generally be stated for Chap. 5 that for cases where pure AC applications are present, a form of representation free of contradictions has been developed and presented, which transforms magnetic circuit arrangements into equivalent electrical circuit arrangements. This allows the simulation of more complex magnetic components without a parallel calculation of the electric circuit and the magnetic field. In the following, a selection of dual switching elements for circuit design is presented. (Table 5.4)

Table 5.4 Building and structural elements with dual performance (transformation reversibly unambiguous)

Element		Dual element	
Designation	Circuit diagram	Circuit diagram	Designation
Mesh			Node
Resistance	R1	G1	Conductance
Capacity	C1	L1	Inductance
Voltage source	E1	I1	Current source
Gyrator	Gy1	Gy1	Gyrator
Ideal transformer	Tr1	Tr1	Ideal transformer
Magnetic resistance	R_{m1}	G_{m1}	Magnetic conductance/inductance for $N = 1$, permeance $\Lambda = G_m$
Magnetomagnetic force (MMF)	$\Theta 1$	$\Phi 1$	Ideal magnetic flux source
Diode	D1	D1	Diode
"Normally off" switch	S1	S1	"Normally-on" switch

References

1. Tellegen, B.D.H.: The gyrator, a new electric network element. Philips Research Report 3: pp. 81–101, April 1948
2. Strutt, M. J. O.: Eindeutige Wahl des Pfeilsinnes für Ströme und Spannungen. Archiv für Elektrotechnik. XIII. Band, Heft 1, 1955
3. Nanokristallines VITROPERM® 500F für stromkompensierte Drosseln -Vergleich zu Ferritwerkstoffen. Präsentation Fa. HAUG, Vertragshändler von VAC
4. VAC Vaccumschmelze: VITROVAC 6025Z: Ringbandkerne für Transduktordrosseln, 1999
5. Mallwitz, R.: GTO-Thyristoren in Hochspannungs-Impulsanwendungen. Diplomarbeit. TU Magdeburg 1993.
6. Friebe, J.: Permanentmagnetische Vormagnetisierung von Speicherdrosseln in Stromrichtern. Dissertation, Universität Kassel, 2013
7. Mecke, H.: Betriebsverhalten und Berechnung von Transformatoren für das Lichtbogenschweißen. Habilitationsschrift. TH Magdeburg 1979
8. Zacharias, P., Kleeb, Th., Fenske, F., Pfeiffer, J., Wende, J.: Controlled Magnetic Devices in Power Electronic Applications. EPE, Warszow 2017

Calculation and Modelling of Linear Magnetic Field Sections in Magnetic Devices

6

6.1 Elementary Magnetic Conductance/Permeance

Up to now, the simple design formula for magnetic conductance $G_m = \Lambda$ of prismatic bodies has been assumed. This is useful for homogeneous magnetic fields. With most inductive components one is confronted with the situation that the shapes of the field lines are neither homogeneous nor obey simple laws of symmetry. Besides, in contrast to the electric current field, there are no real "magnetic insulators". In the case of the electric current field, one speaks of 'conductors' with a conductivity $\sigma > 10^4$ S/m and of insulators with $\sigma < 10^{-8}$ S/m. This is a difference of 12 orders of magnitude. For comparison: copper has a conductivity of $56 \cdot 10^6$ S/m.

This fact leads to the rule that when calculating magnetic fields – especially leakage fields – the parts in the vicinity of the component must be included. Since these are often very complex calculations, procedures have been developed that can be used in certain frequently occurring situations. In general, the problems can always be solved with 3D FEM field calculation programs, but even these are relatively complex and not in focus of this book.

In the following, the problem shown in Fig. 6.1 will be assumed. An EE70 core is provided with an air gap in the middle leg. The excitation windings are applied to the outer legs. The flux generated from the left and right side is guided through the middle leg with a variable air gap. Due to the non-zero magnetic conductivity of the air, magnetic field lines are also formed there. This means that between the upper and lower yoke there is magnetic flux also parallel to the air gap. The maximum yoke conductivity is reached when $\delta = 0$. Then the conductance is essentially determined by the permeability of the middle leg. The minimum magnetic conductance is obtained when the air gap is equal to the yoke distance. The total permeance for a structure can be determined if the magnetic conductances G_m or Λ are combined. A pure parallel connection of the conductances thus results in

© Springer Fachmedien Wiesbaden GmbH, part of Springer Nature 2022
P. Zacharias, *Magnetic Components*,
https://doi.org/10.1007/978-3-658-37206-4_6

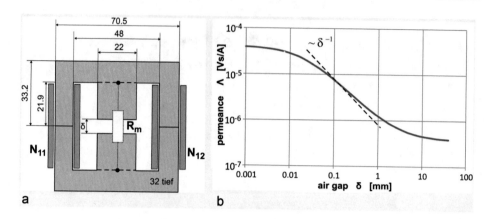

Fig. 6.1 EE70 core with an air gap. (**a**) Dimensional sketch with inserted additional yoke reluctance, (**b**) Dependence of the magnetic conductance between upper and lower yoke on the length of the air gap δ

$$\Lambda_{tot} = \sum_{v} \Lambda_{v}$$

The basic equation for magnetic resistance in a homogeneous magnetic field is

$$R_m = \frac{l}{\mu \cdot A} = \frac{1}{G_m} = \frac{1}{\Lambda}$$

where l: length of the section, A: cross-sectional area of the magnetized section.

For the conductance between any 2 equipotential surfaces of the magnetic vector potential, the integral representation

$$\Lambda = G_m = \frac{1}{R_m} = \mu \int_A \frac{1}{l} dA \approx \sum_{v} \mu \frac{\Delta A}{l}$$

is scheduled. Geometrically ideal boundaries of the field space (e.g. cuboid, spherical segment, spherical shell, etc.) are assumed, which come closest to the true course of the field. Thus the calculation results are only approximations of the actual conductance. A better approximation can only be achieved with considerably more time and calculation effort. This approach is very helpful for many calculation objectives, especially since special, tailored calculation methods also exist for special applications.

Table 6.1 summarizes some guide values for basic forms of the magnetic field, which were taken from the literature.

If you look at magnetic sections of magnetic cores, they serve to guide the magnetic flux as concentrated as possible. The above considerations can also be applied to core shapes or parts thereof, whose form has the "natural" field shape. For some, frequently encountered forms, Table 6.2 shows the graphical representations and the corresponding formulae.

Table 6.1 Magnetic conductance of characteristic regular field sections [1]

The geometry of a linear magnetic segment	Magnetic conductance Λ	Remarks
	$\Lambda = \mu \frac{ab}{c}$	
	$\Lambda \approx \mu \frac{2ab}{\pi(b+c)}$ $\Lambda \approx \mu \frac{a}{\pi} \ln\left(1 + \frac{2b}{c}\right)$	For $c > 3a$ For $c < 3a$
	$\Lambda \approx 0,26 \cdot \mu \cdot b$	$l_m \approx 1,22 \cdot c$ $V_e \approx \frac{\pi}{8} b \cdot c^2$

(continued)

Table 6.1 (continued)

The geometry of a linear magnetic segment	Magnetic conductance Λ	Remarks
	$\Lambda \approx 0,077 \cdot \mu \cdot c$	$l_m \approx 1,22 \cdot c$ $V_e \approx \frac{\pi}{24} \cdot c^3$
	$\Lambda \approx \mu \cdot \frac{a}{4}$	$l_m \approx \frac{\pi}{2}(a+c)$ $V_e \approx \frac{\pi}{24} \cdot \left((c+2a)^3 - c^3 \right)$

Table 6.2 Typical shapes of magnetic core sections

Core sections	magnetic conductance, Inductance Λ	Comment
	$\Lambda = \frac{\mu h}{2\pi} \ln\left(\frac{r_2}{r_1}\right)$ $\Lambda \approx \frac{\mu h}{\pi}\left(\frac{r_2 - r_1}{r_2 + r_1}\right)$	If $\frac{r_2 + r_1}{2} \ll r_2 - r_1$
	$\Lambda = \frac{\mu}{8}\left(r_1 + r_2 - \sqrt{(r_1 + r_2)^2 - 4(r_1 - r_2)^2}\right)$ $\Lambda \approx \frac{\mu(r_1 - r_2)^2}{4(r_1 + r_2)}$	$h = r_2 - r_1$ If $\frac{r_2 + r_1}{2} \ll r_2 - r_1$
	$\Lambda = \frac{2\pi\mu h}{\ln\left(\frac{r_2}{r_1}\right)}$	

(continued)

Table 6.2 (continued)

Core sections	magnetic conductance, Inductance Λ	Comment
	$\Lambda = \mu \dfrac{\pi\left(r_2^2 - r_1^2\right)}{h}$	
	$\Lambda = \dfrac{\mu a b}{h}$	

6.2   Distributed Magnetic Field Quantities Versus Lumped Equivalent Parameters

Especially in magnetic arrangements, where the magnetic flux is conducted from one winding to the next via ferromagnetic sections ("magnetic conductors"), the problem arises that there are practically no magnetic non-conductors. This means that the magnetic line between two conductors (Fig. 6.2) is always connected to a magnetic shunt that is distributed over the entire section. The use of concentrated elements for system description, as explained in Chap. 5, is therefore questionable. How the transmision paramters for cascade connection for the magnetic section is generally designed in such a case and how adequate approximations and their limits can be derived is the focus of this section. Figure 6.2 shows a graphic illustration of the problem.

Since the permeability of air is not zero, in each section 'dx' a portion of the current from the upper conductor is transferred to the lower conductor. dx' can therefore be described by a described differential series magnetic resistance 'dR_m' and a differential transverse magnetic conductance 'dG_m' *(corresponds to the differential permeance '$d\Lambda$')*. Since both quantities are assumed to be linear and frequency-dependent, the complex notation is chosen and a specific, freely selectable frequency is assumed. Thus a largely general representation is achieved. The corresponding equivalent circuit is shown in Fig. 6.3.

Fig. 6.2 Guidance of the magnetic field through two ferromagnetic cable sections of a magnetic network

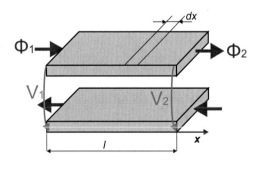

Fig. 6.3 Differential description of a magnetic section according to Fig. 6.2

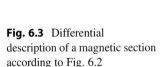

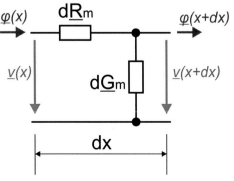

This problem is strongly reminiscent of that of an isotropic double line from the theory of conduction in communications engineering [2]. Accordingly, the structure of its solution can be used and analogous terms can be formulated.

In analogy to the complex characteristic impedance of a line, a magnetic propagation impedance of

$$\underline{Z}_m = \sqrt{\frac{\frac{d\underline{R}_m}{dx}}{\frac{d\underline{G}_m}{dx}}} = \sqrt{\frac{\underline{R}'_m}{\underline{G}'_m}} = \sqrt{\frac{d\underline{R}_m}{d\underline{G}_m}}$$

because differential permeability generally has a complex frequency dependence. The propagation coefficient along the magnetic conduction is accordingly

$$\underline{\gamma}_m = \sqrt{\underline{R}'_m \cdot \underline{G}'_m} = \alpha + j\beta$$

With these parameters, the cascade connection of the electromagnetic transfer matrix can be determined to

$$\begin{bmatrix} V_1 \\ \Phi_1 \end{bmatrix} = \begin{bmatrix} \cosh\left(\underline{\gamma}_m \cdot l\right) & \underline{Z}_m \sinh\left(\underline{\gamma}_m \cdot l\right) \\ \dfrac{1}{\underline{Z}_m} \sinh\left(\underline{\gamma}_m \cdot l\right) & \cosh\left(\underline{\gamma}_m \cdot l\right) \end{bmatrix} \cdot \begin{bmatrix} V_2 \\ \Phi_2 \end{bmatrix}$$

A look at the coefficients shows the following conditions from which a topology with concentrated elements can be derived.

(a) $\underline{V}_1 = \cosh\left(\underline{\gamma}_m \cdot l\right) \cdot \underline{V}_2$
    for $\Phi_2 = 0$
(b) $\underline{V}_1 = \underline{Z}_w \cdot \sinh\left(\underline{\gamma}_m \cdot l\right) \cdot \underline{\Phi}_2$
    for $V_2 = 0$
(c) $\underline{\Phi}_1 = \cosh\left(\underline{\gamma}_m \cdot l\right) \cdot \underline{\Phi}_2$
    for $V_2 = 0$
(d) $\underline{\Phi}_1 = \dfrac{1}{\underline{Z}_m} \sinh\left(\underline{\gamma}_m \cdot l\right) \cdot \underline{V}_2$
    for $\Phi_2 = 0$

The transmission network must therefore be completely symmetrical for the division of magnetic voltage drops and magnetic fluxes. If a T equivalent circuit diagram for the magnetic transmission arrangement is used as a basis, the equivalent transmission equation is obtained according to Chap. 5

Fig. 6.4 Equivalent circuit with concentrated elements

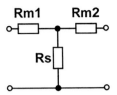

$$\begin{bmatrix} V_1 \\ \Phi_1 \end{bmatrix} = \begin{bmatrix} \dfrac{R_s + R_{m1}}{R_s} & \left(\dfrac{R_{m1}(R_{m2} + R_s)}{R_s} + R_{m2} \right) \\ \dfrac{1}{R_s} & \dfrac{R_s + R_{m2}}{R_s} \end{bmatrix} \cdot \begin{bmatrix} V_2 \\ \Phi_2 \end{bmatrix}$$

The corresponding equivalent circuit diagram is shown in Fig. 6.4.

Here the equality $R_{m1} = R_{m2} = R_m$ is an expression of symmetry.

$$\begin{bmatrix} V_1 \\ \Phi_1 \end{bmatrix} = \begin{bmatrix} \dfrac{R_s + R_m}{R_s} & \left(\dfrac{R_m^2 + 2R_m R_s}{R_s} \right) \\ \dfrac{1}{R_s} & \dfrac{R_s + R_m}{R_s} \end{bmatrix} \cdot \begin{bmatrix} V_2 \\ \Phi_2 \end{bmatrix}$$

With the elements of the main diagonal, you get first for

$$\underline{R}_s = \frac{\underline{Z}_m}{\sinh\left(\underline{\gamma}_m \cdot l\right)} \approx \frac{1}{G'_m \cdot l}$$

According to this

$$\frac{R_m^2 + 2R_m R_s}{R_s} = \underline{Z}_w \cdot \sinh\left(\underline{\gamma}_m \cdot l\right) \text{ be.}$$

A further transformation results in $R_m^2 + 2R_m R_s - \underline{Z}_w^2 = 0$ and therefore for

$$R_m = -R_s \pm \sqrt{R_s^2 + Z_w^2} = -\frac{\underline{Z}_m}{\sinh\left(\underline{\gamma}_m \cdot l\right)} \pm \underline{Z}_m \sqrt{\frac{1}{\sinh^2\left(\underline{\gamma}_m \cdot l\right)} + 1}$$

$$= \frac{\underline{Z}_m}{\sinh\left(\underline{\gamma}_m \cdot l\right)} \left(-1 + \sqrt{1 + \sinh^2\left(\underline{\gamma}_m \cdot l\right)} \right)$$

$$R_m = \frac{\underline{Z}_m}{\sinh\left(\underline{\gamma}_m \cdot l\right)} \left(\cosh\left(\underline{\gamma}_m \cdot l\right) - 1 \right) = \underline{Z}_m \cdot \tanh\left(\frac{\underline{\gamma}_m \cdot l}{2}\right)$$

For the elements of the main diagonal then applies as required,

$$\frac{R_s + R_m}{R_s} = \frac{\dfrac{Z_m}{\sinh\left(\underline{\gamma}_m \cdot l\right)} + \dfrac{Z_m}{\sinh\left(\underline{\gamma}_m \cdot l\right)}\left(\cosh\left(\underline{\gamma}_m \cdot l\right) - 1\right)}{\dfrac{Z_m}{\sinh\left(\underline{\gamma}_m \cdot l\right)}} = \cosh\left(\underline{\gamma}_m \cdot l\right)$$

Based on the Taylor series development for sinh and cosh function, which are as follows

$$\sinh(z) = \frac{e^z - e^{-z}}{2} = z + \frac{z^3}{3!} + \frac{z^5}{5!} + \ldots \, or \, \cosh(z) = \frac{e^z + e^{-z}}{2} = 1 + \frac{z^2}{2!} + \frac{z^4}{4!} + \ldots,$$

so you can $\left(\left|\underline{\gamma}_m \cdot l\right| << 1\right)$ write for low values of 'l':

$$\underline{R}_s = \frac{Z_m}{\sinh\left(\underline{\gamma}_m \cdot l\right)} \approx \frac{Z_m}{\underline{\gamma}_m \cdot l} = \frac{1}{G'_m \cdot l}$$

$$R_m = \frac{Z_m}{\sinh\left(\underline{\gamma}_m \cdot l\right)}\left(\cosh\left(\underline{\gamma}_m \cdot l\right) - 1\right) \approx \frac{Z_m}{\underline{\gamma}_m \cdot l}\frac{\left(\underline{\gamma}_m \cdot l\right)^2}{2} = Z_m \frac{\left(\underline{\gamma}_m \cdot l\right)}{2} = \frac{R'_m \cdot l}{2}.$$

Since the magnetic conductor length is doubled (the halves of the magnetic conductors are combined in the upper and lower conductor in Fig. 6.4), the value for R_m is therefore

$$R_m = \frac{l}{\mu_0 \mu_r A_{Fe}}$$

With an "iron" cross-section of $a \cdot b = 4 \, cm^2$ and a permeability of $\mu_r = 2200$, one obtains for R'_m for example

$$R'_m = 904289 \, \frac{A}{Vs \cdot m}$$

With a distance $d = 2 \, cm$ the value is $G'_m \approx 0.6 \cdot \omega_0 \cdot 2cm/m = 4.8 \cdot 10^{-9} \frac{Vs}{A \cdot m}$.

This corresponds to a value $\gamma = R'_m \cdot G'_m = 0.066 \cdot m^{-1}$ that can be $\left(\left|\underline{\gamma}_m \cdot l\right| << 1\right)$ written for usual values of 'l' and is based on the concentrated parameters in the equivalent circuit diagram Fig. 6.5. Following common practice, the calculation of magnetic arrangements is based on a transverse conductance value for the magnetic field, taking into account the influence of the field between two magnetic conductors (e.g. yoke leakage flux). This has to be taken into account in the structural analysis and development of the magnetic equivalent circuit diagram with concentrated elements.

In the case of cut strip-wound cores or toroidal cores made of soft magnetic material, the issue of transverse conductivity plays a role once again. Here, a long magnetic conductor is

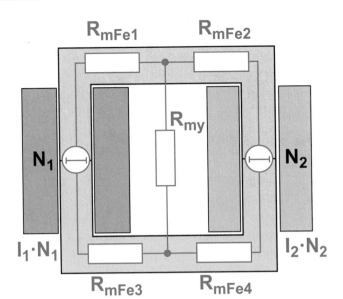

Fig. 6.5 Assumption of a concentrated magnetic resistance R_{my} for the yoke leakage flux of a transformer with strong yoke leakage

wound up. The resulting permeable layers are insulated from each other to reduce eddy currents. As a result, there are no truly closed magnetic field lines within highly permeable material. The insulation by thin layers, however, enables extensive penetration of the material with magnetic flux even at low magnetic force values $I \cdot N$. The magnetic transverse conductance is kept particularly high.

6.3   Inductance of Conductors and Conductor Arrangements

Maxwell's equations can be used to calculate the magnetic fields in and around conductors through which current flows. From the analysis of the field strength distribution, the integral parameter 'inductance' can be determined. According to the definition

$$L = 2\frac{W_{magn}}{I^2} = \frac{2}{I^2}\int_V \frac{BH}{2}\,dV = \frac{2}{I^2}\int_V BH\,dV$$

the inductance is equal to the quotient of the magnetically stored energy and the square of the causing current. As a simple example, the inner inductance of a conductor is calculated below.

Assume a cylindrical conductor of length 'l' and diameter '2R', in which a current flows with constant current density (Fig. 6.6).

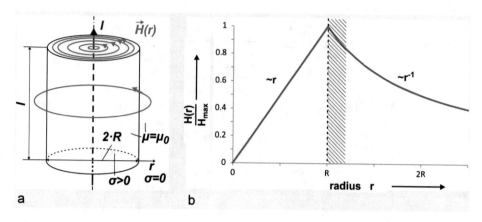

Fig. 6.6 Magnetic field inside and outside a cylindrical conductor. (**a**) geometrical sketch, (**b**) course of the magnetic field strength

The internal magnetic field of the conductor (0 . . .r. . . R) is observed. The magnetic field strength H(r) can then be calculated using Ampére's law:

$$\oint_S \vec{H}(r) \cdot d\vec{s} = H(r) \cdot 2\pi r = \int_A \vec{J} \cdot d\vec{A} = J \cdot \pi r^2 = I(r)$$

$$H(r) = J\left(\frac{r}{2}\right) = \frac{I_0}{\pi R^2}\left(\frac{r}{2}\right)$$

for $0 \leq r \leq R$. The energy content for a conductor of length 'l' is then obtained

$$W_{magn.i} = \int_V \frac{B \cdot H}{2} \cdot dV = l \cdot \int_0^{2\pi} \int_0^R \frac{\mu H^2(r)}{2} \cdot r dr \cdot d\varphi = 2\pi l\mu \int_0^R \frac{I_0^2}{\pi^2 R^4}\left(\frac{r}{2}\right)^2 \cdot r dr = \frac{l\mu}{8\pi}.$$

This value for a cylindric conductor does obviously not depend on the radius/diameter of the conductor.

The inner inductance of a conductor is basically to be added to the inductance of any conductor arrangement. However, the contribution is generally small. It is not dependent on the diameter of the conductor, but on the geometry due to the different field strength distribution. If the conductor itself is ferromagnetic, the contribution can become significant. At higher frequencies the current density distribution changes due to the skin effect. At very high frequencies the current flows only at the surface of the conductor. The result is a reduction of the inner inductance of the conductor. As a result f. e., the inductance of a coaxial cable decreases. If single conductors are stranded, as is usual with HF-litz, the current density distribution created by the skin effect is partly compensated, since the conductors have positions both outside and inside the rope. Consequently, at high frequencies, the inner inductance hardly decreases due to the forced current equal distribution. (A simple twisting of wires does not help here).

At higher frequencies, the magnetic field is forced to the conductor surface and, at the border crossing $\omega \to \infty$, for example, in a coaxial cable, it only covers the space between the inner and outer conductor, which is filled by the magnetic field.

The procedure briefly sketched above is the same as that found in FEM calculations when data on an inductance is generated. In this case, the magnetic field strength is calculated depending on the location. From the energy content of field sections, inductance values can be assigned to them.

An external inductance of a straight conductor is not defined, since a return conductor influencing the external field is always required for a current. Table 6.3 summarizes inductance values for frequently encountered arrangements.

Low inductive line arrangements can be achieved in various ways. Obviously, a coaxial line arrangement has a lower inductance coating than a simple double line of parallel conductors. However, if this double line is *twisted* ("*twisted pair*"), the external magnetic fields of individual line sections compensate each other to a small extent, as they have opposite directions. Twisting improves the transmission properties in particular. The electromagnetic fields of adjacent signal currents can only interfere with short line sections that run parallel in the same direction. The spatial change in direction of the magnetic field section suppresses crosstalk between the wire pairs. The fields of an electromagnetic wave acting on the cable result in signals with the same phase and amplitude on the cores. Since the signal difference of each pair is evaluated at the end of the cable, the interference largely cancels each other out.

The first step towards low-inductance supply lines would be ribbon cables or cable bundles. With HF or pulse loads, however, it must be remembered that high-frequency components of the current are forced to the surface of the sides of the outgoing and return conductors facing each other and to the edges, which increases the losses in particular. This effect is countered by braiding ("braids" and flat stranding) and achieves an effect similar to that of round HF litz.

6.4 Inductance of Windings

If you lay many windings with the same diameter next to each other you get a single layer cylindrical coil. With a homogeneous current distribution, the same effect can be achieved with one turn from a thin strip. Assuming that the length is large against the radius ($l >> R$), a practically homogeneous field is obtained inside the coil and the inductance of the arrangement can be determined by

$$L \approx N^2 \mu \frac{\pi R^2}{l}$$

Table 6.3 Inductance values of typical line arrangements according to [1]

Line arrangement	Inductance	Remarks
	$\Lambda = \frac{\mu l}{8\pi}$	The inner inductance of a conductor
	Parallel cylindrical conductors $\Lambda =$ $\frac{\mu}{\pi} l \left(\frac{1}{4} + \ln\left(\frac{d}{2\sqrt{r_1 \cdot r_2}} \right) \right)$	First summand: Inner inductance of the conductors
	$\Lambda \approx \frac{2\mu}{\pi} l \left(\frac{1}{8} + \ln \frac{c-a}{b} \right)$ $\Lambda \approx \mu l \left(\frac{1}{4} + \frac{a}{b} \right)$	$a \gg \frac{c-a}{2}, a \gg b$ $a \gg b, \frac{c-a}{2} \gg b$
	$\Lambda = \frac{\mu l}{8\pi} \left(1 + 4\ln\left(\frac{r_1}{r_0} \right) \right) +$ $+ \frac{\mu l}{8\pi} \left(\frac{2r_1^2}{(r_2^2 - r_1^2)^2} \left(2\left(\frac{r_2}{r_1} \right)^2 \ln\left(\frac{r_2}{r_1} \right) - \left(\frac{r_2}{r_1} \right)^2 + 1 \right) \right)$	Because of the skin effect in HF, the first in brackets and the last summand disappear

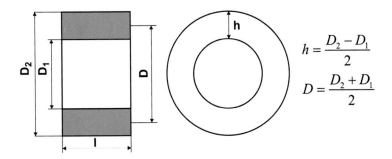

Fig. 6.7 Short cylindrical coil with dimensioning

$$h = \frac{D_2 - D_1}{2}$$

$$D = \frac{D_2 + D_1}{2}$$

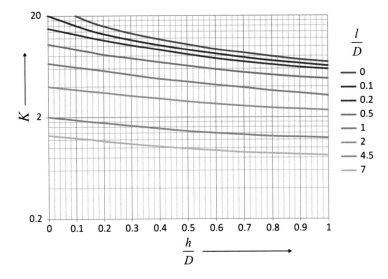

Fig. 6.8 Dependence of the factor 'K' on the ratios h/D and l/D

neglecting the outer Inductance of the magnetic field.

Although this formula is widely used for calculations, it leads to significant deviations for short and multi-layer cylindrical coils because the basic assumptions of the calculation are violated. Figure 6.7 shows such a short cylindrical coil. For h/D →0 a "single-layer" cylinder coil is obtained. For l/D →0 a flat coil is obtained.

According to [1] a general formula for calculating the inductance of such coils is

$$L = \frac{\mu_0}{4\pi} N^2 \cdot D \cdot K\left(\frac{l}{D}, \frac{h}{D}\right)$$

Here K is the evaluation of an elliptic integral. Figure 6.8 shows a graphic representation of the dependence of the "correction factor" K based on [1].

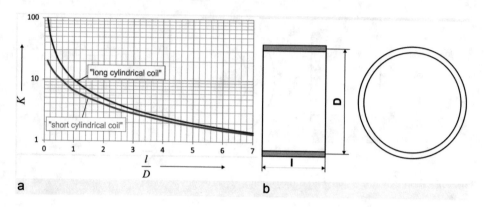

Fig. 6.9 Dependence of the correction factor K for a single-layer cylindrical coil (**a**) Calculation results for (N = 1, h → 0), (**b**) dimensioning of the coil

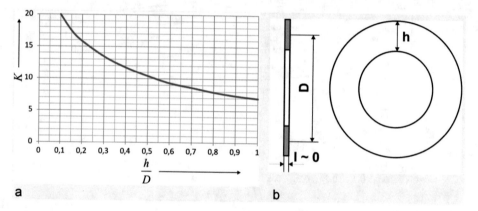

Fig. 6.10 Dependence of the correction factor K for a flat coil. (**a**) Calculation results for (l →0), (**b**) Dimensioning

If one compares this representation with the formula for inductance calculation for a long cylindrical coil (Fig. 6.9a), one sees that the inductance is always calculated too high. The reason is that the magnetic resistance of the space outside the coil is neglected by the formula. The longer the coil, the less this affects the calculation result.

If all windings are arranged in one plane, a flat coil is obtained, to which the formula (for l = 0) can also be applied. The area around the coil is then permeated by an intensive magnetic field. The dependence of the K-factor for this case is shown in Fig. 6.10a. It should be noted that with short coils, high magnetic field strengths penetrate the winding itself and can cause strong eddy current losses due to the proximity effect. This effect can be counteracted by using HF litz wire.

According to these investigations the inductance of a thin ring

$$L = 7\frac{\mu_0}{4\pi}N^2 \cdot D = 0.7 \cdot N^2 \cdot D\left[\frac{\mu H}{m}\right].$$

"Optimization" of Short Cylinder Coils
For practical purposes, short cylinder coils are of greater interest. They will be examined more closely below. An arrangement as shown in Fig. 6.11 is to be regarded as a short cylinder coil. (Deviations from the ideal circular cross-section can be converted to a circular cross-section of the same size). The average coil diameter D and the winding height h result from outer coil diameter D_a and inner coil diameter D_i to.

$$D = \frac{D_a + D_i}{2} \text{ and } h = \frac{D_a - D_i}{2}.$$

The coil length is 'l'. According to [1], the dimensionless parameters ρ and λ can be derived, with which the energy content of the electric field or the inductance L of the arrangement without ferromagnetic core can be described in general:

Fig. 6.11 Short, coreless cylindrical coil with dimensioning

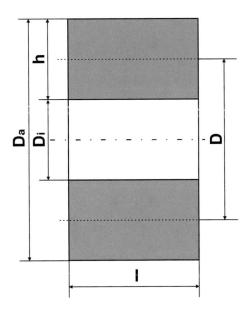

$$L = \frac{\mu_0}{4\pi} \cdot N^2 \cdot D \cdot K(\lambda, \rho) \text{ with } \lambda = \frac{l}{D} \text{ and } \rho = \frac{h}{D}.$$

$K(\lambda, \rho)$ is the result of elliptic integrals from the field calculation and is shown in Fig. 6.8 [1].

With this information, it is easy to determine the expected inductance values of real air-core coils. If coils with a diameter D are to be realized with low ohmic losses, this corresponds to the search for designs with a high ratio of inductance and coil resistance L/R. This means that the above formula can be further written to (geometry-dependent parameters underlined)

$$\frac{L}{R} = \frac{\frac{\mu_0}{4\pi} \cdot N^2 \cdot D \cdot K(\lambda, \rho)}{\rho^* \cdot \frac{\pi \cdot D}{l \cdot h} \cdot N^2} = \frac{\mu_0}{4\pi^2 \cdot \rho^*} \cdot D^2 \cdot \underline{\lambda \cdot \rho \cdot K(\lambda, \rho)} = \frac{\mu_0}{4\pi^2 \cdot \rho^*} \cdot D^2 \cdot \underline{f_0(\lambda, \rho)}$$

$$\frac{L}{R} \cdot \frac{4\pi^2 \cdot \rho^*}{\mu_0 \cdot D^2} = f_0(\lambda, \rho)$$

The function $f_0(\lambda, \rho)$

The ratio L/R can also be considered as a quality characteristic. For $0 < \rho < 0.5$ a degressive increase is recorded, which changes into a linear increase for the range $0.5 < \rho < 1$. (This means that the increase does not decrease further when ρ is enlarged.) The parameter $\rho = 0.5$ can therefore be taken as a 'compromise' in the sense of V. Pareto when searching for coils of high quality.

The magnetically stored energy is $W_L = \frac{L}{2} \cdot I^2$, while the energy dissipated from the coil depends essentially on the surface:

$$P_v = R \cdot I^2 = k^* \cdot A_O \text{ respectively } I^2 = \frac{k^* \cdot A_O}{R}.$$

The total surface of the cylinder coil is calculated to

$$A_O = \pi \cdot (D_a + D_i) \cdot l + 2 \cdot \frac{\pi}{4} \cdot (D_a^2 - D_i^2) = \pi \cdot (D_a + D_i) \cdot l + 2 \cdot \frac{\pi}{4} \cdot (D_a + D_i)(D_a - D_i)$$

$$A_O = 2\pi \cdot h \cdot l + 2\pi \cdot D \cdot h = 2\pi D^2 \cdot \rho \cdot \lambda + 2\pi \cdot D^2 \cdot \rho$$

$$A_O = 2\pi D^2 \cdot (\rho \cdot \lambda + \rho) = 2\pi D^2 \cdot \rho \cdot (\lambda + 1)$$

This makes it possible to formulate for the permanently storable magnetic energy:

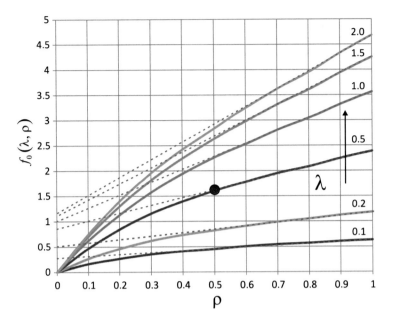

Fig. 6.12 Geometry factor $f_0(\lambda,\rho)$ to determine the "discharge time constant" L/R of a short cylindrical coil as an inductive energy storage

$$W_L = \frac{L}{2} \cdot \frac{k^* \cdot A_O}{R} = \frac{k^*}{2}\frac{L}{R} \cdot 2\pi D^2 \cdot (\rho \cdot \lambda + \rho) = \frac{k^* \mu_0}{4\pi \cdot \rho^*} \cdot D^4 \cdot \lambda \cdot \rho \cdot (\rho \cdot \lambda + \rho) \cdot K(\lambda, \rho)$$

$$W_L = \frac{k^* \mu_0}{4\pi \cdot \rho^*} \cdot D^4 \cdot \lambda \cdot \rho \cdot (\rho \cdot \lambda + \rho) \cdot K(\lambda, \rho) = \frac{k^* \mu_0}{4\pi \cdot \rho^*} \cdot D^4 \cdot f_1(\lambda, \rho)$$

Interesting here is that the (permanently) storable energy of such an air coil is proportional $\sim D^4$. The following applies to the net volume of the coil winding $V = \pi \cdot D \cdot l \cdot h = \pi \cdot D^3 \cdot \lambda \cdot \rho$ and thus to the magnetic energy per net volume

$$\frac{W_L}{V} = \frac{W_L}{\pi \cdot D^3 \cdot \lambda \cdot \rho} = \frac{k^* \mu_0}{4\pi^2 \cdot \rho^*} \cdot D \cdot (\rho \cdot \lambda + \rho) \cdot K(\lambda, \rho) = \frac{k^* \mu_0}{4\pi^2 \cdot \rho^*} \cdot D \cdot f_2(\lambda, \rho)$$

The mean energy density, therefore, increases linearly with the mean diameter D. Figure 6.13 shows the dependence for the geometry factor $f_2(\lambda,\rho)$. It becomes clear that there are no local extreme values in these considerations. Nevertheless, the parameter combination $\lambda = \rho = 0.5$ is often used in the sense of a (Pareto-optimal) compromise. In the diagrams Figs. 6.12 and 6.13 this point is marked. In Fig. 6.14 the proportions of such an "optimum compromise" are shown graphically.

If the diameter of a short cylindrical coil D_C with a winding height h = 0 is compared with the diameter of a short cylindrical coil $D(\lambda, \rho)$ of the same length, a coil with h > 0

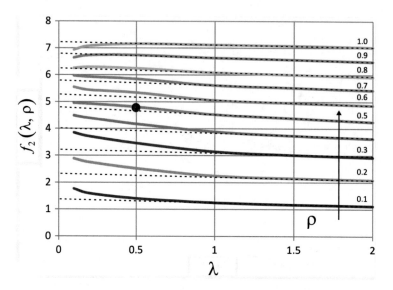

Fig. 6.13 Characteristic value $f_2(\lambda,\rho)$ for the storable energy per net volume as a function of $\lambda = l/D$ and $\rho = h/D$

Fig. 6.14 Proportions of a ("Pareto") optimal short cylinder coil

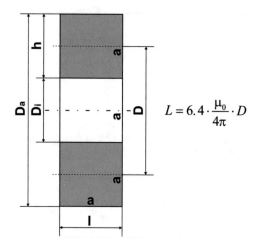

$$L = 6.4 \cdot \frac{\mu_0}{4\pi} \cdot D$$

($\rho > 0$) is always larger than a single-layer cylindrical coil with the same inductance (Fig. 6.15).

For rough calculations, the following formula is often used

$$L^* = \mu_0 \cdot \frac{\pi}{4} \cdot \frac{D^2}{l} \cdot N^2,$$

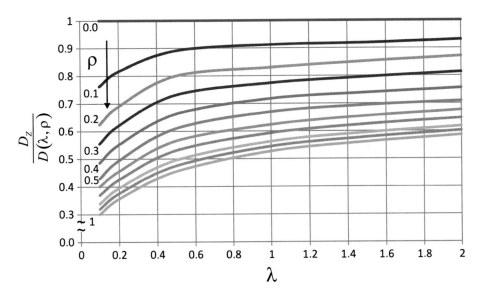

Fig. 6.15 Diameter comparison of a general short cylindrical coil (D(λ,ρ)) with a single-layer ($\rho = 0$) short cylindrical coil of the same length

which applies for an infinitely long chokel $\rightarrow \infty$. The results obtained are usually clearly too large, as shown in Fig. 6.9. The diagrams Figs. 6.15 and 6.16, which show the deviations from this estimate, can be used to correct this as a well-founded estimate. It can also be seen that the contribution of windings near the coil axis to the total inductance is smaller than that of windings far from the axis. Further calculation formulas are given in Table 6.4. For arrangements with core, proceed according to Chap. 5 and assign a corresponding core section to each winding.

6.5   Leakage Inductance Between Windings

The magnetic coupling between windings is described by the leakage inductance. With two magnetically coupled ideal coils (assumed $R = 0$), when a voltage is applied to a winding N_1, part of the flux generated is coupled into winding N_2. According to the law of induction, this induces the open-circuit voltage U_2 there when no load is applied. However, if N_2 is short-circuited, the resulting voltage is identical to zero (because $R_2 = 0$). The previously existing flux in the winding must therefore have been forced out of the winding according to Lenz's rule. The magnetic flux then takes the path through the leakage path. The higher the short-circuit current, the higher the magnetic resistance of the leakage path. The magnetic leakage conductance $G_m = \Lambda$ is the reciprocal value of the magnetic leakage resistance R_m. The values measured on the windings during such short-circuit tests are the leakage inductances related to the respective number of turns $L_{\sigma 1} = N_1^2 \cdot \Lambda$ or $L_{\sigma 2} = N_2^2 \cdot \Lambda$.

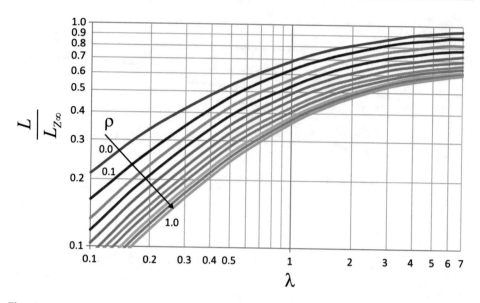

Fig. 6.16 Inductance of a short cylindrical coil in relation to the inductance determined by the calculation formula for a long cylindrical coil

The secondary short-circuit current, which would be infinitely large with an ideal transformer and constant primary voltage, is inductively limited in the real transformer. This is the reason for many applications and designs.

In the following, the principle procedure for determining the leakage inductance in different ways is described. PM cores of type PM62/49 (Fig. 6.17) are used as a demonstration example. There are the two basic possibilities of arranging the primary and secondary winding either in the axial direction one above the other ("disc winding") or in the radial direction one above the other ("tube winding"). The calculation approaches for these two arrangements are shown below.

Magnetic properties (per set)

$\Sigma\, l_{Fe}/A_{FE} = 0.191\ mm^{-1}$.
$l_{Fe} = 109\ mm \quad A_{Fe} = 570\ mm^2$
$A_{min} = 470\ mm^2$
$V_{Fe} = 62,000\ mm^3$
$m \approx 280\ g/set$

The following Fig. 6.18 shows qualitatively the leakage magnetic lekage field between 2 windings in the chambers of a pot core (disk winding), as well as the qualitative course of the leakage magnetic field strength along the z-axis:

From the "yoke" inner side of the core, the magnetic force increases linearly up to a maximum. In the partition wall of the chambers, the magnetic force is almost constant, only to decrease linearly again to zero in the second winding chamber. This would be the case with the secondary short circuit of a transformer with impressed primary voltage:

Table 6.4 Formulas for calculating simple winding arrangements

Winding structure	Inductance/remarks
	Circular ring winding[a] $$\Lambda \approx \mu_0 R\left(\ln\left(\tfrac{8r_1}{R}\right) - 1.75\right)$$
	Short single layer cylindrical coil[a] $$L \approx \tfrac{\mu_0}{4\pi} \cdot N^2 \cdot \tfrac{22.7}{1+2.3\cdot\frac{l}{D}}$$
	Flat toroidal coil[a] $$L \approx \tfrac{\mu_0}{4\pi} \cdot N^2 \cdot \left(6.95 - 6.6675 \ln\left(\tfrac{h}{D}\right)\right)$$
	Short multilayer cylindrical coil $$L \approx \tfrac{\mu_0}{4\pi} \cdot N^2 \cdot K\left(\tfrac{l}{D}, \tfrac{h}{D}\right)$$ (K see Fig. 6.7)
	Short square coil[a] $$L \approx \tfrac{\mu_0}{4\pi} \cdot a \cdot N^2 \cdot \left[0.3448 e^{0.137876\cdot\lambda} - 0.13575 \cdot e^{-\frac{\lambda}{7}}\right]$$ $$\lambda = \tfrac{a}{l}$$

(continued)

Table 6.4 (continued)

Winding structure	Inductance/remarks
	Flat square coil[a] $$L \simeq 33.9 \cdot \frac{\mu_0}{4\pi} \cdot (a+h) \cdot N^2 \cdot e^{-4.418 \cdot \lambda} \approx$$ $$\xi = \frac{h}{a+h}$$

[a]Correction function fitted according to [1]

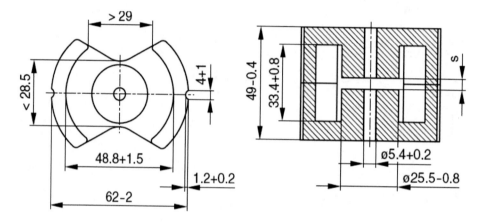

Fig. 6.17 Dimensioned drawing for core PM62/49 according to IEC 61247 especially suitable for power transformers and storage chokes. (Source: TDK/EPCOS 20/)

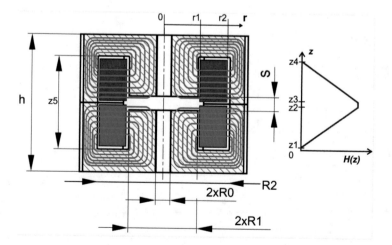

Fig. 6.18 Magnetic leakage field between 2 windings in the chambers of a pot core

$$\Theta_1 = I_1 N_1 = -I_2 N_2 = -\Theta_2$$

The magnetic conductance G_m of a cylindrical element of height dz of the leakage field is

$$dG_m = \frac{2\pi \cdot \mu_0 \cdot dz}{\ln\left(\frac{r_2}{r_1}\right)}$$

If one assumes, for example, a linearly distributed magnetic force θ, which starts at z_1 with zero and increases up to z_2 to the maximum value, one obtains in this range

$$\Theta_1(z) = \frac{N_1 I_1}{z_2 - z_1}(z - z_1).$$

The radial magnetic flux in this element is caused by a magnetic force

$$d\Phi = \Theta(z) \cdot dG_m = N_1 \cdot I_1 \frac{z - z_1}{z_2 - z_1} \cdot \frac{2\pi \cdot \mu_0 \cdot dz}{\ln\left(\frac{r_2}{r_1}\right)}.$$

The induction $B_1(r,z)$ is then consequently $dA = 2\pi r dz$

$$B_1(r, z) = \frac{d\Phi}{dA} = \frac{\Theta(z) \cdot dG_m}{dA} = \frac{\mu_0 \cdot N_1 \cdot I_1}{r \cdot (z_2 - z_1) \cdot \ln\left(\frac{r_2}{r_1}\right)}(z - z_1)$$

This allows the energy stored in a volume element dV to be calculated as

$$dW_1 = \frac{1}{2\mu_0} B_1^2(r, z)dV = \frac{1}{2\mu_0} B_1^2(r, z)2\pi r dr dz$$

Consequently, the following applies to the total leakage field energy of winding N_1

$$\int_V \frac{B_1^2(r, z)}{2\mu_0} dV = \int_{z1}^{z2} \int_{r1}^{r2} \frac{\mu_0^2 \cdot N_1^2 \cdot I_1^2}{2\mu_0 r^2 \cdot (z_2 - z_1)^2 \cdot \ln^2\left(\frac{r_2}{r_1}\right)}(z - z_1)^2 2\pi r dr dz$$

$$\Delta W_1 = \frac{\pi \cdot \mu_0 \cdot N_1^2 \cdot I_1^2}{(z_2 - z_1)^2 \cdot \ln^2\left(\frac{r_2}{r_1}\right)} \int_{z1}^{z2} \int_{r1}^{r2} (z - z_1)^2 \frac{dr dz}{r} = \frac{\pi \cdot \mu_0 \cdot N_1^2 \cdot I_1^2}{(z_2 - z_1)^2 \cdot \ln\left(\frac{r_2}{r_1}\right)} \int_{z1}^{z2} (z - z_1)^2 dz$$

$$\Delta W_1 = \frac{\pi \cdot \mu_0 \cdot N_1^2 \cdot I_1^2}{3 \cdot \ln\left(\frac{r_2}{r_1}\right)}(z_2 - z_1)$$

Analogously, one finds – as explained above – under the assumption that in the case of a short circuit the following applies $I_1 N_1 = I_2 N_2$ to N_2:

$$\Delta W_2 = \frac{\pi \cdot \mu_0 \cdot N_1^2 \cdot I_1^2}{\ln\left(\frac{r_2}{r_1}\right)} (z_3 - z_2) \text{ and } \Delta W_3 = \frac{\pi \cdot \mu_0 \cdot N_1^2 \cdot I_1^2}{3 \cdot \ln\left(\frac{r_2}{r_1}\right)} (z_4 - z_3).$$

This results in the following for the total leakage field energy of this winding arrangement

$$\Delta W_{ges} = \Delta W_1 + \Delta W_2 + \Delta W_3 = \frac{L_\sigma}{2} I_1^2 = \frac{\pi \cdot \mu_0 \cdot N_1^2 \cdot I_1^2}{3 \cdot \ln\left(\frac{r_2}{r_1}\right)} \left[(z_2 - z_1) + 3(z_3 - z_2) + (z_4 - z_3) \right]$$

$$L_\sigma = \frac{2\pi \cdot \mu_0 \cdot N_1^2}{3 \cdot \ln\left(\frac{r_2}{r_1}\right)} \left[(z_2 - z_1) + 3(z_3 - z_2) + (z_4 - z_3) \right]$$

Distance between windings

From this, the A_L value for the standardized characterization of the winding arrangement as a disk winding can be derived as

$$\frac{A_{L\sigma}}{\left[\frac{nH}{Wdg.^2} \right]} = \frac{2\pi \cdot \mu_0 \cdot 10^9}{3 \cdot \ln\left(\frac{r_2}{r_1}\right)} \left[(z_2 - z_1) + 3(z_3 - z_2) + (z_4 - z_3) \right]$$

The procedure is similar for 2 concentric tube windings (Fig. 6.19): Starting from the course of the magnetic force, the spatial dependence of magnetic field strength or magnetic

Fig. 6.19 Magnetic leakage field between 2 windings in the form of coaxial tubes in a PM pot core

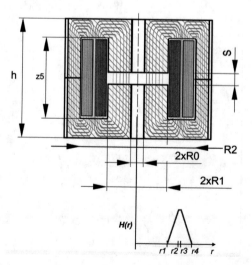

flux density is calculated, the amount of energy stored in the volume for the short-circuit case is determined as an easily calculable special case without consideration of the winding resistances and finally the leakage inductance of the winding arrangement is determined. To demonstrate the effects of a different winding arrangement, this procedure is demonstrated in detail below. The following Fig. 6.19 shows the principle course of the leakage field lines and the qualitative course of the leakage field strength H(r).

An alternating voltage at N_1 is assumed in the event of a short circuit in N_2. The magnetic field strength in the winding N_1 depends on the radius and is in the first range

$$H_1(r) = \frac{I_1 N_1}{l} \frac{r - r_1}{r_2 - r_1}$$

Along the z-axis, the magnetic field strength is constant, so that the following applies to the leakage field energy in the winding Sect. 6.1

$$\Delta W_1 = \int_V \frac{\mu_0}{2} H_1^2 dV = \int_V \frac{\mu_0}{2} H_1^2 l r dr d\varphi = 2\pi \frac{\mu_0}{2(r_2 - r_1)^2} \frac{I^2 N^2}{l} \int_{r1}^{r2} (r - r_1)^2 r dr$$

$$= \frac{\pi \mu_0}{(r_2 - r_1)^2} \frac{I^2 N^2}{l} \int_{r1}^{r2} (r - r_1)^2 r dr$$

$$\Delta W_1 = \frac{\pi \mu_0}{(r_2 - r_1)^2} \left(\frac{IN}{l}\right)^2 \int_{r1}^{r2} \left[(r - r_1)^3 + r_1(r - r_1)^2\right] dr$$

$$= \frac{\pi \mu_0}{(r_2 - r_1)^2} \left(\frac{IN}{l}\right)^2 \left[\frac{(r - r_1)^4}{4} + \frac{r_1(r - r_1)^3}{3}\right]_{r1}^{r2}$$

$$\Delta W_1 = \pi \mu_0 \left(\frac{IN}{l}\right)^2 \left[\frac{3r_2^2 - 2r_1 r_2 - r_1^2}{12}\right]$$

For the other two sections, assuming that primary and secondary current are the same, we find

$$\Delta W_2 = \pi \mu_0 \frac{I^2 N^2}{l} \left(\frac{r_3^2 - r_2^2}{2}\right) \text{ and } \Delta W_3 = \pi \mu_0 \frac{I^2 N^2}{l} \left(\frac{r_4^2 + 2r_3 r_4 - 3r_3^2}{12}\right).$$

The leakage inductance of the arrangement is again obtained over the total stored energy:

$$\Delta W_{ges} = \frac{L_\sigma}{2} I^2 = \pi\mu_0 l \left(\frac{IN}{l}\right)^2 \left[\frac{-(r_2 + r_1)^2 + 2\left(r_3^2 - r_2^2\right) + (r_4 + r_3)^2}{12}\right]$$

$$L_\sigma = \pi\mu_0 \frac{N^2}{6\cdot l} \left[(r_4 + r_3)^2 - (r_2 + r_1)^2 + 2\left(r_3^2 - r_2^2\right)\right]$$

For assumed winding distances to the core and each other of 1.5 mm, the A_L values of the leakage inductance are plotted for the PM core family compared to the A_L value of the core for a single winding without air gap (s = 0).

6.6   Approaches to Determine the Leakage Inductance with Different Winding Arrangements

It is clear from Fig. 6.20 that the leakage inductance between two windings can be strongly influenced by the geometry of the leakage channel. A reduction of the leakage inductance of disk windings or tube windings according to Figs. 6.18 and 6.19 is possible by further interleaving the windings. This is due to the interruption of the concatenation of the leakage flux. A flux leakage part is formed between two adjacent winding parts, which is not linked with other flux leakage parts. The price to be paid for this is a lower filling factor of the winding window, which is caused by additional layer insulation. Maximum nesting is achieved with a bifilar winding. Here, one primary winding and one secondary winding are applied next to each other. The differential voltage between the layers then stresses each winding with its insulation. It should be noted that a low leakage inductance means a

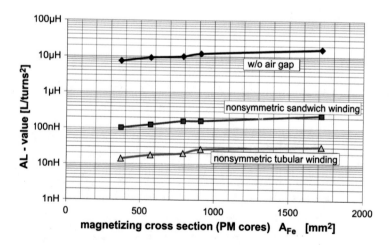

Fig. 6.20 A_L value for PM cores in comparison (top: single winding, air gap s = δ = 0; bottom: leakage inductance for two different winding arrangements)

Fig. 6.21 Winding structure of primary and secondary winding: (**a**) unbalanced, (**b**) balanced

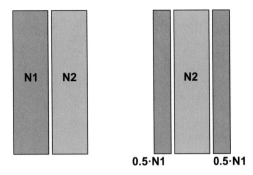

'close' coupling of the coils. However close coupling of the coils also generally means a high capacitance between the windings.

Bifilar windings can also be constructed for different transmission ratios of 1:1 N_1: N_2. With a transmission ratio of 1:6, for example, the full number of turns N_2 is wound continuously and the number of turns N_1 is placed six times on the same core as the secondary winding, whereby the winding segments of N_1 are all connected in parallel. The leakage inductance can be reduced by the distribution of the partial windings in the winding window. Section 10.27 describes a high-voltage impulse transformer for 30 kV with a correspondingly voltage-resistant coaxial HV cable, in which the cable was wound with the number of turns N_2 and the shield as the primary winding was repeatedly disconnected after several turns N_1. These were then connected in parallel to a primary winding. This can be easily achieved, especially with toroidal cores, since the symmetry of the structure results in quite even current distributions to the parallel-connected windings. The method described also shows that not every transmission ratio can be realized in this way.

A simple method to reduce the leakage inductance is the symmetrical division of the windings (Fig. 6.21). By dividing a winding in half, the maximum values of the leakage field strength and thus the values of the magnetic energy density decrease. This simple measure approximately halves the leakage inductance with a symmetrical structure compared to an asymmetrical structure. In the above-mentioned examples Figs. 6.18 and 6.19 we are dealing with asymmetrical structures. The calculated values in Fig. 6.19 can be approximately halved by a symmetrical setup.

If you want to achieve a higher leakage inductance, the enlargement of the leakage channel is a possibility. In the case of a disk winding, as shown in Fig. 6.18, the insulating bars/insulating distances between the windings are widened. For tube windings, appropriate insulating spacers are wound in. The resulting spaces can be used for improved air or oil cooling. Instead of insulating bars, ferromagnetic bars can also be inserted to increase the leakage inductance. It must be taken into account that these materials then also saturate and can themselves be sources of loss.

The existing geometric constraints limit the scope for such measures. If large leakage inductances are required, it is advisable to use yoke leakage (Sect. 6.5). The yoke leakage

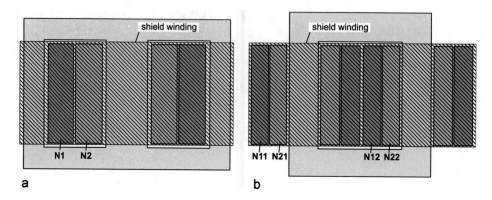

Fig. 6.22 Short-circuit winding to suppress the yoke leakage field in winding arrangements of sheath type (**a**) and core type (**b**)

field occurs in transformers of the sheath type as a parasitic field. It can lead to interference elsewhere through inductive coupling. An effective measure is the application of a short circuit winding for the yoke leakage field (Fig. 6.22) made of copper strip. The yoke leakage field is then largely compensated by an induced counter field of the short-circuit winding. The effects in the outside area (induced interference voltages and interference currents) can be drastically reduced.

The calculation shown only works for closely coupled windings on a core. It requires certain simplifying assumptions, otherwise, the fields cannot be adequately described. More precise results, which can be relied upon, especially for the leakage inductance values, are provided by FEM methods. These can be used to determine the distribution of the magnetic field strength and then, based on this spatially distributed data, to determine the energy content, etc. and finally the leakage inductances.

At the beginning of the last century Walter Rogowski [3] presented a work that allows usable values to be determined with little information using a formula. This formula is still widely used.

$$L_\sigma = \mu_0 \cdot N^2 \cdot l_m \cdot \lambda \cdot k_\sigma\left(\frac{a}{b}\right)$$

Here N: number of turns, l_m: average length of turns, λ: relative leakage conductivity and k_σ is a correction factor for the geometry of the outer contour (Fig. 6.23).

The relative leakage conductivity λ can be determined for the simple coil arrangement of coils wound on top of each other to

$$\lambda = \frac{1}{b}\left(\delta + \frac{a_1 + a_2}{3}\right),$$

with b: winding width, a_1, a_2: winding height of the two coils and δ: the winding distance between N_1 and N_2.

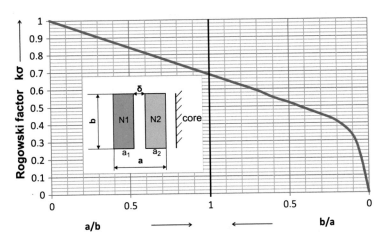

Fig. 6.23 Rogowski factor as a function of the aspect ratio a/b (for b > a) and b/a (for b < a) [3, 4]

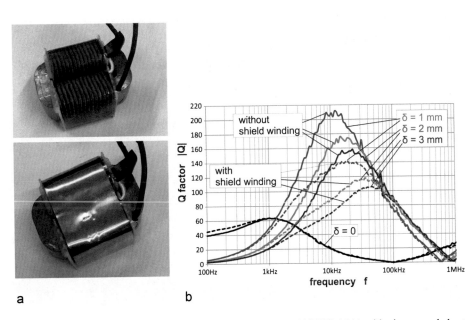

Fig. 6.24 Structure of a choke (40.5 mΩ) made of CC cores (AMCC-16A) with air gap and short-circuit winding for leakage field suppression. (**a**) Structure. (Source: Zacharias 2017), (**b**) Quality Q in comparison

6.7   Design of Transformers and Chokes with Strong Yoke Leakage

Even in the case of chokes with an air gap for setting the inductance, the properties can be influenced by means of shield windings. The inductance of the choke becomes slightly smaller. The currents in the short-circuit winding cause losses which, however, can be easily controlled with appropriate geometry (e.g. distance to the winding and core).

Figure 6.25 shows an equivalent circuit diagram obtained from the morphological observations on the magnetic resistances, which was parameterized via selected points of dependence $Z(\omega)$, such as resonance points, $\omega = 0$, etc. The associated parameters, some of which depend on the air gap δ, are summarized in Table 6.5. Where a parameter determination was not possible, but the presence of the parameter is physically plausible, '-' was entered.

R_{Cu} is the DC resistance of the winding. L_σ symbolizes the unlinked flux. Parallel to this, the resistor R_σ is representative of the associated eddy current losses. L_μ unites the magnetic field in the core, while R_μ reflects the core losses. By inserting an air gap δ, magnetic energy is stored in L_δ, that is, in the air gap and the environment. At the same time, the magnetic field reaches further into the room and causes eddy current losses in the shielding winding but also in the core and, if necessary, on structural parts.

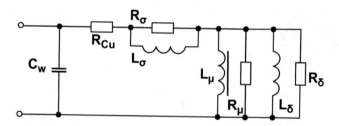

Fig. 6.25 Equivalent circuit diagram for a winding arrangement according to Fig. 6.24, taking into account the skin effect ($R\sigma$) and the eddy current losses in the shield winding R_δ

Table 6.5 Parameters from measured values for the equivalent circuit diagram Fig. 6.25 (m. S.: with shield winding, o. S.: without shield winding)

	$\delta = 0$		$\delta = 1$ mm		$\delta = 2$ mm		$\delta = 3$ mm	
	o. S.	m. S.	o. S.	m. S.	o. S.	m. S.	o. S.	m. S.
R_{Cu}	40.5 mΩ							
L_μ	2.65 mH							
R_μ	43 kΩ/42 kΩ							
L_δ	∞	∞	918 µH	913 µH	471 µH	465 µH	334 µH	329 µH
R_δ	∞	∞	881 kΩ	399 kΩ	188 kΩ	134 kΩ	125 kΩ	76 kΩ
L_σ	2.5 µH							
R_σ	–	–	–	–	–	–	–	–
C_W	95.6 pF							

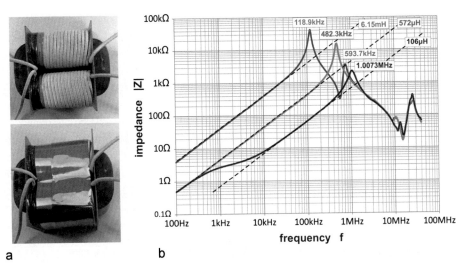

Fig. 6.26 Transformers with strong yoke leakage and one winding (33.6 mΩ) per leg on core AMCC-320 without and with a short circuit for the external leakage field. (**a**) Structure. (Source: Zacharias 2017), (**b**) primary impedance dependencies on frequency: blue: open circuit of N_2, green: short circuit of N_2 without shielding winding, red: short circuit of N_2 with shielding winding

In order to keep the leakage field strength and thus the effects of the proximity effect within the windings low, it is recommended, if possible, to divide a large air gap into several small air gaps connected in series. It should be noted that with an air gap in the core the leakage inductance between the windings is practically unaffected, but the leakage field into the room is.

A transformer with 2 windings is constructed with a similar winding arrangement (Fig. 6.26). In no-load operation, such a transformer has hardly any leakage field into the environment. The magnetic field is concentrated in the core. If one of the two windings is now short-circuited, the magnetic field from this winding is to a certain extent forced out of the winding. The magnetic field lines in the space run between the upper and lower yoke and penetrate the windings and the environment. This leakage field is therefore much more intense in the outer area than the leakage field of an air gap.

If a short-circuit winding is inserted into this field, it is forced out of this winding. Above a certain frequency, the winding acts like a "magnetic non-conductor" on the field, which can no longer propagate through the short-circuited area. This means that the resistance of the leakage field is increased and consequently the leakage inductance is reduced. This can be clearly seen in the red curve in Fig. 6.26. The effect starts at 1-2 kHz and the effective leakage inductance decreases by a factor of about 7.

In Fig. 6.26, the no-load measurement follows the reactance of the magnetizing inductance in the lower frequency range. With the two winding capacitances, primary and secondary form a parallel resonant circuit, which has a resonant frequency at 118.9 kHz. The winding capacitance can be determined from this. In the case of a secondary short circuit without shielding winding, the impedance in the lower frequency range essentially follows the reactance of the leakage inductance until it has a parallel resonance with the primary winding capacitance. At sufficiently low frequencies, the induced current in the shield winding is not yet sufficient to compensate for parts of the leakage field. This is only the case at sufficiently high frequencies (>10 kHz). As a result, the measurable leakage inductance between the primary and secondary windings is significantly lower. However, this effect is associated with losses in the shielding winding, which must be dissipated. In the upper-frequency range, this influence decreases again in relation to other effects. Leakage field-shaping can therefore be used to adapt the leakage inductance to specific applications (taking into account the losses that occur but are usually tolerable). In the case described, the effect of a shielding winding has been described. Similar effects can be achieved by providing the upper and/or lower yoke with a short circuit winding. In both cases, a reduction of the leakage inductance is achieved. An increase in leakage inductance is achieved by placing ferromagnetic material in the leakage path. However, these also cause corresponding magnetization losses.

The described processes are reflected in the Q-factor curves Fig. 6.27 The insertion of a short circuit winding for a part of the leakage flux leads to induced, oppositely acting currents which result in losses. The latter significantly reduces the quality of the winding arrangement. In a higher frequency range, however, this influence is less than in the lower frequency range. It should be noted that it is possible to subsequently adjust the leakage inductance of a transformer with strong yoke leakage. This must be taken into account when installing such transformers in vessels whose metal parts are in the leakage field and on the one hand form loss sources and on the other hand change the leakage inductance. According to Poynting's theory, energy is ultimately transported in this leakage field. Therefore the design of these field areas is of special interest.

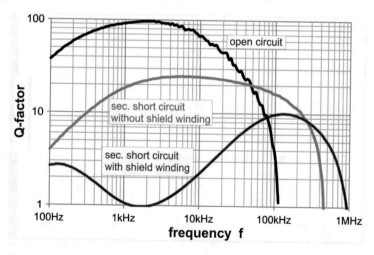

Fig. 6.27 Impedance quality for no-load and short-circuit measurement of the 2-winding transformer according to Fig. 6.26 without and with shield winding

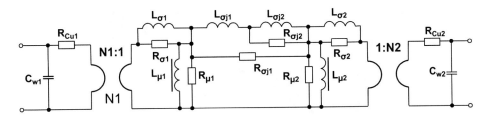

Fig. 6.28 Equivalent circuit diagram after small-signal measurements, taking into account the main influencing factors

Table 6.6 Parameters of the transformer equivalent circuit diagram Fig. 6.28 after measurements (parameters calculated with reference to the primary side)

	Without shield winding	With shield winding
$R_{Cu1} = R_{Cu2}$	33.6 mΩ	
$C_{w1} = C_{w2}$	208 pF	
$L_{\mu1} = L_{\mu2}$	12.3 mH	
$R_{\mu1} = R_{\mu2}$	92.4 kΩ	
$L_{\sigma1} = L_{\sigma2}$[a]	–	–
$R_{\sigma1} = R_{\sigma2}$[a]	–	–
$L_{\sigma j1}$	107 µH	
$R_{\sigma j1}$	20.8 kΩ	
$L_{\sigma j2}$	604 µH	
$R_{\sigma j2}$	∞	1.4 Ω

[a]Not accessible for metrological determination

A 'complete' electrical equivalent circuit diagram for a transformer with yoke leakage and shield winding is shown in Fig. 6.28. The energy of the entire leakage field is concentrated for the most part in the yoke leakage field. The two legs with the windings are largely decoupled by the leakage inductance. In resonant circuits, this can lead to one leg saturating while the other leg is still operating in the 'normal' range. Although the leakage components within the winding are present, they cannot be easily parameterized with the structure shown in Fig. 6.26. For this reason, Table 6.6 does not contain any values where the parameters of the equivalent circuit determined from measurements are summarized.

Figure 6.29 shows the application of these effects on the design example of a transformer 160 V/5 kV@50 kHz with oil insulation, which is part of a resonant circuit concept. The U-shaped bent 3 mm thick aluminium sheet is not part of the insulation concept but part of the magnetic structure. The leakage field of the transformer is shaped to give a certain leakage inductance. It also shields other structural parts from the transformer's leakage magnetic field, thus preventing heating and interference. The reproducibility of the results of this measure is excellent.

Fig. 6.29 Design example of a transformer 160 V/5 kV@50 kHz with oil insulation. (Source: Zacharias 2017)

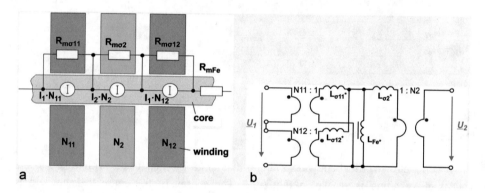

Fig. 6.30 Example of the formation of leakage flux in arrangements with several windings. (**a**) Winding arrangement; (**b**) equivalent electrical circuit

Design Options for the Leakage Inductance of Two Coupled Windings

The leakage inductance describes the coupling between 2 windings. You can choose different analytical description possibilities. However, these are suitable to a varying extent for determining the leakage inductance and for deriving measures for its design. For explanation, a methodology is used which is explained in more detail in Chap. 5. The magnetic force $I \cdot N$ drives a magnetic flux in the core. However, not all the flux remains in the core. Part of it closes the circuit via the air in the winding. This is structurally recorded in Fig. 6.30 by the magnetic leakage resistors $R_{m\sigma}$.

The model shown represents only a part of the dependencies between electrical and magnetic quantities. But it illustrates several possible measures. The effective leakage inductance of the arrangement shown in Fig. 4.30 can be calculated as follows

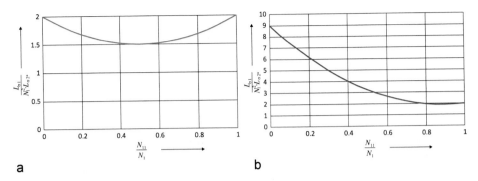

Fig. 6.31 Variation of the leakage inductance $L_{\sigma 1}$ by winding distribution of N_1: (**a**) $L_{\sigma 11*} = L_{\sigma 12*} = L_{\sigma 2*}$; (**b**) $8 \cdot L_{\sigma 11*} = L_{\sigma 12*} = L_{\sigma 2*}$

$$L_{\sigma 1} = \left(\frac{U_{11} + U_{12}}{\omega \cdot I_1} \right) = N_{11}^2 \cdot L_{\sigma 11*} + (N_{11} + N_{12})^2 \cdot L_{\sigma 2*} + N_{12}^2 \cdot L_{\sigma 12*}$$

Several possible measures for setting a certain inductance can be derived from this dependence as rules:

- The leakage inductance $L_{\sigma 1}$ is dependent on the square of the number of turns and can be adjusted via it. A reduction in the number of windings is limited by saturation. An increase in the number of turns means higher winding losses.
- With a constant number of primary windings $N_1 = N_{11} + N_{12}$ and equally large leakage field resistances, so that $L_{\sigma 11*} = L_{\sigma 2*}$, the leakage inductance can be adjusted by distributing the number of windings to the partial windings. With $N_{11} = N_{12}$, this results in a minimum (Fig. 6.31a).

$$L_{\sigma 1} = N_1^2 \cdot \left(L_{\sigma 2*} + \left[\left(1 - \frac{N_{11}}{N_1} \right)^2 + \left(\frac{N_{11}}{N_1} \right)^2 \right] \cdot L_{\sigma 11*} \right)$$

with the minimum $L_{\sigma 1\,\mathrm{min}} = N_1^2 \cdot (L_{\sigma 2*} + 0,5 \cdot L_{\sigma 11*})$
and the maximum $L_{\sigma 1\,\mathrm{max}} = N_1^2 \cdot (L_{\sigma 2*} + L_{\sigma 11*})$

- If $L_{\sigma 11*} \neq L_{\sigma 12*}$ applies due to the winding design, the leakage inductance can be adjusted over a relatively wide range by combining the number of turns.

$$L_{\sigma 1} = N_1^2 \cdot \left(\left(\frac{N_{11}}{N_1} \right)^2 \cdot L_{\sigma 11*} + L_{\sigma 2*} + \left(1 - \frac{N_{11}}{N_1} \right)^2 \cdot L_{\sigma 12*} \right)$$

with the minimum $L_{\sigma 1\,\mathrm{min}} = N_1^2 \cdot \left(L_{\sigma 2*} + \frac{L_{\sigma 11*} L_{\sigma 12*}^2 + L_{\sigma 12*} L_{\sigma 11*}^2}{(L_{\sigma 11*} + L_{\sigma 12*})^2} \right)$

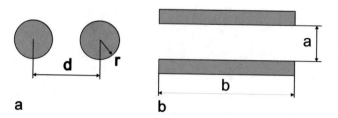

Fig. 6.32 Inductance values of double conductors for small conductor spacings including the internal inductance of the conductors. (**a**) round conductors, (**b**) thin foils

and the maximum $L_{\sigma 1\,max} = N_1^2 \cdot (L_{\sigma 2*} + \max(L_{\sigma 11*}; L_{\sigma 12*}))$

In the example, the flux linkage of the leakage flux of the primary winding is interrupted by inserting the counterflux of N_2. As a result, the leakage inductances of the partial windings add up approximately. This fact can be used if very small leakage inductances are to be achieved. Approximately, the leakage inductance is inversely proportional to the number of partial windings. A maximum subdivision is achieved when the counter-flooding of one winding is achieved by an adjacent winding. If $N_1 = N_2$, the leakage inductance is then equal to the inductance of the wound double cable of the corresponding length. This also represents the achievable minimum. Figure 6.32 shows 2 double cables with the approximate calculation formulae for the inductance at small conductor spacings.

In general, it should be noted that a low leakage inductance also means a close coupling and thus results in a relatively large coupling capacitance between the windings. There is often a misunderstanding that in transformers the leakage inductance can be influenced by inserting air gaps. This is usually not the case. In most designs, the air gap only affects the magnetizing inductance. The leakage inductance is a consequence of the winding arrangement against each other and is practically not or hardly influenced by the reluctance of the core.

With flux branches formed by cores, however, certain coupling factors or leakage inductances can be set. Defined flux division factors are best set with air gaps. You can also influence the leakage inductance by specifically designing the leakage channels, that is, the areas through which the leakage flux is closed.

In the case of closely coupled windings, the leakage inductance can be increased by widening the leakage cross-section or by inserting ferromagnetic materials (Fig. 6.33). However, the leakage field can also be reduced by working specifically with shield constructions in the leakage field. Eddy currents are induced in these conductive materials, which "adjust" the path of the leakage flux and thus increase the magnetic resistance in the leakage path. The applicability of this method is limited by the additional losses that occur at the same time.

Fig. 6.33 Increasing the leakage inductance of closely coupled windings by widening the leakage channel and inserting ferromagnetic material

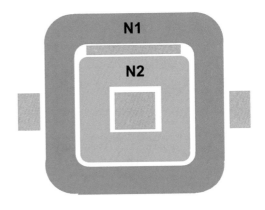

6.8   Determination of Parameters of Inductive Components with Toroidal Cores

Concentric Windings on Toroidal Cores

If primary and secondary windings are wound on top of each other on the entire circumference of a toroidal core, concentric "tubular" windings are obtained (Fig. 6.34).

The magnetizing inductance of the core at a core height 'h' is

$$L = \frac{N^2}{R_m} = N^2 \frac{\mu_r \mu_0}{2\pi} \cdot h \cdot \ln\left(\frac{r_6}{r_5}\right)$$

where N: number of turns, h: core height, r_1: inner diameter, r_2: outer diameter.

Due to the ring shape, the core is magnetized uniformly. As a result, all parts of the core go into saturation practically simultaneously and the magnetization inductance "disappears abruptly" in such a case. This is the reason for the so-called rush effect when such a transformer is switched on: If a transformer is switched on at a sinusoidal voltage and then switched $\omega t = 0$ on, the flux linkage reaches the maximum value $\widehat{\psi}_{max} = 2\frac{\sqrt{2}U_1}{\omega}$ while $\omega t = \frac{\pi}{2}$ only the peak value $\widehat{\psi}_{max} = \frac{\sqrt{2}U_1}{\omega}$ is reached at the time of switch-on. In the first case, depending on the design reserve, the transformer core goes into saturation and the inrush current rises sharply, so that the I^2t value of upstream fuses may be exceeded. These then trip. This effect can be prevented by connecting a switch-on resistor upstream, which is then shunted during normal operation, or an NTC resistor to limit the starting current. Figure 6.35 shows the switch-on situation and the course of the inrush current at different switch-on points as well as the corresponding course of the integral

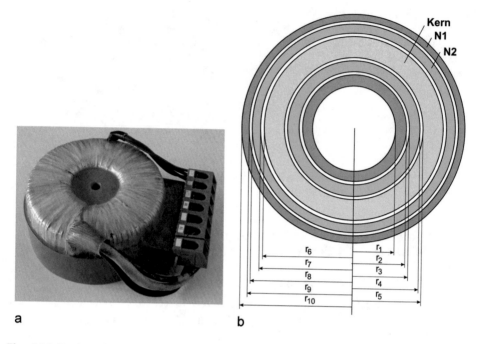

Fig. 6.34 Design of a toroidal transformer 230 V/2x16V/270VA. (Source: Zacharias 2017). (**a**) Photo, (**b**) Dimensional sketch

Fig. 6.35 Switch-on situation and inrush current curve at different switch-on points of a toroidal transformer and the corresponding curve of the integral "I^2t"

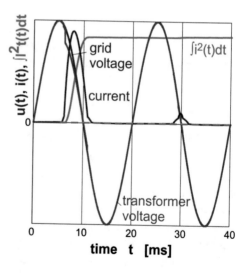

$$I^2 t = \int_0^t i_1^2(t)dt,$$

that as "load integral" represents a triggering measure for fuses. If the maximum load integral "$I^2 t$" of a fuse is exceeded, it trips and the circuit is interrupted. This can be prevented by either connecting an NTC resistor or by switching on a resistor for current limitation, which is shunted by a second switch. The latter does not cause permanent losses but requires an additional switch.

Analogous to Sect. 6.5, the leakage inductance between 2 windings, related to the primary winding, can be determined. The analytical method is relatively complex. An easier way is to use the Rogowski formula

$$L_\sigma = \mu_0 \cdot N^2 \cdot l_m \cdot \lambda \cdot k_\sigma \left(\frac{a}{b}\right).$$

The values for 'a' and 'b' are then and $a = \frac{(r_4 - r_1) + (r_{10} - r_7)}{2}$ $b = 2\pi \frac{r_5 + r_6}{2}$. Due to the slightly different winding heights, the relative conductance λ is calculated with the average values of

$$\lambda = \frac{1}{b}\left(\frac{r_3 - r_2 + r_9 - r_8}{2} + \frac{r_2 - r_1 + r_4 - r_3 + r_8 - r_7 + r_{10} - r_9}{6}\right).$$

Due to the great length of the leakage field lines, very small leakage inductances can easily be realized with toroidal cores. Toroidal transformers, therefore, have a very small internal voltage drop and are suitable for compact power supplies. Similar values for leakage inductances can be achieved with LL, CC, UU and UI core forms. Primary and secondary windings are placed half on top of each other to achieve low leakage inductance values (Fig. 6.36).

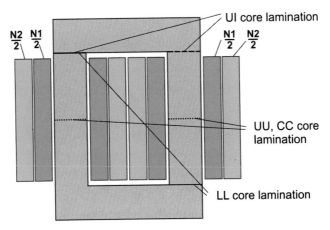

Fig. 6.36 1-phase transformer with low leakage based on LL, CC, UU and UI core sections

If half the primary and half the secondary winding are each applied to one leg and the winding halves are connected in series in pairs, the leakage inductance from the individual leakage inductances of the partial windings is obtained as concentric tube winding from their sum

$$L_{\sigma 1} = L_{\sigma 11} + L_{\sigma 12} = N_{11}^2 L^*_{\sigma 11} + N_{12}^2 L^*_{\sigma 12}$$

The values of the leakage inductances are added here since the leakage fluxes of the partial windings are not interlinked. Since the core shapes deviate from the "natural" shape of the magnetic field, they do not go into saturation uniformly, so that the rush effect is not quite as pronounced as with the toroidal core.

Segmented Windings of Toroidal Cores
In addition to the largely linear chokes and toroidal transformers with concentric windings used in normal applications, other winding designs are also used. The following Fig. 6.37 shows examples with 2, 3 and 4 winding segments. The purpose of these components is the suppression of common-mode interference when connecting single-phase and multi-phase electrical systems.

In the following, an investigation of the structure of the couplings of these windings will be carried out. A symmetrical division of the windings into 2 segments is comparable to the winding arrangement of 2 windings on 2 separate legs of a UU, LL, or UI core. The systematic description is therefore best done analogously via a π replacement schematic diagram. With the additional winding, the complexity of the magnetic and electrical model of the arrangement increases with a three-fold segmentation. Figure 6.38 shows a test setup for a symmetrically wound toroidal core with 3 windings and the associated magnetic equivalent circuit diagram.

Fig. 6.37 Examples of toroidal cores with segmented windings, designed for use as current-compensated chokes for common-mode suppression in high-frequency filters. (Manufacturer: Vacuumschmelze Hanau, Source: VAC 2017)

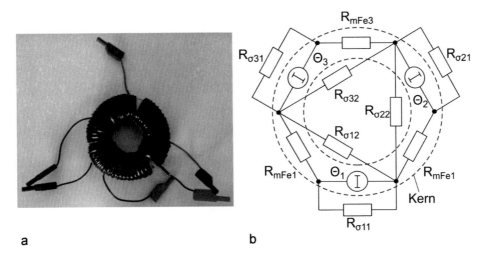

Fig. 6.38 Toroidal core with 3 winding segments. (**a**) Experimental setup, (**b**) associated magnetic equivalent circuit. diagram. (Source: Zacharias 2017)

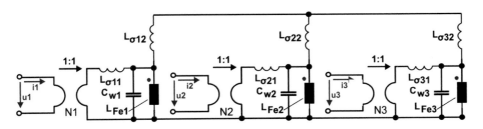

Fig. 6.39 Electrical equivalent circuit diagram of the symmetrically coupled toroidal core choke according to Fig. 6.38

Formally, the equivalent electrical circuit diagram is obtained by graphic circuit inversion and rearrangement of the components, taking into account the effect of the winding capacities as shown in Fig. 6.39.

To determine the parameters of the model, a series of measurements were carried out, the results of which are summarised in Table 6.7.

The number of turns of the windings N_1, N_2 and N_3 are equal. It can be assumed that the magnetizing inductances of L_1, L_2 and L_3 are large compared to the leakage inductances. Despite the asymmetries in the design of the choke, as can be seen in Fig. 6.38, the following measured values with low leakage result (Table 6.8):

If one starts from $L_x \gg L_\sigma$, the following relationships can easily be derived from the electrical equivalent circuit diagram using the average values of the inductance measurements

Table 6.7 Measurement results for no-load and short-circuit tests on a 3-winding arrangement according to Fig. 6.38

Winding	Inductance [H]	Short curcuit of winding					
		L1	L2	L3	L1 + 2	L2 + L3	L1 + L3
L1	1,26E-02	–	8,40E-04	8,51E-04	–	6,30E-04	–
L2	1,23E-02	8,97E-04	–	8,52E-04	–	–	6,72E-04
L3	1,21E-02	9,01E-04	8,45E-04	–	6,68E-04	–	–
L1 + L2	3,77E-02	–	–	8,56E-04	–	–	–
L1-L2	2,61E-03	–	–	8,56E-04	–	–	–
L1 + L2 + L3	7,80E-02	–	–	–	–	–	–
L1 + L2-L3	1,26E-02	–	–	–	–	–	–

Table 6.8 Results of the evaluation of the measurements according to Table 6.7

Inductance	Average value	Effective deviation	Effective deviation
L_x (self-inductance a winding)	12.3 mH	140.5 µH	1.14%
L_σ' at short circuit one winding	864 µH	24.7 µH	2.86%
L_σ'' at short circuit of 2 windings	657 µH	18.9 µH	2.88%

(a) Short circuit of a winding

$$864\mu H = 2(L_{\sigma 11} + L_{\sigma 12})$$
$$L_{\sigma 11} + L_{\sigma 12} = 432\mu H$$

(b) Short circuit of 2 windings

$$657\mu H = L_{\sigma 11} + L_{\sigma 12} + \frac{L_{\sigma 11} + L_{\sigma 12}}{2} = 1{,}5 \cdot (L_{\sigma 11} + L_{\sigma 12})$$
$$L_{\sigma 11} + L_{\sigma 12} = 438\mu H$$

The deviation from the average value of *435 µH* is here $3\mu H \triangleq 0.7\%$. By these measurements both scattered components cannot be separated. The inclusion of the magnetizing inductance in the calculations leads to nonsensical results since these – in contrast to the leakage inductances – are strongly non-linear.

To determine the relationship between $L_{\sigma31}$ and $L_{\sigma32}$, a resonance test was therefore carried out with N_3, while N_1 and N_2 were short-circuited to get additional information. The analysis showed the following results

Measured inductance (10 kHz) $L_{3(KS\ von\ L\sigma31\ und\ L\sigma32)} = 668\ \mu H$
Frequency of the parallel resonance $f_{p.res} = 0.9744\ MHz$
Resonance capacity $C_W = = 41\ pF$

With the last size, C_W results an effective inductance for the parallel resonance to 650 µH. Thus one receives the value of $L_{\sigma31} = 18\ \mu H$.

Analogous statements with small deviations are obtained for the measurements on the other windings. Thus, an adequate symmetrical electrical equivalent circuit diagram can be dimensioned for simulations according to the arrangement in Fig. 6.39:

$L_{Fex} = 36.93\ mH$
$L_{\sigma1x} = 18\ \mu H$
$L_{\sigma2x} = 417\ \mu H$
$C_W = 41\ pF$

According to this, the ratio between $L_{\sigma31}$ and $L_{\sigma32}$ is about 1:23, while the ratio $(L_{\sigma31} + L_{\sigma31}):L_{Fe3} = 1:85$. For simulations, it will generally be sufficient to combine $_{L\sigma1x}$ and $_{L\sigma2x}$ and use them in the Γ equivalent circuit diagram. Alternatively, the leakage inductances can be switched as Y, as shown in Fig. 6.39, or as Δ. In principle, the design is completely symmetrical, which makes the construction interesting for various applications (current compensated chokes, storage chokes for multiphase DC/DC converters).

Due to the symmetry of the arrangement, the non-linearity of the magnetic resistance of the core and the large differences between magnetizing inductances and leakage inductances, combinations of short-circuit and no-load measurements are sometimes misleading. The measurements were performed f. e. at a constant voltage. When windings are connected in series, the reactance of the winding increases approximately proportionally to the $\sim \left(\sum_{i=1}^{k} N_i\right)^{-2}$ square of the number of turns $\sim \left(\sum_{i=1}^{k} N_i\right)^{2} \sim \left(\sum_{i=1}^{k} N_i\right)^{-1}$.

This means that the measurements on the coupled windings of the arrangement are carried out at different states of the magnetic core. The relative permeability of the core is a function of the magnetic field strength. Figure 6.40 shows the dependence for the permeability in normalized form for the case described above as a function of the flux density/ magnetic field strength. If permeabilities are sufficiently high, the values for the leakage inductances are practically independent of the concrete values of the core.

It follows:

Leakage inductances depend primarily on the geometries and the position of the windings. Due to air gaps in the core, leakage inductances of winding arrangements cannot be influenced or only slightly influenced.

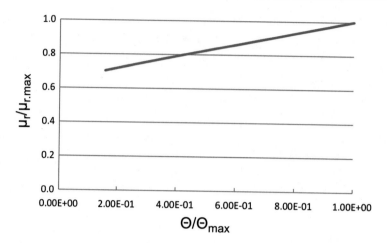

Fig. 6.40 Relative permeability as a function of magnetic force/magnetic field strength for the measurements in Table 6.7

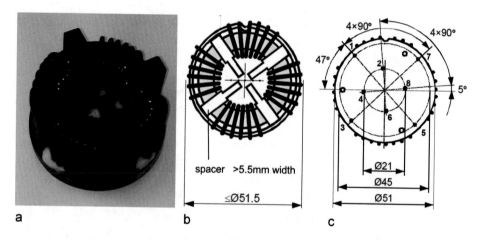

Fig. 6.41 Current compensated choke VAC X401A6123. (Source: Vacuumschmelze Hanau). (**a**) Photo, (**b**) Dimensional drawing top view (**c**) Dimensional drawing bottom view

Figure 6.41 shows the design of a current compensated choke with 4 windings. Following the above findings, a model is to be developed here as well, which enables the simulation of circuits with this component. Taking into account the air leakage paths, the magnetic equivalent circuit diagram shown in Fig. 6.42a can be found. From this, the electrical equivalent circuit diagram Fig. 6.42b can be derived. This gives the structure of the model, but not yet the parameterization.

In the case of symmetry of the structure, the approach is

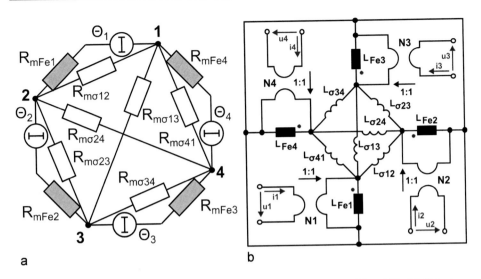

Fig. 6.42 Modeling for a toroidal core with 4 windings according to Fig. 6.41. (**a**) Reluctance model, (**b**) derived electrical model

$$L_{\sigma 1} = L_{\sigma 2} = L_{\sigma 3} = L_{\sigma 4} = L_{\sigma r}\left(\text{with } L_{\sigma r} = N^2 \cdot G_{m\sigma x} = N^2 \cdot \Lambda_{m\sigma x}\right)$$

and

$$L_{\sigma 13} = L_{\sigma 24} = L_{\sigma d}$$

Measurement results at the current compensated choke VAC X401A6123 at 20 kHz result in $L_{Fe} = 3.04$ mH. The resulting leakage inductance between 2 opposite windings is calculated using the model parameters for a model according to Fig. 6.42

$$\frac{2 \cdot \frac{L_{\sigma r}}{2} \cdot L_{\sigma d}}{L_{\sigma r} + L_d} = 4.75 \mu H$$

Accordingly, in the case of a simultaneous short circuit of 2 windings adjacent to one winding, the value is

$$\frac{L_{\sigma r}}{2} = 2.84 \mu H \text{ or } L_{\sigma r} = 5.68 \ \mu H.$$

From this follows for $L_{\sigma d} = 25{,}08$ µH in Fig. 6.42. This means that the coupling of the 4 windings is balanced but *not always* the same in *pairs*, as is the case with a balanced current compensated choke with 3 windings on a toroidal core. It is not possible to convert the square with the diagonal elements into a star-shaped equivalent circuit for all leakage inductances.

If only one winding adjacent to a winding is short-circuited, the measurement yields an average value of 3.76 µH in this case. This corresponds approximately to the datasheet specification ~3.2 µH/100 kHz. However, the value is made up of $L_{\sigma p}$ and $L_{\sigma q}$.

With the model described above, which is based on measurement data and whose parameters can still be set as a function of frequency, the behavior of filter circuits can be determined much more precisely than with only one specification in the datasheet.

If 2 adjacent windings are connected in series, the model shown above changes to the π equivalent circuit diagram valid for 2 windings.

There is the general problem of how far magnetic and electrical symmetry can be designed for simple concentrated windings. For an arrangement with 3 windings on one core, the solution seems to be obvious. The transformation of a star into a triangle and vice versa is possible in the model of reluctances as well as in the model of leakage inductances. If 'n' windings are symmetrically placed on an annular core or another core with its centerline in the plane, a convex n-corner for the reluctance model with 'd' **diagonal** connections between all corner points is inevitably created, where

$$d = \frac{n(n-3)}{2}$$

is. Diagonal connections are available in the plane geometry by definition for n > 3. The number of connections of the structure to the n-corner corresponds to the number 'n = r' of the corners. The total number 'z' of connections results from the sum z = n + d. The following figure shows the relations up to n = 8 (Fig. 6.43).

An n-star has only n = r elements. A mapping of 'n' elements to z > n elements will generally always be possible, as the Kennelly theorem shows. In the opposite direction, this is generally not the case [5], but rather with conditions like

$$G_{m.\lambda\nu} = \frac{G_{m.\lambda} \cdot G_{m.\nu}}{\sum\limits_{\nu=1}^{n} G_{m.\nu}}$$

knotted. In this example, the variables $G_{m\nu}$ are to be the magnetic conductance (permeances), which are each assigned to a point 'ν' of the n-star (see Fig. 6.42). At the same time, $G_{m\nu\lambda}$ describes the permeances of a corresponding n-corner between the points 'ν' and 'λ'. With complete symmetry of the n-star, all $G_{m\nu}$ have the same size. This would also require the same values for all permeances of the n-corner. This contradicts the assumptions made above regarding the geometric design of the winding arrangement. Consequently, the design of truly symmetrical magnetic components with more than 3 phases/windings quickly reaches practical limits.

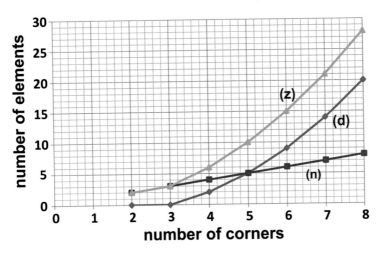

Fig. 6.43 Minimum number of possible connecting elements in an electrical network with 'n' "corners" [2] (n: number of corners of the n-corner, z: total number of connections, d: diagonal connections between the corners)

In a "plane" symmetrical magnetic core arrangement, a symmetrical star would also be expected for a star-shaped reluctance model for the leakage. This leads to a logical contradiction already at n = 4. The leakage field of a uniformly "convex" shaped core, for example, in ring form, does not allow for equal magnetic leakage resistances between adjacent windings and between diagonally arranged winding endpoints. The logical conclusion from this is that such a (uniform) coupling can only be achieved if, for example, the following conditions are available.

The use of a 3-dimensional structure of the core or the use of co-connections and counter-connections of winding components, or a uniform core cross-section is avoided, or additional ferromagnetic paths are specifically inserted to support the desired coupling. If, for example, 4 coupled windings on a toroidal core are postulated with a star-shaped equivalent circuit diagram of the leakage inductances, this structure could be represented in Fig. 6.44a. The transformation into the reluctance equivalent circuit diagram is shown in Fig. 6.44b, while Fig. 6.44c shows a sketch for a possible technical design.

The structure shown in Fig. 6.44a leads to relatively large leakage inductance values with extensive symmetry in the magnetic and electrical range. The magnetic/geometric design shows that it is still relatively easy to insulate the windings from each other. As the electrical equivalent circuit diagram clearly shows, core sections go into saturation individually in the event of saturation phenomena, as is also the case with current-compensated chokes.

An alternative for achieving the electrical equivalent circuit diagram with regard to the leakage with low values of the leakage inductance would be an (e.g. tetra-filar) arrangement of the partial windings around the entire core. As with the 2-winding transformer on a

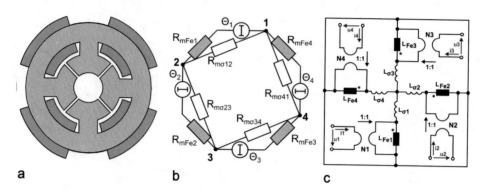

a   b   c

Fig. 6.44 Derivation of a possible structure for an arrangement of 4 coupled windings on a toroidal core with an arrangement in segments for the approximate implementation of the electrical equivalent circuit. (**a**) schematic structure of the core and arrangement of the windings, (**b**) derived reluctance model, (**c**) electrical equivalent circuit

toroidal core in the previous section, saturation would saturate the entire core at once. Depending on the application, this behaviour is more or less desirable. The leakage inductances and winding resistances of these windings are not equal in such a design and disturb the symmetry of the magnetic-electric structure. The equivalent electrical circuit of such an arrangement with non-segmented, closely coupled windings around the entire circumference is shown in Fig. 6.45.

To achieve a very close coupling of 2 coils, different approaches can be used. Concentric or nested windings are obvious. Figure 6.46 shows 2 possibilities with nested windings with the same effective number of turns. In Fig. 6.46a, these 6 turns of 4 parallel wires are applied continuously in the same winding direction around the core. If large voltage differences occur between the start and end of the winding during operation, this can lead to insulation problems. In Fig. 6.47a the result of a calculation of the distribution of the electric field strength for an approximately concentric winding arrangement is shown Fig. 6.46a. These calculations were performed for a 2-winding transformer in a high voltage pulse application. Clearly visible are the excessive field strengths at the ends of the windings.

In Fig. 6.46b these problems are avoided by dividing the entire winding into 2 partial windings with the same number of turns and half the conductor cross-section. One partial winding consisting of 4 parallel wires with the same winding direction is placed in the upper half of the toroidal core, while the second partial winding is placed on the lower half of the core in Fig. 6.46b with the opposite winding direction. When the windings are connected in parallel in pairs, the directions of the fluxes are the same. This also achieves the goal of uniform magnetization of the core. However, the start and end of the windings are far apart, which makes it easier to solve insulation problems. Figure 6.47b shows a practical design using this method for 2 coupled windings on a toroidal core. The function of this arrangement is that of a saturable choke in a pulse forming high voltage network.

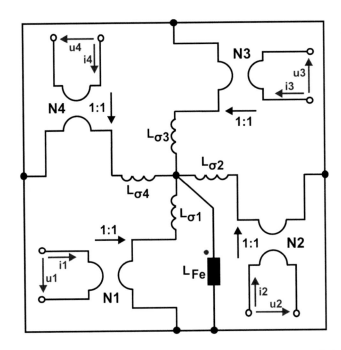

Fig. 6.45 Electrical equivalent circuit diagram of a winding arrangement consisting of 4 closely coupled windings on a toroidal core

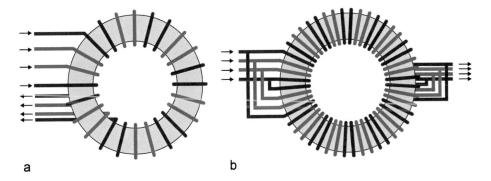

a   b

Fig. 6.46 Possible arrangements of 4 closely coupled windings on a toroidal core. (**a**) Design as a tetra-filament winding with adjacent winding start and ends, (**b**) Segmented division of the windings into two parallel upper and lower parts with different winding directions

This means that voltages of >10 kV occur briefly (<1 µs) between the ends of the component. The core of coiled nanocrystalline magnetic tape with layer insulation was insulated with Teflon. At the same time, Teflon-insulated stranded wire was used for the windings. This winding arrangement and the materials used can be used in oil to permanently overcome the insulation problems.

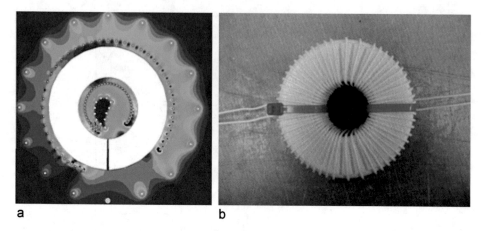

a   b

Fig. 6.47 Comparison of couplings of two windings on a toroidal core. (**a**) Distribution of the electric field strength of superimposed windings with adjacent winding starts and ends, (**b**) Design example of a switching choke consisting of two windings according to the design principle of Fig. 6.46. (Source: Strowitzki 2009 [6])

6.9   Permeability as a Complex Parameter

Complex Permeability of Core Material
When an alternating magnetic field penetrates matter, it results in a coupled electromagnetic field according to Maxwell's equations. Figure 6.48 is used for further analysis. It represents a cylindrical section of material which is penetrated by an alternating magnetic field in the direction of the axis.

If one formulates the correlations (without considering the induced secondary magnetic field), one obtains

$$B(t) = \mu_r \mu_0 \cdot H(t)$$
$$E(r, t) = \frac{-d}{dt} B(t) \cdot \pi r^2 = \frac{-d}{dt} \mu_r \mu_0 \cdot H(t) \cdot \pi r^2$$

For harmonic quantities, these equations can be formulated with phasorr quantities (RMS value phasors are used):

$$\frac{\underline{E}(r, t)}{2\pi r} = -j\omega \underline{B}(t) \cdot \pi r^2 = -j\omega \mu_r \mu_0 \cdot \underline{H}(t) \cdot \pi r^2$$

The electric field results in a electric current field that is composed of a "galvanic" and a dielectric component. The current density is then

Fig. 6.48 Cylindrical body, penetrated by an alternating magnetic field

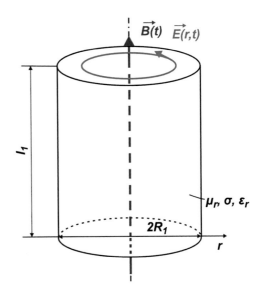

$$\underline{J}(r, t) = (\sigma + j\omega\varepsilon) \cdot \underline{E}(r, t)$$

If one now separates the material influence, one breaks down the idea in Fig. 6.2.1 into a component that is assigned to the vacuum and one that is assigned only to the material, one obtains

$$\underline{B} = (1 + (\mu_r - 1))\mu_0\underline{H} = \underline{B}_0 + \underline{P}$$

In the magnetic field, the equation pairs for the components caused by the vacuum and the components caused by the core material are

$$\underline{B}_0 = \mu_0\underline{H}$$
$$\underline{J}_0 = j\omega\varepsilon_0\underline{E}_0$$

and

$$\underline{B}_0 = (\mu_r - 1)\mu_0\underline{H}$$
$$\underline{J}_0 = (\sigma + j\omega(\varepsilon_r - 1))\varepsilon_0\underline{E}_0$$

$W_{magn0} = \frac{\Phi^2}{L_0} = \frac{I_0^2}{\omega_0^2 C_0}$ which can be $C_0 = \frac{I_0^2 L_0}{4\pi^2 f_0^2 \Phi^2}$ calculated by setting the adjustment condition ω_0.

The arrangement is similar to a cylindrical (*pillbox*) cavity used in microwave technology. (In the present case, however, the cavity is not hollow, but filled with a polarizable material). The E modes ((TM modes), $H_z = 0$, $E_z \neq 0$) form the dual problem there. Here, distributed parameters for the electromagnetic field and its wavelike propagation are assumed. Therefore there are harmonics (modes) in addition to the fundamental oscillation.

The magnetic field is perpendicular to the direction of propagation. Natural frequencies are then

$$f_{mnp} = \frac{c_0}{2\pi\sqrt{\mu_r\varepsilon_r}}\sqrt{\left(\frac{X_{mn}}{r}\right)^2 + \left(\frac{p\pi}{h}\right)^2}$$

m = order of Bessel function first type (m = 0, 1, 2, 3, ...)
n = counter of the zero (n = 2 means: second zero of the Bessel function first type (n = 1, 2, 3, ...))
X_{mn} = x-value of the n-th zero of the Bessel function of the first order type (e.g. X01 = 2.405)
p = number of half wavelengths in axial direction (p = 0, 1, 2, 3, ...)
A cylinder with cylinder radius r = 10 cm and cylinder height h = 11.95 cm with m = 0, n = 1, p = 0 ((010)-mode) has a natural frequency f_0 = 1.146 GHz and is therefore uninteresting for classical power electronic applications. However, the picture changes when ferrites are considered as core material. For cuboid resonators, Lord Rayleigh already in 1896 established the relationship

$$f_0 = \frac{c_0}{2\pi\sqrt{\mu_r\varepsilon_r}}\sqrt{\left(\frac{\pi n_x}{l_x}\right)^2 + \left(\frac{\pi n_y}{l_y}\right)^2 + \left(\frac{\pi n_z}{l_z}\right)^2}$$

specified [7, 8].

For "concentrated field elements", inductance and magnetic flux are given by the geometry to

$$L_0 = \mu_0\frac{\pi r_1^2}{l_1} \text{ and } \Phi = \pi r_1^2 \cdot B.$$

The magnetic and electric field are connected via a gyrator (Fig. 6.49).

A rough estimate of the parameters and the natural frequency is given by the formula

$$f_0 = \frac{1}{2\pi\sqrt{\mu_0\frac{\pi r_1^2}{l_1}\cdot\frac{\varepsilon_0\cdot l_1}{4\pi}}} = \frac{1}{\pi\cdot r_1\sqrt{\mu_0\varepsilon_0}} = \frac{c_0}{\pi\cdot r_1},$$

which is close to the formula for the basic mode.

Fig. 6.49 Connection of
magnetic and electrical
quantities via a gyrator

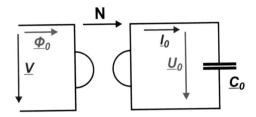

Despite Fig. 6.49, a loss-free parallel resonant circuit is assumed as a simplified model for the vacuum component. In analogy, the core-related part can be represented by a further parallel resonant circuit. Analogously, the associated elements l_1, C_1 and R_1 receive the values

$$L_1 = \frac{(\mu_r - 1)\mu_0 \cdot \pi r_1^2}{l_1}, C_1 = \frac{(\varepsilon_r - 1)\varepsilon_0 \cdot l_1}{4\pi} \text{ and } R_1 = \frac{4\pi}{\sigma l_1}.$$

The natural resonant frequency is accordingly

$$f_{mnp} = \frac{c_0}{2\pi \sqrt{(\mu_r - 1)(\varepsilon_r - 1)}} \sqrt{\left(\frac{X_{mn}}{r}\right)^2 + \left(\frac{p\pi}{h}\right)^2}$$

For ferrites, a rough estimate can be made $\mu_r \approx 1000$. For MnZn ferrites [9] the relative dielectric constant, frequency-dependent (0.1 MHz...100 MHz), indicates a range of $\varepsilon_r = 200{,}000...10{,}000$. According to this information, the range of NiZn ferrites lies in the range 100...12. Accordingly, a natural resonant frequency of a core section with $r_1 = 10$ cm would lie in the range 81 kHz...360 kHz. For NiZn ferrites one would accordingly obtain a natural resonance range of 3.6...10.5 MHz. A displacement current does not take place in metals. However, they have a conductivity σ. With the values L_1 and R_1 a cut-off frequency

$$f_{max} = \frac{R_1}{2\pi \cdot L_1} = \frac{2}{(\mu_r - 1)\mu_0\sigma \cdot \pi r_1^2}$$

The cut-off frequency thus becomes larger in inverse proportion to the conductivity and depends quadratically on the sheet thickness r_1. This results in the equivalent circuit diagram for a core (without taking into account the skin effect due to its opposing magnetic field) in Fig. 6.50. The winding resistance R_s and the winding capacitance C_w are taken into account.

Such an arrangement was realized with PM cores and analyzed metrologically (Fig. 6.51). From the conspicuous resonances, the results compiled in Table 6.9 can be determined.

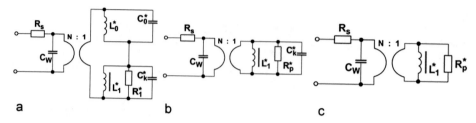

Fig. 6.50 Equivalent circuit for a winding with a lossy core without leakage field. (**a**) "complete" equivalent circuit, (**b**) neglecting the "vacuum component", (**c**) electrical model of a metallic core

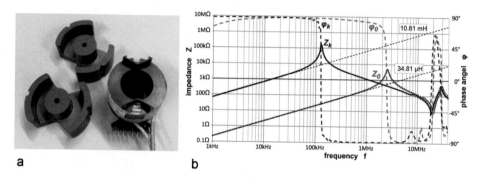

Fig. 6.51 Measured course of impedance and phase of the inductance with and without core (N = 30, core PM74/Mf102), also drawn is the course calculated with the determined substitute parameters (dotted). (**a**) Winding with core half-shells. (Source: Zacharias 2017); (**b**) Measured and recalculated impedance and phase angle as a function of frequency

Table 6.9 Measured resonant frequencies at an inductor without and with core (PM74/Mf102) according to Fig. 6.51a

Resonance type	f_{1x}	f_{2x}	f_{3x}
	Parallel	Series	Parallel
Without core (0)	2.54 MHz	20.83 MHz	29.9 MHz
With core (k)	126.5 kHz	19.09 MHz	28.5 MHz

The two resonant frequencies f_{10} and f_{1k} should behave like the root of the ratio of the two inductors with and without a core. One finds however.

$$\frac{f_{10}}{f_{1k}} = \frac{2.54}{0.1265} = 20.08 \text{ and } \sqrt{\frac{L_{1k}}{L_{10}}} = \sqrt{\frac{10.81 mH}{34.81 \mu H}} = 17.62.$$

As explained above, the deviation could be caused by the "intrinsic" capacity of the core. From f_{10}, an effective winding capacity of

$$C_w = \frac{1}{4\pi^2 \cdot (2.54MHz)^2 \cdot 34.81\mu H} = 112.79pF$$

$$C_w + C_k^* = \frac{1}{4\pi^2 \cdot (126.5kHz)^2 \cdot 10.81mH} = 146.4pF$$

Thus the intrinsic capacity of the core, extrapolated $C_k^* = 33.64pF$ to the number of turns would be 30. The intrinsic core capacity related to one turn is then

$$C_k = N^2 \cdot C_k^* = 30^2 \cdot 33.64\text{pF} = 30.3nF$$

To verify the thesis, a single-layer winding with $N = 18$ was designed. There the control measurement with the same core showed

$$C_w = \frac{1}{4\pi^2 \cdot (6.811MHz)^2 \cdot 8.148\mu H} = 67.01pF$$

$$C_w + C_k^* = \frac{1}{4\pi^2 \cdot (307, 894kHz)^2 \cdot 3,075mH} = 86.9pF$$

The capacitance C_k^* related to one turn is then 6.44 nF. The measurement was checked several times with the same result. The explanation for the difference may be that the difference in winding capacitance when the core is placed is the intrinsic core capacitance plus part of the winding capacitance, which is the operating capacitance over the core. The expected value for C_k for MnZn is in the range

$$C_k \approx \frac{l_{Fe}}{4\pi} \varepsilon_r \varepsilon_0 = \frac{128mm}{4\pi} \cdot (5...100 \cdot 10^4) \cdot 8,86\frac{pF}{m} = 100pF...2nF$$

However, this does not explain the other resonance points of the impedance characteristic Fig. 6.50. These are series and parallel resonances. If we assume that the winding capacitance cannot be represented by a single capacitance, but by two or more that only appeared as a total capacitance at the first resonance due to magnetic coupling, we obtain an equivalent circuit diagram as shown in Fig. 6.52.

In C_1 and C_2, C_{w1}, C_{w2} and C_k are combined proportionally and related to the number of turns N. The sizes L_1, L_σ, R_1 and R_2 also refer to the number of turns N. Here L_σ represents a summary of the two leakage inductance parts $0.5 \cdot L_\sigma^*$ (Γ equivalent circuit diagram) transformed to N. If the leakage inductances $L_{\sigma1}$ and $L_{\sigma2}$ are combined to L_σ, one can write $L_\sigma \ll L_{10}$ for the resonance frequencies under the assumption $_{L\sigma}$:

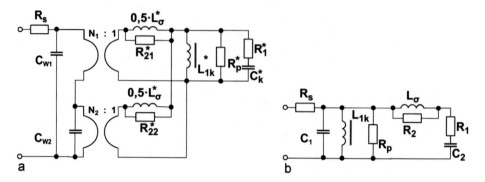

Fig. 6.52 Model for one winding. (**a**) for on a core with 2 separate winding capacities, (**b**) simplified simulation model for the frequency response

$$C_{20}L_\sigma = \frac{1}{4\pi^2 f_{20}^2} \text{ and } \frac{C_{10}C_{20}}{C_{10} + C_{20}}L_\sigma = \frac{1}{4\pi^2 f_{30}^2}.$$

That's...

$$\frac{C_{10}}{C_{10} + C_{20}} = \frac{f_{20}^2}{f_{30}^2} = \left(\frac{20.83MHz}{29.9MHz}\right)^2 = 0.485.$$

Correspondingly, analogous to this, with a built-in core

$$\frac{C_{1k}}{C_{1k} + C_{2k}} = \frac{f_{2k}^2}{f_{3k}^2} = \left(\frac{19,09MHz}{28,5MHz}\right)^2 = 0.4487.$$

This formally results in the following for the capacities sought

$$C_{10} = \frac{f_{20}^2}{4\pi^2 L_{10}f_{10}^2 f_{30}^2} = 54.74pF \text{ and analog } C_{1k} = \frac{f_{2k}^2}{4\pi^2 L_{1k}f_{1k}^2 f_{3k}^2} = 65.7pF.$$

Also, the resistances R_{px} and R_1, which can be estimated from the measurement data, are included:

$$R_{px} = Z(f_{1x}), R_1 = Z(f_{2x})$$

The resistance R_2 was determined by mathematical adjustment of the calculation result to the measured data. Table 6.10 gives an overview of the parameters determined in this way, with which the frequency behavior of the inductance up to 40 MHz in the small-signal range can be simulated.

Table 6.10 Compilation of the determined parameters of the equivalent circuit diagram Fig. 6.52a		Coreless	With core
	L_{1x}	34,81 µH	10.81 mH
	L_s	1006 µH	0.8613 µH
	C_{1x}	54.74 pF	65.7 pF
	C_{2x}	58.05 pF	80.7 pF
	$C_{1x} + C_{2x}$	112.79 pF	146.4 pF
	R_s	26.33 mΩ	26.33 mΩ
	R_1	8 Ω	8 Ω
	R_2	4200 Ω	4200 Ω
	R_p	5100 Ω	196 kΩ

At lower frequencies, the $Z(\omega)$ curves are proportional to the frequency, which results in an ascending straight line on a double logarithmic scale. The following applies to the reactance of an inductor

$$\underline{X}_L = j\omega L = j\omega \frac{\mu_r \mu_0 A}{l}$$

It is known, however, that the impedance of a real inductive component consists of a real part and an imaginary part. The measurement results are therefore interpreted via a so-called complex relative permeability μ_r, which is defined in such a way that the calculation results reflect the usual relationships. The following should therefore be applied

$$\underline{Z}_L = j\omega L + R = j\omega(\mu_r' - j\mu_r'') \frac{\mu_0 A}{l}.$$

The information on the so-called complex permeability can be found in the data sheets on ferromagnetic materials as graphics. Especially μ_r'' is strongly frequency-dependent. It is used to describe the loss properties of the core material during remagnetisation. The losses are mainly – but not only – caused by the electric field that swirls around the magnetic field. These losses are caused by the low conductivity of the core material and the lossy dielectric component of the core. A parallel connection of an electrical loss component to the inductive component would therefore be a physically more appropriate approach to describe the core losses. The eddy current losses due to the proximity effect can also be well mapped with a resistance parallel to the inductor in a component with only one winding. They occur practically transformational in the winding material. In the case of a transformer, proximity losses are best represented by resistors parallel to the leakage inductance. A series connection rather maps the ohmic winding losses of a component. Figure 6.53 shows the complex permeability of two HF ferrites in comparison.

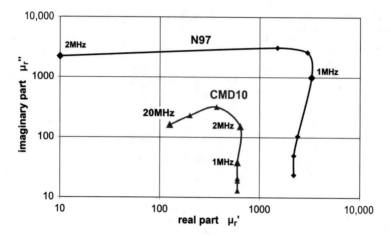

Fig. 6.53 Complex permeability of the MnZn ferrite N97 (EPCOS) and the NiZn ferrite CMD10 as relative permeability in a locus curve on a double logarithmic scale

6.10 Complex Permeability of Winding Material and Structural Parts

In principle, there are no insulators of the magnetic (direct) field. Even with diamagnetic materials, μ_r' differs only slightly from 1. According to Maxwell, in an electric field induced by an alternating magnetic field, a current field with the current density $\vec{J}(t) = \left(\sigma + \frac{d}{dt}\varepsilon\right)\vec{E}(t)$ or in complex notation $\underline{\vec{J}} = (\sigma + j\omega\varepsilon)\underline{\vec{E}}$. This means that a current flows, which in turn results in a magnetic field that counteracts its cause. This effect will be investigated for metals (i.e. $\sigma >> \omega\varepsilon$). This situation exists for many construction materials and winding materials. For quantitative estimation, the arrangement of a conductive, infinite half-space shown in Fig. 6.54 is assumed.

Due to this arrangement and $\frac{\partial H}{\partial x} = 0$ $\frac{\partial H}{\partial y} = 0$. For a sufficiently small ε_r (and thus neglecting the dielectric properties) one can formulate for the diffusion equation of the magnetic and electric field in differential or complex notation [1]

$$\Delta \vec{B} = \mu\sigma \frac{d}{dt}\vec{B} \text{ respectively } \Delta \underline{\vec{B}} = j\omega\mu\sigma \underline{\vec{B}}$$

$$\Delta \vec{E} = \mu\sigma \frac{d}{dt}\vec{E} \text{ respectively } \Delta \underline{\vec{E}} = j\omega\mu\sigma \underline{\vec{E}}$$

The general solution of this second order partial differential equation with the exponential approach is

Fig. 6.54 Conductive infinite half-space in a homogeneous alternating magnetic field

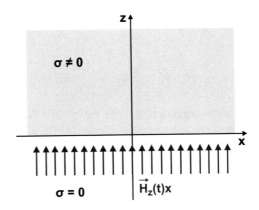

$$B_z = Ae^{\sqrt{j\omega\mu\sigma}\cdot z} + Be^{-\sqrt{j\omega\mu\sigma}\cdot z} = Ae^{\sqrt{\frac{\omega\mu\sigma}{2}}\cdot z}e^{j\sqrt{\frac{\omega\mu\sigma}{2}}\cdot z} + Be^{-\sqrt{\frac{\omega\mu\sigma}{2}}\cdot z}e^{-j\sqrt{\frac{\omega\mu\sigma}{2}}\cdot z}$$

This corresponds to the equation of 2 damped waves propagating in opposite directions. For $z \to \infty$ $B_z \to 0$ must apply and $z = 0$ for $B_z(0) = \mu \cdot H_z(0) = \mu \cdot H_0$. Thus one obtains

$$B_z = \mu H_0 e^{-\sqrt{\frac{\omega\mu\sigma}{2}}\cdot z}e^{-j\sqrt{\frac{\omega\mu\sigma}{2}}\cdot z}$$

If one considers that $rot\,\vec{E} = -\frac{d}{dt}\vec{B}$ or $rot\,\underline{E} = -j\omega\underline{B}$, one obtains for the components of the E-field

$$rot\,\underline{E} = \det \begin{bmatrix} \vec{e}_x & \frac{\partial}{\partial x} & \underline{E}_x \\[2mm] \vec{e}_y & \frac{\partial}{\partial y} & \underline{E}_y \\[2mm] \vec{e}_z & \frac{\partial}{\partial z} & \underline{E}_z \end{bmatrix}$$

$$= \left(\frac{\partial}{\partial y}\underline{E}_z - \frac{\partial}{\partial z}\underline{E}_y\right)\vec{e}_x + \left(\frac{\partial}{\partial z}\underline{E}_x - \frac{\partial}{\partial x}\underline{E}_z\right)\vec{e}_y + \left(\frac{\partial}{\partial x}\underline{E}_y - \frac{\partial}{\partial y}\underline{E}_x\right)\vec{e}_z$$

$$rot\,\underline{E} = \left(\frac{\partial}{\partial x}\underline{E}_y - \frac{\partial}{\partial y}\underline{E}_x\right)\vec{e}_z = -j\omega\underline{B} = -j\omega\mu\underline{H} = -j\omega\mu\underline{H}_z$$

$$\underline{E}(z) = E_0 e^{-\sqrt{\frac{\omega\mu\sigma}{2}}\cdot z}e^{-j\sqrt{\frac{\omega\mu\sigma}{2}}\cdot z}$$

From the course of the electric field strength, the electric current density and thus the power loss density can be determined

$$\frac{dP}{dV} = \sigma E_0^2 \cdot e^{-2\sqrt{\frac{\omega\mu\sigma}{2}}\cdot z}$$

$$P = \int_V \sigma E_0^2 \cdot e^{-2\sqrt{\frac{\omega\mu\sigma}{2}}} z\, dx\, dy\, dz = x \cdot y \cdot \frac{-\sigma E_0^2}{2\sqrt{\frac{\omega\mu\sigma}{2}}} \cdot e^{-2\sqrt{\frac{\omega\mu\sigma}{2}}z}\Big|_0^\infty = x \cdot y \cdot \frac{-\sigma E_0^2}{\sqrt{2\omega\mu\sigma}}$$

Since the magnetic and not the electric field strength was given at the beginning, this dependence still has to be mapped. For this purpose, the law of induction in its differential form is used. If you look at a magnetic field line on the z-axis, it is swirled around by an E-field in the xy-plane. Especially when y = 0, the component E_x becomes maximum and equal to the magnitude of the electric field strength.

$$\frac{\partial}{\partial z}E_x = -j\omega\mu H_z$$

$$\frac{\partial}{\partial z}E_x = -j\omega\mu H_y = \frac{\partial}{\partial z}\left(E_0 e^{-\sqrt{\frac{\omega\mu\sigma}{2}}\cdot z}e^{-j\sqrt{\frac{\omega\mu\sigma}{2}}\cdot z}\right) = -E_0(1+j)\sqrt{\frac{\omega\mu\sigma}{2}}e^{-(1+j)\sqrt{\frac{\omega\mu\sigma}{2}}\cdot z}$$

$$H_0 = \frac{E_0(1+j)\sqrt{\frac{\sigma}{2\omega\mu}}}{j} = \frac{1+j}{j}E_0\sqrt{\frac{\sigma}{2\omega\mu}}$$

$$H_0 = E_0\sqrt{\frac{\sigma}{\omega\mu}}$$

$$E_0^2 = H_0^2\frac{\omega\mu}{\sigma}$$

$$P = x \cdot y \cdot \sqrt{\frac{\omega\mu}{2\sigma}}H_0^2$$

The converted power loss consequently increases with frequency and permeability and decreases with the conductivity of the material. Ferromagnetic parts in the area of the leakage field of a transformer can therefore be quickly heated up to form red-hot parts or partial areas. On the other hand, it is worthwhile to silver-plate the surfaces of conductive parts (windings, wall surfaces of waveguides or cavity resonators) to achieve low damping losses at high frequencies. It is interesting to note that the electromagnetic field in the conductor material decreases rapidly:

$$B(z) = B_o e^{-\frac{z}{s}} \text{ with } s = \sqrt{\frac{2}{\omega\sigma\mu}}.$$

The power loss density decreases proportionally to the square of the electric field strength. The resulting layer thickness for power conversion is therefore only 0.5 s. If you want more than 95% of the possible power to be absorbed in a layer thickness d, you have to change this to

$$d > \frac{3}{2}s = 3\sqrt{\frac{1}{2\omega\sigma\mu}}$$

select. Another approach is to use the material, for example, to attenuate/shield interfering electromagnetic waves. For this purpose, the attenuation of the magnetic or electric field strength by the factor 'k' can be written

$$\frac{E(d)}{E_0} = e^{\frac{-d}{s}} = k$$

Consequently, the material thickness must be $d = -s \cdot \ln(k)$ selected for this purpose.

The effects of both magnetic and electric fields occur in a material. In the following, it is investigated how these effects can be represented together. It is assumed that in a cylindrical material a magnetic field of strength 'H' propagates homogeneously in the axial direction. According to the law of induction, an electric field whirls around it.

The following relations apply

$$H \cdot l = B \cdot A \text{ or and } V = \left(\frac{l}{\mu A}\right) \cdot \Phi = R_m \cdot \Phi = \frac{1}{\Lambda} \cdot \Phi$$

$$\oint E \cdot ds = \frac{d}{dt} B \cdot A$$

$$\underline{Z_L} = \frac{U_0}{I_0} = N^2 \cdot \left(\frac{j\omega\Lambda}{1 + j\omega\frac{L}{R}}\right) = j\omega N^2 \cdot \left(\frac{\Lambda\left(1 - j\omega\frac{\Lambda}{R^*}\right)}{1 + \omega^2\frac{\Lambda}{R^*}}\right)$$

$$= j\omega N^2 \cdot \left(\frac{\frac{\mu_r\mu_0\pi R_1^2}{l_1}\left(1 - j\omega\frac{\mu_r\mu_0\sigma\pi R_1^2}{4\pi}\right)}{1 + \omega^2\left(\frac{\mu_r\mu_0\sigma\pi R_1^2}{4\pi}\right)^2}\right)$$

From this, one can determine the resistance R*.

$$R^* = \frac{U^*}{\int_A J \cdot dA} = \frac{\oint E \cdot ds}{\int_A \sigma E \cdot dA} = \frac{j\omega B \cdot \pi R_1^2}{\int_0^{R1} \sigma\frac{j\omega B \cdot \pi r^2}{2\pi r} \cdot l_1 \cdot dr} = \frac{\pi R_1^2}{\int_0^{R1} \sigma\frac{r}{2} \cdot l_1 \cdot dr} = \frac{\pi R_1^2}{\sigma\frac{R_1^2}{4} \cdot l_1}$$

$$= \frac{4\pi}{\sigma \cdot l_1} \cdot$$

Electric and magnetic field quantities are only connected via Maxwell's equations. The voltage U* corresponds to the voltage U_0 converted to the number of turns $N = 1$. This means that the interaction can also be represented by a parallel connection of Λ and R. The impedance of the component then results in

$$\underline{Z}_L = \frac{\underline{U}_0}{\underline{I}_0} = N^2 \cdot \left(\frac{j\omega\Lambda}{1 + j\omega\frac{L}{R^*}} \right) = j\omega N^2 \cdot \left(\frac{\frac{\mu_r\mu_0\pi R_1^2}{l_1}\left(1 - j\omega\frac{\mu_r\mu_0\sigma\pi R_1^2}{4\pi}\right)}{1 + \omega^2\left(\frac{\mu_r\mu_0\sigma\pi R_1^2}{4\pi}\right)^2} \right)$$

$$\underline{Z}_L = j\omega N^2 \frac{\mu_0\pi R_1^2}{l_1} \cdot (\mu' - j\mu'') = j\omega N^2 \frac{\mu_0\pi R_1^2}{l_1} \cdot \underline{\mu}_r$$

For the quality factor Q, the following applies

$$Q = \frac{\omega \cdot \mu'\mu_0 \frac{\pi R_1^2}{l_1}}{\omega \cdot \mu''\mu_0 \frac{\pi r_1^2}{l_1}} = \frac{\mu'}{\mu''} = \frac{4}{\omega\mu_0\sigma R_1^2}$$

$$\omega Q = \omega \frac{\mu'}{\mu''} = \frac{4}{\mu_0\sigma R_1^2}$$

By means of this representation, a characteristic value for the comparison of core materials can be generated from the material data (Fig. 6.55). The metallic material Vitroperm VP500 corresponds to the assumptions of the metallic cable. Up to a frequency of approximately 30 kHz, the characteristic value μ_r is constant and then increases. It can be assumed that both the electric current field and the magnetic field are forced out of the 17–25 µm thick

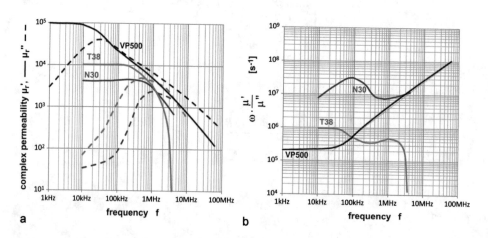

Fig. 6.55 Comparison of real part μ_r' and imaginary part μ_r'' of the permeability of 3 materials. (**a**) Frequency dependence of real part μ_r' and imaginary part μ_r'', (**b**) Ratio of μ_r' and μ_r'' multiplied by the "correction factor" ω [metallic: $\mu_r = 100{,}000$, $\sigma \sim 8.7 \cdot 10^5$ S/m, [9], ferritic: T38, N30)

strips by the eddy currents generated. The characteristic value increases, which corresponds to increased resistance in the parallel equivalent circuit diagram. The reason for this could be the displacement of the magnetic field to the surface of the core strips due to the skin effect. The equivalent curves for the MnZn ferrites T38 and N30 have different behaviour. This indicates that the nature of the losses is different here. The conductivity of the material is only one of several loss sources here. Another important one is the dielectric losses of the ferrite.

When measuring the impedance, the available information U, I and φ can be interpreted as a series or parallel equivalent circuit diagram. For the impedance this means

$$\underline{Z} = \frac{U}{I} = R_s + j\omega L_s = \frac{R_p \cdot j\omega L_p}{R_p + j\omega L_p}$$

By comparing the real and imaginary parts, both equivalent circuit diagrams can be converted into each other for a certain frequency. If the structure of the equivalent circuit diagrams corresponds to physical reality, they can also be used as an approximation in a certain frequency range. Then you can even use equivalent circuit diagrams for simulations. This also means that the modelling with FEM calculations should be carried out most sensibly with definitions of the complex permeability that reflect reality over a large frequency range. If necessary, the large signal properties can be approximated by amplitude-dependent resistors and inductors, to get results closer to reality.

The ratio of the imaginary and real impedance is called quality 'Q'. Therefore the following applies

$$Q = \frac{\text{Im}\{\underline{Z}\}}{\text{Re}\,\{\underline{Z}\}} = \tan{(\varphi)}$$

An inductive component is expected to have an imaginary part of the impedance that is much larger than the real part. Nevertheless, Q values of $Q > 100$ are difficult to achieve in practice. As an illustration, Fig. 6.56 shows the Q(f) curve of the magnetizing inductance of two transformers made of amorphous C-section tape-wound cores in comparison to an ordinary power transformer made of a toroidal tape-wound core of dynamo sheet metal. The resonance points are the result of parallel resonance with the winding capacitances. The maximum quality values Q(f) of the primary winding of the power transformer are 2. As Fig. 6.57 shows, it quickly loses its desired properties at higher frequencies compared to a transformer with an AMCC core.

Not only the core material of an inductive component is permeated by a magnetic field. It must be considered that there are practically no magnetic "non-conductors". The field of a ring-shaped tubular winding is formed in such a way that there is hardly any difference in its shape with or without a core. However, here too, the magnetic field permeates the winding and any core. However, the field lines in the winding are very long in the case of a toroidal core. This means that eddy current losses due to the proximity effect are very low.

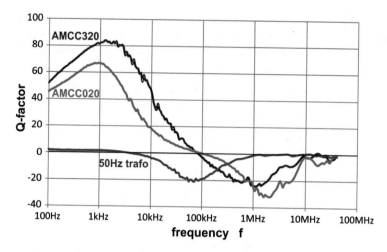

Fig. 6.56 Q(f) curve of the magnetizing inductance of 2 transformers made of amorphous C-section stripwound cores compared to a toroidal transformer made of dynamo sheet metal

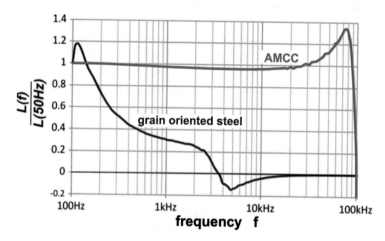

Fig. 6.57 Course of the measured inductance in the no-load operation of transformers with simple dynamo sheet and amorphous material

Outside the ring-shaped component, magnetic fields are usually negligible. They are mainly caused by deviations in geometry from ideally symmetrical conditions.

If an air gap is inserted into a toroidal core, the conditions change. In the area around the air gap, the field lines emerge more strongly and penetrate the environment and thus also the winding material. By dividing an air gap into many small air gaps, as is done in powder magnetic cores, these leakage fields can be geometrically minimized. Cores with a different amount of powder thus act like homogeneous cores with different permeabilities. The proximity effect then plays a subordinate role here as well. With elongated cores, a similar effect can be achieved by dividing an air gap to be inserted into several series-connected cores. This reduces the field strength and the range of the leakage fields. The eddy current losses in the affected winding parts are simultaneously reduced. Eddy current losses can otherwise only be countered by using insulated conductors with a small cross-section. In contrast to the skin effect, this measure also works without stranding the individual wires. In any case, however, winding space is lost due to a lower filling factor.

The leakage fields become stronger and stronger the more a core form deviates from the "natural" form of the magnetic field. In transformers, the magnetic field of a short-circuited winding is "forced" out of this winding and then permeates both the space between the windings and the windings themselves (see Sect. 6.4). The leakage field strengths may be considerable. Even if toroidal cores are used, far-reaching leakage fields can be generated in this way if the toroidal core is wound in segments. The effect is correspondingly strong in grooves in motors. As an example, a bar with a rectangular cross-section in a rotor of an asynchronous machine with a squirrel-cage rotor is considered here (Fig. 6.57). The rotor is the moving part of a transformer. The rotating field of the stator also passes through the rotor. If the rotor has the same rotational frequency as the stator field (synchronism), the frequency of the magnetic field in the rotor is practically zero, since the coordinate system of the rotor rotates synchronously with the stator field. Current forces can only occur when there is a speed difference between mechanical rotor rotation and the stator rotating field frequency. In a standardized form this is called slip 's':

$$s = \frac{\Omega_{stator} - \Omega_{rotor}}{\Omega_{stator}}$$

Consequently, when an asynchronous motor is loaded, the speed drops slightly, so that the alternating magnetic field in the rotor induces a current which applies the motor forces/rotor torque. The frequency $\omega = 2\pi fr$ in the rotor is between 0 (rotor standstill, s = 1) and f_{netz} (synchronous speed, s = 0) during motor operation. The conductor bars distributed on the circumference of the rotor represent parts of the secondary winding of a transformer and are accordingly also penetrated by a magnetic field with variable frequency (see Fig. 6.58). The interaction with the magnetic field of the induced current results in a variable current density distribution J(x) depending on the frequency. Since the distribution of the Lorentz forces is also set up, as a result, the speed-torque characteristic $\Omega(M)$ can also be influenced by the shape of the rotor bars. The example is used here to show that the magnetic interactions with conductor material at sufficient conductor cross section are already significant even at low frequencies.

To simulate this effect of current displacement, a cascade of elements as shown in Fig. 6.59 can be used. In analogy to Sect. 6.2, one can calculate a propagation coefficient

$$\underline{\gamma_i} = \sqrt{j\omega L' \cdot G'} = \sqrt{\omega L' \cdot G'} \cdot e^{j\frac{\pi}{4}} = \sqrt{\frac{\omega L' \cdot G'}{2}}(1+j)$$

determine. The frequency-dependent propagation impedance.

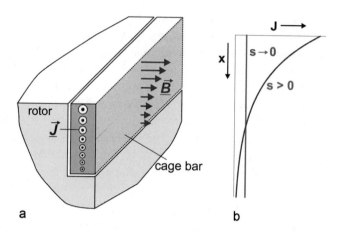

Fig. 6.58 Conductor bar in the squirrel cage rotor of an asynchronous machine. (**a**) with a magnetic field and electric current field at slip s ≠ 0, (**b**) qualitative course of the current density J(x) during operation close to synchronism (s = 0) and at the output of mechanical power (s > 0)

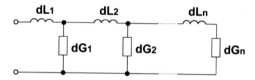

Fig. 6.59 Equivalent electrical circuit diagram to simulate the current distribution in a rotor bar at variable effective frequency $f = s \cdot f_0$

$$Z_i = \sqrt{\frac{j\omega L'}{G'}} = \sqrt{\frac{\omega L'}{2G'}}(1+j) \text{ respectively } \underline{Y}_i = \frac{1}{\underline{Z}_i} = \sqrt{\frac{G'}{j\omega L'}} = \sqrt{\frac{G'}{2\omega L'}}(1-j).$$

which changes $\sim \sqrt{\omega^{-1}}$ with it, which reflects the corresponding proportional penetration depth of the electric current. A discretization as a cascaded conductor, as shown in Fig. 6.58 with different conductance values dG_i also allows the representation of different rod shapes.

By their very nature, eddy currents are opposite in direction to their cause (Lenz's rule). An alternating magnetic field can therefore only penetrate an electrical conductor to a limited extent. The amplitude sounds according to.

$$\frac{H(d)}{H(0)} = e^{-\frac{d}{s}} \text{ with me } s = \sqrt{\frac{2}{\omega\sigma\mu}} = \frac{1}{\sqrt{\pi f \sigma \mu}}.$$

Here f: frequency, σ: conductivity and μ: permeability. A graphic representation of this is shown in Fig. 6.60. In this way, electromagnetic alternating fields can be shielded very

Fig. 6.60 Conductor material
as a shield for electromagnetic
alternating fields

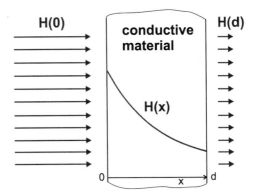

effectively with metallic covers. In the case of strong fields, the associated heat generation must be taken into account. With good conductors such as copper, the penetration depth 's' is low. Therefore the required weights for this are usually low. Ferromagnetic materials should only be used for shielding in weak fields, otherwise, they can produce considerable losses/heat.

The materials considered above, permeated by the alternating magnetic field, are isotropic. This means that their properties are not dependent on the direction in space. Neither the components of the magnetic field nor those of the electric field find different propagation conditions in a homogeneous solid. This changes immediately when the solid is structured. Figure 6.61 shows typical arrangements as found in the windings of an inductive component. The magnetic field shown in blue is perpendicular to the direction of current conduction and thus to one of the geometric axes of a conductor. The electric field therefore only has the same propagation conditions in the direction of this axis as in a homogeneous body. In the case of the round wires Fig. 6.61a,b, the other two spatial directions are limited by the diameter of the wires. An induced electric vortex field is therefore strongly restricted in one dimension. Figure 6.61a and b differ in the packing density of the wires so that in the same volume element invariant (b) more conductor material is influenced by the magnetic field. In Fig. 6.61c foils are used. Here only one spatial direction is influenced by the thickness of the foils insulated from each other. It is obvious that in all the cases shown, the interactions of the material will be quite different if the direction of the magnetic field is parallel to the direction of the current. With round wires, for example, the eddy current fields are then symmetrically limited by the wire cross section. In foil winding, elliptically shaped eddy current fields are obtained.

The magnetic fields which penetrate the winding material can have any direction in space. The different influence in the 3 directions in space can be represented by a tensor in relation $B = B(H)$:

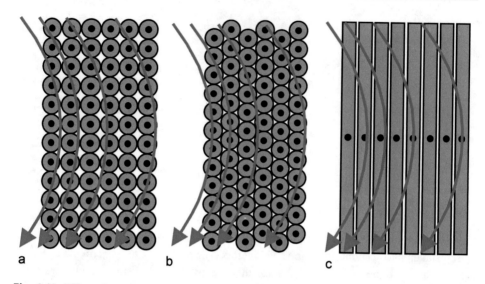

Fig. 6.61 Effect of magnetic fields in different winding constructions. (**a, b**) Winding construction with round wire, (**c**) with foils

$$\vec{B} = \begin{pmatrix} \mu_{rxx} & 0 & 0 \\ 0 & \mu_{ryy} & 0 \\ 0 & 0 & \mu_{rzz} \end{pmatrix} \cdot \mu_0 \cdot \vec{H}$$

In this equation, the properties are different exactly in the directions of the selected coordinate system. If these (main) directions do not match, a corresponding transformation must be performed.

Figure 6.62a shows a schematic diagram of a test arrangement with which winding material in the transverse field is to be investigated. For this purpose 2 U-cores with the air gap are used. In this air gap, only one insulating material is inserted to set the distance, and the material to be investigated is inserted in the other. The yoke leakage is done via outer space. In the case of the conductor material introduced into the air gap, transverse to the direction of the magnetic field, the effect of the eddy current losses of the material is added to the effect of the void space. The corresponding impedances are therefore in series in the electrical equivalent circuit diagram Fig. 6.62b.

It is a symmetrical structure. That is, it's $\underline{Z}_{\mu 1} = \underline{Z}_{\mu 2} = \underline{Z}_{\mu 3} = \underline{Z}_{\mu 4}$, and $\underline{Z}_{\sigma 1} = \underline{Z}_{\sigma 2} \underline{Z}_{\delta 1} = \underline{Z}_{\delta 2}$. For this purpose, the same number of turns is applied to each leg. The windings are connected in series in the same direction. If there is no air gap, the impedance of the windings can be adjusted at the terminals of the windings.

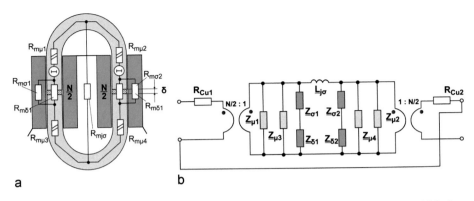

Fig. 6.62 Arrangement for the metrological determination of the complex permeability. (**a**) Mechanical construction and magnetic equivalent circuit diagram; (**b**) Electrical equivalent circuit diagram for the determination of the parameters (see text)

$$\underline{Z}_1(\omega) = R_{Cu1} + R_{Cu2} + N^2 \cdot \frac{\underline{Z}_{\mu1}(\omega)}{4}$$

measure. If you now insert an air gap δ, you measure the impedance at the terminals

$$\underline{Z}_2(\omega, \delta) = R_{Cu1} + R_{Cu2} + N^2 \cdot \frac{\underline{Z}_{\mu1}(\omega)}{4} \, // \left(\frac{\underline{Z}_{\sigma1}(\omega, \delta)}{2} + \frac{\underline{Z}_{\delta1}(\omega, \delta)}{2} \right).$$

This is $\underline{Z}_{\delta1}(\omega, \delta) = j\omega \cdot \frac{\mu_0 \cdot A_{Fe}}{\delta}$ so that it can be $\underline{Z}_{\sigma1}(\omega, \delta)$ determined because the medium in the air gap is simply air. If the structure is sufficiently symmetrical, $L_{j\sigma}$ is not important, because no current will flow in this element. If the air gap is filled with a sample of the winding material, the impedance curve can be $\underline{Z}_{\delta1}^*(\omega, \delta)$ determined by changing the following equation. This can then be used to determine the complex permeability with

$$\underline{\mu}_r(\omega) = \frac{\underline{Z}_{\delta1}^*(\omega, \delta) \cdot \delta}{j\omega \cdot \mu_0 \cdot A_{Fe}} .$$

However, there are limits to the practical application of these considerations. The measurements concern both ferromagnetic and paramagnetic components. The amplitudes or rms values of the current and voltage of the two poles and the phase shift between the two quantities $u_z(t)$ and $i_z(t)$ are available as measured quantities for determining Z. Both quantities are partly non-linearly dependent on the modulation (i.e. they are affected by harmonics) and are noisy. This can lead to considerable deviations caused by measurement errors. In order to at least show the effect of such an approach, the setup shown in Fig. 6.62 was implemented and observations were made on the change in impedance Z(ω) during

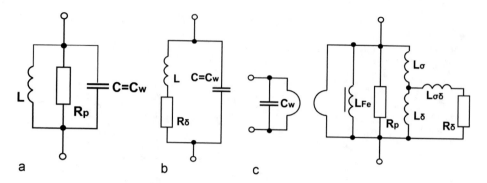

Fig. 6.63 Winding models with varying degrees of detail. (**a**) Pure parallel resonant circuit with X_L// R//X_C. (**b**) Parallel resonant circuit with $(X_L + R)$//X_C, (**c**) approximate illustration of the conditions with winding material in the air gap without ohmic resistance of the winding

measurements on the winding with HF-Litz wire on a CC core of amorphous material with an air gap. For the interpretation, the correlations shown in Fig. 6.63 are used.

The normalized Z-position curves of the arrangement without material in the air gap (Fig. 6.64a) form almost perfect circles through the coordinate origin after subtraction of the winding resistance $R_{Cu} = R_{Cu1} + R_{Cu2}$. The center of these circles is on the real axis. The resonance frequency of a pure parallel resonance depends practically only on the air gap. The 45° frequencies $f_{45°}$, that is, the frequencies at which the phase shift between current and voltage $\varphi = \varphi_u - \varphi_i = 45°$ or $Q = 1$, are in this case symmetrical to the calculated frequency of the parallel resonance f_p with

$$f_p = \frac{1}{2\pi\sqrt{L \cdot C}}$$

The following applies to the amount of the impedance

$$|\underline{Z}(\omega_{45°})| = \frac{Z(\omega_p)}{\sqrt{2}} = \frac{Z(2\pi f_p)}{\sqrt{2}},$$

while because of $Q = 1$ applies

$$R_p = \omega_{45°} \cdot L - \frac{1}{\omega_{45°} \cdot C}.$$

The corresponding equivalent circuit diagram is shown in Fig. 6.63a. It corresponds to the further analysis of Fig. 6.62b assuming $L_{\mu x} > > L_{\sigma x}$ without material in the air gap.

If a conductive material is inserted into the air gap, eddy currents are coupled like in a transformer, which weaken the impressed magnetic field with their magnetic field. As in the case of a transformer with a connected load resistor, this situation can be represented by

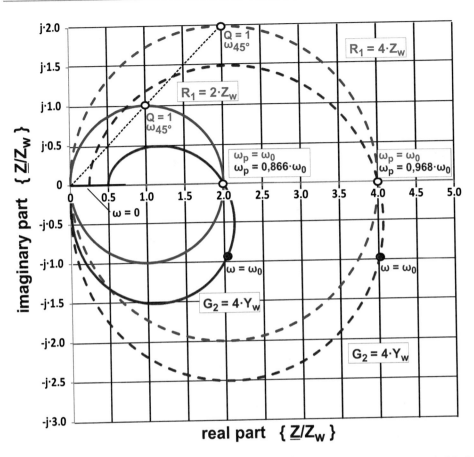

Fig. 6.64 Comparison of the calculated Z-location curves of a "real" parallel resonant circuit (blue) with a parallel resonant circuit with a resistor $R_2 = R_\delta$ in series (red)

an equivalent circuit, which is shown in Fig. 6.62c. The mutual inductance of the two coupled magnetic fields leads via the corresponding coupling factor to a "leakage inductance" $L_{\mu\delta}$. The inductance of the pure air gap is the magnetizing inductance of the transformer $L_{\mu\delta}$. The electrical conductivity of the material leads to a loss resistance R_δ, which represents the losses in the introduced material. If R_δ is sufficiently small, there is no pure parallel resonance. The resonance inductance L is formed from the inductance of the magnetic field around the air gap plus the "leakage inductance" $L_{\sigma\delta}$ of the air gap. The different behaviour of the impedance at variable frequency is shown in Fig. 6.64. The resonance at $\sphericalangle\left(\underline{U}_Z, \underline{I}_Z\right) = 0$ shifts to lower values than ω_0 with the "pure" parallel resonance circuit:

$$\omega_p{}^2 = \frac{1}{L^2C}(L - CR_\delta^2) = \frac{1}{LC}\left(1 - \frac{CR_\delta^2}{L}\right) = \omega_0^2\left(1 - \frac{R_\delta^2}{Z_W^2}\right)$$

with $\omega_0^2 = \frac{1}{LC}$, and $L = L_\sigma + L_{\sigma\delta}$ $Z_W^2 = \frac{L}{C}$.

The imaginary component of $Z(\omega_p)$ disappears so that at this frequency the following is measured

$$\text{Re}\left\{\underline{Z}(\omega_p)\right\} = R^* = \frac{1}{R_\delta^2 + (\omega_p L)^2} \cdot \frac{L^2}{R_\delta C^2} \approx \frac{1}{(\omega_p L)^2 R_\delta} \cdot \frac{L^2}{C^2} \Rightarrow R_\delta = \frac{1}{\omega_p^2 C^2 R^*}$$

The approximation applies if the resistor R_δ is small enough. That means the resistor R_δ is determined with C. C must therefore be known. Since C can be determined as winding capacitance from other measurements (or can also be connected in parallel from outside as a capacitor to change the measuring frequency), R_δ can therefore also be measured not only at one but also at several frequencies. The connected capacitors serve as reference components.

The inductance then results in

$$L = \frac{1}{2\omega_p^2 C} + \sqrt{\frac{1}{4\omega_p^4 C^2} - \frac{R_\delta^2}{\omega_p^2}} \approx \frac{1}{\omega_p^2 C}$$

The analytical representation of the problem repeatedly falls back on approximations for simplification. A general closed solution for determining the parameters does not seem possible. However, numerical algorithms can be derived from these approaches in connection with experiments for the approximate calculation of the parameters.

One starts from the equivalent circuit Fig. 6.63c This is obtained for $\omega_p = 2\pi f_p$ from Fig. 6.61b, if the complex conductance values of $Z_{\mu x}$ complement each other to zero with the connected capacitance or the winding capacitance. Then the value sought from R_δ or the imaginary part of the permeability for the direction used in Fig. 6.61a can be determined step by step.

The inductance $L_0(\delta = 0) = 57.8$ mH was measured. Assuming this as a constant value even at frequencies near 300 kHz, measuring the inductance $L_1 = 3.94$ mH at $\delta = 2 \cdot 2.7$ mm without inserted material for the equivalent circuit Fig. 6.63c gives an inductance value of

$$L^* = \frac{L_0 \cdot L_1}{L_0 - L_1} = 4.23 mH$$

for a pure parallel connection. Known is now the inductance of the air gap L_δ without filling material as

$$L_\delta = \frac{\mu_0 \cdot \Delta x \cdot \Delta y}{2 \cdot \delta} \cdot N^2 = \frac{4\pi \cdot 10^{-7} H/m \cdot 22mm \cdot 50mm}{2 \cdot 2.7mm} \cdot 108^2 = 2.986mH.$$

For an interpretation as a series connection, as in Fig. 6.63c, the inductance $L_\sigma = L^* - L_\delta = 4.23$ mH $- 2.986$ mH $= 1.242$ mH is thus determined. The parallel resistance of an equivalent parallel equivalent circuit determined for the measurement with the air gap is $R^* = 76.29$ kΩ. If the material sample of the winding material is now inserted into the air gap, the resonant frequency f_p and $Z(f_p) = R_{vp} = 20.94$ kΩ. This is further interpreted in the following with the equivalent circuit diagram Fig. 6.63c. The measured effective inductance at the resonant frequency f_p is then 3.45 mH. This gives the inductance for the coreless component

$$L^{**} = \frac{L_0 \cdot 3.45mH}{L_0 - 3.45mH} = 3.67mH$$

Accordingly, for the interpretation as a pure parallel resonant circuit, the apparent parallel resistance is obtained from the measurement

$$R^{**} = \frac{R^* \cdot R_{vp}}{R^* - R_{vp}} = \frac{76.29k\Omega \cdot 20.94k\Omega}{76.29k\Omega - 20.94k\Omega} = 28.86k\Omega$$

For the further procedure, it is best to use the conversion of a series connection into an equivalent parallel connection and vice versa in succession for the frequency of the parallel resonance $f_p = 291.6$ kHz. The conversion formulas are summarized in Table 6.11.

For the conversion of R^{**} into an equivalent series resistance, one then obtains

$$R_s = 28.86k\Omega \cdot \frac{(449\Omega)^2}{(28.86k\Omega)^2 + (449\Omega)^2} = 670\Omega$$

and for the inductance accordingly

Table 6.11 Conversion of elements of a series connection into elements of a parallel connection with equal phase shift and impedance at a certain frequency		Series connection	Parallel connection
	Circuit	R_s L_s	R_p L_p
	Resistance	$R_s = R_p \cdot \dfrac{(\omega L_p)^2}{R_p^2 + (\omega L_p)^2}$	$R_p = R_s \cdot \dfrac{R_s^2 + (\omega L_s)^2}{(\omega L_s)^2}$
	Inductance	$L_s = L_p \cdot \dfrac{R_p^2}{R_p^2 + (\omega L_p)^2}$	$L_p = L_s \cdot \dfrac{R_s^2 + (\omega L_s)^2}{R_s^2}$

$$L_s = (3.67mH - 1.242mH) \cdot \frac{(28.86k\Omega)^2}{(28.86k\Omega)^2 + (4449\Omega)^2} = 2.37mH$$

From this inductance, $L_\sigma = 1.242$ mH is to be subtracted, so that the resulting inductance of a series connection of the inductive part and ohmic part $L_\delta^* = 2.24$ mH. This gives the impedance of the air gap with filling material:

$$\underline{Z}_{\delta.N=1} = \frac{670\Omega + j4345\Omega}{108^2} = 0.05742\Omega + j0.3725\Omega = j\omega\underline{L}$$

$$\underline{L} = \frac{0.05742\Omega}{j\omega} + \frac{0.3725\Omega}{\omega} = (2.033 \cdot 10^2 - j3.134 \cdot 10^{-8})H = \underline{\mu}_r \cdot \mu_0 \cdot \frac{22mm \times 50mm}{2 \cdot 2.7mm}$$

$$\underline{\mu}_r = \frac{(2.033 \cdot 10^2 - j3.134 \cdot 10^{-8})H}{4\pi \cdot 10^{-7}\frac{H}{m} \cdot \frac{22mm \cdot 50mm}{2 \cdot 2.7mm}} = 0.7943 - j0.1224$$

In this way, the complex permeability is formally determined according to its definition. However, the losses in the inserted material depend on the voltage applied to the inductor. This means that, from a physical point of view, a parallel connection of the loss resistance makes sense. However, an approach for further analysis until L_δ, $L_{\delta\sigma}$ and R_δ could not be found. If the air gap filled with winding material is analysed in more detail, eddy current losses occur in it, which are dependent on the induced voltage. This corresponds to a loss resistance R_δ'' parallel to the inductance. From the relationship for the power loss density in the electromagnetic field derived in subchapter

$$P = x \cdot y \cdot \sqrt{\frac{\omega\mu}{2\sigma}}H_0^2$$

the effective conductivity for the penetration of the magnetic field through the air gap can be estimated to

$$\sigma(\omega_p) \cong (x \cdot y)^2 \frac{\omega_p\mu}{2} \cdot \frac{H_0^4}{P^2} \approx (x \cdot y)^2 \cdot \frac{\omega_p\mu}{2} \cdot \frac{\left(\frac{I \cdot N}{2\delta}\right)^4}{(I \cdot N)^4 R_\delta'^2} = \frac{(x \cdot y)^2}{16 \cdot \delta^4} \cdot \frac{\omega_p\mu}{2R_\delta''^2}$$

$$\sigma(\omega_p) \approx \frac{(22 \cdot 50)^2}{16 \cdot 2.7^4} \cdot \frac{2\pi \cdot 291600s^{-1} \cdot 4\pi \cdot 10^{-7}\frac{Vs}{Am}}{2 \cdot 0.058784^2\Omega^2} = 474060\frac{S}{m}$$

According to the assumptions made before, this is the ("effective") conductivity in the xy-plane, when the H-vector points in the z-direction. Assuming a single conductivity in all spatial directions leads to unusable results in total. In the direction of the wire axis, one has a very high conductivity. Transversely to this, there is a spatially very restricted area that can only be traversed by induced currents. The driving electromotive forces act in the plane

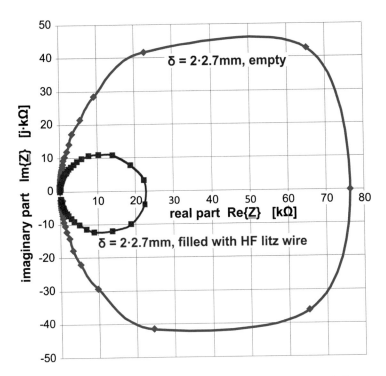

Fig. 6.65 Measured impedance locus curve of an inductance, formed by 2 windings connected in series and a total of 108 turns with 2 air gaps (2.7 mm × 22 mm × 50 mm) unfilled and filled with HF-litz wire 300 × 0.35 CuL wire

with the surface vector in the direction of the magnetic field. This has little to do with the design equation of the magnetic resistance and can only be solved with finely resolved FEM calculations. Figure 6.65 shows the results of the performed measurements of the impedance locus curve $Z(j\omega)$ in graphical form. The corresponding evaluations are summarized in Table 6.12.

With the concept of complex permeability, one can characterize volume units for the propagation of the electric field. This can be used in FEM calculations for the electric field in winding materials. Otherwise, a very large number of meshes is required to map the problems which are only interesting in sum. The derivation carried out above shows that this is a frequency-dependent variable $\mu_r(\omega)$.

Table 6.12 Results of the evaluation of measured locus curves $Z(\omega)$ in Fig. 6.65

	Air gap 2 × (2.7 × 22 × 50) not filled	Air gap 2 × (2.7 × 22 × 50) filled with 2 × (300 × 0.35 CuL × 50 mm) Cu wires
Number of turns	108	
$L(\delta) = 0$	57.8 mH	
Effective capacity C	88.6 pF	
Air gap inductance L_0 2 × (2,7 mm × 22 mm × 50 mm)	$L_0 = 256nH \cdot N^2 = 2.986mH$	
The inductance of the boundary field	1.242 mH	
Parallel resonance-frequency f_p	268.7 kHz	291.6 kHz
Effective inductance L	3.94 mH	3.45 mH
Inductance, core influence excluded	4.23 mH	3.67 mH
Inductance to be assigned to the air gap	2.99 mH	2.428 mH
Characteristic impedance Z_w	6671 Ohm	6241 Ohm
Air gap inductance L_0 2 × (2,7 mm × 22 mm × 50 mm)	$L_0 = 256nH \cdot N^2 = 2.986mH$	
Re$\{Z_{(\omega p)}\}$	$R_p = 76.29$ kOhm	$R^* = 20.94$ kOhm
$R^{**} = \frac{R_p \cdot R^*}{R_p - R^*}$	–	28.86 kΩ
R_δ	–	1486 Ohm
$R_\delta{}'$ for N = 1 (in series)	–	0.05742 Ω
$R_\delta{}''$ for N = 1 (parallel)	–	0.058784 Ω
Conductivity σ_{xy}	0	474·10³ S/m
$\underline{\mu}_r = \mu_r' - j\mu_r''$ at f_p and vertical direction of the magnetic field in the image plane of Fig. 6.62a	1	**0.7943 − j·0.1224**

References

1. Philippow, E.: Taschenbuch Elektrotechnik, Band 1, 1986
2. Rindt, C.: Taschenbuch der Hochfrequenz- und Elektrotechnik, Band 1, S. 331 ff.
3. Rogowski, W., Über das Streufeld und den Streuinduktionskoeffizienten eines Transformators mit Scheibenwicklung und geteilten Endspulen, (Dissertation), VdI, Mitteilung über Forschungsarbeiten auf dem Gebiet des Ingenieurwesens, 1909
4. Mecke, H.: Betriebsverhalten und Berechnung von Transformatoren für das Lichtbogenschweißen. Habilitationsschrift, TH Magdeburg 1079
5. Stadler, A., M. Albach, and S. Chromy, "The optimization of high frequency operated transformers for resonant converters," Proceedings of 11th European Conference on Power Electronics and Applications (EPE2005), 77, Dresden, Germany, Sept., 2005.
6. Strowitzki, C.: Simulation and Design of a high speed Solid State Switch for Excimer Laser. Dissertation Universität Kassel 2009

7. D. M. Pozar: Microwave engineering.. 4. Auflage. J. Wiley, New York 2012, ISBN 978-0-470-63155-3
8. https://mhf-e-wiki.desy.de/Moden_eines_zylindrischen_Hohlraumresonators
9. Ferroxcube: Soft Ferrites and Accessories. Handbook 2013, p.13

Characterization of Inductive Components

7.1   Basics

Inductance is a characteristic quantity for inductive components. Due to the non-linear properties of many magnetic materials, inductance is generally not a constant quantity. However, there are various application situations, which result in correspondingly different specifications for the definition and measurement of inductance. For linear conditions, all these inductance values are the same.

To explain the problem, a real magnetization characteristic curve of a soft magnetic sheet shall be used (Fig. 7.1a). This exhibits a non-linear curve typical for many soft magnetic materials. The relationship $B = \mu \cdot H = \mu_r \cdot \mu_0 \cdot H$ describes a linear relationship between B and H and is basically only applicable here for very low field strengths in the range of the initial permeability μ_i (Fig. 7.1b).

From this characteristic curve, the characteristic curve $\Psi(I)$ can be generated for an inductive component with constant given magnetization cross section A_{Fe}, iron path length l_{Fe} and number of turns N by the following multiplications of the axis designations:

$$\Psi = N \cdot A_{Fe} \cdot B$$

$$I = \frac{H \cdot l_{Fe}}{N}$$

For a sheet of quality M400-50A, both the magnetization characteristic curve (a) and the values of the inductors according to the above definitions (b) for an iron path length of 200 mm, a magnetization cross section of 200 mm^2 and a number of turns of 100 are shown in the following Fig. 7.2. A corresponding core can be produced by a lamination stack of appropriate thickness consisting of circular rings with an inner diameter of 60 mm and an outer diameter of 67.3 mm.

© Springer Fachmedien Wiesbaden GmbH, part of Springer Nature 2022
P. Zacharias, *Magnetic Components*,
https://doi.org/10.1007/978-3-658-37206-4_7

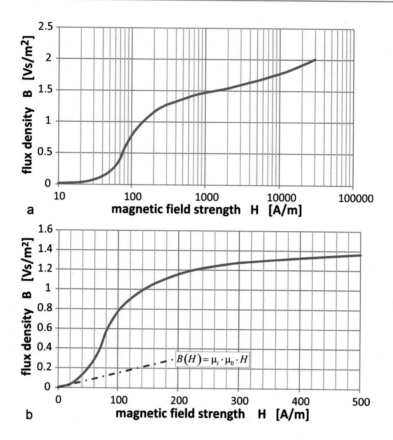

Fig. 7.1 Magnetization characteristic curve of sheet M400-50A (raw data: [1]). (**a**) in the field strength range up to 30kA/m; (**b**) in the initial range up to 500 A/m

Analogous to the relationship between the field sizes B and H, the relationship between the integral component sizes Ψ and I is described by an approximately linear relationship

$$\Psi = L \cdot I$$

The inductance L is a characteristic quantity of the component. Obviously, in the general, non-linear case L is not a constant quantity.

Amplitude Inductance L_a, Amplitude Permeability μ_{ra}

If one considers voltage drop and current at a choke, the general relationship is as follows

$$i_L(t) = \frac{1}{L(i)} \int u_L(t) \cdot dt = \frac{1}{L(i)} \psi(t).$$

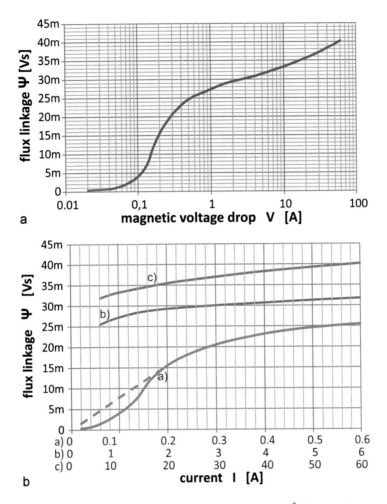

Fig. 7.2 Sheet metal M400-50A (l_{Fe} = 200 mm, A_{Fe} = 200 mm^2, N = 100): magnetization characteristic curve. (**a**) In single-logarithmic representation, (**b**) in linearly divided coordinates with linearization (dashed) by a tangent to the curve running through the coordinate origin (raw data: [1])

For periodic voltage curves with the mean value $\overline{u_L(t)} = 0$, an amplitude inductance L_A can now be defined, where

$$L_A = \frac{\widehat{\psi}}{\widehat{i_L}} \text{ applies.}$$

Based on the design equation for the inductance $L = \mu_r \cdot \mu_0 \cdot N^2 \cdot \frac{A_{Fe}}{l_{Fe}} = \frac{\psi}{i}$, an amplitude permeability

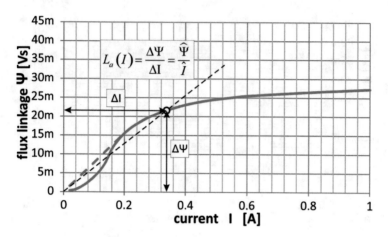

Fig. 7.3 Definition of the amplitude inductance $L_a(I)$

$$\mu_{ra} = \frac{\widehat{\Psi}}{\widehat{I}} \cdot \frac{l_{Fe}}{A_{Fe} \cdot \mu_0 \cdot N^2} = \frac{\widehat{\Psi}}{N \cdot A_{Fe}} \cdot \frac{l_{Fe}}{N \cdot \widehat{I}} \frac{1}{\mu_0} = \frac{\widehat{B}}{\mu_0 \cdot \widehat{H}}$$

can be defined (see Fig. 7.3).

The following Fig. 7.4 shows the curves u(t), ψ(t) and i(t) for a certain working point L_a to illustrate the importance of this size definition. L_a describes the ratio of the maximum values of ψ(t) and i(t). This is the same in both cases despite the different curve shapes.

Differential Inductance L_d, Differential Permeability μ_{rd}
For the voltage induction, the following applies in principle

$$u_L(t) = \frac{d}{dt}\psi(i_L) = \frac{d}{di} \cdot \frac{di}{dt} \cdot \psi(i_L) = \left(\frac{d}{di} \cdot \psi(i_L)\right) \frac{di}{dt} = L_d(i) \cdot \frac{di_L}{dt}.$$

This leads to another value for defining an inductance: the differential inductance L_d, which can be used for small current changes $\Delta I \rightarrow di \rightarrow 0$ to describe the reactions of the component (Fig. 7.5).

In accordance with the considerations made above, an associated differential permeability μ_{rd} can also be defined here:

$$\mu_{rd} = \frac{1}{\mu_0} \cdot \frac{dB}{dH}$$

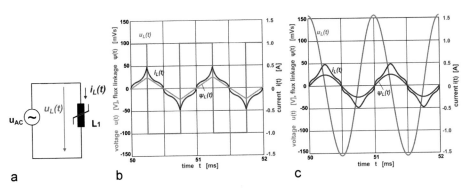

Fig. 7.4 Current curves for the inductance according to Fig. 7.2. (**a**) Circuit diagram, (**b**) for rectangular voltage, (**c**) for sinusoidal voltage with the same maximum flux linkage ψ_{max}

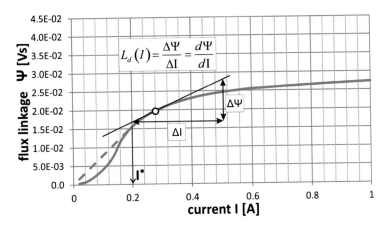

Fig. 7.5 Definition of the differential inductance $L_d(I)$ (I^*...contact point of the tangent to $\Psi(i)$ by the coordinate origin)

The meaning of the terms thus defined is shown below (Fig. 7.6). There, a square-wave alternating voltage is applied to an inductor L1 by a buck converter. The voltage has the mean value '0'. The average current through the inductance can be changed by a variable load resistance. This shifts the operating point on the $\Psi(I)$ characteristic curve and the change in current is dependent on the local/differential inductance L_d. In the vicinity of the maximum of the differential inductance, the current ripple becomes minimal.

Energy Related Induktance L_e

A common characterization of inductive components is done by means of energetic considerations. The energy stored in the magnetic field of a magnetic component is

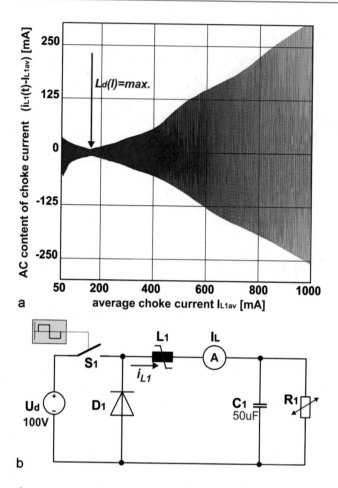

Fig. 7.6 Alternating component of the choke current curve for a buck converter equipped with a smoothing choke according to Fig. 7.2. (**a**) Circuit diagram, (**b**) curve with increasing current (f_{sw} = 50 kHz, R = 1000 Ohm...50 Ohm)

$$W_{magn.tot} = \int_V \left[\int_0^I H(I) \cdot dB \right] \cdot dV.$$

With a linear relationship between B and H with $B = \mu_r \cdot \mu_0 \cdot H = \mu_r \cdot \mu_0 \cdot \frac{N \cdot I}{l}$, the stored magnetic energy is

$$W_{magn.tot} = \int_V \left[\int_0^I H(NI) \cdot dB(NI) \right] dV = \int_V \left[\int_0^I \mu_r \mu_0 \cdot \frac{NI}{l} \cdot \frac{dNI}{l} \right] dV$$

$$= \frac{I^2}{2} \int_V \left[N^2 \frac{\mu_r \mu_0}{l^2} \right] dV = L_{eff} \cdot \frac{I^2}{2}$$

In general, the entire space around the component that contributes to varying degrees to energy storage (core, air gap, leakage field) should be considered. In soft magnetic inductive components, magnetic field strength and magnetic induction are dependent on an electric current. For the non-linear case, an energy-related, current-dependent inductance can therefore be defined with

$$L_{eff}(I) = L_e(I) = \frac{2 \cdot W_{magn.tot}}{I^2} = \frac{2}{I^2} \int_V \left[\int_0^I H(I) \cdot dB \right] \cdot dV = \frac{2}{I^2} \int_0^I I \cdot d\Psi$$

If such an inductor is connected to a charged linear capacitor, oscillations are formed in which the two storage elements alternately exchange their energy contents. As a function of maximum current/maximum voltage, the energy storage capacity of the storage choke changes as shown in Fig. 7.7, so that the natural oscillation frequency in Fig. 7.8 also changes as a function of voltage/current. In addition to the "S-shape" of the magnetization characteristic curve of sheet metal M400-50A, the effect of a tangent to this characteristic curve in the range of small currents is shown in dashed lines in Fig. 7.7.

But the shape of the courses of currents and voltages also changes. For the example introduced above, these relationships are shown in Fig. 7.9. The representation is the result of simulation calculations with a loss-free non-linear inductance.

With natural oscillation, it is interesting to note that the distortion of the voltage, expressed by the THD (*total harmonic distortion*), is significantly lower than that of the current (Fig. 7.10b). Only at relatively high current/voltage amplitudes does the THD_u exceed the value of 8%, which according to IEC is the permitted limit for the mains voltage. The THD_i is then already 30% in this case. This can be advantageously used for test circuits for non-linear inductors.

In summary, it can be stated that it is useful to use different definitions for the inductance of inductive components to characterize them. The measuring method used or applicable for their determination depends on the respective effect to be described. The example used shows that in the general case, the 3 definitions of amplitude inductance examined are different. In the case of a linear relationship between B and H, L_a, L_d and L_e are identical. To illustrate this finding, the $\Psi(I)$ characteristic curve is compared below with the curve of the $L_x(I)$ characteristic (Fig. 7.11). The dashed line shows the course for a sectionally linearized case (in the lower current range, the tangent is laid to the $\Psi(I)$ characteristic by

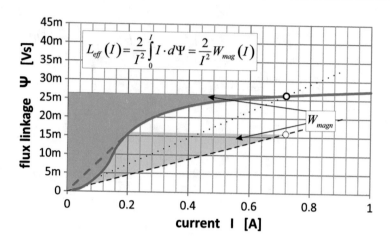

Fig. 7.7 Definition of the amplitude inductance $L_{eff}(I)$ as equivalent "linear" inductance with the same energy content

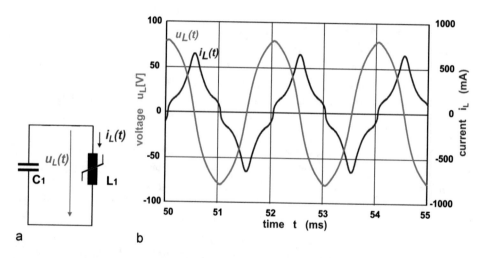

Fig. 7.8 Formation of natural oscillations of a non-linear choke with a linear capacitor ($C = 1.862\ \mu F$, $U_c(t = 0) = 80$ V). (**a**) Circuit, (**b**) u-, i-curves

the coordinate origin). At the point of contact I^* of the $\Psi(I)$ curve in Fig. 7.5, the course at $L_a(I)$ and $L_d(I)$ in Fig. 7.11 merges practically directly into the course for the nonlinear case. Since the course of the magnetization characteristic curve in the range $(0...I^*)$ plays a role in the definition of L_e, the course for the partially linearized case for $I > I^*$ approaches the curve $L_e(I)$.

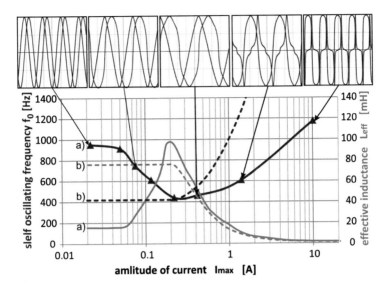

Fig. 7.9 Change of natural oscillation frequency with change of current amplitude by changing $U_c(t = 0)$ and corresponding effective inductance L_{eff}. Above the diagram, the normalized u(t) curves (blue) and i(t) curves (red) are shown for different vibration conditions. (Dotted lines: Results for "linearized curves" as shown in Fig. 7.2b)

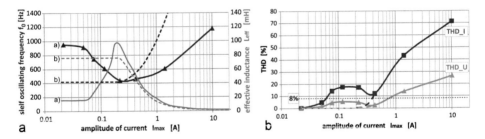

Fig. 7.10 Variation of characteristic values of u(t) and i(t) as well as L_{eff} when changing the current amplitude by changing the capacitor voltage $U_c(t = 0)$ (THD: *total harmonic distortion*): (**a**) for the example characteristic M400-A50, (**b**) for characteristic of (**a**) linearized in the lower current range by a tangent to the characteristic as shown in Fig. 7.2b

7.2   Equivalent Circuits for Inductors

When determining the permeability of a substance, it is generally assumed that it consists of a real part $\mu_r{}'$ and an imaginary part $-j \cdot \mu_r{}''$. When linear conditions are considered, the magnetic permeability of a cylindrical / prismatical magnetic section is

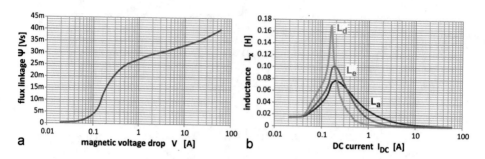

Fig. 7.11 Comparison of the different definitions of inductance using an example. (**a**) Magnetization characteristic curve as $\Psi(V)$ characteristic curve, (**b**) inductance values as a function of operating current 'I' for sheet metal M400-50A ($l_{Fe} = 200$ mm, $A_{Fe} = 400$ mm^2, N = 100)

$$\underline{G}_m = \frac{1}{\underline{R}_m} = (\mu_r' - j \cdot \mu_r'') \cdot \mu_0 \cdot \frac{A_m}{l_m}.$$

The impedance of an inductive component with the number of turns N is accordingly

$$\underline{Z}_{ind} = j\omega \cdot \left(N^2 \underline{G}_m\right) = \omega N^2 \cdot \mu_r'' \mu_0 \cdot \frac{A_m}{l_m} + j\omega N^2 \cdot \mu_r' \mu_0 \cdot \frac{A_m}{l_m}$$
$$= R_s(\omega) + jX_s(\omega). = R_s(\omega) + j\omega L_s(\omega)$$

This definition leads to a series replacement circuit for an inductor. If one considers the nature of the core losses, these (eddy current losses, remagnetization losses) are rather losses coupled into the core by a transformer ($\sim B^2 \sim U^2$). This could be represented in an equivalent circuit with a parallel connection of an inductor and an electrical resistor. Figure 7.12 shows this using the example of the material 3E27 from Ferroxcube [2].

If a toroidal core with a cross section of 1 cm^2 and an average length of 10 cm is made from this material and equipped with 10 turns, the curves shown in Fig. 7.13 can be calculated for the real and imaginary parts of the impedance of the resulting component using the above-mentioned curves.

Each series connection of L_s and R_s can be converted into a parallel connection L_p and R_p if, for a given frequency, the real and imaginary parts of the impedance $\underline{Z}$ are equated in pairs:

$$\underline{Z} = R_s + jX_s = \frac{R_p \cdot jX_p}{R_p + jX_p} \rightarrow R_p = \frac{R_s^2 + X_s^2}{R_s} X_p = \frac{R_s^2 + X_s^2}{X_s}$$

If this conversion is applied to Fig. 7.13 and the corresponding component parameters for series and parallel connection are calculated, the following results are obtained in graphic form. If you want to calculate with parameters that are as independent of frequency as

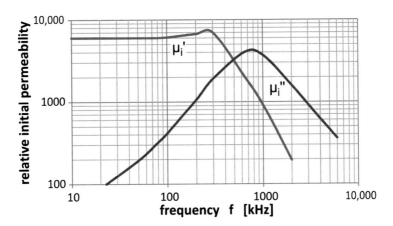

Fig. 7.12 Real part and imaginary part of the relative permeability at H → 0 for the material 3E27 of Ferroxcube

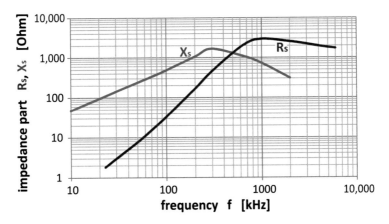

Fig. 7.13 Inductive and resistive impedance components (series replacement circuit diagram) for an inductive component with the data $A_{Fe} = 1$ cm^2, $l_{Fe} = 10$ cm and N = 10

possible, parallel connection is obviously the preferred variant of the equivalent circuit diagram (Fig. 7.14).

In a real inductive component, a serial loss component will also have to be taken into account, since, for example, the ohmic resistance of the winding must be taken into account. Thus, the combination of parallel and series connection shown below would be an adequate equivalent circuit that can be used to characterize inductive components (Fig. 7.15).

For the impedance of the circuit in Fig. 7.15, one gets

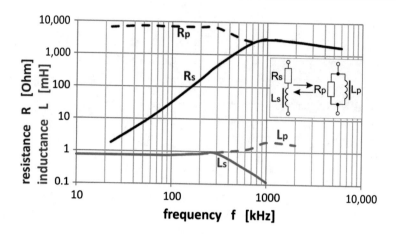

Fig. 7.14 Frequency dependency of the impedance components in series and parallel set circuits

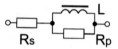

Fig. 7.15 Physically based 2-pole equivalent circuit diagram for an inductive component with one winding, consisting of 3 elements

$$\underline{Z}(\omega) = R_s + \frac{R_p \cdot j\omega L}{R_p + j\omega L} = \frac{R_s R_p + (R_p + R_s) \cdot j\omega L}{R_p + j\omega L}$$

$$= \frac{\left(R_s R_p + (R_p + R_s) \cdot j\omega L\right)(R_p - j\omega L)}{R_p^2 + \omega^2 L^2}$$

The locus curve of this impedance $\underline{Z}(\omega)$ obviously resembles a semicircle with the center $((R_s + 0.5 \cdot R_p); 0)$ and the radius of the length $0.5 \cdot R_p$ (Fig. 7.16).

In general, the arrangement according to Fig. 7.15 is:

$$\underline{Z}(\omega) = \frac{\left(R_s R_p^2 + (R_p + R_s) \cdot \omega^2 L^2 + j\omega L R_p^2\right)}{R_p^2 + \omega^2 L^2}$$

$$\mathrm{Im}\{\underline{Z}(\omega)\} = X(\omega) = \frac{\omega L R_p^2}{R_p^2 + \omega^2 L^2} = \frac{\omega L}{1 + \omega^2 \frac{L^2}{R_p^2}} \rightarrow \max\left[\mathrm{Im}\{\underline{Z}(\omega)\}\right] = X_{\max} = \frac{R_p}{2} \text{ at } \omega$$

$$= \frac{R_p}{L}$$

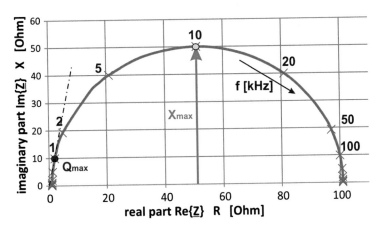

Fig. 7.16 $\underline{Z}$ locus curve with frequency 'f' as parameter for $R_s = 1$ Ohm, $R_p = 100$ Ohm, $L = 1.592$ mH, $\omega = 2\pi f$

$$\text{Re}\left\{\underline{Z}(\omega)\right\} = R(\omega) = \frac{\left(R_s R_p^2 + (R_p + R_s) \cdot \omega^2 L^2\right)}{R_p^2 + \omega^2 L^2}$$

$$Q\{\underline{Z}(\omega)\} = \frac{\text{Im}\{\underline{Z}(\omega)\}}{\text{Re}\left\{\underline{Z}(\omega)\right\}} = \frac{X(\omega)}{R(\omega)} = \frac{\omega L R_p^2}{R_s R_p^2 + (R_p + R_s) \cdot \omega^2 L^2}$$

$$= \frac{\omega L R_p}{R_s R_p + \left(1 + \frac{R_s}{R_p}\right) \cdot \omega^2 L^2}$$

Figure 7.17 shows the corresponding curves for $X(\omega)$, $R(\omega)$ and $Q(\omega)$ in diagrams for the example of an equivalent circuit used in Fig. 7.15.

For small frequencies $\omega \to 0$, the following approximations apply

$$Q\{\underline{Z}(\omega)\} \simeq \frac{\omega L}{R_s},$$

while for high frequencies $\omega \to \infty$, the following applies accordingly

$$Q\{\underline{Z}(\omega)\} \simeq \frac{R_p^2}{(R_p + R_s) \cdot \omega L}$$

The point of intersection of these asymptotes to the quality curve $Q\{\underline{Z}(\omega)\}$ on a double logarithmic scale is at

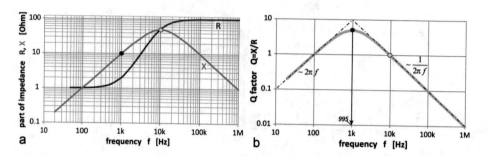

Fig. 7.17 Frequency responses of the equivalent circuit according to Fig. 7.15 assuming frequency-independent components. (**a**) Impedance components X(ω) and R(ω), (**b**) Quality value Q(ω)

$$\frac{R_p{}^2}{(R_p + R_s) \cdot \omega^* L} = \frac{\omega^* L}{R_s} \rightarrow \omega^* = \frac{R_p}{L}\sqrt{\frac{R_s}{R_p + R_s}}$$

and also marks the frequency ω* of the maximum of Q.

For further analysis, an extreme value consideration of the Q(ω) dependence is carried out. If one sets the first derivative after the angular frequency ω to zero, one obtains

$$\frac{d}{d\omega}Q\{\underline{Z}(\omega)\} = \frac{d}{d\omega}\left(\frac{\omega L}{R_s + \left(\frac{1}{R_p} + \frac{R_s}{R_p{}^2}\right)\cdot \omega^2 L^2}\right) = 0$$

For the resulting value of ω*, the quality is then calculated as a maximum

$$\omega^{*2} = \frac{R_s R_p{}^2}{(R_p + R_s)L^2} \rightarrow \omega^* = \frac{R_p}{L}\sqrt{\frac{R_s}{R_p + R_s}}\,(\text{see above}).$$

→ For example, $f(Q_{max}) = 995$ Hz for Fig. 7.15.

When inserted into the imaginary part of Z(ω), this results in

$$\mathrm{Im}\{\underline{Z}(\omega^*)\} = X(\omega^*) = \omega^* L\frac{1 + \frac{R_s}{R_p}}{1 + 2\frac{R_s}{R_p}} \approx j\omega^* L\left(1 - \frac{R_s}{R_p}\right) \approx j\omega^* L$$

This means: The inductance measured at $Q = Q_{max}$ is approximately equal to the "actual inductance" due to the resistance ratios. When inserting ω* into the real part of Z(ω), you get the approximation

$$\mathrm{Re}\,\{\underline{Z}(\omega^*)\} = R(\omega^*) = \frac{2R_s(R_p + R_s)}{R_p + 2R_s} \approx 2R_s\left(1 - \frac{R_s}{R_p}\right) \approx 2R_s$$

The maximum quality $Q_{max}(\omega^*)$ for this arrangement is

$$Q^2_{max} = \frac{R_p{}^2}{4R_s(R_p + R_s)} \quad \text{or}\ Q_{max} = \frac{R_p}{2R_s}\sqrt{\frac{R_s}{R_p + R_s}}$$

(In the example $Q_{max} = \frac{100\ \Omega}{2 \cdot 1\ \Omega}\sqrt{\frac{1\ \Omega}{100\ \Omega + 1\ \Omega}} = 4.975$ or $Q_{max} \approx 0.5\sqrt{\frac{R_p}{R_s}} = 5$)

The relationship between R_s, R_p and Q can be derived from this:

$$R_p = 2R_s\left(Q^2_{max} + Q_{max}\sqrt{Q^2_{max} + 1}\right) \approx 4Q^2_{max} \cdot R_s$$

When measuring the impedance from current, voltage and phase position, the basic elements L and R with their sizes can be determined for the serial or parallel equivalent circuit. For an equivalent circuit diagram consisting of 3 elements, more information is naturally required. For an equivalent circuit diagram, which can be used in a wide frequency range, the procedure shown above is suitable:

1. Determination of the courses $X(\omega)$ and $R(\omega)$ and calculation of $Q(\omega) = X(\omega)/R(\omega)$
2. Determination of Q_{max} and ω^* or $X^*(\omega^*)$ and $R^*(\omega^*)$
3. Determination of the asymptotes at the course $Q(\omega)$ ($q_1(\omega) = k_1 \cdot \omega$ left of Q_{max} and $q_2(\omega) = k_2 \cdot \omega^{-1}$ right of Q_{max}) in double logarithmic scale.
4. Determination of the intersection ω^* of the tangents
5. Calculation of the 3 elements of the equivalent circuit diagram from

$$L \approx \frac{1}{\omega^*} \cdot X^*(\omega^*),\ \ R_s \approx \frac{1}{2} \cdot R^*(\omega^*),\ \ R_p \approx 4Q^2_{max} \cdot R_s$$

7.3   Characterization of Inductive Components by Measuring Sinusoidal Voltage, Current and Their Phase Relation

For a two-pole measurement on an inductive component, only the voltage $u_L(t)$ and $i_L(t)$ are available as characteristic measured variables. Assuming linear conditions, the RMS values U_L and I_L and the phase shift φ are sufficient to characterize the component (Fig. 7.18).

The ideal relationships for the progressions are

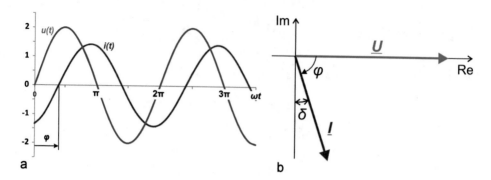

Fig. 7.18 Current and voltage at an electrical two-terminal with inductive component. **(a)** Time courses, **(b)** corresponding phasor image

$$u(t) = \sqrt{2} \cdot U \cdot \sin(\omega t) \quad \text{and} \quad i(t) = \sqrt{2} \cdot U \cdot \sin(\omega t - \varphi)$$

For the complex impedance then applies

$$\underline{Z} = \frac{U}{I} = R_s + jX_s, R_{xs} = Z \cdot \cos\varphi, X_s = Z \cdot \sin\varphi \text{ with } Z = \frac{U}{I}.$$

The inductance is determined during the measurement via interpretation $X_s = \omega L_{xs}$.

It can be assumed that errors occur in the measurement of the individual parameters U, I and ϕ, which affect the errors in the determination of L_{xs} and R_{xs}. The total differential can be used for this purpose. Determination of the maximum measurement error for R using the total differential is as follows

$$\max|\Delta R| = \left| \frac{\partial}{\partial U} \cdot \left(\frac{U}{I} \cdot \cos\varphi \right) \cdot \Delta U \right| + \left| \frac{\partial}{\partial I} \cdot \left(\frac{U}{I} \cdot \cos\varphi \right) \cdot \Delta I \right|$$
$$+ \left| \frac{\partial}{\partial \varphi} \cdot \left(\frac{U}{I} \cdot \cos\varphi \right) \cdot \Delta\varphi \right|$$

or

$$\max|\Delta R| = \left| \left(Z \frac{\Delta U}{U} \cdot \cos\varphi \right) \right| + \left| \left(-\frac{Z\Delta I}{I} \cdot \cos\varphi \right) \right| + \left| (-Z \cdot \sin\varphi) \cdot \Delta\varphi \right|$$

$$\max |\Delta R| = \left|\left(Z\frac{\Delta U}{U} \cdot \cos\varphi\right)\right| + \left(-\frac{Z\Delta I}{I} \cdot \cos\varphi\right) + |(-Z \cdot \sin\varphi) \cdot \Delta\varphi|$$

$$\max \left|\frac{\Delta R}{Z}\right| = \left|\left(\frac{\Delta U}{U} \cdot \cos\varphi\right)\right| + \left(-\frac{\Delta I}{I} \cdot \cos\varphi\right) + |(-\sin\varphi) \cdot \Delta\varphi|$$

$$= \left(\frac{\Delta U}{U} + \frac{\Delta I}{I}\right) \cdot \cos\varphi + \Delta\varphi \cdot \sin\varphi$$

$$\max \left|\frac{\Delta R}{Z}\right| = \left(\frac{\Delta U}{U} + \frac{\Delta I}{I}\right) \cdot \cos\varphi + \Delta\varphi \cdot \sin\varphi$$

$\Delta\varphi$ is with $\Delta\varphi = 2\pi(\Delta t/T)$ expression of the relative error of the time measurement. Especially the angular error has a strong effect on low-loss inductors $X_s >> R_s$. For $\varphi \approx \frac{\pi}{2}$, the following approximation is obtained

$$\Delta R \cong |Z \cdot \Delta\varphi| \cdot \sin\varphi \approx |Z \cdot \Delta\varphi|$$

For the maximum relative measurement error, one then obtains

$$\frac{\Delta R}{R} \cong \frac{|Z \cdot \Delta\varphi| \cdot \sin\varphi}{Z \cdot \cos\varphi} \approx \frac{|Z \cdot \Delta\varphi|}{Z \cdot \cos\varphi} = \frac{|\Delta\varphi|}{\cos\varphi} = \frac{|\Delta\varphi|}{\cos\left(\frac{\pi}{2} - \delta\right)} = |\Delta\varphi|\frac{\sqrt{1 + \tan^2\delta}}{\tan\delta} \approx \frac{|\Delta\varphi|}{\tan\delta}$$

for sufficiently small loss angles δ. If we now go back to the total differential, we obtain for the relative measurement uncertainty of the ohmic component.

$$\max\left(\frac{\Delta R}{R}\right) \cong |\Delta\varphi|\frac{1}{\tan\delta} + \left|\left(\frac{\Delta U}{U}\right)\right| + \left|\left(-\frac{\Delta I}{I}\right)\right| \text{ or } \max\left|\frac{\Delta R}{Z}\right| = \left(\left|\frac{\Delta U}{U}\right| + \left|\frac{\Delta I}{I}\right|\right)$$
$$\cdot \cos\varphi + \Delta\varphi \cdot \sin\varphi$$

If one proceeds analogously for the imaginary part of the impedance, one gets for $\varphi \to 0; \delta \to \frac{\pi}{2}$

$$\max\left(\frac{\Delta X}{X}\right) \cong |\Delta\varphi|\tan\delta + \left|\left(\frac{\Delta U}{U}\right)\right| + \left|\left(-\frac{\Delta I}{I}\right)\right| \text{ or }$$

$$\max\left|\frac{\Delta X}{Z}\right| = \left(\frac{\Delta U}{U} + \frac{\Delta I}{I}\right) \cdot \sin\varphi + \Delta\varphi \cdot \cos\varphi.$$

As an example, a temporal resolution of *0.1%* of the period duration and relative measurement errors of 0.1% for current and voltage measurement are assumed. The time measurement corresponds in radians to 0.00628319 rad. If we also assume a resistance of 1 Ohm in series with an inductance of 0.000015915H (X = 1 Ohm at 10 kHz), the following results

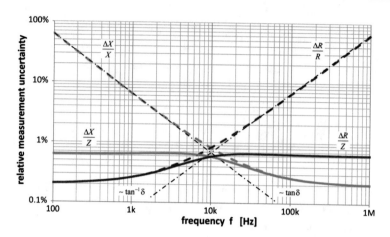

Fig. 7.19 Maximum relative measurement uncertainties (calculated) for the assumptions
(R = 1 Ohm, L = 15.915 µH, for the maximum relative errors $\Delta U/U = \Delta I/I = \Delta t/T = 0.001$

are obtained for the relative measurement uncertainty for the impedance measurement via
the phase difference (Fig. 7.19).

The maximum measurement uncertainty, related to both components $\left|\frac{\Delta R}{R}\right|, \left|\frac{\Delta X}{X}\right|$, reaches a
minimum at $\tan\delta=1$ with this measurement method and otherwise deviates considerably
from this. This means that the smallest measurement errors occur when the real and
imaginary parts of the impedance are of the same order of magnitude. If the measurement
results are related to the total impedance Z, 0.63% is not exceeded over the entire range
under consideration in the example.

Another way of determining impedance can be done by power measurements
(Fig. 7.20). Here, apparent power and active power are measured.

The measurement consists in the formation of the mean scalar product of current and
voltage. The active power is then general:

$$P = \frac{1}{T}\int_{t1}^{t1+T} u(t) \cdot i(t) \cdot dt = \frac{1}{T}\int_{t1}^{t1+T}\left[\sum_{v=1}^{\infty}\sqrt{2}\cdot U_v \cdot \sin\left(v\omega t + \varphi_{uv}\right)\right]$$
$$\cdot\left[\sum_{\mu=1}^{\infty}\sqrt{2}\cdot I_\mu \cdot \sin\left(\mu\omega t + \varphi_{i\mu}\right)\right]\cdot dt$$

If the applied voltage in particular is a pure sine wave voltage, the following applies to the
active power

$$P = U_1 \cdot I_1 \cdot \cos\left(\varphi_{u1} - \varphi_{i1}\right),$$

while a fundamental reactive power Q_1

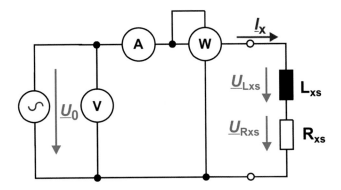

Fig. 7.20 Schematic diagram of a power measurement of an inductive component

$$Q_1 = U_1 \cdot I_1 \cdot \sin\left(\varphi_{u1} - \varphi_{i1}\right)$$

can be defined. Under these conditions, the apparent power is calculated as

$$S = U \cdot I = \sqrt{\int_{t1}^{t1+T} \left[\sum_{\nu=1}^{\infty} 2U_\mu^2 \cdot \sin^2(\nu\omega t + \varphi_\nu)\right] \cdot dt}$$

$$\cdot \sqrt{\frac{1}{T}\int_{t1}^{t1+T} \left[\sum_{\mu=1}^{\infty} 2I_\mu^2 \cdot \sin^2(\mu\omega t + \varphi_\mu)\right] \cdot dt}$$

$$\text{or } S = U \cdot I = \frac{1}{T}\sqrt{\left(\sum_{\nu=1}^{\infty} U_\nu^2\right) \cdot \left(\sum_{\mu=1}^{\infty} I_\mu^2\right)}.$$

As known, there is a relation for sinusoidal voltage $S^2 = P^2 + Q_1^2 + D^2$ where 'D' is the distortion reactive power. For the series equivalent circuit diagram and parallel equivalent circuit diagram, one obtains for the ohmic components

$$R_{xs} = \frac{P}{I_x^2} \text{ or } R_{xp} = \frac{U_0^2}{P}.$$

The quantity $\sqrt{D^2 + Q_1^2} = \sqrt{S^2 - P^2}$ describes a reactive power quantity, whereby reactive power describes an energy exchange between storage units. Accordingly, an effective inductance can be assigned here, which is derived from energetic considerations (see Sec. 7.1)

$$\omega L_{xs.eff} = \frac{\sqrt{S^2 - P^2}}{I_x^2} \text{ or } \omega L_{xp.eff} = \frac{U_0^2}{\sqrt{S^2 - P^2}}.$$

The determination of the loss components in particular is often carried out in this way, as shown in Sect. 7.5. In order to obtain comparable measurements, a sufficiently low harmonic voltage is used. The results depend heavily on the bandwidth and phase response of the measuring amplifiers and power meters used. Electrodynamic power meters can be used very well for line-frequency applications. Both of the basic methods described find practical application in the characterization of materials and have been standardized for different areas of application to ensure comparability.

7.4 Bridge Circuits for Determining the Impedance Components of Inductive Components

With bridge circuits (Fig. 7.21), components can be compared with each other. If the voltages of the bridge elements ($\underline{Z}_1; \underline{Z}_3$) and ($\underline{Z}_2; \underline{Z}_4$) are the same in pairs, the elements are similar, i.e. they can be converted into each other using proportionality factors. Generally, bridge circuits are used for linear elements. This enables the comparison of complex AC impedances in terms of magnitude and phase using sinusoidal voltages. If the bridge voltage is $\underline{U}^* = 0$, $\underline{U}_1 = \underline{U}_3$ and $\underline{U}_2 = \underline{U}_4$ follow automatically. A bridge is also calibrated if $\frac{\underline{U}_1}{\underline{U}_2} = \frac{\underline{U}_3}{\underline{U}_4} \rightarrow \frac{I_1 \cdot \underline{Z}_1}{I_1 \cdot \underline{Z}_2} = \frac{I_3 \cdot \underline{Z}_3}{I_3 \cdot \underline{Z}_4} \rightarrow \frac{\underline{Z}_1}{\underline{Z}_2} = \frac{\underline{Z}_3}{\underline{Z}_4}$.

A logical consequence for the determination of impedance components is the **Maxwell bridge** (Fig. 7.22). Here a linear reference inductance L_1 with series resistance R_1 is compared with an unknown inductance ($L_x; R_x$). In doing so, R_1 and the ratio R_3/R_4 can be changed. Obviously the following applies in the calibrated state

$$\frac{R_1 + j\omega L_1}{R_x + j\omega L_x} = \frac{R_3}{R_4} \rightarrow (R_1 + j\omega L_1) = \frac{R_3}{R_4}(R_x + j\omega L_x) \rightarrow R_x = R_1 \frac{R_4}{R_3} \text{ and } \frac{R_3}{R_4} L_x = L_1 \frac{R_4}{R_3}$$

For the quality, one gets

$$Q_x = \frac{1}{\tan \delta_x} = \frac{\omega L_x}{R_x} = \frac{\omega L_1}{R_1}$$

As L_1, a coreless inductor is best used. The (frequency-dependent) resistance of the winding must be added to the resistance R_1.

For larger measuring ranges or already with low nonlinearities of the core material of L_x, the **Maxwell-Wien bridge** proves to be more advantageous. Here, the dual circuit (parallel connection of C_1 and R_3) for series connection of L_x and R_x is placed in the diagonal position of the bridge (Fig. 7.23). C_1 serves here as a reference component. R_3 is used to

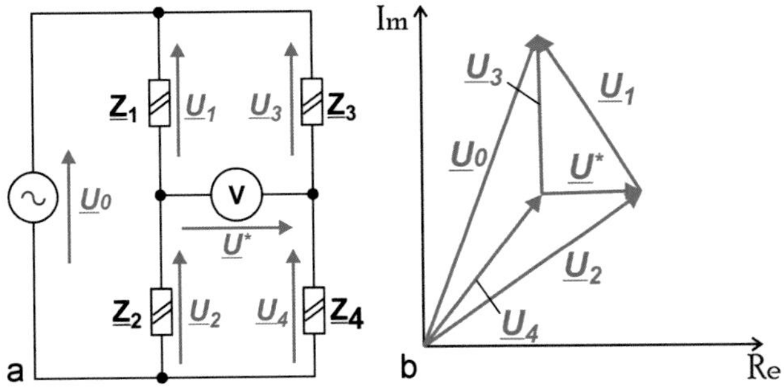

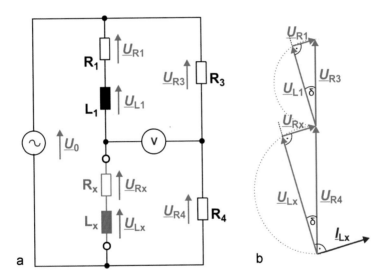

Fig. 7.21 General alternating current bridge. (**a**) Circuit, (**b**) phasor diagram

Fig. 7.22 Structure and function of the Maxwell bridge. (**a**) Principle circuit diagram, (**b**) associated phasor diagram for the case of an adjustment (the voltage measured with the voltmeter is then zero)

match the phase shift of the currents ($\underline{I}_{C1}$; $\underline{I}_{R3}$) to the ($\underline{U}_{Lx}$; $\underline{U}_{Rx}$). The matching condition and finally the determination of the impedance components result from this consideration

$$L_x = R_1 \cdot R_4 \cdot C_1, R_x = \frac{R_1 \cdot R_4}{R_3}, Q_x = \omega \cdot R_3 \cdot C_1.$$

If L_x or R_x is non-linear, the adjustment condition can no longer be fulfilled. The nonlinearities cause harmonics of the currents and voltages in this branch, which makes adjustment impossible, because the other branch does not cause harmonics (Fig. 7.24a). It is possible here to adjust for a minimization of the bridge voltage U^* → min. The sensitivity of the bridge when determining L_x and R_x depends decisively on the size of the minimum.

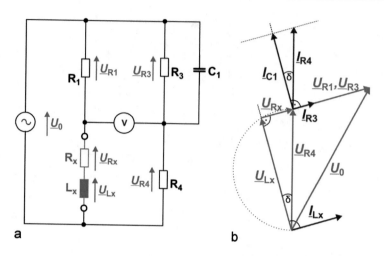

Fig. 7.23 Maxwell-Wien bridge. (**a**) Principle circuit diagram, (**b**) corresponding phasor diagram for the case of an adjustment

The Maxwell bridge differs little from the Maxwell-Wien bridge, where L_x interacts only with a resistor R_1 and not with another inductor (Fig. 7.24b). As an indicator of the minimum, the rectifier average value is more sensitive than the RMS value in the balancing branch.

If you exchange the parallel connection of R_3 and C_1 for a series connection of R_3 and C_1 with the same phase shift and the same amount, you get the Hay bridge (Fig. 7.25).

The adjustment condition is shown in Fig. 7.25b and leads to the relationships:

$$L_x = \frac{R_1 \cdot R_4 \cdot C_1}{1 + \omega^2 R_3^2 \cdot C_1^2}, \quad R_x = \frac{\omega^2 \cdot R_1 \cdot R_3 \cdot R_4 \cdot C_1^2}{1 + \omega^2 R_3^2 \cdot C_1^2}, \quad Q_x = \frac{1}{\tan \delta_x} = \frac{1}{\omega \cdot R_3 \cdot C_1}$$

The adjustment with C_1 and R_3 is somewhat more difficult here than with the Maxwell bridge, since both results L_x and R_x are influenced by it.

Another interesting bridge circuit is the Owen bridge. Here the necessary phase shifts of the second bridge branch are generated by two capacitors (Fig. 7.26).

From $\underline{U}_{R1} = \underline{U}_{C1}$ and $\underline{U}_{Lx} = \underline{U}_{R4}$ follows $L_x = R_1 \cdot R_4 \cdot C_1$,

with which one can derive from $\underline{U}_{Rx} + \underline{U}_{R3} = \underline{U}_{C2}$ and $\underline{U}_{Lx} = \underline{U}_{R4}$ for $R_x = \frac{C_1}{C_2} R_1 - R_3$

The quality factor 'Q' is then $Q_x = \frac{1}{\tan \delta_x} = \frac{\omega R_1 \cdot R_4 \cdot C_1}{\frac{C_1}{C_2} R_1 - R_3}$.

In general, C_1 and C_2 are constant values. The adjustment is performed using R_3 and R_4. The Owen bridge can be used for a wide range of L_x.

A synthesis of the basic ideas of Maxwell-Wien bridge and Hay bridge is represented by the Anderson bridge, the principle circuit of which is shown in Fig. 7.27. The phasor image is designed for the condition

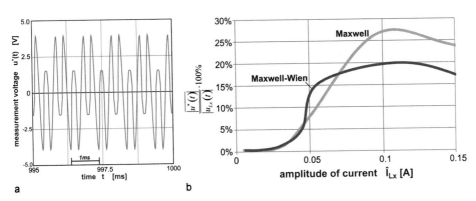

Fig. 7.24 Voltage at adjustment in the bridge branch with non-linear inductance (according to the material example in Sect. 7.1) with a Maxwell bridge $\hat{u}_0 = 35$ V, 1 kHz, $L_1 = 50$ mH. (**a**) Course of the bridge voltage, (**b**) change of the minimum bridge voltage with the amplitude of the choke current

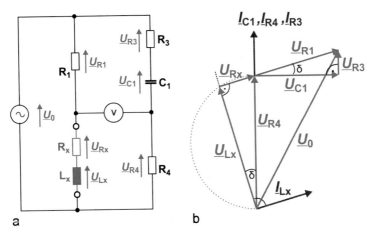

Fig. 7.25 Hay bridge. (**a**) Principle circuit diagram, (**b**) corresponding phasor diagram for the case of an adjustment

$R_3 \| R_4 = \frac{R_3 R_4}{R_3 + R_4} << \sqrt{R_2^2 + (\omega C_1)^{-2}}$. An impedance converter may therefore have to be connected between R_2 and R_3, R_4 in order to exclude the reaction of the RC element on the voltage divider.

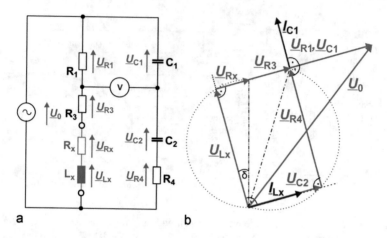

Fig. 7.26 Owen bridge. (**a**) Principle circuit diagram, (**b**) corresponding phasor diagram for the case of an adjustment

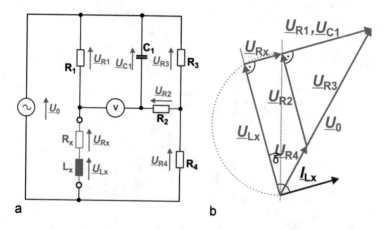

Fig. 7.27 Anderson bridge. (**a**) Principle circuit diagram, (**b**) corresponding phasor diagram for the case of an adjustment

$$1. \quad \frac{U_{Rx}}{U_{R1}} = \frac{U_{R4}}{U_{R3}} \qquad \rightarrow R_x = R_1 \cdot \frac{R_4}{R_3}$$

$$2. \quad \frac{U_{Lx}}{U_{R2}} = \frac{U_0}{U_{R3}} \qquad \rightarrow L_x = \frac{R_1 \cdot R_2 \cdot C_1}{R_3}(R_3 + R_4)$$

$$3. \quad Q_x = \frac{\omega L_x}{R_x} \qquad \rightarrow Q_x = \frac{1}{\tan \delta_x} = \frac{R_2 \cdot \omega C_1}{R_4}(R_3 + R_4)$$

If the phase shifter $R_2 C_1$ has a feedback effect on the voltage divider R_3, R_4, the result is
$L_x = \frac{R_1 \cdot C_1}{R_3}(R_3(R_2 + R_4) + R_2 R_4)$, $Q_x = \frac{1}{\tan \delta_x} = \frac{\omega C_1}{R_4}(R_3(R_2 + R_4) + R_2 R_4)$

With R_2 and R_3, for example, the bridge can be effectively adjusted. It is also well suited for the analysis of resistances with (internal) inductance.

Bridge circuits that use the resonance between the inductor under investigation and a capacitor are also useful. The first bridge to be mentioned here is the series resonance bridge according to Grüneisen and Griebe (Fig. 7.28). If a capacitance C is connected in series with the inductance to be measured in the first branch of the bridge, so that $\omega_0 L_x = \frac{1}{\omega_0 C_1}$, only the loss resistance R_x (for the fundamental oscillation) remains as the resulting resistance. The resistors R_1, R_3 and R_4 are purely ohmic. During the measurement, the bridge is first adjusted with short-circuited C with direct current. Then the bridging of C is removed and changed until resonance is reached and the measuring instrument shows no deflection with alternating current. Alternatively, the frequency can also be changed. The searched components L_x and R_x are then

$$R_x(\omega = 0) = R_1 \cdot \frac{R_4}{R_3}, Q_x = \frac{R_3}{\omega C_1 R_1 R_4}.$$

As the phasor image Fig. 7.28b shows, the component $R_x(\omega_0)$ could also be determined in the resonance case:

$$R_x(\omega_0) = R_1 \cdot \frac{R_4}{R_3}.$$

The prerequisite for this is a sufficiently good linearity of 'L'. The series resonance measuring bridge is a very simple and fast way to determine the effective inductance. However, determining the losses via R_x leads to questionable results for non-linear behavior of L, since losses are generally specified for sinusoidal voltages in order to make the data comparable. The harmonic content may be too large for an accurate statement. Even in the loss-free case, a voltage drop can be observed at the series connection of L and C due to the harmonics. As shown in Fig. 7.29, current and voltage are then orthogonal to each other. The measured voltage drop (without fundamental component) is an expression of distortion reactive power. Only if this is small compared to the voltage drops due to losses, such a measurement is useful.

In the following, the possible uses of a parallel resonance bridge will be investigated. The corresponding circuit is shown in Fig. 7.30. As an interpretation, the parallel connection of the unknown components L_{px} and R_{px} can be used here. Adjustment is done via $\omega = 2\pi f$ and R_3. At parallel resonance, U_{C1} and U_{R4} are in phase. This can be used as an indicator. As above, the components of the impedance can be calculated to

$$L_{px} = \frac{1}{\omega^2 C_1} \text{ or } L_x = \frac{C_1 R_1^2 R_4^2}{(\omega C_1 R_1 R_4)^2 + R_3{}^2}, R_{px} = R_1 \cdot \frac{R_4}{R_3} \text{ or } R_x = \frac{R_1 R_3 R_4}{(\omega C_1 R_1 R_4)^2 + R_3{}^2}$$

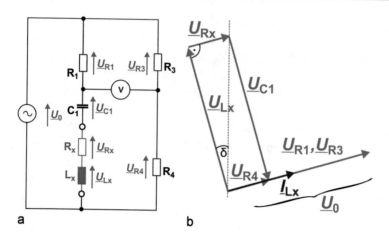

Fig. 7.28 Series resonance bridge according to Grüneisen and Griebe. (**a**) Principle circuit, (**b**) corresponding phasor diagram for the case of an adjustment

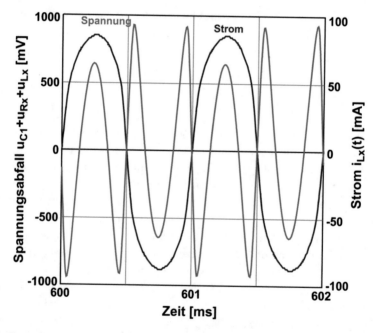

Fig. 7.29 Current and voltage curves on a series resonant circuit with a non-linear inductance according to Sect. 7.1 at resonance (simulation without losses)

$$\text{with } Q_x = \frac{\omega C_1 R_3}{R_1 R_4} ..$$

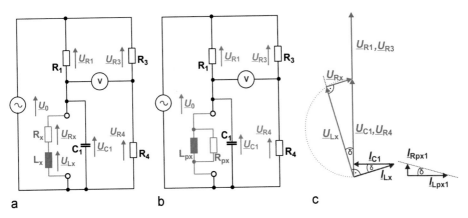

a   b   c

Fig. 7.30 Principle circuit diagram of the parallel resonance bridge. (**a**) With series equivalent circuit diagram, (**b**) with parallel equivalent circuit diagram, (**c**) corresponding phasor diagram for the case of an adjustment

It is interesting to note that due to the current-integrating behaviour of C_1, the voltage harmonics at the connected inductive component are much lower than in the previous example, as Fig. 7.29 shows. This makes it possible to determine the power dissipation in a much larger voltage range than with the other bridge circuits. The circuit thus acts as a "linearizing" impedance transformation. $\underline{U}_{C1} = \underline{U}_{Rpx} = \underline{U}_0 \cdot \frac{R_4}{R_3 + R_4}$ applies to the balanced bridge. For the losses in the inductive component can be written, neglecting the capacitor losses .

This is obviously very easy to determine. The calculation applies exactly only to sinusoidal variables. This condition is much better fulfilled for the bridge with parallel resonance than for the bridge with series resonance or the Maxwell-Wien bridge, as the following figure shows using the example of the inductor used in Sect. 7.1, which was supplemented by a series resistor and a parallel resistor. Since the model is clear in this way, simulation can be used to determine how far the measured values deviate from the actual loss values.

In the simulated measurement sequence for Fig. 7.31, the resonance condition was first set by setting $U_{R1} \rightarrow$ max for the series resonant circuit and $\angle(\underline{U}_{R1}; \underline{U}_{C1}) = 0$ for the parallel resonant circuit.

In the case of a series resonant circuit, the voltage should be equal to the "loss voltage" at the resistor of the series equivalent circuit at the time of calibration. However, compensation of the voltage drops of L_x and C_1 is only made for the fundamental oscillation. The harmonics in the current simultaneously cause larger voltage drops at the inductor and smaller ones at the capacitance. Therefore, $U_{Lx} + U_{Rx} + U_{C1}$ is strongly dominated by harmonics (Fig. 7.29). During bridge adjustment, R_4 is changed. With larger non-linearities, a minimum is hardly detectable, since the fundamental oscillation is compared with a increased harmonic voltage. The measurement uncertainty is thus increasingly large. This is noted in Fig. 7.31. The "measuring points" entered nevertheless are

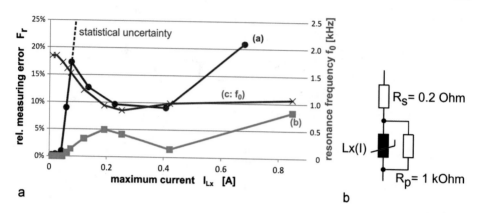

a   b

Fig. 7.31 Measurement error of the losses. (a) When determined via a series resonant circuit measuring bridge according to Fig. 7.28 (red (**a**)), a bridge with parallel resonance according to Fig. 7.30 (green, (**b**)) with corresponding resonance frequency at $C_1 = 0.5$ µF, for a parallel resonance taking into account the series resistance R_s, (**b**) supplemented inductance with the resistances R_s and R_p

based on the evaluation of the determined effective values at resonance, so that $P_v = I_{R1} \cdot U(-u_{Lx}(t) + u_{Rx}(t) + u_{C1}(t))$. The statistically assumed measurement uncertainty is much greater than the "measurement error" resulting from the concrete example.

The situation is similar for the measurement with a series resonant circuit, but the effect is completely different. The current harmonics caused by the inductive component are damped by C_1 in relation to the voltage curve. In bridge balancing, therefore, a relatively weakly distorted fundamental oscillation voltage is compared with the fundamental oscillation. Although an adjustment $U^* \rightarrow 0$ is then no longer possible, an adjustment to $U^* \rightarrow$ min is possible. When using the RMS value $\widetilde{U}^*$, practically the same minimum is obtained for small distortions as when using the rectification average value $\overline{|u^*(t)|}$. For larger distortions, the latter is more suitable for adjustment. This circuit is therefore quite suitable for determining losses in inductors.

An interesting extension of the circuit for practical use is shown in Fig. 7.32, where the combination of the inductance with 2 capacitors causes an internal resonance boost of the voltage at the inductance. This allows the inductance to be tested with higher voltages than the signal generator actually delivers. This is interesting for large signal measurements. The signal generator has to deliver essentially only the power converted into R_{px} plus the power converted into R_1.

Adjustment takes place here as in the bridge according to Fig. 7.30. U_{Lpx} is also measured. When the bridge is adjusted, U_{R1}, I_{R1} and U_{C1} are in phase. Since only R_{px} can cause this active power, the following must apply.

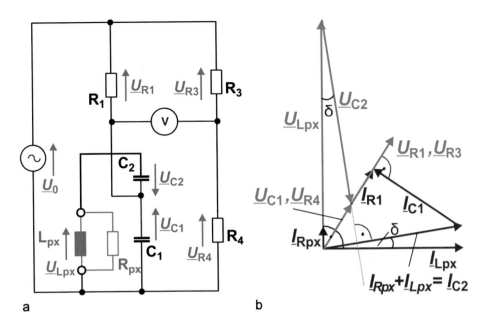

Fig. 7.32 Principle circuit diagram of the parallel resonance bridge with internal resonance voltage boost. (**a**) With parallel equivalent circuit diagram, (**b**) corresponding phasor diagram for the case of an adjustment

$$P_x = I_{R1} \cdot U_{C1} = \frac{U_0^2 \cdot R_3 R_4}{R_1 (R_3 + R_4)^2} = \frac{U_{Lpx}^2}{R_{px}} \rightarrow R_{px} = \left(\frac{U_{Lpx}}{U_0}\right)^2 \frac{R_1 (R_3 + R_4)^2}{R_3 R_4}$$

Since the two-pole, formed by $X_{C1} \| (X_{C2} + X_{Lpx} \| R_{px})$, observed from the outside, is reactive-power-free, an internal reactive power compensation can be assumed. This means:

$$Q_{C1} + Q_{C2} = Q_{Lpx} \rightarrow \omega C_1 U_{C1}^2 + \omega C_2 U_{C2}^2 = \frac{U_{Lpx}^2}{\omega L_{px}} \rightarrow L_{px} = \frac{U_{Lpx}^2}{\omega^2 C_1 \left(\frac{R_4 U_0}{R_3 + R_4}\right)^2 + \omega^2 C_2 U_{C2}^2}$$

This means that to determine the effective inductance, U_{Lpx} and U_{C2} must be determined in addition to the ratio $R_4/(R_3 + R_4)$. If δ is small enough, $U_{C2} \approx U_{Lpx} + U_{C1}$ can also be used. In the past, this circuit has been successfully used time and again to determine losses for chokes and transformers in power electronics.

In general, the use of bridge circuits as shown in Figs. 7.21–7.32 is problematic when analyzing inductors with nonlinear behavior. In particular, the determination of the loss component is often only possible with great error. Further problems occur with measurements over a wide frequency range because the reference elements $R_{1...4}$ and C_1, C_2 are not ideal but have parasitic, band-limiting components.

7.5   Measuring the B(H(T)) Characteristic

The determination of the dynamic characteristic curve B(H(t)) is one of the more frequent problems in the design of inductive components. For this purpose, a number of standardized measuring arrangements have been developed, which aim at the comparability of measurement results and take into account the properties of the materials under investigation. These standards will not be discussed here, but only referred to [3, 4]. Only the basic principles and practical applications in development practice will play a role here. The measurement of the magnetization characteristic B(H) serves to characterize a material or an inductive component or a section of an inductive component. Since the direct measurement of the flux density is associated with the introduction of a B-sensor into the magnetic circuit, which usually changes the magnetic circuit in an impermissible manner, the flux density is usually determined indirectly by voltage induction. The flux linkage Ψ is determined as a function of I. The simpler measurement tasks arise when periodic test signals are used, since the effect can then be investigated in a stationary manner.

If a magnetic section, fulfilled with a homogeneous magnetic field H, with the magnetic length l_{Fe} and the constant cross section A_{Fe} is surrounded by 2 closely coupled coils with the numbers of turns N_1 and N_2, then the following basic relationships are obtained for the case that N_1 is flowed through by a current $i_1(t)$:

$$H(t) = \frac{N_1}{l_{Fe}} \cdot i_1(t) \text{ or } \psi_2(t) = \int u_2(t)dt = N_2 \cdot \varphi(t) = N_2 \cdot A_{Fe} \cdot B(t) B(t) = \frac{\psi_2(t)}{N_2 \cdot A_{Fe}}.$$

From this, as shown schematically in Fig. 7.33, by scaling $\psi(t)$ and $i(t)$ with the geometry data, the dependence B(H(t)) can easily be represented. This leads to a (frequency-dependent) hysteresis curve for B(H).

The voltage drop $u_1(t)$ is composed of the internal ohmic voltage drop and that of the inductive component in an assumed series replacement circuit. It is therefore also possible to write

$$u_1(t) = u_{Lx1}(t) + u_{Rx1}(t) = L_{x1} \frac{d}{dt} i_1(t) + R_{x1} i_1(t)$$

If R_{x1} is known and constant, the self-induction voltage $u_{Lx1}(t)$ can be determined by a circuit that performs the operation

$$u_{Lx1}(t) = u_1(t) - k_1 i_1(t)$$

with $k_1 = R_{x1}$. The signal processing with a digital oscilloscope can be easily realized according to Figs. 7.33 and 7.34.

To determine the commutation new curve or the DC commutation curve, the reaching of a final state must be examined in a single process from the demagnetized state in order to

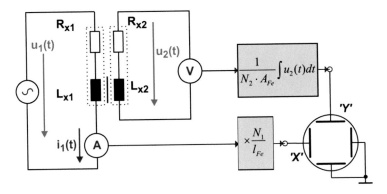

Fig. 7.33 Metrological determination of the $\Psi(I)$ characteristic of an inductor

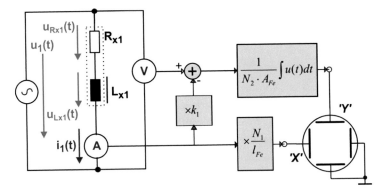

Fig. 7.34 Compensation of the effect of the ohmic resistance in the measuring branch with regard to the voltage time area by a correction factor k_1

generate a measuring point. This results in a relatively complex measurement setup. A good voltage integrator is required. Ballistic galvanometers were used for this in earlier times. Today, it is advisable to carry out the measurement in electronically controlled measuring set-ups with digital evaluation of the measurement results, which allows good compensation of measurement errors.

Determination of the Properties of a Magnetic Branch

An example is the investigation of the operating behaviour of an orthogonally pre-magnetized branch of a magnetic circuit (Fig. 7.35). Here the middle leg is drilled through in the direction of the induction field lines. The magnetization characteristic curve of the drilled section is to be investigated if a direct current pre-magnetization I_d (yellow lines) occurs through the drilled holes. The measurement is performed via the N_3 winding (white) of the middle leg. This is used to measure the small-signal inductance of the middle leg.

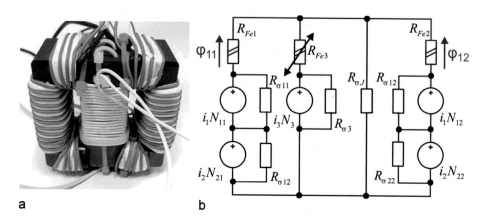

Fig. 7.35 Example of an experimental arrangement for the investigation of the influence of an orthogonal premagnetisation of the central core section with the aid of applied measuring windings. (**a**) Practical setup. (Source: Zacharias 2016), (**b**) magnetic equivalent circuit diagram of the arrangement

The structure consists of two outer legs with two windings each (red: $(N_1 = N_{11} + N_{12})$ and green: $(N_2 = N_{21} + N_{22})$), which are connected in series in pairs. One winding is used for 'energizing', while the other winding remains unloaded and is only used for flux measurement in this leg. The middle, white winding is used for flux measurement in the middle leg. When winding N_1 is connected to an AC voltage source, a current is formed that results in a flux in N_3, which can be measured there by integrating the voltage. The partial windings of N_1 are of the same size and connected in such a way that their fluxes add up in the middle leg. The characteristic curve $\psi_3(i_1)$ changes as a function of the bias current through the yellow wire, which is passed through the middle leg by means of 6 holes parallel to the flux. The higher the bias current, the lower the effective permeability. If the permeability is sufficiently low, an increased flux component through $R_{\sigma j}$ must be expected. The question arises as to the course of the permeance of the middle leg as a function of the bias current through the yellow wire.

If the inductance $L_3(I_d)$ of winding N_3 is measured, it is obtained, as shown in Fig. 7.35b from

$$L_{N3}(I_d) \approx \frac{N_3^2}{R_{Fe3} + R_{\sigma j}//R_{Fe1}//R_{Fe2}} = \frac{N_3^2}{R_{Fe3} + \dfrac{R_{\sigma j} \cdot \frac{R_{Fe1}}{2}}{R_{\sigma j} \cdot \frac{R_{Fe1}}{2}}}$$

Figure 7.36 shows the connection of the windings. The partial windings of N_1 and N_2 are connected in series. N_3 serves to observe the flux in L_{Fe3}. Because the magnetic fields of the DC premagnetization and the AC excitation of N_1 are perpendicular to each other, there is no inductive coupling between the two. The bias winding is therefore not shown. (Due to

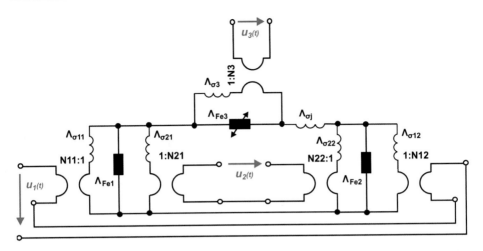

Fig. 7.36 Electrical equivalent circuit diagram of the measurement setup according to Fig. 7.35

the non-linear B(H) characteristic, there is a coupling via the parameters of the differential equations of the law of induction).

For the windings $N_{11} = N_{12} = N_{21} = N_{22} = N_1/2 = N_2/2$ was chosen. Except for the magnetic leakage resistance $R_{\sigma j}$, the drawn leakage resistances are representatives of the unlinked flux components of the windings, which are represented in the electrical equivalent circuit diagram by leakage inductances. During the measurement, practically only winding N_1 is energized.

Windings N_2 and N_3 are connected to integrators with high impedance. The voltage drop at the associated leakage inductances is therefore negligible. Due to the bifilar structure of N_1 and N_2, the total flux

$$\varphi_i(t) = \varphi_{11}(t) + \varphi_{12}(t) = \frac{2}{N_2}\int (u_{21}(t) + u_{22}(t))dt = \frac{2}{N_2}\int u_2(t)dt$$

of winding N_1 can be recorded. With N_3, the flux through the influenced magnetic section of the ferrite core is observed. This must be taken into account in a final evaluation of the separated influence of the premagnetisation on the ferrite core. The leakage inductance $L_{\sigma j}$ related to N_1 can be easily determined by short-circuiting N_3. The flux in N_3 is then practically zero if the conductor cross section of N_3 is chosen large enough. $L_{\sigma j}$ is then obtained by short-circuiting N_3 to

$$R_{\sigma j} \approx \frac{0.5 \cdot i_1 \cdot N_1}{\varphi_2(t)} = \frac{i_1 \cdot N_1}{\frac{2}{N_2} \int u_2(t)dt} = \frac{i_1 \cdot N_1 \cdot N_2}{2 \int u_2(t)dt}$$

$$\Lambda_{\sigma j} = \frac{1}{R_{\sigma j}} = \frac{2 \int u_2(t)dt}{i_1 \cdot N_1 \cdot N_2}.$$

With this parameter and the reasonable assumption that $\Lambda_{\sigma 11} \approx \Lambda_{\sigma 12} \approx \Lambda_{\sigma 21} \approx \Lambda_{\sigma 22} \approx 0$ and $R_{Fe1} = R_{Fe2}$, the value for the inductance determined by a small signal measurement at N_3 is

$$L_3(I_d) = \frac{N_3^2}{R_{Fe3}(I_d) + (R_{\sigma j}//R_{Fe1}//R_{Fe2})} = \frac{N_3^2}{R_{Fe3}(I_d) + \left(R_{\sigma j}//\frac{R_{Fe1}}{2}\right)}$$

From this, the proportion that can be assigned to the middle leg only can be calculated

$$R_{Fe3}(I_d) = \frac{N_3^2}{L_3(I_d)} - \left(R_{\sigma j}//\frac{R_{Fe1}}{2}\right) = \frac{1}{\Lambda_{Fe3}(I_d)}$$

$$L_{Fe3.N3}(I_d) = N_3^2 \cdot \Lambda_{Fe3}(I_d)$$

Figure 7.37b shows results of the small signal measurements at 1 kHz. During premagnetisation, the inductance is measured at winding N_3. By means of the calculation procedure described above, the effect of the unaffected areas of the magnetic circuit is deducted and represented as $L_{Fe3.N3}(I_d)$.

Observation of the Magnetization of a Transformer During Normal Operation in a Switching Power Supply

Using a pulse-width modulated half-bridge circuit, a charger for the periodic charging of high-voltage capacitors was realized, which showed irregularities in the capacitor discharges during longer operation and heating. It was suspected that the core of the HV transformer would go into saturation, although a safety margin was provided. This should be checked by measurement. The principle circuit of the device can be seen in Fig. 7.38 and in detail of the transformer in Fig. 7.39.

The "inner" magnetizing current in the transformer equivalent circuit diagram is

$$i_\mu^*(t) = i_1^*(t) - i_2^*(t) = N_1 \cdot i_1(t) - N_2 \cdot i_2(t) \rightarrow \frac{i_\mu^*(t)}{N_1} = i_1(t) - N_2 \frac{i_2(t)}{N_1}.$$

The current $i_1(t)$ is measured via a current transformer with a transmission ratio of 1:200, while the current $i_2(t)$ is measured via a shunt R_2. An observation of the magnetization state

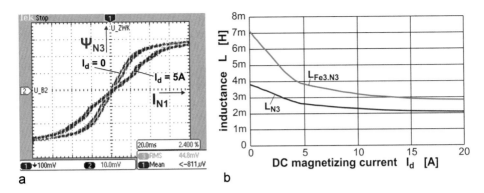

a   b

Fig. 7.37 Measurement results of the above described observation of the influence of DC pre-magnetisation. (**a**) Characteristic curve $\psi_3(I_1N_1)$ at current change. (Source: Zacharias 2016), (**b**) influence of yoke leakage on the measurement result: red - measured with yoke leakage, green - yoke leakage deducted

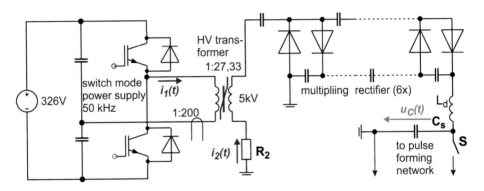

Fig. 7.38 Principle circuit diagram of a capacitor charger consisting of a PWM-controlled IGBT inverter in a half-bridge circuit, transformer and downstream voltage multiplication

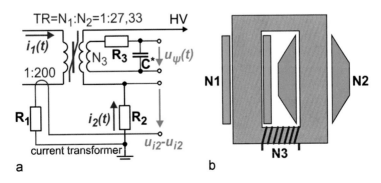

a   b

Fig. 7.39 Measurement setup for determining the magnetization state of the transformer core during operation. (**a**) Principle circuit, (**b**) winding arrangement

of the core requires a determination of the magnetization current. This can be mapped to the voltage drop $u_i(t)$ by appropriate wiring of the measuring equipment and its dimensioning. If the "inner" magnetizing current is transformed to the primary side and measured with a current transformer 1:200 ($R_1 = 20$ Ohm$\rightarrow \Delta U_{R1}/\Delta I_1 = 0.1$ V/A), one obtains

$$u_i(t) = \frac{i_\mu^*(t)}{200 \cdot N_1} R_1 = \frac{R_1}{200}\left(i_1(t) - N_2\frac{i_2(t)}{N_1}\right) = \frac{R_1}{200}i_1(t) - N_2\frac{i_2(t)}{N_1}\frac{R_1}{200} = \frac{R_1}{200}i_1(t)$$
$$- i_2(t)R_2$$

$$\text{with } R_2 = \frac{N_2}{N_1} \cdot \frac{R_1}{200} = 27.33 \cdot \frac{20\,\Omega}{100} = 2.733\,\Omega.$$

Thus, the magnetizing current $i_{\mu 1}(t) \sim u_i(t)$ can be measured with a resolution of 0.1 V/A in relation to the primary side. This quantity is fed to the X-deflection of the oscilloscope.

To determine the induction, an additional winding, not traversed by current, is applied to the core. In the example shown here, we are dealing with a transformer with strong yoke leakage. The auxiliary winding $N_3 = 7$ was arranged as shown in Fig. 7.39b. The induced voltage is then

$$u_{ind3} = N_3\frac{d}{dt}\varphi(t) = N_3 A_{Fe}\frac{d}{dt}B(t)$$

This induced voltage is then passed through a low-pass filter, so that between the Laplace-transformed voltages $\underline{u}_\psi$ and $\underline{u}_{ind}$, the relationship $\underline{u}_\psi = \frac{\underline{u}_{ind3}}{1+j\omega\cdot R_3 C^*}$ exists. For $\omega >> \frac{1}{R_3 C^*}$, you can also write $\underline{u}_\psi \approx \frac{\underline{u}_{ind3}}{j\omega\cdot R_3 C^*}$, which corresponds to an integrating behaviour of the low pass. So the output voltage of the low pass is

$$u_\psi \approx \frac{1}{R_3 C^*}\int u_{ind3}dt = \frac{N_3 A_{Fe}}{R_3 C^*}B(t).$$

The effective core area of the GUTV64 x 40 x 20 is 290 mm^2. To achieve a display scale of 2 V/0.3 T, the relation

$$\frac{1}{2\pi \cdot 50\text{ kHz}} = 3.183 \cdot 10^{-6}s < < R_3 C^* = N_3 A_{Fe}\frac{B(t)}{u_\psi(t)} = \frac{7}{2\text{ V}}0.3\frac{\text{Vs}}{\text{m}^2} \cdot 290$$
$$\times\ 10^{-6}\text{m}^2 = 3.045 \cdot 10^{-4}s$$

apply. At a capacitor $C^* = 33$ nF, a resistor $R_3 = 9227$ Ohm is therefore set. The measurement results are shown in Figs. 7.41a,b. After the discharge of the HV capacitor, an oscillation process obviously occurs on the secondary side, which is determined by the high voltage cascade and the secondary winding. Due to magnetization from the secondary

Fig. 7.40 Measured plot $B(I_\mu)$ with periodic charging of the high-voltage capacitor and set charging pause 0.1 ms. (Source: Zacharias 1994)

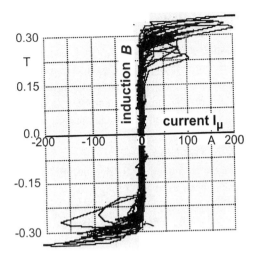

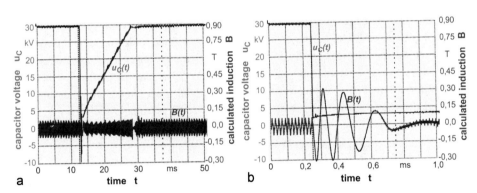

Fig. 7.41 Voltage of the HV capacitor and induction of the transformer core after a discharge with a set charging pause of 0.5 ms. (**a**) Time scale 50 ms, (**b**) time scale 1 ms. (Source: Zacharias 1994)

side, this leads to much higher induction values than those resulting from dimensioning for minimum total losses. This can lead to saturation if magnetization from the primary side is simultaneously applied in the same direction (Fig. 7.40).

The dimensioned current amplitude for the charging operation was approximately 50A. The problem was solved by extending the charging pause and a slower soft start of the inverter's PWM. The effect can be seen in Fig. 7.41. As a result, risky magnetisation of the transformer from both sides was no longer possible even with larger HV capacitors.

7.6   Transformer Measurement Methods

Measurement on Manufactured Magnetic Cores
The determination of the electrical parameters of the electrical equivalent circuit diagram, which is required for the modeling/simulation of electrical circuits, can only ever be carried out concretely due to the complex interactions. The core losses can now be predicted relatively well using FEM calculations based on standarized measurements.

Epstein Test Frame
A standardized method for determining the remagnetization losses of electrical steel sheets that has been used for a long time is the Epstein test frame described below (Fig. 7.42). Here, the losses in electrical steel are measured by means of an electrical measurement as a function of the magnetic polarization. In this process, strips of sheet metal are double-overlapped (i.e. at the beginning and the end) and layered to form a square magnetic conductor of length l_m with cross section A_{Fe}. Around the 4 sides, there are 4 windings, each with an excitation coil and a measuring coil $N_1 = 175$ or 75 and correspondingly N_2, which are each connected in series. Figure 7.42b shows the wiring of the arrangement for measurement. In addition to the main windings, a mutual inductance is inserted to compensate for the "air leakage field" (compensation transformer). It is adjusted so that without inserted core material, the induced voltage VM2 is zero. In physical terms, this means that only the effect of polarization of the material is measured. The measurement results are thus theoretically independent of the cross-sectional area of the coils filled with

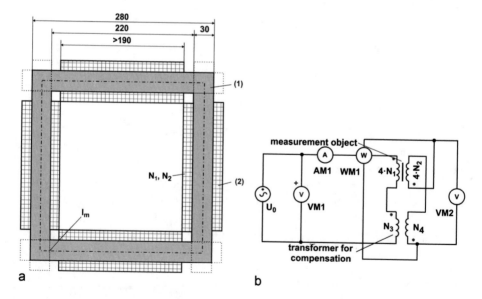

Fig. 7.42 Epstein test frame: (**a**) Design of an Epstein test frame according to IEC 60404-2 and wiring of the windings, (**b**) equivalent circuit of use

sheet metal in their normalization. According to the specifications of the standard, $l_m = 940$ mm for the construction.

With the winding N_1, the magnetic force $4 \cdot N_1 \cdot i_1$ is generated, so that the following applies

$$H(t) = \frac{4N_1 \cdot i(t)}{l_m} \rightarrow B = B_0 + J = \mu_0 H + (\mu_r - 1)\mu_0 H = (\mu_0 + (\mu_r - 1)\mu_0) \frac{4N_1 \cdot i(t)}{l_m}$$

Accordingly, a self-induction voltage u_{ind1} is induced in this winding

$$u_{ind1}(t) = 4N_1 \cdot \frac{d}{dt}\left(\mu_0 \mu_r \frac{4N_1 i(t)}{l_m} A_{Fe} + \mu_0 \frac{4N_1 i(t)}{l_m} A_{air1} \right).$$

A_{Fe} is the cross-sectional area of the coil filled by the sheet to be magnetised and A_{air} is the inner cross section of the coil through which the magnetic field also passes. The induced voltage in the $4N_2$ winding is then

$$u_{ind2}(t) = N_2 \cdot \frac{d}{dt}\left(\mu_0 \mu_r \frac{N_1 i(t)}{l_m} A_{Fe} + \mu_0 \frac{N_1 i(t)}{l_m} A_{air2} \right).$$

If a coreless mutual inductance formed by windings N_3 and N_4 is added in the way shown in Fig. 7.42b, the induced voltages are

$$u_3^*(t) = L_3^* \frac{d}{dt} i_1(t) \text{ and } u_4^*(t) = M_{34}^* \frac{d}{dt} i_1(t).$$

The difference $u_2(t) - u_4^*(t)$ is fed into the voltage path of the wattmeter.

$$u_W(t) = u_2(t) - u_4^*(t) = \left(16\mu_0 \mu_r A_{Fe} \frac{N_2 N_1}{l_m} + 16\mu_0 A_{air1} \frac{N_2 N_1}{l_m} - M_{34}^* \right) \frac{di_1(t)}{dt} \text{ or }$$

$$u_W(t) = \left(16\mu_0 (\mu_r - 1) A_{Fe} \frac{N_2 N_1}{l_m} + 16\mu_0 (A_{air1} + A_{Fe}) \frac{N_2 N_1}{l_m} - M_{34}^* \right) \frac{di_1(t)}{dt}$$

M_{12}^* is calibrated so that $M_{12}^* = 16\mu_0 (A_{air1} + A_{Fe}) \frac{N_2 N_1}{l_m}$. Then the induced (additional voltage), which is only due to the sheet material, is

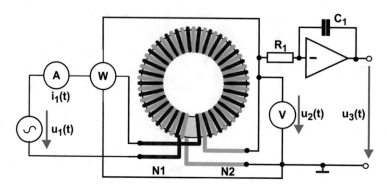

Fig. 7.43 Principle setup for measuring losses and the hysteresis characteristic of ferrite toroidal cores with high frequencies

$$u_W(t) = 16\mu_0(\mu_r - 1)A_{Fe}\frac{N_2 N_1}{l_m}\frac{di_1(t)}{dt} = 16N_2\frac{d}{dt}\frac{\mu_0(\mu_r - 1)A_{Fe}N_1 \cdot i_1(t)}{l_m}$$

$$= 16N_2 A_{Fe}\frac{d}{dt}J(t)$$

The measured voltage of the wattmeter is thus proportional to the "pure" inductive voltage in the excitation winding N_1, which is due to the inserted sheet metal. The measured power provides the losses of the lamination packs. Today's Epstein test frames are designed for about 1 kg of material to be tested, whereas earlier arrangements required 10 kg to be tested. The losses are determined related to mass.

A similar procedure can be used for ferrite materials. Figure 7.43 shows a structure with a similar mode of operation. Here, the two windings N_1 and N_2 are applied bifilar to a toroidal core. The 2nd (unloaded) winding N_2 supplies the voltage signal $u_2(t)$ for the power meter, which must be correspondingly broadband. If you want to look at the hysteresis characteristic, you have to integrate the voltage $u_2(t)$ to $u_3(t)$. Since the measurement procedure is based on the subtraction of values that are large in relation to the measurement result, broadband measuring equipment with small errors for magnitude and phase of the measured quantities current and voltage is required. However, the problem of active power measurement at a voltage sufficiently close to the sinusoidal form is somewhat relaxed, since the harmonics of the current are practically orthogonal to the oscillation of the voltage.

7.7 Calorimetric Methods for Loss Determination

For the measurement of losses in large-signal measurements with non-sinusoidal voltages, electrical measurements can theoretically be used. In practice, however, one encounters massive metrological problems. The "better" an inductive component is (high Q factor), the greater the ratio between reactive power (Q) and active power (P). If a non-sinusoidal voltage is applied to the component, all harmonics of the currents also contribute to the active power. The electrical determination of the losses would therefore have to be very broadband and with very small errors at each measurement point. If it is possible to measure the heat generated per unit of time directly, this problem can be avoided. However, other problems must be solved.

The measuring methods that work in this way are called calorimetric measuring methods. For power loss measurement, methods are used that use the heat capacity of an object and those that use the thermal resistance of an object.

Use of the Heat Capacity
If you heat a body with a certain heat capacity C_{th} with the heat loss, the temperature of this body changes.

$$(\vartheta_2 - \vartheta_1) \approx \frac{P_v \cdot \Delta t}{C_{th}}$$

Consequently, the heat capacity of the heated object is used to determine the power loss. The actual measurements are those of time and temperature differences. The power loss is then calculated as

$$P_v \approx C_{th} \cdot \frac{(\vartheta_2 - \vartheta_1)}{(t_2 - t_2)}$$

Prerequisites for such an approach are

- Known heat capacity
- Uniform heating of the object
- Low heat emission to the environment from the object

The requirement of 'low heat emission' corresponds to a high thermal resistance towards the environment. Actually the temperature curve would correspond to the following relationship

$$\vartheta = P_v \cdot R_{th}\left(1 - e^{-\frac{t}{R_{th} \cdot C_{th}}}\right) + \vartheta_{amb}$$

P_v:	Power loss
C_{th}:	Heat capacity
R_{th}:	Thermal resistance
ϑ_{amb}:	Ambient temperature

A linear course is only obtained at the beginning of the heating process. The measurement time for the application must therefore be significantly shorter than the thermal time constant of the object towards the environment. At the same time, however, the internal compensation processes during the heating of the object must be sufficiently short for the measured temperature to be representative. This method is therefore often only used as a rough measurement. On the one hand, the secondary conditions could only be fulfilled with very simply structured structures. On the other hand, losses, especially in ferrites, are also strongly temperature-dependent. If one wants to assign the losses to a certain temperature, only small changes should be allowed. The measurement errors that occur then combine in an unfavorable way.

The measurement of the temperature rise corresponds approximately to a differentiation of the temperature over time

$$\frac{d\vartheta}{dt} = \frac{P_v}{C_{th}} e^{-\frac{t}{R_{th} \cdot C_{th}}} \approx C_{th} \cdot \frac{(\vartheta_2 - 0)}{(t_2 - 0)} \text{ for } t_2 \gg \tau_{th} = R_{th} \cdot C_{th}$$

Obviously the measured temperature for $t_2 >> \tau_{th}$ is about

$$\vartheta \approx P_v \cdot R_{th} + \vartheta_{amb}$$

This means that if the ambient temperature and the thermal resistance are known, the power dissipation can be measured by measuring the achieved stationary temperature of the object. In the steady-state case, one can assume that the compensation processes have subsided. However, one must wait approximately $3 \cdot \tau_{th}$ after the start to reach such a point in time. Such a method was used to determine the power loss of storage chokes. A calorimeter was made for this purpose and fitted with a total of eight sensors on the inner and outer walls (for the principle structure, see Fig. 7.44).

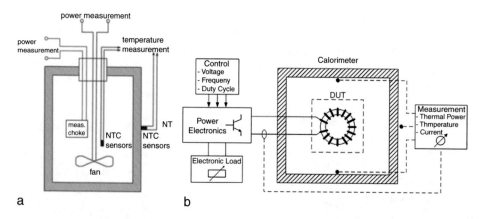

Fig. 7.44 Schematic diagram of a calorimeter for power loss measurements in a storage choke [2]. (**a**) Principle mechanical design, (**b**) wiring of the electrical components. (Source: Kleeb, Dombert Uni Kassel 2016)

In this method, the thermal resistance of the calorimetric box is used to determine the losses. From the consideration of the equation for the temperature curve, it can be seen that the losses in the storage choke are proportional to the derivative of the temperature $d\vartheta/dt$ with respect to time.

To reduce the measurement errors and thus minimize the influence of the second summand, the temperature derivative must be virtually zero. This is possible if a measurement takes long enough. In a period of about 2 to 3 hours, the measurement error would then be 1–2 mW and can be neglected.

The temperature difference was determined by means of four precision temperature sensors in and on the calorimeter. The mean value of the measured values was used in each case. The thermal resistance R_{th} was calculated by reference measurements with different power resistances, which roughly correspond to the power dissipation in the chokes. This results in a calibrated measurement setup with which the self-heating and thus the power loss of inductive components or entire assemblies can be determined. The NTC temperature sensors used in [2] belong to the category of precision NTCs and are advantageously characterized by very small dimensions, a low thermal mass and a considerably higher sensitivity. Furthermore, they have a high impedance and therefore require only a minimal measuring current. The calorimeter including the circuit was placed in a climatic cabinet. The temperature was kept at 25 °C.

Obviously, the measurement results are only based on the power dissipation, the measured values for temperature and various types of error influences. The latter are mainly sufficiently constant ambient temperature, measurement errors in temperature measurement and errors in the determination of the thermal resistance R_{th}. The ambient temperature is a problem in an open laboratory setup, as shown in Fig. 7.44, because constant values must be maintained over several hours. Since a difference of 2 temperatures is measured, the tolerance of the transducers also plays a role. In order to set a certain temperature level in the variant shown in Fig. 7.44, the outdoor temperature / ambient temperature would have to be adjustable.

This leads to a slightly modified structure (Fig. 7.45.), which is however very powerful and can be used in a variety of ways. The calorimeter in Fig. 7.45 consists of 2 chambers. The test sample is located in the inner chamber. This inner chamber is heated up to the set temperature T_{maes}, at which the measurement of the losses is to be carried out. The chamber is surrounded by a thermally insulating, i.e. poorly heat-conducting wall. This wall is lined inside and outside with aluminium sheet. This makes it easy to achieve almost identical temperatures on the inside and outside of the partition wall. Nevertheless, the temperatures are measured at different points and average values are formed for the inside ϑ_i and the outside ϑ_a. The heat flow from the inside to the outside in the stationary state is then

$$P = \frac{dQ}{dt} = \frac{(\vartheta_i - \vartheta_a)}{R_{th}}$$

(Q...heat quantity)

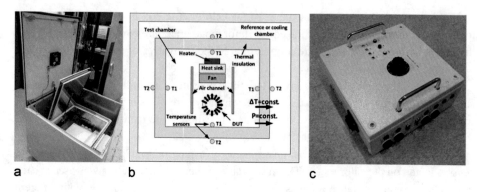

Fig. 7.45 Design of a double-shell calorimeter with controlled reference temperature. (**a**) Open inner and outer chamber, (**b**) basic functional design of the calorimeter, (**c**) control box for manual control and connection of the control computer [2, 5]. Photos: KDEE/University of Kassel. (Source: Kleeb, Dombert/Uni Kassel 2016)

The temperature level ϑ_a in the space between the inner and outer chamber is also controlled by a separate heater. In this way, a constant heat flow through the inner wall is achieved. To ensure that the stationary temperature conditions are set as quickly as possible, both the air in the inner chamber and the air in the space between are circulated. The power consumption of the internal fans has to be measured and is in principle included in the power balance of the interior. The power consumption of the fans changes slightly with the temperature and must be taken into account if necessary. Otherwise, the temperatures in the inner chamber and the intermediate chamber / reference chamber are constant. Therefore, the heat flow from the inner chamber to the reference chamber is also constant.

If an object that produces power loss is now brought into the inner chamber, the equilibrium is initially disturbed. The excess heat causes the temperature to rise and the inner control circuit reduces the heating power to return to the original temperature. In the steady state, the control deviation is zero. The amount by which the heating power had to be reduced corresponds to the power loss of the object. It is recommended to bring the object into the inner chamber already during the heating phase. Otherwise, it is necessary to wait the time for the object to heat up by itself until a sufficiently stable state for reading out the measured values is reached. This compensation method transforms a problem of measuring alternating current losses, which is difficult to measure, into a simple problem of measuring direct current losses.

Figure 7.46 illustrates the effect of the procedure. After increasing the power loss, the heating power requirement decreases in order to maintain the set temperature level. The temperature-dependent controlled heating power decreases. In the new stationary state, the difference to the initial state is the power loss introduced. The set temperature level of the inner chamber is also the ambient temperature of the material at which the measurement is made. Inside an inductive component, the heat conduction processes in the component

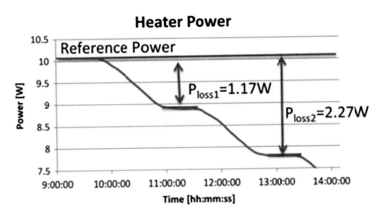

Fig. 7.46 Heat output curve with stepwise change of the operating point of the test sample. (source: Dombert, Kleeb [2, 6])

cause different temperatures depending on the location. This can be counteracted by special designs (e.g. cooling channels between core and winding). To monitor the measuring process, it is useful to insert additional temperature sensors in the measured object. A high sensitivity of the measuring arrangement can be achieved by keeping the thermal resistance R_{th} between the inner and outer chamber high with a small heat capacity C_{th}. The thermal insulation should therefore be made with foamed or porous thermal insulation material that is permanently suitable for the maximum temperature level (for example aero gele). Despite this effort, a time interval of more than 1 h per measured value must be expected. During this time, the conditions must be kept constant.

Since the object to be measured generates its power loss through electromagnetic effects in the case described, the required electrical power must be supplied via electrical lines. In high current applications, another technical problem is encountered. The electrical cables with the necessary cross section of the leads must also be fed through the inner wall, but they represent a shunt for the heat conduction from the inner chamber to the outer chamber and thus a source of error for the measurements. Two measures are to be mentioned here, which eliminate the problem partially or almost completely.

- Reduction of the cross section for the length of the passage through the heat-insulating wall and cooling via the connection points
- Insulated installation of the electrical supply lines through the intermediate chamber and compensation of heat conduction by heating this section to the internal temperature at the output by a third temperature controller. The cross section when passing through the heat insulating wall between the reference chamber and the environment is also reduced to minimize this heating power.

Figure 7.47 shows two of the views of the control program for the calorimeter as shown in Fig. 7.45. Due to the design with 2 chambers, several sensors and fans for air circulation,

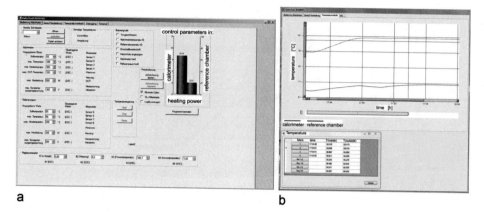

Fig. 7.47 Control of the heating of the test specimen. (**a**) User interface of the control program for the calorimeter, (**b**) measured temperature curve of measuring chamber and reference chamber after insertion of the test sample. (source: Kleeb, Dombert/University of Kassel 2016 [5])

the calorimeter has become a complex device to be controlled. A high degree of flexibility is achieved by the possibility of setting several parameters (Fig. 7.47a). Since the thermal compensation curtains are similar but different for each object to be measured, the temperatures of the inner chamber and reference chamber are also displayed (Fig. 7.47b). In the case shown here, one can see the progression of these temperatures after introducing and starting a new measurement. By opening the chambers, the temperature drops briefly and then rises due to the additional power loss. The actual measurement process takes more than an hour, since a time must pass before a quasi-stationary state is reached. Appropriately tuned control loops lead to this state. The measurement itself is carried out automatically.

References

1. Katalog Stanzwerk AG Unterentfelden
2. Ferroxcube Soft Ferrites and Accessories Handboook 2009
3. DIN EN 60404-2:2009-01: Titel (deutsch): Magnetische Werkstoffe - Teil 2: Verfahren zur Bestimmung der magnetischen Eigenschaften von Elektroband und -blech mit Hilfe eines Epsteinrahmens (IEC 60404-2:1996 + A1:2008); Deutsche Fassung EN 60404-2:1998 + A1:2008
4. DIN EN 10252:1997-0:Titel (Deutsch): Magnetische Werkstoffe - Verfahren zur Messung der magnetischen Eigenschaften von Elektroblech und -band bei mittleren Frequenzen; Deutsche Fassung EN 10252:1997
5. Kleeb, Th.; Dombert, B.; Araújo, S.; Zacharias, P.: Loss Measurement of magnetic components under real application conditions. 15th European Conference on Power Electronics and Applications (EPE), 2-6 Sept. 2013, Electronic ISBN: 978-1-4799-0116-6
6. Kleeb, Th.; Dombert, B.; Araújo, S.; Nöding, Ch. Zacharias, P.: Kalorimeter zur der Verluste von Baugruppen bis 250 W. Interner Bericht. KDEE, Universität Kassel 2016

Apparent Power and Volume at Inductive Components

8

8.1 Basic Considerations Using the Example of Transformers

In this chapter, it shall be investigated which principal factors influence the size and power/ apparent power and in which way. For this purpose, abstract fundamental considerations are used, which do not aim at the calculation of a specific case, but at determining the order of certain influencing factors.

For this purpose, a heat conduction problem is first investigated:

A sphere with radius R is given, in whose interior heat is produced with a certain, constant power density p_V. The heat can only be dissipated via the surface by convection to the environment, which has a temperature T_a. The temperature in the centre of the sphere, where it is maximum in a given case, should not exceed a certain value T_{imax}. The heat is transported from the inside of the sphere to the surface via a thermal conductivity λ, which is assumed to be the same everywhere.

For a given heat transfer coefficient k_o, the external thermal resistance R_{tha} of the arrangement can be described as

$$R_{tha} = \frac{T_A - T_O}{P_{Vtot}} = \frac{k_O}{4\pi R^2}$$

Inside the sphere, a temperature difference results at a spherical shell of thickness dr

$$dT = p_V \cdot \frac{4}{3}\pi r^3 \cdot \frac{dr}{\lambda \cdot 4\pi r^2} = p_V \cdot \frac{rdr}{3\lambda}$$

If you integrate all temperature differences from the centre to the surface, you get

© Springer Fachmedien Wiesbaden GmbH, part of Springer Nature 2022
P. Zacharias, *Magnetic Components*,
https://doi.org/10.1007/978-3-658-37206-4_8

$$T_{i\max} - T_O = p_V \cdot \frac{R^2}{6\lambda}$$

The total power loss produced is

$$P_{Vtot} = p_V \cdot \frac{4\pi R^3}{3}$$

This results in an internal thermal resistance of

$$R_{thi} = \frac{T_{i\max} - T_O}{P_{Vtot}} = \frac{p_V \cdot \frac{R^2}{6\lambda}}{p_V \cdot \frac{4\pi R^3}{3}} = \frac{1}{8\pi R\lambda}$$

The total permissible heat output inside the sphere can then be calculated as

$$P_{Vtot} = \frac{T_{i\max} - T_A}{R_{thi} + R_{tha}} = \frac{T_{i\max} - T_A}{\frac{1}{8\pi R\lambda} + \frac{k_O}{4\pi R^2}} = (T_{i\max} - T_A)\frac{4\pi R^2}{\frac{R}{2\lambda} + k_O}$$

For sufficiently small radii R, the following approximation applies for the permissible power loss

$$P_{Vperm} \sim \frac{R^2}{k_O}, \text{or for the thermal resistance } R_{th} = \frac{T_{\max} - T_O}{P_{v\max}} \sim \frac{1}{R^2}$$

while for larger dimensions, the approximation

$$P_{Vtot} \sim \lambda R \left(\text{or for the thermal resistance } R_{th} \sim \frac{1}{R} \right)$$

can orientate.

With a proportional increase of the dimensions, winding window A_{Cu} and magnetization cross section A_{Fe} each increase $\sim R^2$, while the mean winding length l_w only changes $\sim R$. This should be taken into account by an enlargement factor k_1. With $R = k_1 \cdot R_0$ and a fill factor k_f of the winding space with conductor cross section $A_{Cu} = k_f \cdot A_W$

$$R_{Cu} = \frac{\rho_{Cu} N^2 l_m}{k_f A_W} = \frac{\rho_{Cu} \cdot N^2 \cdot k_2 \cdot l_{m0}}{k_1^2 k_f A_{W0}} = \frac{\rho_{Cu} \cdot N^2 \cdot l_{m0}}{k_1 k_f A_{W0}}$$

The core must not reach saturation and is therefore loaded with a maximum induction B_{max}, whereby the following applies

$$B_{max} = \frac{\sqrt{2}U}{2\pi f \cdot A_{Fe} \cdot N} = \frac{\sqrt{2}U}{2\pi f \cdot k_1^2 A_{Fe0} \cdot N} \quad or \, N = \frac{\sqrt{2}U}{2\pi f \cdot k_1^2 A_{Fe0} \cdot B_{max}}$$

Thus, for a given voltage U, the following applies to the winding resistance

$$R_{Cu} = \frac{\rho_{Cu} \cdot l_{m0}}{k_1 k_f A_{W0}} \cdot N^2 = \frac{\rho_{Cu} \cdot l_{m0}}{k_1 k_f A_{W0}} \cdot \left(\frac{\sqrt{2}U}{2\pi f \cdot k_2^2 A_{Fe0} \cdot B_{max}}\right)^2 = \frac{\rho_{Cu} \cdot l_{m0} \cdot 2 \cdot U^2}{k_1^5 k_f A_{W0}(2\pi f \cdot A_{Fe0} \cdot B_{max})^2}$$

The winding losses are then

$$P_{VCu} = 2 \cdot I^2 \cdot R_{Cu} = 2 \cdot I^2 \cdot \frac{\rho_{Cu} \cdot l_{m0} \cdot 2 \cdot U^2}{k_1^5 k_f A_{W0}(2\pi f \cdot A_{Fe0} \cdot B_{max})^2} = \frac{4 \cdot S^2 \cdot \rho_{Cu} \cdot l_{m0}}{k_1^5 k_f A_{W0}(2\pi f \cdot A_{Fe0} \cdot B_{max})^2}$$

$$< \frac{T_{max} - T_{amb}}{R_{th}}$$

They must be smaller than the dissipative losses at a certain maximum temperature difference. If the core losses are neglected, this results in the maximum apparent power for a design of

$$S < 2\pi f \cdot A_{Fe0} \cdot B_{max} \cdot \sqrt{\frac{\left(\frac{T_{max} - T_{amb}}{R_{th}}\right) \cdot k_1^5 k_f A_{W0}}{4 \cdot \rho_{Cu} \cdot l_{m0}}} \sim k_1^{\frac{5}{2}},$$

where R_{th}, as shown above, is dependent on the growth factor k_2. Using the proportionalities of the above shown dependencies, depending on the size, a proportionality results for the apparent power of $S \sim k_1^{\frac{5}{2}+(1...2)} = k_1^{3.5...4.5}$. If the core volume with the growth factor k_2 is implemented to $V = k_2 \cdot V_0 = k_1^3 \cdot V_0$ or $k_1 = k_2^{\frac{1}{3}}$, the apparent power increases with the volume of $S \sim k_2^{1.167...1.5}$. Figure 8.1 shows the apparent power of transformers based on cut strip cores and shows the dependence of $S \sim V^{1.238}$. Obviously higher power densities can be achieved at higher operating frequencies.

The power density here obviously increases with $\frac{S}{V} \sim k_1^{0.238}$, that is about $\frac{S}{V} \sim \sqrt[4]{k_1}$.

As the formula derived above also shows, the apparent power can be slightly increased with increased frequency under the conditions given. For 400 Hz applications, a lower number of windings is theoretically required by a factor of 9 compared to 50 Hz under the above assumptions of loss. As Fig. 8.2 shows, a factor '6' was set to take the core losses into account.

The current densities actually applied for dimensioning are also interesting, as shown in Fig. 8.3.

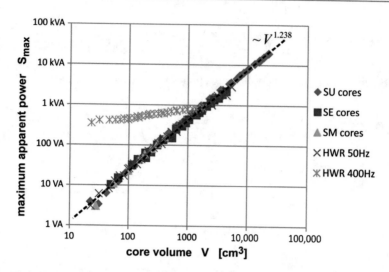

Fig. 8.1 Dependence of the maximum apparent power for cut strip-wound cores with different proportions, according to [1]

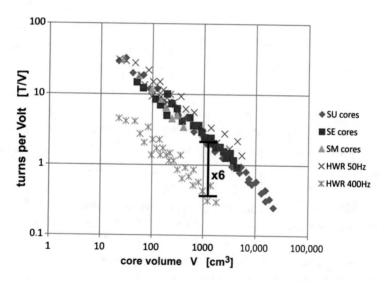

Fig. 8.2 Number of turns set for cut strip-wound cores with different proportions according to [1]

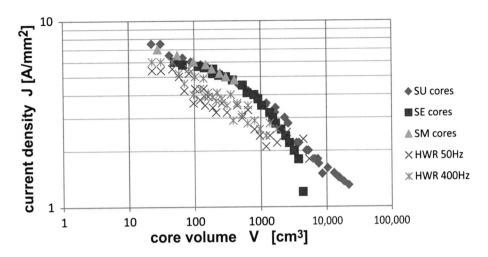

Fig. 8.3 Current densities used for transformer dimensioning with cut strip-wound cores according to [1]

8.2   Size of Optimized Ferrite Chokes and Transformers

For alternating current applications of electromagnetic components, the winding and core losses to be dissipated play a decisive role for the size. Therefore, these interrelationships are to be analysed in more detail here. The total losses result from the sum of the core losses and winding losses. If only the core losses and the ohmic losses at low frequencies are taken into account (i.e. without skin and proximity effect), one obtains N

$$P_V = P_{Vcore} + P_{Vwinding} = V_c \cdot k_e \cdot \left(\frac{\widehat{B}}{B_0}\right)^\alpha \cdot \left(\frac{f}{f_0}\right)^\beta + k_W I^2 \cdot N^2 \frac{\rho \cdot l_{mCu}}{k_f \cdot A_W}$$

V_c = core volume	$\widehat{B}$ = amplitude of induction	f = frequency
N = number of windings	B_0 = reference value (1 T)	f_0 = reference value (1 kHz)
A = material parameters	β = material parameters	k_e = material parameters
A_W = winding cross section l_{mCu} = average winding length	k_f = fill factor	
P = spec. Resistance	k_W = number of windings	
	k_w = 1 or 2	

With $\widehat{B} = \frac{\sqrt{2}U}{\omega \cdot N}$, it follows

$$P_V = V_c \cdot k_e \cdot \left(\frac{\sqrt{2}U}{2\pi \cdot f \cdot N \cdot A_{Fe} \cdot B_0} \right)^\alpha \cdot \left(\frac{f}{f_0} \right)^\beta + k_W I^2 \cdot N^2 \frac{\rho \cdot l_{mCu}}{k_f \cdot A_W} = \xi(U) \cdot \frac{1}{N^\alpha} \cdot f^{\beta - \alpha}$$
$$+ \chi(I) \cdot N^2$$

or the optimization problem

$$P_V = \xi(U) \cdot \frac{1}{N^\alpha} \cdot f^{\beta - \alpha} + \chi(I) \cdot N^2 \Rightarrow \min$$

with $\xi(U) = V_c \cdot k_e \cdot \left(\dfrac{\sqrt{2}U}{2\pi \cdot A_{Fe} \cdot B_0} \right)^\alpha \cdot \left(\dfrac{1}{f_0} \right)^\beta$ and $\chi(I) = k_W I^2 \cdot \dfrac{\rho \cdot l_{mCu}}{k_f \cdot A_W}$.

The condition for a local extreme value lists

$$\frac{dP_V}{dN} = -\alpha \cdot \xi(U) \cdot \frac{1}{N^{\alpha + 1}} \cdot f^{\beta - \alpha} + 2 \cdot \chi(I) \cdot N = 0$$

and finally

$$\text{to } N_{opt} = \left(\frac{\alpha}{2} \cdot \frac{\xi(U)}{\chi(I)} \right)^{\frac{1}{\alpha + 2}} \cdot f^{\frac{\beta - \alpha}{\alpha + 2}}$$

as the optimal solution for N_{opt}. This means that the optimum number of windings decreases with increasing frequency $\sim f^{\frac{\beta - \alpha}{\alpha + 2}}$. For the basic approximations $\alpha = 2$ and $\beta = 1$, this means: $N_{opt} \sim f^{\frac{1}{4}}$.

Using the optimum number of windings leads to the minimum power loss under the assumed conditions.

$$P_{Vopt} = \left(\left(\frac{2}{\alpha} \right)^{\frac{\alpha}{\alpha + 2}} + \left(\frac{\alpha}{2} \right)^{\frac{2}{\alpha + 2}} \right) \cdot \chi(I)^{\frac{\alpha}{\alpha + 2}} \cdot \xi(U)^{\frac{2}{\alpha + 2}} \cdot f^{2\frac{\beta - \alpha}{\alpha + 2}} = \sim f^{2\frac{\beta - \alpha}{\alpha + 2}} \text{ respectively}$$

For the simplified assumption $\alpha = 2$ and $\beta = 1$, this results in the dependency

$$P_{Vopt} = 2 \cdot \sqrt{\chi(I) \cdot \xi(U)} \cdot f^{-0,5} = \sim f^{-0,5}.$$

The maximum power loss of a component with a thermal resistance R_{th} is then reached at

$$P_{Vopt}(U, I) \cdot R_{th} = (T_{\max} - T_{amb})$$

($T_{\max}$... maximum internal temperature, T_{amb}... ambient temperature)

That is, voltage and/or current are set to meet the condition above. This results in 'S' implicitly for the maximum apparent power:

$$\left(\left(\frac{2}{\alpha}\right)^{\frac{\alpha}{\alpha+2}} + \left(\frac{\alpha}{2}\right)^{\frac{2}{\alpha+2}}\right) \cdot \chi(I)^{\frac{\alpha}{\alpha+2}} \cdot \xi(U)^{\frac{2}{\alpha+2}} \cdot f^{2\frac{\beta-\alpha}{\alpha+2}} \cdot R_{th} = (T_{max} - T_{amb})$$

and explicitly with the agreements made

$$S_{opt} = \frac{2\pi \cdot A_{Fe} \cdot B_0 \cdot f_0 \left(\frac{f}{f_0}\right)^{\frac{\alpha-\beta}{\alpha}}}{(V_c \cdot k_e)^{\frac{1}{\alpha}} \cdot \sqrt{\frac{k_W \cdot \rho \cdot l_{mCu}}{k_f \cdot A_W}}} \cdot \left(\frac{\frac{T_{max} - T_{amb}}{R_{th}}}{\left(\frac{2}{\alpha}\right)^{\frac{\alpha}{\alpha+2}} + \left(\frac{\alpha}{2}\right)^{\frac{2}{\alpha+2}}}\right)^{\frac{\alpha+2}{2\alpha}}$$

As can be seen from the formula, even with a known core shape (right factor), the achievable apparent power depends in a complex way on the material. In addition to the frequency, the parameters α and β also influence the result via links to other parameters. If only the frequency influence is considered, the achievable apparent power of a core follows proportionality $S_{opt} \sim f^{\frac{\alpha-\beta}{\alpha}}$ and $S_{opt} \sim \left(\frac{\frac{T_{max}-T_{amb}}{R_{th}}}{\left(\frac{2}{\alpha}\right)^{\frac{\alpha}{\alpha+2}} + \left(\frac{\alpha}{2}\right)^{\frac{2}{\alpha+2}}}\right)^{\frac{\alpha+2}{2\alpha}}$. For the basic approximations of α and β, one can expect $S_{opt} \sim \sqrt{f}$ and $S_{opt} \sim \frac{T_{max}-T_{amb}}{2 \cdot R_{th}}$.

The denominator of the bracket expression has the value 2 in a relatively wide range as shown in Fig. 8.4. For $\alpha = 2$, it is identical to 2. In this case, the two loss components of core and winding are also equal. But also for the relevant range of α, the assumption of equal loss components is a good approximation for finding the loss minimum.

The solution of the optimization task described above requires knowledge of the Steinmetz parameters α, β and k_e. Functions of this kind are planes in double logarithmic 3D representation. In 2D representation with additional parameters, they are parallel straight lines. With satisfactory accuracy over a wide frequency or induction range, however, this only applies to a few materials. Especially power ferrites in the higher frequency range show strong deviations from this approach. As described in Chap. 9, the (multidimensional) value fields can be described with practically any precision by approximation functions, but the optimization problem can then only be solved numerically. This makes it difficult to handle the data.

Therefore, a method is described below, which provides the information for optimum number of turns and maximum apparent power in a relatively simple way.

As an illustration, Fig. 8.5 shows the losses of the materials N27, N87 and N88 specified by the manufacturer TDK/EPCOS.

The above mentioned deviations of the characteristic curves from the assumed linearity are clearly visible. In order to map the frequency dependence in a range as large as possible, the best possible approximation of the Steinmetz parameters is sought at the left and right edges (blue line) of the primary data. A pair of values at constant frequency is selected at

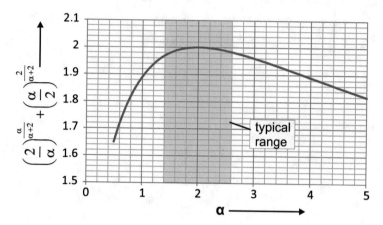

Fig. 8.4 Factor dependence on the material parameter α

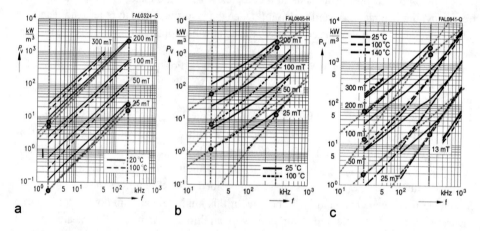

Fig. 8.5 Specific core losses of various materials. (**a**) N27, (**b**) N87 and (**c**) N88 according to TDK/EPCOS with random samples to determine the Steinmetz parameters for 100 °C at low and high frequencies

both edges and supplemented by a further value at the other edge. Because of the partly non-linear dependencies, the courses are "estimated" from tangents to the curves. Thus, the method remains consistent at least in the vicinity of the edges. The 3 Steinmetz parameters on the left and right edges of the value range of the diagrams can be estimated for each material from the two triplets of points (Table 8.1).

From the size $(\alpha - \beta)/\alpha$, it can be seen that the growth of apparent power with frequency differs greatly between materials in the upper and lower frequency range. An assessment of the maximum apparent power can only be given for a specific core shape under certain conditions. In the following, the three materials will therefore be compared using the example of an ETD 49/25/16 (Table 8.2).

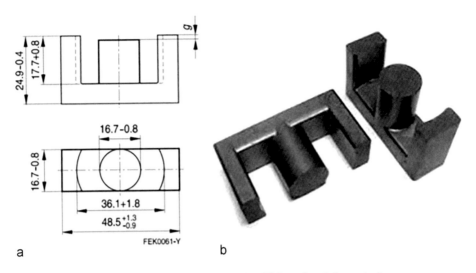

Fig. 8.6 Core type ETD 49/25/16. (Source: TDK/EPCOS product information)

Table 8.1 Steinmetz parameters determined according to the method described for materials N27, N87 and N88

	N27_LF	N27_HF	N87_LF	N87_HF	N88_LF	N88_HF
α	2348	2.15	2885	2.49	2565	2.31
β	1239	1.31	1295	2.35	1302	2.12
kC	94,476	64.15	96,297	0.201	59,189	0.543
$(\alpha - \beta)/\alpha$	0.4726	0.3900	0.5509	0.05708	0.4924	0.08299

Table 8.2 Characteristic values for core ETD 49/25/16 according to Fig. 8.6

Size	Value	Unit	Size	Value	Unit
V_e	24.1	cm^3	T_{max}	100	°C
A_{Fe}	2.11	cm^2	T_{amb}	50	°C
A_w	343	mm^2	f_o	1	1 kHz
l_{Cu}	83.72	mm	B_o	1	1 T = 1 Vs/m^2
R_{th}	8	K/W	ρ_{Cu}	0.01724	Ohm*mm^2/m
k_f	0.4	–	k_w	2	–

The optimization area is limited by the maximum possible induction

$$B = \frac{\sqrt{2}U}{2\pi f \cdot N^* \cdot A_{Fe}} < B_{max} \text{ or } N^* > \frac{\sqrt{2}U}{2\pi f \cdot B_{max} \cdot A_{Fe}}$$

If $N^* = N_{opt}$, then the limit of the optimization area is reached and the following applies to the frequency limit

$$N^* = N_{opt}$$

This condition leads to the implicit relationship

$$f_{min} = \left(\frac{\sqrt{2}S_{opt}(f_{min})}{2\pi \cdot A_{Fe}}\right)^{\frac{2}{2+\beta}} \cdot \left(\frac{1}{B_0}\right)^{\frac{-\alpha}{2+\beta}} \left(\frac{1}{B_{max}}\right)^{\frac{\alpha+2}{\beta+2}} \cdot \left(\frac{2 \cdot k_W \cdot \rho \cdot l_{mCu}}{\alpha \cdot V_c \cdot k_e \cdot k_f \cdot A_W}\right)^{\frac{1}{2+\beta}} \cdot f_0^{\frac{\beta}{2+\beta}},$$

so that the minimum frequency for optimal conditions can be determined.

But it is also clear that

$$V_c \cdot k_e \cdot \left(\frac{B_{max}}{B_0}\right)^{\alpha}\left(\frac{f}{f_0}\right)^{\beta} + k_W I^2 \cdot N^2 \frac{\rho \cdot l_{mCu}}{k_f \cdot A_W} \leq \frac{T_{max} - T_{amb}}{R_{th}}$$

must be fulfilled. This is obtained by changing over to the apparent power S at a given maximum induction B_{max}:

$$S \leq f \cdot B_{max} \cdot \sqrt{\frac{\left(\frac{T_{max} - T_{amb}}{R_{th}}\right) - V_c \cdot k_e \cdot \left(\frac{B_{max}}{B_0}\right)^{\alpha}\left(\frac{f}{f_0}\right)^{\beta}}{k_W \cdot \left(\frac{\sqrt{2}}{2\pi \cdot A_{Fe}}\right)^2 \frac{\rho \cdot l_{mCu}}{k_f \cdot A_W}}}$$

as a limitation. This is an absolute barrier and has nothing to do with an optimum. According to this formula, the maximum apparent power is reached when the core losses are set to zero. Then

$$S \leq f \cdot B_{max} \cdot \sqrt{\frac{\left(\frac{T_{max} - T_{amb}}{R_{th}}\right)}{k_W \cdot \left(\frac{\sqrt{2}}{2\pi \cdot B_{max} \cdot A_{Fe}}\right)^2 \frac{\rho \cdot l_{mCu}}{k_f \cdot A_W}}}$$

is a barrier given for a ferromagnetic core type. On the other hand, the apparent power is zero when.

$$\left(\frac{T_{max} - T_{amb}}{R_{th}}\right) = V_c \cdot k_e \cdot \left(\frac{B_{max}}{B_0}\right)^{\alpha}\left(\frac{f}{f_0}\right)^{\beta} \quad \text{or} f_{min} = f_0 \cdot \left(\frac{(T_{max} - T_{amb})}{V_c \cdot k_e \cdot R_{th}}\right)^{\frac{1}{\beta}}\left(\frac{B_0}{B_{max}}\right)^{\frac{\alpha}{\beta}}.$$

In addition to this (loss-defined) limit, there is the limit independent of induction, which setsthe material at the frequency f_{max} at which the initial permeability μ_i' 'collapses'. These statements are summarized in Table 8.3 for ETD49 and 100 °C for the materials N27, N87 and N88. B_{max} was determined from the point of intersection of the tangent to the magnetization curve in the saturation range with the ordinate. The value f_{max1} corresponds

Table 8.3 Upper and lower optimization limits for the operating frequency of an ETD 49 with 2 windings at 100 °C

	α	β	k_C (kW/m³)	B_{max} @100 °C (T)	f_{min} (kHz)	f_{max1} (kHz)
N27	2.15	1.31	64.15	0.405	12.84	2600
N87	2.49	2.35	0.201	0.375	19.09	1700
N88	2.31	2.12	0.543	0.395	19.39	1400

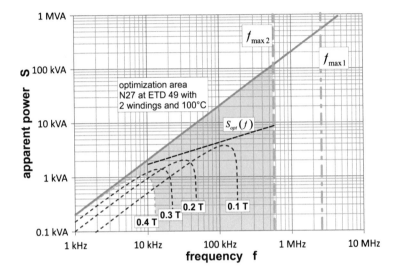

Fig. 8.7 Optimization area for N27 at different maximum flux densities and frequencies using the example of an ETD 49 with marked optimization results $S_{opt}(f)$ for $B_{max} = 0.4$ T

to the frequency at which μ_i' drops to half its value in the lower frequency range. The initial permeability μ_i depends on the temperature. Usually only the curve at 25 °C is available. This was therefore chosen as a parameter. A magnetic component cannot be sensibly operated at the "resonance frequency". For comparison purposes, the value f_{max2} chosen is the one characterized by the relationship $\mu_i' = 20 \cdot \mu_{ii}''$, i.e. the imaginary part of the complex permeability is 5% of the real part. This specification is based on the assumption that the calculation model becomes too inaccurate if the relative losses are even greater. Furthermore, the primary data of the approximation model are usually far away from f_{max1}, so that an extrapolation is not meaningful. The imaginary part of the complex permeability is also strongly dependent on temperature. Since usually only the values for 25 °C are available, these were used for comparison purposes instead of the values at 100 °C, which are important for operation.

Figure 8.7 shows these ratios for N27 graphically. It becomes obvious that a smaller maximum induction results in a massive restriction of the optimization area. The apparent

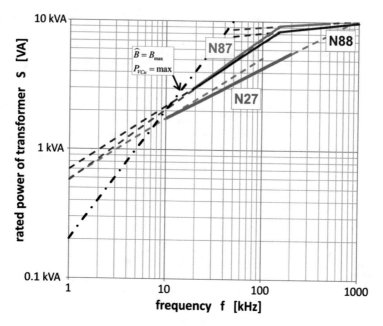

Fig. 8.8 Optimization results for the transmittable power of a transformer based on an ETD 49 with 2 windings at $T_{max} = 100\ °C$, approximated by the tangents at the edges of the optimization area

power is thus limited upwards as well as downwards in the frequency range. The upper frequency limit is set by the core material itself.

The optimization results for the materials N27, N87 and N88 are shown in Fig. 8.8. From the behavior of the power dissipation at the upper and lower frequency limits of the representations in the data sheets, the tangents were placed at the characteristic $S_{opt}(f)$ and the rough curve was compiled from this. With N27, the curve is less bent than with the other two materials. However, the primary data also stop at 200 kHz. At higher frequencies, there is probably also a stronger increase with frequency, which the curves of N87 and N88 suggest. This would cause a "flattening" of the rise of S(f). The curves were drawn with full lines in the frequency domain only as far as primary data is available. The dashed area is thus a range of extrapolated speculation. N87 and N88 are apparently very similar materials in terms of apparent power at 100 °C. However, N88 has its minimum specific power dissipation at 140° and therefore allows a higher apparent power than N87 with the same core, if a maximum temperature of 140 °C is applied. However, the construction and insulation materials used must then also correspond to thermal class F (up to 155 °C) instead of thermal class A (up to 105 °C) or E (up to 120 °C).

Consequently, a quite high apparent power could be achieved if only winding losses were incurred. And the difference between optimum power and maximum power becomes greater and greater the higher the frequency is, as Fig. 8.7 shows. It is therefore worth looking for coreless alternatives for magnetic components in the high frequency range.

The transmittable power at maximum frequency can be analytically formulated as follows

$$
S_{opt.\,max} = \frac{2\pi \cdot A_{Fe} \cdot B_0 \cdot f_0 \left(\frac{f_{max2}}{f_0}\right)^{\frac{\alpha-\beta}{\alpha}}}{(V_c \cdot k_e)^{\frac{1}{\alpha}} \cdot \sqrt{\frac{k_w \cdot \rho \cdot l_{mCu}}{k_f \cdot A_w}}} \cdot \left(\frac{\frac{T_{max} - T_{amb}}{R_{th}}}{\left(\frac{2}{\alpha}\right)^{\frac{\alpha}{\alpha+2}} + \left(\frac{\alpha}{2}\right)^{\frac{2}{\alpha+2}}}\right)^{\frac{\alpha+2}{2\alpha}}
$$

with $B_0 = 1$ T and $f_0 = 1$ kHz.

This limit is fictitious because it is impossible to operate the component at the frequency limit. The formula only contains 'fixed' parameters that characterize the core. The values are useful for a material comparison. As 'Steinmetz parameter', the parameters for high frequencies should be used. If one does this, one obtains the results shown in Table 8.4.

The underlying quality of the primary data used to construct Table 8.4 varies considerably. Both the density of the information data and the ranges of the influencing variables are not uniform. The comparison of materials therefore works well, especially for materials from one manufacturer. For this comparison and for comparison between different manufacturers, the fictitious apparent power at 1 MHz was calculated. To sound out the material limits, the value $S_{opt.max}(f_{max2})$ was calculated and given. The absolute values are not so important for the statements, since wide extrapolations are subject to large uncertainties furthermore there is a mixing of small and large signal parameters. But the relations are interesting. Thus, very high power densities can be achieved at high cut-off frequencies, where $k_f = 0.4$ is by far no longer realistic attainable taking the skin effekt into

Table 8.4 Maximum transmittable apparent power of a transformer with ETD49 core with the above agreements at $T_{max} = 100\ °C$ for different materials

Material	Manufacturer	B_{max} (100 °C) [T]	$S_{opt.\,max}$ (1 MHz) [kVA]	f_{max1} [kHz]	f_{max2} [kHz]	$S_{opt.\,max\,2}$ [kVA]	μ_i (100 °C)	Type
N27	TDK	0.405	10.5	2600	550	8.31	4600	MnZn
N87	TDK	0.375	10.2	1700	500	9.82	4500	MnZn
N97	TDK	0.400	10.8	2000	520	10.4	4000	MnZn
3F4	Ferroxcube	0.335	12.1	6500	2050	11.0	1100	MnZn
3F5	Ferroxcube	0.320	18.5	8000	2200	18.6	750	MnZn
4F1	Ferroxcube	0.240	9.23	80,000	26,000	36.9	140	NiZn
Fi 335	Sumida	0.360	11.1	2300	750	11.4	2100	MnZn
Fi 327	Sumida	0.310	20.5	5500	1300	23.3	1500	MnZn
Fi 212	Sumida	0.300	6.53	92,000	12,000	31.2	105	NiZn
Mf 102	Tridelta	0.360	13.0	2000	400	8.85	3500	MnZn
Mf 106	Tridelta	0.380	13.1	1500	530	11.4	3400	MnZn
Mf 108	Tridelta	0.435	17.5	1800	680	14.6	3700	MnZn

account. Fig. 4.18 shows the frequency dependence of the effective resistance of the winding. NiZn ferrites seem to have advantages here. The low initial permeabilities play only a minor role at these frequencies. The relevant impedances result from

$$X_L = 2\pi f \cdot L = 2\pi f \cdot \frac{\mu \cdot A_e}{l_e} \cdot N^2$$

A_e: magnetization cross section, l_e: magnetization length, μ: permeability, N. number of turns, f: frequency.

At high frequencies, a certain reactance is achieved even with lower permeability. Obviously, the core materials used have a considerable influence on the performance of the magnetic/inductive components. This aspect will be dealt with separately below. In addition to the material characteristics, the size of the component also has an influence on the power loss density.

First of all, for comparison purposes, impedance location curves $\underline{Z}(j\omega)$ are generated from the manufacturers' data sheets for the materials used above (Fig. 8.9). With the initial permeability data, the impedance is

$$\underline{Z} = j\omega N^2 (\mu' - j\mu'') \frac{A_e}{l} = R + jX$$

For the illustrations in Figs. 8.9 and 8.10, the curves of the real and imaginary parts of the permeability of the materials were digitized from the manufacturer's data sheets ($\omega = 2\pi f$). The real and imaginary parts of the impedance, related to one turn, then result in.

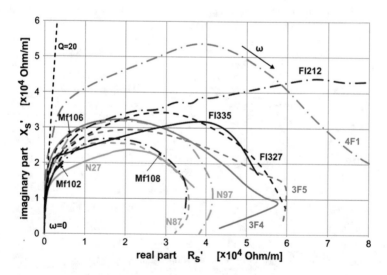

Fig. 8.9 Impedance locus curves of selected ferrite core materials based on the data sheets of various manufacturers. A black dotted line is also drawn, which characterizes a quality of Q = X/R = 20 [2]

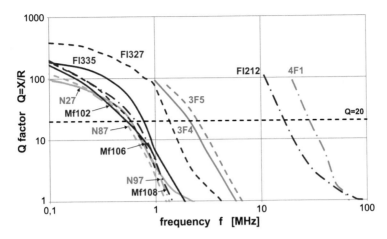

Fig. 8.10 Quality factor Q of selected ferrite core materials from various manufacturers. A black dotted line is also drawn, which characterizes a quality factor of Q = X/R = 20

$$R_s(\omega) = \omega \cdot \mu_0 \cdot \mu_r''(\omega) \cdot \tfrac{A}{l} \text{ or } \mathrm{R}_s'(\omega) = \tfrac{R_s(\omega) \cdot l}{A} = \omega \cdot \mu_0 \cdot \mu_r''(\omega).$$

$$X_s(\omega) = \omega \cdot \mu_0 \cdot \mu_r'(\omega) \cdot \tfrac{A}{l} \text{ or } X_s'(\omega) = \tfrac{X_s(\omega) \cdot l}{A} = \omega \cdot \mu_0 \cdot \mu_r'(\omega).$$

The frequency-dependent Q(ω) course is the ratio of the two components of the impedance

$$Q(\omega) = \frac{X(\omega)}{R(\omega)} = \frac{\mu_r'(\omega)}{\mu_r''(\omega)}$$

and shown in Fig. 8.10. It can be clearly seen that this representation shows a much stronger order of the material data. The order with regard to the maximum frequency with respect to use in power applications can be found here. For comparison, the level Q = 20 is entered in Fig. 8.10.

If you look at the impedance locus curves in Fig. 8.9, you can approximately see the pattern "Circular segment through the coordinate origin". This can be interpreted as a parallel connection of an inductor L_1 with other elements. A parallel connection with a resistor R_1 (see Fig. 8.11) would result in a semicircle through the coordinate origin for the Z locus curve. A capacitor C_1 connected in parallel to this provides a full circle in the value range $0 < \omega < \infty$. The imaginary part of the impedance disappears at the resonant frequency $f_0 = \omega_0/(2\pi)$. With this you can calculate a fictitious capacity for the core to

$$C_1 = \frac{1}{\omega_0^2 L_1}$$

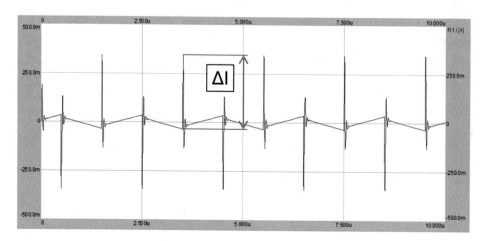

Fig. 8.11 Current curves through a smoothing choke with core when applying a high-frequency rectangular alternating voltage. (Source: Kleeb, Zacharias [3])

If one now considers that with a phase angle of 45°, the total impedance of the imaginary part and the real part of the impedance become equally large, a further model value can be estimated from the condition $X(\omega) = R(\omega)$ (or $\mu' = \mu''$) i.e. at $\varphi = 45°$. If the model for Fig. 8.9 is used as a basis, the estimated value for the resistance R_1 is

$$R_1 \approx \frac{\omega_{45°} \cdot L_1}{\left(1 - \omega_{45°}^2 \cdot C_1 L_1\right)}$$

Another value combination of $[X(\omega); R(\omega)]$ with the previous estimates also provides a value for R_2, which is in series with. Because of the nonlinear dependencies in reality, only the basic behavior can be achieved with relatively rough parameter estimates of the model. The procedure can also be further developed. However, the results become more and more sensitive to the digitization and measurement errors of the primary data. In Fig. 8.12, the described procedure is summarized graphically. It must be remembered that the components are always non-linear components whose properties result from complex relationships. Nevertheless, the model illustrates essential properties of the inductive component.

In experimental investigations on this topic, it is noticeable that in inductors with cores, in contrast to coreless inductors, relatively large current peaks can be observed when applying a switched voltage (Fig. 8.11). This cannot be explained by the winding capacity alone. If we look at the equivalent circuit diagram Fig. 8.12, the influence of C_1 and R_2 becomes clear. In addition, dividing the reluctance of the core into an "air/vacuum portion" of the magnetic properties in parallel with the polarized portion leads to a series connection of L_0 and L_1 in the equivalent circuit. With a voltage jump ΔU, the current amplitude can then be estimated to

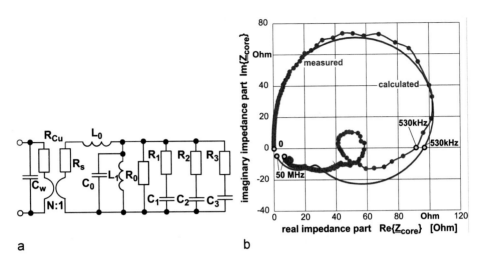

Fig. 8.12 Simple linear equivalent circuit diagram for a PM114/N27/9 turns choke as an example. (**a**) Equivalent circuit diagram, (**b**) impedance locus curves for the core related parts) in comparison of the model according to **a** and the measured data with the parameter estimated values $L_0 = 0.67\ \mu H$, $L_1 = 15.4\ \mu H$, $C_0 = 0.324\ nF$, $C_1 = 3.9\ nF$, $C_2 = 1.7\ nF$, $C_3 = 0.4\ nF$, $R_0 = 5.6kOhm$, $R_1 = 37Ohm$, $R_2 = 222Ohm$, $R_3 = 333Ohm$, $R_s = 3.7mOhm$ ($C_w = 5.77\ pF$, $R_{Cu} = 13mOhm$)

$$\Delta I \approx \frac{\Delta U}{\sqrt{\dfrac{L_0}{C_1 + C_W} + R_2^2}}$$

Relative to the inductance value of the core, $L_0 = L_1/(\mu' - 1)$ is close to zero. The size of the capacity C_1 will be discussed further below.

8.3   Consideration of Dielectric Properties in the Core

If the toroidal core is regarded as a toroidal resonator Fig. 8.13, the corresponding self-inductance can be approximately determined with a correspondingly thin ring and assuming a homogeneous magnetic field inside

$$L_{i0} \approx \frac{\mu_0 \mu_r r_1^2}{2 r_2}$$

The following applies to the induced voltage on a path with radius 'r' concentric around the centre line of the ring.

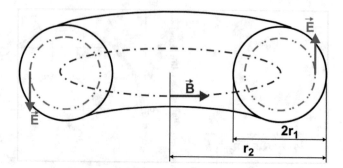

Fig. 8.13 Toroidal core as electromagnetic "cavity" resonator with material filling

$$U_{ind}(r) \approx \omega \int_A \vec{B} \cdot d\vec{A} = \omega \pi r^2 B \text{ with } 0 \le r \le r_1 \text{ and } E(r) \approx \omega \frac{U_{ind}(r)}{2\pi r} = \omega \frac{rB}{2}.$$

Consequently, the maximum energy in the electric field of the ring is approximately.

$$W_{el} = \int_V \frac{dW}{dV} \, dV \approx \int_0^{2\pi} \int_0^{r_1} \frac{\varepsilon E^2(r)}{2} l_{Fe} r \, dr \, d\varphi = \frac{\pi \varepsilon}{16} \omega^2 B^2 r_1^4 l_{Fe}$$

for sinusoidal variables. From this, an "equivalent capacitance" of the ring resonator can be easily calculated:

$$C_1 = \frac{2W_{el}}{U^2(r_1)} \approx \frac{2 \frac{\pi \varepsilon}{16} \omega^2 B^2 r_1^4 l_{Fe}}{\left(\omega \pi r_1^2 B\right)^2} = \frac{\varepsilon}{8\pi} l_{Fe} = \frac{\varepsilon}{4} r_2$$

This gives the resonator's "natural" resonant frequency of

$$f_0 = \frac{1}{2\pi\sqrt{L_{i0}C_1}} = \frac{1}{2\pi\sqrt{\frac{\mu_0\mu_r r_1^2}{2r_2} \frac{r_2\varepsilon_0\varepsilon_r}{4}}} = \frac{\sqrt{2}}{\pi r_1 \sqrt{\mu_0\mu_r\varepsilon_0\varepsilon_r}} = \frac{\sqrt{2}c_0}{\pi r_1 \sqrt{\mu_r\varepsilon_r}}$$

(where $c_0 = 299{,}792{,}458$ m/s is the speed of light in a vacuum). Conversely, the relative permittivity (dielectric constant) of the material can also be determined by measuring the natural resonant frequency of the toroidal core - assuming constant values for μ and ε:

$$\varepsilon_r = \frac{2c_0^2}{\mu_r\left(\pi r_1 f_0\right)^2}.$$

It is interesting that the cores apparently have a natural resonant frequency that depends on their size (expansion transverse to the field direction). This frequency decreases inversely proportional to the size of the toroidal core.

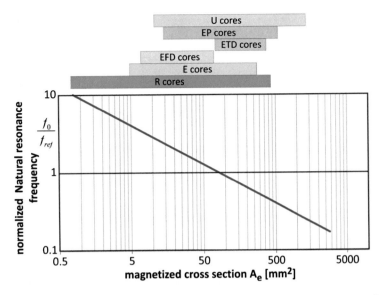

Fig. 8.14 Natural resonant frequency of a core in relation to that of an R34 (82.06 mm²)cores for measuring the core properties (https://www.mag-inc.com/Products/Ferrite-Cores/Ferrite-Shapes)

$$f_0 = \frac{\sqrt{2}c_0}{\pi r_1 \sqrt{\mu_r \varepsilon_r}}$$

Obviously, a component can only be operated as an inductive component below this frequency, if one does not want to use especially this effect. Estimated natural resonant frequencies are shown in Fig. 8.14. Extra losses are to be expected at the resonant frequency in switched applications, since the parallel resonance of the resonant circuit and the serial resonance with L_0 in Fig. 8.12 leads to a "current surge" within the core. For standard measurements of the core properties, the quantity r_1 is known and is, for example, an estimate from the magnetization cross section A_e:

$$r_1 = \sqrt{\frac{A_e}{\pi}} = \sqrt{\frac{83 \text{ mm}^2}{\pi}} = 5.14 \text{ mm}$$

for an R34 core. This provides absolute bounds for the use of frequencies at different core shapes for a given material. The maximum achievable operating frequency is thus also dependent on the magnetization cross section. The calculations of the core losses according to the Steinmetz formula therefore only provide an approximate result without taking this effect into account. One can shift the "natural" resonance frequency of a magnetic branch or circuit to higher values by inserting air gaps. But then you lose inductance with the same number of windings.

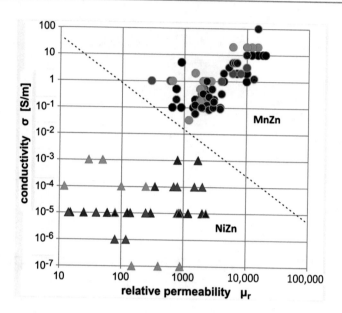

Fig. 8.15 Conductivity and permeability of MnZn and NiZn ferrites from Ferroxcube, TDK, Sumida and Tridelta in comparison

The selected NiZn ferrites stand out clearly from the MnZn ferrites in Fig. 8.15. This is due to their special combinations of material characteristics. Basically, it is noticeable that the conductivities of these ferrites differ by several orders of magnitude (Fig. 8.15). This has an effect on the properties in the electromagnetic alternating field. In the alternating current field, the current density is given by.

$\vec{J} = \sigma \vec{E} + \frac{d}{dt} \varepsilon \vec{E}$ or in complex notation for sinusoidal variables

$$\underline{\vec{J}} = (\sigma + j\omega\varepsilon) \cdot \underline{\vec{E}} = \underline{\kappa} \cdot \underline{\vec{E}}$$

The value of the complex conductivity κ is then

$$|\underline{\kappa}| = \sqrt{\sigma^2 + (\omega\varepsilon)^2}$$

If one applies the formula for determining the thickness 's' of the "skin" on the surface of the body influenced by the electromagnetic field, one obtains

$$d_s = \frac{1}{\sqrt{\pi f \mu \kappa}} = \frac{1}{\sqrt{\pi f \mu \sqrt{\sigma^2 + (\omega\varepsilon)^2}}}$$

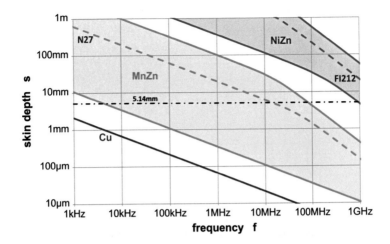

Fig. 8.16 Calculated penetration depth of the magnetic field in MnZn- and NiZn ferrites compared to copper with the reference value 5.14 mm as effective radius of the core cross section of R34 cores. Calculation results for the material representatives N27 and FI212 entered as comparison ($\varepsilon_{rMnZn} = 10^5$, $\varepsilon_{rNiZn} = 10^2$ according to Ferroxcube 2009)

That means, for frequencies where $\omega \cdot \varepsilon > \sigma$ applies, the layer thickness is approximately $s \sim 1/f$. For $\omega \cdot \varepsilon < \sigma$, $s \sim f^{-0.5}$ applies. This results in very different penetration depths of the electromagnetic field for MnZn and NiZn ferrites (Fig. 8.16). While the penetration depths of NiZn ferrites in frequency ranges relevant for power electronics are practically always sufficient to fill the entire core cross section sufficiently evenly with a magnetic field, this is not the case with MnZn ferrites for large cores. This circumstance was also pointed out in [12]. The consequence is probably that extra losses due to the skin effect occur here with non-sinusoidal/switched voltages in the upper frequency range, since the maximum induction is significantly higher than the average induction for a short time. A parallel segmentation of the magnetic conductors should remedy this situation - similar to that of electrical steel in the low-frequency range.

The following conclusions can be drawn from the above:

- Cores with small magnetization cross sections behave differently than those with large magnetization cross sections
- Short magnetic length reduces the internal core capacitances
- At frequencies >1 MHz, NiZn ferrites can be advantageous due to their low conductivity, although they have comparatively lower permeabilities.
- Small cores are more suitable for high frequencies, i.e. splitting the power over several components (modularization) is useful for achieving high power densities. This also favours heat dissipation from the relatively poorly heat-conducting core materials.

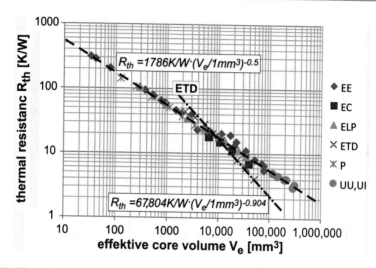

Fig. 8.17 Thermal resistance of ferrite core moulds according to specifications of various manufacturers [2]

- "Parallel connection" of several cores is also more favourable in the upper frequency range for achieving lower losses. Each individual core acts as a separate resonator with only a fraction of the total cross section.
- The Steinmetz formula approximates the dependence of the losses in the core only in the lower frequency range proper. In the upper frequency range, the loss distributions are influenced by the core shape as well as the core size, air gaps and winding distributions.
- At high frequencies (>100 kHz) the used winding material and cooling are highly sgnificant to achieve high reactive power in the case of chokes or transmittable power in case of transformers.

To achieve certain impedance, the permeability can be reduced to a tenth at 10 times the frequency with the same structure. Permeabilities of >100 can therefore be used without any problems at frequencies >10 MHz, even if one considers that the optimum number of turns decreases with $f^{-\frac{1}{4}}$.

For the previous considerations, the ETD49 was used, which stands for the kW power range. If other core shapes are considered, they differ mainly in the thermal resistance R_{th} with regard to the dissipated power loss. Figure 8.17 summarizes the thermal resistances of a number of ferrite core types. Although the entire surface of the component is decisive for heat dissipation to the environment, the R_{th} values in Fig. 8.16 follow a dependency

$$R_{th} \sim \frac{1}{\sqrt{V_e}}.$$

Only the ETD core form deviates significantly from the above dependence.

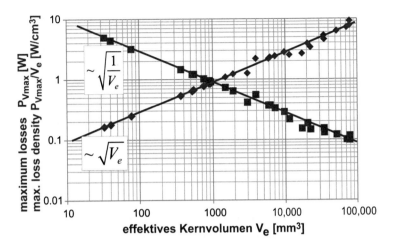

Fig. 8.18 Power dissipation and power density of EE core pairs of different manufacturers (Ferroxcube, Philips, TDK, EPCOS, Tridelta) at $\Delta T = 50$ K

A frequently asked question is the permissible power loss density in the core. An indicator for this is Fig. 8.18. For EE cores, the dissipatable power loss and the corresponding power loss density are shown here. Assuming that winding losses and core losses are approximately equal, power loss densities <100 mW/cm^3 can be assumed for cores with $V_e > 20$ cm^3 for orientation.

8.4   Small Transformers for Mains Frequency

In this section, typical designs of low power line transformers are presented with regard to their size and performance. The mains frequency used is 50/60 Hz for general power supply and 400 Hz in the aircraft industry. The latter is referred to in the corresponding examples. Transformers with toroidal cores are mainly found in the lower power range. The core form best meets the "natural" formation of main and stray fields. This is reflected in its transmission properties. Primary and secondary windings are wound 'concentrically' on each other. This keeps the length of the stray field lines at a maximum and the 'cross section' of the stray field small. As a result, the leakage inductance limiting the short-circuit current can be kept low even with transformers of low rated power. This is a circumstance that must be taken into account when inserting into a system. A connected load is connected to the energy source practically without current limitation when it is switched on. An uncontrolled rectifier with capacitive smoothing causes high inrush currents in this way.

The effects of the rush effect can be even greater when the transformer is switched on with unfavourable phase position. Due to the design, the T-equivalent circuit diagram can be assumed to be a physically adequate equivalent circuit diagram for the electrical behaviour of the transformer. The leakage inductance is very small. If the component is connected to a voltage source $u_1(t)$, an inrush current is generated (without load) according to the relationship

$$i_1(t) = \frac{1}{L_{\sigma 1} + L_\mu(\psi(t))} \int_0^t u_1(t)dt = \frac{1}{L_{\sigma 1} + L_\mu\left(\int_0^t u_1(t)dt\right)} \int_0^t u_1(t)dt$$

The magnetizing inductance itself depends on the effective flux linkage of the core. It becomes practically maximum when the switch-on process takes place at a voltage zero crossing with sinusoidal mains voltage. The voltage-time integral of the input voltage reaches its maximum at the next zero crossing. If this value is greater than the flux linkage when the core is saturated, the magnetizing inductance in the above expression 'disappears' and i_1 assumes very high values. The effect corresponds to that a mains short-circuit by means of a transductor. The $\Psi(i)$ characteristic of the primary winding follows the B (H) characteristic of the core material because of the almost homogeneous field in the annular core. Saturation therefore occurs everywhere at the same time. Since the rush effect can cause fuses to blow, inrush currents are limited by bridgeable resistors or NTC (*soft start*) resistors. The low leakage inductance causes only a low internal voltage drop of the transformer under load, even at low rated power, which is a popular feature. Correspondingly high inrush current peaks can be expected if necessary. The dependency of the rated power on the transformer mass is shown for the usual application range of toroidal transformers for 50 Hz or 400 Hz (Fig. 8.19).

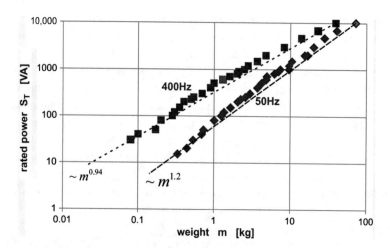

Fig. 8.19 Rated power and mass for toroidal transformers for 50 Hz and 400 Hz. (Data: Information sheet Tauscher Transformatorenfabrik GmbH)

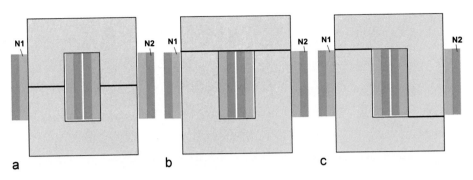

Fig. 8.20 Core shape of the magnetic circuit as core type. (**a**) UU core, (**b**) UI core, (**c**) LL core

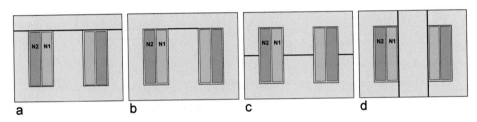

Fig. 8.21 Transformers with magnetic shell-type circuit: (**a**) EI cores, (**b**) M cores, (**c**) EE cores, (**d**) CI cores

The rush effect occurs similarly in transformers with wound primary and secondary windings of the core type, which are discussed below. The basic versions with different sheet metal sections are shown in Fig. 8.20, where the primary and secondary windings are divided in half over 2 legs and connected in series in pairs. A parallel connection of the secondary windings would in principle provide the same leakage reactance values, but leads to balancing currents due to the non-existent identity of the winding coupling and thus results in additional losses. The winding design 'channels' both the main flux and the leakage flux. This makes the electrical behaviour similar to that of a toroidal transformer. If N_1 and N_2 are arranged on separate windings, a structure with large yoke leakage and correspondingly large leakage inductance is obtained.

Another form of the magnetic circuit is the shell-form, for which designs with typical sheet metal sections are shown in Fig. 8.21. In the basic design, there is only one leakage channel, so that a comparatively higher leakage inductance must be expected than with the aforementioned core type, where practically 2 not coupled magnetic leakage inductances with half number of turnsare connected in series. In the case of saturation phenomena, it is mainly the core section enclosed by the excitation winding that goes into saturation, while the remaining section is less affected. As a result, the rush effect is much less pronounced.

Figure 8.22 shows a magnetic core of the shell-type with a coordinate system and characteristic dimensions. These enable an estimation of the relationship between design size and construction performance for a transformer.

Fig. 8.22 Magnetic circuit with coordinate system for estimating the relationship between mechanical size and rated power

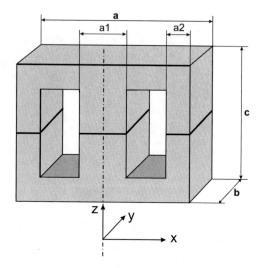

The following must apply to the number of turns: $N \leq \frac{\sqrt{2} \cdot U}{\omega \cdot A_{Fe} \cdot B_{max}}$.
A certain core loss also results in a certain iron loss density, i.e.

$$P_{VFe} = k_{Fe} \cdot V_{Fe}$$

In addition, taking the above condition into account, the winding losses

$$P_{VCu} \leq \frac{2 \cdot I^2 \cdot \rho \cdot l_m}{k_f \cdot A_W} \cdot N^2 \frac{2 \cdot I^2 \cdot \rho \cdot l_m}{k_f \cdot A_W} \cdot \left(\frac{\sqrt{2} \cdot U}{\omega \cdot A_{Fe} \cdot B_{max}} \right)^2 = \frac{4 \cdot S^2 \cdot \rho \cdot l_m}{k_f \cdot A_W \cdot \omega^2 \cdot A_{Fe}^2 \cdot B_{max}^2}$$

Assuming a simple model with a thermal resistance, one obtains a temperature increase compared to the environment of

$$\Delta T_{max} \leq (P_{VFe} + P_{VCu}) \cdot R_{th} = \left(k_{Fe} \cdot V_{Fe} + \frac{4 \cdot S^2 \cdot \rho \cdot l_m}{k_f \cdot A_W \cdot \omega^2 \cdot A_{Fe}^2 \cdot B_{max}^2} \right) \cdot R_{th}$$

$$S = A_{Fe} \cdot B_{max} \sqrt{\frac{k_f \cdot A_W}{4 \cdot \rho \cdot l_m} \left(\frac{\Delta T_{max}}{R_{th}} - k_{Fe} \cdot V_{Fe} \right)} \sim A_{Fe} \cdot \sqrt{A_W} \sim V_{Fe}$$

This means that the apparent power of a component is proportional to the magnetization cross section A_{Fe} under the conditions mentioned. Thus, it is also proportional to the dimension 'b' or $V_{Fe}^{0,667}$. This can be used by simply stacking more sheets. This is used, for example, as standard for the cores of many cores. Figure 8.23 shows an example with EI cores. The thermal resistance is proportional to the surface, i.e. $R_{th} \sim (ab + ac + bc)$.

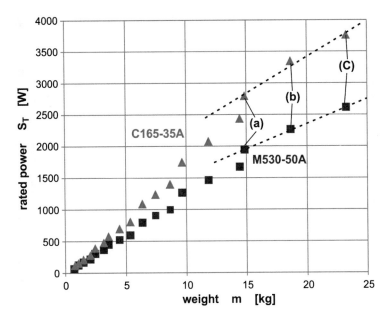

Fig. 8.23 Relationship between rated power and mass for 1-phase transformers with EI sheet metal cuts EI92...231a...c for the materials M530-50A and C165-35A (EI231a, b, c differ only in the stacking height of the sheets)

8.5 Problems of Increasing the Operating Frequency

As Sect. 8.2 shows, an increase in frequency leads to a steady reduction in the maximum usable flux density in ferrites due to the additional core losses. Power dissipation-optimized solutions are far removed from theoretical power without consideration of the core losses.

Assuming a uniform development of the losses in the winding and core, the estimation formula is

$$\frac{P_{vCu}}{V} \approx \frac{I^2 \cdot R}{V} = \frac{I^2 \cdot \rho_{Cu} \cdot \frac{l}{A_{Cu}}}{l \cdot A_{Cu}} = \frac{I^2 \cdot \rho_{Cu}}{A_{Cu}^2} = J^2 \cdot \rho_{Cu} \approx \frac{P_{vFe}}{V}$$

For large transformers, the guideline value for the maximum current density is 2.5 A/mm². This corresponds to a power loss density of about 100 mW/cm³. This value is a proven, conservative value for magnetic components with natural air cooling in the range 300 W...3 kW for ferrites. However, this value should only be taken as an orientation, as it is very strongly linked to the cooling conditions. After all, the rated power depends primarily on whether the associated losses can be permanently dissipated from the component. This is expressed by the thermal resistance R_{th}, which differs greatly for the same

Table 8.5 Selected values of thermal conductivity and saturation induction for typical core materials [5]

Material	Thermal conductivity [W/K/m]	B_{sat} [T]
M 350–50 A	24.5	1.56
M 470–50 A	30.3	1.59
M 600–50 A	33.6	1.62
M 800–50 A	42.4	1.65
MnZn-ferrite	3.5…5	0.3….0.55
NiZn-ferrite	3.5…5	0.2…0.4
Mu-metal	17–19	0.8
VACOPERM 100	17–19	0.74
PERMENORM 5000 H2/V5	13–14	1.55
VACOFLUX 50	30	2.35
$Co_{66}Fe_4Mo_2Si_{16}B_{12}$	9	0.4…1.2
Iron (Fe 99.95%)	80	2.16
Nickel	90.9	0.6
Cobalt	100	1.76

component depending on whether cooling is by natural air flow, forced air cooling or convection in oil. At the core, heat transport is limited by the thermal conductivity of the core material. The thermal conductivity of some materials is listed in Table 8.5.

Table 8.5 shows that the thermal properties of the materials sometimes differ considerably. In particular, ferrites as ceramic materials and electrical sheets differ by a factor of up to 10 at thermal conductivity. In addition, heat conduction is anisotropic even in non-grain-oriented sheets, as the sheets are insulated from each other with an even worse heat-conducting material.

Considering that a heat flow 'P' on an element with a cross section 'A' and thickness 'd' causes a temperature difference ΔT with.

$\Delta T = P \cdot \frac{d}{\lambda \cdot A}$, you get a thermal gradient of $\frac{\partial}{\partial x} T \approx \frac{\Delta T}{d} = \frac{P}{\lambda \cdot A}$. If the core losses can only be emitted via heat conduction (in 2 directions), the difference between maximum temperature and surface temperature is roughly

$$T_{max} - T_O = p_{VFe} \cdot V_{Fe} \cdot \frac{l}{8 \cdot \lambda \cdot A_{Fe}} = p_{VFe} \cdot \frac{l^2}{8 \cdot \lambda}$$

Figure 8.24 shows the temperature increase in the core compared to the surface temperature at different thermal conductivities in comparison. It is also clear here that for larger dimensions, the power density must be reduced so that the temperature does not exceed a specified maximum temperature. Consequently, higher rated power is also associated with lower loss densities. If, however, a different cooling method is used, which, for example, also allows heat to be dissipated via the lateral surfaces, the power loss density

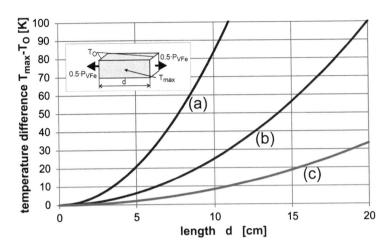

Fig. 8.24 Temperature rise within a core section due to core losses at $p_{VFe} = 200$ mW/cm^3, (**a**) $\lambda = 3$ W/K/m, (**b**) $\lambda = 10$ W/K/m, (**c**) $\lambda = 30$ W/K/m

can be increased. This is also an indication that the power density can be further increased with improved cooling methods. Thresholds that cannot be overcome are explained in Sect. 8.2.

In addition to the core losses, winding losses also occur in magnetic components. Depending on the frequency, only an outer part of the conductor cross section with the skin thickness d_s is used for a high-frequency current. This leads to a simple requirement for the wire diameter

$$d_s = \frac{1}{\sqrt{\pi \mu f \kappa}} \Rightarrow d_{Cu} < 2 \cdot d_s = \frac{2}{\sqrt{\pi \mu f \kappa}}$$

A further limit is represented by possible eddy current (proximity) losses. These occur mainly in the stray field that passes through winding parts perpendicular to the winding direction (Fig. 8.25).

To estimate the influence of eddy current losses, a comparison was made with effectiveness calculations for inductive heating (Fig. 8.25). The shielding effect of interspersed winding parts for adjacent parts was not considered here. Figure 8.26 shows that the eddy current losses decrease very quickly when the conductor diameter falls below a certain limit value. This also agrees with our own observations. The definition of a maximum limit for the wire diameter is arbitrary here and is a compromise on size and cost. According to Fig. 8.26, the threshold lies at $d_{Cu} = 3.04 \cdot s$. For example, a requirement to achieve 5% of the possible proximity effect losses leads to $R_{Cu}/s < 0.55$ or $d_{Cu} < 1.1 \cdot s$. In this way, a dimensioning requirement can be formulated. As with the skin effect, the reduction of the diameter leads to a reduction of the cross section, which must be compensated by connecting several conductors in parallel in order to obtain the desired resistance. It must

Fig. 8.25 The air field of a coil
penetrates the winding and
causes eddy current losses

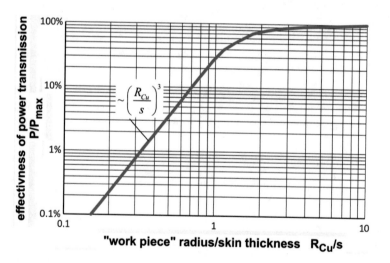

Fig. 8.26 Effectiveness of magnetic energy coupling for induction heating of a cylindrical part (according to: [6])

be remembered that the resistance of an HF strand is higher than that of solid wire and direct current even with the same cross section. The many partial conductors only allow a more even current distribution over the total cross section by stranding. In any case, the available winding space with stranded conductors is less well utilized. The insulation of the single wires as well as the twisting and stranding lead to a decrease of the filling factor for the winding space. This drop can be partially counteracted by profiling the cross-sectional contour.

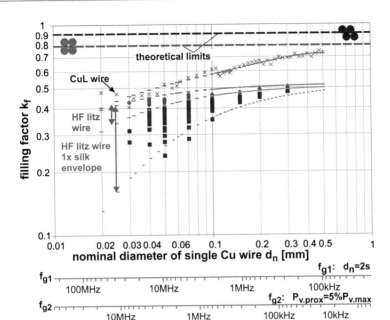

Fig. 8.27 HF litz wires with different single wire diameters and achievable filling factors of the winding space in comparison with the usage frequencies resulting from the skin effect and proximity effect. (Data: Electrisola, source: Wendt and Zacharias [13])

Figure 8.27 shows the achievable fill factors for different strands with different single core thicknesses and different overall cross sections. The different frequencies result once from the condition that the wire diameter is equal to twice the layer thickness s. The other assumption is based on the assumption that the proximity losses according to Fig. 8.25 are a maximum of 5% of the maximum possible. It can be seen that small wire thicknesses lead to low fill factors. This certainly has to do with the insulating layer. The varnish normally used cannot be applied in any desired thickness.

The volume advantage resulting from frequency increase is partly compensated by the effects described above, so that real reductions in component volumes can only be achieved by appropriately adapted new designs. The ferromagnetic core has the task of bundling the magnetic flux. A field scattering into space is thus reduced. As seen above, however, the ferromagnetic core material limits the maximum induction by means of the loss density, which increases rapidly with induction. This also has a limiting effect on the size. In very high frequency ranges of communications engineering, components with magnetic cores are also not used for flux concentration but for other purposes such as mixing, splitting, damping, etc. Therefore, the question has to be asked whether it is possible to equip components without an active core and what the consequences are.

With individual inductors, it is important to accommodate a magnetic field of a certain energy in as little space as possible. The core usually helps here by giving preference to certain field lines so that the energy content of the external field remains low. In alternating

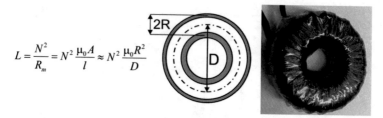

$$L = \frac{N^2}{R_m} = N^2 \frac{\mu_0 A}{l} \approx N^2 \frac{\mu_0 R^2}{D}$$

Fig. 8.28 Example of a toroidal coil without ferromagnetic core. (Source: Zacharias 2017)

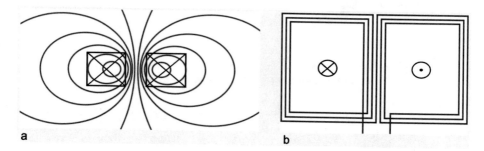

Fig. 8.29 Coreless coil arrangements as examples of coreless coil arrangements. (**a**) Solenoid coil, (**b**) flat gradiometer coil

current applications, the remagnetization of the core causes losses that must be dissipated in addition to the winding losses. The first question is the most compact design of a coreless arrangement. The obvious choice would be a toroidal coil (Fig. 8.28). Here the core functions only as a mechanical carrier for the windings and consists of a paramagnetic, non-conductive material. The field lines propagate mainly within the toroid. Ideally, an external magnetic field does not exist. This results in only minimal external influences. Due to the long field lines, the eddy currents caused in the winding are also low. But at the same time, the 'magnetic conductor' is relatively long and has a small cross section A. This means that the magnetic resistance is high and the achievable inductances are quite low.

Higher inductance values are obtained with the same external dimensions using a solenoid shape (= flat short cylindrical coil, Fig. 8.29a). The magnetic field represents, so to speak, the shell-type of the component. The toroid is an example of the 'core type' of the magnetic field. If the winding of the toroid is arranged in a plane (Fig. 8.29b), the shape of the magnetic field is only partially determined by the coil. There is still a concatenation of a large proportion of flux. However, this spreads out in space. Due to the large area interspersed with field lines, a relatively low reluctance is obtained. This results in a greater inductance than with the toroid.

If inner diameter = winding height = winding width = a, then the approximate formula for the inductance with D = 2·a is

$$L = 6.7 \cdot 10^{-7} \frac{H}{m} \cdot N^2 \cdot D$$

The problem is, as with the solenoid, that the magnetic field generated extends far into space. This requires shielding measures. The effect of this problem will be briefly assessed below. A magnetic component of an alternating magnetic field hitting a conductor surface perpendicularly causes eddy currents that are directed in the opposite direction to the cause. The field in the material is partially compensated by this. The magnetic field strength on the surface facing away from the cause is then

$$H_a = H_i \cdot \exp\left(\frac{-d}{\delta}\right)$$

with
  Shielding of the alternating magnetic field

Magnetic field strength inside the shielding H_i
Magnetic field strength outside the shielding H_a
Thickness of the shielding d
Damping constant = skin penetration depth δ

  The skin layer thickness or damping constant for a conductive material is

$$\delta = \sqrt{\frac{2}{\omega \cdot \mu \cdot \kappa}} = \frac{1}{\sqrt{\pi \cdot f \cdot \mu \cdot \kappa}},$$

with f: frequency, μ: permeability and κ: conductivity.
  A damping to $0.01\% = 10^{-4}$ results in, for example
$H_a = H_i \cdot \exp\left(\frac{-d}{\delta}\right) = 10^{-4} \cdot H_i$.
  This means for the layer thickness of the shield a minimum of

$$d = -\delta \cdot \ln\left(\frac{H_a}{H_i}\right) = 4 \cdot \delta.$$

Figure 8.30 shows this relationship for AlMg3 and shielding attenuation of 10^{-4} and 10^{-6}. Table 8.6 shows the electrical conductivity of selected metals at a temperature of 300 K.
  Above 100 kHz, the required shielding thickness of aluminium falls below 3 mm. From this point on, coreless inductors can also become interesting. Hard aluminium (e.g. AlMg3) is less conductive than e.g. almost pure, but soft (electrical) aluminium. In this illustration, this pushes the limits almost imperceptibly upwards. Such shields could also be used as small oil containers. The insulating oil then transports the winding heat very effectively to the surface of the vessel.

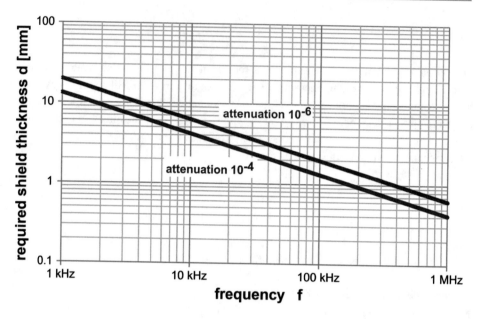

Fig. 8.30 Minimum shielding thickness for aluminium AlMg3 with attenuation 10^{-4} or 10^{-6} times

Table 8.6 Electrical conductivity of selected metals at a temperature of 300 K [7]

Material	κ in S/m	μ_r
Silver	$61.35 \cdot 10^6$	1
Copper	$\geq 58.0 \cdot 10^6$	1
Gold	$44.0 \cdot 10^6$	1
Aluminium	$36.59 \cdot 10^6$	1
Brass	$\sim 14.3 \cdot 10^6$	1
Iron (96% Fe, 4% Si)	approx. $10.02 \cdot 10^6$	400...8000
Stainless steel (1.4301)	$\sim 1.4 \cdot 10^6$	<1.3
μ-metal (76% Ni, 17% Fe, 5% Cu, 2% Cr)	$1.82 \cdot 10^6$	12,000...45,000...(300,000)

The natural resonant frequency of all inductive components is the limiting factor for their use. Although this cannot be avoided by appropriate winding technology (basket winding, honeycomb winding, cross winding, etc.), it can be reduced (Fig. 8.31). The leakage inductance between two closely coupled windings hardly depends on the permeability of the core but only on the arrangement of the windings relative to each other. Bifilar windings lead to very small leakage inductances. Sufficient magnetizing inductances can be achieved at high frequencies.

However, regulatory requirements also have a massive impact on the size and use of frequency ranges. As an example, a single-stage filter is examined below, which is intended to dampen push-pull harmonics towards the mains connection. Figure 8.31 serves to

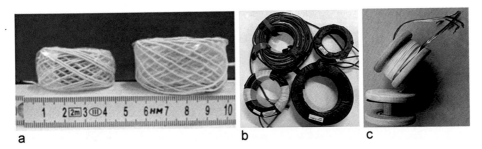

Fig. 8.31 Different designs of coreless windings. (**a**) Windings in cross winding technology. (Source: Wendt [4]), (**b**) solenoid coils, (**c**) 1:1 transformer from a double line ($L_\mu = 120$ μH, $L_\sigma = 415$ nH, $f_{res} = 3$ MHz). (Source: Zacharias 2017)

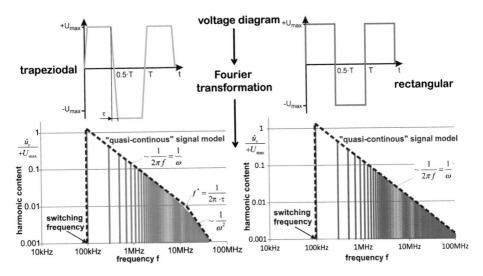

Fig. 8.32 Origin of line spectra in switch mode power converters (SMPC). (Source: Zacharias [8])

explain the problem. A switched-mode power supply with a DC link voltage of 360 V generates more or less steep voltage pulses depending on the switching speed of the semiconductor switches used. Slower switching causes the spectral power density to decrease in the upper frequency range. However, the low-order oscillations are not affected. For further considerations, therefore, the effect of increasing the switching frequency with ideal switches should be investigated. Due to their ever decreasing losses, these enable ever higher switching frequencies. Since field-controlled switches are involved, topologies and control methods will be used, which limit the voltage slope at switch-on (e.g. ZVS). But a large part of the problems of electromagnetic compatibility already arise in the range of the frequency of the fundamental oscillation. In the following, the treatment of the problem assumes that the envelope of all possible harmonics can be represented by a limit curve inversely proportional to the frequency (Fig. 8.32).

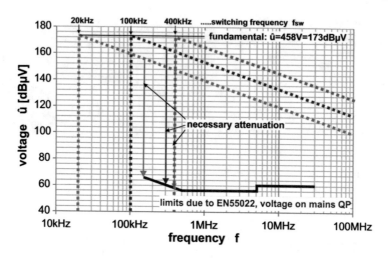

Fig. 8.33 Shift of the envelope of a quasi-continuous signal model of the harmonics of a square wave signal compared to the limits of the quasi-peak measurement specified by EN55022 when assuming a DC link voltage of 360 V. (Source: Zacharias [8])

For grid connection, guidelines have long existed that limit both the harmonic content of voltages and currents at the grid connection point (EN 55022). One such limit is shown in Fig. 8.33. The voltages are entered in $dB_{\mu V}$, as is usual for these considerations. The envelope curve of the line spectrum of the primary interference voltage decreases ~1/f and should serve as an indicator of the considerations. It becomes clear that very different attenuations are required, depending on how close the fundamental frequency is to the first limit at 150 kHz. At 150 kHz, the regulated range definitely begins.

If you increase the switching frequency, you can easily calculate the required attenuation of the filter. The dependence on the frequency then has the appearance of a staircase curve as shown in Fig. 8.34. For orientation, the limit values of Fig. 8.33 are shown dotted. The curve ends at 30 MHz, because that is where the regulated range ends again.

For the sake of simplicity, it is assumed for further considerations that it is a single-stage LC filter. This leads to a voltage attenuation of

$$\frac{U_1}{U_2} = \left|1 - \omega^2 LC\right| \approx \omega^2 LC = (2\pi f)^2 LC$$

If $V_L \sim L$ and $V_C \sim C$ are now applied to the volumes of L and C, $k \cdot \sqrt{LC}$ can also be taken as the geometric mean value V* of the two volumes. For this, you can also write

$$V^* = \sqrt{V_L V_C} \sim k \cdot \sqrt{LC} = k \cdot \frac{\sqrt{A_U}}{2\pi f}$$

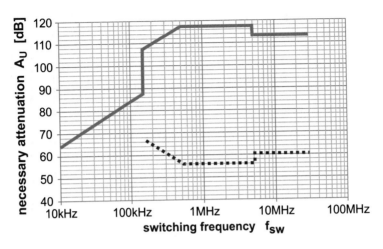

Fig. 8.34 Required attenuation of a filter with variable switching frequency of the converter. (Source: Zacharias [8])

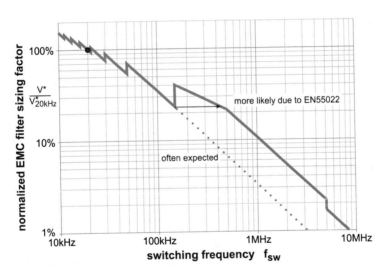

Fig. 8.35 Volume indicator for EMC filters normalized to a value at 20 kHz. (Source: Zacharias [8])

Applying this relationship to Fig. 8.34, one obtains a simplified basic curve for the change in volume with frequency (Fig. 8.35). The curve is normalized to a volume at a switching frequency of 20 kHz. Although it is only an estimation, it shows the problems. It is then advisable to keep the switching frequency either below 150 or above 500 kHz, otherwise the expected reduction in volume will not occur. The effect of other regulators can be investigated in a similar way.

8.6 Transmission and Distribution Transformers of Power Supply Systems

In the technology for the transmission and distribution of energy, high efficiencies at reasonable costs are crucial. As Fig. 8.36 shows, efficiency tends to increase with increasing rated power. The transmitted efficiency of a transformer is

$$P = U_{nom} \cdot I \cdot \cos \varphi.$$

Assuming (voltage-dependent) core losses P_0 and current-dependent winding losses P_k, the demand for maximum efficiency can be formulated as follows:

$$\eta = \frac{U_{nom} \cdot I \cdot \cos \varphi}{U_{nom} \cdot I \cdot \cos \varphi + P_0 + \left(\frac{I}{I_{nom}}\right)^2 \cdot P_k} \Rightarrow \max$$

From this follows after differentiation

$$P_0 - \left(\frac{I}{I_{nom}}\right)^2 \cdot P_k = 0$$

Maximum efficiency therefore occurs when winding and core losses are equal to:

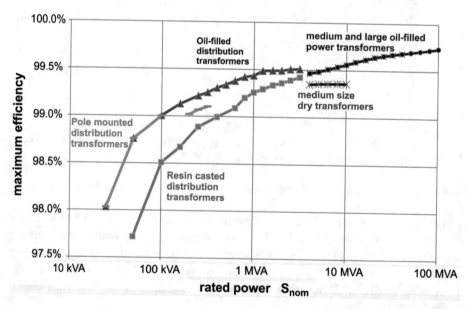

Fig. 8.36 Maximum efficiency of a transformer as a function of the rated equation using the data from [10]

$$\left(\frac{I}{I_{nom}}\right)^2 \cdot P_k = P_0,$$

that is: at partial load according to the condition $\frac{I}{I_{nom}} = \sqrt{\frac{P_0}{P_k}}$.

The maximum efficiency at $\cos\varphi = 1$ is

$$\eta_{max} = \frac{S_{rT} \cdot \sqrt{\frac{P_0}{P_k}}}{S_{rT} \cdot \sqrt{\frac{P_0}{P_k}} + 2 \cdot P_0} = \frac{\frac{S_{rT}}{P_k} \cdot \sqrt{\frac{P_0}{P_k}}}{\frac{S_{rT}}{P_k} \cdot \sqrt{\frac{P_0}{P_k}} + \frac{2 \cdot P_0}{P_k}}.$$

The performance range in which operation most frequently occurs in the respective application is decisive for dimensioning. According to [9], for distribution transformers up to 3150 kVA, the maximum efficiencies are achieved at 27–40% (~33% → $P_k \approx 9 \cdot P_0$) of the rated power. The higher the rated power, the higher the underlying load to be supplied. When the rated power is reached, the efficiency would then be

$$\eta = \frac{S_{rT}}{S_{rT} + 10 \cdot P_0}$$

In medium- and high-power transformers for power distribution and power transmission, additional cooling systems are integrated for heat dissipation, which become active depending on the load. Here, in addition to the losses of the transformer itself, those of the cooling system must also be taken into account. The measurements required to determine efficiency must be carried out using a reliable, accurate and reproducible method. This includes the generally accepted measurement methods. The calculation of the peak efficiency index (PEI) for power transformers is based on the ratio of the transmitted power minus the electrical losses to the transmitted power of the transformer.

$$PEI = 1 - \frac{2(P_0 + P_{C0})}{S_{rT} \cdot \sqrt{\frac{P_0 + P_{C0}}{P_k}}}$$

P_0 = no-load losses at rated voltage and rated frequency at the measured tap
P_{C0} = electrical power of the cooling system at idle
P_k = measured losses at rated current and frequency at the measured tap corrected according to reference temperature according to EN 60076-2
S_{rT} = rated power of the transformer on which PVSC is based
If the efficiency requirements [10] for transformers in the range 10 kVA...100 MVA are plotted in a diagram, Fig. 8.35 is obtained. It is clear that transformers are energy converters

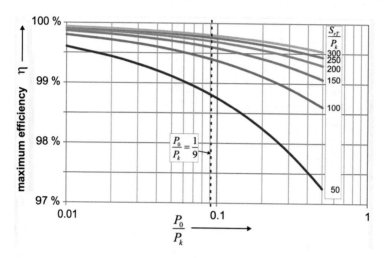

Fig. 8.37 Maximum efficiency as a function of the ratio no-load losses/short circuit losses (P_0/P_k) and the ratio rated power/short circuit losses S_{rT}/P_k

with extremely high efficiency. The high efficiencies that can be achieved (Fig. 8.36) are not comparable with small transformers in the kVA range or below (Fig. 8.37).

High power to be transmitted also requires a correspondingly high input of iron and copper, which leads to correspondingly high weights. In larger transformers, the costs for insulation and cooling also contribute to the weight. In accordance with the preliminary remarks, the underlying growth drivers are to be analysed here. The losses in a two-winding transformer for mains frequency consist of the core losses and the winding losses, as shown above. The core losses are essentially dependent on the voltage, which is why they are also referred to as no-load losses P_0. The winding losses are mainly ohmic losses at 50 Hz and can be mapped using $P_k = 2 \cdot R_w \cdot I^2$. Here, '$R_w$' is the winding resistance of one of the two windings with the current 'I'. The winding resistance depends on the number of windings N, the mean winding length, the size of the winding window A_w and the fill factor k_f:

$$R_w = \frac{\rho \cdot N^2}{k_f \cdot A_w}.$$

With the same core, the winding losses can be set above N and k_f. The losses can be calculated from $P_v = P_0 + P_k$. It was explained above that and why transformers for energy transmission and distribution are designed in such a way that the maximum efficiency is achieved at approximately 1/3 of the nominal current. The maximum power to be dissipated occurs at the rated power and then roughly causes the losses

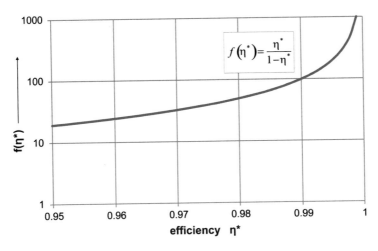

Fig. 8.38 Effect of the choice of efficiency on the rated power via f(η*)

$$P_{v\,max} = P_0 + P_k = 10P_0 = 10k_e \cdot V_e.$$

The rated power therefore follows approximately the following relationship

$$S_{rT} = \frac{10 \cdot P_0 \cdot \eta^*}{(1 - \eta^*)} = \frac{10 \cdot k^* V_e \cdot \eta^*}{(1 - \eta^*)} = 10 \cdot k^* V_e \cdot f(\eta^*)$$

For a given efficiency, the rated power as well as the approximate power loss is proportional to the volume. The efficiency at rated current η* is a variable that can be adjusted within limits. It can thus be used to influence the power loss and the rated power (Fig. 8.38).

If, on the other hand, one considers that the dissipatable power is proportional to the surface area and thus approximately proportional to $\sqrt[3]{V_e^2}$, this means that the rated power can increase primarily by increasing the efficiency. This also applies analogously to the mass. Figure 8.39 summarizes the rated power of oil-insulated transformers for the application under consideration. The determined exponent of the growth of the mass m_e with the power S_{rT} is in the range of 0.75, so that the rated power for this design can be estimated to.

$$m_e = 18\ \mathrm{kg} \cdot \left(\frac{S_{rT}}{1\ \mathrm{kVA}}\right)^{0.752}\ \text{or}\ S_{rT} = 1\ \mathrm{kVA} \cdot \left(\frac{m_e}{18\ \mathrm{kg}}\right)^{1.3294}\ \text{or}\ \frac{S_{rT}}{m_e} = 1\ \frac{\mathrm{kVA}}{\mathrm{kg}}$$
$$\cdot \left(\frac{m_e}{18\ \mathrm{kg}}\right)^{0.3294}.$$

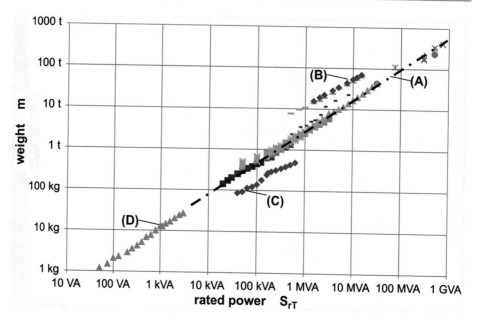

Fig. 8.39 Dependence of the mass of power transformers from different manufacturers on the rated power (**a**) oil-insulated transformers for power transmission and distribution, (**b**) MV cast resin transformers, (**c**) LV dry-type transformers (3~), (**d**) single-phase LV dry-type transformers

A special feature is the determination of the referred short-circuit voltage u_k or the leakage inductance of the transformer. The higher the rated power of a transformer, the higher u_k is selected. The referred short-circuit voltage u_k is a measure of the relative size of the leakage reactance. A larger value means a small short-circuit current. The reason for larger values at higher rated power is the better controllability of the magnetic forces in the transformer and the connected equipment in the event of a short circuit due to the current-limiting effect of the leakage inductance of the transformer. While in the distribution system, the transformers are equipped with $u_k = 4...6\%$, u_k is up to 16% for high-power transformers (see Fig. 8.40).

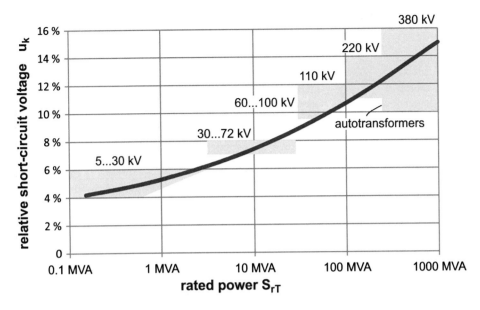

Fig. 8.40 Assignment of rated power and relative short-circuit voltage in distribution and transmission transformers with typical voltage levels according to [9]

8.7 Windings of Copper or Aluminium

Copper is the most widely used conductor material in electrical engineering. But it is relatively expensive and heavy. For a long time now, there have therefore been efforts to replace copper with aluminium. However, this is encountering a number of difficulties due to the material properties. Physical parameters relevant from experience are summarised in Table 8.7 below.

For a comparison, the resistance values at 100 °C should be used, as this is more likely to correspond to the actual application:

- Cross section of Al with the same resistance: 146%
- Mass of Al with the same resistance 44%
- Price relation of Al/Cu with same resistance 44.8% +/−7.4%

This potentially results in a weight advantage of −56% and a price advantage of about −45% for aluminium. With a weight proportion of 30...50% for copper windings in the total weight, the influence on the price is noticeable. Nevertheless, copper will continue to dominate for a long time to come, even for general applications in electrical engineering.

Although aluminium is quite ductile, i.e. plastically deformable, it has lower strength and conductivity than copper and breaks more easily. When exposed to air, aluminium quickly coats itself with a hard, heat-resistant oxide layer, which is not electrically

Table 8.7 Comparison of physical parameters of (unalloyed) copper and aluminium

	Formula sign	Copper (Cu)	Aluminium (Al)	Unit	Al:Cu
Atomic number	Z	29	13	–	–
Specific resistance	ρ_{20} ρ_{100}	0.01754 0.02385	0.02653 0.03481	$\Omega\cdot mm^2/$ m	1.51 1.46
Density	γ_{20}	8.92	2.7	g/cm^3	0.3
Molar mass	M	63.55	26.98	g/mol	–
Thermal conductivity	λ_{20}	394	237	W/ $(m\cdot K)$	0.601
Specific heat capacity	c_w	385 3434.2	900 2430	Ws/ $(kg\cdot K)$ Ws/ $(dm^3\cdot K)$	2338 0.7076
Melting temperature	T_m	1084 1357	660 933	°C K	$\Delta T_m =$ 424 K
Heat of fusion	W_m	13.3 844.6	10.79 291.1	kWs/ mol kWs/g	0.81 0.3447
Young's modulus	E	100...130	70	GPa	0.54...0.7
Tensile strength	σ_m	200...250	90	MPa	0.36...0.45
Elongation at break	A	4.5	60	%	13.3
Electrochemical potential		+ 0.522 (Cu^+) + 0.345 (Cu^{2+})	−1.66	V	–

conductive and therefore makes contacting more difficult. Contacting is possible by means of clamp, crimp and welded connections. The latter also includes the bonding of Al wires on metallized chip surfaces. In principle, soldering is possible with suitable fluxes and solders both as hard soldering and soft soldering, but is largely uncommon in electrical engineering. Due to the surface oxidation of aluminium in an oxygen environment, particular difficulties arise here. The electrochemical potential of −1.66 V identifies aluminium as a "base metal". This is reflected in material combinations with a considerable tendency to corrosion. The contact of joints with electrolytes should therefore be avoided. Condensation in connection with salt water mist, for example, should be avoided at all costs. The use of aluminium alloys in a humid maritime climate is problematic and may require special tests.

It is true that there are a large number of alloys that change the mechanical properties strongly in the direction of steel and also have a positive influence on corrosion resistance. In general, however, it can be said that the attractive good properties in the conduction of electricity and heat in both copper and aluminium can only be achieved with high degrees of purity. Alloys practically always have negative consequences in terms of thermal and electrical properties. Therefore, a separate check of the required and possible properties

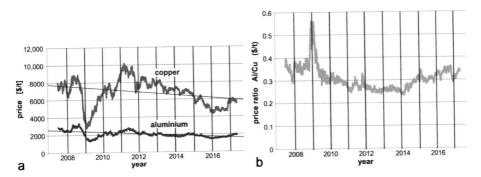

Fig. 8.41 Comparison of copper and aluminium prices per metric tonne over a period of 10 years (8 May 2007 to 8 May 2017) (**a**) Prices in US dollars with trend lines, (**b**) development of the price ratio [11]

must be carried out for each application. In contrast to copper, aluminium tends to flow over a long period of time. This means that the material gives way over time under high pressure. Thus, initially fixed connections can gradually become loose, which may require regular checks. A test of connection technology according to the IEC 61238-1 test standard provides safety here.

Table 8.7 shows a material comparison of aluminium and copper. The density of aluminium is only 30% of the density of copper. The specific resistance of aluminium is about 50% higher than that of copper. A metallic resistor made of aluminium would therefore have only 45% of the weight of a copper resistor with the same resistance value. This is a design advantage that is particularly noticeable in mobile applications where there is less material to move.

But there are also big differences in prices. Figure 8.41 shows the price development on the world market 2007...2017. The price ratio of aluminium to copper in this period was thus 1:3...1:4, which represents a strong incentive to overcome the constructive difficulties in the use of aluminium as indicated above.

References

1. Data: Karl Schupp AG, Zöllikerberg/CH
2. Data of manufacturers: TDK/EPCOS, Ferroxcube, Sumida, Tridelta
3. Kleeb, Th.; Zacharias, P.: Hybrid and coreless magnetics for future power electronics. Uni Kassel, project report to ECPE 2017)
4. Wendt, M.: Ermittlung der Verlustleistungen in einem Synchron-Tiefsetzsteller mit Niedervolt GaN-HFETs (Determination of power dissipation in a synchronous buck converter with low-voltage GaN HFETs). Dissertation, Universität Kassel 2015
5. C. D. Waelzholz (CDW), Kaschke, Vacuumscmelze, S. Shrestha, Kaye and Laby: Table of Physical and Chemical Constants
6. Conrad, H.; Krampitz, R.: Elektrotechnologie, Verlag Technik Berlin 1983

7. Lexikon der Physik. Spektrum Akademischer Verlag, Heidelberg 1998
8. Zacharias, P.: Power Semiconductors – Trends in System Integration and Future Application Aspects. NPERC-J Workshop: Breath and smart Application of Power Electronics System Integration, Tokyo 2016
9. Oswald, B. R.: Vorlesung Elektrische Energieversorgung I, Skript Transformatoren 2005 (Lecture Electric Power Supply I, Script Transformers 2005), Leibniz-Universität Hannover
10. EU Regulation: No. 548/2014 of the EU Commission on the implementation of the Ecodesign Directive 2009/125/EC
11. http://www.finanzen.net/rohstoffe/
12. TDK: Large Size Ferrite Cores for high Power. January 2013
13. Wendt, M.; Zacharias, P.: Einsatz von GaN-Transistoren in DC-DC-Wandlern kleiner Leistungen (Use of GaN transistors in low-power DC-DC converters). GaN-Workshop ETG 2013

Approximation of Empirical Characteristic Curves

9

9.1 Dimensional Analysis for Ordering Influencing and Resulting Values Among Each Other

Dimensional analysis was developed in the nineteenth century to establish relationships between physical quantities that occur in a physical phenomenon under investigation. It is based on the analysis of the units of measurement and dimensions of these quantities and ultimately on the insight that mathematical functions for describing the structure of relationships refer to dimensionless numerical ranges. It is extremely useful for ordering influencing variables whose direct effect is difficult or practically impossible to grasp analytically. If a complete or over-determined set of influencing variables is present, there is thus a chance that dimensional analysis will provide a set of dimensionless parameters with which the empirically obtained data can be described in a largely generally valid way. In this way, the data is grouped without knowing the actual analytical context. This is particularly helpful in order to find meaningful standardization possibilities for measurement data to extend the validity of statements [3].

The method is therefore traditionally much better known and used in more empirically oriented disciplines of physics and engineering sciences such as fluid mechanics or nonlinear solid state mechanics than in electrical engineering. During the time when analog computers were used as process computers, the method also experienced a certain "boom" in electrical engineering. Because of the available technical basis of analog process computers in the years 1960...1980, their physical limitations forced standardization not only with the aim of dimensionlessness, but also with the aim that the result should be within the display range (maximum output voltage) of the computer. Digital signal processing with floating-point arithmetic makes these boundary conditions obsolete. Nevertheless, before evaluating empirical data, or even better: before planning the experiments leading to it, one is well advised to arrange the influencing parameters in such a way that

© Springer Fachmedien Wiesbaden GmbH, part of Springer Nature 2022
P. Zacharias, *Magnetic Components*,
https://doi.org/10.1007/978-3-658-37206-4_9

the results can be generalized. This is the main reason for the following considerations, because it allows a greater degree of generality to be obtained from empirical results.

The dimensional analysis is based on the requirement that an equation expressing the sought-after relationship must be independent of the units used for input and output variables. This requirement corresponds to the demand for equal dimensions of the terms on the right and left side of an equation. A physical quantity Y is represented as a product with its unit of measurement as unit as follows:

$$Y = z \cdot [Y]$$

These are

z   numerical value as an expression of the quantity of Y and
[Y] unit of measurement to characterize the quality of the size Y

The unit of measurement of a general physical quantity can be expressed in the following way:

$$[Y^*] = L^l \cdot M^m \cdot T^t \cdot$$

Y*   derived (secondary) physical quantity
L, M, T   units of measurement of the basic sizes
l, m, t   rational exponents (dimensions)

As basic quantities, a number of physical quantities have been agreed upon according to recognised expediency, from whose units of measurement all other units of measurement of other quantities can be composed according to the above equation. The binding SI unit system (SI: *Système international d'unités*) prescribes the basic quantities listed in Table 9.1 with their units of measurement. By means of the conservation laws of physics, the units of measurement of all agreed quantities of physics can practically be represented from these.

For example, the units for energy W, magnetic field strength H and electric voltage U *can be* written as follows:

$$[W] = \mathrm{kg} \cdot \mathrm{m}^2 \cdot \mathrm{s}^{-2}$$

$$[H] = \mathrm{A} \cdot \mathrm{m}^{-1}$$

$$[U] = \mathrm{kg} \cdot \mathrm{m}^2 \cdot \mathrm{s}^{-3} \cdot \mathrm{A}^{-1}$$

The term dimension, which is actually determined geometrically, is also used in an extended sense for the basic sizes. It is assumed that the dimension of a basic size relative

Table 9.1 The basic physical quantities with their units in the SI system

Physical variable	Formula symbol	Unit of measurement	Abbreviation
Length	l	Meter	m
Mass	m	Kilogram	kg
Time	t	Second	s
Electric current	I	Ampere	A
Temperature	T	Kelvin	K
Luminous intensity	I_v	Candela	cd
Quantity of substance	N	Mol	mol

to itself is equal to 1 and that it does not depend on other sizes. If the dimension of a considered size does not change when using a different scale for a basic size, it has the dimension '0' in this basic size. If it is clear for a measured quantity under investigation to which quantities it is related, but the form of these relationships is not known, dimensional equations can be set up in which the unit of measurement of the output quantity is shown on the left and a product of powers of the units of measurement of the influencing quantities is shown on the right. Since for each basic unit of measure the dimension on the right and left side of the equation must be the same, a linear system of equations is obtained to determine the exponents of the product approach shown above.

The following example is given:

$$\text{The power loss density } [p_V] = 1\,\frac{W}{m^3} \; = \; 1\,\frac{V \times A}{m^3} \; = \; 1\,\frac{kg}{m \; \times \; s^3}$$

of a soft magnetic material is dependent on the following quantities of the electromagnetic field during periodic magnetization

- magnetic field strength $[H] = 1\dfrac{A}{m}$

- induction B with $[B] = [\|\mu\| \cdot H] = 1\,T = 1\dfrac{Vs}{m^2} = 1\dfrac{kg}{A \cdot s^2}$

- permeability $[\mu] = 1\dfrac{V \cdot s}{A \cdot m} = 1\dfrac{kg \cdot m}{A^2 \cdot s^2}$

- frequency,repetition rate $[f] = 1\ s^{-1}$

- electric field strength $\left[E = \dfrac{\partial \psi}{dt}\right] = 1\dfrac{V}{m} = 1\dfrac{kg \cdot m}{A \cdot s^3}$

- electrical conductance $[\sigma] = 1\dfrac{A}{V \cdot m} = 1\dfrac{A^2 \cdot s^3}{kgm^3}$

- permittivity $[\varepsilon] = 1\dfrac{As}{Vm} = 1\dfrac{A^2 s^4}{kg \cdot m^3}$

- characteristic core-length dimension $[d] = 1\ m$

The electric field strength results from induction of the alternating magnetic field in the core and is therefore not considered separately in the following - just like the magnetic induction. The possibilities and limitations of this approach are to be demonstrated by a largely formal approach, i.e. one that is equipped with minimal primary information. The equality of the units of measurement can be expressed as follows for an (absolutely arbitrary) product approach

$$[p_V] = [H]^{\alpha 1} \cdot [\mu]^{\alpha 2} \cdot [f]^{\alpha 3} \cdot [\sigma]^{\alpha 4} \cdot [\varepsilon]^{\alpha 5} \cdot [d]^{\alpha 6}$$

If the basic parameters listed in Table 9.1 are selected, this equation results in the following system of equations by equating the dimensions on the right and left side for each basic parameter unit

$$
\begin{array}{c}
m: \\
kg: \\
s: \\
A:
\end{array}
\begin{bmatrix}
-1 & 1 & 0 & -3 & -3 & 1 \\
0 & 1 & 0 & -1 & -1 & 0 \\
0 & -2 & -1 & 3 & 4 & 0 \\
1 & -2 & 0 & 2 & 2 & 0
\end{bmatrix}
\begin{bmatrix}
\alpha_1 \\ \alpha_2 \\ \alpha_3 \\ \alpha_4 \\ \alpha_5 \\ \alpha_6
\end{bmatrix}
=
\begin{bmatrix}
-1 \\ 1 \\ -3 \\ 0
\end{bmatrix}
$$

The solution using the Gaussian algorithm to reduce the number of equations of the system results in the following procedure.

First you get the initial system

	a_1	a_2	a_3	a_4	a_5	a_6	β_Y	$\sum k_i$
m :	-1	1	0	-3	1	-1	-6	
kg :	0	1	0	-1	0	1	0	
s :	0	-2	-1	3	0	-3	1	
A :	1	-2	0	2	0	0	3	$+1\,Gl.1$

The goal of a triangular coefficient matrix |a| is approached in the first step by adding the first equation to the last equation, and so on.

After the first step, you get

	a_1	a_2	a_3	a_4	a_5	a_6	B_y	$\sum k_i$
m :	-1	1	0	-3	-3	1	-1	-6
kg :	0	1	0	-1	-1	0	1	0
s :	0	-2	-1	3	4	0	-3	1 $+2Gl.2$
A :	0	-1	0	-1	-1	1	-1	-3 $+1\,Gl.2$

After step 2, the final form of the system of equations is obtained

	a_1	a_2	a_3	a_4	a_5	a_6	B_y	$\sum k_i$
m :	-1	1	0	-3	-3	1	-1	-6
kg :	0	1	0	-1	-1	0	1	0
s :	0	0	-1	1	2	0	-1	1
A :	0	0	0	-2	-2	1	-0	-3

Since the system of equations has two equations less than variables, two of the exponents α_i are freely selectable.

The magnetization curve B(H) already shows the energy expended per time with repeated remagnetization as power loss

$$p_V \sim H \cdot B \cdot f \sim \mu \cdot H^2 \cdot f \text{ respectively}$$

$$\alpha_1 = 2$$
$$\alpha_2 = 1$$
$$\alpha_3 = 1$$

This results in the following equations for determining the remaining exponents/matrix coefficients

$$-2 + 1 - 3 + 0 - 3\alpha_5 + \alpha_6 = -1 \quad \text{or} \quad -3\alpha_5 + \alpha_6 = 3$$
$$0 + 1 - 1 + 0 - \alpha_5 + 0 = 1 \quad \text{or} \quad \alpha_5 = -1 \quad \text{or, as the result of which} \quad \alpha_6 = 0$$
$$0 + 0 + 1 - \alpha_4 + 2\alpha_5 + 0 = -1 \quad \text{or} \quad 2\alpha_5 = -2 + \alpha_4$$
$$0 + 0 + 0 - 2\alpha_4 + 2\alpha_5 + 0 = -2 \quad \text{or} \quad 2\alpha_5 = -2 + 2\alpha_4$$

The last two equations seem to contradict each other. Although the first two equations do not contradict each other, they contradict practical experience, at least in the case of metallic materials. Nevertheless, there is obviously a certain order of the influencing variables, which can be used directly. The approach shown above

$$\alpha_1 = 2$$
$$\alpha_2 = 1 \quad \text{or} \, p_V \sim H \cdot B \cdot f \sim \mu \cdot H^2 \cdot f \sim \frac{B^2}{\mu} f$$
$$\alpha_3 = 1$$

is widely used as a Steinmetz formula for calculating magnetic losses in electrical machines. The functional product approach selected for all influencing variables is obviously not achievable in its entirety with these selected variables. This can have 2 main reasons:

- The number of influencing factors is insufficient and/or
- The structure of the chosen approach is not adequate to the physical laws [1, 2, 6].

The second reason is, for example, if the target variable under investigation is the result of the sum of several effects. This is the case with magnetic losses, for example. The decisive advantage and the reason for the widespread application of the dimensional analysis method are the low effort with which even simple qualitative models can be achieved. The largest application of dimensional analysis has been in such areas of physics (hydraulics, aerodynamics, astrophysics, etc.), where a strict solution of problems often encounters significant difficulties, especially due to the large number of influencing variables [4, 5, 8].

From the system of equations created in the example, it becomes clear that for 'r' used basic quantities, there are nevertheless only a maximum of 'r' independent *units of measurement*. This means that all other units of measurement can be represented by power products of the units of these 'r' quantities. In other words, the rank of the dimension matrix (see preceding example) determines the number of independent units (in addition to formal independence, [6, 7] also assumes independence for physical reasons, if, for example, the interactions between individual groups of quantities can be physically excluded).

Based on this, further formal possibilities of dimensional analysis will be investigated. For this purpose, a modified representation of the functional dependence is assumed:

$$f(x_1, \quad x_2, \quad \ldots \quad x_r, \quad \ldots \quad x_{n-1}, \quad x_n, \quad y, \;) = 0$$

Here the quantities $x_1...x_r$ are the selected physical quantities with the independent units of measurement. According to the above findings, it is possible to agree/construct a system of units of measurement on the basis of these, which then - with reference to the assumed functional dependence - has the following form:

$$[x_1] = A_1$$
$$[x_2] = A_2$$
$$\ldots\ldots\ldots$$
$$[x_r] = A_r$$
$$[x_{r+1}] = \prod_{j=1}^{r} A_j^{\alpha_{j(r+1)}}$$
$$\ldots\ldots\ldots$$
$$[x_n] = \prod_{j=1}^{r} A_j^{\alpha_{jn}}$$
$$[y] = \prod_{j=1}^{r} A_j^{\alpha_{j0}}$$

Consequently, a dependence of a quantity y on n variables $x_1...x_n$ can also be written in the following form:

$$f\left(z_1 \cdot A_1, \; \ldots\ldots z_r \cdot A_r, \; z_{r+1} \cdot \prod_{j=1}^{r} A_j^{\alpha_j(r+1)}, \; \ldots, \; z_n \cdot \prod_{j=1}^{r} A_j^{\alpha_j(n)}, \; z_0 \cdot \prod_{j=1}^{r} A_j^{\alpha_j(0)} \right) = 0$$

Since such an equation must also remain true when using other scales for the independent units of measurement, the representation

$$f\left(\beta_1 \cdot z_1 \cdot A_1, \ldots \ldots \beta_1 \cdot z_r \cdot A_r, z_{r+1} \cdot \prod_{j=1}^{r} \{\beta_j \cdot A_j\}^{a_j(r+1)}, \ldots, z_n \cdot \prod_{j=1}^{r} \{\beta_j \cdot A_j\}^{a_j(n)}, \right.$$

$$\left. z_0 \cdot \prod_{j=1}^{r} \{\beta_j \cdot A_j\}^{a_j(0)}\right)$$

$$= 0$$

is also possible. If you now choose specifically for $\beta_i = (z_i \cdot x_i)^{-1}$ under the condition that $x_i \neq 0$, then

$$f\left(1, 1, 1, \ldots.1, z_{r+1} \cdot \prod_{j=1}^{r} \left\{\frac{A_j}{z_j A_j}\right\}^{a_j(r+1)}, \ldots, z_n \cdot \prod_{j=1}^{r} \left\{\frac{A_j}{z_j A_j}\right\}^{a_j(n)}, z_0 \cdot \prod_{j=1}^{r} \left\{\frac{A_j}{z_j A_j}\right\}^{a_j(0)}\right)$$

$$= 0 \text{ or}$$

$$f\left(1, 1, 1, \ldots.1, z_{r+1} \cdot \prod_{j=1}^{r} \left\{\frac{1}{z_j}\right\}^{a_j(r+1)}, \ldots, z_n \cdot \prod_{j=1}^{r} \left\{\frac{1}{z_j}\right\}^{a_j(n)}, z_0 \cdot \prod_{j=1}^{r} \left\{\frac{1}{z_j}\right\}^{a_j(0)}\right) = 0$$

or after reversing the substitution made at the beginning

$$f\left(1, 1, 1, \ldots.1, \frac{x_{r+1}}{\prod_{j=1}^{r} x_j^{a_j(r+1)}}, \ldots, \frac{x_n}{\prod_{j=1}^{r} x_j^{a_j(n)}}, \frac{y}{\prod_{j=1}^{r} x_j^{a_j(0)}}\right) = 0$$

This generalization of the procedure applied at the beginning only to size y leads to the Π theorem of Edgar Buckingham (* 8 July 1867 Philadelphia; † 29 April 1940 Washington, D.C.) [8]:

"If there is a relationship between m parameters in the form

$$f(x_1, x_2, x_3, \ldots x_m) = 0$$

then an equivalent form of the relationship can be found with k dimensionless product complexes Π_i:

$$f(\Pi_1, \Pi_2, \Pi_3, \ldots \Pi_m) = 0$$

The number k is determined by

$$k = m - r$$

Here are

m: number of parameters q_i in the equation $f(x_1, x_2, x_3, \ldots x_m) = 0$ and
r: largest possible number of dimensionally independent variables from the set of influencing variables.

The effect of the Π theorem is demonstrated below using the example of core losses of a magnetic material, which was already considered at the beginning.

$$f(p_V, H, \mu, f, \sigma, \varepsilon, d) = 0$$

The solution of the Gaussian algorithm already provides that the first 4 columns or measurement units are linearly independent of each other. If we now write the Gaussian algorithm in Table 9.2, we obtain with "separation of the independent" variables.

From this, the necessary exponents for the Π complexes are obtained directly:

$$\Pi_{P_V} : \alpha_{(0)} = 1 \quad \alpha_4 = 0 \rightarrow \alpha_3 = 0 \rightarrow \alpha_2 = 1 \rightarrow \alpha_1 = 2$$

This means in other words

$$\Pi_{P_V} = \frac{p_V}{\mu \cdot H^2 \cdot f} \quad \text{or} \quad p_V \sim \mu \cdot H^2 \cdot f$$

Analogously you get

$$\Pi_\varepsilon = \frac{\varepsilon \cdot f}{\sigma}$$

This expression is more familiar to the reader after an expansion of the fraction with dimensionless numbers, which dimensional analysis cannot provide.

$$\Pi_\varepsilon = \frac{\varepsilon \cdot 2\pi f \cdot \frac{A}{l}}{\sigma \cdot \frac{A}{l}} = \frac{Y_C}{G_C} = \frac{R_C}{Z_C} = \tan(\delta_C)$$

Consequently, the dielectric losses in the core material are described here. These are of particular importance for HF ferrites, for example. While MnZn ferrites usually have a relatively low DC conductivity, NiZn ferrites are almost perfect insulators. However, they do have dielectric losses. One would have to write therefore more exactly

Table 9.2 Application of the Gaussian algorithm

	α_1	α_2	α_3	α_4	α_5	α_6	α_7	$\sum k_i$
m :	-1	1	0	-3	-3	1	-1	-6
kg :	0	1	0	-1	-1	0	1	0
s :	0	-2	-1	3	4	0	-3	1
A :	1	-2	0	2	2	0	0	3
m :	-1	1	0	-3	-3	1	-1	-6
kg :	0	1	0	-1	-1	0	1	0
s :	0	-2	-1	3	4	0	-3	1
A :	0	-1	0	-1	-1	1	-1	-3
m :	-1	1	0	-3	-3	1	-1	-6
kg :	0	1	0	-1	-1	0	1	0
s :	0	0	-1	1	2	0	-1	1
A :	0	0	0	-2	-2	1	0	-3

$$\sigma = \omega \cdot \mathrm{Im}\{\underline{\varepsilon}\} \text{ with } \underbrace{\lim_{\omega \to 0}}(\omega \cdot \mathrm{Im}\{\underline{\varepsilon}(\omega)\}) = \sigma_{DC}$$

Finally, the same way you get for d:

$$\Pi_d = d\sqrt{\sigma \cdot f \cdot \mu} \ \text{ or } \ d \sim \frac{1}{\sqrt{\sigma \cdot f \cdot \mu}}$$

The penetration depth of the current or magnetic field in the skin effect is equivalent and is typically

$$s = \frac{1}{\sqrt{\pi \cdot \sigma \cdot f \cdot \mu}}$$

and can thus be seen as an interpretation of the Π complex found.

In the Π theorem, no difference is made between the sizes $(x_1...x_n, y)$. Furthermore, it only implies that a certain number of dimensionless Π complexes finally form the sought-after connection. With regard to the application, the first thing to note is the order of the selected variables, which can have an influence on the result of a calculation.

If, for example, the specific power dissipation p_V is included in the group of the first 4 variables, the Gaussian algorithm (as expected) shows that there is a linear dependence between the dimensions of p_V, H, f and μ. The Π theorem is based on the non-existence of such a relationship to be mapped. Therefore, the order must be changed and/or new influencing variables must be searched for in order to get closer to a physically founded correlation. The same applies to empirical relationships that can be generalized.

The sequence of the approach used at the entrance provided

	H	μ	f	σ	ε	d	p_V	Σk_i
	α_1	α_2	α_3	α_4	α_5	α_6	β_Y	
$m:$	-1	1	0	-3	-3	1	-1	-6
$kg:$	0	1	0	-1	-1	0	1	0
$s:$	0	0	-1	1	2	0	-1	1
$A:$	0	0	0	-2	-2	1	0	-3

You can now change the order, so that, for example, the following image is created at the last step of the algorithm

	H	ε	f	d	μ	σ	p_V	Σk_i
	α_1	α_5	α_3	α_6	α_2	α_4	$\alpha_{(0)}$	
$m:$	-1	-3	0	1	1	-3	-1	-6
$kg:$	0	-1	0	0	1	-1	1	0
$s:$	0	0	-1	0	2	-1	1	1
$A:$	0	0	0	1	-2	0	-2	-3

With this solution, further Π complexes can be determined, which could be the basis for the description of the examined context:

$$\Pi_\mu = \frac{1}{\mu \cdot \varepsilon}\sqrt{\frac{H^3}{d \cdot f}} \qquad \Pi_{Pv(2)} = \frac{\varepsilon \cdot H^{3,5}}{p_V \cdot d^{0,5} \cdot f} \qquad \Pi_\sigma = \frac{f \cdot \varepsilon}{\sigma}$$

The first term provides a Π complex, which is more likely to be assigned to the high-frequency technology. The second term also provides a dependency for $p_V(H, \varepsilon, d, f)$, which remains difficult to interpret as long as the pure (reduced) product is considered. Since permittivity plays a role, the losses must be those resulting from the dielectric behaviour of the core material. This would concern the imaginary component of ε. One would therefore have to extend the complex $\Pi_{Pv}(2)$ with f/f, etc.. The size d, for example, which was assumed here as an example, does not necessarily have to have the same meaning in the different complexes. In dimensional analysis, only influencing variables with different dimensions can be directly distinguished. If it is clear from experience/experiments that certain preferred influencing variables with the same dimension are essential for a description, these can be included without any problems in the table of the Π theorem. A systematic investigation of the possible links also then provides the possible Π complexes.

Despite all the - mathematically very logical - analytical connections, it must be stated that the method presented is itself part of an empirical process. There is no guarantee that the correlations are complete. On the other hand, however, if influencing variables are determined, the "non-solvability" of the systems of equations provides indications of missing variables, but not which ones. If, for example, relative variables already occur within the dependencies, these act like dimensionless variables during the investigation and have no effect in the result. This may be illustrated by such dimensionless quantities as remanence flux density B_r vs. saturation flux density B_r/B_{SAT} or electric field strength vs. breakdown field strength E/E_D. Such relations cannot be clearly found by the described analysis.

Nevertheless, the dimensional analysis is a very helpful method. With the help of the Π theorem, one can examine and order the units of measurement of target variables and influencing variables very effectively. By combining the variables into dimensionless product complexes (Π complexes), it is possible to reduce the number of variables to be investigated. Formal algebraic as well as heuristic methods are required to apply the theorem to a number of secondary or derived variables. The mathematical principles of the Π theorem are explained in full [4, 6, 7].

By applying the method, it is possible to generalize experimental results or to transfer experimental results from a model to a real object. The Π theorem is connected to the similarity theory and is a central theorem of modeling [6, 9]. The method has no means for finding or checking the necessary influencing variables as well as the functional structure required for this. This requires the use of other sources. It is therefore not a completely independent method. It finds its broad practical application in the formulation of analytical relationships, e.g. all arguments of function expressions must be dimensionless.

9.2   Linear Regression Analysis

For easier writing, single and multiple row data fields (matrices) are marked with underlined letters in this chapter. The usual matrix operators are applied to these.

Properties of magnetic materials, such as the magnetization characteristic curve(s) and specific core losses are usually not fully analytically describable. Therefore, simulations and calculations of losses often require the approximation of empirical data. But also for the comparison of materials or components, an empirical modelling is often advantageous.

As a rule, the aim is to describe the data sets, which are inevitably subject to measurement errors, by simple analytical functions that can be continuously differentiated at least once. The simplest mathematical tool is the linear regression analysis of the primary data. Here, the measured data are approximated by a linear combination (function series) of selected functions (basic functions). The selection is made in such a way that its basic properties are as close as possible to those of the measured data. Otherwise, very long function series are obtained. The general procedure is described below and demonstrated with simple examples.

Let y be a measured quantity, which was determined in $i = 1...n$ samples and which depends for each i on a number of m influence quantities $\underline{X} = (x_{i1}; x_{i2} ... x_{im})$. It is postulated that each measured value y_i is composed of the sum of k known functions for which, however, the "weighting factors" are still unknown, i.e.

$$y_i = \sum_{v=1}^{k} a_v f_v(x_{i1}; x_{i2}; \ldots x_{im}) = \sum_{v=1}^{k} a_v f_v(\underline{x}_i)$$

In matrix writing, this connection looks as follows for a single-column vector $\underline{Y}$:

$$\underline{Y} = \sum_{v=1}^{k} a_v f_{vi} = \begin{bmatrix} f_{11} & f_{12} & \cdots & f_{1k} \\ f_{21} & f_{21} & \cdots & f_{2k} \\ \cdots & \cdots & \cdots & \cdots \\ f_{n1} & f_{n1} & \cdots & f_{nk} \end{bmatrix} \cdot \begin{bmatrix} a_1 \\ a_2 \\ \cdots \\ a_k \end{bmatrix} = \underline{F} \cdot \underline{A}$$

The most frequently used criterion for the design of the quality of the approximation is the minimization of the sum of the error squares according to Carl Friedrich Gauss (* 30 April 1777 Braunschweig; † 23 February 1855 Göttingen). Analytically expressed, this means

$$\left(\sum_{i=1}^{n} \left(y_i - \sum_{v=1}^{k} a_v f_{vi} \right)^2 \right) = (\underline{Y} - \underline{F} \cdot \underline{A})^T \cdot (\underline{Y} - \underline{F} \cdot \underline{A}) = \sigma_1^2 \Rightarrow \min!$$

By multiplying out, you get

$$\left(\underline{Y}^T \cdot \underline{Y} - (\underline{F} \cdot \underline{A})^T \cdot \underline{Y} - \underline{Y}^T \cdot \underline{F} \cdot \underline{A} + \underline{F}^T \cdot \underline{A}^T \cdot \underline{F} \cdot \underline{A} \right) = \sigma_1^2 \Rightarrow \min!$$

or by grouping and differentiating according to all coefficients a_v

$$\frac{\partial}{\partial a_v} | (\underline{Y}^T \cdot \underline{Y} - 2\underline{F}^T \cdot \underline{A}^T \cdot \underline{Y} + \underline{F}^T \cdot \underline{F} \cdot \underline{A}^T \cdot \underline{A}) = \sigma_1^2 \Rightarrow \min! \quad \forall a_v$$

$$-2\underline{F}^T \cdot \underline{Y} + 2 \cdot \underline{F}^T \cdot \underline{F} \cdot \underline{A} = 0 \text{ or } \underline{F}^T \cdot \underline{F} \cdot \underline{A} = \underline{F}^T \cdot \underline{Y}$$

From this follows for $\underline{A}$ in general terms, i.e. independent of the selected system of basic functions:

$$\underline{A} = \left(\underline{F}^T \cdot \underline{F} \right)^{-1} \cdot \left(\underline{F}^T \cdot \underline{Y} \right)$$

An example with 2 influencing factors x_1 and x_2 and the following functional approach for a 2-dimensional polynomial of first order shall clarify this, here very formalized approach. The advantage of the matrix notation is the easy transferability to any functional system and finally also the key for the optimized statistical design of experiments for the determination of $\underline{A}$, described in the following chapter.

Table 9.3 Fictitious samples for measured values $y_i(x_{1i}; x_{2i})$ from a test design

Nr.	y	x1	x2	f1	f2	f3	f4
1	2.06	0	0	1	0	0	0
2	4.41	4	4	1	4	4	16
3	2.31	3	2	1	2	3	6
4	1.25	4	1	1	1	4	4
5	3.03	1	3	1	3	1	3
6	1.19	2	0	1	0	2	0
7	1.73	2	1	1	1	2	2
8	3.52	1	4	1	4	1	4
9	2.66	1	2	1	2	1	2
10	0.59	3	0	1	0	3	0

Table 9.3 for $\underline{Y}$ contains the values.

The description approach chosen for these data is

$$y = a_1 f_1(x_1, x_2) + a_2 f_2(x_1, x_2) + a_3 f_3(x_1, x_2) + a_4 f_4(x_1, x_2) = a_1 + a_2 \cdot x_1 + a_3 \cdot x_2 + a_4 \cdot x_1 \cdot x_2.$$

This corresponds to a well-founded assumption for the fundamental course of the dependence, which can only be determined with unavoidable statistical measurement errors.

The expression generating the artificial sample is therefore for the demonstration of the algorithm's mode of operation

$$y = 2 + 0.3 \cdot x_1 - 0.5 \cdot x_2 + 0.2 \cdot x_1 \cdot x_2 + 0.2 \cdot z_v,$$

where zv is a random number with the property

$$z_v \in (0; +1)$$

so that a "real systematic" mean error of 0.08 has been inserted compared to the "real" data with a dispersion around this mean of 0.065. These properties result analytically formulated:

$$E[z_1, \ldots z_{10}] = 0.08 \text{ and } \sigma[z_1, \ldots z_{10}] = 0.065$$

To give an idea of the model function used, a graphical representation is shown in Fig. 9.1. The error sizes are not shown here because they are only generated at the sampling points. Figure 9.1 does not show the true relationship, but rather the probable course of the searched function if the assumptions are correct.

The expression for $\underline{F}^T \cdot \underline{F}$ is then

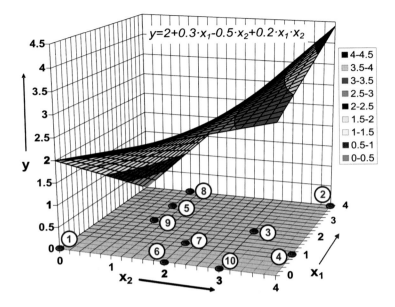

Fig. 9.1 Graphical representation of the function to be approximated used for the calculation example with marked locations and sequence of the sample values

$$\begin{bmatrix} 1 & 1 & 1 & 1 & 1 & 1 & 1 & 1 & 1 & 1 \\ 0 & 4 & 2 & 1 & 3 & 0 & 1 & 4 & 2 & 0 \\ 0 & 4 & 3 & 4 & 1 & 2 & 2 & 1 & 1 & 3 \\ 0 & 16 & 6 & 4 & 3 & 0 & 2 & 4 & 2 & 0 \end{bmatrix} \cdot \begin{bmatrix} 1 & 2 & 3 & 6 \\ 1 & 1 & 4 & 4 \\ 1 & 3 & 1 & 3 \\ 1 & 0 & 2 & 0 \\ 1 & 1 & 2 & 2 \\ 1 & 4 & 1 & 4 \\ 1 & 2 & 1 & 2 \\ 1 & 0 & 3 & 0 \end{bmatrix} = \begin{bmatrix} 10 & 17 & 21 & 37 \\ 17 & 51 & 37 & 111 \\ 21 & 37 & 61 & 111 \\ 37 & 111 & 111 & 341 \end{bmatrix}$$

The matrix inverse $\left(\underline{F}^T \cdot \underline{F}\right)^{-1}$ then has the value

$$\begin{bmatrix} 10 & 17 & 21 & 37 \\ 17 & 51 & 37 & 111 \\ 21 & 37 & 61 & 111 \\ 37 & 111 & 111 & 341 \end{bmatrix}^{-1} = \begin{bmatrix} 0.765 & -0.239 & -0.267 & 0.082 \\ -0.239 & 0.142 & 0.081 & -0.047 \\ -0.267 & 0.081 & 0.134 & -0.041 \\ 0.082 & -0.047 & -0.041 & 0.023 \end{bmatrix}$$

The multiplication of the matrices $\underline{F}^T \cdot \underline{Y}$ gives

$$
\begin{bmatrix} 1 & 1 & 1 & 1 & 1 & 1 & 1 & 1 & 1 & 1 \\ 0 & 4 & 2 & 1 & 3 & 0 & 1 & 4 & 2 & 0 \\ 0 & 4 & 3 & 4 & 1 & 2 & 2 & 1 & 1 & 3 \\ 0 & 16 & 6 & 4 & 3 & 0 & 2 & 4 & 2 & 0 \end{bmatrix} \cdot \begin{bmatrix} 2.06 \\ 4.41 \\ 2.31 \\ 1.25 \\ 3.03 \\ 1.19 \\ 1.73 \\ 3.52 \\ 2.66 \\ 0.59 \end{bmatrix} = \begin{bmatrix} 22.75 \\ 53.73 \\ 46.39 \\ 121.37 \end{bmatrix}
$$

From which, finally, by performing the operation $\underline{A} = (\underline{F}^T \cdot \underline{F})^{-1} \cdot (\underline{F}^T \cdot \underline{Y})$ the vector of the coefficient matrix for the "estimator" gives for $\underline{A}$:

$$
\begin{bmatrix} 0.765 & -0.239 & -0.267 & 0.082 \\ -0.239 & 1.142 & 0.081 & -0.047 \\ -0.267 & 0.081 & 0.134 & -0.041 \\ 0.082 & -0.047 & -0.041 & 0.023 \end{bmatrix} \cdot \begin{bmatrix} 22.75 \\ 53.73 \\ 46.39 \\ 121.37 \end{bmatrix} = \begin{bmatrix} 2.128 \\ 0.246 \\ -0.482 \\ 0.23 \end{bmatrix}
$$

If the error size is freed from its mean value, i.e. $E[z_1, \ldots z_{10}] = 0$, the equations for the multiplication of the matrices $\underline{F}^T \cdot \underline{Y}$ and the determination of the coefficient vector $\underline{A}$ *are*

$$
\begin{bmatrix} 1 & 1 & 1 & 1 & 1 & 1 & 1 & 1 & 1 & 1 \\ 0 & 4 & 2 & 1 & 3 & 0 & 1 & 4 & 2 & 0 \\ 0 & 4 & 3 & 4 & 1 & 2 & 2 & 1 & 1 & 3 \\ 0 & 16 & 6 & 4 & 3 & 0 & 2 & 4 & 2 & 0 \end{bmatrix} \cdot \begin{bmatrix} 1.99 \\ 4.34 \\ 2.24 \\ 1.18 \\ 2.96 \\ 1.12 \\ 1.66 \\ 3.45 \\ 2.56 \\ 0.52 \end{bmatrix} = \begin{bmatrix} 22.02 \\ 52.48 \\ 44.89 \\ 118.72 \end{bmatrix}
$$

or

Table 9.4 Comparison of the approximation results with error sizes with existing and disappearing mean

Coefficient	Default	Calculated with E[z] = 0.08	Calculated with E[z] = 0
a_1	2	2.128	2.052
a_2	0.3	0.246	0.246
a_3	−0.5	−0.482	−0.481
a_4	0.2	0.23	0.229
Mean deviation of the model of the "original" in $x_1 \in (0; 4)$ and $x_2 \in (0; 4)$		0.216	0.163
Mean deviation of the model of the "original" on the bases		0.217	0.155

$$\begin{bmatrix} 0.765 & -0.239 & -0.267 & 0.082 \\ -0.239 & 0.142 & 0.081 & -0.047 \\ -0.267 & 0.081 & 0.134 & -0.041 \\ 0.082 & -0.047 & -0.041 & 0.023 \end{bmatrix} \cdot \begin{bmatrix} 22.02 \\ 52.48 \\ 44.89 \\ 118.72 \end{bmatrix} = \begin{bmatrix} 2.052 \\ 0.246 \\ -0.481 \\ 0.229 \end{bmatrix}$$

A comparison of the "actual" and the calculated coefficients results in values close to each other (Table 9.4) despite the inserted error sizes with a non-zero mean value (see above):

Subtracting the mean value of this sample $E[\underline{Y}]$ from the measured sample $\underline{Y}$, one obtains with $E[\underline{Y}] = 2.275$ and

$$\underline{Y}^* = \underline{Y} - \begin{bmatrix} 1 & 0 & .. & 0 & 0 \\ 0 & 1 & .. & 0 & 0 \\ .. & .. & 1 & .. & .. \\ 0 & 0 & .. & 1 & 0 \\ 0 & 0 & .. & 0 & 1 \end{bmatrix} \cdot E[\underline{Y}]$$

The values in Table 9.3 change accordingly in Table 9.5:

Since the subtraction of $E[\underline{Y}]$ also subtracts the mean value of the error quantity $E[\underline{Z}]$, the result is independent of the latter. By "centering" the data set $y(x_1; x_2)$ in this way, the structure of the functional approach to description is not changed in principle. It only reads in this example

$$y^* = y - E[\underline{Y}] = a_1^* f_1(x_1, x_2) + a_2^* f_2(x_1, x_2) + a_3^* f_3(x_1, x_2) + a_4^* f_4(x_1, x_2) = a_1^* + a_2^* \cdot x_1 + a_3^* \cdot x_2 + a_4^* \cdot x_1 \cdot x_2$$

or

Table 9.5 Table of values with fictitious measurement results reduced by the mean value

Nr.	$y^*(x1,x2)$	x1	x2	f1	f2	f3	f4
1	−0,22	0	0	1	0	0	0
2	2.14	4	4	1	4	4	16
3	0.04	2	3	1	2	3	6
4	−1.03	1	4	1	1	4	4
5	0.76	3	1	1	3	1	3
6	−1.09	0	2	1	0	2	0
7	−0.55	1	2	1	1	2	2
8	1.25	4	1	1	4	1	4
9	0.39	2	1	1	2	1	2
10	−1.69	0	3	1	0	3	0

$$y = \left(a_1^* + E[\underline{Y}]\right) + a_2^* \cdot x_1 + a_3^* \cdot x_2 + a_3^* \cdot x_1 \cdot x_2$$

The numerical execution of the operations to obtain the coefficients $\left(a_1^*, a_2^*, a_3^*, a_4^*\right)$ yields the equations for the multiplication of the matrices $\underline{F}^T \cdot \underline{Y}$ and the determination of the coefficient vector $\underline{A}$

$$
\begin{bmatrix}
1 & 1 & 1 & 1 & 1 & 1 & 1 & 1 & 1 & 1 \\
0 & 4 & 2 & 1 & 3 & 0 & 1 & 4 & 2 & 0 \\
0 & 4 & 3 & 4 & 1 & 2 & 2 & 1 & 1 & 3 \\
0 & 16 & 6 & 4 & 3 & 0 & 2 & 4 & 2 & 0
\end{bmatrix}
\cdot
\begin{bmatrix}
-0.22 \\ 2.14 \\ 0.04 \\ -1.03 \\ 0.76 \\ -1.09 \\ -0.55 \\ 1.25 \\ 0.39 \\ -1.69
\end{bmatrix}
=
\begin{bmatrix}
0 \\ 15.12 \\ -1.39 \\ 37.32
\end{bmatrix}
$$

and

$$
\begin{bmatrix}
0.765 & -0.239 & -0.267 & 0.082 \\
-0.239 & 0.142 & 0.081 & -0.047 \\
-0.267 & 0.081 & 0.134 & -0.041 \\
0.082 & -0.047 & -0.041 & 0.023
\end{bmatrix}
\cdot
\begin{bmatrix}
0 \\ 15.12 \\ -1.39 \\ 37.32
\end{bmatrix}
=
\begin{bmatrix}
-0.182 \\ 0.28 \\ -0.492 \\ 0.205
\end{bmatrix}
$$

The mathematical model for the primary data is then

$$y(x_1; x_2) = E[\underline{Y}] - 0.182 + 0.28 \cdot x_1 - 0.492 \cdot x_2 + 0.205 \cdot x_1 \cdot x_2 \text{ or}$$

$$y(x_1; x_2) = 2.093 + 0.28 \cdot x_1 - 0.492 \cdot x_2 + 0.205 \cdot x_1 \cdot x_2$$

The "effective" deviation, i.e. the RMS value of the error, of the model from the "original" is now only $\sigma = 0.062$ instead of 0.215. It is clear from the example that "centering" the data set $\underline{Y}$ when calculating the coefficients for the approximation function results in significant advantages with regard to model accuracy or error suppression. The operation $\underline{Y} = \underline{F} \cdot \underline{A}$ can be seen as a image/projection of $\underline{Y}$ into the 4-dimensional vector space spanned by $\underline{f}_1 \cdots \underline{f}_4$. The single-column matrix $\underline{A}$ contains the scalar products that are generated by this projection. Figure 9.2 shows a visualization of this approach for the plane spanned by $\underline{f}_2$ and $\underline{f}_3$. Usually, only selected properties of the reality to be reflected can be depicted in an image. The vectors of $\underline{f}_2$ and $\underline{f}_3$, with their 10 components each, span a subspace that is no longer directly graspable by the human imagination. If only the vectors $\underline{f}_2, \underline{f}_3$ and $\underline{Y}$ are considered in their subspace, each vector is characterized by a magnitude and an angle in relation to the other vector. The magnitude of a vector is given by the scalar product with itself:

$$\left| \underline{f}_2 \right| = \sqrt{\underline{f}_2^T \cdot \underline{f}_2} = 7.14 \text{ and } \left| \underline{f}_3 \right| = 7.81 \text{ and } |\underline{Y}| = 7.787$$

At the same time, the angles between $\underline{f}_2, \underline{f}_3$ and $\underline{Y}$ result in

$$\sphericalangle(\underline{f}_2; \underline{f}_3) = \frac{180°}{\pi} \arccos\left(\frac{\underline{f}_2^T \cdot \underline{f}_3}{|\underline{f}_2| \cdot |\underline{f}_3|} \right) = 48.4° \; \sphericalangle(\underline{f}_2; \underline{Y}) = 19.4° \; \sphericalangle(\underline{f}_3; \underline{Y}) = 42.5°$$

By "centering" the measured value vector $Y \rightarrow Y^*$, its position relative to the basis vectors changes, as shown in Fig. 9.2.

This space will be further analysed in the following using the selected example. Analogously, for other angles between the vectors, you get

$$\sphericalangle(\underline{f}_1; \underline{f}_2) = 41.2° \quad \sphericalangle(\underline{f}_1; \underline{f}_3) = 31.8° \quad \sphericalangle(\underline{f}_1; \underline{f}_4) = 50.7°$$

$$\sphericalangle(\underline{f}_3; \underline{f}_4) = 41.5° \quad \sphericalangle(\underline{f}_2; \underline{f}_4) = 32.7°.$$

The coordinate system used for the projection is therefore not orthogonal. A component of the measurement vector $\underline{Y}$ in a spatial direction of the 4-dimensional subspace affects several components of the "spanning" vectors of $[\underline{f}_1, ... \underline{f}_4]$ in any case in the projection. Thus, it can happen that even error quantities with their components can have multiple effects on components of the model function. *As a result, the error of the model function compared to the "true" function may not be smaller than the statistical measurement*

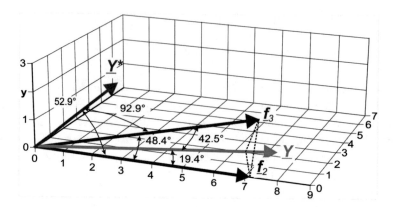

Fig. 9.2 Graphical representation of the magnitudes and angular relationships between the vectors $\underline{f}_2, \underline{f}_3$ and $\underline{Y}$ and the "centered" measured value vector $\underline{Y}*$

error, although this is assumed to be a fact in the general use of regression analysis. If the matrix $\underline{F}^T\underline{F}$ is different from zero only on the main diagonal, a complete orthogonal system of basis vectors is given. This offers some advantages in the calculation. The way to such base vectors will therefore be discussed several times in this chapter.

The structure of the functional approach is not changed by orthogonalization. It is now achieved as follows:

So far, the column vectors $[\underline{f}_1...\underline{f}_4]$ of the matrix $\underline{F}$ - as shown above - form a non-orthogonal group of vectors.

$$\underline{F} = \begin{bmatrix} \underline{f}_1 & \underline{f}_2 & \underline{f}_3 & \underline{f}_4 \end{bmatrix}$$

In principle, these represent the "coordinate system"onto which the vector of the measured values $\underline{Y}$ is mapped. Let us first look at the $(\mathrm{f}_1; \mathrm{f}_2)$ plane. In this plane, the two vectors $\mathrm{f}_1, \mathrm{f}_2$ enclose an angle of $41.5°$ (see Sect. 9.2) (Fig. 9.2, Fig. 9.3).

If one takes as starting point the column vector $\underline{f}_1 = \underline{f}_1^*$ for the transformation of the column vector $\underline{f}_2$ to the orthogonal column vector $\underline{f}_2^*$. With the help of $\underline{f}_2$, a vector with $\underline{f}_2^* \perp \underline{f}_1^*$ is constructed by the linear combination of the two vectors $\underline{f}_1^*$ and $\underline{f}_2$:

$$\underline{f}_2^* = b_{21} \cdot \underline{f}_1^* + \underline{f}_2$$

For orthogonality, the following must apply

$$\underline{f}_1^{*T} \cdot \underline{f}_2^* = b_{21} \cdot \underline{f}_1^{*T} \cdot \underline{f}_1^* + \underline{f}_1^{*T} \cdot \underline{f}_2 = 0$$

For the factor b_1, the value can be derived from this:

Fig. 9.3 Visualization of the method for orthogonalizing an "oblique" coordinate system

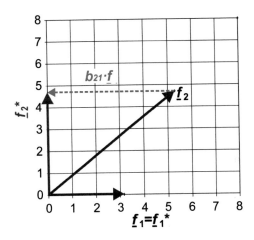

$$b_{21} = -\frac{f_1^{*T} \cdot f_2}{\left| f_1^* \right|^2}$$

The same procedure is used for the next vectors. The two vectors f_1^* and f_2^* are now already orthogonal. The vector f_3 is corrected to this property with the approach

$$f_3^* = b_{31} \cdot f_1^* + b_{32} \cdot f_2^* + f_3,$$

where the scalar products $f_1^{*T} \cdot f_3^*$ and $f_2^{*T} \cdot f_3^*$ must be identical zero to fulfill the set condition. These properties simultaneously provide the solutions for the searched coefficients:

$$f_3^{*T} \cdot f_1^* = b_{31} \cdot f_1^{*T} \cdot f_1^* + b_{32} \cdot f_2^{*T} \cdot f_1^* + f_3^T \cdot f_1^* = 0 = b_{31} \cdot f_1^{*T} \cdot f_1^* + 0 + f_3^T \cdot f_1^*$$

$$\Rightarrow b_{31} = -\frac{f_3^T \cdot f_1^*}{\left| f_1^* \right|^2}$$

$$f_3^{*T} \cdot f_2^* = b_{31} \cdot f_1^{*T} \cdot f_2^{*T} + b_{32} \cdot f_2^{*T} \cdot f_2^* + f_3^T \cdot f_2^* = 0 = 0 + b_{32} \cdot f_2^{*T} \cdot f_2^* + f_3^T \cdot f_2^*$$

$$\Rightarrow b_{32} = -\frac{f_3^T \cdot f_2^*}{\left| f_2^* \right|^2}$$

Proceed analogously with f_3. Finally, the following coefficients are obtained for the given example

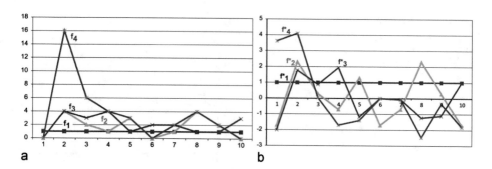

Fig. 9.4 Graphical representation of the components of the base vectors f_i. (**a**) Before orthogonalization and f^*_i, (**b**) after orthogonalization. (The values belonging to each other were only connected to each other to document the togetherness. A physical-technical background does not exist here)

$$b_{21} = -1.7 \qquad b_{31} = -2.1 \qquad b_{41} = -3.7$$
$$b_{32} = -0.05882 \quad b_{42} = -2.17647$$
$$b_{43} = -1.8112$$

These provide the orthogonalized vectors of the basic functions that span the space for the measured values to be mapped. Figure 9.4 visualizes the change by simply plotting the value of the function against the atomic number of the component.

The experimental design with the orthogonalized basic functions now looks as follows (Table 9.6):

The analytical form of the modified basic functions f^* now looks as follows:

$$f^*_1 = f_1 = 1$$
$$f^*_2 = -1.7 \cdot f^*_1 + f_2 = -1.7 + x_1$$
$$f^*_3 \quad = \quad -2 \cdot f^*_1 - 0.05882 \cdot f^*_2 + f_3 = -2 - 0.05882 \cdot (-1, 7 + x_1) + x_2$$
$$\quad = \quad -1.88235 - 0.05882 \cdot x_1 + x_2$$
$$f^*_4 = b_{41} \cdot f^*_1 + b_{42} \cdot f^*_2 + b_{43} \cdot f^*_3 + f_4 = -3,7 + b_{42} \cdot f^*_2 + b_{43} \cdot f^*_3 + x_1 \cdot x_2$$
$$f^*_4 = \begin{aligned} &-3.7 - 2.17647 \cdot (-1.7 + x_1) - 1.81119 \cdot (-1.88235 - 0.05882 \cdot x_1 + x_2) \\ &+ x_1 \cdot x_2 \end{aligned}$$
$$f^*_4 = 3.4093 - 2.069936 \cdot x_1 - 1.81119 \cdot x_2 + x_1 \cdot x_2$$

If one carries out the steps described above to determine the optimal description of the data set with these basic functions, one obtains

$$y(x_1; x_2) = E[\underline{Y}] + 1.923 \cdot 10^{-3} \cdot f^*_1 + 0.63 \cdot f^*_2 - 0.133 \cdot f^*_3 + 0.196 \cdot f^*_4 \text{ or}$$

Table 9.6 Experimental design with the orthogonalized basic functions

Nr.	$y^*(x1,x2)$	x1	x2	f1*	f2*	f3*	f4*
1	−0.22	0	0	1	−1.700	−2.000	3.622
2	2.14	4	4	1	2.300	1.765	4.098
3	0.04	2	3	1	0.300	0.882	0.049
4	−1.03	1	4	1	−0.700	1.941	−1.692
5	0.76	3	1	1	1.300	−1.176	−1.399
6	−1.09	0	2	1	−1.700	0.000	0.000
7	−0.55	1	2	1	−0.700	−0.059	−0.070
8	1.25	4	1	1	2.300	−1.235	−2.469
9	0.39	2	1	1	0.300	−1.118	−0.329
10	−1.69	0	3	1	−1.700	1.000	−1.811

$$y(x_1;x_2) = 2.038 + 0.282 \cdot x_1 - 0.488 \cdot x_2 + 0.196 \cdot x_1 \cdot x_2$$

The comparison of the calculated models with the "original" specification leads to Table 9.7.

Also, the mean value of a random variable equally distributed in an interval is not equal to zero in a finite sample. Therefore, the distinction between $E[z] \neq 0$ and $E[z] = 0$ has a demonstrative character even if systematic measurement errors could be avoided. As Table 9.2 shows, if the initial situation is the same, the way in which the data are treated has a significant effect on the quality of the approximation if the deviations at the grid points of the sample are taken as a basis. The best results are provided by orthogonalized basic functions - related to the experimental design - which can be generated from a basic approach.

The orthogonalization does not change the structure of the approximation problem. The change refers to the construction of an orthogonal coordinate system for the freely selected experimental design $\underline{F}$. This enables only the optimal approximation within the experimental design together with the selected approximation function of this method. This procedure avoids the multiple consideration of uncontrolled error components and thus leads to a minimum description error of the model at the supporting points of the experimental plan. Thus, the most favorable utilization of the *a-priori* information for the parameterization of the model function is achieved. The orthogonality for each basis vector f_i from the experimental design is not transferred to the combination of the approximating functions. Under certain boundary conditions, it is approximately valid for a complete experimental design in which all possible combinations of $(x_1...x_m)$ occur (with equidistant change intervals). The basis for this is the boundary transition from summation to integration.

The minimum of the sum of the error squares has generally established itself as a quality criterion. It goes back to the definition of the distance between two points according to Euclid (third century B.C.) and, in the form used, consists of the individual errors

Table 9.7 Composition of the calculation results for the example without and with centering of the primary data set (*A sample z from a large normally distributed population $\underline{Z}$ with the mean value E $(\underline{Z}) = 0$ practically always provides $E[\underline{z}] \neq 0$, which for normally distributed values z with $\frac{1}{\sqrt{n}}$ strives towards zero)

Coefficient	Default	Calculated with E $[\underline{Z}] = 0.08$ without centring	Calculated with E $[\underline{Z}] = 0^*$ without centring	Calculated with *centered* data set *and* E $[\underline{Z}] = 0.08^*$	Calculated with *centered* data set and *orthogonal* basic functions *and* E $[\underline{Z}] = 0.08^*$
a_1	2	2.128	2.052	2.093	2.038
a_2	0.3	0.246	0.246	0.28	0.282
a_3	−0.5	−0.482	−0.481	−0.492	−0.488
a_4	0.2	0.23	0.229	0.205	0.196
Effective deviation of the model from the "original" in $x_1 \in (0;4)$ and $x_2 \in (0;4)$		**0.216**	**0.163**	**0.094**	**0.039**
"Effective" deviation of the model from the "original" at the support points		**0.217**	**0.155**	**0.065**	**0.052**

$$\Delta_i y := y_i - a_1 f_{1i} - a_2 f_{2i} - a_3 f_{3i} - a_4 f_{4i}$$

together, which are squared and added up, so that you get

$$\Delta^2 Y = \sum_{i=1}^{n} \Delta_i^2 y := \sum_{i=1}^{n} \left(y_i - a_1 f_{1i} - a_2 f_{2i} - a_3 f_{3i} - a_4 f_{4i}\right)^2 .$$

A notation in matrix form returns this quantity as the scalar product of the single-column vectors $(\underline{Y} - \underline{F} \cdot \underline{A})$ with itself

$$(\underline{Y} - \underline{F} \cdot \underline{A})^T \cdot (\underline{Y} - \underline{F} \cdot \underline{A}) = \Delta^2 Y \Rightarrow \text{ min}$$

This implicitly presupposes actually an orthogonal coordinate system in a metric space. The basic functions f_i, which span the space and are available as vector functions with discrete or continuous values, generally do not fulfill the requirement of orthogonality. The procedure described above creates an orthogonal reference system of vectors $\underline{f}_{-v}^*$, so that the Euclidean distance definition can be used optimally.

Compared to other distance definitions, such as

$$\Delta^* Y = \sum_{i=1}^{n} \Delta_i^* y := \sum_{i=1}^{n} |y_i - a_1 f_{1i} - a_2 f_{2i} - a_3 f_{3i} - a_4 f_{4i}| \Rightarrow \min$$

the Euclidean distance definition has the advantage that the resulting quality functionals can be continuously differentiated at least once and thus the extreme value investigations are comparatively simple. Other distance definitions also have their specific applications. For example, in a city, the Euclidean distance definition for two points has little sense. If the streets meet at right angles to each other, and these streets are also partially designated as one-way streets, only the distances that can really be covered can be added up to one distance. In contrast, the Euclidean distance describes the distance between two points as the crow flies. Depending on the type of locomotion, the specification of the minimum distance between two points is very different. The most appropriate distance definition for an optimization task depends on the properties of the data and the optimization task or the optimization goal itself. In general, one works with metric spaces with *Minkowski distances*. It is also possible to map non-metric data to metric spaces, as it often happens in sociology and pharmacy [10]. gives a very good introduction and overview to the application. This literature reference is particularly important under the aspect that material batches, material designations etc. for magnetic materials rarely contain a quantitative order, which is a prerequisite for a metric space. They are designations that characterize a quality. In the simplest case, numbering can be used to create a one-to-one mapping to a metric space and then to perform an analysis of the data.

Of course, there will be no connection that describes the arbitrary position of the material in any list in terms of an atomic number (material) or the like. However, it is possible to extract functions and associated coefficients from a data set, which on the one hand are the same for the entire measurement data, as the basic functions and coefficients spanning the vector space, which are characteristic for each material (the coefficient vectors). Mathematically, basis vectors and coefficient vectors are treated equally. However, both have - as indicated above - different physical meanings. These possibilities are discussed in more detail in Sect. 9.5.

9.3 General Approach as Linear Combination of Influences

The above approach to data analysis aims at minimizing the mean square error of the absolute errors. The properties of the measurement method or the objective of the measurement task may make the minimization of the relative description error appear to be a quality criterion in empirical modeling. The procedure is similar in many respects to the previous ones. It is therefore only briefly outlined here. Here, too, a measurement value sequence $\underline{Y}$ (with all elements $\neq 0$) is to be represented by a linear combination of basic functions $\underline{F} \cdot \underline{A}$. For example, the coefficient vector should contain n = 4 components. The approximation is then

$$\underline{Y} \approx \underline{F} \cdot \underline{A}$$

The vector of absolute description errors is formed by the difference between measured values and the calculated values of the model functions.

$$\underline{\Delta Y} \approx \underline{Y} - \underline{F} \cdot \underline{A}$$

In determining the relative error, each individual difference is related to the value y_i. The measured values are used as the "true value" in this context, since no other primary information is available. The individual (relative) error size can then be expressed by the relationship

$$\Delta_i y = \frac{y_i - \underline{F_i} \cdot \underline{A}}{y_i}$$

In matrix notation, the error vector with the relative description errors looks like this

$$\underline{\Delta Y}^{**} \approx (diag(\underline{Y}))^{-1} \cdot (\underline{Y} - \underline{F} \cdot \underline{A})$$

Here too, the sum of the error squares should be minimized. The corresponding optimization criterion results in

$$\Delta^2 Y^{**} = \underline{\Delta Y}^{**T} \cdot \underline{\Delta Y}^{**} = \left((diag(\underline{Y}))^{-1} \cdot (\underline{Y} - \underline{F} \cdot \underline{A}) \right)^T \cdot \left((diag(\underline{Y}))^{-1} \cdot (\underline{Y} - \underline{F} \cdot \underline{A}) \right)$$
$$\Rightarrow \min$$

or

$$\Delta^2 Y^{**} = \left(\left((diag(\underline{Y}))^{-1} \cdot \underline{Y} \right)^T - \left((diag(\underline{Y}))^{-1} \cdot \underline{F} \cdot \underline{A} \right)^T \right)$$
$$\cdot \left(\left((diag(\underline{Y}))^{-1} \cdot \underline{Y} \right) - \left((diag(\underline{Y}))^{-1} \cdot \underline{F} \cdot \underline{A} \right) \right)$$
$$\Rightarrow \min$$

To determine an extreme value of this function, one differentiates between the individual coefficients. With $n = 4$, one then obtains a linear system of equations of fourth order.

$$\frac{\partial}{\partial_v}|\Delta^2 Y^{**} = \frac{\partial}{\partial_v}\left(\begin{array}{l}\left((diag(\underline{Y}))^{-1}\cdot\underline{Y}\right)^T\cdot\left((diag(\underline{Y}))^{-1}\cdot\underline{Y}\right)-2\cdot\left((diag(\underline{Y}))^{-1}\cdot\underline{Y}\right)^T\cdot\left((diag(\underline{Y}))^{-1}\cdot\underline{F}\cdot\underline{A}\right)+\\ +\left((diag(\underline{Y}))^{-1}\cdot\underline{F}\cdot\underline{A}\right)^T\cdot\left((diag(\underline{Y}))^{-1}\cdot\underline{F}\cdot\underline{A}\right)\end{array}\right)$$

$$= 0$$

$$,\forall v = 1\ldots 4$$

In analogy to the first example, the differentiation according to all elements of the coefficient vector results in $\underline{A}$ and the conversion of the equation to $\underline{A}$ finally:

$$-2\cdot\left((diag(\underline{Y}))^{-1}\cdot\underline{Y}\right)^T\cdot\left((diag(\underline{Y}))^{-1}\cdot\underline{F}\right)+2\cdot\left((diag(\underline{Y}))^{-1}\cdot\underline{F}\right)^T\cdot\left((diag(\underline{Y}))^{-1}\cdot\underline{F}\right)\cdot\underline{A}=0$$

$$\underline{A}=\left(\left((diag(\underline{Y}))^{-1}\cdot\underline{F}\right)^T\cdot\left((diag(\underline{Y}))^{-1}\cdot\underline{F}\right)\right)^{-1}\cdot\left((diag(\underline{Y}))^{-1}\cdot\underline{Y}\right)^T\cdot\left((diag(\underline{Y}))^{-1}\cdot\underline{F}\right)$$

The effect of the quality function for minimizing the relative error is comparatively unclear, since only the measured values are available as reference values for the relative error. However, these are composed of the "truth" of the physical context plus the error quantity unknown in the individual case. In the best case, the value of the variance is known for measured quantities. The current value of the random error quantity is naturally not known. A standardization to the measured quantity is therefore problematic. If the real values are not known, the formulation of this quality function can also become meaningless.

In the modelling of data, there is also the task of presenting in a simple way the clear but complex results of a mathematical analysis. This concerns, for example, the results that are only implicitly available in the form of extensive transcendental algebraic equations. The same applies to integral equations or elliptic integrals. Here, the "truth" and thus the reference value for the normalization can be determined with any degree of accuracy when determining the relative error. The quality functional for minimizing the relative descriptive error then has a real basis for the optimal adjustment even with relatively large differences in the values of $\underline{Y}$.

Weighting to take into account the error characteristics is often a useful way, since additional primary information from the measurement can also be processed here. Depending on the selected measuring ranges and measuring methods, the measurement results can thus be taken into account to a greater or lesser extent. The interpretation of the results of this approach is much more reliable. Furthermore, the weighting factors can be designed in such a way that outliers are automatically suppressed during measurements. This is especially true when evaluating large amounts of data and when one is sure of the type of relationship to be described. The basic approach to the procedure with weighted description errors is only hinted at here. For further details, please refer to the relevant literature [11, 12]. If one keeps the previous notation for the description of an empirical data set and introduces the single-column matrix Γ with the weights γ_i of the individual errors for each measuring point, the approximation problem can be formulated as follows:

$$(diag(\underline{\Gamma}) \cdot (\underline{Y} - \underline{F} \cdot \underline{A}))^T \cdot (diag(\underline{\Gamma}) \cdot (\underline{Y} - \underline{F} \cdot \underline{A})) \Rightarrow \min$$

The treatment of this problem is analogous to that for minimizing the relative and is not presented again separately here.

$$\underline{A} = \left((diag(\underline{\Gamma}) \cdot \underline{F})^T \cdot (diag(\underline{\Gamma}) \cdot \underline{F})\right)^{-1} \cdot (diag(\underline{\Gamma}) \cdot \underline{Y})^T \cdot (diag(\underline{\Gamma}) \cdot \underline{F})$$

Models for Logarithmic Measurement Data

In the case of non-negative empirical data to be represented by model functions, which extend over a wide range of orders of magnitude, it is often appropriate to relate the model functions to the logarithm of the measured data:

$$y_i^* = \ln\left(\frac{y_i}{[y]}\right) \approx \sum_{j=1}^{k} a_j f_{ji} \text{ or } \underline{Y}_i^* \approx \underline{F} \cdot \underline{A}$$

$[y]$ contains the physical dimension of the measured values $\underline{Y}$, since the logarithm function requires a dimensionless argument. Thus, the procedure already described above can be applied to the data set modified via logarithmization. The following quality criterion is obtained

$$\Delta^2 Y^* = \sum_{i=1}^{n} \left(\ln\left(\frac{y_i}{[y]}\right) - \sum_{j=1}^{k} a_j f_{ji}\right)^2 = (\underline{Y}^* - \underline{F} \cdot \underline{A}) \cdot (\underline{Y}^* - \underline{F} \cdot \underline{A}) = (\underline{Y}^* - \underline{F} \cdot \underline{A})^2$$

$$\Rightarrow \min$$

The effect of this function is similar to the minimization of the relative description error, but without the preference of the measured values in $\underline{Y}$.

Because of the properties of the logarithmic function (the real function $ln(y)$ is defined only for $y > 0$), this form of mathematical modelling has the limitation $y_i > 0$ compared to the approach presented at the beginning of this chapter and the formal approach of minimizing the relative error, which only includes the requirement $y \neq 0$. Furthermore, the approach to the description of the original data changes fundamentally. The logarithmization of the data with the aim of a better approximation represents an intermediate illustration that works as follows (illustration based on 2 influence quantities x_1 and x_2:

$$y^* = \ln\left(\frac{y}{[y]}\right) \approx a_0 + \sum_{v=1}^{k} a_v \cdot f_v(x_1, x_2)$$

The quantity a_0 is again a constant as in the example chosen at the beginning. The methods demonstrated here can also be applied to logarithmic data, including the construction of orthogonal basis functions. When centering logarithmic data, the subtraction of the arithmetic mean of the data in the image area is

$$\overline{y^*} = \frac{1}{n}\sum_{i=1}^{n} y_i^*$$

a division of the original data by the geometric mean of the measured data

$$\widehat{y^*} = \sqrt[n]{\prod_{i=1}^{n} y_i^*}$$

If especially the approximation function for intermediate mapping has the structure of a linear combination of products (e.g. for two variables x_1 and x_2), then this approach has the following effects:

$$y^* = \ln\left(\frac{y}{[y]}\right) \approx a_0 + \sum_{v=1}^{k} a_v \cdot f_v(x_1) \cdot g_v(x_2)$$

In the original area, the function of $y(x_1, x_2)$ is then expressed as

$$y \approx \exp\left(a_0 + \sum_{v=1}^{k} a_v \cdot f_v(x_1) \cdot g_v(x_2)\right) \cdot [y]$$

$$y \approx \exp(a_0) \cdot \exp\left(\sum_{v=1}^{k} a_v \cdot f_v(x_1) \cdot g_v(x_2)\right) \cdot [y] \text{ or}$$

$$y \approx \exp(a_0) \cdot \prod_{v=1}^{n} \left(\exp(f_v(x_1) \cdot g_v(x_2))\right)^{a_v} \cdot [y]$$

In the original area, the approach with the logarithmic measurement data thus creates a structurally completely different relationship. The approximation of the logarithmic data is performed with the same means as described at the beginning of this section. The situation described is present, for example, in the description of losses for ferromagnetic cores.

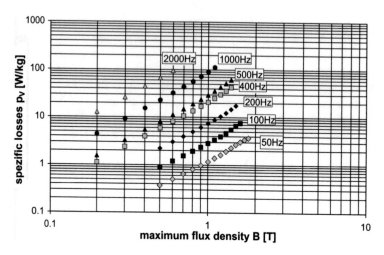

Fig. 9.5 Average remagnetization losses $p_V(B_{max};f)$ for the grain-oriented *PowerCore® 330-35AP* sheet from *ThyssenKrupp Steel as a* function of maximum induction and frequency for sinusoidal, symmetrical magnetization with the mean value $E[B] = 0$ and averaging over magnetization parallel to the rolling direction (0°) and transverse to the rolling direction (90°)

The effect is described below using an example. Figure 9.5 shows a graphic representation of the dependence of the specific remagnetization losses for a grain-oriented PowerCore® 330-35AP sheet from *ThyssenKrupp Steel*. The graph uses directly the values of the data sheet data.

The material 330-35AP is anisotropic. Thus, for example, the amplitude permeabilities in the rolling direction differ in relation to the transverse rolling direction by the factor

$$\frac{\widetilde{\mu}(0\,^{\circ})}{\widetilde{\mu}(90\,^{\circ})} = 0.9|_{1.9\ T}\ldots 0.41|_{0.9\ T}$$

This is also reflected in the specific losses of the material. Figure 9.6 shows for 50 Hz the ratio of the remagnetization losses in the rolling direction relative to the transverse direction.

In the double logarithmic scale of Fig. 9.5, the measured values belonging to one parameter appear to lie on a straight line with the same gradient. Therefore, the approach is obvious to describe the logarithmic data set by a plane in 3-dimensional space, which is spanned by $\lg\left(\frac{B}{[B]}\right)$ and $\lg\left(\frac{f}{[f]}\right)$ as well as $\lg\left(\frac{p_V}{[p_V]}\right)$:

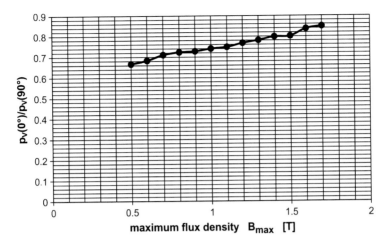

Fig. 9.6 Ratio of the remagnetization losses at 50 Hz in the rolling direction ($0°$) relative to a magnetization transverse to the rolling direction ($90°$) for the material *PowerCore® 330-35AP* from *ThyssenKrupp Steel* (according to data sheet data)

$$\lg\left(\frac{p_V}{[p_V]}\right) = a_0 + a_1 \cdot \lg\left(\frac{\widehat{B}}{[B]}\right) + a_2 \cdot \lg\left(\frac{f}{[f]}\right) \text{ or } \lg\left(\frac{p_V}{1\,\frac{W}{kg}}\right) = a_0 + a_1 \cdot \lg\left(\frac{\widehat{B}}{1\,T}\right) + a_2$$

$$\cdot \lg\left(\frac{f}{1\,s^{-1}}\right)$$

In the original data area, the formula for the specific remagnetization losses is thus obtained

$$p_V = 10^{a_0} \cdot \left(\frac{\widehat{B}}{1\,T}\right)^{a_1} \cdot \left(\frac{f}{1\,s^{-1}}\right)^{a_2} \cdot \left(1\,\frac{W}{kg}\right)$$

This is the well-known Steinmetz formula (Carl August Rudolph Steinmetz: * 9 April 1865 in Breslau; † 26 October 1923 in Schenectady, NY, USA) for approximating the specific losses of magnetic materials. The use of the logarithmic data sheet values for 330-35AP provides the following values after centering and regression analysis

$$a_0 = -2.38 \qquad a_1 = 1.842 \qquad a_2 = 1.432$$

The model formula is then

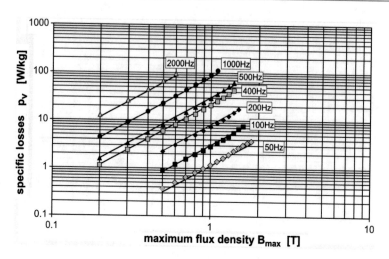

Fig. 9.7 Common presentation of measurement data and model function for the average re-magnetization loss $p_V(B_{max};f)$ for the grain-oriented sheet *PowerCore® 330-35AP* (see also Fig. 9.5)

$$p_V = 4.167 \frac{\text{mW}}{\text{kg}} \cdot \left(\frac{\widehat{B}}{1\ \text{T}}\right)^{1.842} \cdot \left(\frac{f}{1\ \text{s}^{-1}}\right)^{1.432}$$

If the determined model function is now drawn together in a diagram (Fig. 9.7), one notices on the one hand a very good agreement of the approach with the data. This speaks for the quality of the approach with regard to the reflection of the basic properties of the measured data. On the other hand, the deviations of the measured data from the model function consist less in a statistical scattering around it, but rather in area-wise tendentious deviations.

These (relative) deviations, shown in Fig. 9.8, suggest by their shape that the Steinmetz approach must be extended for smaller description tolerances or larger ranges of validity of the model. This is discussed in Sect. 9.5.

9.4   Planning of Experiments

Already in the planning phase of experiments, it is possible to influence the effectiveness of later data analysis procedures, but also the scope of the necessary experimental work. Experimental design in a broader sense can be understood to include all measures that aim to ensure certain comparable experimental conditions, to reduce the time and financial expenditure for the performance of the experiments, and to ensure their favorable evaluation and repeatability. Thus, the conception of a suitable experimental arrangement, the selection or, if necessary, the construction of equipment or methods for data compression

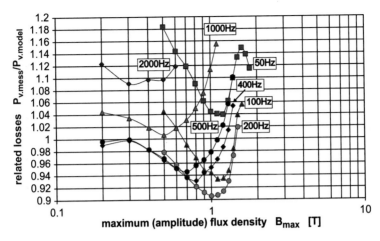

Fig. 9.8 Representation of the relative deviations of measured data and model for the average remagnetization loss $p_V(B_{max};f)$ for the grain-oriented sheet *PowerCore® 330-35AP*

and evaluation, as well as the appropriate delimitation of the field of interest fall within this area. The possibilities for this are extremely diverse, since they depend on the test object and the test objective. In the field of mathematical statistics, statistical design of experiments has emerged as the methodological tool for selecting the most favourable experimental conditions from a given field of interest. Due to the achievable economic effects, statistical design of experiments has already become an integral part of research technology. In the context of this thesis, therefore, only the basic features of possibilities and existing difficulties of the application of this interesting methodology will be discussed. Detailed descriptions are contained in [12–17]. The main aim is to show the potential for rationalising the collection and evaluation of empirically obtained process and material data.

The statistical design of experiments is based on the work of the English mathematician Sir Ronald Aylmer Fisher (* 17 February 1890 in London, England; † 29 July 1962 in Adelaide, Australia) [18], who found that a rational design of experiments does not yield less gain with respect to the characteristics to be determined than an optimized processing of the measured data. When processing empirically obtained data, the requirement is usually to represent a measured quantity y after n measurements quantitatively as a dependence of various factors $x_1...x_m$. The aim of this modelling is often a later optimisation of these influencing factors. The influencing variables or factors can be both qualitative and quantitative. Qualitative factors are for example different types of material, different material batches, shapes of magnetic cores, structure of winding materials and similar. Quantitative factors are measurable physical and technical quantities. Each of these factors assumes different levels in the course of the experiments, which are either quantitatively measurable or are defined by a code (metric). Formally this can be expressed by a further index of the quantities x_i. Thus, the symbol x_{ij} characterizes the level that the factor

x_j (j = 1...m) takes on in the experiment. If the selected factor combinations are compiled in a table, the design $\underline{V}$ for the experiment is obtained. In matrix notation, this results in

$$\underline{V} = \begin{bmatrix} x_{11} & x_{12} & x_{13} & \cdots & x_{1m} \\ x_{21} & x_{22} & x_{23} & \cdots & x_{2m} \\ x_{31} & x_{32} & x_{33} & \cdots & x_{3m} \\ \cdot & \cdot & \cdot & \cdots & \cdot \\ x_{n1} & x_{n2} & x_{n3} & \cdots & x_{nm} \end{bmatrix}$$

The number of levels to be investigated or to be examined is not necessarily the same for each factor x_j, as shown in the matrix, but generally varies from influencing factor to influencing factor. Each of these factors x_j can therefore generally be assigned a number of levels of n_j. If each of the factors x_j (j = 1...m) has a number of n_j levels, the following value results for the number of experiments Z_v when performing experiments of all possible level combinations m

$$Z_V = \prod_{j=1}^{m} n_j$$

Such a design is called a complete factorial design. It can be seen that the number of measurements increases very quickly with the number of influencing factors considered. For example, if all $n_j = n$, there are $Z_V = n^m$ combinations. By applying the dimensional analysis (Sect. 9.1), the number of parameters to be investigated can be reduced. The use of complete factorial experimental designs involves a considerable number of measurements. Complete factorial designs are used especially when nothing is known about the properties of the target y to be investigated. If a model to be completed is known or can be assumed after knowledge of basic properties of the data, another approach is possible.

The following considerations are based on Fig. 9.9, which was also used in the previous section. The dependence of a quantity y on the influencing variables $x_1...x_m$ is determined experimentally according to a test plan $\underline{V}$, in which the level combinations of the influencing variables are compiled. In Fig. 9.9, the selected level combinations of two influencing variables x_1 and x_2 are symbolized by points in the $(x_1; x_2)$ plane and the sequence given in the plan $\underline{V}$ by numbered points.

In the statistical design of experiments for linear regression problems, it is assumed that the vector of the measured values (observation vector) $\underline{Y}$ with i (i = 1...n) components can be represented by linear combinations of $k \le n$ vectors $\underline{f_i} = (f_{i1}, f_{i2}...f_{ik})$. (The symbolic marking of vectors is carried out in the following in accordance with the marking of matrices by underlined formula symbols, since they can be represented as (k; n) matrix). Each of the elements of the vector $\underline{f_{iv}}$ is thereby linked to the influencing variables $x_1...x_m$ of

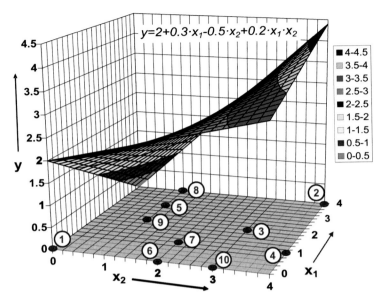

Fig. 9.9 Determination of the dependency of a quantity y on two influencing variables by random measurements for different combinations of x_1 and x_2 with indication of the sequence of the measurements

the experiment via a known functional relationship. Then we arrive at the general representation for one element of the observation vector

$$y_\mu \cong \sum_{\nu=1}^{k} f_\nu(x_{1\mu}; x_{2\mu}; \ldots x_{m\mu}) \cdot a_\nu = \begin{bmatrix} f_{\mu 1} & f_{\mu 2} & f_{\mu 3} & \cdots & f_{\mu k} \end{bmatrix} \cdot \begin{bmatrix} a_1 \\ a_2 \\ a_3 \\ \cdots \\ a_k \end{bmatrix}$$

or in notation as matrix multiplication

$$\underline{Y} = \begin{bmatrix} \underline{f_1}(x_{11}; x_{21}; \ldots x_{m1}) \\ \underline{f_2}(x_{12}; x_{22}; \ldots x_{m2}) \\ \cdots \\ \cdots \\ \underline{f_p}(x_{1p}; x_{2p}; \ldots x_{mp}) \end{bmatrix}^T \cdot \begin{bmatrix} a_1 \\ a_2 \\ . \\ . \\ a_k \end{bmatrix} = \begin{bmatrix} f_{11} & f_{12} & f_{13} & \cdots & f_{1k} \\ f_{21} & f_{22} & f_{23} & \cdots & f_{2k} \\ f_{31} & f_{32} & f_{33} & \cdots & f_{3k} \\ .. & .. & .. & .. & .. \\ f_{p1} & f_{p2} & f_{p3} & \cdots & f_{pk} \end{bmatrix} \cdot \begin{bmatrix} a_1 \\ a_2 \\ . \\ . \\ a_k \end{bmatrix} = \underline{F} \cdot \underline{A}$$

To build the mathematical model, it is necessary to determine the elements of the coefficient vector $\underline{A}$. As will be shown below, this problem leads to a linear system of equations with k equations. To solve this, an observation vector $\underline{Y}$ with $p \geq k$ elements must be determined. The minimum number of experiments is therefore $p_{min} = k$. For all experiments, different level combinations have to be chosen. For the most favourable selection of the combinations of the influencing parameters from the complete experimental design, it is necessary to investigate the scatter/variance when determining the vector $\underline{A}$. For this purpose, it is assumed that the quantity to be measured can be completely described by the given model, but that certain errors φ_μ are superimposed on the measured values. The error quantity has the mean value zero ($E(\underline{\Phi}) = 0$) and the variance

$$\sigma_\Phi^2 = E\left[(\underline{\Phi} - E[\underline{\Phi}])^2\right] = D^2[\underline{\Phi}],$$

which is independent of the combination x_{ij}. According to [13], this results in the property.

$$E\left[\underline{\Phi} \cdot \underline{\Phi}^T\right] = \sigma_\Phi^2 \cdot \underline{E} \ (E : \text{unit matrix}),$$

This corresponds to an equal distribution of the variances in all directions of the base vectors spanning the measurement vector. To investigate the effects of the error size Φ on the determination of the unknown parameter vector $\underline{A}$, the following approach is used. The observation vector $\underline{Y}$ is composed of the "true" quantity $\underline{Y}^*$ and the error quantity $\underline{\Phi}$:

$$\underline{Y} = \underline{Y}^* + \underline{\Phi} \simeq \underline{F} \cdot \underline{A}$$

If, in order to determine the unknown parameter vector A, minimization of the sum of the quadratic deviations S^2 is required, the following results are obtained

$$S^2 = (\underline{Y} - \underline{F} \cdot \underline{A})^T \cdot (\underline{Y} - \underline{F} \cdot \underline{A})$$
$$= \underline{Y}^T \cdot \underline{Y} - \underline{A}^T \cdot \underline{F}^T \cdot \underline{Y} + \underline{A}^T \cdot \underline{F}^T \cdot \underline{F} \cdot \underline{A} - \underline{Y}^T \cdot \underline{F} \cdot \underline{A}$$
$$= \underline{Y}^T \cdot \underline{Y} - \underline{A}^T \cdot \left(2 \cdot \underline{F}^T \cdot \underline{Y} - \underline{F}^T \cdot \underline{F} \cdot \underline{A}\right) \Rightarrow \min$$

The system of equations for determining the optimum coefficient vector $\widetilde{\underline{A}}$ is obtained by differentiating the above equation according to all elements from $\underline{A}$ to

$$\underline{F}^T \cdot \underline{F} \cdot \widetilde{\underline{A}} = \underline{F}^T \cdot \underline{Y}$$

If for the rank, the matrix $\underline{F}^T F$ applies

$$Rg\left(\underline{F}^T \cdot \underline{F}\right) = k \leq n,$$

the system of equations can be solved unambiguously and one receives for $\widetilde{\underline{A}}$

$$\tilde{\underline{A}} = \left(\underline{F}^T \cdot \underline{F}\right)^{-1} \cdot \underline{F}^T \cdot \underline{Y}$$

From this equation, because of the linearity in $\tilde{\underline{A}}$ for the error of determination of $\tilde{\underline{A}}$, the relation can be given

$$\tilde{\underline{A}} - \underline{A}^* = \underline{A}_\Delta = \left(\underline{F}^T \cdot \underline{F}\right)^{-1} \cdot \underline{F}^T \cdot \underline{\Phi}$$

(projection of the base vectors on the error size Φ). From this, the covariance matrix Γ of the so estimated parameter vector $\tilde{\underline{A}}$ is calculated:

$$\underline{\Gamma} = \left[\left(\underline{F}^T \cdot \underline{F}\right)^{-1} \cdot \underline{F}^T \cdot \underline{\Phi}\right] \cdot \left[\underline{\Phi}^T \cdot \underline{F} \cdot \left(\underline{F}^T \cdot \underline{F}\right)^{-1}\right]$$

This must be equivalent to

$$\underline{\Gamma} = \sigma_A^2 \cdot \left(\underline{F}^T \cdot \underline{F}\right)^{-1} = \tilde{\underline{A}}_\Delta \cdot \tilde{\underline{A}}_\Delta^T$$

The error in the determination of $\tilde{\underline{A}}$ can be influenced by selecting different experimental conditions $(x_{1j}...x_{mj})$ on which the shape of the matrix $(F^T F)^{-1}$ depends. In statistical design of experiments, therefore, the selection of a given number of measurement points from a complete plan is carried out in such a way that certain conditions with respect to the matrix $(F^T F)^{-1}$ are fulfilled.

For a clear formulation of the optimality criteria, the above equation is appropriately further transformed:

$$\tilde{\underline{A}}_\Delta^{-1} \cdot | \quad \underline{\Gamma} = \sigma_A^2 \cdot \left(\underline{F}^T \cdot \underline{F}\right)^{-1} = \tilde{\underline{A}}_\Delta \cdot \tilde{\underline{A}}_\Delta^T$$

$$\sigma_A^2 \cdot \tilde{\underline{A}}_\Delta^{-1} \cdot \left(\underline{F}^T \cdot \underline{F}\right)^{-1} = \tilde{\underline{A}}_\Delta^{-1} \cdot \tilde{\underline{A}}_\Delta \cdot \tilde{\underline{A}}_\Delta^T = \tilde{\underline{A}}_\Delta^T \quad | \cdot \left(\underline{F}^T \cdot \underline{F}\right) \cdot \tilde{\underline{A}}_\Delta$$

$$\sigma_A^2 = \tilde{\underline{A}}_\Delta^T \cdot \left(\underline{F}^T \cdot \underline{F}\right) \cdot \tilde{\underline{A}}_\Delta$$

This expression describes a k-dimensional ellipsoid in the space spanned by $\underline{A}$, which with a certain probability contains the scatter range of the calculated vector $\tilde{\underline{A}}$ around the "true" value. This ellipsoid (concentration rotation ellipsoid) is the basis of many optimality criteria for statistical experimental design [14, 15]. The most frequently used optimality criterion is the D-optimality [15, 16]. A design is called D-optimal if the value of the determinant

$$\det\left(\left(F^T F\right)^{-1}\right) \Rightarrow \min$$

This is equivalent to minimizing the volume of the concentration ellipsoid of the defect size. However, this is only one possible measure for assessing the quality of the approximation. Another criterion requires the minimization of the mean half-axis length of the k-dimensional concentration ellipsoid, which is expressed by minimizing the trace of the square matrix $(\underline{F}^T\underline{F})^{-1}$:

$$Sp\left(\left(F^T F\right)^{-1}\right) \Rightarrow \min.$$

For a square matrix, the term *trace* as the sum of its main diagonal elements is defined as the sum of the elements of the main diagonal [22]. An experimental design that meets this requirement is called A-optimal.

Further optimality criteria, which refer to certain properties of the concentration ellipsoid, are contained in [15, 16].

Further possibilities for an optimized experimental design (G-optimality) are obtained by demands on the model adaptation itself. The dispersion of the model value $\tilde{y}$ around the true value y* can be determined according to [17] by

$$D^2\left(\tilde{y}\right) = E\left[\left\{\left(\tilde{y}(\underline{x})\right) - \underline{y}^*(\underline{x})\right\}^2\right] = \sigma^2 \cdot \underline{f}^T(\underline{x}) \cdot \left(\underline{F}^T \cdot \underline{F}\right)\underline{f}(\underline{x})$$

as a variance function that depends on the vector of the influence quantities $\underline{x}$.

In addition to the criteria mentioned above, orthogonality and rotatability play a role as properties of experimental designs. The calculation of the coefficient vector in equation

$$\underline{\tilde{A}} = \left(\underline{F}^T \cdot \underline{F}\right)^{-1} \cdot \underline{F}^T \cdot \underline{Y}$$

is particularly easy to determine if the matrix $\underline{F}^T\underline{F}$ and thus also its inverse matrix diagonal shape. An experimental design that leads to such a matrix is called orthogonal [22]. With orthogonal designs, all elements of the coefficient vector $\underline{\tilde{A}}$ are determined independently of each other. Each element of $\underline{\tilde{A}}$ is calculated by dividing the corresponding element of $\underline{F}^T\underline{Y}$ by the corresponding element of the diagonal of $\underline{F}^T\underline{Y}$. Orthogonality cannot be achieved for all model functions and experimental designs. For polynomials of a higher degree than model functions, for example, orthogonal polynomials would have to be constructed beforehand - a procedure that is rarely used [19, 21], but that is to be pointed out here, since the effort is comparatively small. In Sect. 9.5, however, orthogonal development functions are extracted from the measured primary data themselves. Nevertheless, it should be pointed out here that the construction of a mathematical-statistical experimental design is not possible without investment of time and is only worthwhile for larger test series or relatively clear structural relationships.

It is possible to determine a new system of model functions Ψ *for* a given experimental design and the matrix $\underline{F}^T\underline{F}$ known with it by a transformation, for which the matrix $\Psi^T\Psi$ is of the diagonal type. The basic idea is the following: By linear combinations of the functions contained in $\underline{F}$, a new function matrix Ψ is created. That is:

$$\underline{\Psi} = \underline{F} \cdot \underline{C}$$

with $\underline{C}$ as transforming matrix. The approximation problem is then

$$\underline{Y} = \underline{\Psi} \cdot \underline{B}$$

where $\underline{B}$ is the new coefficient vector to be determined. The transforming matrix $\underline{C}$ is now to be determined so that

$$\underline{\Psi}^T \cdot \underline{\Psi} = \underline{C}^T \cdot \underline{F}^T \cdot \underline{F} \cdot \underline{C}$$

has a diagonal shape. This problem can always be solved because the matrix $\underline{F}^T\underline{F}$ is symmetrical [11]. Matrix $\underline{C}$ consists of columns of the eigenvectors of $\underline{F}^T\underline{F}$. This 'model transformation' offers the possibility to reduce the calculation effort for frequently recurring modeling tasks, e.g. recalibration of sensors and model updates in adaptive systems.

A design is called rotatable if the variance function, when the design $\underline{V}$ is rotated around the central point $\underline{X}_0$, depends only on the distance $|X - X_0|$. Rotatable designs are obtained by using the corners and center of a regular polyhedron as test points. They are particularly useful for spherical areas of interest and are preferred for higher order polynomial models.

The numerical solution of the optimization tasks set up is generally complex. If the complete experimental design contains N possible experiments, then are

$$z = \frac{N!}{n!(N-n)!}$$

possibilities for the selection of an experimental design with n experiments. Determining the optimal experimental design by checking all possibilities becomes very complex if N is large. In the literature, the possibilities of special search procedures and programs are therefore given, which result in a considerable reduction of the calculation time [15]. For polynomial models, standardized experimental designs can be taken from the literature, so that the sometimes considerable numerical effort of the design is not necessary [13].

The experimental effort can be reduced to a large extent by using experimental designs constructed according to statistical criteria. The design requires the assumption of a sufficiently flexible model. By far the largest part of the investigations in this field relates to linear combinations of parameter-dependent functions (or, translated into the language of geometry: parameter-dependent vectors). The optimal experimental design of models that contain at least one non-linear parameter a is very complicated in both theoretical and numerical treatment. Only in special cases, it is possible to specify an optimal design from the outset. As a rule, the experimental design is changed step by step starting from a certain number of experiments. The numerical determination of the most favourable conditions for the next experiment may require the use of somewhat more complex calculations with predetermined search algorithms. The experimenter must check here that the effort for the design does not exceed the effort for the experimental work.

In the case of technical problems, the focus is usually on recording the effects of factors quantitatively with sufficient accuracy and at the lowest possible cost. In many cases, no analytical model is available for this purpose. For this reason, a linear combination of functions should, if possible, be used accordingly when selecting the model because of the difficulties in further processing the problem. (In part, non-linear approaches can also be traced back to this form). Since the selection of a suitable model is to be regarded as a key problem of experimental information retrieval, the following section is devoted to the related questions. The planning of the sample size, special conditions for mixing experiments (e.g. for testing the compositions of different magnetic conductors in series) will not be dealt with here. In this respect, reference is made to the extensive literature [14, 15, 24].

9.5   Determination of an Optimized Functional System for the Approximation of Measurement Data from Random Samples of the Measurement data

When working on tasks in the field of magnetic components, a wide range of variations of the characteristic influencing parameters must often be taken into account. As a result, polynomial approaches, for example, do not meet the requirements for model adequacy. Even the description of the losses for the grain-oriented *PowerCore® 330-35AP/Thyssen-Krupp Steel* sheet shows that the simple formula according to Steinmetz is not sufficient for larger ranges of values of the influencing parameters. For an adequate description of the dependencies, it is necessary to select problem-adapted functions with which a simple and flexible model can be constructed (also taking into account material influences/ compositions or designs).

The descriptive mathematical model in its most general form can often be determined from physical relationships or from random measurements. Aids can be additional useful or plausible assumptions or specifications for a further treatment of the modelling task. Following [23, 19] introduces a concept how an optimized functional system can be extracted from empirical data sets by a transformation to describe the technical facts. This is assumed in the following section.

The methodology is explained, with the help of which a problem-adapted system of functions can be determined for different tasks and focal points of the model development. Since the primary information is primarily available in the form of discrete measured values, a point-by-point calculation of the most favourable approximating functions is assumed as the objective. The presented methodology is based on the representation of a measurement data set as points of a metric space, whose distances from each other and with respect to a coordinate system are examined. The aim is to uncover the internal structure of the measurement dataset and to trace it back to linear vector functions. Due to the few prerequisites that have to be met, the procedure is largely universally applicable.

The derivation of data analysis rules can be done from different starting points. In the present work, a form is chosen, which, in the opinion of the author, is appropriate to engineering, pragmatic and in accordance with practical experience. The characteristics of the determined model functions are discussed afterwards.

First of all, there is a data set $\underline{Y}$ determined according to a *complete* experimental design, whose elements depend on two varied influencing factors 'x' and 'z'. Summarized in Table 9.8, the measurement results y_{ij} and the associated test conditions x_i and z_j are presented in the following form:

For technical data sets of this type, it can usually be assumed that adjacent columns or rows of a table have similarities among themselves, which result from the determined relationship between the target variable and the influencing parameters.

Starting from this consideration, the approach for the *inner mathematical structure* of the descriptive model is a symmetrical product form, which - so to speak - postulates a "symmetrical proportionality

$$y_{ij} \approx a_i \cdot c_j \text{ or } \underline{Y} = \underline{A} \cdot \underline{C}^T = \underline{A} \bigotimes \underline{C}.$$

This expresses the similarity assumed above in symbolic notation. According to [22], the vector function $\underline{A} \cdot \underline{C}^T$ is called a *dyadic product* in the following. The notation $\underline{A} \bigotimes \underline{C}$ expresses the same relation, but is called tensor product in this notation. Since in most cases of the relationships to be described here, the treatment of the variables as tensors has no physical justification, the term dyadic product is used in the following. This is intended to provide a simpler overview of the matrix operations carried out. For an optimal description of the product approach, the minimization of the variance of the approximation error is required:

$$\sum_{i=1}^{n} \sum_{j=1}^{m} \left(y_{ij} - a_i \cdot c_j \right)^2 \Rightarrow \min$$

It should be noted that the problem - based on the required properties of the model functions - can also be treated from the point of view of vector analysis or mathematical statistics [10, 23].

Table 9.8 Structure of a complete experimental design with two influencing factors

		Influencing factor $z \rightarrow$					
		z_1	z_2	z_3	z_4	......	z_m
Influence	x_1	y_{11}	y_{12}	y_{13}	y_{14}	......	y_{1m}
factor x	x_2	y_{21}	y_{22}	y_{23}	y_{24}	......	y_{2m}
$\downarrow$	x_3	y_{31}	y_{32}	y_{33}	y_{34}	......	y_{3m}
	....	....	....	....	....	....	....
	x_n	y_{n1}	y_{n2}	y_{n3}	y_{n4}	......	y_{nm}

By differentiating according to the unknown values a_i and c_j, the necessary condition for the desired minimum is obtained

$$a_i = \frac{\sum_{j=1}^{m} y_{ij} \cdot c_j}{\sum_{j=1}^{m} c_j^2} \quad \text{and} \quad c_j = \frac{\sum_{i=1}^{n} y_{ij} \cdot a_i}{\sum_{i=1}^{n} a_i^2}$$

After the mutual insertion and rearrangement of the elements of this system of equations, the eigenvalue problem arises for the determination of the vector $\underline{A}$ (or analogously for $\underline{C}$)

$$\lambda \cdot a_\mu = \sum_{i=1}^{n} a_i \sum_{j=1}^{m} y_{\mu j} \cdot y_{ij} \forall \mu = 1 \ldots n$$

or its abbreviated form

$$\lambda \cdot \underline{A} = \underline{Y} \cdot \underline{Y}^T \cdot \underline{A} \text{ or } \lambda \cdot \underline{C} = \underline{Y}^T \cdot \underline{Y} \cdot \underline{C}$$

with

$$\lambda = \sum_{i=1}^{n} a_i^2 \sum_{j=1}^{m} c_j^2$$

This eigenvalue problem has at most $r = min(m, n)$ eigenvalues different from zero λ_v and as many assigned eigenvectors $\underline{A}_v$ and $\underline{C}_v$. It follows that the measurement dataset $\underline{Y}$ can be completely described by a sum of dyadic products (dyad):

$$\underline{Y} = \sum_{v=1}^{r} \underline{A}_v \cdot \underline{C}_v^T$$

According to the derivation, each member of this series meets the above requirement for minimizing the sum of the quadratic deviations.

If the eigenvectors $\underline{A}_\nu$ and $\underline{C}_\nu$ are normalized using the Euclidean norm, so that $\left|\underline{A}_\nu^*\right| = \left|\underline{C}_\nu^*\right| = 1 \forall \nu = 1 \ldots r$ applies, this leads to the following form of representation

$$\underline{Y} = \sum_{\nu=1}^{r} \sqrt{\lambda_\nu} \cdot \underline{A}_\nu^* \cdot \underline{C}_\nu^{*T}$$

This will be used in the following section to discuss some properties of the approximating vectors $\underline{A}$ and $\underline{C}$. The eigenvalue problem corresponds to the problem arising in the derivation of the discrete Karhunen-Loeve transform from the theory of stochastic processes [25] (Kari Karhunen (1915–1992), Michel Loève (1907–1979)). The eigenvectors $\underline{A}_\nu$ and $\underline{C}_\nu$ therefore have similar properties as the basis vectors of the Karhunen-Loeve evolution. This means that the 'amount of information' is concentrated on a few links of the Karhunen-Loeve evolution according to the size λ_ν. From the shown derivation - especially if the data set has been centered or if $E[\underline{Y}] = 0$) applies - it follows directly that

$$\sum_{\nu=1}^{r} \lambda_\nu = \sum_{i=1}^{n} \sum_{j=1}^{m} y_{ij}^2 = D^2(\underline{Y}) = S_0^2$$

where S_0^2 is the variance of the data set $\underline{Y}$. Each pair of eigenvectors $\underline{A}_\nu$ and $\underline{C}_\nu$ contributes to this variance with a certain proportion λ_ν. The entropy function introduced in [26] for an approximation function $\eta \approx y$

$$H(\eta) = \sum_{\nu=1}^{\nu \leq r} \frac{\lambda_\nu}{S_0^2} \ln \left(\frac{\lambda_\nu}{S_0^2}\right)$$

assumes a small value and reaches the smallest value for a centered data set [25, 26]. Analogous to the entropy term according to Claude Elwood Shannon (* 30 April 1916 in Petoskey, Michigan; † 24 February 2001 in Medford, Massachusetts), $H(\eta)$ characterizes the "indeterminacy" of the model. It assumes the greatest value when the proportions of all members of the dyad series are equal and the information of the data set is therefore distributed equally over all members of the series. $H(\eta)$ tends towards zero if the information is concentrated in one or a few links of the series. In this way, different model approaches can be compared.

The effect of the procedure is demonstrated below. As an example, the dependence of the amplitude permeability μ_a for the ferrite N95 at the temperatures 25 °C and 120 °C is assumed (Fig. 9.10).

Fig. 9.10 Course of the amplitude permeability of ferrite N95 over the maximum induction B at 25 °C and 120 °C. (Source: TDK 2006)

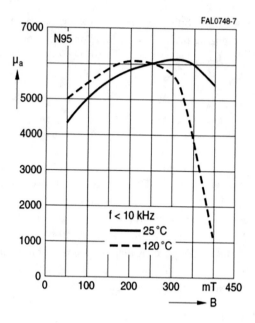

The objective is to express the measured value matrix μ_a using the interpolation points B = 50 mT; 100 mT... 400 mT by their eigenvectors $\underline{A}_v(\vartheta)$ and $\underline{C}_v(B)$ after separation of the mean value of $E[\mu_a] = 5182$ according to the following relationship in such a way that

$$\underline{\mu_a} = 5182 + \sum_{v=1}^{2} \underline{A}_v \cdot \underline{C}_v^T$$

The eigenvalue problem describing this is accordingly when using the grid points B = 50 mT; 100 mT...400 mT:

$$\lambda \cdot \underline{A} = \underline{\mu}_a^* \cdot \underline{\mu}_a^{*T} \cdot \underline{A} = \begin{bmatrix} 3.5461 \cdot 10^6 & 8.0331 \cdot 10^4 \\ 8.0331 \cdot 10^4 & 2.0491 \cdot 10^7 \end{bmatrix} \cdot \underline{A}$$

This returns the eigenvalues $\lambda_1 = 2.0491 \cdot 10^7$ and $\lambda_2 = 3.546 \cdot 10^6$ with the values set to the amount 1

$$\text{normalized eigenvectors } \underline{A}_1 = \begin{bmatrix} 4.7411 \cdot 10^{-3} \\ 1 \end{bmatrix} \underline{C}_1 = \begin{bmatrix} 0.057 \\ -0.061 \\ -0.145 \\ -0.194 \\ -0.184 \\ -0.107 \\ 0.31 \\ 0.89 \end{bmatrix}$$

$$\text{and } \underline{A}_2 = \begin{bmatrix} 1 \\ -4.7411 \cdot 10^{-3} \end{bmatrix} \underline{C}_2 = \begin{bmatrix} 0.477 \\ 0.112 \\ -0.157 \\ -0.327 \\ -0.438 \\ -0.487 \\ -0.43 \\ -0.12 \end{bmatrix}$$

The composition of the measured data thus formally results in

$$\underline{\mu}_a = 5182 \pm \sqrt{\lambda_1} \cdot \underline{A}_1 \cdot \underline{C}_1^T \pm \sqrt{\lambda_1} \cdot \underline{A}_2 \cdot \underline{C}_2^T = 5182 + \underline{\mu}_{a1}^* + \underline{\mu}_{a2}^*$$

Except for the sign, the coefficients are determined for the individual links in the series.

After inserting the eigenvalues into the output equation and normalizing the eigenvectors $\sqrt{\lambda}$, the equation

$$\underline{\mu}_a = 5182 - 4527 \cdot \underline{A}_1 \cdot \underline{C}_1^T - 1883 \cdot \underline{A}_2 \cdot \underline{C}_2^T$$

yields the desired relationship. A common representation of the measured values and the model values is shown in Fig. 9.11. The pointwise representation of the output data sequence [$\mu_a(120\ °C)$; $\mu_a(25\ °C)$] was interpolated (blue). The blue curve can be constructed point by point from the two orthogonal vectors $\underline{\mu}_{a1}^*$ (orange) and $\underline{\mu}_{a2}^*$ (green). The components of the base vectors $\underline{A}_1$, $\underline{A}_2$ depend only on the temperature and those of the base vectors $\underline{C}_1$, $\underline{C}_2$ only on the maximum induction. This supports a separate evaluation of the influencing factors.

Figure 9.11 illustrates the decomposition of the initial data set into 2 orthogonal data sets. The descriptive vector functions can also be represented in relation to the influencing factors. While the vector $\underline{A}$ here is only 2 values, the vectors $\underline{\mu}_{a2}^* = \sqrt{\lambda_2} \cdot \underline{A}_2 \cdot \underline{C}_2^T$ and

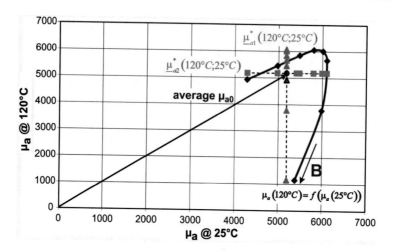

Fig. 9.11 Splitting the data set $\underline{\mu}_a(120\,^\circ C; 25\,^\circ C)$ into two orthogonal vector functions plus the mean value of the data set (material N95, EPCOS)

$\underline{\mu}^*_{a1} = -\sqrt{\lambda_1} \cdot \underline{A}_1 \cdot \underline{C}^T_1$ clearly shows how the original functions can be reconstructed from them (Fig. 9.12).

As a further example, the approximation of the temperature dependence in Fig. 9.13 is to be used, since this is a complete value matrix Y with more than 2 influence levels in each parameter.

If these empirically determined dependencies are decomposed according to the pattern described above, a maximum of 4 pairs of basic functions are obtained. The associated eigenvalues for the centered data set are summarized in the vector $\underline{\lambda}$.

$$\underline{\lambda} = \begin{bmatrix} 37.245 \\ 0.399 \\ 3.91 \cdot 10^{-3} \\ 1.88 \cdot 10^{-4} \end{bmatrix}$$

If the eigenvectors are determined and normalized in such a way that their magnitude is equal to 1, one initially obtains a very confusing system of functions. A weighting of the eigenvectors with the root of the corresponding eigenvalue contributes to the clarity and at the same time contains the information of the amount of influence of a corresponding vector/function pair. Figures 9.14 and 9.15 contain the results of the corresponding data analysis, performed on values read from the diagram in Fig. 9.13.

With the extracted functions shown, the original data can be reconstructed. The analytical context is

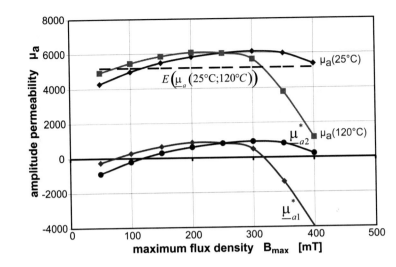

Fig. 9.12 Representation of the output data together with the pointwise extracted basic functions

Fig. 9.13 Representation of the temperature dependence of the core losses of N95. (source: TDK 2006) at 100 kHz

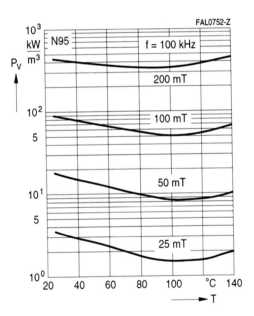

$$1\ g\left(\frac{P_V}{kW \cdot m^{-3}}\right) = 1.439 + \sum_{v=1}^{4} A_v\left(\frac{B}{mT}\right) \cdot C_v\left(\frac{\vartheta}{{}^\circ C}\right)$$

Since the third and fourth eigenvalues are extraordinarily small compared to the first two eigenvalues, the series can be broken off after the second element without much loss of

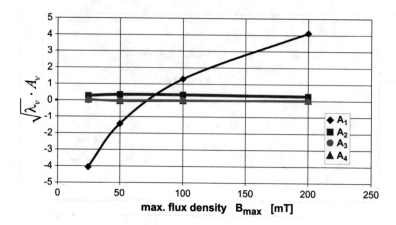

Fig. 9.14 Representation of the induction-dependent influencing functions of the core losses of N95 (EPCOS) at 100 kHz

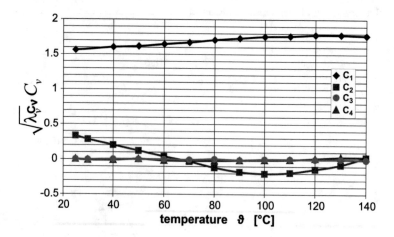

Fig. 9.15 Representation of the temperature-dependent influence functions of the core losses of N95 (TDK 2006) at 100 kHz

information. Figure 9.16 shows the original data reconstructed with these two members of the series.

Often only part of the information is available in a complete experimental design. The following example is intended to investigate the possibilities of extrapolation from such areas of investigation to others with less primary information and to show their possibilities and limitations. The power loss density for the ferrite N95 (EPCOS) as a function of frequency and induction amplitude under sinusoidal control is used as an example (Fig. 9.17). Here, the values of the empirically obtained dependence $P_V(B;\ \vartheta)$ are not available for all combinations of the influencing variables. For a first analysis, therefore, an

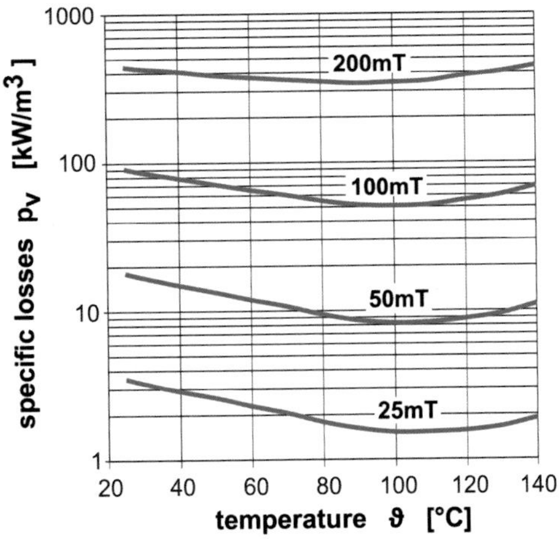

Fig. 9.16 Representation of the temperature-dependent influence functions of the core losses of N95. (source: TDK 2006) at 100 kHz, reconstructed with the first two pairs of basic functions $\{A_1(B); C_1(\vartheta)\}$ and $\{A_2(B); C_2(\vartheta)\}$

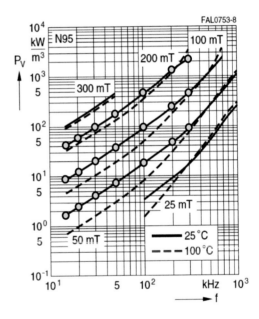

Fig. 9.17 Power loss density for ferrite N95 (TDK 2006) as a function of frequency and induction amplitude for sinusoidal control with a "complete experimental design" for the primary information

area was selected that allows a complete "experimental design". This is marked in the diagram.

The approximation method is first applied to the area in Fig. 9.17 for which a complete data set ("experimental design") can be generated (yellow dots). This means that the dependent variable y is

$$y = \lg\left(\frac{P_V(f,\ B)}{\frac{kW}{m^3}}\right)$$

The individual levels of the influencing factors (f = 15 kHz, 20 kHz, 30 kHz, 50 kHz, 100 kHz, 200 kHz, 300 kHz and B = 50 mT, 100 mT, 200 mT) are known. The mean value of the observed dependence lg(y(f, B)) is $E(\underline{Y}) = 1.6916$. The eigenvalue problem for the centered values $\underline{Y}^*$ already described above leads to the 3 eigenvalues

$$\begin{bmatrix} \lambda_1 \\ \lambda_2 \\ \lambda_3 \end{bmatrix} = \begin{bmatrix} 7.784 \\ 6.808 \\ 1.014 \cdot 10^{-3} \end{bmatrix}$$

λ_3 apparently deviates significantly from the first two eigenvalues, which indicates a significantly lower influence of the associated eigenvectors. Figure 9.18 shows the values of the eigenvectors weighted with the root of the associated eigenvalues as a function of the induction for better illustration. The influence of $\underline{A}_3$ can obviously be neglected.

A similar representation can also be made with regard to frequency as an influencing variable, as shown in Fig. 9.19. The approximation of these 1-dimensional functions is much easier with suitable analytical functions than with multidimensional functions. In the chosen example, 2 pairs of functions are sufficient, as the significant differences in the magnitude of the eigenvalues suggest. The easily comprehensible example assumes a complete data set Y. As already shown in [20], the described procedure can also be extended to incomplete value matrices using weighting factors. At the same time, an extension to more than 3 influencing variables is possible. However, the procedure for determining the eigenvectors then changes somewhat. Such approaches are discussed in the following section. The simple example selected here for demonstration purposes contains only a part of the primary information from Fig. 9.17, which also shows a significant influence of temperature as a further parameter.

In the following, the approximation function is to be assembled from the results obtained so far. If the series are broken off after 2 terms, then one receives

$$y = E[\underline{Y}] + \sqrt{\lambda_1} \cdot a_1\left(\lg\left(\frac{B}{0.1\ T}\right)\right) \cdot c_1\left(\lg\left(\frac{f}{1\ kHz}\right)\right) + \sqrt{\lambda_2} \cdot a_2\left(\lg\left(\frac{B}{0.1\ T}\right)\right)$$
$$\cdot c_2\left(\lg\left(\frac{f}{1\ kHz}\right)\right)$$

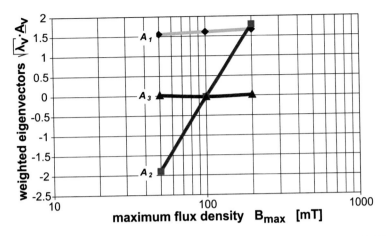

Fig. 9.18 Values of the eigenvectors $\underline{A}_1, \underline{A}_2, \underline{A}_3$ as a function of the induction (weighted with the root of the corresponding eigenvalues)

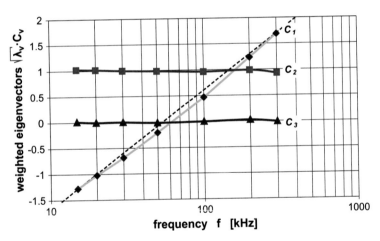

Fig. 9.19 Values of the eigenvectors $\underline{C}_v$ as a function of frequency (weighted with the root of the corresponding eigenvalues)

$$y = 1.6916 + 2.79 \cdot a_1 \left(\lg\left(\frac{B}{0.1\ \text{T}}\right)\right) \cdot c_1\left(\lg\left(\frac{f}{1\ \text{kHz}}\right)\right) + 2.609 \cdot a_2\left(\lg\left(\frac{B}{0.1\ \text{T}}\right)\right)$$

$$\cdot c_2\left(\lg\left(\frac{f}{1\ \text{kHz}}\right)\right)$$

An approximation of the discrete data of the basis vectors by polynomials as basis functions results

$$a_1\left(\lg\left(\frac{B}{0.1\ \text{T}}\right)\right) \approx 0.578 + 0.0664 \cdot \left(\lg\left(\frac{B}{0.1\ \text{T}}\right) - 0.1657 \cdot \lg\left(\frac{B}{0.1\ \text{T}}\right)\right)$$

$$a_2\left(\lg\left(\frac{B}{0.1\ \text{T}}\right)\right) \simeq -0.0096 + 2.3486 \cdot \left(\lg\left(\frac{B}{0.1\ \text{T}}\right) - 0.049 \cdot \lg^2\left(\frac{B}{0.1\ \text{T}}\right)\right)$$

$$c_1\left(\lg\left(\frac{f}{1\ \text{kHz}}\right)\right) \simeq -1.0837 + 0.4008 \cdot \left(\lg\left(\frac{f}{1\ \text{kHz}}\right) + 0.2832 \cdot \lg^2\left(\frac{f}{1\ \text{kHz}}\right)\right)$$

$$c_2\left(\lg\left(\frac{f}{1\ \text{kHz}}\right)\right) \approx 0.424 - 0.0271 \cdot \lg\left(\frac{f}{1\ \text{kHz}}\right)$$

With this information, the function series can be calculated from its summands. put together

$$y_0 = E[\underline{Y}] = 1.691610^{E[\underline{Y}]} = 10^{y_0} = 10^{1.6916} = 49.159$$

$$y_1(B;f) = \sqrt{\lambda_1} \cdot a_1\left(\lg\left(\frac{B}{0.1\ \text{T}}\right)\right) \cdot c_1\left(\lg\left(\frac{f}{1\ \text{kHz}}\right)\right)$$

$$10^{y_1(B;f)} = 0.01788 \cdot \left(\frac{B}{0.1\ \text{T}}\right)^{-0.1997 + 0.033 \cdot \lg\left(\frac{B}{0.1\ \text{T}}\right)} \times$$

$$\cdot \left(\frac{f}{1\ \text{kHz}}\right)^{0.6463 + 0.07425 \cdot \lg\left(\frac{B}{0.1\ \text{T}}\right) + 0.1803 \cdot \lg\left(\frac{f}{1\ \text{kHz}}\right) + 0.0206 \cdot \lg\left(\frac{f}{1\ \text{kHz}}\right) \cdot \lg\left(\frac{B}{0.1\ \text{T}}\right) - 0.0303 \cdot \lg\left(\frac{f}{1\ \text{kHz}}\right) \cdot \lg^2\left(\frac{B}{0.1\ \text{T}}\right) - 0.0.01223 \cdot \lg^2\left(\frac{B}{0.1\ \text{T}}\right)}$$

$$y_2(B;f) = \sqrt{\lambda_2} \cdot a_2\left(\lg\left(\frac{B}{0.1\ \text{T}}\right)\right) \cdot c_2\left(\lg\left(\frac{f}{1\ \text{kHz}}\right)\right)$$

$$y_2(B;f) = 0.976 \cdot \left(\frac{f}{1\ \text{kHz}}\right)^{0.00068}$$

$$\cdot \left(\frac{B}{0.1\ \text{T}}\right)^{2.598 - 0.1273 \cdot \lg\left(\frac{B}{0.1\ \text{T}}\right) - 0.166 \cdot \lg\left(\frac{f}{1\ \text{kHz}}\right) + 0.00814 \cdot \lg\left(\frac{f}{1\ \text{kHz}}\right) \cdot \lg\left(\frac{B}{0.1\ \text{T}}\right)}$$

Then the description function in the original area results in

$$\frac{P_V}{\left[\frac{kW}{m^3}\right]} = 10^{E[\underline{Y}]+y1(B;f)+y2(B;f)} = 49.159 \cdot 0.01788 \cdot \left(\frac{B}{0.1\ T}\right)^{-0.1997+0.033\cdot\lg\left(\frac{B}{0.1\ T}\right)} \times$$

$$\cdot\left(\frac{f}{1\ kHz}\right)^{0.6463+0.07425\cdot\lg\left(\frac{B}{0.1\ T}\right)+0.1803\cdot\lg\left(\frac{f}{1\ kHz}\right)+0.0206\cdot\lg\left(\frac{f}{1\ kHz}\right)\cdot\lg\left(\frac{B}{0.1\ T}\right)-0.0303\cdot\lg\left(\frac{f}{1\ kHz}\right)\cdot\lg^2\left(\frac{B}{0.1\ T}\right)-0.0.01223\cdot\lg^2\left(\frac{B}{0.1\ T}\right)} \times$$

$$\cdot 0.976 \cdot \left(\frac{f}{1\ kHz}\right)^{0.00068} \cdot \left(\frac{B}{0.1\ T}\right)^{2.598-0.1273\cdot\lg\left(\frac{B}{0.1\ T}\right)-0.166\cdot\lg\left(\frac{f}{1\ kHz}\right)+0.00814\cdot\lg\left(\frac{f}{1\ kHz}\right)\cdot\lg\left(\frac{B}{0.1\ T}\right)}$$

Further summary finally results in

$$\frac{P_V}{\left[\frac{kW}{m^3}\right]} = 0.8579$$

$$\cdot\left(\frac{f}{1\ kHz}\right)^{0.64698+0.07425\cdot\lg\left(\frac{B}{0.1\ T}\right)+0.1803\cdot\lg\left(\frac{f}{1\ kHz}\right)+0.0206\cdot\lg\left(\frac{f}{1\ kHz}\right)\cdot\lg\left(\frac{B}{0.1\ T}\right)-0.0303\cdot\lg\left(\frac{f}{1\ kHz}\right)\cdot\lg^2\left(\frac{B}{0.1\ T}\right)-0.0.01223\cdot\lg^2\left(\frac{B}{0.1\ T}\right)}$$

$$\times \cdot\left(\frac{B}{0.1\ T}\right)^{2.3983-0.124\cdot\lg\left(\frac{B}{0.1\ T}\right)-0.166\cdot\lg\left(\frac{f}{1\ kHz}\right)+0.00814\cdot\lg\left(\frac{f}{1\ kHz}\right)\cdot\lg\left(\frac{B}{0.1\ T}\right)}$$

Figure 9.20 shows the calculation results compared to the original data. In the data area from which the primary data was taken, a very good match can be seen. Naturally, the further one moves away from the area of primary information, the worse the match becomes. On the one hand, the detail of the representation should show that it is possible to determine functions for multidimensional dependencies of material parameters. The proposed procedure leads pointwise to functions that are only dependent on one influencing variable. They can either be approximated by analytical functions or represented by spline interpolation or similar approaches as single or multiple differentiable functions. This makes a numerical evaluation in optimization calculations relatively simple. On the other hand, it must always be clear that the real a-priori information can never be replaced by such model formations. An interpolation is allowed if one is sure that the function between two points is monotonous. Functional values outside the range of influence parameters investigated by measurement are speculations, the more one moves away from the range of the primary data field, the less meaningful they are.

The described procedure is used to examine a set of acquired measurement data for its internal relationships. A system of optimal (with regard to residual error) linear vector functions in parameter representation is determined. The components of these vectors represent the *most favourable development functions* belonging to the corresponding influencing factor *pointwise*. By interpolation (or renewed approximation with analytical functions), the course of the sought-after development functions is obtained. If a representative data set has been analyzed, the entire class of dependencies can be approximated by these functions with little effort [20]. The methodology shown here using a very simple example can in principle also be extended to a number of influencing factors >2 with an incomplete experimental design. This is necessary, for example, for the additional recording of temperature influences, dependencies on the spatial direction or for the characterization of different materials or material batches. The solution steps required for this are described in the following chapter.

Fig. 9.20 Comparison of the approximation function for the losses in N95 at 25 °C with the primary information from the manufacturer EPCOS, a "complete test plan

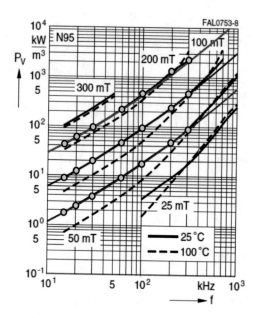

9.6   Properties of the Approximating Basic Vectors

For further work with the eigenvectors determined, knowledge of a number of their properties is advantageous. This section therefore summarizes the most important properties.

As can be seen from [22], the eigenvectors belonging to an influence factor are orthogonal to each other because of the symmetry of $\underline{Y}^T\underline{Y}$ or $\underline{Y}.\underline{Y}^T$, i.e. for the scalar products.

$$A_v^T \cdot A_\mu = 0 \ \text{ or } \ C_v^T \cdot C_\mu = 0 \forall \mu \neq v$$

From this relationship, a relationship between the eigenvalues and the sum of all squares of the measured values can be derived [20, 22]:

$$\sum_{i=1}^{n}\sum_{j=1}^{m} y_{ij}^2 = \sum_{i=1}^{n}\sum_{j=1}^{m} \left(\sum_{v=1}^{r}\sqrt{\lambda_v}\cdot a_{vi}\cdot c_{vj}\right)^2 = \sum_{v=1}^{r}\lambda_v$$

The eigenvalue belonging to a member of the approximating dyad thus indicates the share of this member in the sum of the measured value squares and thus reflects its influence within the data set or model.

From the previous section, it became clear that, in principle, all values of the matrix $\underline{Y}$, including the measurement errors contained therein, are reconstructed by the development presented if all 'r' members of the series are applied. In this context, the question arises at which point the development can be stopped without losing a substantial part of the information. Assuming completely independent random errors (ideal case), they form an r-dimensional point cloud in the feature space with the same variance in each dimension. In the presented approximation method, the total variance is proportionally projected into each dimension of the spanned orthogonal vector space so that it can be written:

$$\sigma^2_{total} = \sum_{\nu=1}^{r} \sigma^2_\nu$$

Mixed products with the "error-free, true" eigenvectors mean that the errors in the calculated eigenvectors are preserved proportionally in an attenuated form. A denser "measurement network" therefore leads to a smaller deviation of model and a-priori information even if measured values are determined only once, but the number of measured values in one spatial direction is greater than 'r'. Vectors with eigenvalues of higher order - or smaller magnitude - then practically only reflect error components, so that it makes sense to neglect them above a certain magnitude ν [23].

If the development is broken off after the k-th summand, the mean square error of the approximation results in

$$\overline{\Delta y^2} = \frac{1}{n \cdot m} \left(\sum_{i=1}^{n} \sum_{j=1}^{m} \left(y_{ij} - \sum_{\nu=1}^{k} \sqrt{\lambda_\nu} \cdot a_{\nu i} \cdot c_{\nu j} \right)^2 \right) = E\left[\underline{Y}^2\right] - \frac{1}{n \cdot m} \sum_{\nu=1}^{k} \lambda_\nu$$

Where this value comes into the order of magnitude of the variance of the observed measurement error, the development should be stopped. Otherwise the determined links would only serve to reconstruct the measurement errors. Important for the evaluation of the approximation method is the investigation of the effects of statistical measurement errors. For this purpose, an estimation is made in the following. It is assumed here that the (n, m) matrix of the measured values $\underline{Y}$ is composed of the matrix of the actual dimension $\underline{Y}^*$ and a matrix $\underline{\Phi}$ of error sizes, so that the following applies

$$\underline{Y} = \underline{Y}^* + \underline{\Phi}$$

By developing into dyadic products $\underline{A}_\nu \cdot \underline{C}_\nu^T$, the space spanned by centered data set $\underline{Y}$ can be represented as the sum of $r = min(n, m)$ orthogonal eigenspaces [22], because

$$E\left[\left(\left(\underline{A}_v \cdot \underline{C}_v^T\right) \cdot \left(\underline{A}_v \cdot \underline{C}_v^T\right)\right) \cdot \left(\left(\underline{A}_v \cdot \underline{C}_v^T\right) \cdot \left(\underline{A}_v \cdot \underline{C}_v^T\right)\right)^T\right] = \left(\underline{A}_v \cdot \underline{C}_v^T\right) \cdot \left(\underline{A}_\mu \cdot \underline{C}_\mu^T\right) = 0 \forall v \neq \mu$$

If the elements of the error matrix Φ_{ij} are random and independent errors with constant variance σ_φ^2, then the development of Φ as dyad also generates 'r' orthogonal eigenspaces of Φ. Because of the assumptions made, the eigenvalues belonging to each subspace or their expected values are ideally equal $\lambda_v = \frac{n \cdot m}{r} \cdot \sigma_{\varphi.total}^2$. This means that the dispersion of Φ is evenly distributed over all eigenspaces. If the matrix of the undistorted measured values $\underline{Y}$ can be represented as the sum of 'k' dyadic products, the influence of Φ on the approximate determination of the dyad of $\underline{Y}^*$ would result from the projection of Φ onto $\underline{Y}^*$. Since all eigenvalues of $\Phi^T\Phi$ should ideally be of the same magnitude due to the requirements made for the error size, each of the 'k' eigenspaces can be assigned a portion of the dispersion of Φ in the amount of $\sigma_{\varphi*}^2 = \frac{\sigma_\phi^2}{r}$. According to these considerations, a probable mean error for each element of the development can be defined as an expected value analogous to [22].

$$\sqrt{\sigma_\varphi^2} = \frac{\sigma_\varphi}{\sqrt{r}}$$

With the preceding formula, it is possible to estimate the probable error of the calculated model if the variance of the measurement errors is known. At the same time, this relationship makes it possible to define a termination criterion for a series development more precisely. Applying the above statements, the variance of the residual error of a data set for the idealized behavior of the residual error results in the value

$$\sigma_{\varphi.rest}^2 = \left(E\left[\underline{Y}^2\right] - \frac{1}{m \cdot n} \sum_{v=1}^{k} \lambda_v\right) = \frac{r-k}{r} \cdot \sigma_\varphi^2.$$

The data set can then be decomposed at most into $r = min(n, m)$ dyadic products. In other words: If the eigenvalues λ_v with $v = (k + 1)...r$ are in the order of magnitude $\frac{\lambda_v}{m \cdot n} \approx \frac{\sigma_\varphi^2}{r}$, the series expansion can be terminated at 'k'.

9.7   Quality Functionalities for the Approximation

The quality criterion for an approximation task can be selected under different aspects. The objective of the approximation, the accuracy requirements, the measurement method used for data acquisition, the type of measurement errors, etc. require the application of quality criteria that take these conditions into account. In the considerations of the previous sections, the minimization of the Euclidean distance between the vector of the measured values Y and that of the model values H was used as a quality functional for the approximation to be carried out:

$$\|\underline{Y} - \underline{H}\| = \sqrt{\sum_{i=1}^{n} (Y - H)^2} \Rightarrow \min$$

This quality criterion corresponds to that on C. F. Gauss method of least squares. With this method, the best linear approximations can be determined for normally distributed error quantities, as can be shown by the maximum likelihood method.

Other definitions of the distance between Y and H may also be appropriate for the formulation of the criterion for optimal fit. They have different meanings for application to technological data. A selection of the most important criteria is compiled below with a description of their mode of operation.

(a) The above requirement minimises the variance of the description error$\sigma^2 = D^2 [\underline{Y} - \underline{H}]$. It is often useful to consider these errors differently weighted with weighting factors γ_i. This is how the claim is created:

$$\sum_{i=1}^{n} \gamma_i (y_i - \eta_i)^2 \Rightarrow \min$$

The non-negative weighting factors γ_i can be used, for example, to take into account the different variance of the measurement errors for the individual points of the experimental design or to highlight parameter areas of greater interest compared to others. Also a different density of levels, which has the same effect as a different weighting of measured values, can be compensated in this way. If $\gamma_i = y_i^{-2}$ is set, a minimization of the mean relative error is achieved by this requirement. With a known system of development functions and unknown parameter vector $\underline{A}$, the system of equations is obtained

$$\underline{F}^T \cdot \underline{\Gamma} \cdot \underline{Y} = \underline{F}^T \cdot \underline{F} \cdot \underline{A}$$

with

$$\underline{\Gamma} = \begin{bmatrix} \gamma_1 & 0 & \cdots & 0 \\ 0 & \gamma_2 & \cdots & 0 \\ \cdots & \cdots & \cdots & \cdots \\ 0 & 0 & \cdots & \gamma_n \end{bmatrix}$$

The determination of optimized approximating parameter vectors by means of the weighted quadratic errors discussed in this section only leads to an eigenvalue problem if $\gamma_{ij} = \rho_i \cdot \psi_j$ applies. It then has the form

$$\lambda \cdot \underline{A} = \underline{\Psi} \cdot \underline{Y} \cdot \underline{Y}^T \cdot \underline{P} \cdot \underline{A}$$

These are

$$\underline{P} = \begin{bmatrix} \rho_1 & 0 & \cdots & 0 \\ 0 & \rho_2 & \cdots & 0 \\ \cdots & \cdots & \cdots & \cdots \\ 0 & 0 & \cdots & \rho_n \end{bmatrix} \text{ and } \underline{\Psi} = \begin{bmatrix} \psi_1 & 0 & \cdots & 0 \\ 0 & \psi_2 & \cdots & 0 \\ \cdots & \cdots & \cdots & \cdots \\ 0 & 0 & \cdots & \psi_m \end{bmatrix}$$

Such an eigenvalue problem cannot be formulated for any weighting factors γ_{ij} (e.g. for minimizing the relative error). In this case, special algorithms are required to solve the approximation task.

(b) A quasi minimization of the mean relative error occurs when, before applying the criterion of minimizing the sum of the quadratic errors, the transformation

$$y^* = \ln\left(\frac{y}{y_0}\right)$$

is carried out for the measured values y_i. This simultaneously influences the model structure, because

$$y_i = y_0 \cdot \exp\left(\eta(A; x_i)\right)$$

This modified criterion can be applied if changes in y_i have to be approximated over several orders of magnitude [19].

(c) Minimisation of the maximum deviation of the model from the measured data shall be achieved by

$$\max_{i=1\ldots n} |y_i - \eta(\underline{A}; x_i)| \Rightarrow \min$$

(Tschebyschew approximation). This approximation method evaluates unfavorable cases more sharply than an averaging criterion (e.g. case a)). Since the maximum distance between the measured values and the approximation function is minimized, so-called outliers can have a considerable influence on the result. For this reason, and because of the relatively high numerical effort required to determine the coefficients sought, this method is only of limited suitability for practical tasks in technology.

(d) Randomly strongly deviating values (outliers) are evaluated as weaker than by a) and
b), if the distance between two vectors is the so-called city block distance [10]. The
criterion for minimizing the distance between $\underline{Y}$ and $\underline{H}$ derived from this results in

$$\|\underline{Y} - \underline{H}\| = \sum_{i=1}^{n} |y_i - \eta(\underline{A}, \; x_i)|$$

The determination of the free parameters a_i of η with this quality criterion leads to a
nonlinear system of equations, since the differentiation after an a_μ results in the occurrence
of the sign function 'sign':

$$\sum_{i=1}^{n} sign(y_i - \eta(\underline{A}, \; x_i)) \cdot \left(\frac{\partial \eta}{\partial a_\mu}\right) = 0 \forall \mu = 1 \ldots m$$

The practical execution of an approximation that satisfies this requirement encounters
numerical problems. Possible solutions are given by algorithms of nonlinear optimization
or stochastic approximation [1, 22]. The measures for the distance between two points in
space presented under a, c and d are special cases of the Minkowski distances. These and
other examined distance definitions are contained in [10]. The gain in robustness of the
approximation against randomly large errors is often disproportionate to the increase of the
calculation effort.

(e) An attempt is therefore made to increase the robustness of the parameter determination
by calculating special weighting factors. The weighting factors are calculated
depending on the description errors of a previous approximation from $\underline{H}$ to $\underline{Y}$. For the
determination of the components of the vector $\underline{A}$, this method requires the repeated
solution of the system of equations [27]. The weighting factors occurring in this system
γ_i are then calculated by an expression of the type

$$\gamma_i^{(2)} = \frac{\zeta\left(\frac{1}{s}\left(y_i - \eta(\underline{A}; x_i)\right)\right)}{y_i - \eta(\underline{A}; x_i)} \cdot s$$

With the weighting function ς, errors of the previous approximation are evaluated
depending on their size. If the distribution function of the occurring errors is known, ς
can be determined exactly. For the case of normally distributed errors, $\gamma_i = 1$. The
procedure then leads to the method of least squares. If the distribution of the error size is
unknown, various empirical functions are used as weighting function ς [27]. One possibility is to use the so-called Andrews function:

$$\zeta(x) = \begin{cases} \sin\left(\dfrac{x}{a}\right) f\ddot{u}r |x| < a \cdot \pi \\ 0 f\ddot{u}r |x| \geq a \cdot \pi \end{cases}$$

The parameters α and s are used to influence the robustness of the approximation. These methods can be used especially for the approximation of data sets with a relatively small size, where the effect of random measurement errors was not attenuated by multiple measurement and averaging.

It becomes clear that with the selection of the quality criterion for an approximation to be carried out, a certain range of variation of the decisions for the user is given. Based on his experience and process knowledge and in accordance with the objective of the modeling, he must select the most appropriate variant. The possibilities for approximation presented above on the basis of their quality criteria are illustrated in the following simple example with regard to their effects. A set of ten points is to be approximated by a straight line (Fig. 9.21). The different effect of the quality criteria on the definition of the searched straight line becomes visible from the graphic representation.

The values $y(x = 3)$ and $y(x = 6)$ have a considerable influence on the course of the straight line, which was determined by minimizing the mean relative error (b) and the largest absolute error (c), respectively. The determination of the straight line using the method Air least squares (a) and the robust version (d) modified according to [27] is much less sensitive to these strongly deviating values. For the selection or formulation of a suitable quality criterion for the approximation, both the objective of the approximation and certain properties of the error size (e.g. size of the variance, the expected maximum error, the dependence of the variance on influencing factors, etc.) must therefore be taken into account.

9.8   Algorithms for the Numerical Solution of the Approximation Tasks

The possibilities of approximating empirical data sets by linear combinations of given functions, which have been discussed so far in Sect. 8.2, lead to linear systems of equations of the form

$$\underline{F}^T \cdot \underline{F} \cdot \underline{A} = \underline{F}^T \cdot \underline{Y}$$

in the determination of the unknown parameter vector $\underline{A}$. For the solution of the system of equations, sufficiently known algorithms such as the Gaussian algorithm, matrix operations or - in the case of higher-order systems of equations - iterative procedures can be used. A selection of solution methods is compiled in [22], while a detailed explanation of solution methods for linear systems of equations is given in [29]. If the same experimental design is used several times with the same approximation functions, the evaluation of experimental

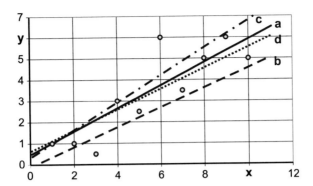

Fig. 9.21 Approximation of ten points by regression lines using different quality criteria (**a**) method of least squares, (**b**) minimization of the mean relative error, (**c**) minimization of the maximum model error, (**d**) approximation by means of the robust method with s = 0.75 and the Andrews function with a = 1.339 according to [27]

results is simplified by the calculation of the inverse matrix $(F^T F)$. If this is known, the system of equations can be solved by simple matrix multiplication. The determination of the most favourable development functions for an approximation by means of an approach according to Sect. 8.4 leads to an eigenvalue problem for symmetrical matrices

$$\lambda \cdot \underline{A} = \underline{Y} \cdot \underline{Y}^T \cdot \underline{A}$$

$$\text{or} \left(\underline{Y} \cdot \underline{Y}^T - \lambda \cdot \underline{E} \right) \cdot \underline{A} = 0$$

$\underline{E}$...unity matrix.

Since the matrix $\underline{Y} \cdot \underline{Y}^T$ is symmetrical, all eigenvalues are real and in this particular case also positive. Furthermore, it is known that all eigenvectors of $\underline{Y} \cdot \underline{Y}^T$ are orthogonal [22].

In principle, it is possible to derive from the necessary condition for the existence of zero different eigenvectors

$$\det \left(\underline{Y} \cdot \underline{Y}^T - \lambda \cdot \underline{E} \right) = 0$$

to determine the magnitude of the eigenvalues λ_ν by solving the characteristic equation. The associated homogeneous systems of equations

$$\left(\underline{Y} \cdot \underline{Y}^T - \lambda_\nu \cdot \underline{E} \right) \cdot \underline{A}_\nu = 0$$

must then be solved and the eigenvectors $\underline{A}_\nu$ normalized. For large formats of the matrix $\underline{Y} \cdot \underline{Y}^T$, this direct method has the decisive disadvantage that small errors in the calculation of the characteristic equation can sometimes have considerable effects on the calculation of the eigenvalues λ_ν. Iteration methods are therefore more favourable in such cases [28]. Such a solution method is the vector iteration according to Richard von Mises (* 19 April 1883 in Lviv, Galicia; † 14 July 1953 in Boston, Massachusetts, USA). The eigenvector with the largest eigenvalue λ is calculated. Based on this method, an algorithm according to [20] tailored to the approximation problem is presented and briefly explained.

For a clearer spelling, the agreement $\underline{Y} \cdot \underline{Y}^T = \underline{K}$ is made where $\underline{K}$ is a symmetrical matrix of type (n;n). The eigenvalue problem is then:

$$\underline{K} \cdot \underline{A}_\nu = \lambda_\nu \cdot \underline{A}_\nu \, \forall \nu = 1 \ldots n$$

Each vector $\underline{A}^*$ of the n-dimensional vector space used as an initial approximation can be interpreted as a linear combination of the (in this case orthogonal) eigenvectors of $\underline{K}$, i.e.

$$\underline{A}^* = \sum_{\nu=1}^{n} b_\nu \cdot \lambda_\nu \cdot \underline{A}_\nu$$

The eigenvectors $\underline{A}_\nu$ are ordered according to the size of their eigenvalues and normalized to the amount 1, so that $\underline{A}_l$ has the largest eigenvalue.

The selected initial approximation $\underline{A}^*_{(0)}$ is multiplied by $\underline{K}$ from the left and the obtained result $\underline{A}^*_{(1)}$ is considered as a new approximation solution:

$$\underline{K} \cdot \underline{A}^*_{(0)} = \sum_{\nu=1}^{n} b_\nu \cdot \lambda_\nu \cdot \underline{A}_\nu = \underline{A}^*_{(1)}$$

If this process is repeated μ times, the following results after the last step

$$\underline{A}^*_{(\mu)} = (\underline{K})^\mu \cdot \underline{A}^*_{(0)} = \sum_{\nu=1}^{n} b_\nu \cdot \lambda_\nu^\mu \cdot \underline{A}_\nu = \lambda_1 \cdot \sum_{\nu=1}^{n} b_\nu \cdot \left(\frac{\lambda_\nu}{\lambda_1} \right)^\mu \cdot \underline{A}_\nu$$

Since λ_1 is, as agreed, the largest eigenvalue, we are striving with

$$\lim_{\mu \to \infty} \left(\frac{(\underline{K})^\mu \cdot \underline{A}^*_{(0)}}{b_1 \cdot \lambda_1^\mu} \right) = \underline{A}_1$$

against the corresponding eigenvector $\underline{A}_l$. The practical execution of the method consists in the following rule

$$\underline{A}^*_{(i+1)} := \underline{K} \cdot \frac{\underline{A}^*_{(i)}}{\left| \underline{A}^*_{(i)} \right|}$$

The calculation is aborted if a given standard of the difference of 2 successive solutions is smaller than a given limit ε, i.e. $\left| \underline{A}^*_{(i+1)} - \underline{A}^*_{(i)} \right| < \varepsilon$.

After calculating the eigenvector $\underline{A}_i$, the original approach is then used to solve the approximation task further, the associated eigenvector $\underline{C}_i$ is calculated and the difference between the "primary" data set $\underline{Y}$ and the created approximation with the largest eigenvalue as starting point $\underline{Y}^*$ is created for the next step:

$$\underline{Y}^* := \underline{Y} - \underline{A}_1 \cdot \underline{C}_1^T$$

Then the eigenvector $\underline{A}_2$ with the largest eigenvalue is determined for $\underline{K}^* = \underline{Y}^* \cdot \underline{Y}^{*T}$, and so on. In this way, all required eigenvectors can be determined with their eigenvalues for an approximation. The convergence speed of the iteration method depends mainly on the ratios of the magnitude of the eigenvalues.

Analogous to the method explained above, a simple algorithm can be derived for the special matrices $\underline{K} = \underline{Y} \cdot \underline{Y}^T$, in which both vectors $\underline{A}_1$ and $\underline{C}_1$ are determined simultaneously. The advantage of this calculation method is in particular that it can be modified for more than two influencing factors as well as that it can be extended for experimental designs, which have a reduced number of measured values compared to a complete factorial experimental design.

It was shown that a given matrix of measured values can be represented as a series of dyads

$$\underline{Y} = \sum_{v=1}^{n} \sqrt{\lambda_v} \cdot \underline{A}_v \cdot \underline{C}_v^T$$

The eigenspaces spanned by the dyadic products of the basis vectors are orthogonal to each other. From this, a simple numerical method for the determination of the basis vectors can be derived based on the von Mises method [20]:

An initial approximation $\underline{A}^*_{(1)}$ for the eigenvector $\underline{A}_1$ can be represented as a linear combination of the eigenvectors $\underline{A}_v$ (while $\|\underline{A}_v\| = 1$), so that $\underline{A}^*_{(1)}$ can be represented by

$$\underline{A}^*_{(1)} = \sum_{v=1}^{n} b_v \cdot \underline{A}_v$$

If the transposed measured value matrix $\underline{Y}^T$ is multiplied by the initial approximation $\underline{A}^*_{(1)}$, the first approximate solution of $\underline{C}_1$

$$\underline{C}^*_{(1)} = \underline{Y}^T \cdot \underline{A}^*_{(1)} = \left(\sum_{v=1}^{n} \sqrt{\lambda_v} \cdot \underline{C}_v \cdot \underline{A}_v^T \right) \cdot \left(\sum_{v=1}^{n} b_v \cdot \underline{A}_v \right)$$

$$= \left(\sum_{v=1}^{n} \sqrt{\lambda_v} \cdot b_v \cdot \underline{C}_v \right)$$

The procedure is repeated to determine the second approximate solution of $\underline{A}_1$ by the corresponding multiplication of $\underline{Y}$ by $\underline{C}^*_{(1)}$. This results in

$$\underline{A}^*_{(2)} = \underline{Y} \cdot \underline{C}^*_{(1)} = \sum_{v=1}^{n} \sqrt{\lambda_v} \cdot b_v \cdot \underline{A}_v$$

Consequently, for the μ-th approximate solutions for $\underline{A}_1$ and $\underline{C}_1$, the relationships are given:

$$\underline{A}^*_{(\mu)} = \lambda_1^{\frac{\mu-1}{2}} \cdot \sum_{v=1}^{n} \left(\sqrt{\frac{\lambda_v}{\lambda_1}} \right)^{\mu-1} \cdot b_v \cdot \underline{A}_v \text{ and } \underline{C}^*_{(\mu)} = \lambda_1^{\frac{\mu-1}{2}} \cdot \sum_{v=1}^{n} \left(\sqrt{\frac{\lambda_v}{\lambda_1}} \right)^{\mu-1} \cdot b_v \cdot \underline{C}_v$$

If λ_1 is the largest eigenvalue, the following quotients consequently strive asymptotically against the searched eigenvectors.

$$\lim_{\mu \to \infty} \left(\frac{\underline{A}^*_{(\mu)}}{\left| \underline{A}^*_{(\mu)} \right|} \right) = \underline{A}_1 \text{ and } \lim_{\mu \to \infty} \left(\frac{\underline{C}^*_{(\mu)}}{\left| \underline{C}^*_{(\mu)} \right|} \right) = \underline{C}_1$$

The practical iteration rule for calculating the eigenvectors is

$$\underline{C}^*_{(i)} := \underline{Y}^T \cdot \frac{\underline{A}^*_{(i)}}{\left| \underline{A}^*_{(i)} \right|} \to \underline{A}^*_{(i+1)} := \underline{Y} \cdot \frac{\underline{C}^*_{(i)}}{\left| \underline{C}^*_{(i)} \right|} \to \underline{C}^*_{(i+1)} := \underline{Y}^T \cdot \frac{\underline{A}^*_{(i+1)}}{\left| \underline{A}^*_{(i+1)} \right|}$$

The calculation is terminated if

$$\left\| \frac{\underline{A}^*_{(i+1)}}{\left| \underline{A}^*_{(i+1)} \right|} - \frac{\underline{A}^*_{(i)}}{\left| \underline{A}^*_{(i)} \right|} \right\| < \varepsilon \text{ and/or } \left\| \frac{\underline{C}^*_{(i+1)}}{\left| \underline{C}^*_{(i+1)} \right|} - \frac{\underline{C}^*_{(i)}}{\left| \underline{C}^*_{(i)} \right|} \right\| < \varepsilon$$

is lower than a given barrier ε. Both the maximum norm and the Euclidean norm can be used as a standard. As you can easily convince yourself, the iteration rule stated above includes the alternate application of the formulas for the calculation of the components of the eigenvectors. If all components of the initial approximation $A_{(1)}$ are chosen to be of the same size (e.g. all $a_\nu = 1$), the result after the first cycle of calculation is identical with the algorithm presented in [19] as an extended mean value method.

The modification of the iteration procedure for more than two influencing variables can be easily derived and has the following appearance, for example for a product approach $y_{ijk} = a_i \cdot b_j \cdot c_k$:

$$a_i^{(\mu+1)} = \left[\sum_{j=1}^{m} \left(b_j^{(\mu)}\right)^2 \sum_{k=1}^{p} \left(c_k^{(\mu)}\right)^2 \right]^{-\frac{1}{2}} \sum_{j=1}^{m} \sum_{k=1}^{p} y_{ijk} \cdot b_j^{(\mu)} \cdot c_k^{(\mu)}$$

For the other sought-after components of $\underline{B}$ and $\underline{C}$, the determination formulae are formed analogously to this formula. If it is necessary to calculate further eigenvectors with lower eigenvalues, the difference between the measured values and the model results is formed and the calculation method is applied to the matrix of description errors. This is continued until the desired accuracy of the model is achieved.

A particular problem is the determination of function values of general development functions when using *incomplete* experimental designs. In the case of a product approach, an approximation requires minimizing the variance of the description error in the form

$$\sum_{i=1}^{n} \sum_{j=1}^{m} \gamma_{ij} \left(y_{ij} - a_i \cdot c_j\right)^2 \Rightarrow \min$$

Here is

$$\gamma_{ij} = \begin{cases} 1, & \text{if } y_{ij} \text{ is a measured value} \\ 0, & \text{if } y_{ij} \text{ is not a measured value} \end{cases}$$

The necessary conditions for reaching the required minimum are obtained by differentiating between a_i and c_j to

$$a_i = \left(\sum_{j=1}^{m} \gamma_{ij} \cdot c_j^2 \right)^{-1} \cdot \sum_{j=1}^{m} \gamma_{ij} \cdot y_{ij} \cdot c_j \quad \text{and} \quad c_i = \left(\sum_{i=1}^{n} \gamma_{ij} \cdot a_i^2 \right)^{-1} \cdot \sum_{i=1}^{n} \gamma_{ij} \cdot y_{ij} \cdot a_i$$

It is not possible to convert these receivables into an eigenvalue problem. However, the previous considerations can be used as the basis of an algorithm for the iterative calculation of the approximating vectors.

For this purpose, starting from an initial approximation, the components of the vectors $\underline{A}$ and $\underline{C}$ are calculated in a stepwise approximation by alternately applying the above equations. Here, too, the idea of the projection method for the solution of systems of equations is the theoretical basis for the procedure. In formal matrix notation, the above iteration rule looks like this.

$$\underline{A}_{v(i)} = \left(\underline{e}_A \div \left(\underline{\Gamma}^T \cdot \overrightarrow{\left(\underline{C}_{v(i)} \cdot \underline{C}_{v(i)} \right)} \right) \right) \cdot \left(\overrightarrow{(\underline{\Gamma} \cdot \underline{Y})} \cdot \underline{C}_{v(i)} \right) \rightleftarrows \underline{C}_{v(i+1)}$$

$$= \left(\underline{e}_C \div \left(\underline{\Gamma} \cdot \overrightarrow{\left(\underline{A}_{v(i)} \cdot \underline{A}_{v(i)} \right)} \right) \right) \cdot \left(\overrightarrow{(\underline{\Gamma} \cdot \underline{Y})}^T \cdot \underline{A}_{v(i)} \right)$$

$\underline{e}_A$ is a 1-column matrix with the same number of matrix elements as A, where all elements are identically 1. The same applies analogously to $\underline{e}_C$.

The expressions $\overrightarrow{\left(\underline{C}_{v(i)} \cdot \underline{C}_{v(i)} \right)}$ or rather $\overrightarrow{\left(\underline{e}_A \div \underline{M}_{v(i)} \right)}$ denote the vectorized matrix operations of a division or multiplication, where - e.g. in contrast to the usual matrix multiplication - only the matrix elements with the same indices are affected. During numerical treatment in corresponding programs, it is often not possible to enter the complete formula as shown above. The operations in the brackets are then to be processed one after the other, starting with the innermost one.

The procedure is subsequently applied to the data in Fig. 9.22, which represents an "incomplete" experimental design. For a network of.

$$B_{max} = 25,50,100,200,300 \text{ mT}$$

and

$$f = 15,20,30,50,100,200,300,500,1000 \text{ kHz}$$

a data record is created at the possible positions

$$y_{ij}^* = \lg \left(\frac{P_V \left(B_{max\,(i)}; f_{(j)} \right)}{\frac{kW}{m^3}} \right)$$

from which the mean value 1.8086 is subtracted (Fig. 9.22). This gives a centered data field $\underline{Y}$ and the corresponding weighting matrix $\underline{\Gamma}$. If primary information of a matrix element of $\underline{Y}$ is not available, the corresponding element of $\underline{\Gamma}$ is set equal to 0. Otherwise, this element is set equal to 1.

The data fields to be approximated now have the following content:

Fig. 9.22 Selection of the "measurement points" in an incomplete factorial experimental design

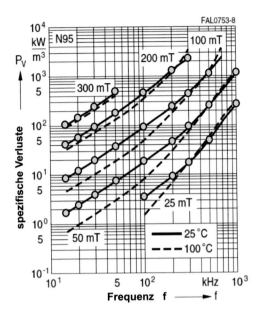

$$
\underline{Y} = \begin{bmatrix}
0 & -1.594 & -0.832 & -0.225 & 0.150 \\
0 & -1.439 & -0.692 & -0.069 & 0.318 \\
0 & -1.219 & -0.496 & 0.150 & 0.563 \\
0 & -0.935 & -0.238 & 0.421 & 0.835 \\
-1.284 & -0.561 & 0.150 & 0.835 & 0 \\
-0.871 & -0.160 & 0.563 & 1.287 & 0 \\
-0.612 & 0.150 & 0.848 & 1.520 & 0 \\
-0.121 & 0.576 & 1.209 & 0 \cdot & 0 \\
0.576 & 1.197 & 0 & 0 & 0
\end{bmatrix} \approx \underline{A} \cdot \underline{C}^T; \quad \Gamma
$$

$$
= \begin{bmatrix}
0 & 1 & 1 & 1 & 1 \\
0 & 1 & 1 & 1 & 1 \\
0 & 1 & 1 & 1 & 1 \\
0 & 1 & 1 & 1 & 1 \\
1 & 1 & 1 & 1 & 0 \\
1 & 1 & 1 & 1 & 0 \\
1 & 1 & 1 & 1 & 0 \\
1 & 1 & 1 & 0 & 0 \\
1 & 1 & 0 & 0 & 0
\end{bmatrix}
$$

The iterations lead to a continuous reduction of the differences between the approximate solutions, as shown in Fig. 9.23.

Figure 9.24a and b show how - starting from an initial approximation with identical elements for $\underline{A}$ and $\underline{C}$ - the solution vectors (here represented by their connected elements) asymptotically approach a course. For the representation, the solution vectors are

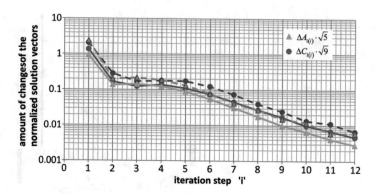

Fig. 9.23 Relative changes of the first pair of solution vectors for which the given example was used in relation to a solution vector with equally distributed components (solid line: root of the variance of the changes; broken line: maximum single deviation)

normalized to $\left|\underline{A}_{1(i)}\right| = 1$ and $\left|\underline{C}_{1(i)}\right| = 1$, respectively. This is also recommended during the iteration calculation for a better overview.

Figures 9.25 and 9.26 show the first 3 pairs of solution functions determined by this method, representing 99.87% of the variance of the primary data (Table 9.9).

The solution vectors obviously no longer have the property of orthogonality. By the effect of the weighting matrix Γ, the "remainder data set" has generally also a mean value different from zero. One can subtract this before the determination of a new pair at solution vectors from the remainder data set and add to the mean value specified above at the beginning. This leads to a more effective development of the function series. In the case study shown here, however, this procedure was omitted and only the residual data set was transferred to the next solution vector pair $(\underline{A}_v; \underline{C}_v)$ (Figs. 9.25 and 9.26) by iteration.

For the calculation of the logarithmic primary data set, the approach

$$Y^* \approx E[\underline{Y}^*] + \underline{A}_1 \cdot \underline{C}_1^T + \underline{A}_2 \cdot \underline{C}_2^T + \underline{A}_3 \cdot \underline{C}_3^T$$

is used. With the model approach

$$y_{ij} \approx a_i + c_j$$

it is assumed that the variable under investigation is the sum of individual effects of influencing variables. The search for the best approximation according to this approach leads - as already shown at the beginning of the section - to a linear system of equations of the form

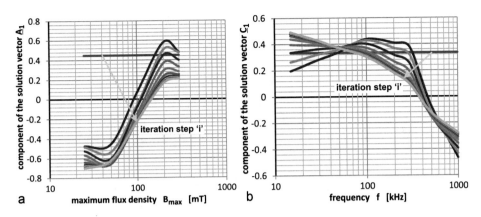

Fig. 9.24 Development of a pair of solution vectors from an initial approximation (all components of a start solution vector equal). (**a**) Iteration course for $\underline{A}_1$, (**b**) iteration course for $\underline{C}_1$

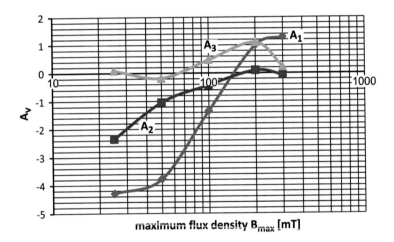

Fig. 9.25 Pointwise determination of solution functions $A_{1...3}\,(B_{max})$

$$\underline{K} \cdot \underline{A} = \underline{Z}$$

for the determination of one of the development vectors $\underline{A}$. To solve this system of equations, the iteration method described below, which is similar to the previous one, is suitable. The matrix $\underline{K}$ is broken down into the unit matrix $\underline{E}$ and the matrix $\underline{K}^*$:

$$\underline{K} = \underline{E} + \underline{K}^*$$

The matrix $\underline{K}^*$ has special properties that can be used in methods for solving the system of equations. From this approach, it can be derived that the sum of all elements of a row vector $\underline{k}_v^*$ is always 1:

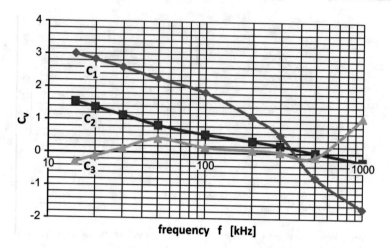

Fig. 9.26 Pointwise determination of solution functions $C_{1\ldots3}(f)$

Table 9.9 Distribution of the variance proportion and eigenvalues of the first 3 development functions according to Figs. 9.25 and 9.26

Serial number	Variance σ_ν^2	Proportion of total variance	σ_ν	Fictitious proportion of eigenvalue $\sqrt{\lambda_\nu}$
1	0.6751	62.97%	0.8216	37.1540
2	0.3777	35.23%	0.6146	6.8100
3	0.0178	1.66%	0.1335	1.5020
Total:	1.0720		1.0354	

$$\sum_{i=1}^{n} |\underline{k}_{\nu i}^{*}| = \sum_{i=1}^{n} \left(\sum_{j=1}^{m} \gamma_{ij} \right)^{-1} \cdot \sum_{j=1}^{m} \left(\gamma_{\nu j} \cdot \gamma_{ij} \cdot \left(\sum_{i=1}^{n} \gamma_{ij} \right)^{-1} \right)$$

From the triangle inequality, the following results for the amounts of the row vectors $\underline{k}_\nu^{*}$

$$|\underline{k}_\nu^{*}| \leq 1$$

The amount of $\underline{k}_\nu^{*}$ is always less than 1 if at least one of the two influence factor levels selected for the measurement of a quantity is used in a further sample measurement. Under this condition, a simple algorithm can be formulated to solve the system of equations. The $(\mu + 1)$th approximation of $\underline{A}$ is given by

$$\underline{A}_{(\mu+1)} := \underline{Z} - \underline{K}^* \cdot \underline{A}_{(\mu)}$$

For the solution of the system of equations $\underline{A} = \underline{A}_{(\infty)}$ exactly the following equation applies

$$\underline{A}_{\infty} = \underline{Z} - \underline{K}^* \cdot \underline{A}_{(\infty)}$$

To prove the convergence of the calculation method, the difference between $\underline{A}_{(\mu+1)}$ and $\underline{A}_{(\infty)}$ is examined:

$$\left| \underline{A}_{(\mu+1)} - \underline{A}_{(\infty)} \right| = \left| \underline{Z} - \underline{K}^* \cdot \underline{A}_{(\mu)} - \left(\underline{Z} - \underline{K}^* \cdot \underline{A}_{(\infty)} \right) \right| = \left| \underline{K}^* \cdot \underline{A}_{(\infty)} - \underline{K}^* \cdot \underline{A}_{(\mu)} \right|$$

$$= \left| \underline{K}^* \cdot \left(\underline{A}_{(\infty)} - \underline{A}_{(\mu)} \right) \right|$$

Since the magnitudes of the line vectors of $\underline{K}^*$ are less than 1, the difference between the approximate solutions and the final solution becomes smaller with each cycle. The differences $\Delta a_v^{(\mu)}$ between the components of the approximate solution and the final solution decrease according to the relationship.

$$\Delta a_v^{(\mu)} := \left| k_v^* \right|^{\mu - 1} \cdot \Delta a_v^{(1)} \quad \text{with} \quad \left| k_v^* \right| = \frac{1}{\sqrt{2}} \cdots \frac{1}{\sqrt{n}}$$

The specified calculation rule for $\underline{A}$ can be further simplified for the numerical version. If you write down this rule for a component $a_v^{(\mu+1)}$ of $\underline{A}_{(\mu+1)}$, you get the modified rule after some transformations.

$$a_v^{(\mu+1)} := \left(\sum_{j=1}^{m} \gamma_{ij} \right)^{-1} \cdot \left(\sum_{j=1}^{m} \gamma_{ij} \left(y_{ij} - c_j^{(\mu)} \right) \right) \quad \text{and} \quad c_j^{(\mu)} := \left(\sum_{i=1}^{n} \gamma_{ij} \right)^{-1}$$

$$\cdot \left(\sum_{i=1}^{n} \gamma_{ij} \left(y_{ij} - a_i^{(\mu)} \right) \right)$$

The iteration rule consists in the alternate application of these two equations. As an initial approximation, the components of one of the two solution vectors are set to zero. The calculation of the approximate solutions is repeated cyclically until their changes are sufficiently small. This iteration method can be easily applied to problems with more than two influencing variables [20]. Figure 9.27 contains a graphical representation of the "measured values" reconstructed from the basic functions shown in Figs. 9.25 and 9.26 for the primary data set in Fig. 9.22.

The quantitative modelling of interactions between physical and derived technical quantities using empirical results can be divided methodically into the areas of test

Fig. 9.27 Comparison of original values and model values in the approximation of the data according to Fig. 9.22

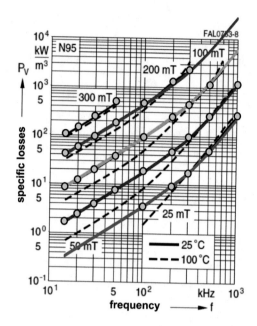

preparation and data analysis of test results. In the phase of test preparation, dimensional analysis offers possibilities of qualitative modelling as well as the ordering and summarising of influencing variables. Statistical design of experiments can significantly reduce the experimental effort required to develop a model of the effects of influencing factors on the investigated variable. Statistical design of experiments in the narrower sense includes methods for the preparation of optimal experimental designs. Optimality criteria used refer to the variance of the results and reduce the effects of random errors. Orthogonal designs make data analysis particularly easy numerically. However, the preparation time required is high. For recurring measurement problems, this effort is worthwhile.

The aim of experimental design and data analysis is to establish quantitative relationships between the parameters of interest with minimum effort and high reliability. Of particular importance is the linear approximation of a measurement data set with functions. A fundamental problem in this context is the adequacy of the obtained model, since by choosing an unfavourable system of basic functions, considerable deviations of the model from reality can occur between the investigated measured values.

An effective possibility for the selection of suitable basic functions or for their design is given by specifying the structure of the relationship between the influencing parameters. Using statistical data analysis, the most suitable basic functions for an approximation are determined point by point. Thus, the most favourable basic functions can be determined by interpolation or new approximation. The problem of multidimensional approximation is thus reduced to one-dimensional problems. This makes it much easier to select the most favorable development functions or analytical approximation functions. For existing analytical models, you can transform them to determine the optimal development functions for development into a series of functions.

Depending on the modeling task and the properties of the error quantities, different quality functionals are used for an approximation. With the help of special methods, robust criteria can be formulated against randomly large errors ("outliers"). The algorithms required for a numerical treatment of the described approximation tasks are presented separately. The proposed solution methods are specially tailored to the approximation problem and can be applied with little numerical effort. The central problem of the choice of a suitable model for the empirical modeling of complicated interrelationships becomes easier to solve with the described approach. It is characterized by extensive universality and relatively low computational effort.

References

1. Werner, J.: Vorlesung über Approximationstheorie. Universität Göttingen 1984
2. Stadler, S.: Messtechnische Bestimmung und Simulation der Kernverluste in weichmagnetischen Materialien. Dissertation. Universität Erlangen–Nürnberg 2009
3. Heck, C.: Magnetische Werkstoffe und ihre technische Anwendung. 2. Auflage, Hüthig-Verlag 1975
4. Bridgeman, P. W.: Dimensional Analysis. Yale University Press New Haven 1932
5. Van Driest, E.: On Dimensional Analysis and the Presentation of Data in Fluid Flow Problems. Journal of Applied Mechanics. Transactions of the ASME Ser. A (1946)
6. Зуев, К. И.: Основ⬚ теории подобия. издательство ВлГУ, Владимир 2011 (Sujev, K, I.: Grundlagen der Ähnlichkeitstheorie. Verlag WlGU, Wladimir 2011
7. Huntley, H. E.: Dimensional Analysis. McDonald and Co. London 1953
8. Buckingham, E.: On Physically Similar Systems. Illustrations of the Use of Dimensional Equations. Physical Reviews (1914) 4, pp. 345
9. Седов ⬚. И.: Метод⬚ подобия и размерности в механике. изд. Наука, 1981, (Sedov, L. I.: Methoden der Ähnlichkeit und Dimensionen in der Mechanik (russ.). 9. Auflage. Verlag Nauka Moskau 1981
10. Hartmann, W.: Geometrische Modelle zur Analyse empirischer Daten. Akademie-Verlag Berlin. 1979, 256 S.
11. Serfling, R. J.: Approximation Theorems of Mathematical Statistics. JOHN WILEY & SON 2002
12. Johnson, L. N.; Leone, F. C.: Statistics and Experimental Design. Band 2, John Wiley & Sons, New York 1977, 516 S. Kap. 19, S. 410-440
13. Бродский В.⬚., Бродский ⬚.И. и др.: Таблиц⬚ планов ⬚ксперимента. Для факторн⬚х и полиномиальн⬚х моделей. Справочное издание. Металлургия, 1982 (Brodskij, V. S. u. a.: Tabellen von Experimentierplänen für Faktor- und Polynommodelle (russ.). Nachschlagewerk. "Metallurgija" Moskau 1982)
14. Асатурян, В. И. Т⬚ория планирования ⬚ксперименов - учебное пособие. Радио и связь, Москва 1978 (Asaturjan, V. I.: Theorie der Planung von Experimenten - Lehrbuch. "Radio i svjaz" Moskau 1978
15. Bandemer, H.; Bellmann, A.: Statistische Versuchsplanung. Teubner Verlagsgesellschaft Leipzig 1976
16. Bandemer, H. u. a.: Optimale Versuchsplanung. Akademie-Verlag, Berlin

17. Rasch, D.; Herrenhöfer, G.: Statistische Versuchsplanung. Deutscher Verlag der Wissenschaften, 1982

18. Fisher, R. A.: Statistical Methods for Research Workers. University of London, Oliver and Boyd, London 1934

19. Zacharias, M.; Zacharias, P.: Beitrag zur Optimierung von Technologien. messen-steuern-regeln 27(1984)6, S. 302-305

20. Zacharias, P.: Elektrophysikalische und Elektrochemische Abtragsverfahren - ihre technische Entwicklung und empirische Modellierung. Habilitationsschrift TH Magdeburg 1985

21. Michaelis, B.: Einführung in zusammengesetzte Messgrößen - ein Konzept zur Messdatenreduktion. messen-steuern-regeln 26(1983)4, S. 167-171

22. Bronstein, I. N.; Semendjajew, K. A.; Musiol, G.; Muehlig, H.: Taschenbuch der Mathematik. Harri Deutsch Verlag 2008

23. Michaelis, B.: Zusammengesetzte Messgrößen und ihre Anwendung. Diss. B/Habilitationsschrift, TH Magdeburg 1980

24. Storm, R.: Wahrscheinlichkeitsrechnung, Mathematische Statistik, Statistische Qualitätskontrolle. Fachbuchverlag Leipzig

25. Reck, H.: Karhunen-Loeve-Transformation, ein Verfahren der Signalverarbeitung. Nachrichtentechnik-Elektronik 29(1979)5, S. 186-188

26. Winkler, G.: Stochastische Systeme - Analyse und Systeme. Akademische Verlagsgesellschaft Wiesbaden 1979

27. Launer, R. L.: Robustness in Statistics. Academic. New York – London. 1979

28. Kiesewetter, H.; Maess, G.: Elementare Methoden der numerischen Mathematik. Akademie-Verlag Berlin 1974

29. Bronshtein, I.N.: Handbook of Mathematics. Published October 1st 2007 by Springer (first published January 1st 1976)

Application Examples

10

10.1 Shunts for Current Measurement

A very common task where the effect of self-induction can be observed is to measure the current by measuring a voltage drop. Figure 10.1 shows an example of such a situation. On a conductor section with radius R_1, length l and resistivity ρ, a voltage drop is caused by the current i(t). This voltage drop is composed of the ohmic voltage drop, the voltage drop at the inner inductance of the conductor and the induced voltage in the conductor loop with the area $A = a{\cdot}l$, so that one can write

$$u^*(t) = u(t) + \frac{d}{dt}\left(\int_A \vec{B} \cdot \vec{dA} \right) = i \cdot \rho \cdot \frac{l}{\pi R_1^2} + L_i \frac{di}{dt} + \frac{d}{dt} \int_{R_1}^{R_1+a} \frac{\mu_0 \cdot i}{2\pi \cdot r} \cdot l \cdot dr$$

Thus, the measured voltage is composed of an ohmic and an inductive voltage drop. In the example shown, there is a compensation of the effect of the internal inductance and the external magnetic field.

$$u^*(t) = i \cdot R + \frac{\mu_0 \cdot l}{2\pi}\left(\frac{1}{4} + \ln\left(\frac{R_1 + a}{R_1} \right) \right) \cdot \frac{di}{dt} = i \cdot R + L_{eff} \cdot \frac{di}{dt}$$

In complex notation, the proportionality is thus obtained by using the inner inductance L_i and the outer inductance L_a

$$\underline{u}^* = \left(R + j\omega L_{eff} \right) \cdot \underline{i} = \left(R + j\omega(L_{i0} + L_a) \right) \cdot \underline{i}$$

© Springer Fachmedien Wiesbaden GmbH, part of Springer Nature 2022
P. Zacharias, *Magnetic Components*,
https://doi.org/10.1007/978-3-658-37206-4_10

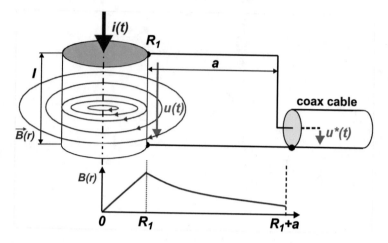

Fig. 10.1 Measurement of an AC voltage on a straight section of the conductor

L_a depends on the cable routing of the measuring connections and is practically not dependent on the frequency. The direct current resistance R_{W0} results from the design formula for electrical resistance of

$$R_{W0} = \frac{l}{\sigma \cdot A} = \frac{l}{\sigma \cdot \pi \cdot R_a^2}$$

L_{i0} is for cylindrical conductors

$$L_{i0} = \frac{\mu \cdot l}{8\pi}$$

If only the resistance and the internal inductance L_{i0} are taken into account, the impedance Z_{W0} is

$$Z_{W0} = \sqrt{R_{W0}^2 + (\omega L_{i0})^2} = \sqrt{\left(\frac{l}{\sigma \cdot \pi \cdot R_a^2}\right)^2 + \left(\omega \frac{\mu \cdot l}{8\pi}\right)^2}$$

From a cut-off frequency f_0 onwards, the inductive component predominates. Where f_0 is equal to

$$f_0 = \frac{4}{\sigma \cdot \mu \cdot \pi \cdot R_a^2}$$

The internal inductance of a conductor depends on the frequency via the skin effect or the current distribution in the conductor. To estimate the effect, proceed as follows. The thickness s of the conductor through which the current flows is given by

$$s = \frac{1}{\sqrt{\pi \cdot f \cdot \sigma \cdot \mu}},$$

where

f...   Frequency
σ...   Conductivity
μ...   Permeability

are. It is further assumed that only one layer s, which is dependent on the frequency f, is conductive and carries the entire current. Then the resistance R_W of the resulting "pipe" changes at $s < R_a$ according to

$$R_W \approx \frac{l}{\sigma \cdot \pi \left(R_a^2 - (R_a - s)^2 \right)} = \frac{l \cdot \sqrt{\pi \cdot f \cdot \sigma \cdot \mu}}{\sigma \cdot \pi \cdot \left(2 \cdot R_a - \frac{1}{\sqrt{\pi \cdot f \cdot \sigma \cdot \mu}} \right)}$$

The resistance, therefore, increases approximately with frequency. From the point of intersection of the tangents to the curve $R_W(f)$ Fig. 10.5, the cut-off frequency f_1 is

$$f_1 = \frac{4}{\pi \cdot R_a^2 \cdot \sigma \cdot \mu}$$

Above this frequency, the resistance increases due to the skin effect $f_0 = f_1$. Minimizing the internal inductance is one of the methods to create a broadband shunt. Possibilities for this are demonstrated below using two simple arrangements. The first arrangement shows the basic structure of a tubular resistor body through which current flows axially. The end faces are contacted with good conductive material. This provides a practically homogeneous current flow field in the resistor body at sufficiently low frequencies. This means that the current density J in the tubular jacket can be assumed to be constant.

The current density 'J' results from

$$J = \frac{I}{\pi \left(R_a^2 - R_i^2 \right)}$$

The magnetic field lines are arranged in a circle around the symmetry axis. The application of Ampere's law yields for the magnetic field strength 'H' at distance 'r' from the axis of symmetry

Fig. 10.2 Structure of a tubular
resistor through which current
flows axially with indicated
contacts at the end faces

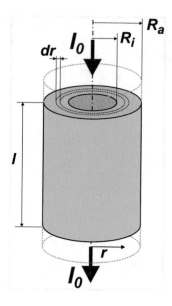

$$\oint_{s} \vec{H}(r) \cdot d\vec{s} = I(r) = \frac{I_0 \cdot \left(r^2 - R_i^2\right)}{\left(R_a^2 - R_i^2\right)}$$

$$H(r) = \frac{I(r)}{2\pi r} = \frac{I_0 \cdot \left(r^2 - R_i^2\right)}{2\pi r \cdot \left(R_a^2 - R_i^2\right)}$$

From the definition for the inductance in connection with energy, the following can be
derived for the internal inductance $_{Li}$

$$W_{Lia} = \frac{1}{2} L_{ia} \cdot I_0^2 = \int_{V} \frac{\mu}{2} \cdot H^2(r) \cdot dV = \frac{\mu}{2} \cdot \left(\frac{I_0}{2\pi \cdot \left(R_a^2 - R_i^2\right)}\right)^2 \int_{V} \left(\frac{\left(r^2 - R_i^2\right)}{r}\right)^2 \cdot l \cdot 2\pi r \cdot dr$$

$$L_{ia} = \mu \cdot \frac{l}{2\pi \cdot \left(R_a^2 - R_i^2\right)^2} \int_{V} \left(\frac{\left(r^2 - R_i^2\right)}{r}\right)^2 \cdot r \cdot dr$$

The inner inductance of the arrangement according to Fig. 10.2 is then

$$L_{ia} = \frac{\mu \cdot l}{2\pi \cdot \left(R_a^2 - R_i^2\right)^2} \left(\frac{R_a^4 - 4R_a^2 R_i^2 + 3R_i^4}{4} + R_i^4 \ln\left(\frac{R_a}{R_i}\right)\right)$$

The ohmic resistance R_{sha} in the axial direction is

Fig. 10.3 Structure of a tubular resistor through which current flows radially with indicated contacts on the sheath surfaces

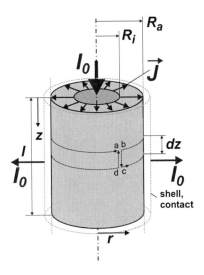

$$R_{sha} = \frac{l}{\sigma \cdot \pi \left(R_a^2 - R_i^2\right)}.$$

This corresponds to an "inner" cut-off frequency f_0 of the shunt of

$$\tau = \frac{1}{2\pi f}$$

$$f_{0a} = \frac{1}{2\pi\tau} = \frac{R_{sha}}{2\pi L_{ia}} = \frac{\left(R_a^2 - R_i^2\right)}{2\pi \cdot \sigma \cdot \mu \cdot \left(\dfrac{R_a^4 - 4R_a^2 R_i^2 + 3R_i^4}{4} + R_i^4 \ln\left(\dfrac{R_a}{R_i}\right)\right)}$$

The same basic arrangement can also be used to construct a resistor arrangement with a radial current flow field (Fig. 10.3). Here the current I_0 is supplied via an internal contact and flows through the resistor body over a length 'l' from the inside to the outside.

Figure 10.3 shows the integration path 's' for the application of flood law. Ampere's law provides

$$\oint_s \vec{H}(r,z)\cdot d\vec{s} \;=\; \int_a^b \vec{H}(r,z)\cdot d\vec{s} + \int_b^c \vec{H}(r,z)\cdot d\vec{s} + \int_c^d \vec{H}(r,z)\cdot d\vec{s} + \int_d^a \vec{H}(r,z)\cdot d\vec{s}$$

$$= \; dI = \frac{I_0}{l}dl$$

$$2\pi r \cdot H(r,z) + 0 - 2\pi r \cdot H(r,z+dz) + 0 = dI = \frac{I_0}{l}dz$$

$$2\pi r \cdot H(r,z) - 2\pi r \cdot \left(H(r,z) + \frac{\partial H(r,z)}{\partial z}dz \right) = \frac{I_0}{l}dz$$

$$- 2\pi r \cdot \frac{\partial H(r,z)}{\partial z}dz = \frac{I_0}{l}dz$$

$$\frac{\partial H(r,z)}{\partial z} = \frac{-I_0}{2\pi rl}$$

For reasons of symmetry H(r,z) has only one component in the (r, φ) plane. At z ≥ l the field strength H is zero because of the included current of zero. For an arrangement as shown in Fig. 10.3 the following relationship can be found

$$H(r,z) = \frac{I_0}{2\pi rl}(l-z) = \frac{I_0}{2\pi r}\left(1 - \frac{z}{l}\right)$$

As in the previous example, the internal inductance of the electric current field in the resistance material can be determined from this.

$$L_{ir} = \frac{\mu}{I_0^2}\int_V H^2(r,z)dV = \frac{\mu}{4\pi^2 l^2}\int_V \frac{(l-z)^2}{r^2}2\pi r dr dz = \frac{\mu}{2\pi l^2}\int_0^l \int_{R_i}^{R_a} \frac{(l-z)^2}{r}dr dz$$

$$L_{ir} = \frac{\mu}{2\pi l^2}\int_0^l \int_{R_i}^{R_a} \frac{(l-z)^2}{r}dr dz = \frac{\mu l}{6\pi}\cdot \ln\left(\frac{R_a}{R_i}\right)$$

With a single-sided current supply, the current density in the cylindrical inner conductor in the arrangement shown in Fig. 10.3 decreases linearly from the maximum value to zero. This results in a field characteristic inside the inner conductor of approximately

$$H(r,z) \approx I_0 \cdot \frac{r}{2\pi R_i^2}\cdot \left(1 - \frac{z}{l}\right)$$

At low frequencies the inner inductance of the inner conductor Lii of length 'l' is added to the total inner inductance:

$$L_{ii} \approx \frac{\mu l}{24\pi}.$$

At high frequencies, this proportion disappears quickly due to the skin effect. Therefore this part will be neglected in the following. The ohmic resistance is

$$R_{shr} = \frac{1}{2\pi \cdot \sigma \cdot l} \ln\left(\frac{R_a}{R_i}\right)$$

This corresponds to an "inner" cut-off frequency f_0 of the shunt of

$$\tau = \frac{1}{2\pi f}$$

$$f_{0r} = \frac{1}{2\pi\tau} = \frac{R_{shr}}{2\pi L_{ir}} = \frac{3}{2\pi \cdot \sigma \cdot \mu \cdot l^2}$$

While the inner cut-off frequency of the axial cylindrical resistance does not depend on the length 'l', it is independent of the radius R_a for the radial resistance. To achieve high cut-off frequencies, the length must be kept very short.

If one equates the two expressions f_{0a} and f_{0r} for the cut-off frequency, one obtains the geometric condition for the same cut-off frequency of both designs.

$$l_{radial} = R_{a\_axial}$$

$$\cdot \sqrt{\frac{3}{2\left(1 - \left(\frac{R_{i\_axial}}{R_{a\_axial}}\right)^2\right)}} \cdot \left(\frac{1 - 4\left(\frac{R_{i\_axial}}{R_{a\_axial}}\right)^2 + 3\left(\frac{R_{i\_axial}}{R_{a\_axial}}\right)^4}{4} + \left(\frac{R_{i\_axial}}{R_{a\_axial}}\right)^4 \ln\left(\frac{R_{a\_axial}}{R_{i\_axial}}\right)\right)$$

If the respective inner inductances are normalized to the inner inductance of a cylindrical conductor, the following is obtained

$$\frac{L_{ir}}{L_{i0}} = \frac{\frac{\mu l}{6\pi} \cdot \ln\left(\frac{R_a}{R_i}\right)}{\frac{\mu l}{8\pi}} = \frac{4}{3} \cdot \ln\left(\frac{R_a}{R_i}\right)$$

In Fig. 10.4 axial and radial shunts are compared. Figure 10.4a shows the inner inductances of the two variants normalized to the inner inductance of a cylindrical conductor in the axial current direction. It can be seen on the one hand that the axial design provides lower internal inductances and on the other hand that thin-walled designs ($R_i \rightarrow R_a$) make the internal inductance very small. The magnitude of the resistance R_{sh} in both designs depends in different ways on geometrical factors. Despite the behavior shown in

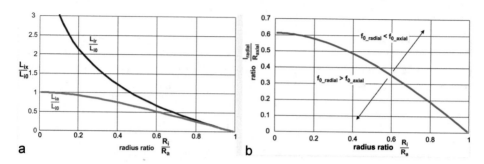

Fig. 10.4 Comparison of axial and radial design of measuring shunts. (**a**) Inner inductance of both designs normalized to the inner inductance of a straight cylindrical conductor of the same length, (**b**) Limit of the geometry factors for the same cut-off frequency

Fig. 10.4a, b shows a limit of the geometry factors for reaching the same cut-off frequency. This gives the user freedom of design for broadband measurement shunts.

To illustrate the behavior of measurement shunts at high frequencies, the change in impedance components at high frequencies is now investigated. The internal inductance of the cylindrical conductor becomes smaller due to the skin effect. Assuming a tube of wall thickness 's' containing the entire current, the corresponding estimate of the inner inductance for s ≥ 0.5·Ra follows the term

$$L_i = \frac{\mu l}{2\pi \cdot \left(R_a^2 - (R_a - s)^2\right)^2}$$
$$\cdot \left(\frac{R_a^4 - 4R_a^2(R_a - s)^2 + 3(R_a - s)^4}{4} + (R_a - s)^4 \ln\left(\frac{R_a}{R_a - s}\right)\right)$$

The layer thickness 's' of the flowing current is

$$s = \frac{1}{\sqrt{\pi \cdot f \cdot \sigma \cdot \mu}}$$

At the same time, the effective resistance increases due to the reduction in cross-section with

$$R_{sh} = R_{sh0} \cdot \frac{R_a^2}{\left(R_a^2 - (R_a - s)^2\right)}$$

From this information, the frequency-dependent impedance components of the shunt can be calculated. Figure 10.5 is inserted to illustrate these relationships. The cut-off frequency

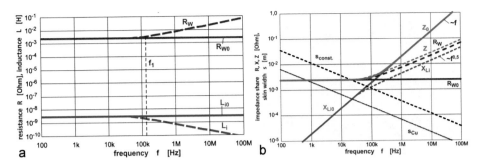

Fig. 10.5 Characteristic values of cylindrical shunts made of constantan. (**a**) Changes in impedance, (**b**) frequency dependence of resistance and internal inductance ($\sigma = 1.96 \cdot 10^6$ S·m (constantan), $R_a = 2$ mm, $l = 6$ cm)

does not change here due to the skin effect. The increase in impedance with frequency is lower. For comparison, s is entered for copper.

The calculations carried out so far assume a cylindrical conductor of length l and with the outer diameter R_a. The cut-off frequency is proportional~ $\frac{1}{R_a^2}$. The question arises whether it is not possible to achieve a smaller internal inductance with a flat arrangement. For this purpose, an estimation of the internal inductance is made from the assumption of rectangular field lines to show the influence of the ratio of length and width. The following expression deviates at a/b = 1 from the value for a cylindrical conductor, which is due to the deviations in the assumptions of the field condition (rectangular geometry) during integration. However, at b/a ≪ 1 or a/b ≫ 1, these approximate, so that the conclusions appear justified.

$$L_i\left(\frac{a}{b}\right) \approx \frac{\mu l}{8}\left(\frac{1}{\left(1+\frac{b}{a}\right)^2}\cdot\frac{b}{a} + \frac{1}{\left(1+\frac{a}{b}\right)^2}\cdot\frac{a}{b}\right)$$

Figure 10.6 shows this dependence in a standardized form.

For strong deviations of the aspect ratio b/a from one, the inductance L_i decreases proportionally to a/b or b/a. Flat designs of measuring resistors are therefore advantageous. Table 10.1 shows the resistivity of various common resistance alloys.

When measuring a current, it is important to meet several targets. On the one hand, the measured current should correspond to the actual current (via the illustration rule) with as little error as possible. On the other hand, the insertion of a current sensor should change the 'actual' arrangement to be investigated as little as possible. In reality, however, each insertion of a shunt into the current path A-B leads to a change, since the geometry of the current path is usually changed. Therefore, the equivalent circuit diagram shown in Fig. 10.7 is introduced for a shunt. The actual shunt resistance is formed from its resistance R_{sh} plus the internal inductance L_i including a possible external inductance L_{al} in series.

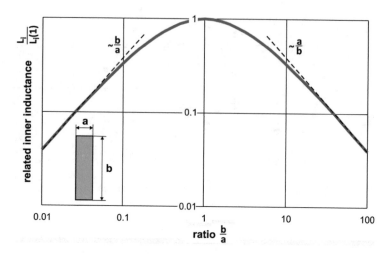

Fig. 10.6 Estimation of the influence of the ratio of width 'b' and height 'a' on the size of the internal inductance

Table 10.1 Specific resistances of resistance alloys in comparison with copper

Material	Specific resistance [Ohm-m]
Manganese	4.30E-07
Constantan	5.10E-07
Isaohm	1.32E-07
Zeranin	2.90E-07
Copper	1.73E-08

L_{a1} results, for example, if the measuring voltage $u_{mess}(t)$ is not tapped directly at the ends of the resistor body. If sufficiently unloaded, the external inductance L_{a2} is not important. The voltage induction shown in Fig. 10.1 is the result of the mutual induction via the mutual inductance 'M'. The increase in inductance in the measuring circuit symbolizes L_{a3}. Due to the self-induction voltage at this element, the voltage $u_{AB}(t)$ increases in relation to $u_{mess}(t)$. R_{a3} is a resistor that represents the attenuation due to induced eddy currents (skin effect, proximity effect).

The geometric size of the shunt resistor was based on the heat to be dissipated. For a pure pulse operation, the temperature increase during a pulse plays an important role. During an impulse of the duration t_1, the amount of heat R_{sh}

$$W_{th} = \int_0^{t_i} R_{sh} \cdot i^2(t) \cdot dt$$

recorded. This leads to a temperature increase of

Fig. 10.7 Electrical equivalent circuit diagram for a shunt resistor inserted in a measuring circuit for current measurement

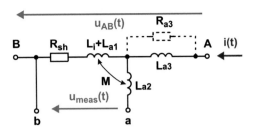

$$\Delta T = \frac{W_{th}}{m \cdot c_w} = \frac{W_{th}}{\rho \cdot c_w \cdot V}$$

with

ρ...   Density
σ...   Conductivity
c_w...   Specific heat
V...   Volume
l...   Length of the resistance
A...   Cross-section of the resistance

For a resistor in the form of a 'prism' with a cross-section A and length l, one obtains the dependence

$$\Delta T \leq \frac{\int\limits_{0}^{t_1} i^2(t)\,dt}{\rho \cdot \sigma \cdot c_w \cdot A^2}$$

The resistors for such an operation should therefore have the largest possible cross-section to ensure a low-temperature rise. For continuous operation it must be taken into account that the unavoidable heat loss is dissipated via the surface by convection and heat conduction and via the electrical connections by heat conduction (Fig. 10.8). The total power loss is the sum of the components.

$$P_V = I^2 \cdot R_{sh} = P_{V\_K1} + P_{V\_K1} + P_{V\_L} + P_{V\_W}$$

For each heat path, a separate thermal resistance can be defined as the ratio of the temperature increase on this path at a certain power loss with

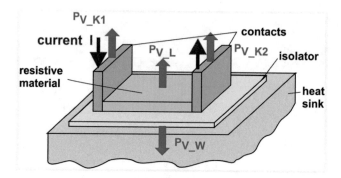

Fig. 10.8 Dissipation of the heat loss of a resistor via the surface by convection and heat conduction and via the electrical connections by heat conduction

$$R_{th\_x} = \frac{\Delta T_x}{P_{V\_x}}$$

The combination of the thermal resistances of the individual paths leads to a temperature increase that must be limited.

$$\Delta T = P_v \cdot \left(R_{th\_K1} // R_{th\_K2} // R_{th\_L} // R_{th\_W} \right) = \frac{P_v}{\frac{1}{R_{th\_K1}} + \frac{1}{R_{th\_K2}} + \frac{1}{R_{th\_L}} + \frac{1}{R_{th\_W}}}$$

If we now assume that each of these thermal resistances is defined by a heat path with a certain effective cross-section A_x and the corresponding properties for heat transport, we can apply the following formula for each thermal resistance

$$R_{thx} = \frac{1}{k_x \cdot A_x}$$

This gives us the connection

$$\Delta T = \frac{P_v}{k_{K1} \cdot A_{K1} + k_{K2} \cdot A_{K2} + k_L \cdot A_L + k_W \cdot A_W}.$$

To achieve small temperature changes, the cross-sections for heat transport must therefore be kept large. The thermal design is ancillary to the measurement. Smaller temperature changes only affect the measurement uncertainty. Larger temperature changes destroy the resistance R_{sh}. Shunts for continuous currents of up to 15kA, for example, are used in the operational measurement of electrical networks. The nominal voltages are standardized **60**, 75, 100, **150** and 300 mV (preferred sizes in bold). A continuous current of 1000 A causes

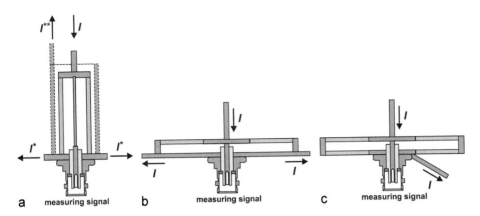

Fig. 10.9 Coaxial shunt resistors for current measurement with cylindrical basic structure in sectional view. (**a**) Axial current conduction, (**b**) radial current conduction, (**c**) radial current conduction in the resistor, but additional external inductance $_{La1}$ due to the current conduction

losses of 60 W at 60 mV, which must be permanently dissipated by suitable cooling measures.

Since the internal inductance is frequency-dependent due to the skin effect, frequency compensation of the measurement error can only be achieved within limits. For high current pulses, short shunts with large width and small thickness provide a path in this direction if the feedback effects on the measuring circuit are acceptable. Therefore there is no generally valid solution. In the following, some selected designs with different requirements for power connections and space are listed. Figure 10.9a shows a coaxial shunt. With this design, external inductances L_{a1} and L_{a2} are practically avoided. However, depending on the "original" current conduction, that is, the best possible low-inductive current conduction before inserting the shunt, L_{a3} can cause considerable interference. A lower overall height is made possible by the design shown in Fig. 10.9b. In addition to the inner inductance, a small proportion of an outer inductance L_{a1} is added here, which results from the insulation distance of the resistor from the GND electrode. Figure 10.9c shows only apparently a similar construction. Here, the current conduction encloses a ring-shaped magnetic field, which leads to a corresponding proportion of external inductance of type L_{a1}.

Figure 10.10 shows two further designs with a coaxial connection for the voltage signal, where the measurement signal is picked up via a coaxial connector. In case Fig. 10.10a it is a measuring resistor, which is arranged as a layer between the two conductors of a current path. The ground connection of the coaxial socket is connected to one conductor. The measuring lead is connected isolated by the resistor layer. If the measuring line is radially shielded (shown here only in sectional view), only the internal inductance is to be expected during measurement. The external inductance L_{a3} when inserted into the circuit depends on the design specifications of the circuit under test. Figure 10.10b also has a coaxial connection. It is interesting here that the shunt resistor can be pulled out of the circuit

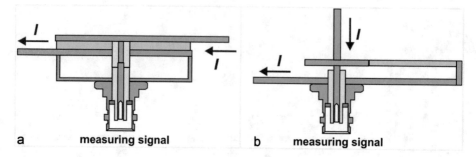

Fig. 10.10 Shunt resistors with coaxial measurement connection. (**a**) Flat construction for insertion between 2 busbars and avoidance of the inductance component L_{a1}, (**b**) Insertion of a flat, wide resistor in a circuit with earth fault

under investigation. This usually allows a very low inductance L_{a3} to be set. At the same time, the value of L_{a1} is increased by the current flow. This can be counteracted if the insulation distance between the resistor and the GND electrode is small and a maximum possible width of the resistor is selected. Laminated structures are suitable here, which can also absorb the magnetic force effects at high currents well. This minimizes the cross-section for the magnetic field in the resulting 'turn' while maximizing the length of the field lines.

Figure 10.11 shows two basic structures for resistors on an insulating plate as a carrier. Depending on the resistance value, the resistor materials to be used, the tolerances to be ensured and the power dissipation to be controlled, different designs are used. The intimate connection with the carrier substrate enables good heat conduction over the substrate if the substrate itself is a good heat conductor. Ceramics with good thermal conductivity (e.g. Al_2O_3, AlN) are therefore often used as carriers. Figure 10.11a shows a design in which the current is passed through the resistor body in a meandering pattern. The resulting magnetic flux is partially chained. Therefore the self-inductance of the shown structure is relatively high. If, however, the rear side is metallized or is mounted on a heat sink with good electrical conductivity, the magnetic flux generated causes eddy currents. The arrangement then acts as a loaded transformer. The inductance of the transformer then still effective for the resistor/shunt corresponds to the leakage inductance of the transformer. The arrangement in Fig. 10.11b has a much lower inductance than variant a. Here too, the inductance becomes even smaller if there is good conductive material on the back of the carrier. At first glance, the integration into PCB assemblies may seem to be a good idea. However, it requires resistance materials that can be integrated into the production process of printed circuit boards.

Figure 10.12 shows two versions for shunts in circuits with high continuous currents according to DIN 43703. The resistor bodies consist of alloys inserted into the connection lugs for the current to be measured. Two additional connections for the measuring leads of the voltmeter are provided to fix defined conditions for measurement. To achieve low inductance and high mechanical stability with simultaneous effective cooling, the resistor

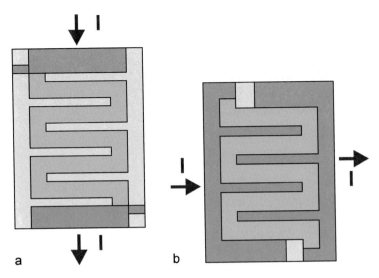

Fig. 10.11 Designs for shunts, which are applied to an insulating, usually ceramic carrier. (**a**) for higher resistance values, (**b**) for low resistance values

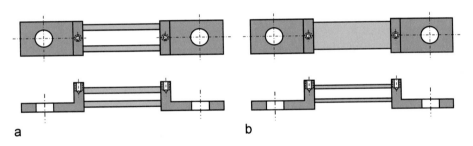

Fig. 10.12 Shunts in circuits with high continuous currents according to DIN 43703. (**a**) Resistor body in bar form, (**b**) Resistor body in band form

material is designed in rod or strip form. This increases the surface area which is important for cooling. At the same time, individual current paths are pulled apart. This has a similar effect to widening the conductor and leads to smaller inductors. When measuring large currents, one must remember that the nominal voltage drop should have a certain value at the nominal current. This means

$$U_{nom} + \Delta U = \sqrt{R_{sh}^2 + (\omega L)^2} \cdot I_{nom} = R_{sh} \cdot I_{nom} \cdot \sqrt{1 + \left(\frac{\omega L}{R_{sh}}\right)^2} = U_{nom} \cdot \sqrt{1 + \left(\frac{\omega L}{\frac{U_{nom}}{I_{nom}}}\right)^2}$$

$$1 + \frac{\Delta U}{U_{nom}} = \sqrt{1 + \left(\frac{\omega L}{\frac{U_{nom}}{I_{nom}}}\right)^2}$$

$$2\frac{\Delta U}{U_{nom}} \approx \left(\frac{\omega L}{\frac{U_{nom}}{I_{nom}}}\right)^2$$

From the maximum tolerance ΔU, the nominal current I_{nom} and the requirement for the maximum frequency f, the requirement for the inductance 'L' results

$$L \le \frac{1}{2\pi f} \cdot \frac{U_{nom}}{I_{nom}} \cdot \sqrt{2\frac{\Delta U}{U_{nom}}}$$

This leads to the demand for small inductors even at high currents and low frequencies.

Resistors are offered in a wide range of designs. In principle, the statements made apply here as well. Resistors are often wound on a carrier to achieve the resistance and power values. To reduce the increased inductive component, the carrier can be wound bifilar. The beginning and end of the resistor are thus next to each other, which must be taken into account in the insulation. The resistor thus has the inductive properties of a short-circuited double cable. The fact that induced voltages can be coupled with and against each other by counter-induction also makes it possible in principle to compensate the frequency response of shunts inductively, which will only be mentioned here.

10.2   Current Transformers for Current Measurement

Short-circuited transformers can be used as sensors for the potential-free measurement of currents. A core without an air gap is used for this purpose. The current to be measured is allowed to flow through winding N_1. In an ideal transformer, the current $I_2 = (N_1/N_2) \cdot I_1$ is induced there when N_2 is short-circuited. In the ideal case, therefore, one obtains an image of the current I_1 provided with a proportionality factor. For a description of the problem, Fig. 10.13 is used first. The magnetic field passes through a ferromagnetic core and a region of space outside the core. This applies to both winding N_1 and N_2. The result is the T equivalent circuit diagram of a transformer. Winding N_1 has resistance R_1 and winding N_2 has resistance R_2. If N_1 is small, the frequency response of the impedance is affected by the winding capacitance of N_2. A measuring device or a load resistor R_B is connected to the output (N_2) as a load to map the current.

By rearranging elements, the simplified T equivalent circuit diagram in Fig. 10.14 is obtained. The coupling capacitance between the primary and secondary windings is not

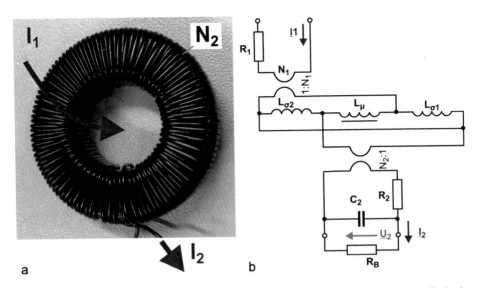

a   b

Fig. 10.13 Design of a current transformer. (**a**) Toroidal core with winding. (source: Zacharias 2017) (**b**) Derived electrical equivalent circuit diagram

Fig. 10.14 Equivalent circuit diagram of a simple current transformer with a toroidal core referred to as the secondary side ($N_1 = 1$)

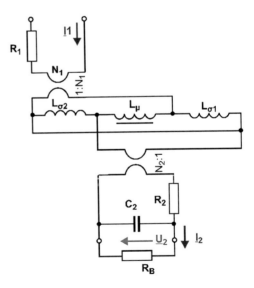

shown. The inductance $L_{\sigma 1}$ plays a subordinate role for the function of the converter, just like the resistor R_1.

Regarding the equivalent circuit diagram Fig. 10.2, the ratio of the secondary and primary current is

$$\frac{\underline{I_2}}{\underline{I_1}} = \frac{N_1}{N_2} \cdot \frac{1}{\frac{(L_\mu + L_{\sigma2}) + R_2 R_B C_2}{L_\mu} + \frac{1}{j\omega L_\mu}\left(R_2 + R_B\left(1 - \omega^2\left(L_\mu + L_{\sigma2}\right)C_2\right)\right)}$$

In the ideal case the load resistance $R_B = 0$, then the limiting case for the current transmission ratio is

$$\frac{\underline{I_2}}{\underline{I_1}} = \frac{N_1}{N_2} \cdot \frac{1}{\frac{(L_\mu + L_{\sigma2})}{L_\mu} + \frac{R_2}{j\omega L_\mu}} = \frac{N_1}{N_2} \cdot \frac{1}{\left(1 + \frac{L_{\sigma2}}{L_\mu}\right) + \frac{R_2}{j\omega L_\mu}}$$

In the lower frequency range, the current transfer factor approaches zero. The cut-off frequency, from which an application as a current transformer can take place, is apparently reached at $\omega(L_\mu + L_{\sigma2}) = R_B$ and is thus

$$f_{mess} \geq \frac{R_2}{2\pi\left(L_\mu + L_{\sigma2}\right)}$$

$$\underline{Z}_\mu = \frac{j\omega L_\mu}{1 + j\omega \frac{L_\mu}{R_p}}$$

One aim of the current measurement is to cover as wide a frequency range as possible. This means that R_2 should be as small as possible and L_μ as large as possible. Current transformers therefore usually also have a core with correspondingly high permeability. In the equivalent circuit diagram shown in Fig. 10.2, no upper cut-off frequency would occur at $R_B = 0$. If a ferromagnetic core is used, it has various losses, which can be represented by a resistor R_p in parallel with L_μ. The transformation ratio for the current is then

$$\frac{\underline{I_2}}{\underline{I_1}} = \frac{N_1}{N_2} \cdot \frac{1}{\left(1 + \frac{L_{\sigma2}}{L_\mu} + \frac{R_2}{R_p}\right) + j\omega\left(\frac{L_{\sigma2}}{R_p} - \frac{R_2}{\omega^2 L_\mu}\right)}$$

For very small values of ω applies here approximately:

$$\frac{\underline{I_2}}{\underline{I_1}} \sim \frac{N_1}{N_2} \cdot \frac{1}{\left(1 + \frac{L_{\sigma2}}{L_\mu} + \frac{R_2}{R_p}\right)} \cdot \frac{\omega L_\mu}{R_2} = \frac{N_1}{N_2} \cdot \frac{1}{\left(1 + \frac{L_{\sigma2}}{L_\mu} + \frac{R_2}{R_p}\right)} \cdot \omega\tau_1$$

For very large values of ω, the following approximation applies accordingly

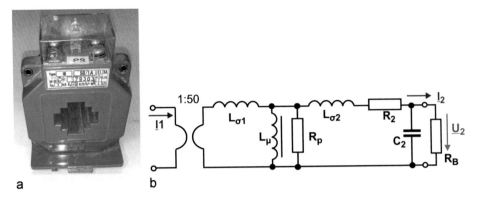

Fig. 10.15 Current transformer for mounting on busbars. (**a**) Current transformer 50/1A. (Source: Zacharias 2017), (**b**) corresponding electrical equivalent circuit diagram (**b**) ($L_{\sigma2} = 19.5 \ \mu H$, $R_2 = 0.11$ Ohm, $C_2 = 12.2$ pF)

$$\frac{I_2}{I_1} \sim \frac{N_1}{N_2} \cdot \frac{1}{\left(1 + \frac{L_{\sigma2}}{L_\mu} + \frac{R_2}{R_p}\right)} \cdot \frac{1}{\omega \frac{L_{\sigma2}}{R_p}} = \frac{N_1}{N_2} \cdot \frac{1}{\left(1 + \frac{L_{\sigma2}}{L_\mu} + \frac{R_2}{R_p}\right)} \cdot \frac{1}{\omega \tau_2}$$

With the knowledge of a few parameters, the frequency response of a current transformer in a short circuit can be determined. An example is shown in Fig. 10.15, where the electrical equivalent circuit diagram was determined for a current transformer via the impedance curve. This was used to estimate the current transfer factor in the case of a short circuit and with a changed load R_B. Due to the toroidal tape core used, the inductance is relatively highly frequency-dependent. The inductance decreases with the effective permeability as a result of the eddy currents in the core material. The eddy currents displace the magnetic field from the inner area of the strips so that the observed permeability decreases. At the same time, the loss resistance R_p, which represents the eddy current losses, increases due to the skin effect. These correlations were used for parameter estimation for the equivalent circuit diagram Fig. 10.15b. The characteristic curve $L(\omega)$ can be mapped $L(f) \approx 48$ mH $\cdot (1 + f/450$ Hz$)^{-0.657}$ very well with the correlation. The first resonance is a parallel resonance with $f_1 = 4.185$ MHz. From this, $L(f_1) = 115.5 \ \mu H$ is used to determine the "responsible" capacitance $C_2 = 12.2$ pF. The frequency change of L_μ and R_p (Fig. 10.16a) leads to changes in the current transfer ratio. This is shown in a standardised form in Fig. 10.16b. It becomes clear that in the upper-frequency range the bandwidth of the current transformer is mainly determined by the core material. In the lower range, the terminating resistance (burden) of the current transformer plays an important role. To achieve a certain accuracy class, a maximum resistance R_B is therefore specified for a current transformer. A good approximation for the description of the current transformation ratio is

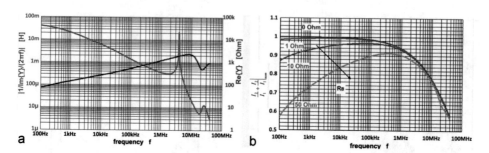

a   b

Fig. 10.16 Measured course of the imaginary part, interpreted as inductance, of the parallel connection of a reactance with a parallel resistance of the current transformer according to Fig. 10.15, (**b**) calculated change of the current transmission ratio with changed load RB

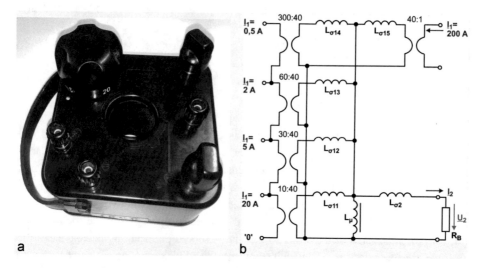

a   b

Fig. 10.17 Laboratory current transformer with switchable transmission ratio (0.5; 1; 2; 5; 20A: 5 A, $N_2 = 40$). (**a**) View. (Source: Zacharias 2017), (**b**) Basic circuit diagram of the inductive components

$$\frac{\underline{I_2}}{\underline{I_1}} \approx \frac{N_1}{N_2} \cdot \frac{1}{\left(1 + \frac{L_{\sigma 2}}{L_\mu}\right) + \frac{R_2 + R_2}{j\omega L_\mu}}$$

Figure 10.17 shows a switchable laboratory current transformer. There are various possibilities to choose from. The measuring ranges are set by a variable transmission ratio by switching over from N_1 with fixed N_2. For the measuring ranges 0.2...20A, the windings of N_1 are applied to the toroidal core. For higher currents, the conductor can also be inserted once or several times through the central hole in the sensor. The diagram of the inductive components can be seen in Fig. 10.17b. The frequency responses of the impedances of this current transformer are much more complex than those of the one with one winding Fig. 10.15. The reason for this is the unloaded windings, which produce additional parallel and series resonances when measuring, which changes the impedance. Switching is done on the primary side of the transducer. This means that the current transfer behavior practically does not change in the different measuring ranges at low frequencies.

Even with air coils, the value of R_p is finite, since eddy current losses are to be expected in the vicinity of the magnetic field. For the short-circuit case of the secondary winding, the effect of the winding capacity is largely eliminated. Therefore the highest upper cut-off frequency is reached here. The maximum use of the bandwidth is possible with a suitable trans-impedance amplifier. After the low-impedance decoupling of a sufficiently low current, an impedance adjustment to the transmission of the measurement signal can be made.

For meter movements an upper measuring frequency is not relevant.. When measuring a current $i_1(t)$, it is usual to map it to a voltage $u_2(t)$. The technically simplest possibility is to simply terminate the arrangement with a resistor R_B, at which the voltage drop is measured. If broadband measurements are to be made here, R_B must be selected equal to the characteristic impedance Z_w of the connected measuring cable. Coaxial cables for measurement purposes usually have a characteristic impedance of 50 Ohm. Broadband current transformers such as Pearson probes (Fig. 10.18) therefore have a 50 Ohm resistance already built-in.

The transformation ratio from a current I_1 to a voltage $\underline{U}_2$ is, with the specifications of Fig. 10.2 when terminated with a resistor RB

Fig. 10.18 Pearson current sensor 0.01 V/A, Z = 50 Ohm, I_{pmax} = 50 kA, Δf = 0.25 Hz...4 MHz. (Source: Zacharias 2017)

$$\frac{\underline{I_2}}{\underline{I_1}} = \frac{N_1}{N_2} \cdot \frac{1}{1+j\omega C_2 R_B} \cdot \frac{j\omega L_\mu}{j\omega\left(L_\mu + L_{\sigma2}\right) + R_2 + \frac{R_B}{1+j\omega C_2 R_B}}$$

This equation states for high values of ω a course of the current transmission ratio proportional to $1/(\omega C_2 R_B)$. This means that the upper limit frequency is further restricted. C_2 increases with the increasing number of turns N_2.

The previous observations were made for the small-signal range to show basic relationships. In general power supply technology, current transformers are often used for which certain operating conditions apply, such as maximum terminating resistance (load), insulation voltage and frequency range. Using the definitions in Fig. 10.14, for the section of a conductor with an inserted current transformer the impedance

$$\underline{Z}_1 = N_1^2 \cdot \frac{j\omega L_\mu R_p}{R_p + j\omega L_\mu} \cdot \frac{\frac{R_2}{N_2^2} + \frac{R_B}{N_2^2\left(1+j\omega C_p R_B\right)} + j\omega L_{\sigma2}}{\frac{j\omega L_\mu R_p}{R_p + j\omega L_\mu} + R_2 + \frac{R_B}{1+j\omega C_p R_B} + j\omega L_{\sigma2}} \approx N_1^2 \cdot \left(\frac{R_2 + R_B}{N_2^2} + j\omega L_{\sigma2}\right)$$

This impedance may cause high voltage drops which may disturb the device under test (DUT). To minimize the problem, R_B must therefore always be selected as small as possible.

To avoid saturation above a maximum permissible induction B_{max}, the following must apply

$$\frac{\sqrt{2}\cdot \underline{I_1} \cdot \underline{Z}_1}{N_1 \cdot \omega \cdot A_{Fe}} \approx \frac{\sqrt{2}\cdot \underline{I_1}}{\omega \cdot A_{Fe}} N_1 \cdot \left(\frac{R_2 + R_B}{N_2^2} + j\omega L_{\sigma2}\right) \leq B_{max}$$

This inequality provides the condition for a maximum load resistance (load) R_B to

$$R_B \leq N_2^2 \cdot \left(\frac{\omega \cdot B_{max} \cdot A_{Fe}}{\sqrt{2}\cdot \underline{I_1} \cdot N_1} - j\omega L_{\sigma2}\right) - R_2$$

This is not an operating condition, but a barrier for R_B! These correlations mean that a missing termination can be dangerous for the current transformer and the system under investigation. Very high voltages may then be induced at the output of the current transformer, combined with strong heating. The insertion impedance of a current transformer is lowest for the DUT if the output N_2 is short-circuited. An open output results in a high impedance, which may change the function of the DUT.

10.3 Rogowski Coils for Current Measurement

Rogowski coils, named after the electrical engineer Walter Johannes Rogowski (* 7 May 1881 in Obrighoven; † 10 March 1947 in Aachen) are widely used as broadband current sensors [1]. They have several great advantages:

- Galvanically isolated measurement.
- Can be inserted into a system at a later date.
- The circuit does not need to be disconnected for the measurement.
- Spools are usually very thin and flexible so that the space requirement is very low.
- Low impact on the measured object.
- Coil size and measuring range are largely independent of each other.
- Are practically not overloadable and therefore very robust in operational measurement technology.
- Relatively high bandwidth of the measuring frequency possible.
- High linearity in the mapping of the measurement signal.
- DC components do not affect the measurement results.

However, it can be disadvantageous that DC measurements are impossible and single pulse measurements are therefore problematic. As with current transformers based on the transformer principle, the upper cut-off frequency is lower than that of coaxial shunts.

They consist of a flexible coil with the number of turns 'N', which is placed around a current-carrying conductor (Fig. 10.19). This creates a closed circuit around the conductor. Assuming a constant coil cross-section of 'A' and a winding density dN/dl along the path 's', the partial voltage induced per unit length 'dl' is

$$\Delta u_{ind} = \frac{d}{dt} B \cdot A \cdot \frac{dN}{dl} \cdot \Delta l = \frac{d}{dt} \mu \cdot H \cdot A \cdot \frac{dN}{dl} \cdot \Delta l$$

From this, the dimensioning can be derived using the following equation

$$u_{ind}(t) = \frac{d}{dt} \oint_s B \cdot A \cdot \frac{dN}{dl} \cdot ds = \frac{d}{dt} \oint_s \mu \cdot H \cdot A \cdot \frac{dN}{dl} \cdot ds = \mu \cdot A \cdot \frac{dN}{dl} \cdot \frac{d}{dt} i(t) = \mu \cdot A \cdot \frac{N}{l}$$

$$\cdot \frac{d}{dt} i(t)$$

The following applies

$$\oint_s d\vec{s} = l$$

Fig. 10.19 Arrangement of a
Rogowski coil along a path
s around a current path

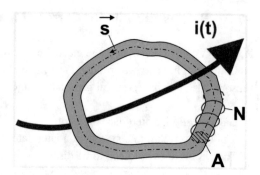

Both Ampère's circuital law and the law of induction are used to generate the induced voltage with the Rogowski coil. The measured voltage is thus not dependent on the current density distribution but only on the total current. In the simplest case, such a coil can be produced by winding a plastic tube and connecting the ends with a pin. No core is used in Rogowski coils. The non-linear and frequency-dependent properties of the core material therefore do not influence the measurement result.

The induced voltage depends on the derivative of the current with respect to time. This means that to measure the current as a value directly, the signal must still be integrated. In principle, this task can easily be solved by a downstream integrator. However, an integrator also integrates offset voltages (to theoretically infinitely large values), so that the measurement of constant quantities with this measuring method is impossible. The lower bandwidth of measurable frequencies is thus limited. The upper bandwidth is limited, among other things, by the winding capacitance. Each of the windings has a capacity to the adjacent winding and against an infinite counter electrode. If an equivalent circuit diagram is formed from this, a parasitic effective capacitance is obtained parallel to the coil output (Fig. 10.20)

$$C_p^* = C_p + \frac{C_{e1} \cdot C_{e2}}{C_{e1} + C_{e2}}$$

If necessary, the input capacitance of the connected measuring system must be added to this.

A further, easily determinable variable is the internal resistance R_i of the coil. Thus the output voltage $\underline{U}_{20}$ of the unloaded Rogowski coil is equal to

$$\underline{U}_{20} = \underline{U}_{ind} \cdot \frac{\frac{1}{j\omega C_p^*}}{R_i + \frac{1}{j\omega C_p^*} + j\omega L_{\sigma 2}} \approx \underline{I}_1 \cdot \frac{j\omega N_2^2 L_\mu}{1 - \omega^2 N_2^2 (L_\mu + L_{\sigma 2}) C_p^* + j\omega C_p^* R_i}$$

If this voltage is integrated via an integrator with a time constant τ_1 (Fig. 10.3), assuming that $I_2 = 0$ for the output voltage U_a of the integrator, one obtains

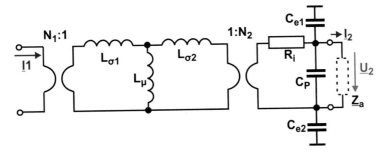

Fig. 10.20 Electrical equivalent circuit diagram of a Rogowski coil

$$\underline{U}_a = \frac{1}{j\omega\tau_1} \cdot \underline{U}_{20} = \frac{\underline{I}_1}{\tau_1} \cdot \frac{N_2^2 L_\mu}{1 - \omega^2 N_2^2 (L_\mu + L_{\sigma2}) C_p^* + j\omega C_p^* R_i}$$

This means that at a sufficient distance from the frequency

$$f_0 = \frac{1}{2\pi\sqrt{N_2^2 (L_\mu + L_{\sigma2}) C_p^*}}$$

the output voltage $\underline{U}_a$ follows proportionally the current $\underline{I}_1$ except for a phase error. This observation makes it possible to identify important factors influencing the transmission characteristics of the measuring system. The larger L_μ is and the smaller τ_1 is, the higher the output voltage U_a is. It can be used up to near the frequency f_0. C_p^* depends on the number of turns but also on other properties of the Rogowski coil. Therefore, considerable know-how is realized in the constructional design. A relatively high winding resistance R_i is advantageous, as the resonance superelevation of the output signal is reduced. In principle, this allows many turns of a thin wire to achieve a high inductance L_μ. The design of the Rogowski coil with a downstream integrating amplifier specially adapted to its construction characterizes the measuring system. By connecting a connecting cable between the Rogowski coil and the measuring amplifier, the assumption $I_2 = 0$ in the derivation shown above is already violated. The actual transmission network to be considered is more complicated. Systems with very high bandwidths are therefore very complex, although the basic structure (Fig. 10.21) still follows simple rules.

The characteristics of the amplifier are matched to the Rogowski coil itself as well as to the connected test lead. Its task is to compensate as far as possible for the parasitic properties of the frequency response. But this is only possible with limited success since the HF model of the Rogowski coil used here only represents part of the actual properties. Consequently, an amplifier with a high input impedance will be connected to the output of the Rogowski coil to put little strain on the voltage provided. After the measured value has been processed and the measurement signal has been integrated, an impedance converter is

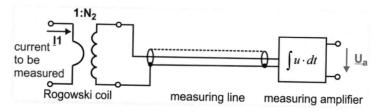

Fig. 10.21 Basic structure of a current measuring system with Rogowski coil

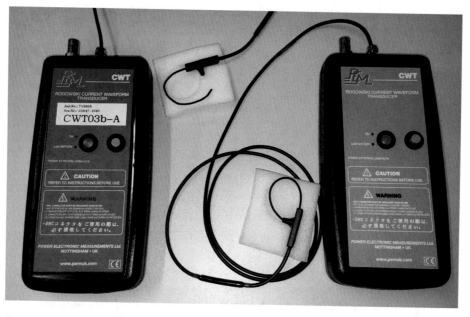

Fig. 10.22 Commercial broadband measuring systems with Rogowski coil with connected test leads and measuring amplifiers 100 mV/A (I_{peak} = 60 A) and 5 mV/A (I_{peak} = 1200 A). (Source: Zacharias 2017)

used to provide a signal for subsequent measurement stages. Figure 10.22 shows two commercial systems for current measurement according to the described principle for different measuring ranges.

The high input impedance of the subsequent amplifier has the disadvantage that the Rogowski coil as a sensor is very sensitive to rapid voltage changes in the environment, which are coupled into the surface of the coils via parasitic coupling capacitances.

Pearson probes, which operate as current transformers with one load, are surrounded for such purposes with a shield that does not form a secondary short circuit to the measuring winding and which shunts du/dt interference to GND. When such a measure is applied to a Rogowski coil, the value of C_p is increased dramatically, resulting in a significant further reduction of the upper cut-off frequency. Nevertheless, a similar measure would be very

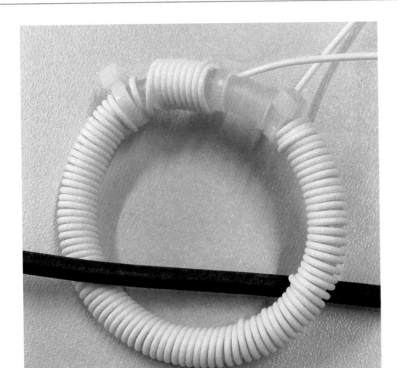

Fig. 10.23 Design of a Rogowski coil with a miniature coaxial cable. (source: Zacharias 2017)

helpful, since voltage gradients of >200 V/ns are measured in switched arrangements, for example, with SiC and GaN switches (see e.g. [2]).

Therefore the following arrangement was experimented with. The aim was to generate a symmetrical measuring signal with an inserted shield for the symmetrical derivation of capacitive interference signals against GND. For this purpose, a miniature coaxial cable with the following data was used to construct the Rogowski coil.

Outside diameter	$D_a = 1.17$ mm
Impedance	$Z = 50$ Ohm
Capacity coating	$C' = 85$ pF/m
Inductance coating	$L' = 212$nH/m
Resistance of the inner conductor	$R_i' = 2.05$ Ohm/m
Resistance of the outer conductor	$R_a' = 0.27$ Ohm/m

With this, a Rogowski coil was realized, whereby the inner conductor of the coaxial cable was used as a measuring coil and the shield conductor for the shielding. The dimensions of the arrangement shown in Fig. 10.5 are

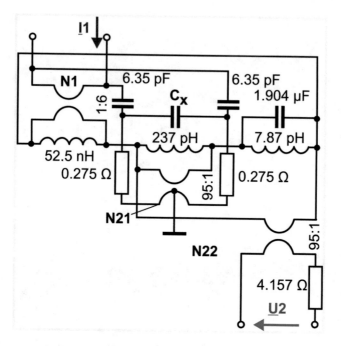

Fig. 10.24 Experimentally determined an electrical equivalent circuit diagram for a Rogowski coil based on a miniature coaxial cable (see Fig. 10.23)

Number of turns	95
Diameter of the winding support	5 mm
Inner diameter of the Rogowski coil	40 mm
Outer diameter of the Rogowski coil	54 mm

From the geometrical arrangement, the equivalent circuit diagram shown in Fig. 10.6 can be derived, which was parameterized by corresponding test measurements. The coaxial cable used was converted with its parameters L and C to the effect of one turn. These were represented as concentrated parameters. To compare the design approaches, alternatives were also measured:

- Rogowski coil with aluminium shield electrode
- Straight coaxial cable of equal length (2034 mm)

Interesting observations were made during the measurements to determine the model parameters. First of all, it becomes clear that an Al shield electrode significantly increases the parasitic winding capacity of the shield winding and thus the measuring winding (Fig. 10.25). The result is a reduction in the upper cut-off frequency of the coil inductance. If the coaxial cable had been 2034 mm long, a measured capacitance between the inner and outer conductor of 173 pF would have been expected. However, 211 pF (0.4...4 MHz) was

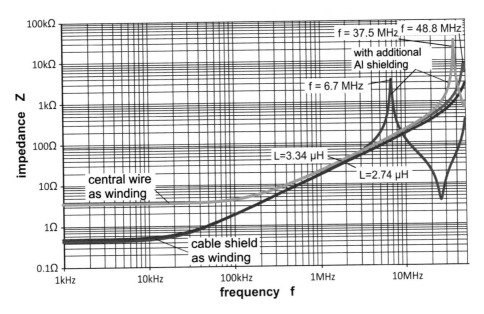

Fig. 10.25 Impedance curve of the Rogowski coil formed from the outer conductor of a coaxial cable with and without additional shield electrode (C_x in Fig. 10.35. 3: with shield electrode: $C_x = 5.84$ pF; without shield electrode: $C_x \sim 2.28$ pF)

measured in the installed state. The natural resonant frequency of the mini coaxial cable of this length is 23.8 MHz. The effect of an overall shield increases the effective capacitance more than the coaxial shield of a cable.

10.4   Design Aspects of Chokes

Chokes with Air Gap

The basic structure of an inductive component consists of an unbranched magnetic circuit with a winding for magnetization. This winding can be arranged around a core section without an air gap or around one with an air gap. Figure 10.26 shows the magnetic and electrical equivalent circuit diagram of such an arrangement, neglecting leakage flux. Figure 10.27 shows a corresponding arrangement for which the field distributions were determined.

Such an air gap can be achieved, for example, by shortening one leg of UU ferrite cores. With suitable layering, such an arrangement can also be built up from sheet metal (Fig. 10.27). This was used in [3] to investigate some fundamental questions.

If we first look at the case without air gap, a 3D calculation gives us the flux density distribution (Fig. 10.28). It becomes clear that an increase in field strength occurs at the

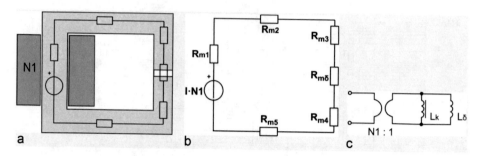

Fig. 10.26 Unbranched magnetic circuit with air gap. (**a**) Construction sketch, (**b**) magnetic equivalent circuit, (**c**) electrical equivalent circuit

Fig. 10.27 Construction of a simple inductor from a laminated UI core for computer experiments. (Source: Rhode [3])

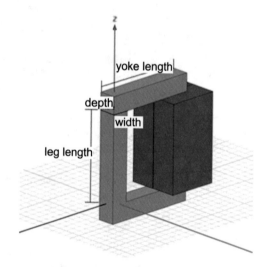

inner edges and a weakening at the outer edges. The flux practically remains in the core. This basic statement changes when an air gap is inserted. To a large extent, the flux remains in the core, but it also emerges massively in the vicinity of the air gap. For this purpose, the flux density distribution in the core (a) and on the core surface (b) was calculated. Due to the finite permeability of the core and the low permeability of the air, there is a stray magnetic field in the space, but its flux density is very low compared to the core.

At the moment when an air gap is inserted into the magnetic circuit, as in Fig. 10.4, the flux density distribution in the core changes significantly. The reason for this is the low permeability of the air, which quickly turns $R_{m\delta}$ into significant magnetic resistances, even when the δ is small. The "magnetic resistance of the outer space" is then connected in parallel to this, so that flux lines emerge on the surface of the ferromagnetic core. This emergence of field lines is particularly intense near the air gap (Fig. 10.29). Therefore, the greatest stray field strengths are to be expected near an air gap.

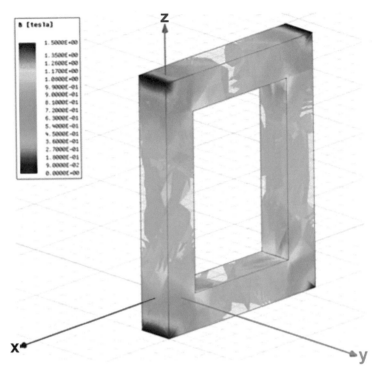

Fig. 10.28 Magnetic flux in the core without air gap when magnetized with one winding. (Source: Rhode [3])

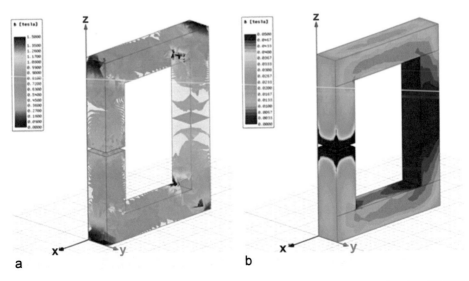

a   b

Fig. 10.29 Magnetic flux of a core with 3-mm air gap. (**a**) Inside, (**b**) on the surface in a 3D-FEM calculation. (Source: Rhode 2017 [3])

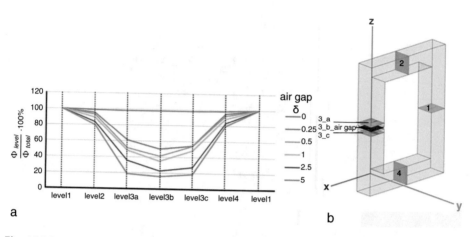

Fig. 10.30 Flux components in the core (not around the leg with air gap when winding). (**a**) The measuring positions, (**b**) for air gaps of different sizes. (Source: Rhode 2017 [3])

In order to reduce the effect of this stray field, e.g. on windings and structural parts, the air gap δ can be divided in such a way that it is distributed over the core with the same overall length. The influenced volume of the surrounding area will then be reduced.

The results of the calculations summarized in Fig. 10.30 also lead to this statement. As the air gap increases, the amount of flux entering the outer space increases rapidly. Even very small air gaps are sufficient for this. This is also interesting for transformer arrangements with large yoke scattering. A short-circuited second turn at the position of the present air gap would have a similar effect. However, the resulting counter-magnetization results in a largely symmetrical field image.

The calculation results are shown in a different way in Fig. 10.31. Shown is the flux fraction that really goes through the air gap, i.e. within the extended body edges of the core in relation to the total flux. This is done as a function of the air gap length. Due to the low permeability of the air, flux lines very quickly push around the air gap. As a result, the flux guided through the air gap decreases rapidly with increasing air gap length. At $\delta = 0.02 \cdot A_{Fe}^{0.5}$, the flux in the air gap is already only 40%. The numerical values here are not constant values because they depend on the core geometry. But they are an indication of orders of magnitude. Thus, if the air gap is divided into many small sections, the formation of stray fields is counteracted. This is used, for example, for metal powder cores.

Such a breakout of the magnetic field lines from the iron core can be largely prevented by arranging the winding around the air gap. This is shown in Fig. 10.32. While a strong stray field reaching into the space is created when the air gap is arranged outside the winding (a), this is not the case when the air gap is arranged in the winding (b). This is not shown in an unbranched magnetic equivalent circuit diagram. However, the FEM calculation shown is also only a relatively rough approximation, since only one current density is

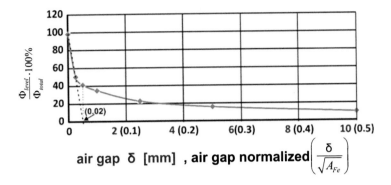

Fig. 10.31 Magnetic flux in the air gap as a function of the air gap length

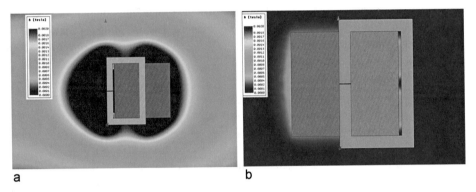

a   b

Fig. 10.32 Magnetic flux distribution in the vicinity of the core, when the winding is *not* located above the air gap (**a**) and when the winding is located *above* the core in the same core (**b**). (Source: Rhode 2017 [3])

assigned to the cross-sectional area of the winding and a shielding effect due to proximity effects is not shown. This could be considered by a complex permeability, which also maps the conductivity of the winding (depending on the material and geometry of the individual conductors). However, this has not been realized for the creation of the images. It was only about the field propagation in the external space.

Conclusions

- Air gaps and sharp buckling in the contour lead to strong "breaking out" of the magnetic field lines from the ferromagnetic core and cause a strong stray field even at greater distances.

- If the winding is arranged around the air gap, significantly lower remote effects can be observed. However, a stronger magnetic field in the winding is to be expected. This results in corresponding losses due to the proximity effect.
- The stray field can be reduced if the contours of the core follow the "natural" course of the characteristic curves. As a result, toroidal cores, for example, have significantly lower stray fields.
- A division of a total air gap into several shorter lengths leads to an increase of the flux in the air gap relative to the total flux. This is also a measure to reduce the external stray field.

Possibilities of Low-Capacity Windings

Low-capacitance windings are important for the applications of inductive components in the higher frequency range. Due to the operating capacitance, a parallel resonance initially occurs when the frequency is increased. Above this resonance, the component behaves capacitively and loses its effect as an inductive component. The operating capacitance of a winding can be caused by this resonance:

- electric field between the windings
- electric field between the layers of a winding
- electric field between winding and earth potential
- electric field between windings

This determination of causes also provides the approaches for influencing the respective proportions. It is interesting to see how these capacitance components add up to finally form an electrical equivalent circuit diagram suitable for simulations. First, the effect of the capacitance between the windings is considered (Fig. 10.8). The capacitance of a wire, a wire winding, exists in principle even without another wire. A wire loop also has a capacity against an infinitely distant counter electrode. In the example of a single-layer cylindrical coil on which Fig. 10.8 is based, the operating capacitance of a winding is therefore derived from the capacitances between the turns and those of the turns against a far-distant counter-electrode (Fig. 10.33).

The coupling of the windings results in

$$C^* = \frac{C_W}{N} + \frac{2N+1}{N} C_e \quad L^* = N^2 \cdot L_m$$

$$f_{res} = \frac{1}{2\pi \cdot \sqrt{N^2 \cdot L_m \cdot \left(\frac{C_W}{N} + \frac{2N+1}{N} C_e\right)}} = \frac{1}{2\pi \cdot \sqrt{N \cdot L_m \cdot (C_W + (2 \cdot N + 1)C_e)}}$$

This shows that the earth capacitance can have a greater influence on the resonance capacity than the capacity between the windings. How this influence turns out obviously

Fig. 10.33 Model of an inductance with winding capacitance. (**a**) Schematic composition of the measured apparent capacitance C* of a transformer from the capacitances of individual windings C_w of a single-layer winding and the capacitance to earth C_e, (**b**) simple model for $f < f_{res}$

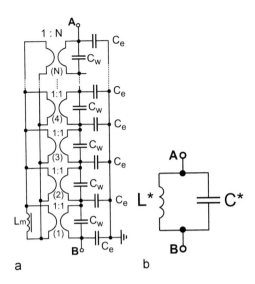

depends on the design of the winding and the ratio of the sizes C_e and C_w. The formulas above are not intended to calculate the winding capacity exactly, but only to investigate the influencing factors. It is not possible to give a generally valid formula for this because of the complexity. A transferability of the statements is given also for multi-layer windings. Figure 10.34 shows the effective "individual capacitors" for a coil arrangement with several layers and several slices of a profiled conductor connected in series.

The following qualitative statements can be derived

- The smaller the surface areas facing each other in individual turns and the greater the distance between the turns, the lower the C_w
- The higher and the narrower a multi-layer winding is, the lower the C_w
- The filling factor of a winding tends to decrease when the intrinsic capacitance decreases. This runs counter to the fact that
- C_e is the lower, the more compact an air core coil is
- A metallic and grounded core increases C_e

In addition, ferrite cores have an intrinsic capacitance that depends on the magnetization cross section. In order to find an RF-capable coil design, various approaches are pursued, which are briefly described below (Figs. 10.35 and 10.36).

In case of resonance, the imaginary part of the impedance disappears. The maximum achieved corresponds to the resistance of an LCR parallel resonant circuit. A pure inductance L* in parallel to a capacitance C* with series resistance R_s with the resonant frequency f_{res} provides the effective resonant capacitance C* to

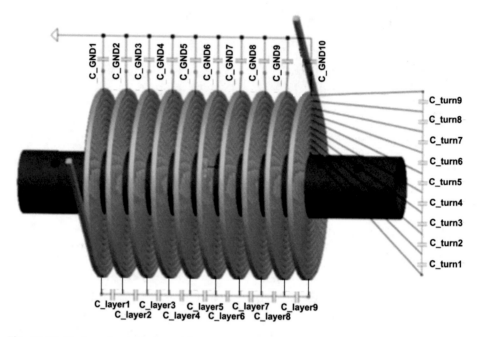

Fig. 10.34 Design of a choke from 400 μH, consisting of winding discs with identification of the capacitive effects. (Source: Zhang [4])

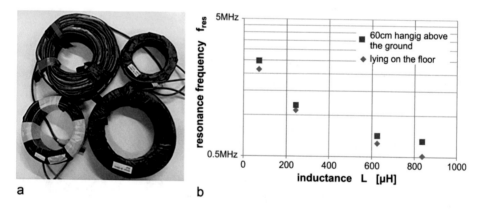

a   b

Fig. 10.35 Different coreless inductors for the double pulse test of semiconductor switches. (**a**) View. (Source: Zacharias 2017), (**b**) natural resonant frequency (parallel resonance) at different useful positions (**b**)

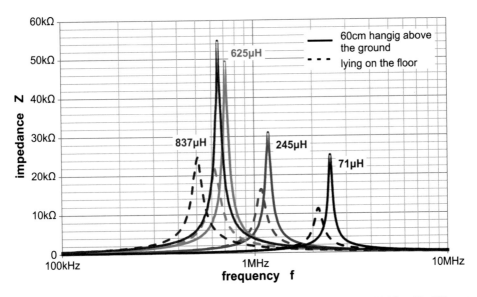

Fig. 10.36 Change in the resonance properties of the coils according to Fig. 10.35a with different mounting positions

$$C^* \approx \frac{1}{4\pi^2 f_{res}^2 L^*}$$

The quality of the oscillating circuit is then

$$Q = \frac{R_p}{2\pi f_{res} L^*}$$

Converted into the corresponding series resistance for the same quality, you get

$$R_s = \frac{2\pi f_{res} L^*}{Q} = \frac{(2\pi f_{res} L^*)^2}{R_p} = R_{s.Cu} + R_{s.Diel}$$

In this measurement or calculation, the influence of the line resistance is far exceeded by the dielectric losses of the insulation material. A "classic" winding form is the multi-layer winding, as shown in Fig. 10.37. Here the winding width is wound in several layers one above the other. In the "normal" layer winding shown, the start of the new layer is placed at the end of the previous one. The field strength in the layer insulation increases linearly over the winding width. If you always start on the same side of the winding, it is called a rectified winding. This has a constant field strength in the winding insulation. The winding capacity to be assigned here is somewhat lower than with normal layer winding. The return of the

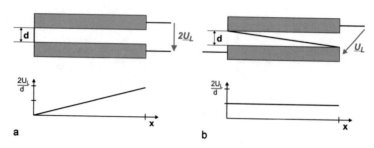

Fig. 10.37 Wiring of winding layers. (**a**) Normal layer winding, (**b**) rectified layer winding **b** with corresponding field strength distribution over the layer width

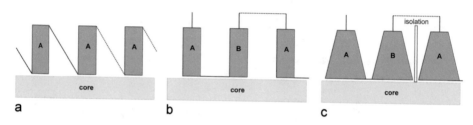

Fig. 10.38 Winding diagrams for disc spools. (**a**) Disc spools with the same winding direction (A), (**b**) disc spools with alternating winding direction in pairs (A-B), (**c**) trapezoidal spools with alternating winding direction in pairs

wire to the start side of the winding, which leads to local increases in field strength, results in a reduction of the filling factor. It is also possible to return the wire to the starting position outside the winding. This significantly increased effort leads to an increase of the filling factor, because the wire can be positioned between 2 wires underneath (orthocyclic winding). In this way, the highest filling factors are achieved (theoretical limit for round wire: 0.907). At the same time, however, the winding capacity tends to increase as the windings get closer together. At higher application frequencies, these opposing effects must be taken into account and whether the increased expenditure is justified. This can only be clarified by test windings in a specific case.

The fact that the resulting winding capacities become smaller the higher the ratio of winding height to winding width (series connection of layer capacities) leads to disc windings as a further characteristic (Fig. 10.38). The variants Fig. 10.38a and b differ in the application mainly in their connection technology. With variant 'A', the individual winding discs are produced by starting with a winding with the same winding direction 'A' again and again. The beginning of the winding is therefore always at the bottom of the winding. A connection is made, as shown, by the upper winding end of the preceding disc with the winding start of the following disc. This procedure can lead to insulation problems at higher voltages.

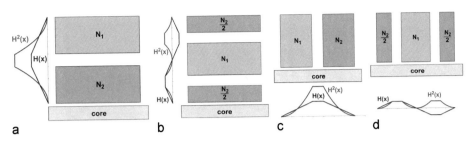

Fig. 10.39 Symmetrical and asymmetrical layer winding (**a, b**) and symmetrical and asymmetrical disk winding (**c, d**)with qualitative representation of the course of the stray field strength H(x). The energy density of the stray field is ~H₂(x)

An alternative is to use an alternating winding direction (right–left–right–..., A–B–A–...), as shown in Fig. 10.38b and c. Variant c) additionally shows the use of a trapezoidal winding design, which is only possible due to the alternating winding sense and allows better mechanical stability and better utilization of the winding space. Two discs A-B can be wound in a winding chamber with insulating walls.

With the different winding types, the leakage inductances between the windings of coupled coils are also influenced. Figure 10.39 shows, in addition to these considerations, layer windings and disc windings in symmetrical and asymmetrical designs, although this subchapter is only dedicated to simple chokes. Those can be achieved by cores without air gaps by counter-orientation of winding parts.

By segmenting and nesting the primary and secondary windings into each other, a reduction of the leakage inductance between the primary and secondary side is achieved. The reason is the reduction of stray field energy due to this measure. Each nesting leads to an interruption of the concatenation of the stray flux. The leakage inductances of individual part-winding pairs simply add up as a result:

$$L_s = \sum L_{si} = \sum \frac{N_i^2}{R_{m.si}}$$

(N_i: number of partial turns, R_{m.si}: proportionate scattering reluctance)

By such measures, symmetrical winding structures (see Fig. 10.39), for example, achieve significantly lower leakage inductances compared to asymmetrical structures, whereby, depending on the layer insulation, the leakage inductances are a factor of 2 or more. Figure 10.40 shows the stringing together of several disc windings of small width and large height to achieve a low winding capacity. In transformer designs, these can also be connected alternately or in groups to achieve specific leakage inductances.

A minimum value of flux linkage is achieved when the corresponding coupled winding is placed next to each winding (bifilar winding, Fig. 10.41). The resulting stray field is then the magnetic field between the windings, comparable to the problem of the magnetic field

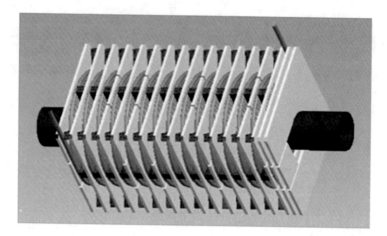

Fig. 10.40 Design of a coreless inductor (400 μH) with series-connected disc coils and insulating bars between the windings. (Source: Zhang 2011 [3])

Fig. 10.41 Bifilar winding design as a way to achieve minimum scattering between primary and secondary winding

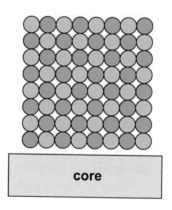

of a double line. This minimizes the linkage of the windings with the leakage flux. This also leads to minimal control inductances.

A close magnetic coupling always means that there is a very small distance between the conductors of different windings. This means a correspondingly increased coupling capacity.

The capacitance between two electrodes becomes smaller the greater the distance and the smaller their surface area. This basic connection results in a further design type for low-capacitance coils. If windings only come close to each other at crossing points, their mutual capacitance also becomes very small. In [4], such approaches were tested to realize coreless coils of high quality for metrological filter applications. The results are so-called honeycomb coils (Fig. 10.42). In the version shown, a choke for a network simulation for EMC measurements was used. A design with high continuous current carrying capacity and high resonant frequency was sought.

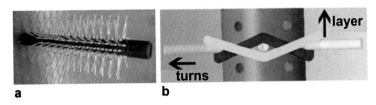

Fig. 10.42 Construction of a coil with disc windings in honeycomb/basket winding technique. (**a**) Carrier construction, (**b**) winding scheme to minimize the internal capacities of each disc. (Source: Zhang 2011 [4])

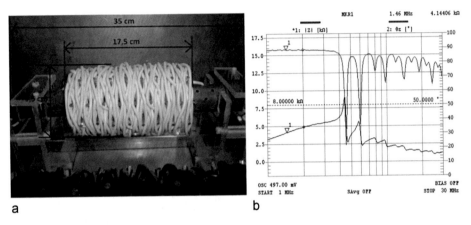

Fig. 10.43 Implementation of an inductance for 20 A/400 μH. (**a**) Design of an inductor of 16 disk coils with 10 turns each, which reaches a total inductance of 398 μH due to the magnetic coupling, (**b**) with amplitude and phase response of the impedance $\underline{Z}$. (Source: Zhang 2011 [4])

The shown design in honeycomb winding technology has a low copper filling factor of the winding space, but can be cooled well. Figure 10.42 shows the design of the coil carrier with removable mandrels as winding aid. Figure 10.43 shows the realized inductance with the frequency response for impedance and phase.

The realized coil has inductive properties up to f > 4 MHz. The first parallel resonance with capacitive components of the component is at 4.2 MHz. Several resonance points are connected to it. The resonance points cause strong deviations in the impedance. These should be avoided. For this reason, damping resistors have been inserted in each winding disc with the aim of achieving the most uniform impedance possible over the widest possible frequency range. Such a condition is shown in Fig. 10.44. It is clear that the impedance curve can be smoothed, but the result is a much lower Q-factor (Q = ωL/R).

Figure 10.45 shows the results of a heating test with half and full nominal current. A thermal camera was used for this. Due to the many gaps in the disc and honeycomb winding, the temperatures can be controlled with air self-cooling even at 10 A (a) and 20 A (b). In a grid simulation, self-heating is mainly achieved by mains frequency current, i.e. at

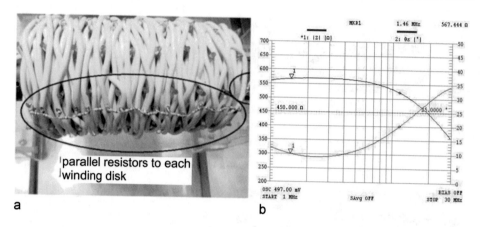

a   b

Fig. 10.44 Avoiding resonance peaks of the impedance $\underline{Z}$. (**a**) Damping of the natural oscillations of each disk, (**b**) frequency and phase response of $\underline{Z}$ no longer show resonance peaks - at the price of a lower impedance. (Source: Zhang 2011 [4])

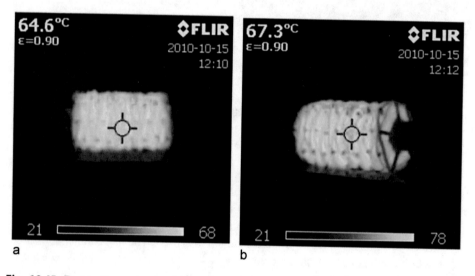

a   b

Fig. 10.45 Due to the many gaps in the disc and honeycomb winding, the temperatures can be controlled with air cooling even at 10 A (**a**) and 20 A (**b**). (Source: Zhang [4])

50/60 Hz. Therefore, stranded wire is already suitable as conductor material. To "smooth" the impedance curve, parallel resistors are used anyway to increase attenuation. The effect is comparable to that of the skin effect with thick cables.

A different situation occurs if chokes are to be used in the high-frequency range in switch-mode power supplies for smoothing or if an HF transformer with low intrinsic losses is to be implemented. The magnetic field of the high-frequency fundamental

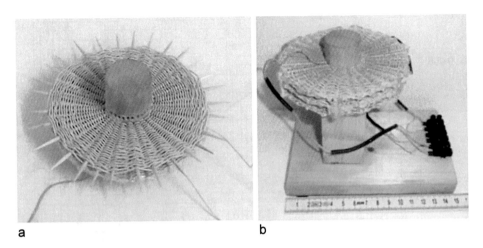

Fig. 10.46 Honeycomb wound coils ("basket coils"). (**a**) Single coil on wooden core or wooden carrier: single disc, (**b**) 3 disc windings in honeycomb winding technology with the possibility of series connection and magnetic coupling. (Source: Wendt [5])

oscillation is then also the main cause of self-heating due to the proximity effect or the skin effect. In such cases, the use of high-frequency litz wire is useful. However, the problem of the winding capacity and the associated natural resonant frequency cannot be solved by using RF litz wire. Above the natural resonance, the coil behaves more like a capacitance than an inductor. To keep this internal capacitance small, the basket winding technique already shown in Fig. 10.43 can also be used. In Fig. 10.46, this is used as the basis for disc windings. Here too, the aim was to achieve the highest possible natural resonant frequencies at high inductance values.

It becomes clear that the realization of large inductance values at high natural resonant frequencies in a component is difficult. It is possible to use a series connection of several inductors with lower values but high natural resonant frequencies or to use a ferromagnetic core or another winding construction, which allows a more compact design. A ferromagnetic core means that the amplitude-dependent losses of the component become dependent on the temperature-dependent core losses. This relationship cannot be separated from the amplitude influences. Rod core chokes use this design approach. The core material must be suitable for the frequency range. Figure 10.47 shows the different usable frequency ranges of the basket coils in Fig. 10.46.

In honeycomb winding technology, relatively high inductance values can be achieved with low values of winding capacity. Up to the mH range, natural resonances at >1 MHz can easily be achieved. The space requirement may be unfavourable. A constructive solution can be the use of the cross-wound winding technique (Fig. 10.48). Even with this technique, which requires appropriate devices for its use, a low self-capacitance of the windings is achieved by the fact that the windings only rest on each other at short sections. The arrangement of the windings makes it possible to realize narrow, high windings, which

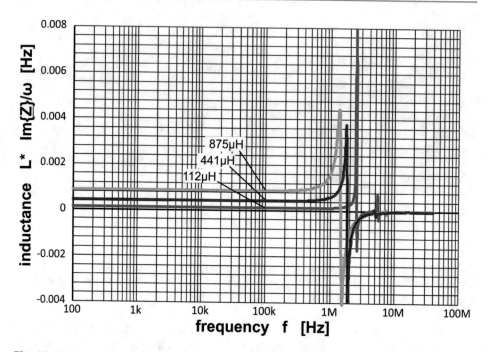

Fig. 10.47 Measured inductances of the 3 windings of the arrangement in Fig. 10.21b (lowest coil only (blue), lowest + middle coil in series (brown), all three coils in series (green))

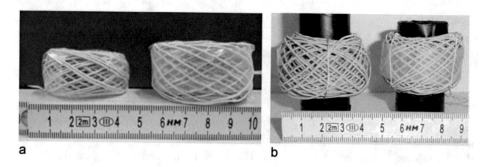

a   b

Fig. 10.48 Cross-wound coils with different inductance values. (**a**) Self-supporting, (**b**) on carrier tube. (Source: Wendt [5])

are self-supporting. Fixation of the structure is possible simply by impregnation with lacquer.

Due to the relatively good chaining of the fluxes, very compact coreless inductors can be achieved, which are interesting for many metrological applications. Figure 10.49 shows measurement results for the impedance of coils based on this design principle. Inductances

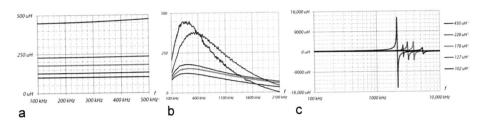

Fig. 10.49 Frequency dependence of the inductance of the coreless coils under investigation. (**a**) In the frequency range 100–500 kHz, (**b**) quality 'Q' **b**, (**c**) natural resonance of the air coils of the air coils under investigation (L is the impedance of the impedance analyzer interpreted by the impedance analyzer as inductance). (Source: Wendt [5])

Table 10.2 Thermal properties of some relevant materials

	Density	Thermal conductivity	Specific heat	Thermal conductivity
Unit of measurement	g/cm^3	W/m/K	Ws/g/K	10^{-6} m^2/s
Ferrite	4...5	3...8	0.5...0.9	~0.1...0.3
Aluminium	2.7	237	0.888	98.8
Iron	7.86	81	0.452	22.8

of approximately 0.5 mH have been realized, which can be used as inductive memories up to approximately 1.5 MHz.

Coreless Inductors

At higher frequencies, one encounters the problem that the possible modulation becomes lower and lower if one does not want to exceed a certain power loss density in the core material. If necessary, one has to provide core geometries that can be cooled well. Above a certain frequency, a core tends to become problematic, since it limits the maximum flux density not through saturation, but primarily through the dissipatable power loss.

The thermal properties of some relevant materials make this clear (Table 10.2):

This raises the question of whether, at a certain frequency, it is better to be served without a core than with a core, because

- the winding can be cooled better and
- as a result, a higher current density can be used at the same maximum temperature of the winding material and the component tends to become more compact
- passive liquid cooling (e.g. oil) would further improve the compactness

A disadvantage would be the stronger stray field in the vicinity of the coreless component, which could require shielding measures. This will be considered below. Basic types of magnetic components are choke and transformer. A double coil ($N_1 = N_2$) or a 1:1 transformer (Fig. 10.50) is used as an example. In order to achieve a close coupling, the

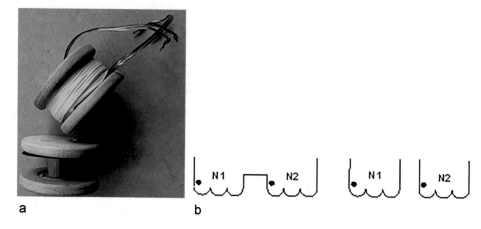

Fig. 10.50 Coupling with bifilar construction. (**a**) 2 bifilar wound coils mounted on a wooden coil carrier. (Source: Zacharias 2017), (**b**) Connection in series or separately, ($N_1 = N_2 = 65$ turns)

winding was designed with a double line bifilar. Wood was used as the coil carrier. The magnetic field between the partial coils is therefore formed in the same way as with a double line.

Figure 10.51 shows the measured frequency response of the inductance for series connection of the windings and a single winding. Due to the selected design of the bifilar winding of two partial windings, there is a close capacitive coupling of the partial windings when the windings are connected in series. This leads to a high effective capacitance C_{eff} when the windings are connected in series. If only one winding is measured, the magnetic coupling to the other winding cannot be switched off. The electric equivalent circuit thus maps also a non-loaded winding. The behaviour therefore reflects that of two magnetically coupled parallel resonant circuits, each with the same winding inductances and winding capacitances.

The frequency responses of the (apparent) inductors $L = \text{Im}\{Z\}/\omega$ of the series connection and the coupling of the bifilar windings result in the following characteristic parameters for the design.

Number of windings	$N_1 = N_2 = 65$
Inductance of the series connection	$L_1 + L_2 = 0.486$ mH @ 10 kHz
Resonant frequency with series connection	$f_{res} = 468$ kHz
Effective resonant frequency	$C_{eff.par} = 238$ pF
Measured inductance of the single coil	124 µH @ 100 kHz (idle of the second winding $f_{res} = 2.07$ MHz, $C_{eff} = 47.8$ pF)
	$L_\sigma = 8.64$ µH @ 100 kHz (short circuit of the second winding)

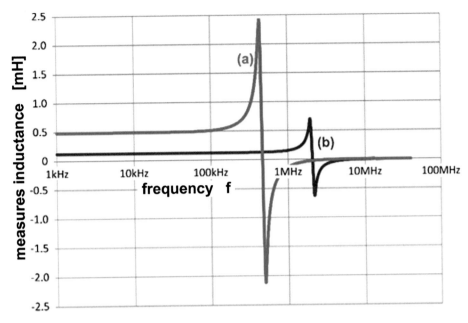

Fig. 10.51 Frequency responses of the (apparent) inductances $L = \mathrm{Im}\{Z\}/\omega$ of the series connection and coupling of the bifilar windings $N_1 = N_2 = 65$, $L_1 + L_2 = 0.486$ mH @ 10 kHz, $f_{res} = 468$ kHz, $C_{eff} = 238$ pF (**a**) and measured inductance of the single coil 124 µH @ 1 00 kHz at no-load of the second coil. Winding $f_r = 2.07$ MHz, $C_{eff} = 47.8$ pF (**b**) leakage inductance $L_\sigma = 8.64$ µH @ 100 kHz

Energy is transferred in the electromagnetic field between the windings. The leakage inductance describes the energy content of the magnetic field between the windings during energy transfer. The coupling capacitance between the windings results from the capacitance coating between the lines for the selected design. From this it follows that these quantities can also be determined from the properties of the wound double cable. For the setup shown in Fig. 10.25, the amplitude response and phase response of the admittance of the coiled double line were measured. The results of the measurements are shown in Fig. 10.52 for the line in open circuit and in short circuit. Characteristic resonance frequencies are determined for both measurements. The first resonance frequency at no load corresponds to a parallel resonance. Below this frequency, the line behaves like a capacitance of 90.2 pF. The resonant frequency corresponds to the reciprocal propagation time of the signal from the input to the output at one quarter of the wavelength of the signal on the line. At higher frequencies, additional resonant frequencies are added, corresponding to the reflection of the voltage wave at the open end of the line. From the increase in admittance, the coupling capacitance C_k between the lines can be easily determined. If the line is short-circuited, the diagram in Fig. 10.52b is obtained. Here the line behaves like an inductance of 8.4 µH for frequencies $f < 4 \cdot c/\lambda$. The values of capacitance and inductance together result in the resonant frequency, which appears here as the series resonance of C_k and L_σ. The higher resonant frequencies result from the

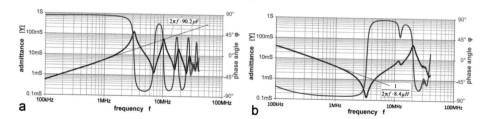

Fig. 10.52 Amplitudes and phase response of the admittance of the double line used. (**a**) At no load, (**b**) short circuit

Fig. 10.53 Electrical equivalent circuit diagram for the coreless transformer according to Fig. 10.24 with parameters estimated from the frequency responses without capacitance to earth

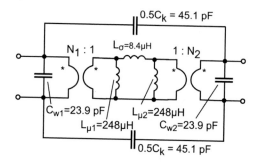

superposition of the waves between the lines (*differential mode*) and those between the line and the ground potential (*common mode*).

In the electrical equivalent circuit diagram Fig. 10.53 for the 1:1 transformer, it is noticeable that the coupling capacitance C_k between the windings is greater than the effective winding capacitance $C_{eff} = C_{w1} + C_{w2}$. It should be noted that the electrical equivalent circuit diagram is only an approximation and can only be used meaningfully below the first resonant frequency. Above the first resonance frequency, it is noticeable that the wave properties of energy propagation become effective in this frequency range.

The minimum leakage inductance is achieved with bifilar winding and then corresponds to the inductance coating of the double line. Table 10.3 summarizes the values of 3 frequent line arrangements.

A high energy density is achieved with short multilayer cylindrical coils (solenoid coils), where winding height and winding length are approximately equal and the average diameter of the coil is about twice the winding height. There is no real minimum size here. A problem in the application is the unbundled magnetic field that extends far into space. It penetrates surrounding construction parts and sometimes causes unwanted interactions. The winding is also permeated by the magnetic field with high field strengths, as shown in Fig. 10.54.

It can be seen that in all 3 cases examined, on the one hand the magnetic field strength reaches far into the room and on the other hand the highest field strengths are reached in the area of the inner edge. If there are conductor materials with corresponding thicknesses

Table 10.3 Elementary line arrangements inductance and capacitance coating

	Inductance coating L' [H/m]	Capacity coating C' [F/m]	Basic structure
Coaxial line	$\dfrac{\mu_r \mu_0}{2\pi} \cdot \ln\left(\dfrac{D_a}{D_i}\right)$ $= 0{,}2\dfrac{\mu H}{m} \cdot \ln\left(\dfrac{D_a}{D_i}\right)$	$\dfrac{2\pi \cdot \varepsilon_r \varepsilon_0}{\ln\left(\dfrac{D_a}{D_i}\right)}$	
Dual line	$\dfrac{\mu_r \mu_0}{\pi}\left(\dfrac{1}{2} + \ln\left(\dfrac{2D_2 - D_1}{D_1}\right)\right)$ $\approx 0{,}4\dfrac{\mu H}{m}\left(\dfrac{1}{2} + \ln\left(\dfrac{2D_2 - D_1}{D_1}\right)\right)$	$\dfrac{\pi \cdot \varepsilon_r \varepsilon_0}{ar\cosh\left(\dfrac{D_2}{D_1}\right)}$	
Double stripe line	$\approx \mu_r \mu_0 \left(\dfrac{a}{b}\right) h$ $\approx 1{.}2567\dfrac{\mu H}{m}\left(\dfrac{a}{b}\right)$	$\approx \varepsilon_r \varepsilon_0 \cdot \dfrac{(b+c)}{h}$	

D_a diameter of the inner conductor, D_i diameter of the outer conductor

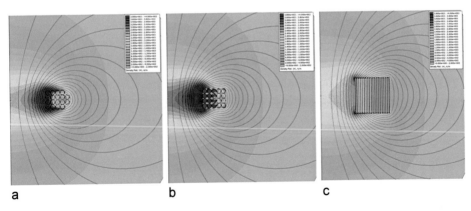

a   b   c

Fig. 10.54 Calculation results for the magnetic field strength of a solenoid coil ("Brooks" coil) 10 µH, FEM simulation parameters: $I_{dc} = 5$ A, $I_{ac} = 4.08$ A 1 MHz. (**a**) HF-wire (345 × 0.071 mm) 18 turns, (**b**) solid wire (1.32 mm) 20 turns, (**c**) foil turn (10.5 × 0.1 mm) 20 turns. (Source: Kleeb [6])

there, strong eddy current losses occur, which heat up the coil strongly in this area. HF litz wire (e.g. variant a) is particularly helpful in avoiding this. In outdoor applications, the field effect can be reduced or eliminated by shielding measures.

The shielding of a metallic shield for an alternating magnetic field can be easily calculated using the following specifications:

Magnetic field strength inside the shielding H_i
Magnetic field strength outside the shielding H_a
Thickness of the shielding   d
Damping constant = skin penetration depth   s

$$H_a = H_i \cdot \exp\left(\frac{-d}{s}\right) \text{ with } s = \frac{1}{\sqrt{\pi \cdot f \cdot \sigma \cdot \mu}}$$

Example:

A required damping to $0.01\% = 10^{-4}$ results in $H_a = H_i \cdot \exp\left(\frac{-d}{s}\right) = 10^{-4} \cdot H_i$

$$\text{Or } \exp\left(\frac{-d}{s}\right) = 10^{-4}.$$

This means for the layer thickness of the shield a minimum of

$$\left(\frac{-d}{s}\right) = \ln\left(10^{-4}\right) = -4 \cdot \ln(10) \text{ or } d = 4 \cdot \ln(10) \cdot s = 9{,}21 \cdot s$$

The skin layer thickness or damping constant for a conductive material is $s = \sqrt{\frac{2}{\omega \cdot \mu \cdot \sigma}} = \frac{1}{\sqrt{\pi \cdot f \cdot \mu \cdot \sigma}}$, with f: frequency, μ: permeability and σ: conductivity.

The electrical conductivity of selected metals at a temperature of 300 K is given in Table 10.4.

Table 10.4 Conductivity and relative permeability for selected materials (*e.g. 1998 Spektrum Akademischer Verlag, Heidelberg*)

Material	σ in S/m	μ_r
Silver	$61.35 \cdot 10^6$	1
Copper	$\geq 58.0 \cdot 10^6$	1
Gold	$44.0 \cdot 10^6$	1
Aluminium	$36.59 \cdot 10^6$	1
Brass	$\sim 14.3 \cdot 10^6$	1
Iron (96% Fe, 4% Si)	Approx. $10.02 \cdot 10^6$	400...8000
Stainless steel (1.4301)	$\sim 1.4 \cdot 10^6$	<1.3
μ-metal (76% Ni, 17% Fe, 5% Cu, 2% Cr)	$1.82 \cdot 10^6$	12,000...45,000...(300,000)

This allows the layer thickness/attenuation constant for different materials to be determined as a function of frequency (Fig. 10.56). Aluminium is particularly interesting as a construction material. From approximately 50 kHz, the layer thickness/penetration depth of the magnetic field is only less than 0.37 mm. In the higher frequency range, a strong shielding effect can be expected even with relatively thin walls. However, the use of ferromagnetic materials for shielding in strong magnetic fields leads to high losses and is therefore used primarily in weak magnetic fields. Ferrites with their low conductivity are an exception. Here the effect is limited by the saturation induction. Hence ferrite Materials are mainly used to shield subsystems in case of EMC problems.

It should not be forgotten that the shielding effect of the metal is an interaction of the eddy currents generated in the shield with the generating field. These eddy currents are generated by transformer via the stray field and cause losses. However, these losses decrease rapidly as the distance between the screen and the coil increases. At the same time, heat can easily be dissipated via a metallic shield. If the shield is designed as an oil container, for example, about three times the current density could be used at the same maximum temperature, so that the volume can also be reduced by this effect. However, the thermal expansion of the oil must be prevented by design.

In summary, these measures can be summarised as follows

- the inductance of the coil decreases with a smaller distance between coil and screen,
- the type of shielding (see Fig. 10.55) may have a considerable influence on the size of the inductance at small distances, and
- stray fields are always to be expected in the case of incomplete encapsulation.

Figure 10.56 shows the shielding effect of aluminium as a function of frequency.

Above 100 kHz, the required shielding thickness of aluminium falls below 3 mm even with attenuation values of 10^{-6}. Above 100 kHz, coreless inductors could also become interesting from this perspective. Hard aluminium (e.g. AlMg3) is less conductive than e.g. almost pure, but soft (electrical) aluminium. This pushes the limits in this illustration almost imperceptibly upwards.

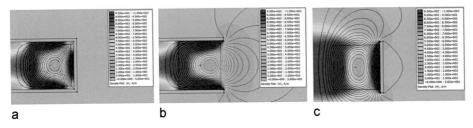

a   b   c

Fig. 10.55 Field strength distribution of a short solenoid coil with different shielding: (**a**) complete enclosure, (**b**) shielding plates arranged above and below the coil, (**c**) cylindrical shielding arrangement. (Source: Kleeb [6])

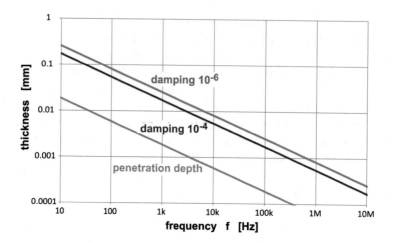

Fig. 10.56 Attenuation constant and minimum thickness of the shielding for aluminium with an attenuation of 10^{-4} or 10^{-6} times

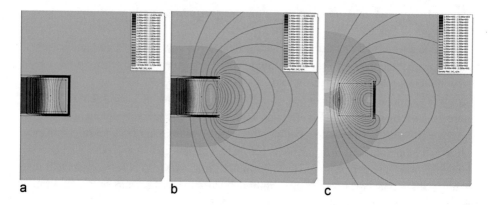

Fig. 10.57 Field calculations for a short solenoid coil with ferromagnetic shielding material: (**a**) complete shielding, (**b**) shielding of the shield surface, (**c**) shielding of the jacket surface. (Source: Kleeb [6])

Another shielding variant is achieved with ferromagnetic material. Ferrites are used here to avoid high losses. Figure 10.57 shows the effect of such shielding measures with short solenoid coils.

In summary, the calculations on the influence of ferrite shielding give the following results

- The inductance becomes higher the smaller the distance between shield and winding is.
- The effect of a cylindrical screen is less than that of an axial screen.

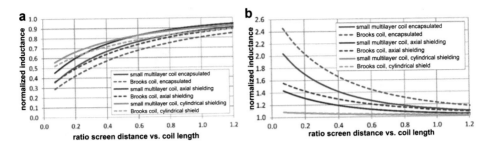

Fig. 10.58 Change in inductance of a short solenoid (Brooks) coil. (**a**) With metallic shielding with copper, (**b**) with ferromagnetic shielding. (Source: Kleeb [6])

- The coil shape has less influence on the result than with a metallic shield.
- The ferrites generate additional losses, depending on the flux density in the material.
- In general, a weaker shielding effect than with metallic shielding can be expected.

The results for both screen types are summarized graphically in Fig. 10.58 below.

The ferromagnetic shielding was investigated because it is positioned in a range of relatively low magnetic field strengths, so that formally low loss densities can be expected at a location that can be easily cooled. The RF shielding effect of ferrite is significantly lower than that of copper or aluminum. Due to the material properties, however, certain minimum strengths, greater than those of metals, must be expected in any case. With such shielding, which is based on shaping the magnetic field, the inductance increases with a smaller shielding distance.

10.5   Controllable Inductors

Approaches to Change the Inductance Via Mechanical Intervention

Passive components are usually accepted as constant. However, variable components are used for both capacitors and inductors. Variable capacitors include in particular variable capacitors and capacitance diodes. With variable capacitors, the active capacitor surface is changed mechanically, whereas with capacitance diodes the change in capacitance of a diode operated in the reverse direction with a changed voltage is used. The latter example is a very small-signal application. Similar applications are also known for inductors.

Especially obvious is the change of inductance by changing the effective magnetic resistance and or the positive and negative coupling of winding parts. Figure 10.59a shows a possible example of this approach. In the so-called variometer (Fig. 10.59b, c), the magnetization axes of two coils are twisted against each other, resulting in a variable counter-inductance. If N_1 and N_2 are connected in series with the inductors L_1 and L_2, the following relationship is obtained

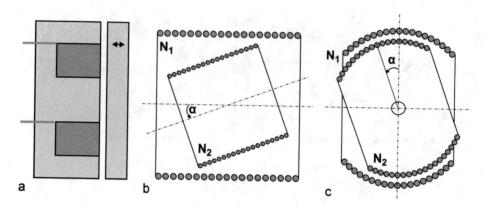

Fig. 10.59 Change in the inductance of an arrangement. (**a**) by changing the effective reluctance by means of, (**b**), (**c**) by changing the mutual inductance of 2 windings

$$L_{ges} = L_1 + L_2 \pm k \cdot \sqrt{L_1 \cdot L_2}$$

This means that inductance values can be infinitely adjusted over a relatively wide range of values. As coreless versions, these components can also be used in the large signal range. Basically, an AC or three-phase asynchronous motor with stator and rotor winding is an analog arrangement with a ferromagnetic core and air gap. Depending on the rotor position, the inductance value of the windings connected in series or parallel can be changed. Because of the much higher inductance values and currents, much higher energies are stored here. A change in the energy content can therefore be associated with considerable force effects. For such an application, the rotor would have to be fixed mechanically, for example, by a worm gear. This then absorbs the starting torque of the 'asynchronous motor'.

Also obvious is the change in the number of turns of an inductor by changing the number of turns. Figure 10.60 shows the principally possible arrangement for changing the active number of turns of a coil via sliding or rolling contacts. This arrangement is used for apparently "infinitely variable" variable transformers. In reality, there is an integer change in the effective number of turns. This option is used in particular for variable transformers.

The examples of mechanical modification possibilities are given here for the sake of completeness. Since the control options of inductive components have been used in power engineering for a very long time. The focus will be on the possibilities and applications of electrically variable inductors.

Electronic Imitation of a Variable Inductance

Inductances can be generated from their dual component, the capacitance, by means of a gyrator (Fig. 10.3). An ideal gyrator is a linear two-port in whose chain matrix only the secondary diagonal is occupied:

Fig. 10.60 Change in the number of turns of an inductor by changing the number of turns with rolling contact

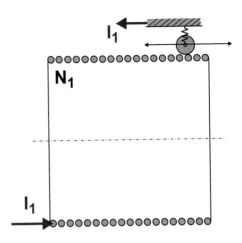

$$\begin{bmatrix} u_2 \\ i_2 \end{bmatrix} = \begin{bmatrix} 0 & R_g \\ \dfrac{-1}{R_g} & 0 \end{bmatrix} \cdot \begin{bmatrix} u_1 \\ i_1 \end{bmatrix}$$

The gyration resistance R_g represents the selectable factor, which influences the inversion. The ideal gyrator is a lossless passive two-port. It is the counterpart of the ideal transformer in the ideal components. The input impedance of Fig. 10.61 is thus calculated as

$$\underline{Z}_e = \frac{R_g^2}{\underline{Z}_e} = j\omega C \cdot R_g^2 = j\omega L^* \text{ with } L^* = C \cdot R_g^2$$

The input terminals of the circuit behave like an inductor, although no such component is included in the circuit.

Gyrators for this application can only be generated by active electronic circuits. Figure 10.62 shows a circuit example from the literature.

For the conditions and $R_1 = \frac{R_2 \cdot R_4}{R_3}$ $R_7 = \frac{R_4 \cdot R_6}{R_5}$ obtain the transfer equations

Fig. 10.61 Using a gyrator to provide an inductor by using a capacitor

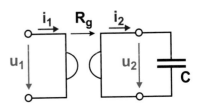

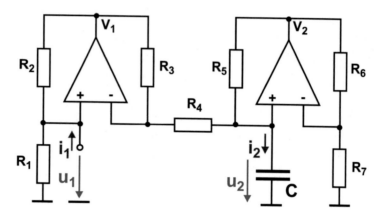

Fig. 10.62 Connection of a gyrator according to [7, 8] with C-termination to provide an inductance

$$i_1 = 0 \cdot u_1 + \frac{R_3}{R_2 R_4} \cdot u_2$$

$$i_2 = \frac{R_5}{R_4 R_6} \cdot u_1 + 0 \cdot u_2$$

Then

$$i_1 = 0 \cdot u_1 + \frac{R_3}{R_2 R_4} \cdot u_2$$

$$\underline{u}_1 = \frac{R_4 R_6}{R_5} \cdot \underline{i}_2 = \frac{R_4 R_6}{R_5} \cdot j\omega C \cdot \underline{u}_2 = \frac{R_2 R_4^2 R_6}{R_3 R_5} \cdot j\omega C \cdot \underline{i}_1$$

This corresponds to the behaviour of an inductor with the value $L = \frac{R_2 R_4^2 R_6}{R_3 R_5} \cdot C$. Since the above-mentioned secondary conditions still exist for a changeable inductance, several resistors would have to be changed in parallel to comply with these conditions. The circuit is therefore only suitable for a single constant inductance value. This can reach very high Q-values. As a rule, the same precision resistors are to be used for resistors $R_1 \ldots R_7$. Thus, for example, with $R_1 = R_2 = \ldots R_7 = R_g = 10$ kOhm and $C = 1$ µF an inductance of $L = 100$ H is obtained. The high Q-values for capacitances are reflected here in the inductance values. Interestingly, the circuit manages without an inductive component. This leads to a very low tolerance of the (virtual) properties.

However, since at least one active electronic element is always required for such circuits, this element must be able to handle the entire apparent power. These and related circuits are therefore only used in communications engineering/signal processing [7]. The ability to store electrical energy required for power engineering is missing here. For a controllable or adjustable inductance, it would be advantageous if this could only be achieved by changing one of the component parameters. A corresponding circuit

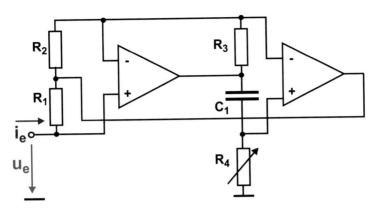

Fig. 10.63 Gyrator circuit for generating a virtual adjustable inductance

(Fig. 10.63) can be found in [9]. A variable resistor R_4 makes it possible to change the inductance represented by the circuit. For the reasons mentioned above, an application is only possible in signal processing. Limits result not only from the energetic conditions but also from the demand for closely toleranced values of the resistors, the voltage limitations of the operational amplifiers and their parasitic properties.

The inductance of the circuit measured against GND as shown in Fig. 10.5 is determined by the relationships

$$Z_e = \frac{U_e}{I_e} = \frac{R_1 \cdot R_3 \cdot R_4}{R_2} \cdot j\omega C_1$$

$$L = \frac{R_1 \cdot R_3 \cdot R_4}{R_2} \cdot C_1$$

Switched Impedance to Replace a Controllable Impedance

The question arises on how to realize a variable or electronically adjustable inductance in the large signal range for power engineering purposes. The "chopper resistance" in drive power converters can serve as a model for this. There, the energy conducted into the inverter when a drive is braked is converted into heat in a periodically switched resistor. At a certain voltage, the power converted in the resistor is

$$P_R = D \cdot \frac{U^2}{R} = \frac{U^2}{R^*}$$

This means that the apparent resistance R^* is inversely proportional to the duty cycle D. A simple switching on and off is out of the question for inductors. If the circuit were to be interrupted, very high voltages would occur at the switch. Therefore a way must be

Fig. 10.64 Principle of current control for the series connection of an inductive and a capacitive storage device

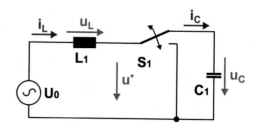

found to prevent this at all costs. The circuit principle is shown in Fig. 10.64, where the current is either conducted to a capacitor C or past it. In an ideal, infinitely fast switch, the voltage at the switch is always defined and finite.

The capacitor voltage then results from the integration of the current components. If we consider the average capacitor current i_C and the total voltage u_{ges}, we obtain $\bar{i}_C = D \cdot i_L$ and for the voltage drop

$$u^* = D \cdot u_C = \frac{D}{C_1} \cdot \int i_C \cdot dt = \frac{D^2}{C_1} \cdot \int i_L \cdot dt$$

This gives you an adjustable "virtual" capacitor of the size $C^* = \frac{C_1}{D^2}$. If this is connected via an inductor to a sinusoidal voltage source with a frequency $\omega = 2\pi \cdot f$ significantly lower than the switching frequency, the current

$$\underline{I}_L = \frac{U_0}{j\left(\omega L - \frac{1}{\omega C^*}\right)} = \underline{I}_L = \frac{U_0}{j\left(\omega L - \frac{D^2}{\omega C_1}\right)} = \frac{U_0}{j\omega L^*}$$

If one interprets the formed two-terminal as inductance L*, a variable inductance of

$$L^* = L - \frac{D^2}{\omega^2 C}$$

When $\omega L - \frac{D^2}{\omega C} = 0$ the reactance disappears. A series resonance occurs. Below the corresponding frequency, the two-pole behaves like a capacitance, above like an inductance. This circuit is suitable for use in alternating current applications, even in large-signal applications, if the generating sources provide sinusoidal signals. The technical design of the switch must be such that the circuit is not interrupted at any time, as with a normal mechanical changeover switch. Figure 10.65 shows an example of this.

The capacity limits the maximum switch voltage. At the same time, it is an oscillating system that is also excited by the frequency of the switching voltage with an underlying change in pulse width. The frequency of the switching voltage and should not be an integral multiple of the "basic frequency". Figure 10.66 shows the dual arrangement to Fig. 10.65.

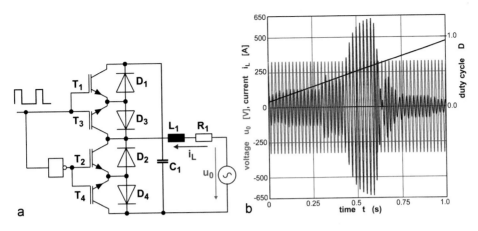

Fig. 10.65 Virtual variable inductance (1). (**a**) clocked "tunable series resonant circuit" with inductive and capacitive working range L = 10 mH, C = 253 μF, R = 0.5 Ohm, ω = 2π·50 Hz, f_{sw} = 70 kHz, U_{0max} = 326 V, (**b**) current and voltage characteristics with variable duty cycle D (b)

The circuit variants shown in Figs. 10.65 and 10.66 are strictly speaking variants for controlling the reactance of a two-pole at a frequency below the switching frequency. Control of the parameter 'inductance' does not occur here. An application therefore only makes sense if the reactance control for a certain frequency is important. The interconnection of the elements results in a second-order system which, in contrast to first-order systems, can oscillate with a simple inductance, so that the effect, for example, in switched systems, differs from a "real" variable inductance.

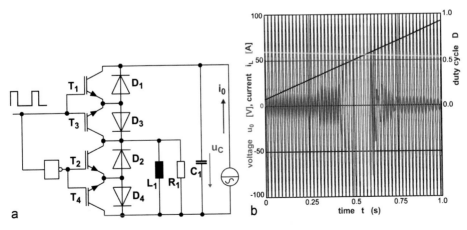

Fig. 10.66 Virtual variable inductance (2). (**a**) tunable parallel resonant circuit with inductive and capacitive working range L = 10 mH, C = 253 μF, R = 79 Ohm, ω = 2π·50 Hz, f_{sw} = 70 kHz, i_{0max} = 1 A, (**b**) current and voltage characteristics with variable duty cycle D

Possibilities of Electrical Control of the Inductance

First of all, it should be noted that if there is proportionality between induction and magnetic field strength (linearity), it is not possible to change the inductance by superimposing two field quantities. The superposition principle, that is, the separate observation and additive superposition of all effects, is based precisely on linearity. A "pivoting" of characteristic curves by linear combinations of fields is thus not possible, only a "shifting".

If an inductor with a ferromagnetic core is connected to an AC voltage, the B (H) characteristic of the material results in characteristic curves I(Ψ). If one looks at the amplitudes of current I_a and flux linkage Ψ_a, one obtains in the loss-free case

$$I_a = H \cdot l = f\left(\frac{U_a}{\omega} \cdot N\right) = f\left(\frac{\sqrt{2} \cdot U}{\omega} \cdot N\right) = f(\Psi_a)$$

The $I_a(\Psi_a)$ characteristic is an odd function. From the descriptive corresponding power series of the form

$$I_a = f(\Psi_a) = \sum_{v=1}^{n} a_v \left(\frac{\Psi_a}{\Psi_0}\right)^{2v-1} \quad \text{with}$$

$$\psi(t) = \int u_L(t) \cdot dt = \int \hat{u}_L \cdot \sin(\omega t) \cdot dt = -\frac{\sqrt{2} \cdot U_L}{\omega} \cdot \cos(\omega t)$$

is obtained for the current course

$$i_L(t) = -\sum_{v=1}^{n} a_v \cdot \left(\frac{\sqrt{2} \cdot U_L}{\omega \cdot \Psi_0}\right)^{2v-1} \cdot \cos^{2v-1}(\omega t)$$

$$i_L(t) = -\left(a_1 \cdot \left(\frac{\sqrt{2} \cdot U_L}{\omega \cdot \Psi_0}\right) \cdot \cos(\omega t) + a_2 \cdot \left(\frac{\sqrt{2} \cdot U_L}{\omega \cdot \Psi_0}\right)^3 \cdot \frac{1}{4}[\cos(3\omega t) + 3\cos(3\omega t)] + \ldots\right)$$

This means that harmonics can occur in the currents depending on the coefficients a_v and the modulation Ψ_a. For the realization of a variable characteristic $\Psi(I)$, it must therefore be ensured that the proportionality $I = \Psi/L$ with low error applies to as large a portion of the characteristic as possible.

For further considerations, an isotropic material with a magnetization curve B(H) is assumed. The magnetic field strength is a vectorial quantity in 3D space and, together with the magnetization characteristic curve, represents the relationship

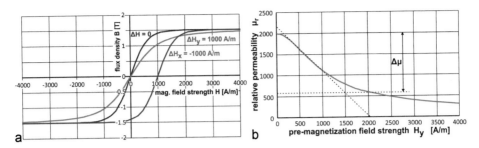

Fig. 10.67 Influence on the magnetization characteristic. (**a**) by homogeneous premagnetization parallel and orthogonal to the 'main direction', (**b**) change of permeability for $H_x = 0$ assuming homogeneous fields and orthogonal superposition

$$\vec{B} = \vec{B}\left(\vec{H}\left(\vec{r}\right)\right) \text{ with and } \vec{r} = x \cdot \vec{i} + y \cdot \vec{j} + z \cdot \vec{k} \quad |r| = \sqrt{x^2 + y^2 + z^2}$$

here. The vectors $\vec{i}$, $\vec{j}$ and $\vec{k}$ thereby span a Cartesian coordinate system. To a field strength in the i-direction, a superimposed field strength can be oriented in the i-direction or perpendicular to it (orthogonal) in j- or k-direction. Figure 10.67a shows the calculated effect of such a measure by an example. A superposition in the x-direction leads to an asymmetrical characteristic curve of the magnetic conductor due to a shift in the characteristic curve. Orthogonal premagnetization produces a symmetrical characteristic which can be deformed under the influence of the premagnetization. The rise at the zero crossings of B (H) can be interpreted as small-signal permeability. In the example shown, a change by approximately 2/3 is technically feasible by orthogonal bias. It should be noted that here the superposition of homogeneous H-fields is used as a secondary condition.

For practical applications, one often wants symmetrical characteristic curves with controlled reluctances. A simple arrangement for this task is shown in Fig. 10.68, where a reluctance section is formed by connecting 2 sections pre-magnetized in opposite directions in parallel. This means that the magnetic voltage drop over both parallel sections is the same. The total magnetic flux results from the sum of $\Phi = \Phi_1 + \Phi_2$. Superimposing the main field results in a non-linear, symmetrical magnetization characteristic curve as shown in Fig. 10.68b. The voltages induced in the winding for premagnetization by the variable flux $\Phi(t)$ cancel each other out partially (and with $I_d = 0$ also completely) if both parts of the winding have the same size. The shape of the characteristic curve changes considerably during premagnetisation. For operation at the fundamental frequency, the 'amplitude inductance' is particularly important, that is, the quotient of the maximum values of the total flux Φ and the magnetic voltage drop V.

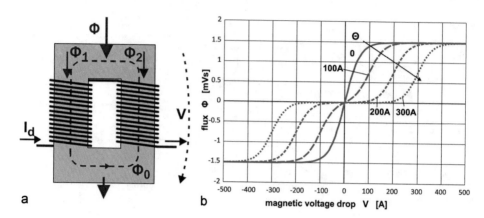

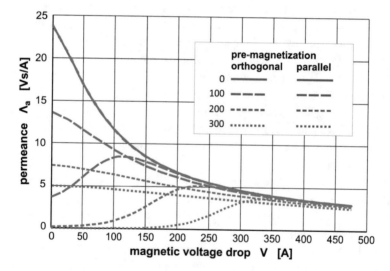

Fig. 10.68 Parallel and antiparallel superposition of the main field and DC premagnetization field. (a) Structure of a controllable reluctance in core form, (b) Change of the magnetization curve at a premagnetizable reluctance of the core type ($l_{FE} = 10$ cm, $A_{Fe} = 5$ cm^2) from the fictitious material used in Fig. 10.67 at different flooding of the premagnetization

Fig. 10.69 Comparison of the permeance values for parallel and orthogonal premagnetization

$$\Lambda_a = \frac{\max(\Phi)}{\max(V)}$$

The calculated dependencies are based on the assumption of homogeneous magnetic fields. For compact magnetic cores, this can be assumed at least approximately for the main field (Fig. 10.69). In the case of parallel magnetization, this secondary condition is also usually fulfilled approximately. For orthogonal premagnetization, the condition requires two

Fig. 10.70 Superposition of a homogeneous main field with an inhomogeneous "control field

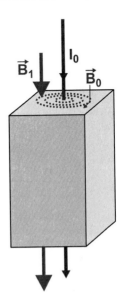

perpendicularly crossing core sections, each with a homogeneous field. Although this is feasible in principle, it is not trivial in terms of production technology.

If inhomogeneous fields for superposition with homogeneous fields are considered, some additional possibilities arise. Such a state can be achieved, for example, by drilling through a core section in the direction of the main field lines and crossing it with a conductor to generate the auxiliary field. The magnetic field lines of the auxiliary field then swirl around the conductor and are perpendicular to the field lines of the main field at all points (Fig. 10.70). If several conductors with different orientations are inserted, it is possible, for example, to achieve that the surroundings of a conductor become saturated depending on the current. This process can be used to estimate the effectiveness of the measure. For an estimation of the diameter D of the saturated homogeneous core material, one can approximately write

$$\frac{\pi}{4}d^2 \cdot J_{d.\,max} \approx I \approx \pi D \cdot \frac{B_{sat}}{\mu_{rFe}\mu_0} \text{ respectively } D \approx \frac{\mu_{rFe} \cdot \mu_0}{4} \cdot \frac{J_{d.\,max}}{B_{sat}} \cdot d^2$$

d...   wire diameter,

$J_{d.max}$...   permissible current density in the conductor for premagnetisation

D...   diameter of the saturated area

If one assumes for an estimation that a jump $\mu_{rFe} \to 1$ occurs when saturation is reached in the material, the limit for the dependence of the permeance for small bias currents I_d can be written as

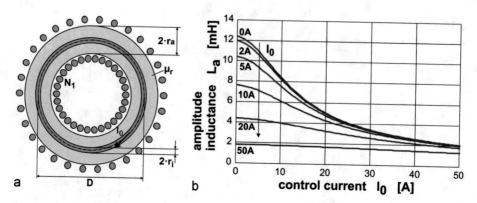

Fig. 10.71 Modulation of inductance type A. (**a**) Possible construction of inductance with a circular core cross-section with radius r_a, an inserted conductor (r_i) with current I_0 and a number of turns N_1, (**b**) Calculation of the dependence of inductance L_a on winding current I_1 ($N_1 = 100$, $D = 0.2$ m, $r_i = 2.5$ mm, $r_a = 10$ mm, $\mu_r = 2000$, $B_{sat} = 0.4$ T)

$$\Lambda_{Fe}(I_d) \approx \frac{\mu_0}{l_{Fe}} \left(\mu_{rFe} \cdot A_{Fe} - (\mu_{rFe} - 1) \cdot \left(\frac{\mu_{rFe}\mu_0}{B_{sat}} \right)^2 \cdot \frac{I_d^2}{4\pi} \right)$$

The formula applies under the assumption that the magnetizing current I_d is parallel to the magnetic flux of the main field. With these preliminary considerations, 2 basic concepts can be followed.

Type A: Modulation of the Magnetization Cross-section A_{Fe} or the Saturation Flux

This concept is followed when the bias current is parallel to the main flux. Due to the partial saturation of the cross-section, only a part of the original cross-section is available for the main flux with high permeability. Thus, two reluctances are connected in parallel - one with low permeability and one with high permeability. The low permeability reluctance shows only minor further saturation phenomena, while the high permeability reluctance still has the original material properties. In the electrical equivalent circuit diagram, this is a series connection of two inductors. For the same current Id, the flux linkages of the two inductors add up. As an example, a fictitious cylindrical core is used here, which is axially traversed by a conductor with the current I0. The distance to the centerline is 'r'. The core with the mean diameter D is provided with a number of turns N_1 so that the arrangement in Fig. 10.71 is obtained schematically. For the homogeneous core material, the relationship

$$B = B_{sat} \cdot \tanh \left(\frac{\mu_r \cdot \mu_0 \cdot H}{B_{sat}} \right)$$

For the dependence of the field strength, the following can be derived

$$H(r) \approx \sqrt{\left(\frac{I_0}{2\pi r}\right)^2 + \left(\frac{N_1 \cdot I_1}{\pi \cdot D}\right)^2}$$

For the magnetic flux in the z-direction, the physical laws provide the relationship

$$\Phi = \int_A \vec{B} \cdot d\vec{A} =$$

$$\Phi_1 = \begin{array}{c} 2\pi \cdot B_{sat} \cdot \displaystyle\int_{r_1}^{r_a} \tanh\left(\frac{\mu_r \cdot \mu_0}{B_{sat}} \sqrt{\left(\frac{I_0}{2\pi \cdot r}\right)^2 + \left(\frac{N_1 \cdot I_1}{\pi \cdot D}\right)^2}\right) \\ \cdot \cos\left(\arctan\left(\frac{2 \cdot I_1}{D \cdot I_0} r\right)\right) \cdot r \cdot dr \end{array}$$

The characteristic curve for the ratios of the amplitudes of flux linkage Ψ_1 and current I_1 provides an inductance value of winding N_1

$$L_a(I_{1a}, I_0) = \frac{N_1 \cdot \Phi_{1a}(I_{1a}, I_0)}{I_{1a}}$$

A graphical representation of this dependency is shown for a dimensioning example in Fig. 10.71. It can be seen that the superposition of a homogeneous field with an inhomogeneous "control field" provides a similar dependence of the amplitude inductance L_a on the current as a superposition of a homogeneous main and control field (see Fig. 10.70).

Type B: Variable Virtual Air Gap
Here only a small part of the reluctance penetrated by the main flux is influenced by the field of the premagnetisation (Fig. 10.72). The effect is similar to an air gap installed transversely to the direction of flux, which can be varied in length. The saturation properties of the core material thus dominate the characteristic curve. Since the uninfluenced parts and the virtual air gap are in series, they appear as parallel inductors in the electrical equivalent circuit diagram. With the same flux linkage, the currents add up. The changes in the flux density are essentially retained so that the loss density according to the Steinmetz formula should also remain approximately constant. Depending on the material, however, the premagnetisation itself also influences the power loss density. This dependence can only be determined experimentally. The effect of the arrangement presented is similar to the effect of a controlled reluctance section shown in Fig. 10.10. Differences exist in the concatenation of the fluxes of the individual conductors.

Fig. 10.72 Pre-magnetization
of the section δ of a reluctance
branch parallel and antiparallel
to the main field B_1

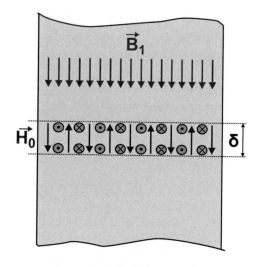

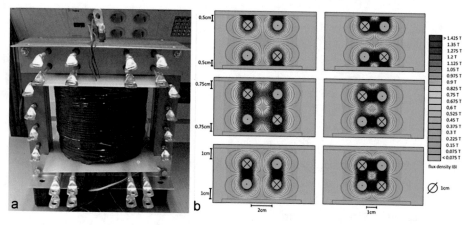

Fig. 10.73 Magnetic setup for demonstrating various possibilities of superimposing magnetic fields:
(**a**) Laboratory setup, (**b**) Simulation results for the distribution of induction at different orientations
of the control currents (Fenske [10])

Examples of Controllable Inductors and Transformers

In the case of laminated cores, control options can be easily added by providing the core
laminations with holes that are used for the conductors to generate the necessary flux.
Figure 10.73 shows such an arrangement. Depending on how the holes are assigned to the
current directions, there are different dependencies of the flux linkage on the current and
thus different control characteristics. For experimental purposes, the control currents were
passed through compact conductors. For possible applications, it is advisable to conduct
the current through several windings and to design the core laminations differently so that it
is possible to work with prefabricated winding parts.

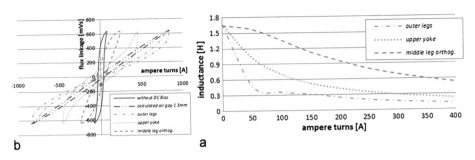

Fig. 10.74 Concept of the virtual air gap. (**a**) Possibilities of changing the magnetization curve with different options of premagnetization according to Fig. 10.73, (**b**) Small-signal inductance at different locations of the premagnetization as a function of their flux flow (Fenske [10])

Figure 10.16 shows a variety of magnetization characteristics with different design approaches. Figure 10.74b makes it clear that the penetration of the control signal can be very different for different design approaches. The effect of the orthogonal magnetization of the center leg is not surprising. Due to the structure of the core made of sheets insulated against each other, the magnetic field lines of the control current must repeatedly cross intervals with a relative permeability of $\mu_r = 1$. The necessary flux and the magnetic force to achieve the required flux densities for a control system are correspondingly high.

For the purely orthogonal premagnetisation of the central leg, it must be taken into account that the core consists of metal sheets insulated from each other. When a current flows in the direction of the main flux, it is surrounded by a magnetic field which must penetrate the insulation layers between the laminations several times. The effect of these insulating layers corresponds to the effect of a distributed air gap. To achieve a magnetizing effect, a very high magnetic force is required. In Fig. 10.73a, the middle leg is provided with an axial conductor with which these relationships can be investigated metrologically. For sheet metal, therefore, the superposition of the control field with the main field as shown in Fig. 10.72 is of particular interest. The arrangement according to Fig. 10.75 allows a series of experiments with this design approach. The main flux is driven by the long winding on the left leg. It must pass through two controllable sections and an air gap. The series connection of the sections in the equivalent magnetic circuit diagram becomes a parallel connection of a linear (air) inductor and two controllable sections. By connecting the sections in series with the air gap, a linearisation of the characteristic curve is achieved. In this configuration, the controlling flux is generated by winding a wire that carries the control current. Due to the geometrically relatively large areas of influence of the control coils, partial saturation of the core material causes a stray field to "break out" into the environment. This stray field can be very disturbing since it does not have a sinusoidal time course due to the saturation processes causing it.

Figure 10.76 shows a "classic" form of a magnetic amplifier. The main flux (alternating flux) is generated in the middle leg and conducted to the outer legs. There it is

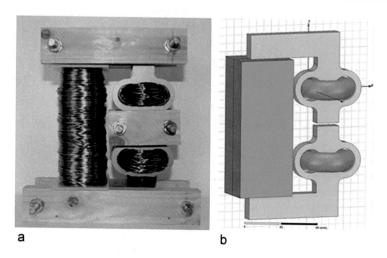

Fig. 10.75 Experimental design of throttle with a fixed air gap and 2 virtual variable air gaps: (**a**) assembled design, (**b**) design diagram (Fenske [10])

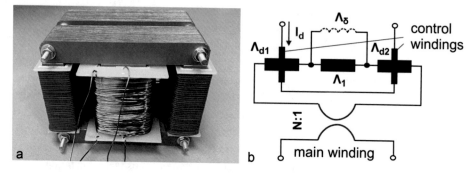

Fig. 10.76 Magnetic amplifier. (**a**) Setup with premagnetization via the two outer legs parallel or antiparallel to the main flux, (**b**) electrical equivalent circuit diagram (Fenske [10])

superimposed with a direct flux in each direction. This results in a non-linear controllable inductance. Use is based either on the control of the current by the variable impedance or the voltage transformation of a further coil on the middle leg. For current control, the aim is usually to influence the effective current within as far apart limits as possible. For this purpose, cores with high permeability without an air gap are used. To control the induced voltage in a second winding of the middle leg, the main winding can be connected in series with another (non-controllable) choke. In this way, a controllable inductive voltage divider is obtained. The voltage drop of the controllable magnetizing inductance is then induced in the additional winding and appears there as controllable voltage. The complex internal resistance of this AC voltage output depends on the set impedance of the parallel-connected inductors of the voltage divider. An air gap in the middle leg is therefore used to increase the output current.

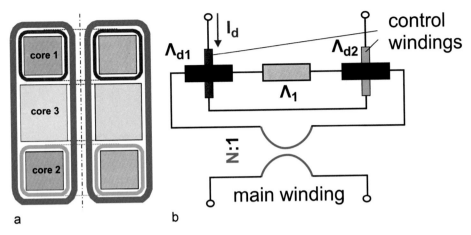

Fig. 10.77 Magnetic amplifier. (**a**) Structure of 3 toroidal cores, (**b**) associated magnetic equivalent circuit diagram

Magnetic amplifiers in the form shown in Fig. 10.76 also lead to relatively strong stray fields. These can be significantly reduced by using toroidal cores (Fig. 10.77). Highly permeable cores made of amorphous or nanocrystalline material or highly permeable ferrites can then be used for the feed areas of the core material. The energy in the component is absorbed by a third core, which must therefore have lower permeability.

Figure 10.77 shows the basic structure and the electrical equivalent circuit diagram. There are 3 separate inductors. Due to the chosen design, this arrangement has hardly any stray fields in case of symmetrical construction. Although the design approach of the component is completely different, the electrical equivalent circuit diagram and thus the function is largely the same.

In contrast to laminated cores, ferrite cores are elements made of a material that is isotropic to the magnetic field. The field propagation has no preferential directions, as is the case with insulated sheets or a texture after rolling. A basic structure is shown in Fig. 10.78, where the inductance of a pot core is controlled by generating an annular flux in the center leg by means of a current through the center hole, which overlaps with the main flux in the axial direction. This changes the magnetization characteristic curve (Fig. 10.78b).

The problem of large-signal measurement of the magnetization characteristic curve was solved here as follows. A DC magnetic force $I_{Bias} \cdot N_{st}$ was impressed for premagnetization. A current I_{DC} with superimposed measuring current I_{AC} was passed through the main winding around the middle leg. This allows the differential inductance

$$L_{diff}(I_{DC}) = \frac{d\Psi(I_{DC})}{dI_{DC}} \approx \frac{\partial L(I_{DC})}{\partial I_{DC}} \cdot \Delta I = \frac{\partial L(I_{DC})}{\partial I_{DC}} \cdot I_{AC},$$

measured at I_{DC} by a small signal measurement. The magnetization characteristic curve Fig. 10.20b is then obtained by integration

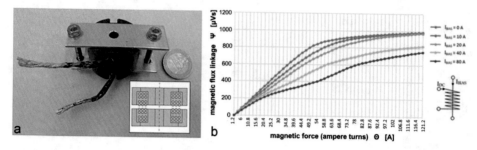

Fig. 10.78 Use of a P36 ferrite core for the construction of an inductance adjustable by orthogonal pre-magnetization. (**a**) Construction (the central opening is used for the conductors for pre-magnetization, (**b**) change in magnetization characteristic curve with different pre-magnetization flux (Pfeiffer 2017 [10])

$$\Psi(I_{DC}) = \int\limits_0^{I_{DC}} L_{diff}(I_{DC})dI_{DC}$$

With different methods of ablation, even brittle ferrites can be processed. Some design approaches of laminated cores can thus be transferred to ferrites. Figure 10.79 shows E70 core pairs made of a power ferrite with a maximum frequency of 2 MHz with holes for a pre-magnetization of core areas similar to Fig. 10.73. The holes for the lead for the pre-magnetization were made in the middle leg in the direction of the main flux (right)

Fig. 10.79 Experimental setup for the investigation of the orthogonal (left, type A) and mixed (right, type B) premagnetization (yellow wire) of the center leg of an EE70 ferrite core with different windings for the measurement of induced voltages. (source: Zacharias 2017)

Fig. 10.80 Simplified electrical
equivalent circuit diagram of the
superstructures shown in
Fig. 10.79a

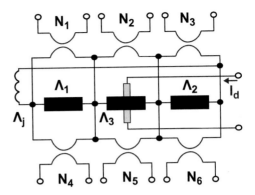

and across it (left). This allows the effective inductance to be adjusted via a direct current
through the yellow lines in a technically sensible ratio of 3:1. Larger setting ranges are
possible. However, relatively high magnetic forces are required in this design.

In the right structure, the main field and control field are perpendicular to each other.
With constant permeability, there can be no mutual interference. This means that there
would also be no induced voltages in the control winding. One can indeed formulate
separate differential equations for each field direction. However, these have a connection
via the material parameters. These are then differential equations with the non-linear
coupling of the coefficients of the equations. Since the control current changes the energy
content of the influenced volume, this can also be observed at the other winding. The
coupling of the windings is therefore nevertheless given, but not as directly as with the law
of induction due to the "integrating" behaviour of the magnetic material. In both cases, an
inductor must therefore be connected in series with the control winding in the simplest case
to achieve a constant control current. Care must be taken that the natural resonant
frequency of this choke is far enough above the operating frequency to avoid interactions
with the main flux. In Fig. 10.79, in addition to the yellow control winding, other windings
of different colours can be seen. These are used to measure the properties of the core
sections and to demonstrate various functions.

Figure 10.80 shows the electrical equivalent circuit diagram of the structure of
Fig. 10.79a. There are 2 windings on each leg. They are used to supply current to and
measure the induced voltage on each leg without current. Λ_j is the permeance of the yoke
scattering. With this setup, various experiments on flux redirection by reluctance change
are possible, for example, for the setup of transformers with controllable output voltage or
the realization of variable inductors.

During the metrological investigations it becomes clear that permeability is not simply a
fixed material coefficient, but an integral descriptive means for the interaction of material
components with small grain size with an externally applied magnetic field. After switching
off a DC pre-magnetization, accommodation effects can be observed even in the range of
10...100 kHz for a ferrite with a preferred use. This means that although the inductance
changes rapidly with the disappearance of the control field, it does change to a large value

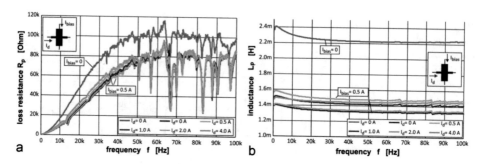

Fig. 10.81 Real part (**a**) of the impedance measurements interpreted as a parallel circuit L_p//R_p and L_p of this equivalent circuit (**b**) of an orthogonally pre-magnetized ferrite section (see Fig. 10.79, measured values: step size $\Delta f = 25$ Hz, bandwidth 10 Hz, (Küster [11])

close to the rest value. However, the change then takes several seconds to minutes to reach this value. The behaviour is similar to that of a very viscous liquid. In high-frequency applications, these processes are practically irrelevant. In very-low-frequency applications, however, they must be taken into account. If one examines the influences of the premagnetisation in small frequency steps, one finds that even in the useful frequency range of the ferrite, the resistance R_p in the parallel equivalent circuit of the impedance shows local minima. This is accompanied by local reductions in the quality factor Q of the component.

Figure 10.81a shows the real part of the impedance measurement interpreted as resistance R_p in the parallel circuit diagram L_p//R_p. Local dips of this value at certain frequencies can be seen. Although these are only small-signal measurements, this means in principle that the core losses increase at these frequencies. An operating frequency at these values should therefore be avoided. Both the position and intensity of the resonance points can be influenced by orthogonal premagnetization. One interpretation of these measurements could be that there are clusters of certain grain sizes of the ferrite material. Due to their magnetic and mechanical properties, these particles, as mechanical micro-oscillators, carry out oscillations that become visible as resonances. The parallel inductance L_p calculated from the data is largely constant and independent of frequency and shows slight reductions in the inductance value at the points with a minimum of the parallel resistance of the equivalent circuit. This is a further indication of resonance phenomena at this point in the material.

10.6   DC/DC Converter with 2 Chokes Without Coupling on One Core

Storage chokes are used in practically all power converters. In many cases, there are variants in which 2 or more magnetic storages are used, which are used alternately. This requires a relatively large installation space. The windings can indeed be dimensioned in such a way that they are not overloaded in the time average. However, the magnetic circuit

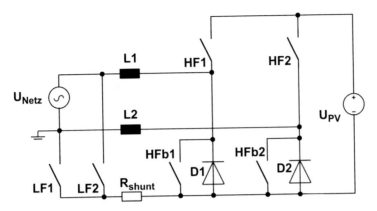

Fig. 10.82 Circuit example for forcing sinusoidal mains currents at $\cos\varphi = \pm 1$ at very high efficiency with 2 uncoupled smoothing chokes [12–14]

must always be designed in such a way that it is possible to store the maximum amount of energy required. The magnetic circuit thus sets fixed basic conditions for the design. The problem could be partially mitigated if the same magnetic circuit could be used for both chokes. In this case, however, magnetic coupling takes place, which, via the law of induction, causes additional voltage loads in the respectively inactive circuit parts (Fig. 10.82). The figure shows a topology that can feed electrical energy from a DC voltage source into the electrical AC grid without the need for a transformer. For this purpose, switches HF1 and HF2 are alternately operated as step-down converters. When the polarity is changed, there is a switchover between the two step-down converters with LF1 and LF2. HFb1 and HFb2 are operated as synchronous rectifiers for diodes D1 and D2. The inductors L1 and L2 are thus only operated with the sinusoidal current for one half-wave each and must not be magnetically coupled. This is avoided by a separate construction of L1 and L2, which leads to an increase in the required construction volume. The achievable efficiencies of this circuit are peak values compared to other topologies.

At $\cos\varphi = 1$, the mains supply is provided alternately by the separate step-down converters formed by the elements (HF_1, L_1, D_1 and LF_1) and formed by the elements (HF_2, L_2, D_2). Magnetic coupling with mutual inductance $M > 0$ for $L_1 \leftrightarrow L_2$ would induce a voltage in the other (unused) inductance, which would increase the voltage load of the (open switches). The considerations aimed to use the same core or magnetic structure as energy storage for both chokes and at the same time to avoid the undesirable or harmful induction voltages in the other choke in order to achieve a more compact structure of the entire circuit, which is essentially determined by the magnetic components. The achievable savings in installation space and material as well as the number of components can also lead to cost reductions. The above problem analysis results in a solution approach:

- The magnetic circuit is used alternately by two circuit parts with at least one winding and is driven with (usually the same) maximum magnetically stored energy.
- The magnetic material should be controlled as uniformly as possible with regard to its limit values (saturation induction, power loss density).
- The windings must be arranged on the core in such a way that the resulting induced voltages in the respective unused winding complement each other to zero so that no additional disturbing voltages occur.

Solution: 2 Two Non-Coupled Storage Chokes Using a Magnetic Core

Table 10.5 shows two basic solutions for this task. Here, consideration is given to common or available core designs. Variant B assumes a magnetic circuit basically divided into 2 parts (MM-cut or EI-cut), which consists of 3 legs. On each of the 3 legs, there are winding parts that are connected in a way that is appropriate for the purpose. The magnetic flux of the middle winding passes through the two outer windings and induces voltages there when it changes. However, these voltages cancel each other out again because of the interconnection of these windings and because of the symmetry of the structure. On the other hand, when current flows through the two outer windings, two magnetic fluxes are generated, which cancel each other out in opposite directions due to the symmetry of the structure in the middle winding and therefore cannot be the cause of the voltages induced there. The corresponding derived design equations for air gap and inductance are shown in Fig. 10.82.

Variant A differs essentially in that the other basic forms of magnetic cores used practically result in a basic division into 4 parts. The unavoidable air gaps between the components cause a decoupling of the two magnetic circuits to the right and left of the axis of symmetry. The equations for the magnetic circuits are thus slightly modified. Nothing changes in the mode of operation described above.

The special feature of the arrangement is that the magnetic core material is practically fully utilised by both windings in both functional phases and that, due to the electrical connection of the windings, the induced voltages compensate each other in one functional phase and the magnetic fluxes complement each other to (practically) zero in the other functional phase due to the connection of the magnetic circuits of the partial windings. This prevents an induced voltage in the overall winding.

Variant A

The complete magnetic equivalent circuit diagram of variant A is shown below (Fig. 10.83).

The reluctances R_{m01} and R_{m02} represent the small air gaps between the 4 U-cores. If the distance between the cores is very small, the reluctance of such gaps becomes small. In the ideal case (when the cores lie flat on their sides), this can be considered a "magnetic short

Table 10.5 Possible designs of storage chokes on a core with coupling factor k = 0

	Common core storage chokes with compensated coupling	
	Variant A	Variant B
Core types	UU cores, UI cores (magnetically decoupled)	EE cores, EI cores
Winding structure	 * Start of winding	 * Start of winding
Claims, Specifications	$L_1 = L_2$, M = 0, Magnetization cross sections outside: A_{Fe}, inside: $2A_{Fe}$ Same maximum flux density in all legs at $I_1 = I_2 = I$ Alternative use of N_1 and N_2.	
Magnetic equivalent circuit diagram	 $$B = \frac{N_1 \cdot I}{\mu_0 \cdot 4\delta_1} = \frac{N_2 \cdot I}{\mu_0 \cdot (\delta_1 + \delta_2)} = \frac{N_2 \cdot I}{\mu_0 \cdot 2\delta}$$	 $$B = \frac{N_1 \cdot I}{\mu_0 \cdot 2\delta_1} = \frac{N_2 \cdot I}{\mu_0 \cdot \delta_1}$$
Number of windings	$N_{11} = N_{12} = \frac{N_1}{2}$ $N_1 = 2 \cdot N_2$	$N_{11} = N_{12} = \frac{N_1}{2}$ $N_1 = 2 \cdot N_2$
Winding window	$A_W = \frac{2 \cdot I \cdot N_2}{k_f \cdot J_{zul}}$	$A_W = \frac{2 \cdot I \cdot N_2}{k_f \cdot J_{zul}}$
Air gap	$\delta_1 = \delta_2 = \delta = \delta^*$	$\delta_2 = 2\delta_1$
Inductors	$L_2 = L_2 = \frac{N_2^2}{\mu_0 \cdot 2\delta} A_{Fe}$	$L_2 = L_1 = \frac{N_2^2}{\mu_0 \cdot \delta} A_{Fe}$

circuit". If the parasitic air gaps cannot be neglected, they can be considered as decoupling between the two cores, as shown in Fig. 10.84.

Since the reluctances R_{0x} are very large, the two magnetic circuits can be considered as separate magnetic circuits as shown in Fig. 10.84. In variant B the magnetic fluxes are parallel in the middle leg. If we assume direct fluxes, then the flux in the right outer leg is twice as large as in the middle, while the flux in the left leg is only half as large. This must be taken into account when dimensioning the cross-sections with regard to saturation. From all conditions together, one can deduce that $\frac{N_1}{2} = N_2$. Since the two halves of the primary winding are connected in series, the resulting primary inductance is the sum

Fig. 10.83 Magnetic
equivalent circuit diagram of
variant A from Table 10.5

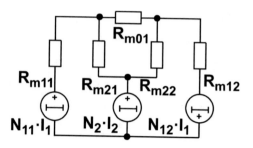

Fig. 10.84 Reduced magnetic
equivalent circuit diagram of
choke variant A

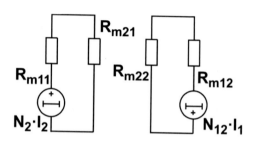

$$L_1 = L_{11} + L_{12} = \frac{\mu_0 \cdot N_1{}^2 \cdot A_{Fe}}{4 \cdot (\delta_1 + \delta_2)}$$

The inductance of the secondary winding is also:

$$L_2 = \frac{\mu_0 \cdot N_2{}^2}{(\delta_1 + \delta_2)} \cdot A_{Fe}$$

The inductors are the same. Due to the independence of the two magnetic circuits, energy storage is also possible in the middle air gaps when winding N1 is active.

Variant B

In variant B the two winding parts of N_1 are arranged in such a way that the resulting magnetic flux in the middle leg is zero. This means that no voltage is induced in the second winding N_2. Consequently, no energy is stored in the air gap δ_2. In this case, the middle leg of the EE core does not contribute to energy storage. The effective series connection of the two windings N_{11} and N_{12} is considered here. The two air gaps δ_1 add up because they are also connected in series, while δ_2 is not taken into account. When winding N_2 is active, the total resulting induced voltage of winding N_1 is zero because both halves are connected against each other. However, the resulting flux in the external leg is not zero and consequently, energy is stored distributed in all air gaps. L1 and L2 should be equal. This means

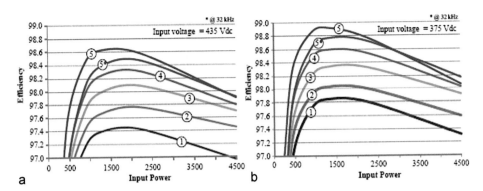

Fig. 10.85 Measured efficiency curves for a test setup of the circuit according to Fig. 10.82 at the intermediate circuit voltages (**a**) 375 V, (**b**) 435 V. (source: Araújo [12])

$$L_1 = \mu_0 \cdot A_{Fe1} \cdot \frac{(N_{11} + N_{12})^2}{2\delta_1} = \frac{N_2^2}{\frac{2\delta_1}{2\mu_0 \cdot A_{Fe1}} + \frac{\delta_2}{\mu_0 \cdot A_{Fe2}}} = L_2$$

For $A_{Fe2} = 2 \cdot A_{Fe1}$, this results in

$$\frac{(N_{11} + N_{12})^2}{2\delta_1} = \frac{N_2^2}{\delta_1 + \frac{\delta_2}{2}} \text{ respectively } \frac{(N_{11} + N_{12})^2}{N_2^2} = \frac{4\delta_1}{2\delta_1 + \delta_2}$$

There are various possibilities to formulate further objectives here, such as minimum total losses, equal losses for both windings, $N_{11} + N_{12} = N_2$, etc. The last approach lists for example

$$\delta_2 = 2\delta_1$$

This topological approach leads to very low overall losses or a very high efficiency since practically all components can be individually optimized for their maximum efficiency. Results have been reported in several publications, for example in [1, 2]. Figure 10.85 contains measurement results for efficiencies obtained at different intermediate circuit voltages for different combinations of transistors and diodes made of silicon and silicon carbide.

Table 10.6 gives an overview of the component combinations used.

Table 10.6 In Fig. 10.85: tested component combinations [12]

	T1–T2	T3–T4	D1–D2
1	Trench IGBT 1st G 25A/1200 V	Trench IGBT 1st G 30A/600 V	Stealth Diode 30A/1200 V
2	Trench IGBT 1st G 25A/1200 V	2nd G CoolMOS 38A/650 V	Stealth Diode 30A/1200 V
3	Trench IGBT 1st G 25A/1200 V	2nd G CoolMOS 38A/650 V	SiC Diode 20A/1200 V
4	Trench IGBT 2nd G 25A/1200 V	2nd G CoolMOS 38A/650 V	SiC Diode 20A/1200 V
5	SiC D-MOSFET 20A/1200 V	2nd G CoolMOS 38A/650 V	SiC Diode 20A/1200 V

10.7 Controllable Harmonic Absorber Filters to Reduce Harmonic Mains Currents

Power converters and non-linear loads lead to current harmonics, which in turn lead to voltage harmonics. The harmonic currents injected into the network can be reduced by injecting currents in the opposite phase. The odd harmonics (3., 5., 7., 9.) are of particular importance here. The harmonics which can be divided by 3 cancel each other out partially or completely in a 'Dy' switching group of the transformer from medium voltage to low voltage or in the case of symmetrical loads, the harmonics of the fifth and seventh order (250 Hz and 350 Hz) are of particular interest. Since the harmonics are practically forced by the load, the load can be considered as a current source for the harmonics [15–18].

$$Z_{SK} = R_S + j\left(\nu\omega(L_0 + L_S) - \frac{1}{\nu\omega C}\right)$$

The current which is then injected into the mains in a 1-phase connection can be calculated according to the current divider rule as follows

$$\frac{\underline{I}_{L1}}{\underline{I}_1} \approx \frac{\underline{Z}_{SK}}{\underline{Z}_{L1} + \underline{Z}_N + \underline{Z}_{SK}}$$

The losses caused by this are in the series resistance of the suction circuit for the nth harmonic:

$$P_{vin} = I_n^2 \cdot R_S$$

With exact resonance and $R_S = 0$, the currents would be limited and no losses would occur.

If for the sake of simplification, only the reactive components of the mains impedance and the components of a harmonic absorber are taken into account, the following is obtained

$$\frac{\Delta f}{f_0} = k$$

$$\omega = (1 \pm k) \cdot \nu \cdot \omega_0 = \frac{(1 \pm k)}{\sqrt{(L_0 + L_S)C}}$$

By exact tuning to the frequency of the harmonics, it is theoretically possible to zero the counter of the expression, so that almost complete elimination of the harmonics should be possible. In reality, this is prevented by at least 2 factors:

- The effective resistance R_S of the series resonant circuit is not zero. This means that the current level cannot fall below a minimum reduction.
- The mains frequency is not constant but can fluctuate within a certain tolerance range during normal operation. The normal range, which is set with positive or negative control energy, is 49.8 ... 50.2 Hz. This corresponds to a normal fluctuation range of +/− 0.4%. The corresponding detuning changes the effect of the harmonic absorber s, which are set to a fixed frequency.
- At the same time, the current through RS causes a power loss $P_v = I_{SK}^2 \cdot R_S$, which accordingly causes heating. The guarantee of a maximum current must be secured by constructive measures.

The harmonic absorber s (Fig. 10.86) are all set to frequency. This means that they represent a capacitive load on the grid at the base frequency and cause capacitive

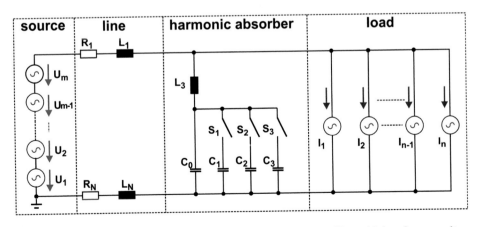

Fig. 10.86 Schematic structure of a single-phase harmonic absorber filter with interference voltage sources on the mains side and interference current sources on the consumer side

reactive power in the grid. For a clearer calculation, the following definitions are introduced.

Relative frequency change $\quad k = \frac{\Delta f}{f_0}$

or via the angular frequency $\quad \omega = (1 \pm k) \cdot \nu \cdot \omega_0 = \frac{(1 \pm k)}{\sqrt{(L_0 + L_S)C}}$

In relation to the resonance point, the frequency fluctuation in the network can reach a maximum impedance change of

$$\frac{\Delta Z}{R_S} = \sqrt{1 + \left(\frac{(\nu\omega_0(1 \pm k))^2(L_0 + L_S)C - 1}{\nu\omega_0(1 \pm k)CR_S} \right)^2} \approx \sqrt{1 + \left(0.008...0.016 \cdot \frac{Z_w}{R_S} \right)^2}$$

$$\approx \left(1 + 0.004...0.008 \cdot \frac{Z_w}{R_S} \right)$$

This means that the smaller R_s, the greater the change in impedance with frequency fluctuations and the greater the fluctuations in the derived harmonic currents. But the mains voltage also contains harmonics. With a filter tuned to the mth harmonic, the losses caused in R_S are approximately maximum

$$P_{vum} \leq \frac{U_m^2}{R_S}$$

This means that with $R_S \to 0$ an adjustable harmonic absorber filter is desirable, with which the current consumption can be adjusted by slightly shifting the resonance frequency. Standard is a switching of capacitors as shown in Fig. 10.86. Continuously adjustable power capacitors are certainly not technically possible here. However, inductivities adjustable by pre-magnetization are in principle suitable for continuously following a setpoint value for a maximum current. For this purpose, the impedance characteristic of a harmonic absorber is shown in a standardized form in Fig. 10.87.

$$\frac{\underline{Z}(\omega)}{Z_w} = \frac{R_S}{Z_w} + j\left(\frac{\omega}{\omega_0} - \frac{\omega_0}{\omega} \right)$$

Obviously, high effectiveness of control measures can be achieved near the resonance with small changes of the resonance frequency.

Assuming the operating frequency is to be the fifth harmonic $\omega 5$, the ratio of R_s/Z_w is 0.005 and the natural resonant frequency is to be changeable by a Wer from $\Delta\omega_0$ from a value $\Delta\omega_{0max}$. The ratios of ω_{0max}/ω_5 and $\Delta\omega_0/\omega_5$ are then variable dimensioning variables for a control range of the impedance Z. If the ratio of maximum and minimum of $Z(\omega, \Delta\omega)$ is compared, a statement about the sensitivity of the control measure is obtained:

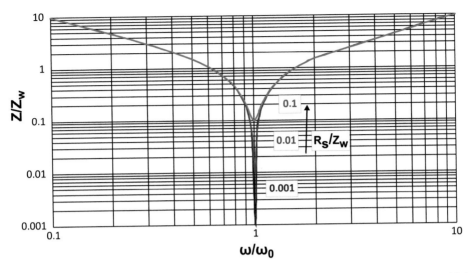

Fig. 10.87 Impedance curve $Z_w(\omega/\omega_0)$ for the relative series resistance $R_s/Z_w = 0.001; 0.01$ and 0.1

$$\frac{Z_{max}(\omega)}{Z_{min}(\omega)} = \frac{\sqrt{\left(\frac{R_S}{Z_w}\right)^2 + \left(\frac{1}{\frac{\omega_{0\,min}}{\omega_5} + \frac{\Delta\omega}{\omega_5}} - \left(\frac{\omega_{0\,min}}{\omega_5} + \frac{\Delta\omega}{\omega_5}\right)\right)^2}}{\sqrt{\left(\frac{R_S}{Z_w}\right)^2 + \left(\frac{\omega_5}{\omega_{0\,min}} - \frac{\omega_{0\,min}}{\omega_5}\right)^2}}$$

Figure 10.88 shows examples of the results of this calculation. For example, with a control range of 4% for the resonance frequency, an impedance ratio of approximately 1500% can be achieved. This numerical example is used for subsequent calculations and derivations.

For an application-related consideration (1-phase) the following conditions are assumed in the following

Frequency of the network harmonic	$f_5 = 250$ Hz
Mains inductance	$L_1 + L_N = 0.8$ mH
Reactance	$X_L(250$ Hz$) = 1.257$ Ohm
Attenuation of the harmonic to	5%

Setting the range of the resonance

Frequency of the filter	4%
Rs/Zw	0.005

The maximum attenuation is obtained with resonance frequency = mains frequency closed:

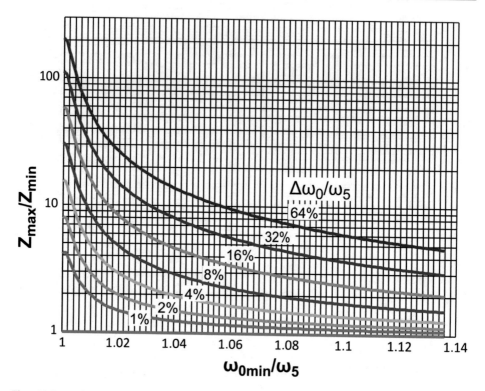

Fig. 10.88 Influence of the relative resonance frequency and the range of adjustment of the resonance frequency on the ratio of maximum and minimum impedance of the harmonic absorber

$$\frac{I_{L1}}{I_1} = 0.05 \approx \min \left| \frac{\underline{Z}_{SK}}{\underline{Z}_{L1} + \underline{Z}_N + \underline{Z}_{SK}} \right| \approx \frac{R_S}{\sqrt{R_S^2 + (5 \cdot \omega(L_1 + L_N))^2}}$$

This allows the series loss resistance to be estimated at

$$R_S = \frac{0.05 \cdot 5 \cdot \omega_{Netz}(L_1 + L_N)}{\sqrt{1 - 25 \cdot 10^{-4}}} \approx 0.05 \cdot 5 \cdot \omega_{Netz}(L_1 + L_N) = 62.8 \ \text{mOhm}.$$

For the characteristic impedance of the harmonic absorber, this gives the relationship

$$Z_w \leq \frac{R_S}{0.005} = 12.566 \, \text{Ohm}.$$

At 250 Hz the capacitor may have at most this apparent resistance, that is,

Fig. 10.89 Division of the controllable choke into a fixed and a variable component

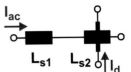

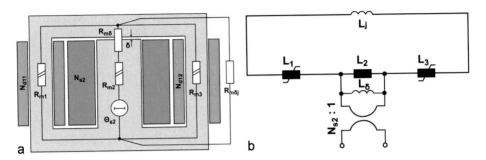

Fig. 10.90 Design of a magnetic amplifier with a magnetic equivalent circuit with the alternating current winding N_{s2} (red), the two direct-current windings $N_{d1} + N_{d2}$, the air gap δ and the iron cross-section $_A A_{FE2} = 2 \cdot A_{Fe1} = 2 \cdot A_{Fe3}$ (**a**) and electrical equivalent circuit (**b**) (assumption $\mu_{rFe} \rightarrow \infty$)

$$C \geq \frac{1}{5 \cdot 2 \ \pi \cdot 50 \ \text{Hz} \cdot 12.566 \ \text{Ohm}} = 50.66 \ \mu F$$
$$C = 2 \cdot 33 \ \mu F = 66 \ \mu F$$

Thus $Z_w = 9.65$ Ohm. If now the resonance frequency is to be adjustable by 4%, the maximum resonance frequency is 260 Hz. This results in maximum and minimum inductance as target values for $L_{max} = 6.141$ mH and $L_{min} = 5.677$ mH. A control range of 0.43 mH is therefore required. How can this control range be achieved?

In the interest of a low tax burden and low additional network distortions, it makes sense to divide the controllable choke into a fixed and a variable component (Fig. 10.89).

In the following, the design of a controllable choke by a *magnetic amplifier* (transductor) is considered. Figure 10.90 shows the basic design of such a component with an EI sheet metal cut with an air gap in the middle leg. The middle winding is the AC winding whose effective inductance is to be controlled. The two outer windings are connected in series and have the same number of turns. This cancels out the direct current flow in the middle leg. The voltages induced in the resulting winding N_d for premagnetisation cancel each other out at $I_d = 0$. The two outer legs have two $\Psi(I)$ characteristics shifted by the premagnetising direct current and are connected in parallel with the magnetic resistance of the yoke scattering. In the electrical equivalent circuit diagram, this leads to a series connection of two non-linear and one linear inductance. The inductance of the yoke scattering L_j can be easily determined by measuring the inductance for the core without the air gap when both premagnetising windings are short-circuited.

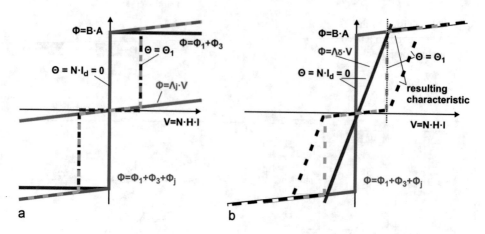

Fig. 10.91 Characteristic curve control. (**a**) Characteristic curve of the series connection of the pre-magnetized core sections, (**b**) Resulting characteristic curve of the magnetic amplifier including the air gap [19]

The superposition of the two nonlinear outer sections with the yoke scattering is shown schematically in Fig. 10.91a. The same current flows through all 3 elements connected in series. Therefore, the fluxes must be added to determine the total characteristic curve for the inductances (or magnetic conductance values) related to the number of turns 1. For a better overview, $\mu_{rFe} \ll 1$ is assumed because of $\mu_{rFe} \to \infty$. A saturation flux Φ_S results from the saturation induction and the iron cross-section A_{Fe} . Figure 10.91a shows the resulting characteristic curve with and without premagnetisation. The characteristic curve of the air gap is essentially connected in parallel to this characteristic curve. This means that with the same flux linkage / same flux, the currents add up. This results in the characteristic curve shown in Fig. 10.91b. By approximating the characteristic curves with straight-line sections, approximation formulas can be easily derived to calculate the resulting inductance.

For the dependency of the inductance of the magnetic amplifier, one obtains approximately

$$L_{magamp}(I_d) \approx \frac{N_{s2}^2 \cdot B_{sat} \cdot A_{Fe1}}{\frac{B_{sat}}{\mu_0} \cdot \delta + (N_{d1} + N_{d3}) \cdot I_d}$$

This inductance designates the value Ψ/I at the upper inflection point of the characteristic curve. An inductance adjustable between 0...100% cannot exist in this way. Maximum and minimum inductance are when the saturation flux is reached.

$$L_{magamp.\,max} \approx \frac{N_{s2}^2 \cdot B_{sat} \cdot A_{Fe1}}{\frac{B_{sat}}{\mu_0} \cdot \delta} \quad \text{and} \quad L_{magamp.\,min} \approx \frac{N_{s2}^2 \cdot B_{sat} \cdot A_{Fe1}}{\frac{B_{sat}}{\mu_0} \cdot \delta + (N_{d1} + N_{d3}) \cdot I_{d\,max}}$$

This results in the ratio of both to

$$\frac{L_{magamp.\,max}}{L_{magamp.\,min}} \approx 1 + \mu_0 \frac{(N_{d1} + N_{d3}) \cdot I_{d\,max}}{B_{sat} \cdot \delta}.$$

Another approach is to set a maximum current at $I(B_{sat} \cdot A_{Fe})$. The maximum inductance is the same as in the example above. For a very low yoke leakage inductance $L_j \rightarrow 0$, the "amplitude inductance" at a constant maximum current amplitude corresponding to the saturation current changes according to

$$\left. \frac{L_{min}}{L_{max}} \right|_{I_s = max} \approx \left(1 - \frac{N \cdot I_d}{\frac{B_{sat}}{\mu_0} \cdot \delta} \right)$$

This relationship can therefore continue to be used for dimensioning in order to be able to make an economic breakdown as shown in Fig. 10.89. If this type of controllable inductance is used in harmonic absorber s, harmonics are also used due to the non-linearities of the characteristic curve. Naturally, these are odd multiples of the resonant frequencies. By connecting the controllable small inductor in series with a larger linear inductor as shown in Fig. 10.89, the non-linear component is kept small. The generation of harmonics is illustrated below with an example. The example started above is continued. The basic values were

Maximum inductance $L_{max} = 6.141$ mH
Minimum inductance $L_{min} = 5.677$ mH

Setting range from 0.43 mH
It applies and $L_s = L_{s1} + L_{s2}$ max $(L_{s2}) = 3 \cdot$ min (L_{s2}) . From this follows the conditions

$$6.141 \ \text{mH} = L_{s1} + 3 \cdot \text{min} \,(L_{s2})$$
$$5.677 \ \text{mH} = L_{s1} + \text{min} \,(L_{s2})$$

For the minimum of the adjustable inductance you get $\min(L_{s2}) = 0.215$ mH. Consequently, the inductance must be $L_{s2}(I_d = 0) = 0.645$ mH. The fixed inductance is then $L_{s1} = 5.496$ mH. The dimensioning should be done for a maximum current amplitude of 15 A. At 250 Hz this corresponds to a reactive power of 114 Var for the controllable choke. To determine the inductance of the air gap, a yoke leakage inductance of 50 µH is assumed as given. The "pure air gap inductance" is then $L_0 = 595$ µH. The characteristic for a transductor-controlled inductance built up with this has then approximately the appearance

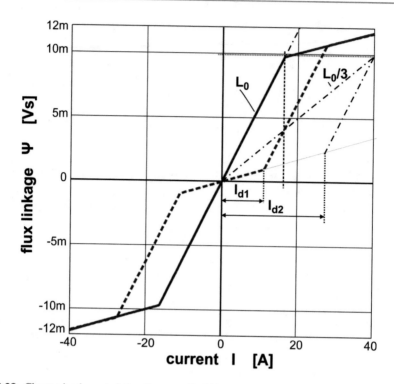

Fig. 10.92 Change in characteristic of a controlled inductor ($L_0 = 595\ \mu H$, $L_j = 50\ \mu H$) [19]

shown in Fig. 10.92. If you only want to achieve a change in inductance to one third at a constant maximum current amplitude, a much smaller current (I_{d1}) is sufficient than if you want to achieve this inductance value at a constant voltage amplitude/flux linkage. The exact goal must be taken into account when formulating the design specifications.

Figure 10.93 shows the controllable filter in its function at 250 Hz. At control current $I_d = 0$, both current and voltage are practically undistorted. When a control current flows, a region of low differential inductance is inserted in the region of the zero-crossing of the characteristic curve. The differential reactance of the entire harmonic absorber is reduced and the voltage at L_{s2} is lower. At the same total voltage, the other components take over the difference and the current increases slightly. However, the resulting current distortion remains very small.

By connecting a largely linear choke in series with a significantly smaller controllable choke, it is possible to achieve a controllable, practically linear choke for the purpose described. The described procedure is only intended to describe the principle. The influence of the core material used was not considered here. The core losses lead to an apparent increase of Rs at the corresponding frequency. For the core materials, therefore, those with the lowest possible core losses should be selected.

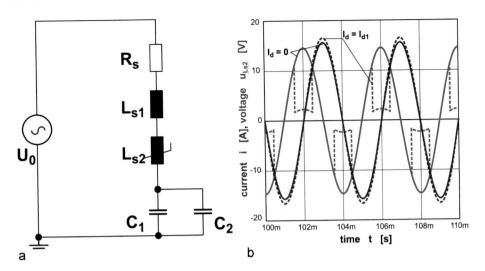

Fig. 10.93 Structure of a controllable harmonic absorber filter. (**a**) Series resonant circuit with controllable inductance, (**b**) Current characteristics and voltage drops at the controllable inductance L_{s2} at control current $I_d = 0$ and $I_d = I_{d1}$. (Maximum total inductance Lmax = 6.141 mH, minimum total inductance L_{min} = 5.677 mH, C_1 = 47 µF, C_2 = 3.3 µF) [19]

10.8   Coupled Inductors for Interleaved Operation

Phase Shifted Operation of 2-Phase DC/DC Converters

Phase-shifted operation of DC/DC converters allows currents to be split between several switching stages. The individually switched current results from the total current divided by the number of phases 'p'. The phase shift is usually 360°/p. Smaller currents allow larger commutation inductances with the same overvoltages at the switches. The staggered switching results in 'virtual' recharging of the buffer capacity with p times the frequency, so that this capacity can be reduced. In principle, magnetically uncoupled and magnetically coupled inductors can be used for the phase-shifted operation. With magnetically coupled inductors, the leakage inductance is used as the storage inductance, while the magnetizing inductance is used for coupling [20]. Figure 10.94 shows a setup in which an attempt was made to come as close as possible to the goal of a symmetrical arrangement. Three toroidal cores are used. In the middle is a highly permeable core of nanocrystalline material. The two cores on the sides are powder magnetic cores. Figure 10.94b, c shows the magnetic and the electrical equivalent circuit diagram. This is obviously a technical realization of the T equivalent circuit diagram. It is interesting to note that the functions of storage and coupling are each assigned their own materials. These can therefore be optimized for the respective task.

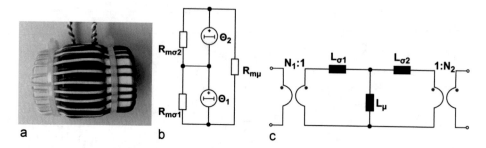

Fig. 10.94 Design of two magnetically coupled storage chokes. (**a**) from 3 toroidal cores. (source: Zacharias 2017), (**b**) magnetic equivalent circuit, (**c**) electrical equivalent circuit

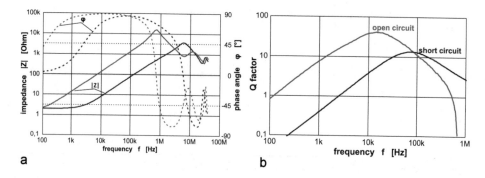

Fig. 10.95 Curves of impedance |Z|, phase angle ϕ (a) and quality factor Q (b) ($L_\mu \approx 2.13$ mH, $L_\sigma \approx 61$ µH)

The arrangement shown is a challenge in terms of production technology. However, the design induces a pure ring field in all 3 cores. Although the leakage inductances are comparatively large, there is practically no leakage field in the area around the component. Magnetic symmetry can also be achieved in other ways, as shown in this subchapter. However, the field lines emerge at least partially from the cores. Figure 10.95 shows the measurement results for Fig. 10.94. for operation as a transformer with secondary open circuit and short circuit. The curves reflect the material properties of the cores used. No other goal was pursued in the realization apart from the demonstration of the principle.

If such a choke arrangement is used in a buck converter with 2 phases and a phase shift of 180°, the circuit shown in Fig. 10.96 is obtained. Due to the magnetic coupling, the choke also acts as an autotransformer. This means that if the outputs of the two bridge branches (T_1; T_2) and (T_3; T4) are at different potentials, the average potential is formed at the unloaded output. This is equivalent to inserting the third level in a step-down converter. This property can also be used with DC/AC converters. Two half-bridges operating in parallel via magnetically coupled chokes, phase-shifted, are perceived on the output side as a 3-point circuit. For easier classification of the calculation results, a transformer equivalent circuit diagram transformed to 1:1 is used.

Fig. 10.96 Circuit of a 2-phase interleaved buck converter with magnetically coupled output chokes

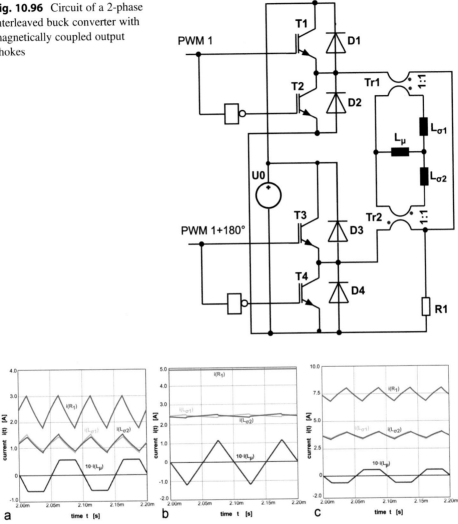

Fig. 10.97 Current characteristics in the circuit according to Fig. 10.3 at different duty cycles. (**a**) D = 0.25, (**b**) D = 0.5, (**c**) D = 0.75

The shown circuit was operated in a simulation with 10 kHz. The detailed modelling allows us to investigate the load of the individual components. Figure 10.97 shows simulation results for D = (0.26; 0.5; 0.75). It becomes clear that a current ripple of 20 kHz, twice the switching frequency, occurs at load resistor R_1. This means that each switch generates switching losses at 10 kHz, although a 20 kHz ripple can be observed at the output. This requires a much smaller smoothing capacitor in parallel to R_1 if further smoothing is necessary. For D = 0.5 the current ripple at the output practically disappears. The figure shows the calculated currents through the model inductors $L_{\sigma1}$, $L_{\sigma2}$ and L_μ.

Table 10.7 Comparison of uncosupled and coupled chokes by characteristic parameters

Parameter	$L_1 = L_2 = 1$ mH - Uncoupled -	$L_{\sigma 1} = L_{\sigma 2} = 1$ mH - Coupled with 10 mH
$\Delta\Psi_{max}(N_1) = \Delta\Psi_{max}(N_2)$	$\frac{U_0 \cdot T}{4}$ with D = 0.5 f = f_{sw}	$\frac{U_0 \cdot T}{4}$ with D = 0.5 f = f_{sw}
$\Delta\Psi_{max}(L_1) = \Delta\Psi_{max}(L_2)$	$\frac{U_0 \cdot T}{4}$ with D = 0.5 f = f_{sw}	$\frac{U_0 \cdot T}{16}$ with D = 0.25 f = $2 \cdot f_{sw}$
$\Delta\Psi_{max}(L_3)$	–	$\frac{U_0 \cdot T}{8}$ with D = 0.5 f = f_{sw}
$\Delta\Psi^2_{max}(L_1) + \Psi^2_{max}(L_2) \cdot f$	$\frac{U_0^2 \cdot T}{8}$	$\frac{U_0^2 \cdot T}{32}$
$\Delta\Psi^2_{max}(L_3) \cdot f$	–	$\frac{U_0^2 \cdot T}{64}$
$f(I_{R1}) =$	$2 \cdot f_{sw}$	$2 \cdot f_{sw}$
$\Delta I_{R1} = 0$ for	D = 0.5	D = 0.5
$Max(\Delta I_{L1})$	$\frac{U_0 \cdot T}{4 \cdot L_1}$ for D = 0.5	$\frac{U_0 \cdot T}{16 \cdot L_1}$ for D = 0.25

Fig. 10.98 Cascading of two 2-phase systems to a symmetrical 4-phase system to generate a 5-level system and further reduction of the current ripple

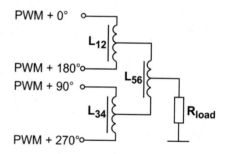

Table 10.7 compares the simulation and calculation results of uncoupled and magnetically coupled smoothing chokes.

If the Steinmetz formula is taken into account, the parameter is $\Delta\Psi^2_{max} \cdot f$ a measure of losses in the building element. With the coupled choke, you have 3 cores instead of 2. However, this potentially results in a lower power loss density. This is an interesting optimization task for the core design. The high currents can be assigned to a core material with high saturation flux density and low permeability. To what extent the sum of the core losses is then lower also depends on the core shape. On the other hand, the power loss can be distributed over 3 cores. A possible larger surface area then allows more heat to be dissipated.

A cascading of 2 interleaved step-down converters (Fig. 10.98) allows the described effects to be used even further. In [21] one finds a multitude of conversion possibilities for combining transformers for DC and AC voltage generation. Virtually the circuit appears like a 5-level circuit. Thus the output ripple becomes zero at D = 0; 0.25; 0.5; 0.75 and 1. The coupled choke L_{56} at the output can be dimensioned much smaller.

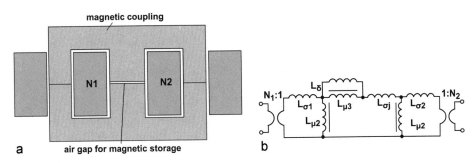

Fig. 10.99 Design of two magnetically coupled chokes with 2 E cores and an air gap. (**a**) basic design, (**b**) electrical equivalent circuit diagram

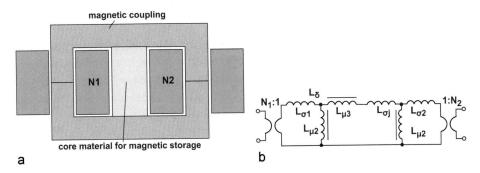

Fig. 10.100 Design of two magnetically coupled chokes with 2 U-cores and a middle leg made of a material with low permeability. (**a**) basic structure, (**b**) electrical equivalent circuit diagram

The coupled chokes in Fig. 10.94 were originally developed to minimize the external stray fields and do not correspond to the most commonly used design. Further examples are shown below. Figure 10.99 shows the most common arrangement because it is easy to carry out. The effective storage inductance is formed from the series connection of yoke leakage inductance $L_{\sigma j}$ and air gap inductance L_δ. $L_{\sigma 1}$, $L_{\sigma 2}$ contribute less than 10% to the storage inductance. A disadvantage is a partly strong stray field into the environment of the component. This leads to eddy current losses in the area of the air gap and the yoke and disturbances due to induced voltages. Wide air gaps can be divided into several smaller air gaps in series to reduce the stray fields in the windings. This, however, increases the manufacturing effort. It is problematic if - as is the normal case in the application described - direct current premagnetisation is present. This premagnetisation "shifts" the magnetisation characteristic so that the useful bipolar magnetisation stroke for coupling the choke becomes smaller. The hoped-for advantage is thus often not achieved to the extent expected.

Because of the disturbing effects of the air gap, the hybrid construction in Fig. 10.100 is used. The strong stray fields in the vicinity of the air gap, which lead to eddy current losses in the windings, are eliminated. This is achieved by using a low-permeable material in the

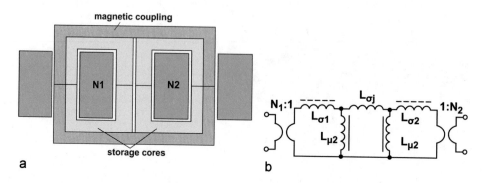

Fig. 10.101 Design of two magnetically coupled chokes with 2 U-core pairs made of different materials in pairs, each for magnetic coupling and storage, and a middle leg made of a material with low permeability. (**a**) basic structure, (**b**) electrical equivalent circuit diagram

middle leg. In the electrical equivalent circuit diagram, the corresponding inductor is connected in series with the yoke leakage inductor. This means that a relatively strong yoke leakage field still exists around the core. Here, too, there is the problem of premagnetisation of the outer magnetic circuit for coupling with direct current. This results in a reduction of the usable induction swing for magnetic coupling.

Many of the disadvantages of the above variants can be avoided by using the design shown in Fig. 10.101. This structure represents the principles presented in Fig. 10.94. However, due to the angular core shapes, a stray field, represented by $L_{\sigma j}$, is induced into the environment. Since, due to the cores used, this special structure is usually used for the winding stray fields $L_{\sigma 1} \approx L_{\sigma 2} > L_{\sigma j}$, the energy content of the yoke stray field is relatively low. A decisive difference, however, is the RF magnetization of the highly permeable material for the magnetic coupling of N1 and N2. The output current in the application according to Fig. 10.101 is ideally composed of two pulsating direct currents of equal magnitude. Due to the orientation of the windings, the fluxes in the connecting core cancel each other out. If $L_{\sigma j} \gg L_{\sigma 1} \approx L_{\sigma 2}$ applies, the asymmetrical flux and thus the DC pre-magnetization is also very low here. Since the additional stray paths for the windings are built-in through the inner cores, the stray fluxes required for energy storage are driven there. Only a very slight premagnetisation occurs. The larger $L_{\sigma j}$, the greater the asymmetry and thus the pre-magnetisation of the "coupling core". If a relatively large inductance is inserted at this very point, as in Fig. 10.99 and 100, the premagnetisation is much higher. By designing the cores as UU pairs, a low-cost version can be achieved by combining ferrite and powder cores. In principle, a modular system is possible in which even ceramic cores can be combined with cut tape-wound cores (e.g. CC with UU cores). In principle, no additional air gaps are required. In the coupling core, a small air gap can be used to compensate for the effects of DC premagnetization due to asymmetries. The yoke leakage $L_{\sigma j}$ can be kept small using the methods described in Sect. 10.4.

Larger air gaps inserted in cores cause field lines to exit from the core surface. In laminated cores or cut strip cores, this causes eddy current losses in the core itself. But also

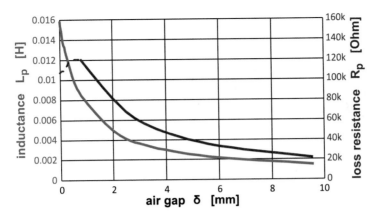

Fig. 10.102 Measured dependency of inductance Lp and loss resistance R_p on the air gap δ (core distance) for an AMCC320–40 core

in the wound material, additional eddy current losses are caused by the transverse field. The consequence of this is a parallel resistance in the equivalent circuit for the impedance of the choke which decreases as the air gap increases (Fig. 10.102).

For small air gaps, no clear trend can be seen in the present case. According to the above-mentioned assumption that additional losses occur via the effect of the exiting field lines, R_p would have to aim for a high limit value.

It should be noted that the equivalent circuit shown in Fig. 10.94 can be approximated as a physical copy of reality by means of various design measures. The equivalent circuit is characterized by a coupling inductance that can be assigned to one core and two storage inductors that can be assigned as leakage inductors to two other cores. The term leakage inductance is precisely connected with the assumption that the corresponding flux of two windings is not linked to each other via the core. If one carries out an analysis of the load quantities of the τ equivalent circuit, one obtains the result shown in Fig. 10.103 for the comparison of a buck converter (BC) with a storage choke L_d, a two-phase BC in *interleaved* mode with separate storage chokes L_d and a two-phase BC with magnetically coupled chokes L_d for the change in flux linkage $\Delta\psi$ as load quantity.

It becomes clear that the flux linkage as a load quantity is the same for the storage chokes of a BC and an *interleaved* operated BC. Because the energy to be stored in *interleaved* mode is smaller for the same ripple of the output current, the space requirement of the chokes is lower in the latter case. With magnetic coupling, the flux linkage or flux is divided among the cores used. Only the coupling core is loaded with the maximum flux at $D = 0.5$. The storage cores are thus only subjected to the difference between paths a) and b) in Fig. 10.103. This means greatly reduced remagnetization losses in these compared to the variant with separate chokes. With an ideal setup, the DC premagnetisation is in $L_{\mu=}$ 0. In reality, asymmetries in the setup or the remaining yoke leakage (see Fig. 10.8) have an effect here. In this way, powder cores with high saturation induction for L_σ and relatively

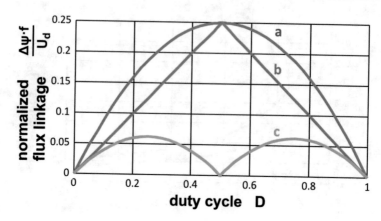

Fig. 10.103 Change in the relative flux linkage $\Delta\psi\cdot f/U_d$ as a load quantity for the inductive components in the BC. (**a**) $\Delta\psi$ to L_d for a simple BC and 2-phase BC with separate chokes and a BC with coupled choke according to Fig. 10.94, (**b**) $\Delta\psi$ to L_μ, c) $\Delta\psi$ to L_σ

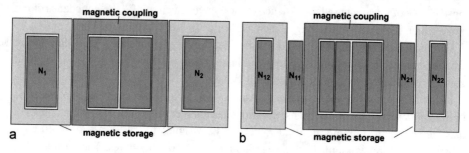

Fig. 10.104 Design variants for largely separate treatment of DC flux and AC flux. (**a**) simple rearrangement of the variant Fig. 10.101a, (**b**) winding subdivision for an additional degree of design freedom

high remagnetization losses can be combined with ferrite cores with low remagnetization losses.

Figure 10.104a shows an equivalent design to Fig. 10.101 with the same electrical equivalent circuit. The combination of cores of different materials is facilitated by this arrangement. The solution shown in [22] is similar in structure. However, the leakage flux is conducted too through the coupling core so that it is also loaded with the DC flux. This makes it impossible to separate the load sizes for different materials for storage and coupling.

Decoupling the number of turns and the inductance of the different cores is made possible by the principle of proportional chaining of the turns with the cores shown in Fig. 10.104b. The principal dependence of the flux changes is shown in Fig. 10.101.

A problem to be solved for all described core forms is that, under certain circumstances, small differences in control and operation can lead to an unbalance of the phase currents

Table 10.8 Comparison of a simple buck converter (BC) with a BC with separate storage chokes and one with magnetically coupled storage chokes

	BC original	BC *interleaved*	BC *interleaved* with autotransformer
Circuit			
Maximum current ripple in output, at	$\Delta I_{Last} = \frac{U_d}{4 \cdot L_d \cdot f_{sw}}$ $D = 0.5$	$\Delta I_{Last} = \frac{U_d}{8 \cdot L_d \cdot f_{sw}}$ $D = 0.25, D = 0.75$	$\Delta I_{Last} = \frac{U_d}{8 \cdot L_d \cdot f_{sw}}$ $D = 0.25, D = 0.75$
Ripple current frequency	f_{SW}	$2 \cdot f_{SW}$	$2 \cdot f_{SW}$
Maximum change in flux linkage L_d Maximum flux linkage change L_k	$\Delta \psi_{Ld} = \frac{U_d}{4 \cdot f_{sw}}$ 100% -	$\Delta \psi_{Ld} = \frac{U_d}{4 \cdot f_{sw}}$ 100% -	$\Delta \psi_{Ld} = \frac{U_d}{16 \cdot f_{sw}}$ 25% $\frac{U_d}{4 \cdot f_{sw}}$ 100%
Maximum current through Storage choke L_d	$I_{Ld.\,max} > I_0 + \frac{U_d}{8 \cdot f_{sw} \cdot L_d}$ 100%	$I_{Ld.\,max} > \frac{I_0}{2} + \frac{U_d}{8 \cdot f_{sw} \cdot L_d}$ > 50%	$I_{Ld.\,max} > \frac{I_0}{2} + \frac{U_d}{32 \cdot f_{sw} \cdot L_d}$ > 50%
Magnetically stored energy with the same ripple current ΔI_{Last}	$W_{magn} \approx \frac{L_d}{2} I_0^2$ 100%	$W_{magn} \approx \frac{L_d}{2} \left(\frac{I_0}{2}\right)^2$ >25%	$W_{magn} \approx \frac{L_d}{2} \left(\frac{I_0}{2}\right)^2$ >25%

[20]. Since the difficulty of the solution increases with the number of phases, the majority of applications will probably be limited to 2 phases.

Comparison of 1- and 2-Phase Buck Converters with and Without Magnetic Coupling

To compare the different circuit concepts, different parameters are used for the same switching frequency f_{sw} and the same DC voltage U_d (Table 10.8):

- Load direct current I_0
- Choke current ripple ΔI_{Last}
- Frequency of the output current f_{sw}
- Modification of the river linkage $\Delta \psi$
- Maximum choke current $I_{Ld.\,max}$
- Maximum magnetically stored energy W_{magn}

a   b

Fig. 10.105 Toroidal cores as the basis for the construction of magnetically coupled chokes. (**a**) Toroidal cores for generating leakage inductance, (**b**) Core for magnetic coupling. (Source: Zacharias 2017)

Multiphase Symmetrically Coupled Magnetic Memories

The question of symmetrical arrangements is always in the air. Often one sees core arrangements that are to be linearized and symmetrized with introduced air gaps. However, since the air gaps are introduced into the cores for magnetic coupling, but the energy storage takes place in the stray field, this approach is not appropriate. A truly complete magnetic symmetry of a magnetic arrangement also requires a symmetrical arrangement in the magnetic equivalent circuit. Therefore, the following section discusses an approach on how to build symmetrical multiphase systems from toroidal cores. The toroidal cores for storage (brown) and magnetic coupling (blue) are shown in Fig. 10.105. Please note that the stored energy in the cores is higher at lower permeabilities.

First, a symmetrical three-phase arrangement is discussed (Fig. 10.106). The blue core is used for magnetic coupling. The windings are applied in such a way that each comprises only one storage core and the coupling core. In this way, the construction principle of Fig. 10.94 is approximately implemented. In the magnetic equivalent circuit diagram, the magnetizing inductance L_μ is formed by the parallel connection of 3 individual inductors. In reality, these sections are connected in parallel by small star-shaped stray inductors. The star point is not accessible. In the case of equal direct currents through the windings, the sum cannot become zero even in the transformed case. There are also possibilities to arrange windings in the magnetic star point area. These could be used, for example, connected in delta, in order to extinguish current harmonics that can be divided by 3. For alternating current, a symmetrical flux distribution and mutual extinction can be achieved, as the magnitude and phase can be adjusted. This does not work for direct current. In

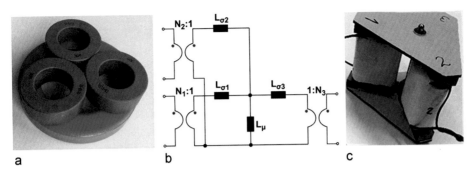

Fig. 10.106 Design example for a symmetrical 3-phase coupled choke. (**a**) basic core arrangement. (source: Zachariah 2017), (**b**) electrical equivalent circuit diagram for (a), (**c**) alternative symmetric approach of the temple-type without a center leg with different equivalent circuit. (source: Liu 2013 [20])

principle, compensation is only possible with an even number of phases. With 3-phase chokes in 3-phase DC controllers, the resulting flux is driven into the stray field. Basically, the same problem exists as with the 3-pulse star type rectifier on a Yy transformer. The DC magnetization of one leg quickly brings the transformer to saturation. To avoid saturation of the coupling core, one or more air gaps can be inserted into the coupling core. Then the energy to be stored is stored in the magnetizing inductance and not in the leakage inductance. Direct and alternating currents are not separated. This effect can also be achieved more easily. Figure 10.106c shows an experimental setup for the design that then emerges, which is reminiscent of the so-called temple-type of a 3-phase transformer. Without an air gap and a centre leg, all "iron" inductors are connected in series in the electrical equivalent circuit of this design and the energy is essentially stored in the yoke leakage inductance, that is, in the environment. To prevent this, air gaps can be inserted into the legs and a middle leg can be added. This must then absorb the total flux (alternating and direct component). The spatially symmetrical structure loses its sense in this case since the symmetrization is mainly done by the air gaps.

Finally, Fig. 10.107 shows two further approaches to how a magnetic coupling of 4 chokes for a 4-phase DC/DC converter can be implemented. If the opposite toroidal storage cores are equipped with voltages applied to the windings offset by 180°, these fluxes can cancel each other out in pairs in the coupling core. The DC flux can thus be largely kept away from this core. The energy is stored as in Fig. 10.94 in the stray field of the arranged storage rings. A symmetrical 4-phase system with a phase shift of 90° to each other is required. Virtual, additional voltage levels of the buck converter also occur here. However, the expenditure is higher than for the variant in Fig. 10.98 combined with Fig. 10.101.

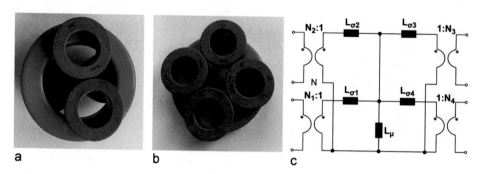

Fig. 10.107 Design examples for a symmetrical 4-phase coupled choke. (**a**) principle core arrangement with storage rings on both sides, (**b**) principle core arrangement with storage rings on one side. (source: Zacharias 2017), (**c**) electrical equivalent circuit diagram

Further Coupling Possibilities

If you go back to the structure shown in Fig. 10.94, you will notice that the function of the equivalent circuit diagram Fig. 10.94c is also represented by the equivalent circuit diagram shown in Fig. 10.108. The physical realization, which is shown in Figure 10.108b consists of an autotransformer 1: 0.5 with the magnetizing inductance L_μ with windings as closely coupled as possible and the two storage chokes with one winding of the autotransformer in series. This is a very simple construction, with which the same effects are achieved as with the arrangements shown in Fig. 10.94 and Fig. 10.104. One of the advantages is the complete decoupling of the required number of windings, which makes dimensioning much easier by decoupling the dimensioning targets for the individual components.

The structure of Fig. 10.108 can be regarded as a basic building block for *interleaved* driven buck converters with even-numbered phases, which are magnetically coupled in pairs. For example, 4- or 6-phase converters can be easily realized by combining the 180°el phase-shifted phases with the storage chokes in series with an autotransformer. If all terminals 'C' are then connected, a balanced 4- or 6-phase system is obtained. For each pair the dimensioning relationships shown in Fig. 10.10 apply.

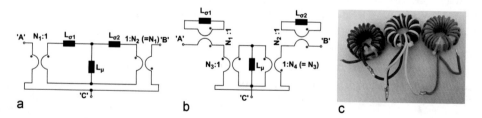

Fig. 10.108 Design of magnetically coupled storage chokes for *interleaved* operation. (**a**) Basic scheme according to Fig. 10.94, (**b**) scheme for separated cores, (**c**) exemplary design with toroidal cores of iron powder and ferrite. (source: Zacharias 2017)

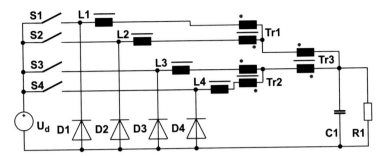

Fig. 10.109 Design of a 4-phase buck converter for *interleaved* operation with cascaded coupled chokes

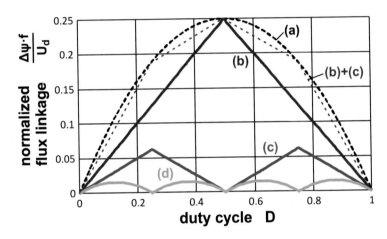

Fig. 10.110 Dependence of the change in the flux linkage on the duty cycle D. (**a**) $\Delta\psi$ with uncoupled storage chokes, (**b**) $\Delta\psi$ with coupling transformers Tr1 and Tr2, c $\Delta\psi$ with coupling transformer Tr3, d) $\Delta\psi$ with storage chokes L1...L4 according to Fig. 10.109

A further reduction of the load magnitude of flux linkage $\Delta\psi$ can be achieved if cascading is carried out as shown in Fig. 10.96. Figure 10.109 shows a version of this approach. The switches S1 and S2 as well as S3 and S4 each operate a 2-phase system in pairs offset by 180°. S1 and S3 are 90°el interleaved. A quarter of the total load current flows through each choke, as in normal *interleaved* operation. The ripple of the current at the load is the same as in normal *interleaved* operation.

However, if one examines the flux linkage of the individual components, one arrives at results that are summarized in Fig. 10.110. As reference line (a) the HF change of the flux linkage at the smoothing chokes of the simple uncoupled *interleaved* operation, which is also valid for the basic circuit of the buck converter, is entered. The characteristic (b) shows the normalized HF change of the flux linkage at the transformers Tr1 and Tr2, while (c) reflects the corresponding load of Tr3. If both dependencies (b) and (c) are added, the

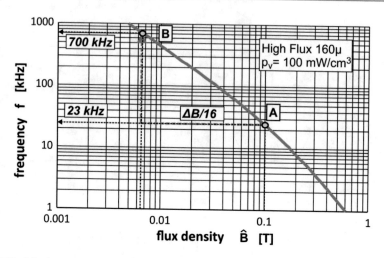

Fig. 10.111 Maximum frequency at loss density of 100 mW/cm³ powder core material 'High Flux 160 µ' at 4-phase *interleaved* operation; **(a)** without magnetic coupling, **(b)** with cascaded coupling

total HF flux linkage $\Delta\psi$ is obtained, with which the coupling inductors are loaded. For the smoothing inductors L1...4, the height of the RF flux linkage as defined in (d) remains, depending on the duty cycle. It is easy to see that the distance between this curve is even greater than in the variant described in Fig. 10.103 between (c) and (a).

This means a further reduction of the remagnetization losses. Consequently, the range of application of ferromagnetic powder cores is extended. For example, the manufacturer [23] gives the following relationship for the specific losses for the powder core material 'High Flux 160 µ'.

$$\frac{P_v}{\left[\frac{mW}{cm^3}\right]} = \left(\frac{\widehat{B}}{[0.1\ T]}\right)^{2.104} \cdot \left(2.117\frac{f}{[kHz]} + 0.1131\left(\frac{f}{[kHz]}\right)^{1.899}\right)$$

From this, for maximum specific core losses of, for example, 100 mW/cm³, the dependence of the maximum frequency on the peak value of the periodic induction with sinusoidal excitation, as shown in Fig. 10.111, can be derived. The material is given as 1.5 T as the level of the saturation induction. If one assumes here for an *interleaved* working buck converter without magnetic coupling a flux change load of $\Delta B = 0.2$ T, one obtains a maximum frequency of approximately 23 kHz. If a 4-phase buck converter with cascade coupling according to Fig. 10.109 is used, the alternating flux load is reduced by a factor of 1/16. This results in a loss density of 100 mW/cm³ at 700 kHz, which is assumed to be permissible.

The described measures of magnetic coupling can significantly reduce the magnetic alternating flux. Thus, the advantage of the high saturation induction of powder cores can be used even at high frequencies. With the cascaded magnetic coupling of 'p' phases, the alternating magnetic flux $\Delta\varphi \sim p^{-2}$ decreases:

$$\frac{\Delta\psi \cdot f}{U_d} = \frac{1}{4 \cdot p^2}$$

The core losses then play a smaller role in the dimensioning. At the same time, it must be taken into account that the total volume is increased by the additional coupling cores. The respective optimum must be sought here.

The following conclusions can be drawn from the preceding considerations:

1. The pure *interleaved* mode has advantages with regard to the size of the winding with a larger number of phases for the same current ripple with regard to the winding goods. Since the direct current is divided into 'p' phases, the total energy $W_{magn} = p \cdot \frac{1}{2} \cdot \frac{L}{p} \left(\frac{I_d}{p}\right)^2 = \frac{L}{2}\left(\frac{I_d}{p}\right)^2$ is to be magnetically stored in separate chokes.
2. With the same inductance L per phase, the ripple of the output current $\Delta I_{Last} \sim p^{-1}$.
3. With an even number of phases p, magnetic coupling can be realized in which the magnetic DC flux in the coupling core is zero.
4. The circuit with autotransformer is particularly interesting due to the comparatively low reactive power load of the inductive memory components. This is a measure of the non-ohmic losses.
5. In 4-phase and multi-phase systems, the cascaded coupling of paired stages leads to a further reduction in the alternating flux load of the storage reactors.
6. The capacitor in parallel with the load in the output experiences p times the frequency of a switch with the same current ripple. You can divide its size by p for the same voltage ripple.
7. If required, the concept also works bidirectionally and can then be combined to form inverters as usual.
8. In the case of very fast switching switches, such as SiC or GaN, the concept allows partial currents of several chips operating in parallel to be balanced at high switching speeds without the need to reduce the partial load for safety reasons as in simple parallel operation. Parasitic inductances cause less overvoltages at low currents than at high currents in relation to a switch.
9. Optimum phase number is according $p = 2^k$.

A further interesting application at the end of this systematic representation is the topology of a "*serial interleaved*" driven buck converter shown in Fig. 10.112. The voltage load of the switches is halved here. At high voltages, this is an advantage with regard to the semiconductor losses that occur. With the *duty cycle* D = 0...1 the load voltage can be changed between 0... U_d.

Fig. 10.112 *"Serial interleaved"* driven buck converter with reduced voltage load on the semiconductor switches

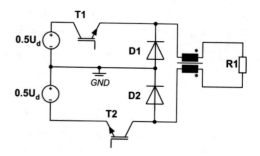

The magnetizing inductance of the two coupled windings forms a high impedance for common-mode interference. If the voltage source Ud is grounded symmetrically, a symmetrical triangular leakage current of

$$\widehat{I}_{GND} = \frac{U_d}{64 \cdot f_{sw} \cdot L_\mu}$$

The leakage current has the switching frequency of a switch. With correspondingly large capacitive AC impedances to GND, the resulting leakage current is correspondingly lower.

An approach presented in [21] in different variations is the use of coupled coils with a transformation ratio different from 1. The potential level resulting from the symmetry of the autotransformer in the middle of the feeding potentials then breaks down into two levels. As shown in Fig. 10.112, this inserts an additional level in the generated output voltage (see Fig. 10.113a). The winding ratio N_2/N_1 can be used to adjust the level of the voltage levels. A modified sine-triangle modulation for the pairs of switches T_1/T_2 and T_3/T_4 allows the adjustment of voltage values between the levels. This provides similar advantages as with

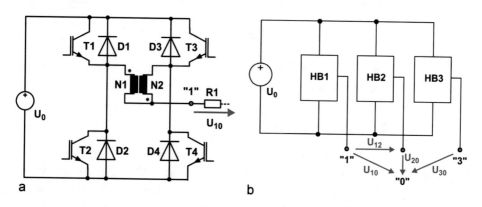

Fig. 10.113 Arrangement of a 3-phase inverter consisting of half-bridge modules (HB) with coupled coils with different numbers of turns. (**a**) Arrangement of a half-bridge module (HB), (**b**) Arrangement of a 3-phase inverter

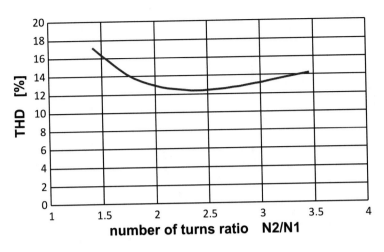

Fig. 10.114 Dependence of the harmonic content THD of the line to line voltage on the number of turns N_2/N_1

multilevel converters. The coupled windings act as a transformer in a first approximation. With appropriate leakage, this can also be used to store or smooth the current.

Figure 10.113 shows how a 3-phase bridge rectifier can be constructed from the half-bridges. While the voltages to the neutral point U_{x0} have only 2 additional levels, the line to line voltages U_{xy} have additional levels because the switching processes take place at different times. The proportion of harmonics depends on the turns ratio N_2/N_1. If the harmonics of the line to line voltage or the THD are minimized, the optimum value for the transmission ratio is obtained

$$opt\left(\frac{N_2}{N_1}\right) \approx \sqrt{6}$$

This leads to the value $THD_{op} = 12.4\%$ even without additional modulation and smoothing. In Fig. 10.114 the corresponding dependence $THD(N_2/N_1)$ is shown.

Figure 10.115 shows the curves of line to star point voltage and line to line of an inverter constructed according to Fig. 10.113b. If the dispersion of the windings is adjusted accordingly, output currents with a low harmonic content can be achieved without additional modulation. The cores of the coupled windings must carry a lower magnetic flux than without the "built-in" stages. If additional pulse width modulation is used, the material must also have low remagnetization losses. This approach is therefore particularly suitable for amorphous and nanocrystalline core materials.

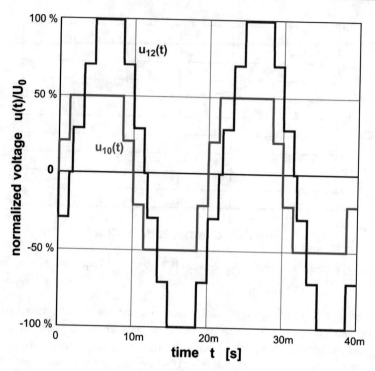

Fig. 10.115 Curves of the line to starpoint voltage $u_{10}(t)$ and the line to line voltage $u_{12}(t)$ for a circuit according to Fig. 10.20b at N2/N1 = 2.45, coupling factor k = 0.9999 (simulation)

10.9 DC/DC Converter with Reduced Voltage Stress at the Switches

Buck converters and *boost converters are among* the basic components in power electronics. They are used for energy transfer and voltage adjustment between source and load. They are hard to beat in terms of simplicity and efficiency.

When using photovoltaic solar systems, the maximum open-circuit voltage of the solar generators is 1500 V. At a given power, the current to be transmitted and thus the required cable cross-section becomes smaller and smaller as the system voltage increases. For larger systems, high voltages are therefore preferred. The basic circuits of the above-mentioned power converters require semiconductor switches which are able to switch the maximum of the occurring currents and voltages. In power supply applications, converters are expected to have a service life of >20 years. In order to minimize the statistical failure due to unavoidable cosmic radiation or its secondary influences, a maximum continuous voltage of 2/3 of the nominal voltage is generally permitted. This means that at a maximum input voltage of 1500 V a required nominal voltage of 2250 V is required for the semiconductor switches and diodes.

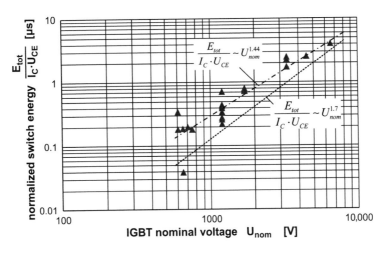

Fig. 10.116 Normalized switching losses per switching operation for silicon IGBTs of the 2nd - fifth generation. (Source: Infineon, Information data on IGBT modules, 2018)

This has a strong impact on the losses to be controlled in such a converter. As Fig. 10.116 shows, the normalized switching losses of the IGBTs in question increase disproportionately with the nominal voltage. This means that two switches of a lower voltage class connected in series may achieve lower losses than a single switch of a higher voltage class (Fig. 10.117b).

When 2 transistors and diodes are connected in series, care must be taken to ensure that the voltage is divided evenly in the static blocking or blocking case, but also during switching. The measures to achieve this usually consume energy themselves. With n elements connected in series, the power dissipation analysis provides the following data for the dissipation of the semiconductor elements for the transistors P_{VT} and the diodes P_{VD}

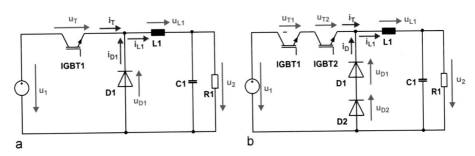

Fig. 10.117 Reduction of voltage stress per switch. (**a**) Basic circuit of a buck converter, (**b**) Use of a series connection of 2 IGBTs of a lower voltage class

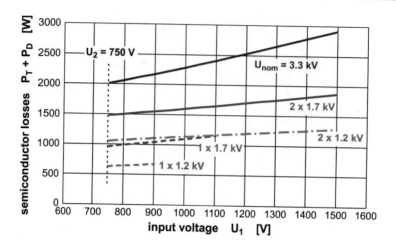

Fig. 10.118 Semiconductor losses in a buck converter (300A, 2 kHz) when using semiconductors with different nominal voltages depending on the possible continuous input voltages U_1 (1200 V: single and 2x in series, 1700 V: single and 2x in series, 3300 V: single)

$$P_{vT} = n \cdot \left(I_2 \cdot U_{TT} + I_2^2 \cdot r_{fT}\right) \cdot \frac{U_2}{U_1} + n \cdot f_{sw} \cdot I_2 \cdot \frac{U_1}{n} \cdot \left(\frac{E_{tot}}{I_{nom} \cdot U_{ref}}\right)$$

$$P_{vD} = n \cdot \left(I_2 \cdot U_{TD} + I_2^2 \cdot r_{fD}\right) \cdot \left(1 - \frac{U_2}{U_1}\right) + n \cdot f_{sw} \cdot I_2 \cdot \frac{U_1}{n} \cdot \left(\frac{E_{rec}}{I_{nom} \cdot U_{ref}}\right)$$

U_T threshold voltage
I_2 load current
r_f differential resistance
f_{sw} switching frequency
E_{tot} Switch-on energy + switch-off energy of the IGBT at datasheet conditions
E_{rec} reverse recovery energy of the diode at datasheet conditions

As the equations show, the switching losses practically do not change when several elements are connected in series, while the forward losses add up. The transition to a different voltage class of semiconductor switches must offer advantages in both areas. Since threshold voltage U_T and differential resistance r_f are different for switches of different voltage levels, the advantage must come from the switching losses.

Figure 10.118 clearly shows how great the advantages of a series connection of 2 switching elements can be. A simple combination of IGBT+diode of voltage class 1200 V and 1700 V is not sufficient to block/disable an input voltage of 1500 V permanently. For most silicon switches, it is assumed that above 2/3 of the nominal voltage a strong increase in statistical failures due to cosmic radiation or its secondary particle radiation sets in. This would result in maximum continuous voltage loads of 800 V for 1200 V switches and 1130 V for 1700 V switches as reference values. When two elements

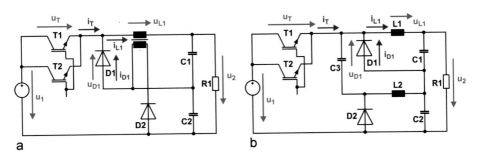

Fig. 10.119 Buck converter with modified voltage transmission to reduce the voltage load on the switch at high output voltages [24]. (**a**) with 2 coupled storage chokes, (**b**) with capacitive energy transmission to the lower buck converter path via C_3

are connected in series, the conduction losses increase, but the switching losses remain practically constant. With semiconductors of voltage class 3300 V, no series connection is necessary, but the losses are by far the greatest. Based on real module data, (fictitious) 600 A modules were uniformly assumed. At 1500 V, when using 2 modules of voltage class 1200 V in series, only 43% of the variant 1 x 3300 V with a DC line of 225 kW is required. The reduction of the voltage load on the switches can therefore lead to significant improvements. Non-generated losses also do not need to be dissipated. The series connection of 2 transistors and diodes requires measures for even voltage distribution in the switching and static cases. For the transistors, two potential-separated drivers are required. This is not the case with a parallel connection of transistors.

If you use the circuit according to Fig. 10.119, you have the situation that the switch $S_1 = T_1 + T_2$ only has to switch a part of the voltage U_1, but a higher current. The two magnetically coupled chokes force the voltages at C_1 and C_2 to be equal. The voltage is divided symmetrically between C_1 and C_2. If the circuit is designed so that $C_1 = C_2$, the output voltage divides symmetrically between the two capacitors. S_1 must then switch a higher load current. If necessary, the switch S_1 can be designed as a parallel connection of 2 transistors. The double load current, however, is far from being reached even at the rms value. This results in significant loss savings compared to the version with semiconductors in series. For the duty cycle 'D', the requirement that the DC voltage at a choke must be zero results in the relationship

$$D = \frac{\frac{U_2}{U_1}}{2 - \frac{U_2}{U_1}}.$$

Figure 10.119a shows a variant with a coupled choke. If the number of turns is equal and the windings are closely coupled, an equal voltage is forced at C_1 and C_2. The leakage inductance of the coupled windings causes overvoltages. These must be limited by snubbers. Circuit variant b) does not have this problem. In Fig. 10.119b C_3 is used for energy transmission. In both cases, the maximum voltage at the transistors is roughly

$$u_{T.\,max} \approx U_1 - \frac{U_2}{2} \le \frac{2}{3} \cdot U_{nom}$$

Under the usual assumption for the voltage design of semiconductors regarding long lifetimes, this leads to the condition for the maximum input voltage

$$U_1 \le \frac{2}{3} \cdot U_{nom} + \frac{U_2}{2}$$

U_{nom}	$U_{1.max}$ @ $U_2 = 750$ V	$U_{1.max}$ @ $U_2 = 1100$ V
1200 V	1275 V	1450 V
1700 V	1500 V	1500 V

Assuming that all components in Fig. 10.119b are ideal and the circuit works in non-stop operation, two phases can be distinguished. The boundary condition in the steady-state is that the voltages on all capacitors are equal to half the output voltage. First, in phase 1 the switch $S_1 = T_1 + T_2$ is closed. The input current is divided between two circuits. One comprises the winding L_1 and the load $_{R1}$ and the second C_3, L_2, C_2. The diodes D_1 and D_2 are blocked during phase 1 and the energy is stored in the chokes L_1 and L_2 and capacitors C_2 and C_3. The current through capacitor C_1 is negligible.

In the second phase, the switch S_1 is open. The polarity of the voltage drops at the two choke windings L_1 and L_2 changes and the inductive accumulators release part of their energy. This leads to the switching through of D_1 and D_2. Another requirement to be met is that capacitors C_2 and C_3 have the same capacitance. As a result, the instantaneous voltage on both capacitors has the same value and the blocking of diodes D_1 and D_2 occurs simultaneously. Characteristic curves of currents and voltages of this circuit are shown in Fig. 10.120.

The curves in Fig. 10.120 show that the energy exchange between the storage systems used is quite complex. In addition to the control times, the shape of the currents also changes depending on the control parameter 'D'. The effectiveness of the circuit shown is underlined by Fig. 10.121. For the intermediate circuit voltages 750 V and 1100 V, the current loads of the transistor switch were calculated and compared for the variants series connection and "parallel connection". The designation "parallel connection" was put in quotation marks for better differentiation. A parallel connection is not justified with the appropriate dimensioning. Depending on the impedances present, the current load that occurs is obviously only insignificantly greater than the load current. The energy exchange between the inductive and capacitive storage devices acts as a power matching of source and load.

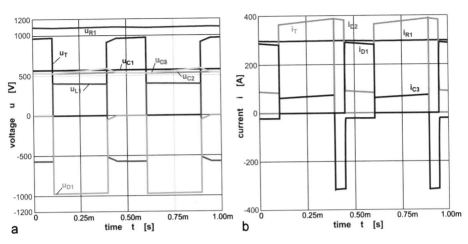

Fig. 10.120 DC/DC converter according to Fig. 10.119b. (**a**) Voltage curves, (**b**) Current curves (conditions $U_1 = 1500$ V, U2 $= 1100$ V, $L_1 = L_2 = 10$ mH, $C_1 = C_2 = 4$ mF, $C_3 = 400$ μF, $f_{sw} = 2$ kHz, $I_{Last} = 300$A)

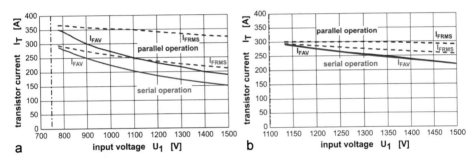

Fig. 10.121 Current stress of the transistors with a series connection according to Fig. 10.2b and the "parallel connection" according to Fig. 10.4b at a load current of 300 A, (**a**) Output voltages 750 V, (**b**) 1100 V

Figure 10.122 compares the resulting losses at an assumed output voltage of 1100 V for the variants series connection and "parallel connection". The significantly lower losses can be seen with the last variant (B). Their voltage dependence of 'B' is somewhat stronger than that of 'A', but both variants are better in their voltage class. To avoid $D = 0$, the minimum input voltage was chosen slightly above the nominal output voltage of 1100 V in this case. The disadvantage is that this converter is a higher-order system and tends to oscillate. This problem can be solved by state control.

The cascading method can also be used with step-up converters. With step-up converters in the basic circuit, the switch must be able to block the highest occurring voltage and carry a high current. Voltage ratios of >5 are therefore possible in principle but lead to a technically questionable design. Cascading results in a DC/DC controller with an

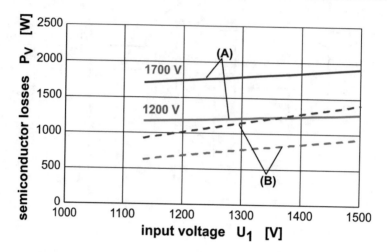

Fig. 10.122 Semiconductor losses of the circuit according to Fig. 10.4b with a load current of 300A at 1100 V at a switching frequency of 2 kHz

extended operating range. The function is similar to a voltage multiplying rectifier. A simple example is shown in Fig. 10.123.

Capacitor C_3 also serves here as a coupling between the two inductors L_1 and L_2. In the output state (T_1 blocks permanently) C_1 is charged to the input voltage. If T_1 now becomes conductive, the state of charge of the memories L_1, L_2 and C_3 changes. These take energy from U_1 and C_1, which they transfer to the capacitors C_1 and C_2 during the blocking phase of T_1. In a state of equilibrium, the voltages are as follows

$$U_{C1} = \frac{1}{1-D} \cdot U_1 \text{ and } U_{C2} = \frac{D}{1-D} \cdot U_1.$$

Where 'D' is the duty cycle of switch T_1 which switches with the period T_{sw}:

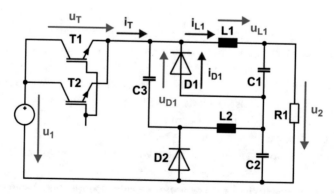

Fig. 10.123 Cascaded step-up converter for generating larger output voltages with low switch load

Fig. 10.124 Cascading of
3 step-up converters to extend
the voltage setting range

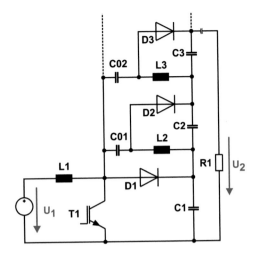

$$D = \frac{t_{on}}{T_{sw}}.$$

The load voltage results from the sum of the two voltages to

$$U_{R1} = U_{C1} + U_{C2} = \left(\frac{1}{1-D} + \frac{D}{1-D}\right) \cdot U_1 = 1 + \frac{2D}{1-D} \cdot U_1.$$

As Fig. 10.124 shows, the operating principle can also be used by adding further stages. The load voltage results accordingly in the general case for 'n' additional stages to

$$U_{R1} = \sum_{\nu=1}^{n} U_{C\nu} = \left(1 + \frac{(n+1) \cdot D}{1-D}\right) \cdot U_1$$

The transistor only needs to block the maximum voltage at C_1. It must, however, ensure that enough energy is stored in L_1 in its conducting phase so that in the blocking phase this energy is sufficient to recharge the output-side capacitors connected in parallel in this respect. In this way, transformerless DC/DC converters with a high voltage ratio can be easily implemented. At high switching frequencies, these approaches have advantageous aspects. The increasing order of the system may be a problem for control fast changing power flow.

10.10 Double Pulse Test for Semiconductor Switching Elements

The double pulse test is usually used for the dynamic characterization of semiconductors. This simple test method does not require complicated control, circuit design, loads or cooling and ensures a flexible and fast setup. In the double pulse test a special, particularly simple but very frequently occurring commutation situation is simulated. It corresponds to the *low side* switch of a half-bridge inverter. The device under test (DUT) is located in the lower voltage branch as switch S_1 (*low side*, Fig. 10.125). This makes it possible to measure voltage and current with ground reference close to the DUT, which minimizes the measurement effort and possible sources of error and justifies the choice of the arrangement. Depending on the switch to be characterized, the load is connected either between the lower voltage branch or the upper voltage branch and the center tap [25, 26].

If the actively switching semiconductor (here MOSFET) is to be evaluated as a test object, the load is connected to the upper voltage branch. Switch S_2 (*high side*) remains unactuated so that only the freewheeling diode remains effective. This arrangement corresponds to the real application. Modules with a half-bridge configuration for the switches and diodes are used as described above. In the case of individual switches, these must be supplemented with the diode with which they must cooperate for commutation. The information about the current and voltage characteristics and resulting losses are only valid for the examined pair of switch and diode. An inductive load L_s is connected opposite the positive pole. S_1 is the examined switch. To avoid feedback effects, it must be ensured that the semiconductor switch is controlled with potential separation. This requires electrical isolation between the driver circuit and its supply of control signals, as well as the operating voltage. Often the galvanic isolation is implemented in the driver circuit itself.

Fig. 10.125 Equivalent circuit diagram for a measurement setup for a double pulse test on an IGBT switch S1 with storage choke L_s with parasitic elements [27]

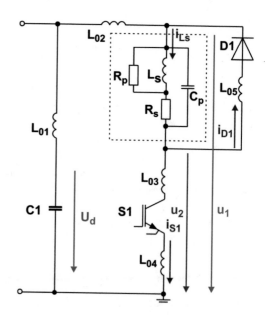

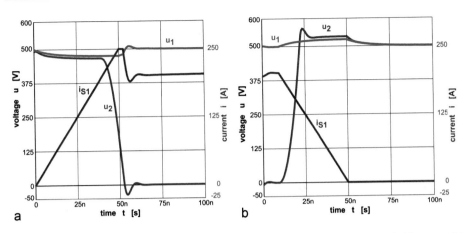

Fig. 10.126 Current and voltage curves during the double pulse test. (**a**) Switching on, (**b**) Switching off i_{S1}

Figure 10.126 shows the currents and voltages that are relevant for the observation of switching operations. During the current commutation, the current alternates between the active switch S_1 and the passive switch D_1. The following applies in the switching interval

$$i_{S1} + i_{D1} = i_{L1} \approx const.$$

From this it can be concluded that the changes in current in switches S_1 and D_1 behave in opposite directions:

$$\frac{di_{S1}}{dt} + \frac{i_{D1}}{dt} \approx 0$$

Taking into account the signed direction of the current, the current change di/dt is therefore the same for the entire commutation circuit. During the commutation phase, a voltage drop ΔU_{LK} occurs across the DC link inductance L_K, caused by the steep switching current edge di_K/dt:

$$\Delta U_{LK} = L_K \cdot \frac{di_k}{dt}$$

The inductive voltage drop Δu_{LK} in the switching process adds itself to the voltage curve of the active switch with the appropriate sign. Using the example of an active switching on of S_1 with simultaneous passive switching off of D_1, it is shown how the DC link inductance affects the switch voltage U_1 during current commutation. Assuming a negligible voltage drop across D_1, the voltage U_{S1}

$$u_{S1} = U_d - \Delta u_{LK} = U_d - L_k \cdot \frac{di_{S1}}{dt} = U_d - (L_{01} + L_{01}) \cdot \frac{di_{S1}}{dt}$$

where Ud is the voltage at the DC link capacitor. The inductive voltage drop can be determined from the measurement of U_{S1} and the inductance of the entire commutation circuit L_K can be determined from the change in current. In addition to the DC-link inductance, the inductance components of the passive switches involved in the switching process are also involved in the measured inductive voltage drop. To determine the actual DC link inductance L_K, these inductance components must be determined and subtracted from the total inductance of the commutation current circuit.

The inductance components of the diode consist of the parasitic track and bond wire inductors that connect the semiconductor chips to the switch terminals on the DC link. To determine this leakage inductance, the switch voltage U_{S1} at the switch terminals to the DC link is measured during the switch-on process of S_1. Figure 10.126, for example, shows the current and voltage curves at switch S_1 during the switching process. The inductive voltage drop at $(L_{01} + L_{02})$ not only causes the switch voltage to drop when switching on but also causes the connection voltage u_1 to drop during the switching process.

$$u_1 = U_d - \Delta U_{Lk} = U_d - L_k \cdot \frac{di_{s1}}{dt} = U_d - (L_{01} + L_{02}) \cdot \frac{di_{s1}}{dt} \text{ respectively } L_k = \frac{\Delta U_{Lk}}{\frac{di_{s1}}{dt}}$$

This gives you a possibility to estimate the inductance of the commutation circuit. The simulated example Fig. 10.126 shows the switching on and off of 20 A with a limited commutation slope. The excess current at the end of the commutation process is due to the parasitic winding capacitance of the storage choke L_s used (in this example 150 μH / 80 pF). Typical values in the range 10...50 nH were assumed for the parasitic inductances. When switching off, these lead to overvoltages at the switch. The parasitic inductances in the commutation circuit should therefore be kept as small as possible. Depending on the shape of the housing, this is limited. The measured losses are therefore not only dependent on the selected pair of switches S_1 and diode D_1, but also on the design and connection to the voltage intermediate circuit $(C_1 + L_{01} + L_{02})$ and the storage choke L_s used. The parasitic elements drawn in can be determined by various tests. Please note that the parasitic inductors in Fig. 10.125 are loss-free. If the real resistances are added, practically the same results are obtained in the simulation. A significant increase of these series resistances is not physically justified. The increase in resistance due to the skin effect is an expression of losses generated by transformers. The corresponding loss resistances are to be arranged parallel to the inductance. Higher current steepnesses then also result in higher losses. In order to map a wider frequency range with its penetration depth, a chain conductor adapted to the geometry of the conductor would have to be arranged to reproduce the impedance. Without taking the losses into account, the equivalent circuit diagram Fig. 10.125 oscillates strongly. In order to be able to examine and characterize the semiconductors close to the dielectric strength even at high operating voltages, the

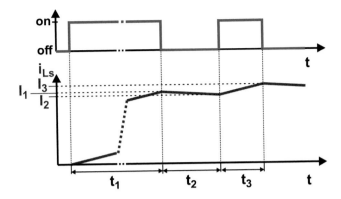

Fig. 10.127 Control signal for the active switch in double pulse mode and current through L_s during the double pulse test

overvoltage at the active switch must be as low as possible. This results in the requirement for a low-inductance DC-link design, which must be implemented for the test application. In the double pulse method, the active semiconductor is driven with two pulses. It is switched on and off twice (Fig. 10.127). The setting of the first pulse length t_1 determines the desired current level I_1 for switching off the active switch, which should be present at the beginning of commutation:

$$I_1 = \frac{U_d \cdot t_1}{L_s}$$

If S_1 is switched off, the current $i_{Ls}(t)$ decreases according to the relationship

$$i_{Ls}(t) = I_1 \cdot \exp\left(\frac{-(t-t_1)}{\tau}\right) = \frac{U_d \cdot t_1}{L_s} \cdot \exp\left(\frac{-(t-t_1)}{L_s} \cdot R_s\right).$$

In the time t_2 after switch-off, the current in L_s changes by the amount

$$\Delta I_{Ls} = i_{Ls}(t_1 + t_2) - i_{Ls}(t_1) = I_1 \cdot \left(\exp\left(\frac{-t_2}{\tau}\right) - 1\right) = \frac{U_d \cdot t_1}{L_s} \cdot \left(\exp\left(\frac{-t_2}{L_s} \cdot R_s\right) - 1\right)$$

Assuming $\tau \ll t_2$ one can break off a Taylor series after the first term and obtain

$$\Delta I_{Ls} \approx -\frac{U_d \cdot R_s \cdot t_1 \cdot t_2}{L_s^2} \text{ respectively } \frac{\Delta I_{Ls}}{i_{Ls}(t=t_1)} \approx -\frac{R_s \cdot t_2}{L_s}$$

The loss resistance R_s but also the current R_p of the storage inductance representing the eddy current losses must be kept very small. Then it is possible to realize relatively long turn-off times t_2 without the current iLs changing significantly during this time. If switch S_1 is then switched on again, practically the same current flows as when switching off at t_1. Thus, a pair of related switch-on and switch-off curves for currents and voltages is generated for each measuring cycle. From these, the losses per switching operation can be determined with high-resolution and high-frequency scanning oscilloscopes.

Coreless coils are used as storage inductance L_s in the measurement setup. Depending on the level of the DC link voltage and the current level to be achieved, the inductance can range from a few µH to several 100 µH. Figure 10.125 already indicates that in the HF range investigated, the current storage choke represents a parallel resonant circuit $L_s//C_p//R_p$. The use of storage choke with core makes the considerations unnecessarily complicated.

Since a high current usually flows in L_s when the device is switched on, a large amount of energy is required within a short time. This cannot usually be covered by the high-voltage source connected to the DC link. To ensure that the voltage on the DC link does not drop too much due to the energy requirement of the load, capacitors on the DC link must briefly cover the required energy requirement. This leads to a further requirement that is placed on the design of the DC link: The total capacitance on the DC link must be sufficiently large to prevent the DC link voltage from collapsing too much. If a maximum voltage dip from ΔU_d to C_1 is specified, the following condition applies to C_1

$$C_1 \geq \frac{U_d(t=0)}{2 \cdot L_s \cdot \Delta U_d}(t_1 + t_3)^2$$

Figure 10.4 shows a measurement setup for a double pulse test. Control and sources for supplying the DC intermediate circuit and for heating the test objects are embedded in a safety concept by interlocks. For common-mode suppression of the test setup against the environment, all supply lines for the power supply are provided with current-compensated chokes. The DC intermediate circuit consists of a series of different capacitors. On the one hand, these should provide the required pulse energy and on the other hand ensure a low commutation inductance L_K. Electrolytic capacitors are therefore connected in parallel with film capacitors and ceramic capacitors at different distances from the test object. The capacitors with the highest natural resonant frequency are placed closest to the commutation branch to achieve a low commutation inductance $L_{01} + L_{02}$. To have a low capacitive impedance for the DUT over a wide frequency range, capacitors with extremely low self-inductance are placed near the switch. To be able to use a DC link for different applications, the structure in Fig. 10.128a is modular. A miniature switching cell, whose layout is adapted to the switch, is connected in parallel with a DC link that functions as an energy storage device. In order to achieve a reliable statement about the losses in the application, the layout in the vicinity of the switch must correspond to that of the application. The storage coil used in this case is shown in red in the upper left corner. To ensure protection against contact and explosion, hinged covers with limit switches are used for monitoring the test stand.

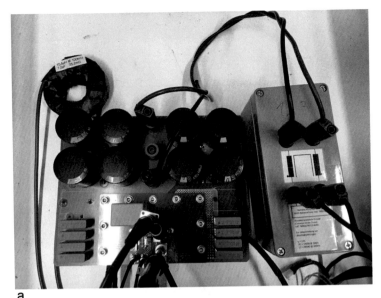

a

b

Fig. 10.128 Setup of a device for the double pulse test of GaN transistors (**a**) Switching cell, (**b**) Overall setup. (Source: Zacharias 2018)

The properties of a storage choke of 425 µH at 100 kHz and 22 pF and $R_s = 0.17$ ohms are shown in Fig. 10.129. It is a coreless solenoid choke (red). The natural resonance is 1.5 MHz. Up to 1 MHz, there is a clean inductive behaviour ($\varphi = 90°$).

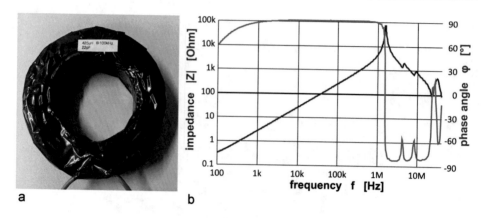

Fig. 10.129 Short coreless solenoid coil for measurements using the double pulse principle. **(a)** Design. (source: Zacharias 2018), **(b)** Frequency and phase response of the impedance of this coil

10.11 Possibilities of Influencing Leakage Inductances

In the case of inductive components with several windings, their satisfactory function is usually tied to a certain tolerance range for the leakage inductance when the coils are coupled. Although this topic is addressed from different perspectives in the present book, this subchapter provides an overview of different methodological approaches for the targeted design of leakage inductances [28].

Definition

In the literature, the terms leakage field and leakage inductance are used. While the term leakage inductance is precisely defined, the term leakage field is not used. In the sense of "the leakage field", the term leakage field also refers to the magnetic field that breaks out of the intended magnetic core and penetrates the environment, thus causing its own effects. The main reason for this is the fact that there are no magnetic non-conductors / insulators as with the electric current field. The best electrical conductor is silver with a conductivity of $61 \cdot 10^6$ S/m. Materials with a conductivity of $<10^{-8}$ S/m are considered non-conductors. The ranges of conductors and non-conductors are thus separated by 12...14 powers of ten in the order of magnitude. The comparable magnitude of relative permeability in a vacuum is 1 and for the best ferromagnetic materials in the order of $10^5...10^6$. Every magnetic arrangement thus has a field that scatters into the environment.

In contrast, the leakage inductance with its leakage field is a quantity that describes the coupling between two coils (Fig. 10.130). Formally, the coupling factor is a quantity that describes the flux in a winding in relation to the total flux generated. Consequently, two windings also have the two coupling factors k_{21} (winding N_1 winding $\rightarrow N_2$) and k_{21} (winding N_2 winding $\rightarrow N_1$). The coupling factor 'k' is the geometric mean of the two coupling factors.

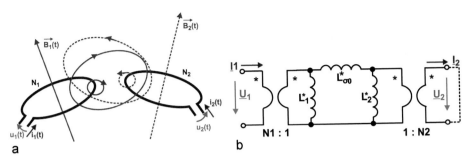

Fig. 10.130 Coupling two coils N_1 and N_2: (**a**) 2 windings with implied flux components, (**b**) electrical equivalent circuit

$$k = \sqrt{k_{12} \cdot k_{21}}$$

with

$$k_{12} = \frac{\varphi_{21}}{\varphi_{11}} \text{ and } k_{21} = \frac{\varphi_{12}}{\varphi_{22}}$$

The extra measurable (self-) inductance for each coil is L_1 and L_2. The mutual inductance of both coils is the geometric mean of both values combined with the coupling factor:

$$M = k\sqrt{L_1 \cdot L_2} = \sqrt{k_{12} \cdot k_{21} \cdot L_1 \cdot L_2}$$

The four-pole equations of a loss-free transformer are thus

$$u_1 = L_1 \cdot \frac{di_1}{dt} \pm M \cdot \frac{di_2}{dt}$$

$$u_2 = \pm M \cdot \frac{di_1}{dt} + L_2 \cdot \frac{di_2}{dt} \text{ or in complex notation } \begin{bmatrix} u_1 \\ u_2 \end{bmatrix} = \begin{bmatrix} j\omega L_1 & \pm j\omega M \\ \pm j\omega M & j\omega L_2 \end{bmatrix} \cdot \begin{bmatrix} i_1 \\ i_2 \end{bmatrix}$$

The minus sign applies to the counting directions used in Fig. 10.30.

This equation system is represented by the π equivalent circuit diagram (Fig. 10.131). The parameters occurring here are the (magnetizing) inductances or magnetic conductance $L_1^* = \Lambda_1 L_2^* = \Lambda_2$ and the leakage inductance or leakage conductance $L_\sigma^* = \Lambda_\sigma$. The designations used indicate the equivalence of two points of view. L_x^* is an inductance quantity of the electrical equivalent circuit diagram transformed to the number of turns 1. Λ_x is the magnetic conductance of the associated magnetic equivalent circuit diagram. Due to the interactions in the magnetic circuit, the individual model components cannot be

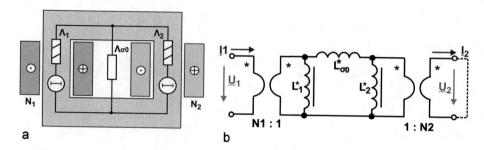

Fig. 10.131 Design of (**a**) transformer with a loose coupling (**b**) π-Equivalent circuit diagram for 2 magnetically coupled coils

measured separately. As is usual for the determination of four-pole parameters, conditions can be formulated which enable metrological determination.

a) Output side open circuit	
Measuring of	
Input impedance	*voltage transfer ratio*
$\underline{Z}_{e0} = j\omega N_1^2 \left(L_2^* // \left(L_{\sigma 0}^* + L_2^* \right) \right) = j\omega L_1$	$\frac{U_2}{U_1} = \frac{L_2^*}{L_{\sigma 0}^* + L_2^*} \cdot \frac{N_2}{N_1}$
With the approximation	
$\underline{Z}_{e0} \approx j\omega N_1^2 \left(L_1^* // L_2^* \right)$ if $L_\sigma^* << L_1^*, L_2^*$	$\frac{U_2}{U_1} \approx \frac{N_2}{N_1}$
b) Output side short circuit	
Measuring of	
Input impedance	*current transfer ratio*
$\underline{Z}_{eK} = j\omega N_1^2 \left(L_1^* // L_{\sigma 0}^* \right)$	$\frac{I_{2K}}{I_1} = \frac{L_1^*}{L_{\sigma 0}^* + L_1^*} \cdot \frac{N_1}{N_2}$
With the approximation	
$\underline{Z}_e \approx j\omega N_1^2 L_{\sigma 0}^*$ if $L_{\sigma 0}^* << L_1^*, L_2^*$	$\frac{I_{2K}}{I_1} \approx \frac{N_1}{N_2}$
c) Input side open circuit	
Measuring of	
Output impedance	*voltage transfer ratio*
$\underline{Z}_{a0} = j\omega N_2^2 \left(L_2^* // \left(L_{\sigma 0}^* + L_1^* \right) \right) = j\omega L_2$	$\frac{U_1}{U_2} = \frac{L_1^*}{L_{\sigma 0}^* + L_2^*} \cdot \frac{N_1}{N_2}$
with the approximation	
$\underline{Z}_{a0} \approx j\omega N_2^2 \left(L_1^* // L_2^* \right)$ if $L_{\sigma 0}^* << L_1^*, L_2^*$	$\frac{U_1}{U_2} \approx \frac{N_1}{N_2}$
d) Input short circuit	
Measuring of	
Output impedance	*current transfer ratio*
$\underline{Z}_{aK} = j\omega N_2^2 \left(L_2^* // L_{\sigma 0}^* \right)$	$\frac{I_{1K}}{I_2} = \frac{L_2^*}{L_{\sigma 0}^* + L_2^*} \cdot \frac{N_2}{N_1}$
with the approximation	
$\underline{Z}_e \approx j\omega N_2^2 L_{\sigma 0}^*$ if $L_{\sigma 0}^* << L_1^*, L_2^*$	$\frac{I_{1K}}{I_2} \approx \frac{N_2}{N_1}$

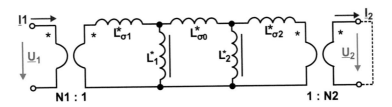

Fig. 10.132 Equivalent circuit diagram of a transformer with "strong yoke leakage" taking into account the unlinked fluxes in the windings

Thus 8 measurements are possible to parameterize the model. Problematic is the non-linearity of the core. With ferromagnetic cores, the measurements must be carried out at the same magnetization state for all measurements. This is usually impossible. Therefore, different approximation methods are used. The mentioned approximations are only examples. In general, the π equivalent circuit diagram can also be converted to a τ equivalent circuit diagram. π and τ equivalent circuit diagrams are only approximate images of physical reality. The physical reasons for both are usually present at the same time in a structure with different weighting. The equivalent circuit diagram itself is already an approximation of the physical reality. Because of the possibility of transformation, both variants can be found in the literature. The τ equivalent circuit diagram reflects closely coupled coils physically better justifiable than the π equivalent circuit diagram. The latter represents loosely coupled windings or pairs of windings with strong yoke leakage physically better. It applies approximately $L_{\sigma T1} + L_{\sigma T2} \approx L_{\sigma \pi}$ to the leakage inductances as shown in the τ and π equivalent circuit diagram.

Except for special constructions, the magnetic field of the leakage inductance is formed in the air or paramagnetic material (Fig. 10.131). Thus the leakage inductance becomes a linear component. If the unlinked flux components in the windings are also taken into account, Fig. 10.132 is obtained. A common approximation is to use the approximation $\mu_r \rightarrow \infty$ when estimating the leakage inductance for the ferromagnetic parts of their permeability or magnetic conductivity. This simplifies many calculations. If finite values are used for the magnetizing inductances, it must be taken into account during parameterization that the inductances concerned are generally not constant.

As a rule, the effects of the non-linked fluxes are combined to form an effective leakage inductance $L^{*}_{\sigma.eff} = L^{*}_{\sigma 1+} L^{*}_{\sigma 0+} L^{*}_{\sigma 2}$. At the same time the relation $L^{*}_{\sigma} \gg (L^{*}_{1}, L^{*}_{2})$ is assumed. This results in a simplified equivalent circuit diagram for short circuit proximity (Fig. 10.133). For a given AC voltage at one winding, the energy content of L_{σ} is at its maximum when the other winding is short-circuited.

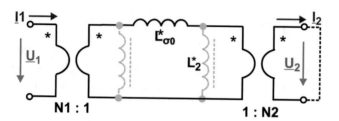

Fig. 10.133 Simplified electrical equivalent circuit diagram for short circuit proximity

$$W_{\sigma.\,max} \approx \frac{N_1^2 L_{\sigma.eff}^*}{2} I_1^2 = \frac{N_1^2 \Lambda_{\sigma.eff}}{2} I_1^2$$

$$W_{\sigma.\,max} \approx \frac{N_2^2 L_{\sigma.eff}^*}{2} I_2^2 = \frac{N_2^2 \Lambda_{\sigma.eff}}{2} I_2^2$$

Derived from this, the following can be written for the leakage inductances measured from the primary or secondary side

$$L_{\sigma.eff1} = N_1^2 L_{\sigma.eff}^* = N_1^2 \Lambda_{\sigma.eff} = N_1^2 \frac{\mu \cdot A_{\sigma.eff}}{l_{\sigma.eff}}$$

$$L_{\sigma.eff2} = N_2^2 L_{\sigma.eff}^* = N_2^2 \Lambda_{\sigma.eff} = N_2^2 \frac{\mu \cdot A_{\sigma.eff}}{l_{\sigma.eff}}$$

This clearly shows the possibilities of influencing the effective leakage inductance:

– Effective cross-section for the propagation of the leakage field
– Effective length of the leakage field lines
– Material properties in the leakage channel
– Number of windings/flux linkage

The possibilities of influence with associated design approaches are discussed below (Fig. 10.134).

Corresponding measures for winding pairs with loose coupling/weak yoke leakage are shown in Fig. 10.135. The main cause for the storage of energy in the leakage field is here the field that forms between the upper and lower "yoke". Apart from the length of the windings, the winding structure plays a minor role here. Since the leakage field lines close mainly in the space between the windings, the width 'b' of the yoke plays a role. But leakage field lines also emerge from the side surfaces and the upper and lower sides of the yoke. The available area in these directions therefore also plays a role (dimensions a and c). The geometry of the selected core has great importance for the leakage inductance.

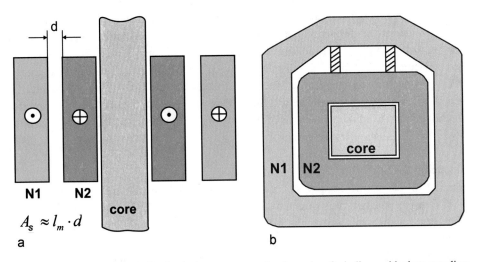

Fig. 10.134 Change of the effective leakage cross-section for pairs of windings with close coupling. (**a**) Reinforced winding insulation, (**b**) Insertion of additional areas for the leakage cross-section by means of spacers, which can also be used for cooling purposes. (lm: average winding length)

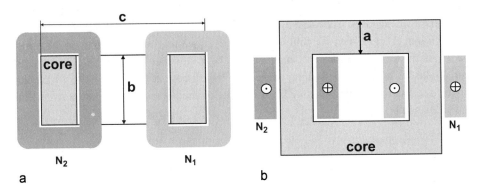

Fig. 10.135 Schematic diagram of a transformer with strong yoke leakage. (**a**) Sectional view, (**b**) side view

Influencing the Effective Cross-Section for the Propagation of the Leakage Field

The magnetic conductance Λ_σ for the leakage field increases as the areas penetrated by the leakage field lines become larger. Figure 10.134 shows examples of closely coupled windings. The leakage field is essentially formed between the windings. The value Λ_σ can be influenced by a more or less thick winding insulation (Fig. 10.134a). Limits are set here by the winding insulation. The leakage field of a pair of windings is not only formed between the windings but rises in a winding starting from zero up to the maximum value at the edge of the winding gap. The field strength remains almost constant in the gap between the windings, only to be reduced to zero again as a result of the counter-flooding of the other winding. The windings thus also contribute to the energy content of the leakage field.

Thus the leakage inductance does not become zero even if the distance between the two windings were zero. A further reduction would only be possible if further measures were taken (see below). Figure 10.134b shows one possibility of increasing the leakage conductance by artificially increasing the distance between the windings/the leakage field cross-section. To do this, an electrically insulating spacer, for example, made of plastic, is inserted into the space between the windings. This measure can also be used to improve the cooling of the windings or sections of the windings. With the latter, spacers are inserted between layers of an otherwise compact winding.

Influencing the Spreading Path Length

Figure 10.136 shows two examples of how to influence the spreading path length $l\sigma$. In the case of Fig. 10.136a, two coils wound on top of each other are visible. If we assume that the space between the coils is mainly a carrier of the leakage field, the coil length $l_{\sigma 2}$ corresponds to the length of the leakage field. Short coil pairs (disc windings) therefore have a higher leakage conductance Λ_σ and thus a higher leakage inductance L_σ than long coil pairs (tube windings). If the windings are arranged on different legs - as is the case with arrangements with large yoke leakage - the coil length also plays a role in the leakage conductance. As a rule, however, the smallest possible dimensions should be used. Shorter coils increase the magnetic yoke conductance because of the area for the leakage field lines exiting the coil increases. The same winding window area can also be accommodated in a core with longer legs. Then the following applies to two tube windings on one leg each: The longer the legs of the core, the smaller the leakage inductance. Coil arrangements with strong yoke leakage / loose coupling often form a very complex leakage field, which can only be quantitatively determined by FEM calculation methods. At this point, only general tendencies can be given.

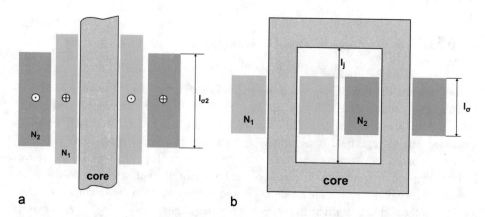

Fig. 10.136 Influencing the spreading length L_σ via the design of winding arrangement. (**a**) for windings with close coupling, (**b**) with loose coupling

Adjustment of Leakage Inductance via Material Properties

The leakage inductance can also be influenced by the permeability (magnetic conductivity). For example, by introducing ferromagnetic material into the leakage channel, the leakage conductance Λ_σ and thus the leakage inductance L_σ can be increased. Figure 10.137 shows examples of this. In Fig. 10.137a, for example, it is assumed that a foil with ferromagnetic properties is arranged in the space between the windings. Such foils usually have relative permeabilities in the range $\mu_r = 5...150$. In the electrical equivalent circuit diagram, an inductance representing the properties of the inserted core must then be entered in addition to the inductance of the "air space" of one winding (Fig. 10.137b). It should be noted that all ferromagnetic materials have saturation properties and generate additional core losses. It is also possible to only partially guide the leakage flux via a magnetic conductor, as shown in Fig. 10.138a. Figure 10.138b shows the corresponding electrical equivalent circuit diagram. One part of the leakage inductance now contains a part with and without a core. Depending on the winding conditions, the leakage inductance can be easily adjusted

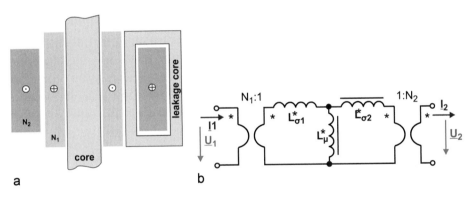

Fig. 10.137 Use of ferromagnetic material in the leakage field to design the leakage inductance when a winding is completely enclosed. (**a**) Winding structure, (**b**) equivalent circuit diagram

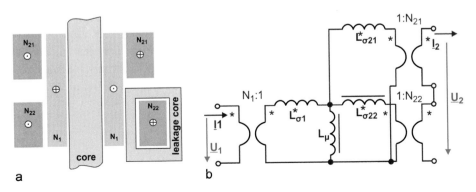

Fig. 10.138 Use of ferromagnetic material in the leakage field to design the leakage inductance when a winding is partially enclosed: (**a**) Winding structure, (**b**) equivalent circuit diagram

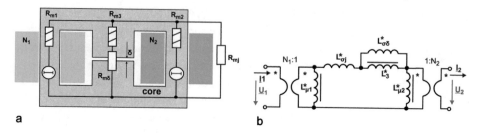

Fig. 10.139 Reinforcement of the yoke leakage inductance by a middle leg with an air gap. (**a**) geometrical structure, (**b**) electrical equivalent circuit diagram

to the desired value, provided it is possible in principle. This approach allows a detailed adjustment of the leakage inductance to the requirements of the component. It should be noted that the introduction of ferromagnetic material into the leakage channel leads to non-linear leakage inductances.

In the leakage field, energy is transferred $\vec{S} = \vec{E} \times \vec{H}$ between the coils via the vector field named after John Henry Poynting. All materials in this field can potentially cause losses and heat up. Simulations of the electrical equivalent circuit diagram provide an overview of how strongly the materials are loaded and which field strengths and perspective losses are achieved. Since all ferro- and ferrimagnetic materials exhibit saturation phenomena and more or less strong remagnetization losses, these properties can also be found in the leakage channels equipped with them. These properties must be taken into account when dimensioning.

The design approach shown in Fig. 10.139 is suitable for coil arrangements with strong yoke leakage, which experience has shown to have significantly higher leakage inductances (approx. 10...30 times) than coil arrangements with low yoke leakage. Here, the two coils are arranged on different legs of the magnetic core. The leakage flux is essentially formed as a yoke leakage flux between the windings. The conductance for the yoke flux leakage increases if an additional leg is inserted for it. With an air gap 'δ' inserted into this, the leakage conductance Λ_σ and thus the leakage inductance L_σ can precisely adjusted. Figure 10.139b also shows the corresponding electrical equivalent circuit diagram. It must be pointed out that with such an arrangement, influence is only dominant with relatively small air gaps. The equivalent circuit diagram shows that the leakage conductance Λ_σ is formed by

$$\Lambda_\sigma = \frac{R_{m\delta} + R_{m3} + R_{mj}}{(R_{m\delta} + R_{m3}) \cdot R_{mj}}$$

This means that the maximum value of the leakage coefficient is reached according to this formula for $\delta = 0$ with

$$\Lambda_{\sigma.\,\max} = \frac{R_{m3} + R_{mj}}{R_{m3} \cdot R_{mj}}$$

If you make the air gap very large ($\delta \to \infty$), then the sum of ($R_{m\delta} + R_{m3}$) $\to \infty$. This means a minimum conductance value of

$$\Lambda_{\sigma.\,\min} = \frac{1}{R_{mj}}$$

This leakage conductance is geometrically defined by the core and cannot be undercut. In many cases, the air gap will be smaller than

$$\delta < \frac{1}{3}\sqrt{A_\sigma}$$

because above this value only small changes in leakage inductance can be achieved with large changes in the air gap 'δ'. A completely different effect can be achieved by inserting electrically conductive material into the leakage channel. If one places there well electrically conductive materials with $\mu_r \approx 1$, eddy currents are induced in these materials. According to Lenz's rule, these are directed against the cause. Materials with a thickness greater than 3 times the penetration depth of the alternating current, that is,

$$d > 3s = \frac{3}{\sqrt{\pi \cdot f \cdot \mu \cdot \sigma}}$$

are therefore in the figurative sense a magnetic non-conductor for correspondingly high frequencies in the corresponding area. Through the effect of eddy currents, they shield the areas behind them from magnetic field lines. At the price of somewhat increased losses, the field of existing transformers with strong yoke leakage can also be shaped in this way afterward. The leakage inductance is reduced in this way. The final leakage inductances effective during operation can only be measured when the inductive component is installed in its final position. The maximum load in the inserted material only occurs at the maximum current or secondary short circuit.

On the other hand, this also means that one must expect a corresponding self-heating of electrically conductive components that are positioned in the leakage field of a transformer. Mounting material and housing should therefore preferably conduct poorly or not at all and have a relative permeability of 1. Oil containers or housings are made of metal for various reasons. Here, care must be taken to ensure that they are not ferromagnetic materials, as these are associated with sometimes considerable additional losses. Figure 10.11 shows a coil arrangement with strong yoke leakage and forming of the yoke leakage field with an aluminium profile. At an operating frequency of 50 kHz and a conductivity $\sigma_{Al} = 45.45$ S/

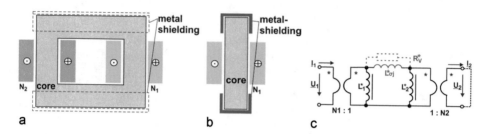

Fig. 10.140 Leakage field formation by electrically conductive mounting parts of a transformer with strong yoke leakage: principle construction front view (**a**), side view (**b**), electrical equivalent circuit diagram

m as well as a relative permeability $\mu_r = 1$, a minimum wall thickness of >1 mm results for the aluminium profile. The generated opposing field makes the aluminium profile impermeable to the original leakage field, whose field lines emerge at the core surface. The original leakage inductance can thus be reduced to about 1/3 (Fig. 10.140).

Influencing the Leakage Inductance Via the Flux Linkage of the Windings Involved
From the derivation of the principal influencing variables for the leakage inductance, it can be seen that in the basic formula there is a quadratic dependence $L_\sigma \sim N^2$. This is only present if all windings include all flux lines. The flux linkage is then maximum. If each winding only includes its own flux, the resulting inductance of the series connection of 'N' windings is only $L \sim N$. This relationship can be used as a basis for further design approaches to leakage inductance. For example, if a voltage is impressed on a transformer, the minimum number of windings is fixed for a given core. The saturation induction alone determines how many windings are required to prevent the core from saturating, to prevent the magnetizing current from becoming too high and to keep core losses within acceptable limits. After a decision for close or loose coupling, the leakage inductance would be fixed in the basic structure. This is not the case, however, if one considers that the principles described above can also be applied to partial winding pairs. Figure 10.141 shows winding

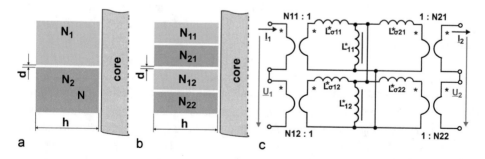

Fig. 10.141 Winding subdivision. (**a**) Simple winding pair of the primary and secondary winding, (**b**) subdivided windings (asymmetrical division), (**c**) associated electrical equivalent circuit diagram

sections of a transformer. Two transformers with the same number of windings are compared. One has only a primary and secondary winding as a disk winding [29]. In the other transformer, the primary and secondary windings are divided in half. The leakage flux can be adjusted by the measure 'd'. The windings are connected in pairs in such a way that the magnetizing fluxes of the winding components are aligned equally. That is, the corresponding induced voltages are added. Compensation of the magnetic fluxes occurs in pairs in the event of a short-circuit of an entire winding.

The electrical equivalent circuit diagram in Fig. 10.141c illustrates this. For illustration purposes, it is assumed that the leakage conductance of a pair of partial coils is In case of two-disc windings it is assumed that the following approximation applies to the leakage conductance

$$l_\sigma = h$$
$$A_\sigma = l_m \cdot d$$

(h: winding height, d: winding distance, lm: average winding length)

Without the nesting shown, if both windings were divided in half, the leakage inductance would be about

$$L_{\sigma 1}^* = N_1^2 \cdot \Lambda_\sigma$$

If, however, the winding is interlaced as in Fig. 10.141, only half of the turns are interlinked with half of the flux at any one time and the approximate approach for pairs of windings with equal spacing yields

$$L_{\sigma 1} = L_{\sigma 1a} + L_{\sigma 1b} = \left(\frac{N_1}{2}\right)^2 \cdot (L_{\sigma 11} + L_{\sigma 12}) + \left(\frac{N_1}{2}\right)^2 \cdot (L_{\sigma 11} + L_{\sigma 12})$$

$$L_{\sigma 1} = 2 \left(\frac{N_1}{2}\right)^2 \cdot \Lambda_\sigma = N_1^2 \cdot \frac{\Lambda_\sigma}{2}$$

The series connection of winding components with unlinked fluxes allows the resulting leakage inductance to be specifically influenced. The following principles can be derived from this

a) *The winding division into winding pairs on one leg*

A distinction is made here between symmetrical and asymmetrical winding subdivisions (Fig. 10.141a and b). If one imagines a winding short-circuited in such an arrangement, one obtains the principle curve of the field strength due to the primary and secondary currents shown in Fig. 10.142. If one has the course of the leakage field strength, the energy content of the leakage inductance results in

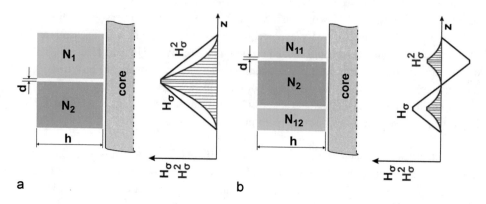

Fig. 10.142 Winding subdivision - principles in comparison (1). (**a**) Asymmetrical winding subdivision, (**b**) symmetrical winding subdivision in disc windings with the qualitative progressions of H (z) and H^2(z)

$$W_\sigma = L_{\sigma 1} \frac{I_1^2}{2} = \frac{\mu_0}{2} \int_V H^2(x) \cdot dV = L_{\sigma 2} \frac{I_2^2}{2}$$

$$L_{\sigma 1} = \frac{\mu_0}{I_1^2} \int_V H^2(x) \cdot dV$$

The area under the function H^2(x) is larger in the asymmetrical case than in the symmetrical case. So the simple change in the winding distribution unsymmetrically → symmetrical leads to a reduction of the leakage inductance. Figure 10.143 shows how this approach can also be implemented for tube windings.

The use of symmetrical winding structures instead of asymmetrical structures leads to a significant reduction of the leakage inductance by a factor of 3...4 with a small loss of the surface of the winding window in reality.

Another approach to reduce leakage inductance is to increase the number of partial windings. Here, a minimum in leakage inductance is achieved if, in the event of a short-circuit of one winding, a further conductor is arranged in parallel to each conductor, in which the opposite flux is generated. With 1:1 transformers, this is referred to as bifilar windings. Here, a double conductor is wound up. The leakage inductance then corresponds to the line inductance of this double line. Integer transformation ratios can be achieved by dividing the conductor with the higher current into several partial conductors, each of which has an equal current in the opposite direction to the other.

b) *Series connection of closely and loosely coupled winding components*

If you want to realize a transformer arrangement with a certain load characteristic, you have the possibility of using closely coupled and loosely coupled winding components at the

Fig. 10.143 Winding subdivision - principles in comparison (1). (**a**) Unsymmetrical (**b**) symmetrical winding subdivision for tube windings

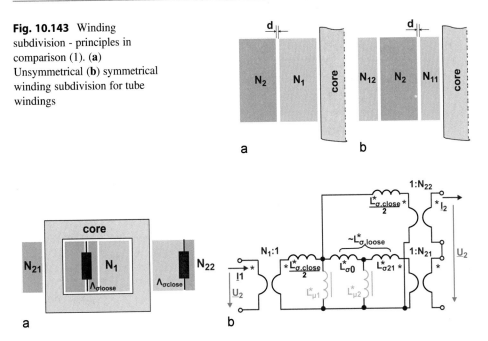

Fig. 10.144 Combination (series connection) of winding components with close and loose coupling. (**a**) Basic arrangement, (**b**) Electrical equivalent circuit diagram [28]

same time (Fig. 10.144). If, as in this example, you have only one primary winding and two secondary windings in series, these can be loosely or closely coupled at the same transformation ratio $N_1 : N_2$ to influence the output current at the same output voltage. The values of the achievable leakage inductance according to Fig. 10.143 can then be set within the following limits

$$N_1^2 \cdot \Lambda_{\sigma.close} < L_{\sigma 1.eff} < N_1^2 \cdot \Lambda_{\sigma.loose} \text{ respectively } N_2^2 \cdot \Lambda_{\sigma.close} < L_{\sigma 2.eff} < N_2^2 \cdot \Lambda_{\sigma.loose}$$

Together with the other design approaches, there is considerable scope for designing the leakage inductance. To give a simple idea of the scope, an estimation is given below. The numbers of turns N_1 and N_2 are assumed to be given and constant. If we now assume as an approximation that the leakage inductance is composed of two winding parts with 2 constant leakage conductance values, we can assume for the permeances using the identity $L^* = \Lambda_\sigma = (R_{m\sigma})^{-1}$, $L^*_\mu \to \infty$ and short circuited N_1 according to Fig. 10.144)

$$\Lambda^*_{\sigma.close} \approx \frac{U_{22}}{\omega \cdot N_{21} \cdot \underline{I}_{22}} \text{ and } \Lambda^*_{\sigma.loose} + \frac{\Lambda^*_{\sigma.close}}{2} \approx \frac{U_{21}}{\omega \cdot N_{21} \cdot \underline{I}_{21}}$$

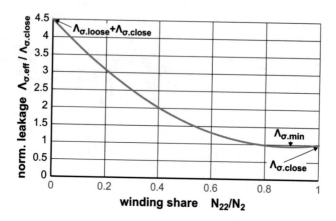

Fig. 10.145 Example of the effect of a winding distribution of a secondary winding on the close and loose coupling to the primary winding with a ratio of the leakage values $\Lambda_{\sigma.lose}/\Lambda_{\sigma.fest} = 4$

The resulting effective leakage conductance $\Lambda_{\sigma.eff}$ is then a function of the turns ratio N_{22}/N_2

$$L^*_{\sigma2.eff} = \left(\frac{N_{21}}{N_2}\right)^2 \cdot L^*_{\sigma.loose} + \left(1 + \left(1 - \frac{N_{21}}{N_2}\right)^2\right)\frac{L^*_{\sigma.close}}{2} = \left(\frac{N_{21}}{N_2}\right)^2 \cdot \Lambda^*_{\sigma.loose}$$

$$+ \left(1 + \left(1 - \frac{N_{21}}{N_2}\right)^2\right)\frac{\Lambda^*_{\sigma.close}}{2}$$

When applied to the arrangement shown in Fig. 10.142b, the two leakage coefficients are equal. The resultant leakage conductance is $L_{\sigma.eff} = 0.5 \cdot N_1^2 \cdot \Lambda_\sigma$ - as derived there - with a number of turns ratio $N_1/N_{11} = 0.5$. Figure 10.145 shows the dependence of the effective leakage conductance for $\Lambda_{\sigma.loose}/\Lambda_{\sigma.close} = 4$. Thus the formula for the effective leakage conductance $\Lambda_{\sigma.eff}$ can be derived as follows

$$L^*_{\sigma2.eff} = \left(1 - \frac{N_{21}}{N_2} + 4,5\left(\frac{N_{21}}{N_2}\right)^2\right)\Lambda^*_{\sigma.close}$$

At the same output voltage or voltage ratio very different output impedances can be set. To illustrate the relationship for the combination of closely and loosely coupled winding components, Fig. 10.145 shows a corresponding dependence of the effective leakage permeance according to Fig. 10.144.

According to these considerations, the realization of the winding design of a transformer can be realized experimentally supported and without FEM calculation. The total number of turns $N_2 = N_{21} + N_{22}$ results from the loss minimization for the transformer and the required transmission ratio. It is also necessary to determine the leakage permeance values $\Lambda_{\sigma.fest}$ and $\Lambda_{\sigma.lose}$. According to Fig. 10.144, these result for $L_\mu \to \infty$ for very large μ_r to

$$L_{\sigma.close} \approx N_{22}^2 \Lambda_{\sigma.close}$$
$$L_{\sigma.loose} \approx N_{21}^2 (\Lambda_{\sigma.loose} + 0,5 \cdot \Lambda_{\sigma.close})$$

With this approximation, the first approximate values of $\Lambda_{\sigma.close}$ and $\Lambda_{\sigma.loose}$ can be determined via short-circuit tests. For this purpose, two test windings are carried out with a material that is well suited for winding. A transformer leg is wound with a layer for the primary winding. The two test windings for a loose and a closely coupled secondary winding are also each wound as a single-layer winding in the arrangement shown in Fig. 10.16. The first approximate values for the required permeances can then be determined from the corresponding short-circuit tests (secondary measurement + primary short-circuit).

These values represent the first approximations of an experimental step-by-step approximation to the searched result. The next step is to carry out the calculation shown at the beginning of this section to achieve the division of the number of windings N_2 between the two partial windings N_{21} and N_{22}. Using these values, a test winding is again carried out with a material that is as easy to process as possible - but this time in such a way that the available winding space is filled, as shown in Fig. 10.16. As winding material different from the final structure is generally used, it is essential to maintain the turn ratio N_{22}/N_{21}. Using the new approximations of $\Lambda_{\sigma.close}$ and $\Lambda_{\sigma.loose}$, the short-circuit measurements lead to slightly different results for N_{21} and N_{22} and typically provide useful results for realizing the desired leakage inductance $L_{\sigma eff}$. The procedure described here leads to a fast experimental approximation to a target value for the leakage inductance.

c) *Co and counter connection of winding components*

If one has the possibility of distributing the windings to different leakage conductance values, another possibility of adapting the leakage inductance to the requirements arises by using the possibilities of connecting and counter-connecting individual winding components. In principle, the voltages of the individual winding components are added to the output voltage. However, since the voltage drop can be added "in phase" or 180° out of phase, the design goal for a transformer with a certain voltage ratio U_2/U_1 and a short-circuit current I_{k2} to

$$U_2 = \frac{N_{21.mit}}{N_1} U_1 - \frac{N_{21.geg}}{N_1} U_1 + \frac{N_{22.mit}}{N_1} U_1 - \frac{N_{22.geg}}{N_1} U_1$$

Fig. 10.146 Transformer with
increased dispersion due to
shifted winding arrangement in
the winding window

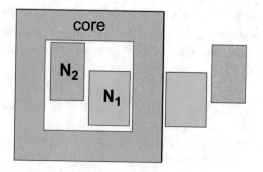

The winding components with loose coupling N21 and with close coupling are divided
up again so that individual winding components can be connected in series and opposition.
The required leakage inductance results approximately from the secondary open-circuit
voltage U_{20} and secondary short-circuit current I_{2k} to

$$Z_2 \approx \frac{U_{20}}{I_{2k}} \approx \omega \cdot N_2^2 \cdot \Lambda_{\sigma.eff}$$

$$L_{\sigma2} \approx \Lambda_{\sigma.lose}\left(N_{21.mit}^2 + N_{21.geg}^2\right) + \Lambda_{\sigma.fest}\left(N_{22.mit}^2 + N_{22.geg}^2\right)$$

It becomes clear that this significantly extends the scope for designing the leakage
inductance. It should be noted, however, that the open circuit voltage of the transformer
is affected and each winding is also a source of losses. With the combined method of
connecting and counter-connecting individual winding components, the total number of
turns is always increased. This is reflected in additional losses. Therefore, one will mainly
use these possibilities if one gains an advantage over a variant with an additional choke.
The relationships shown reflect the basic relationships and are suitable for estimating
calculations. For a more precise calculation, a higher effort has to be made.

d) *Free design of the space for the leakage field*

In transformers with a large winding window, the leakage inductance can also be
influenced by the arrangement of the winding parts in the winding window. This is not a
new design approach, but rather a mixed design approach (Fig. 10.146). The leakage
inductance can also be influenced by the distances and areas of the winding cross-sections
used. Older designs of welding transformers are based, for example, on shifting windings
against each other in the winding window.

Simplified calculation methods or - for recurring similar tasks - tailored size equations
are used to predict the leakage inductances. Calculation methods with finite elements
usually require 3-dimensional calculations if no special symmetry properties of the field
can be used to achieve satisfactory accuracy of the result. There are numerous examples
and instructions in the literature [30–33].

A simple experimentally supported procedure, which provides fast and mostly suffi-ciently accurate results, is described at the end of this subchapter. Since the leakage inductance describes the relationship between two windings and the core normally plays no or a subordinate role, it is also possible to make a (possibly reduced or enlarged) model of the planned winding structure or parts of it. For this purpose, an arrangement of the windings according to scale is necessary. For the design of the windings, an easily workable material (if necessary highly flexible cable) is used. Instead of the core, a ferromagnetic material (sheet metal, stacks of ferrite cuboids, etc.) is used. For such an arrangement, the leakage inductances can be easily measured by combined short-circuit and open-circuit tests. With 1:1 models, the values determined can be used directly as information values for the later realization. In the case of true-to-scale models, a conversion to the real size of the object must still be carried out. In this way, one can go very quickly from a design to a prototype or compare different design approaches.

In the practical design of a transformer, the following procedure has proven to be useful:

1. selecting a suitable core
2. optimization of the total losses to determine the number of turns
3. checking the perspective temperature increase with known Rth
4. select another core if necessary or change the cooling mode
5. test winding for the metrological determination of the leakage permeances via short-circuit and no-load tests for fixed and loose coupling of the coils
6. determination of the winding distribution of closely and loosely coupled winding components.

These successive activities may have to be performed several times depending on the interim results. It also follows from the structural investigations carried out in this subchapter that the leakage inductance is in most cases completely unaffected by an air gap introduced into the air gap of the core. The leakage inductance describes the coupling between two coils. In the case of closely coupled coils on a core, an air gap does not affect the leakage inductance of the arrangement. This is similarly the case for the leakage inductance in arrangements with strong yoke leakage. Only if the magnetic resistance of the leakage field is directly influenced by an air gap the leakage inductance can be influenced.

10.12 Single-Phase Transformers for Voltage Adjustment and Insulation

With the help of electromagnetic induction, energy can be transferred via the electromag-netic field. This is described by the Poynting theorem. The Poynting vector $\vec{S} = \vec{E} \times \vec{H}$ characterizes both direction and "flux density" of the energy in VA/m^2. It is important to note here that both the magnetic and the electric fields are always involved in energy transfer. This applies to the electrical energy transfer in general and the electromagnetic transformer in particular. In the latter, the energy is transmitted in the leakage field, where the power densities can assume very high values. At the same time, the primary and secondary sides can be isolated from each other ($I_{DC} = 0$) and the voltages at the input and output can be set differently by selecting the appropriate number of primary and secondary windings. The conversion of the two forms of energy can be carried out with very high efficiency since very low-loss materials are used for conducting the current (therefore copper conductors in combination with good insulators are suitable) as well as for guiding the magnetic field. The energy transport from one winding to the other is, however, affected by properties of the space between the windings. The effectiveness of the actual transmission is strongly influenced by the losses in this gap. This must be taken into account when designing the interspace. The first thing to be mentioned are eddy current losses in electrically conductive construction parts between the windings or their leakage field [34–36].

The piezo transformer uses a different energy combination for energy transmission. Here the linear expansion of an insulator is used when an electric field is applied. Mechanical oscillations are caused and accelerations and movements occur inside the material. The piezo-transformer, therefore, represents an electrically excited oscillating mechanical system. Since the piezoelectric effect is reversible, an electrical voltage can be assumed elsewhere, the magnitude of which can be determined by the design of the mechanically vibrating transformer. The energy transport is similar to a sound wave through a solid body. The losses that occur are essentially internal friction losses. Naturally, the heat loss is difficult to dissipate from insulators. For this reason, piezo transformers with comparatively low efficiencies are limited to low power levels. They are used more widely, for example, in surface acoustic wave filter assemblies, where narrow-bandedness and stability rank behind efficiency.

The simple voltage increase in the electromagnetic transformer is achieved by the effect of counter-induction. There is no such effect in the electric field. Therefore, energy can indeed be transferred through the alternating electric field if capacitors are inserted in the forward and return conductors. However, the voltage at an active resistance as a load will always be smaller than the generator voltage. Nevertheless, generator and load are galvanically isolated from each other. The higher the frequency, the smaller the coupling capacitance can be chosen. Generator and load are galvanically isolated.

In the following subchapters different arrangements are compared with each other, which are suitable for energy transmission and show one or more of the following characteristics of transformers at high efficiency:

- Bidirectional power transmission
- Voltage adjustment
- DC-isolation of input and output.

Assume a core with a certain magnetization area A_e and a winding cross-section A_w. At a certain voltage and frequency as well as an optimal current density Jo, the number of turns of a transformer results in

$$N_1 = \frac{\sqrt{2}U_1}{\omega \cdot A_e \cdot B_{max}} \text{ or/and } N_2 = \frac{\sqrt{2}U_2}{\omega \cdot A_e \cdot B_{max}} \quad U_2 = \frac{\omega \cdot A_e \cdot B_{max} \cdot N_2}{\sqrt{2}}.$$

Assuming the same fill factor k_f for both sides, the following must apply to the winding window

$$k_f \frac{A_w}{2N_1} = A_{Cu} = \frac{I_1}{J_O} \text{ respectively } I_1 = k_f \frac{A_w}{2N_1} J_O$$

The apparent power is the same in both windings, so

$$S_1 = S_2 = U_1 \cdot I_1 = \frac{1}{2\sqrt{2}} \omega \cdot A_e \cdot B_{max} \cdot k_f \cdot A_w \cdot J_O$$

This formula provides the product of A_e and A_w as a figure of merit (FOM) for the performance of a core as a rough assessment variable.

$$FOM_{Kern} = A_e \cdot A_w$$

This value is often used for a rough selection of cores. However, it does not take into account the fact that heat dissipation can also vary greatly between different designs. Therefore a subsequent detailed consideration of the heating situation is necessary. The rated power of an S_T transformer is the average value of primary and secondary apparent power:

$$S_T = \frac{1}{2} \left(\sum_{\mu=1}^{m} I_{1\mu} \cdot U_{1\mu} + \sum_{v=1}^{n} I_{2v} \cdot U_{2v} \right)$$

The voltage is usually to be regarded as an impressed/pre-set value. The effective value of the currents depends on the connected loads. Particularly in the case of converter transformers, there is a large difference between the effective power to be transmitted and the required rated power.

If the requirement for insulation of the primary and secondary side is dropped first and only the low-loss voltage transformation in magnetic systems is used, this leads to an autotransformer (Fig. 10.147).

If the same core as above is used as an autotransformer, the winding N_1 is connected to the impressed voltage U_1. Winding N_2 should be used here to increase the voltage to U_2 so that $U_a = U_1 + U_2$. This arrangement is often used with rolling or sliding contacts in variable transformers. Insulating variable transformers are rather rare. For low power ratings, autotransformers with sliding contacts with an upstream insulating 1:1 transformer are often used for insulating variable transformers. The apparent power on the input and output side of the autotransformer is again the same.

Fig. 10.147 Circuit diagram of
an autotransformer

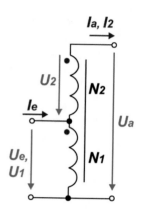

$$S_e = I_e \cdot U = (I_2 + I_1) \cdot U_1 = S_a = I_2 \cdot U_a = I_2 \cdot (U_1 + U_2)$$
$$I_2 U_1 + S_1 = I_2 U_1 + S_2$$

It follows that $S_1 = S_2 = S_T$ still applies, where S_T is the apparent power transmitted by transformer/inductive means.

$$\frac{S_T}{S_a} = \frac{I_2 \cdot U_2}{I_2 \cdot (U_1 + U_2)} = \frac{U_2}{(U_1 + U_2)}.$$

With $U_a = U_1 + U_2$ then follows $\frac{S_T}{S_a} = 1 - \frac{U_e}{U_a}$ for $U_a > U_e$ and analogously $\frac{S_T}{S_a} = 1 - \frac{U_a}{U_e}$ for $U_a < U_e$.

Figure 10.148 summarizes these considerations in a diagram. It can be seen that significant savings can be achieved in the rated power of transformers and thus in their size at transformation ratios in the vicinity of $U_a/U_e = 1$. If space-saving is the only factor that matters, then transmission ratios $U_a/U_e = 0.3 \ldots 3$ are common.

If the transformer galvanically separates the primary and secondary side by 2 separate windings, the entire apparent power must be transmitted via the transformer and thus determines the rated power/size of the transformer. For mains transformers, the voltage is predetermined, while the current is determined by the load. Table 10.9 illustrates this influence. A resistive load is used as a reference. The specified required rated power does not take the efficiency of a transformer into account. For transformers of low power, deviations result for this reason. The different rated powers only result from the current forms caused by the load that deviate from the sinusoidal form. With L-smoothing of the load current, the transformer currents are almost rectangular. With C-smoothing, the capacitor is only recharged near the voltage maximum. This significantly increases the form factor of the current and thus the required rated power of the transformer. Factor 1.4 is only an orientation value. It is set as a function of the leakage inductance of the transformer and the size of C.

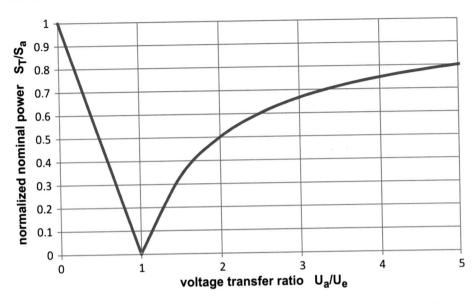

Fig. 10.148 Required rated power S_T of a transformer relative to the transmitted power S_a in an isolating transformer as a function of the voltage ratio

Table 10.9 Increase of rated power at different loads

Designation	Load scheme	Rated power/active power S_T/P
Bridge rectifier + resistive load		1.0
Bridge rectifier + ohmic-inductive load, smooth DC		~1.1
Bridge rectifier + resistive-capacitive load, smooth DC, typical leakage inductance		>1.4 (depending on mains + transformer impedance)

However, real transformer structures are not only characterized by magnetic links. Depending on the application, parasitic elements also play a role. On the one hand, each winding also has a winding capacity. This becomes electrically effective on the other side

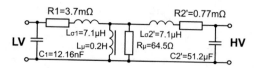

Fig. 10.149 Example of a 1-phase equivalent circuit diagram for the undervoltage side for a 1 MVA transformer (20 kV/0.27 kV) under investigation, taking into account the winding capacities determined from the undervoltage side

with the transmission ratio to the square. For example, with a real winding capacity of 17 pF on the high-voltage side and a transformation ratio N_2/N_1 in a 3 kW oscillating circuit inverter with a switching frequency of 50 kHz, a capacity of

$$C_2^* = 160^2 \cdot 17 \ pF = 25600 \cdot 17 \ pF = 0.435 \ \mu F$$

and was thus of the same order of magnitude as the resonance capacitor of the circuit.

Another example (Fig. 10.149) is the 1-phase HF equivalent circuit diagram determined for a 1 MVA transformer Dy5 (referred to as the primary side). The capacitance was determined by exciting resonances between the leakage inductance and the winding capacitances. This makes one independent of the magnetization state of the core. The square of the transformation ratio determines the effective size of the actual winding capacitance at the terminals.

The winding capacity must be taken into account, especially at higher frequencies and in pulse applications. It cannot be reduced at will. Its size depends on the winding design and the insulation materials used. Although various formulae are available for calculating the winding capacitances, experience shows that these formulae have often been published without the boundary conditions used at the time of their creation. They provide orientation values. Therefore only qualitative information can be given:

- For a given number of windings, the narrower or higher the winding, the lower the winding capacitance.
- For a given winding width, it is useful to divide the winding width into several chambers to achieve small winding capacities, so that several "disc windings" are connected in series.
- Layer windings can be carried out in such a way that the layers are wound right to left and back. Or after each winding layer from left to right to left the next layer starts at the beginning (z-form). The latter variant has a lower winding capacity but is possibly problematic from an insulation point of view.
- Adjacent windings result in a greater capacity than crossing windings.
- A higher fill factor usually also results in a higher winding capacitance.
- From older publications and devices, special winding techniques are also known in this context for low-capacitance windings, which, however, usually require corresponding

Fig. 10.150 Equivalent circuit
diagram of a 2-winding
transformer without loss
representatives and shield
winding

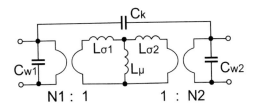

N1 : 1   1 : N2

devices/machines: Basket winding, star winding, honeycomb winding, cross winding technology. These allow extremely low winding capacities with correspondingly small winding widths in self-supporting winding constructions without coil formers. Usually theses methods are combined with lower fill factors.

For an estimation of the winding capacity of a single-chamber winding, see, for example, [37].

$$C_w = \frac{0.12 \cdot \varepsilon_r \cdot l_w \cdot l_m}{h_w \cdot \left(1 - \sqrt{\frac{4 \cdot k_f}{\pi}}\right)}$$

ε_r - relative dielectric constant of wire insulation, k_f - fill factor

More detailed considerations on the subject of winding capacity can be found in [38] for different variants of winding design. Here the tendencies of the effects of measures are clearly explained. According to the experience of the author, the measurement of test windings on appropriate devices is the only reliable way to determine the winding capacity. It is best determined by means of the inductance and resonant frequency of the coreless winding.

If the windings are arranged on top of each other, there is also a capacitive coupling C_k between them. This coupling can lead to problems on the other side in case of transient common-mode voltage changes on one side. Capacitive 'crosstalk' is a current that is transmitted via the dielectric, that is, without any transformer effect. To prevent this, a shielding winding can be placed between the windings and connected to the core or ground. The currents are then diverted to the ground. In principle, close magnetic coupling of the windings also results in a higher coupling capacity. A simple equivalent circuit diagram with concentrated elements can thus be derived for a magnetic transformer Fig. 10.150. Other capacitive effects do exist. However, the figure reflects the dominant quantities.

10.13 Transformers for Multiphase Voltage Systems

As multi-phase voltage systems, 3-phase systems in particular are to be mentioned from the application. They are used to provide and transmit electrical energy. Depending on the distances to be covered, the voltage levels in the transmission and distribution networks must be changed and, if necessary, the loads on the network strings and phases must be equalized. Appropriate transformers are used for this purpose. An ideal, 3-phase symmetrical voltage system is characterised by 3 voltage sources of the same amplitude and frequency with a phase shift of 120° each, which are connected in star (Y) or delta (D). The former can be designed as a system with a loaded star point (4-wire system) or with an unloaded star point (3-wire system). Delta-connected sources always supply a 3-wire system. 3-wire systems can be converted into each other using Y-D transformation. The design and construction of corresponding transformers fill many books and will only be treated here as an overview of some important aspects.

The 1- and 3-phase network configurations have historically developed from different conditions and are therefore not uniform worldwide. Also, the use of 50 Hz or 60 Hz as grid frequency is distributed mixed over the countries and historically. In the 1940s, frequencies between 25 Hz and 100 Hz were common worldwide. Of these, 50 Hz and 60 Hz are left. In Japan, for example, the electrification of Westinghouse (60 Hz) and Siemens (50 Hz) was started from two sides in the nineteenth century. The difference is based on the procurement of generators from Germany by AEG in 1895 for Tokyo, which supplied 50 Hz, and by General Electric from the USA in 1896 for Osaka, which supplied 60 Hz. As a result, today practically half of the country is electrified on a 50 Hz basis and the other half on a 60 Hz basis. The two systems are connected via 4 DC couplings of 300 kVA each. Even with the same nominal voltage, the operating frequency has far-reaching consequences for electrical equipment and systems supplied to these supply areas. The magnetic flux linkage is lower at 60 Hz than at 50 Hz. As a consequence, electrical machines designed for 60 Hz may go into saturation at 50 Hz. The same applies to transformers. On the other hand, with the same leakage inductance, the leakage reactance at 60 Hz is 20% higher than at 50 Hz. The transmittable power is correspondingly lower.

Besides, the speed of connected synchronous and asynchronous motors is frequency-dependent. With speed-independent load torque or - as in the case of fans - speed-dependent load torque, the mechanical and thus the absorbed electrical power increases. In the past, this led to correspondingly complex tests for the respective area of use of the devices and systems. With the spread of *Power Factor Controllers* (PFC) with a DC intermediate circuit, these problems are elegantly solved today with the help of power electronics. When high tolerable voltage ranges are implemented, only the different connectors play a role for the application. It must be taken into account that the power electronics are designed in such a way that the nominal power can be provided. Therefore, the current consumption automatically increases at lower mains voltages. Here we will discuss configurations of the low-voltage network and possible consequences for the transformer design. Table 10.10 shows that there is a great variety of different nominal voltages for 1-phase and 3-phase electrical low-voltage networks.

In the nineteenth century, electrification started in the USA and Germany. As a result, systems with a frequency of 60 Hz and 50 Hz became widespread worldwide. In some countries, such as Japan and Saudi Arabia, both frequencies are in use in the respective areas (Fig. 10.151).

Table 10.10 Nominal voltages in low-voltage distribution networks, worldwide (Source IEA)

Single-phase connection (rated voltage [V])	Three phase connection (rated voltage [V])
100	120
110	190
120	200
125	208
127	240
220	277
230	380
240	400
	415
	440

Fig. 10.151 Nominal voltages and mains frequencies in their worldwide use according to IEA (green: 220–240 V / 50 Hz, yellow: 220–240 V / 60 Hz, blue: 100–127 V / 50 Hz, red: 100–127 V / 60 Hz (source: IEA)

3-Wire Networks

A three-wire *split-phase electric power* system is a power supply system commonly used in North America for single-family homes and small businesses (up to about 100 kW), but also in Australia, New Zealand and Canada (Fig. 10.152). The operational return conductor on the medium voltage side is the grounding. Historically, the configuration is derived from Th. A. Edison's three-wire DC system. The voltages U_{L1N} and U_{L2N} are in phase opposition to the neutral 'N' (phase shift 180°el). The chained voltage U_{L1L2} can also be taken, which is twice as high as the line voltage compared to the neutral conductor. 3-phase rotating field machines can not be operated with this. In the distribution system, the

Fig. 10.152 Three-wire system for 1-phase *electric power* transmission (**a**) (*split-phase electric power*) and phasor diagram of the voltages (**b**) in US colour coding

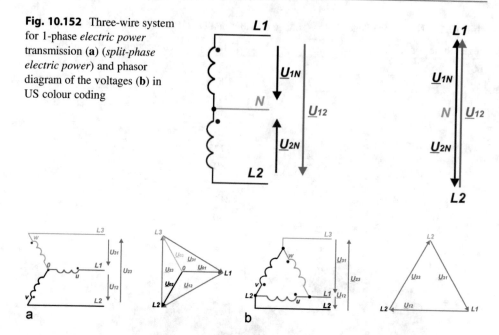

Fig. 10.153 Transformer for an insulated 3-wire network for 3-phase power transmission. (**a**) with unearthed star point, (**b**) in delta connection

transformer is then connected on the high voltage side either to one conductor and the neutral point or, in the case of an unearthed medium voltage network, to two-phase conductors.

3-phase energy distribution networks with only 3 conductors are based on symmetrical 3-phase voltage systems. In Germany, these network configurations were previously also used in low-voltage distribution networks. In a corresponding 3 ~ 220 V network, the windings were connected in delta. This results in a symmetrical 3-wire network (Fig. 10.153). An earth fault exists only in a high-impedance manner via parasitic capacitances and insulation resistances. Against the earthed protective conductor, all possible voltages between 0 and 220Vac can therefore be measured (Fig. 10.153a). Otherwise, unearthed 3-wire systems are mainly found in medium and high voltage networks. A single-pole earth fault of a consumer must be monitored by an insulation voltage. This can be used to set up an IT system (french: *Isolé Terre*) (see also below. It can be seen here that the interconnection of the same windings in Y and D configuration results in 3-phase systems with a phase shift of 30°el.

A 3-wire network can also be operated with 2 single-phase transformers (V circuit, Fig. 10.154). On the output side, 3-phase alternating voltage is again generated in a 3-wire network. The circuit is preferably used for operating consumers with a symmetrical load, for example, drives, and for smaller outputs in the lower kVA range.

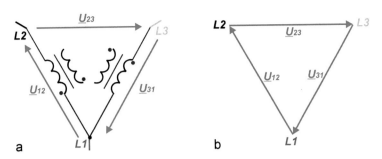

Fig. 10.154 V-circuit for generating a 3-phase network with 2 single-phase transformers. (**a**) Winding circuit, (**b**) Phasor diagram

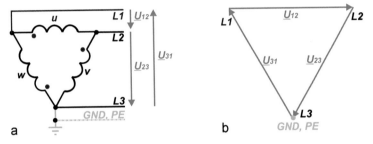

Fig. 10.155 Example of a *corner-grounded delta system*. (**a**) Winding connection, (**b**) Phasor diagram of the voltages

   In principle, it is also possible to connect one of the 3 conductors to the earth potential to transmit 3-phase energy. This is used in some industrial networks (*phase-grounded triangular network, corner grounded system*). It can be found in the USA, for example, and is shown in Fig. 10.155.

4-Wire Networks

The networks derived from a 3-phase system lead to 4-wire networks for the distribution networks in many regions. Here, too, several configurations must be taken into account when planning and implementing devices and systems. The dominant configuration in Europe is a $3 \sim 400$ V network with a neutral point (Fig. 10.156, TN network). This means that the voltages are available on 3 phases compared to the neutral conductor as a 1-phase power supply and $3 \sim 400$ V as a 3-phase interlinked voltage system. The common vector group of a low voltage distribution transformer is Dy5. The neutral point is then also the loadable neutral conductor 'N' of the supply system. The voltages of conductors L1, L2 and L3 measured against this neutral point all have the same magnitude and a phase shift of $120°$el. The 3-phase *wye-connected system* used in the USA corresponds to this configuration.

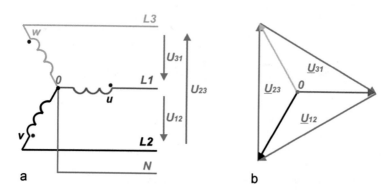

Fig. 10.156 3-phase network with star point earthing (*wye-connected system*). (**a**) Winding connection, (**b**) Phasor diagram of the voltages

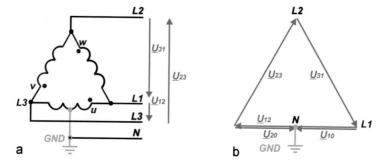

Fig. 10.157 4-wire system with split phase (4 wire split-phase system, *delta 4 wire, high leg delta, red leg delta system*). (**a**) Winding connection, (**b**) p diagram of voltages

A form widely used in North America and Japan is a mixed form of electrical power transmission in the low-voltage distribution network as a 4-wire network with a *split phase* (*split phase*, Fig. 10.157a). This is an extension of the single-phase three-conductor network for three-phase alternating current and is known as *delta 4 wire, high leg delta* or also *red leg delta system*. This type of low voltage supply is unusual in Europe. The advantage of this design is that no changes are required for the single-phase three-wire connections in the 120 V / 240 V range that are partly historically common in the American region. A three-phase feed from the medium-voltage network is required to supply the local transformers. An existing 3-wire single-phase installation can be created by introducing a further phase with 2 additional single-phase transformers. These are comparatively small changes.

This configuration is obtained by connecting the low voltage windings of a 3-phase transformer in the delta. One winding is provided with a centre tap. This results in a neutral conductor 'N' whose potential is not in the centre of the phasor system. L1-N-L2 form the

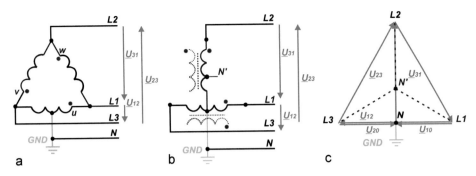

Fig. 10.158 Formation of a 4-wire network with split-phase from a symmetrical 3-phase 3-wire network with a 3-phase transformer in Δ circuit. (**a**) Winding circuit with 3-phase transformer, (**b**) Use of 2 single-phase transformers with tappings (Scott circuit) and attainable N′, (**c**) Phasor image of the voltages

3-wire system for single-phase power transmission described above. Besides, there is the conductor connection L3. This gives the voltages

$$\underline{U}_{L1N} = U \cdot e^{j0°}$$
$$\underline{U}_{L2N} = U \cdot e^{-j180°}$$
$$\underline{U}_{L1N} = \sqrt{3} \cdot U \cdot e^{j90°}$$

For U = 120 V, for example, this gives U_{L1N}=U_{L2N} = 120 V and U_{L3N} = 208 V (Fig. 10.158).

The higher voltage of U_{L3N} is the basis for the designation *High Leg*. To avoid connection errors and according to article 110.15 in NFPA 70, this connection must be marked in orange or red (*red leg delta*).

A disadvantage of the high-leg delta system, as with the single-phase three-conductor network, is that the outer conductors in the medium-voltage network are loaded very unevenly, which can lead to increased unbalanced loads. Distribution of the single-phase loads to all three phase conductors is not possible in this system and a skew load can only be compensated over several local transformer stations. Because of these practically unavoidable unbalanced loads, transformers in North America are usually designed as three individual single-phase transformers. In the case of house connections, these single-phase transformers can be seen hanging from the line masts.

Historically, this has resulted from a 2-phase network of two voltage systems that are 90°el out of phase with each other. These are due to early generator types of the nineteenth century. This voltage system also forms a rotating phasor system and allows the operation of rotating field machines like the usual 3-phase machines. A mathematical connection

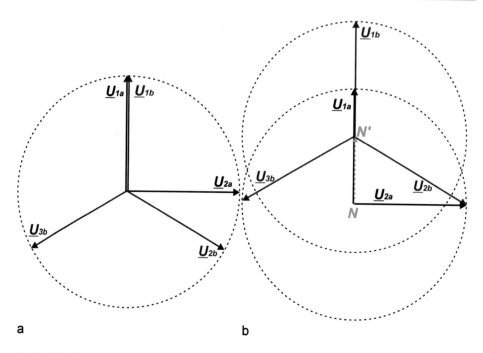

Fig. 10.159 Scott transformation. (**a**) Structure of a 2-phase and a 3-phase voltage system with equal phasor length, (**b**) Star point shift in a way that the conductor voltage $\underline{U}_{23}$ is in phase with $-\underline{U}_{2a}$

between these two systems is created by the Scott transformation. First of all, a symmetrical 2-phase voltage system is given:

$$\underline{U}_{1a} = U \cdot e^{j90°}$$
$$\underline{U}_{2b} = U \cdot e^{j0°}$$

A 3-phase voltage system can be constructed with the same phasor length (Fig. 10.159a):

$$U_{1b} = U \cdot e^{j90°}$$
$$U_{2b} = U \cdot e^{j90° - 120°}$$
$$U_{3b} = U \cdot e^{j90° - 240°}$$

If you shift the 3-phase system so that the phasor of the conductor voltage $\underline{U}_{23}$ is parallel to $\underline{U}_{2a}$, you get a 3-phase system with a "zero system" in the form

$$U_{1b} = \frac{1}{2} U \cdot e^{j90°} + U \cdot e^{j90°}$$

$$U_{2b} = \frac{1}{2} U \cdot e^{j90°} + U \cdot e^{j(90° - 120°)}$$

$$U_{3b} = \frac{1}{2} U \cdot e^{j90°} + U \cdot e^{j(90° - 240°)}$$

This transformation can also be carried out with a so-called Scott transformer so that you can easily get a 3-phase system from a 2-phase system and vice versa. Figure 10.159 shows the phasor images for both systems in comparison as well as the winding arrangement of a Scott T transformer, which connects the 2-phase system with the 3-phase system. In principle, the arrangement can also be used further as a symmetrical 3-phase system with a neutral point if a transformer is provided with a winding tap N′. It should be noted that for the voltages $U_{1N} = U_{2N} = 2 \cdot U_{1N}$ und $U_{3N}:U_{1N} = \sqrt{3}$. Compared to 3-phase systems, true 3-phase systems require less conductor mass for the same power and have therefore displaced the latter from the application. Furthermore, even with balanced loads, the phase currents do not add up to zero, so that the neutral conductor must always carry a considerable current. In machines for 2-phase systems, the individual fluxes and currents do not add up to zero even with balanced loads. This results in magnetostriction and torsional forces that make the machines appear loud.

As described above, transformers whose insulated windings can be regarded as AC voltage sources isolated from each other provide the basis for single- and multi-phase distribution systems for electrical energy. Different treatment of the neutral conductor 'N' and the protective conductor 'PE' connected to earth potential GND leads to different installations. Depending on the conditions and supply objectives, different types of networks have developed, which will be described briefly here. This is not intended to claim to be a complete description of all the subtleties. The aim is a rough overview. The graphic examples for a better understanding are based on a 4-wire system based on a symmetrical 3-phase system with brought out star point.

TT Networks
In the TT system, a point is directly earthed and the conductive parts in the network that are not live during operation are connected to earth electrodes that are independent of the earth electrodes of the supply system. The TT system was formerly known as "protective earthing". The generator neutral point at the generator is directly earthed and is carried as neutral 'N'. On the consumer side, the bodies of the systems are connected to the protective conductor PE. The protective conductor is directly connected to the system earth. The TT system differs from the TN system (see below) in the direct line connection between the system and operational earthing. Residual current devices (RCDs) are used for contact protection. Every insulation fault causes an interruption of the power supply, but the failure is limited to the faulty circuit (Fig. 10.160).

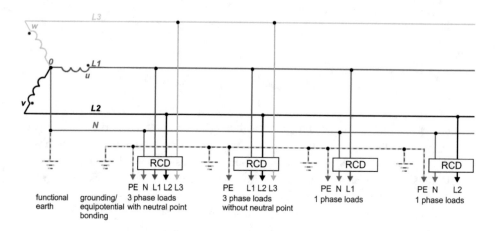

Fig. 10.160 Schematic diagram of a TT system (overcurrent fuses not shown)

TN System

The protective and earth connection is a direct conductor connection between supplier and consumer, whereas in the TT-Net the protective and earth connection between supplier and consumer only exists via the earth. Otherwise, the above statements on the structure of the TT-Net apply. A further distinction is made between TN-C and TN-S systems.

TN-C System (French: Terre Neutre Combiné)

Here, the neutral conductor 'N' and the protective conductor 'PE' are jointly routed as 'PEN'. The 'PEN' must therefore not be interrupted by switching or protective devices (DIN VDE 0100-410:2007-06) in order not to interrupt the protective effect. The connection and earthing of touchable conductive parts and the neutral conductor are mandatory. Interruption when the first fault occurs by means of an overcurrent protection device (circuit breaker, miniature circuit breaker or fuses). Contact protection can only be achieved by RDC sockets individually. This type of network is also problematic in a building because there are potential differences between different protective conductor terminal points due to the voltage drops on the current-carrying neutral conductor, which is also used as a protective conductor in this installation. This can lead to considerable interference voltages for the operation of, for example, audio equipment, measuring instruments, etc.

TN-S System (French: Terre Neutre Séparé)

In a TN-S system, separate neutral and protective conductors are routed from the transformer to the consumers (see Fig. 10.161).

A TN-S system is safer than the TN-C system. The problems that can result from an interrupted PEN conductor do not occur here, and protection is much better guaranteed. The TN-S system is mainly used for new installations in buildings. The house connection

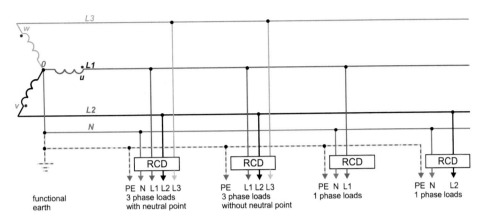

Fig. 10.161 Schematic diagram of a TN-S network (overcurrent fuses not shown)

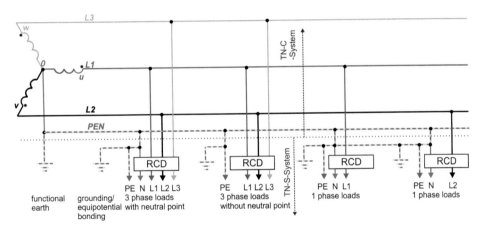

Fig. 10.162 Schematic diagram of a TN-C-S network (overcurrent fuses not shown)

to the local network transformer is connected via a TN-C network, resulting in a TN-C-S system (Fig. 10.162). The transition from a TN-C to a TN-S system is signaled with a separate blue line for the neutral conductor. From the generator to the consumer, a TN-C-S system must not be followed by a new TN-C system.

IT System (French: Isolé Terre)

This is an isolated supply system. In the IT system, all active parts are isolated from earth or a point is connected to earth via an impedance (Fig. 10.163). Touchable parts which are not live during operation must also be connected to the earth. A first fault is detected and reported by an *Insulation Monitoring Device* (IMD: IEC 61557), if necessary connected to a *Surge Protective* (SPD: *Surge Protective Device*), together with mandatory fault location

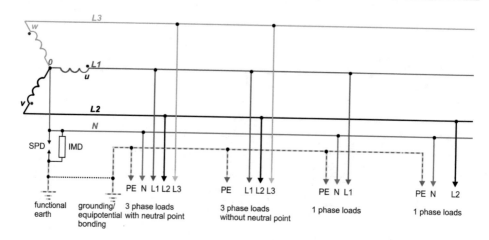

Fig. 10.163 Basic structure of an IT system

and elimination (DIN VDE 0100–410). In the event of a second fault, overcurrent protection devices (e.g. circuit breakers, miniature circuit breakers or fuses) are used to interrupt the circuit.

In principle, 3-phase energy transmission can also be transmitted via 3 single-phase transformers. However, in these applications, the transformers are usually magnetically combined. In principle, energy transmission is a matter of close magnetic coupling. This means here that typically the low voltage winding is first placed on the core and then the high voltage winding. Between the windings and between the winding and the core there is usually a shielding winding that is connected to the core or the housing.

3- and 5-legged designs are used as cores. While in the case of voltage unbalances the magnetic fluxes must close via the air path from yoke to yoke, in the case of the 5-leg transformer the magnetic return is via the outer legs. Figure 10.164a–c shows the basic design and the associated electrical equivalent circuit diagram. In symmetrical voltage systems, the sum of the magnetic fluxes is practically zero, so that the outer legs are hardly loaded at all. In the case of strong unbalanced loads or mains faults, however, the sum of the fluxes is not equal to zero, so that the outer legs of the 5-leg transformer are stressed for this purpose. Due to the high inductance $L_{J'}$ of the magnetic flux because of the outer core legs, this transformer type behaves similar to 3 single-phase transformers, each of which has its magnetic flux. The yoke inductance L_j contain a magnetic core for 5 leg transformers. It is used, for example, for coupling transformers of vector group Yy0.

How the total of six connections on each side of a three-phase transformer are interconnected is determined by the vector group. The usual connections are star and delta connections, which can be combined in principle on both sides as required. With the Z-connection, winding parts of different legs are connected in series.

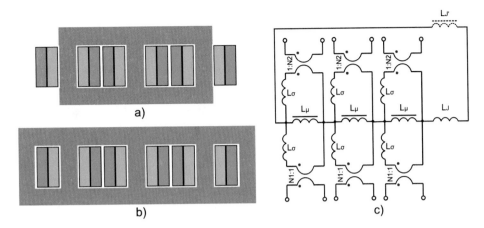

Fig. 10.164 Three-phase transformer versions. (**a**) Three-leg transformer, (**b**) five-leg transformer, (**c**) electrical equivalent circuit diagram with yoke spread L_J and inductance of the outer legs for the 5-leg version $L_{J'}$ (green: low voltage winding, red: high voltage winding)

This results in different phase shifts between the outer conductor voltages of the overvoltage and undervoltage side, which cannot be only 0° or 180° as in single-phase transformers. In these cases, the transformation ratio is expressed by a complex factor which also includes the phase shift. For this reason, when several three-phase transformers are operated in parallel, the vector group must be taken into account, which contains the circuit configurations on the high voltage side (capital letter), low voltage side (small letter), information on the star point brought out and the phase shift in multiples of 30°el.

Circuit configuration of the winding:

Y, y   Star connection
D,   Delta connection
Z,   Zigzag circuit
n   Detailed neutral point
k   Number

Example: **Dyn5**
D = higher voltage winding in delta connection
y = lowervoltage winding in star connection
n = brought out star point (neutral conductor)
5 = Phase shift between high and low voltage is: $5 \cdot 30° = 150°$

Phase shifts between 0° and 180° can be generated by combining winding parts. In power engineering, the switching groups summarized in Table 10.11 have been established as standards.

Table 10.11 Switching groups according to EN61558 / VDE0570

Code	Switching group	Phasor image	Winding-interconnection	Load capacity of the sec. Neutral point
0	Dd0			–
	Yy0			10%
	Dz0			100%
5	Dy5			100%
	Yd5			–

(continued)

			100%	Yz5
6			–	Dd6
			10%	Yy6
			100%	Dz6
11			100%	Dy11

Table 10.11 (continued)

Code	Switching group	Phasor image	Winding-interconnection	Load capacity of the sec. Neutral point
	Yd11			–
	Yz11			100%

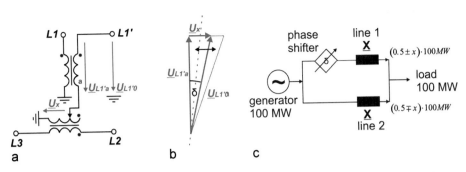

Fig. 10.165 Principle of a phase-shifting transformer. (**a**) Principle structure, (**b**) Phasor diagram, (**c**) Use as the element for load flow control

Phase-Shifting Transformers

Obviously, in alternating current systems, the sum of separate voltage drops can also cause a phase shift of voltages. In power supply systems this is achieved with so-called phase-shifting transformers. Figure 10.165 shows the underlying principle for a 4-wire system (L1, L2, L3, N). A voltage $\underline{U}_x$ is added to a transformed voltage $\underline{U}_{L1'a}$, which is derived transformerically from the chained voltage $\underline{U}_{L2L3}$ and is thus perpendicular to $\underline{U}_{L1'a}$. This results in a voltage that is out of phase with the voltage $\underline{U}_{L1'a}$. The simultaneously increased amplitude can generally be neglected at small angles δ. If one wants to keep a constant amplitude at larger angle changes, this change in size must be compensated somehow. One speaks then of transverse controllers. Variable transformers for voltage change without phase change are called longitudinal controllers.

An essential effect of a phase rotation of the voltage vector inserted in a network is the change of the load flow distribution. Even with 2 parallel lines, the current distribution can be changed by inserting a phase shifter (Fig. 10.165c). This effect is used for load flow control in power supply networks.

A disadvantage of the arrangement in Fig. 10.165a is the fact that the "cross transformer" for $\underline{U}_{L2L3}$ requires high dielectric strength. This can be circumvented with the design shown in Fig. 10.166a. A 60°el out-of-phase voltage is added (in Fig. 10.165 Δφ = 90°). This also causes a phase shift but also leads to a further extension of the voltage phasor. This arrangement is called an phase-angle controller. The insulation problems are thus largely defused. By combining two phase-angle regulators with opposite voltage ratios, pure cross regulators can be obtained. This type of variable transformers makes it possible to control the load flow in lines according to both size and sign. This also means that in AC networks it is possible to reduce a given load flow to zero.

Insulating Transformers

Insulating transformers electrically separate different energy supply areas. This means that apart from a low parasitic coupling capacity, no electrical coupling between the areas is

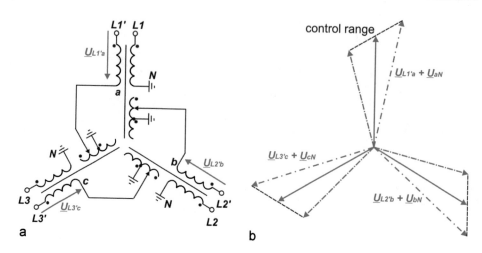

Fig. 10.166 Schematic diagram of an phase-angle control transformer for power supply networks. (a) Principle winding connection, (b) with the associated phasor diagram of the voltages

provided. Signals and energy are transmitted magnetically in single-phase or multiple phases. At the same time, a level adjustment can be carried out. Depending on the application, there are special requirements for the insulation design combined with special test specifications. Obvious examples of applications are, for example, transmitters for measuring current or voltage in electrical networks. In addition to the exact mapping of the variable to be measured to a measurement signal, particularly reliable and durable insulation is one of the main tasks of these elements. The electrical insulation of 2 systems in communication technology is often achieved by magnetic coupling with an appropriate insulation design. These components have the task of eliminating the effects of potential differences between ground points in transmitted signals, which are also caused by uncontrollable currents with their voltage drops when the galvanic coupling is used. An alternative to this is an optoelectronic coupling point, which allows much greater insulation distances to be bridged via optical fibers. However, it does not permit the transmission of significant energy simultaneously with the signal transmission. Also, the temperature limits of optoelectronic semiconductors are much tighter than those of magnetic couplers and the expected service life is much shorter than with magnetic couplers.

Insulating transformers can be integrated into integrated circuits even in the smallest of spaces. Coreless flat coils are used for coupling, which are insulated from each other by extremely homogeneous SiO_2. Due to the small distance between the coils relative to the diameter, coupling factors close to 1 can be achieved.

10.14 Transmission Line Transformers (TLT)

A special design of transformers can be realized for high frequencies and pulse applications. These use the propagation properties of electromagnetic waves in lines and are therefore also called propagation time transformers. The overview of important designs of this technology is therefore briefly preceded by an explanation of the wave propagation of electromagnetic energy in a double line. It provides the terms used to explain the functions of the various designs. In order to describe the fundamental problems involved in modeling dynamic conduction processes, the analytical description of a homogeneous isotropic double conduction should first be preceded. As elements of classical electrical engineering, the means of description are common knowledge, so that only the essential steps and intermediate results will be presented here (Fig. 10.167).

With the introduced parameters for the resistance per length R', inductance per length L', capacitance per length C' and conductance per length G', the initial boundary value problem, generally known as telegraph equation, results after application of Kirchoff's laws:

$$\frac{\partial^2}{\partial t^2} \cdot \begin{bmatrix} u \\ i \end{bmatrix} + \left(\frac{R'}{L'} + \frac{G'}{C'} \right) \cdot \frac{\partial}{\partial t} \begin{bmatrix} u \\ i \end{bmatrix} + \frac{R' \cdot G'}{L' \cdot C'} \cdot \begin{bmatrix} u \\ i \end{bmatrix} = \frac{1}{L' \cdot C'} \cdot \frac{\partial^2}{\partial x^2} \begin{bmatrix} u \\ i \end{bmatrix}$$

With stationary sinusoidal excitation, this problem can be transformed into a normal boundary value problem with complex coefficients after a Laplace transformation. It is always assumed that the line parameters R', L', G' and C' are constant, which is not the case in practice. Especially at high frequencies, the effects of the skin effect, for example, must be taken into account for the parameters R' and L'. Characteristic parameters for a line are in this treatment (approximations for R, G → 0):

- Propagation coefficient $\gamma = \sqrt{(R' + j\omega L') \cdot (G' + j\omega C')} \approx j\omega \sqrt{L' \cdot C'}$
- Characteristic impedance $\underline{Zw} = \sqrt{\frac{R' + j\omega L'}{G' + j\omega C'}} \approx \sqrt{\frac{L'}{C'}} \approx \sqrt{\frac{L' \cdot l}{C' \cdot l}} = \sqrt{\frac{L}{C}}$
- Phase coefficient $\beta = \sqrt{0.5 \cdot \left[\omega^2 L' C' - R' G' + \sqrt{(R'^2 + \omega^2 L'^2) \cdot (G'^2 + \omega^2 C'^2)} \right]} \approx$

$\omega \cdot \sqrt{L' C'}$

The phase velocity or propagation velocity of a wave can be found at

Fig. 10.167 Element of a homogeneous isotropic double line

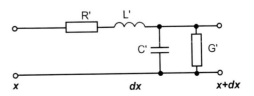

$$v = \frac{\omega}{\beta} \approx \frac{1}{\sqrt{L'C'}},$$

which is on a wavelength of

$$\lambda = \frac{2\pi}{\beta} \approx \frac{2\pi}{\omega \cdot \sqrt{L'C'}}$$

and a term of

$$\Delta t = \frac{l}{v} = l \cdot \sqrt{L'C'}$$

corresponds. If we look at any point 'x' on the line (Fig. 10.167), we obtain the cascade form of the four-pole equations for the stationary excitation in the image area:

$$\begin{bmatrix} U_x \\ I_x \end{bmatrix} = \begin{bmatrix} \cosh \underline{\gamma} \cdot x & -Zw \cdot \sinh \underline{\gamma} \cdot x \\ -\dfrac{1}{Zw} \cdot \sinh \underline{\gamma} \cdot x & -\dfrac{1}{Zw} \cdot \sinh \underline{\gamma} \cdot x \end{bmatrix} \cdot \begin{bmatrix} U_1 \\ I_1 \end{bmatrix}$$

$$\begin{bmatrix} U_x \\ I_x \end{bmatrix} = \begin{bmatrix} \cosh \underline{\gamma} \cdot (l-x) & -Zw \cdot \sinh \underline{\gamma} \cdot (l-x) \\ \dfrac{1}{Zw} \cdot \sinh \underline{\gamma} \cdot (l-x) & \cosh \underline{\gamma} \cdot (l-x) \end{bmatrix} \cdot \begin{bmatrix} U_2 \\ I_2 \end{bmatrix}$$

These equations result in $x = l$ (above) for the special cases:

$$\begin{bmatrix} U_2 \\ I_2 \end{bmatrix} = \begin{bmatrix} \cosh \underline{\gamma} \cdot l & -Zw \cdot \sinh \underline{\gamma} \cdot l \\ -\dfrac{1}{Zw} \cdot \sinh \underline{\gamma} \cdot l & -\dfrac{1}{Zw} \cdot \sinh \underline{\gamma} \cdot l \end{bmatrix} \cdot \begin{bmatrix} U_1 \\ I_1 \end{bmatrix}$$

and for x = 0 (lower term)

$$\begin{bmatrix} U_1 \\ I_1 \end{bmatrix} = \begin{bmatrix} \cosh \underline{\gamma} \cdot l & -\underline{Zw} \cdot \sinh \underline{\gamma} \cdot l \\ \dfrac{1}{Zw} \cdot \sinh \underline{\gamma} \cdot l & \cosh \underline{\gamma} \cdot l \end{bmatrix} \cdot \begin{bmatrix} U_2 \\ I_2 \end{bmatrix}$$

A double line is therefore a transmission element that itself has an impedance $\underline{Z_w}$ and in which there is a phase shift or time delay between input and output. This is described analytically by the above equations. In the case of transformers, it should be remembered that a third line often plays a role as well, to which the same ratios apply with respect to the two lines already mentioned. This gives a total of 3 wave channels, each with its own impedance and wave propagation speed. These parameters can be influenced according to

Table 10.12 Coaxial cable data RG58-50JF, (Source: Radio Frequencies System)

Type	RG58-50JF
Characteristic impedance	$Z_W = 50\ \Omega$
Capacity coating	$C' = 100\ pF/m$
Maximum DC voltage	$U_{dmax} = 15\ kV$
Maximum HF	$U_{HFmax} = 2.5\ kV$
Cable cross-section (inside)	$R_1' = 39\ m\Omega/m$
Cable cross-section (inside)	$R_2' = 15\ m\Omega/m$
Outside diameter	5 mm

the equations described above. The characteristic impedances of individual coaxial lines lie between 30 Ω and 85 Ω. The default values of measuring and antenna cables are 50 Ω and 75 Ω. The characteristic impedance of twin cables is generally higher due to the lower capacitance and is in the range 200...400 Ω with standard values 220 Ω, 240 Ω and 300 Ω. In power electronics, *laminated busbars* are often used in commutation circuits. With a conductor width of 'b' and a conductor spacing of 'd', the characteristic impedance of a dielectric with ε_r and μ_r is

$$Z_{W.lbb} \approx \frac{d}{b}\sqrt{\frac{\mu_r\mu_0}{\varepsilon_r\varepsilon_0}} = \frac{d}{b}\sqrt{\frac{\mu_r}{\varepsilon_r}}376.6\ \Omega.$$

The meaning of the parameters is to be shown based on the manufacturer's specifications for a coaxial cable (Table 10.12):

The following parameters can be obtained from this:

Inductance coating: $L' \approx Z_W^2 \cdot C' = (50\ \Omega)^2 \cdot 100\ pF/m = 250\ nH/m$

Phase velocity: $\nu \approx \frac{1}{\sqrt{L'C'}} = \frac{1}{\sqrt{250\cdot10^{-9}\cdot100\cdot10^{-12}}}\frac{m}{s} = 2\cdot10^8\frac{m}{s}$,

Wavelength: $\lambda = \frac{2\pi}{\beta} \approx \frac{2\pi}{\omega\cdot\sqrt{L'C'}} = \frac{1}{f\cdot\sqrt{L'C'}} = \frac{2\cdot10^8\frac{m}{s}}{f}$

For an illustrative example, a 1:5 impulse transformer is to be realized (Fig. 10.168). The number of windings shall be:

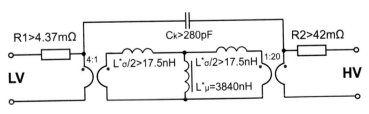

Fig. 10.168 Expected values for the equivalent circuit diagram of a 4:20 transformer with coaxial cable winding on a T80 × 20 × 50 toroidal core

$$N_1 = 4 \text{ and } N_2 = 20 \text{ or } N_2/N_1 = 5$$

A toroidal core T80 × 20 × 50 ($A_{Fe} = 295$ mm^2, $l_{Fe} = 197$ mm, $\mu_r = 2000$) is to be used as core. The approximate length of one winding is then

$$l_W = 2 \cdot ((80 - 50 + 10) + (20 + 10))mm = 140mm$$

The cable inductance of one winding is 35 nH.

The reluctance of the core is

$$R_{m\mu} \approx \frac{193 \text{ mm}}{2000 \cdot 4\pi \cdot 10^{-7}\frac{Vs}{Am} \cdot 295 \text{ mm}^2} = 2{,}60{,}313\frac{1}{H} \text{ respectively } \rightarrow A_L$$

$$\approx 3.84 \ \mu H/Wdg^2$$

The resistance of the windings (4 times screened and 5 times connected in parallel) is

$$R_1 = \frac{4 \cdot 140 \text{ mm}}{5} \cdot 39 \ m\Omega/m = 4.368 \, m\Omega \ \text{ or } R_2 = 20 \cdot 140 \ mm \cdot 15 \ m\Omega/m = 42 \ m\Omega.$$

With a capacitance of 100 pF/m, the coupling capacitance between primary and secondary winding is $C_k > 20 \cdot 0.14$ m $\cdot$ 100 pF/m $= 280$ pF.

The measured parameters to be expected are therefore

sec. Short circuit	$L_{\sigma 1} > 560nH$,	Primary Short circuit	$L_{\sigma 1} > 14 \ \mu H$,
sec. Idle	$L_{\mu 1} > 61.5 \ \mu H$,	sec. Idle	$L_{\mu 1} > 1537 \ \mu H$,
Coupling capacitance	$C_k = 280$ pF		

(between primary and secondary shorted windings).

Similarly, small leakage inductances can be achieved with bifilar windings. For this purpose, the respective parts of the other winding are always magnetically connected in parallel to a 'continuous winding'. The leakage inductance achieved here is relatively small and is quickly reached and exceeded on the primary side by the size of the supply inductance, so that appropriate measures must be taken: For example, wide, closely spaced conductors, coaxial current conduction, etc. In general, one must assume that a small leakage inductance/close magnetic coupling is associated with a comparatively large coupling capacitance between the windings. Even if the design described is not a line transformer in the true sense of the word, it is clear how to deal with the variables. In the following, the most important types of line transformers will be presented. The line types are divided into coaxial lines and symmetrical double lines. These behave differently as lines in principle. The structure of coaxial lines ensures that the external magnetic field disappears. The two conductors (inner and outer) cannot behave symmetrically to GND.

This is different with a double line and is taken into account or used in the following arrangements [39, 40].

Circuits with Coaxial Lines

Pulse Transformer for Voltage Multiplication

If you consider a line as a dead time element, where an applied signal appears at the output after a delay, a transformer can be constructed from this basic consideration. One connects lines in parallel on the input side and series on the output side. Figure 10.3 shows such an arrangement for 4 parallel line connections. The 4 line impedances are connected in parallel on the input side and series on the output side. The resulting impedances are also the values for optimum power matching / zero reflection. TLT1 is connected to GND on both sides so that a pure 2-port is given. This is not the case with lines TLT2...4.

Due to the raiseable potential on the output side, a high output voltage is achieved for a short time. To achieve the effect described in the schematic diagram Fig. 10.169, the wave propagation in the channel against GND is impeded by plugged toroidal cores by drastically reducing the propagation speed. At the same time, the impedance of this channel is significantly increased, so that the load on the associated output in this channel is reduced. With TLT4 the toroidal cores experience a higher flux load than with TLT2. Therefore more cores must be used to prevent saturation. The higher the output voltage and the longer the pulse duration, the higher the flux to be conducted or the magnetisation cross-section to be used. The cores should be arranged as close as possible to the end of the cable. The magnetization causes losses so that the cores must be cooled at higher repetition rates. With

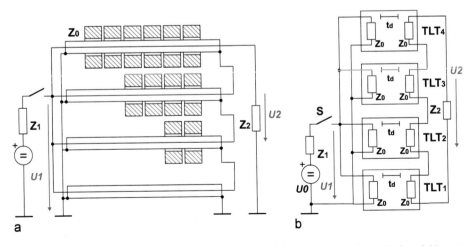

Fig. 10.169 Cable transformer (running time transformer) for pulse generation with four-fold output voltage, (**a**) circuit diagram with toroidal cores for delaying the outer wave channel, (**b**) simplified equivalent circuit diagram (**b**, l...cable length)

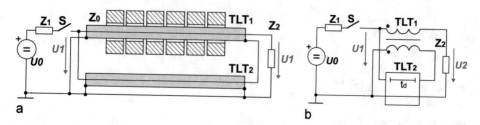

Fig. 10.170 Unun transformer with voltage amplification 2. (**a**) basic design, (**b**) simplified equivalent circuit diagram

the interconnection, a voltage amplification of 4 is achieved [41–42]. The circuit is used to generate short pulses. The transmittable pulse energy depends on the length of a cable or its capacitance:

$$W_{puls} = 4\frac{U_1^2}{Z_o} \cdot \Delta t = 4\frac{U_1^2}{\sqrt{\frac{L'}{C'}}} \cdot l \cdot \sqrt{L' \cdot C'} = 4 \cdot U_1^2 \cdot l \cdot C' = 4 \cdot U_1^2 \cdot C$$

The transmitted power is obviously controlled by the voltage U_1.

From a systematic point of view, the above-mentioned transformer is fed from a source to GND and supplies a load which is also switched to GND. Accordingly, this is called an *unun transformer* (*unbalanced to the unbalanced transformer*).

Unun Transformer with Coaxial Lines
The same principle of the above impulse transformer is used for HF transformers for supplying earthed loads from earthed sources with the appropriate voltage ratio (Fig. 10.170). This results in an impedance transformation as with any real transformer. Source and sink with different impedance can thus be matched to each other for optimum power transmission (power matching). Here too, ferrite cores (as above) serve to delay/suppress wave propagation in the channel against GND.

Balun Transformer with Coaxial Cable
An analog transformer for matching a load symmetrically to GND to a source referenced to GND at a gain of 2 is shown in Fig. 10.171. Of course, the direction of action can also be reversed with such transformers. Unun and balun transformers are used in a wide variety of applications in RF and RF technology. For example, such transformers can be used to connect symmetrical antenna terminals to subsequent coaxial lines. From the equivalent circuit diagram, it is clear that this arrangement involves two 1:1 transformers which are connected on the output side in the antiphase so that double the output voltage is achieved.

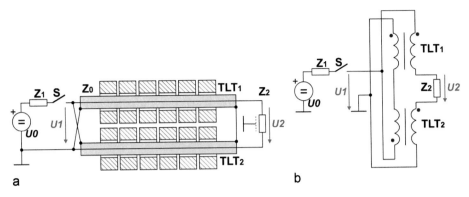

Fig. 10.171 Balun transformer for a voltage amplification of 2 and symmetric load connection, (**a**) basic structure, (**b**) simplified equivalent circuit diagram **b**

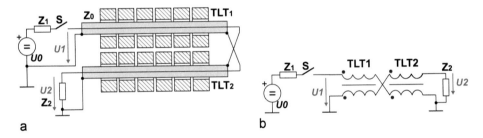

Fig. 10.172 Inverting Unun transformer for a voltage amplification of 1. (**a**) basic structure, (**b**) simplified equivalent circuit diagram

Inverting Unun Transformer

If a line is loaded with a subsequent line whose connections are reversed as in Fig. 10.172, the output signal is inverted. Here too, the ferrite cores impede the wave propagation in the wave channel against GND, so that for sufficiently high frequencies the equivalent circuit diagram can be shown with discrete transformers.

Dynamic Voltage Doubler (Blümlein Circuit, Blümlein Generator)

This circuit is often found in *pulsed power* applications. Due to the construction (one of the two lines is grounded), it can also be realized with double conductors or ribbon conductors. The constructive arrangement is shown in Fig. 10.173a. The simplified equivalent circuit diagram mainly reflects the terminal behavior of the circuit. In pulse applications, the center conductor is first charged to the voltage U_0. Then line 1 is short-circuited by the switch S so that a wave is started to the short-circuited end of the circuit. After reflection at this end, a voltage with a reversed sign builds up at the output of line 1, so that ideally $U_2 = 2 \cdot U_0$.

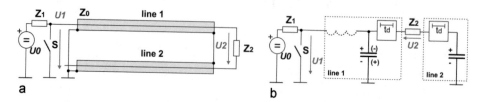

Fig. 10.173 Dynamic voltage doubler (Blümlein circuit) in a pulse application. (**a**) Basic structure, (**b**) equivalent circuit to explain the function

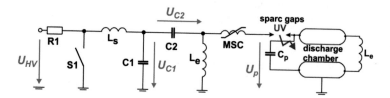

Fig. 10.174 Blümlein circuit with discrete components in a simplified pulse laser application

The circuit is used to generate short pulses. To supply a gas discharge path of nitrogen lasers with extremely short pulses as in N_2 lasers, laminated ribbon lines can be used, for example, where several are connected in parallel to reduce the characteristic impedance. Triggered spark gaps are usually used as switches at higher voltages. Thyratrons are used for lasers with lower dynamic requirements such as excimer lasers or CO_2.

Due to the low impedance of the gas discharge path, which is arranged instead of Z_2, a true power matching is usually not possible and after the first pulse, there is another of lower amplitude, which is due to reflections of excess energy at the ends of the line. The components used must be designed for the corresponding loads. For lasers of higher energy, it is not possible to store the transmitted energy in lines. Therefore, the Blümlein circuit with concentrated components is used here (Fig. 10.174). At the beginning the two capacitors C_1 and C_2 are charged in parallel to the same voltage UHV. When the switch S_1 closes, the voltage at C_1 oscillates above L_s and changes the sign. Two negative voltages are then connected in series at C_1 and C_2. When the *magnetic switch circuit* (MSC) is saturated when the sum voltage reaches its maximum value, a rapid charge transfer of $C_1 + C_2 \rightarrow C_p$. starts via the residual inductance. The fast charging via the spark gap generates hard UV radiation which pre-ionizes the gas in the discharge chamber or conditions it for the subsequent volume discharge. When the maximum voltage at C_p is reached, a volume discharge begins in the discharge chamber. This allows a laser to be pumped. The dynamic voltage doubling by the oscillation of the LC subcircuit $L_s + C_1$ occurs in the same way as the reflection at the end of line 1 in the circuit shown in Fig. 10.173. By connecting the voltage at C1 and C2 in series with the same polarity after oscillation.

The capacity of the lines is simulated by capacitors. The inductances of the lines form the inductances of the circuit. Since the gas-discharge path of the laser is on GND on one side, the components are arranged in such a way that during the charging phase of the two capacitors C1 and C2, this discharge path is bridged by a large inductance L_e (order of magnitude ~100 µH). This inductance does not play a role during the pulse process, just like L_e parallel to the discharge path. The properties of line 1 are simulated by the line inductance of the C1-L1-S resonant circuit in conjunction with a saturable choke MSC (*magnetic switch circuit*). As the voltage at C1 is reversed, a voltage is built up via MSC, the voltage-time area of which finally leads to saturation of MSC. MSC then acts like a switch that connects C1 and C2 in series-parallel to the gas discharge path of the laser. Ideally (by pre-ionization using UV or X-ray radiation), the discharge path breaks down homogeneously throughout the entire discharge chamber at a certain voltage and absorbs the energy of the two capacitors C1 and C2.

Transformers with Symmetrical Double Lines

Balun Transformer

Figure 10.175 shows a balun transformer with a double line. Input and output impedance remains the same as the line impedance. This means that even the voltage does not change in a loss-free case. The ferromagnetic cores result in an "HF insulation" of load Z_2 against GND. The lower the frequencies to be transmitted with this effect, the longer the cable must be. If necessary, it must be inserted several times through the cores.

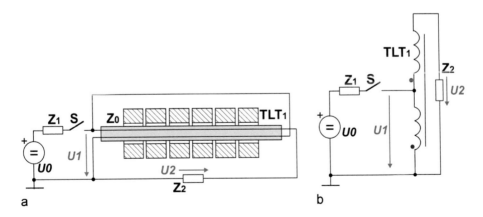

Fig. 10.175 Balun transformer with a voltage ratio of 1 with a double line. (**a**) basic design, (**b**) simplified equivalent circuit diagram

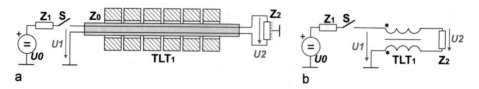

Fig. 10.176 Unun transformer with a voltage ratio of 2 with a double line. (**a**) basic structure, (**b**) simplified equivalent circuit diagram

Unun Transformer

The transmission of an unbalanced voltage to an unbalanced input of a receiver with a voltage ratio of 2 allows the circuit shown in Fig. 10.176. It works similarly to an autotransformer.

Finally, a simple possibility of modeling lines for simulation purposes is pointed out. This is the D'Alembert's principle known from school. It describes a wave as two partial waves propagating in opposite directions and independent of each other. At the beginning and at the end of the 'line' there are interactions with the environment via the characteristic impedance. Something like this can easily be represented by two dead time elements with characteristic impedances, which are switched against each other. These are actually available in most common electrical simulation programs. Basic properties such as propagation time, reflection and energy dissipation at the ends can be easily mapped and simulated in this way.

10.15 Transformers with Controllable Output Voltage

In addition to insulation, an important function of circuits is the adaptation of the voltage level to an application. In this subchapter, an overview of the different possibilities of a transformer for a variation of the output voltage shall be given. As an example, a general illustration of coupled coils (Fig. 10.177) as a transformer with strong yoke leakage shall be considered. Besides, the corresponding magnetic and electrical equivalent circuit diagrams are shown, which provides a quick overview of the conditions. The results of the observations can be transferred to transformers with the close coupling of primary and secondary windings.

The flux components of a winding that are not coupled to the core are generally very small so that the inductors $L_{\sigma 1}$ und $L_{\sigma 2}$ in the equivalent circuit diagram are also very small in relation to the other inductors. If a voltage $\underline{U}_1$ is applied to the winding N1, the voltage $\underline{U}_2$ is given by the two-pole equation (without taking into account the ohmic model components)

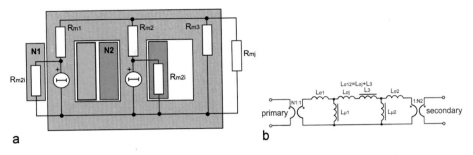

Fig. 10.177 Transformer with equivalent circuit diagrams, taking into account different leakage paths. (**a**) magnetic circuit, (**b**) electrical equivalent circuit diagram

$$\underline{U}_2 \approx \frac{N_2}{N_1} \cdot \frac{L_{\mu2}}{L_{\mu2} + L_{\sigma12}} \cdot \underline{U}_1 - j\omega \frac{L_{\sigma12} L_{\mu2}}{L_{\sigma12} + L_{\mu2}} \cdot N_2^2 \cdot \underline{I}_2$$

This means that

Open circuit voltage $\underline{U}_{20} \approx \frac{N_2}{N_1} \cdot \frac{L_{\mu2}}{L_{\mu2} + L_{\sigma12}} \cdot \underline{U}_1$

Short circuit current $\underline{I}_{2K} \approx \frac{1}{N_1 N_2} \cdot \frac{\underline{U}_1}{j\omega L_{\sigma12}}$

Ideal apparent power

$$S_{T.0} \approx U_{2.0} \cdot I_{2.K} = \frac{1}{\left(1 + \frac{L_{\sigma12}}{L_{\mu2}}\right)} \cdot \frac{\left(\frac{U_1}{N_1}\right)^2}{\omega L_{\sigma12}}$$

If you want to change the open-circuit voltage ($I_2 = 0$) you can do this by changing

1. N_2
2. $L_{\sigma12}$
3. $L_{\mu2}$

Change of the Secondary Number of Turns N_2

In theory, N_2 can be used to change the output voltage without changing the ideal apparent power. In practice, the transmittable current I_2 is limited by the cross-section of the secondary winding material. The no-load output voltage changes proportionally to N_2. The available output current is, in a first approximation, proportional $\sim N_2^{-1}$. This principle is used for variable transformers of various designs. The number of windings is changed via a movable contact (switch groups, sliding contact, rolling contact). It is important during the design process that the circuit is not interrupted from one position to the other, as otherwise electric arcs are induced. Figure 10.178 shows a typical toroidal variable transformer using this principle. The number of secondary windings is changed by a movable roller contact.

Fig. 10.178 1kVA- variable transformer as autotransformer with a toroidal core and roller contact to the voltage tap (Source: Zacharias 2017)

Adjustment of the Output Voltage by Changing the Magnetising Inductance

This control method is also based on an inductive voltage divider ($X_{\mu2}/(X_{\mu2} + X_{\sigma12})$), with which the open-circuit voltage is changed. Since the general rule is $L_\mu > L_\sigma$, the influence on the short-circuit current is not as great as that caused by changing the leakage reactance. However, the magnetizing current increases when the output voltage is reduced, which limits the range of applications. Here, too, the reluctance of an element of the equivalent circuit diagram must be made variable. This can be achieved, for example, by premagnetisation. Figure 10.179 shows a ferrite transformer with orthogonal premagnetisation of the middle leg by an additional (direct) current (yellow power). This allows the effective inductance to be changed continuously.

Adjustment of the Output Voltage by Changing the Leakage Inductance

This control method is based on an inductive voltage divider ($X_{\mu2}/(X_{\mu2} + X_{\sigma12})$) with which the open-circuit voltage is changed. At the same time, however, the output impedance of the two-pole is changed in such a way that a lower output voltage results in a lower output current. The magnetizing current in the open circuit remains largely unaffected. The leakage can be designed via a variable leakage reluctance. This can be achieved by suitable premagnetization or a variable air gap that can be adjusted mechanically (e.g. by means of piezo translators or similar). Figure 10.180 shows an example of a mechanically adjustable

Fig. 10.179 EE70 ferrite core pairs with the variable reluctance of the center leg through inserted holes for premagnetization (yellow) parallel to the main flux and transverse to the main flux (Source: Zacharias 2017)

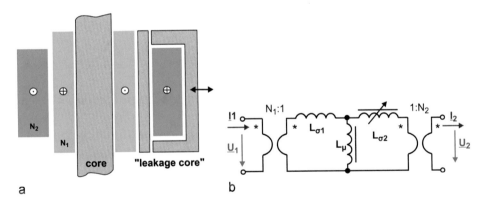

Fig. 10.180 Transformer with adjustable output voltage. (**a**) adjustable leakage reactance, (**b**) electrical equivalent circuit diagram

leakage reactance for a transformer of the type "close couplings": The flux leakage is changed by an adjustable air gap of an additional U-I core (Fig. 10.181).

Inserting a Capacity to Change the Controlled Range

The output impedance of a transformer in the T equivalent circuit diagram is approximately the same $\underline{Z}_2 \approx j\omega(L_{\sigma 1} + L_{\sigma 2})N_2^2$ at an open-circuit output voltage $\underline{U}_{20} \approx \frac{N_2}{N_1} \cdot \frac{L_{\mu 2}}{L_{\mu 2} + L_{\sigma 12}} \cdot \underline{U}_1$.

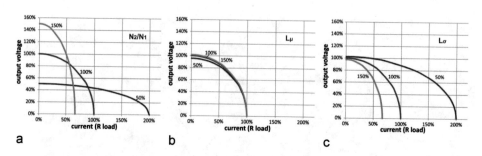

Fig. 10.181 Comparison of the output voltage control options discussed. (**a**) with the change of the transmission ratio N2/N1, (**b**) change of the magnetizing inductance L_μ, (**c**) change of the leakage inductance to the steps 50%, 100% and 150% with the assumption of the ratio $L_{\sigma100\%}: L_{\mu100\%} = 1:20$

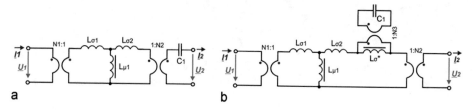

Fig. 10.182 Changing the output impedance and thus the setting range of a transformer by inserting a capacitor C1. (**a**) Increasing the output current, (**b**) decreasing the output current

The output current is approximately $\underline{I}_{2K} = \frac{\underline{U}_{20}}{\underline{Z}_2} = \frac{\underline{U}_{20}}{j\omega(L_{\sigma1}+L_{\sigma2})N_2^2}$.

If energy is transferred at a fixed frequency, it is possible to influence the output current retroactively, that is, with a given transformer design. For this purpose, a capacitor is inserted into the construction, as shown in Fig. 10.182. Variant Fig. 10.182a, with $(\omega C_1)^{-1} < \omega(L_{\sigma1} + L_{\sigma2})$, allows increasing the output current by reducing the effective output impedance

$$\underline{I}_{K2} \approx \frac{\underline{U}_{20}}{j\left[\omega(L_{\sigma1} + L_{\sigma2})N_2^2 - \frac{1}{\omega C_1}\right]} = \frac{j\omega C_1 \cdot \underline{U}_{20}}{1 - \omega^2(L_{\sigma1} + L_{\sigma2})N_2^2 C_1}.$$

The current carrying capacity of the transformer and capacitor, as well as the voltage carrying capacity of the capacitor, set the essential limits here. Switchable capacitors increase the control range via the output impedance.

Figure 10.182b shows one possibility for reducing the maximum output current specified by the design. Here the output impedance is given by $\underline{Z}_2 \approx j\omega\left(L_{\sigma2} + \frac{L_\sigma^*}{1 - \omega^2 L_\sigma^* N_3^2 C_1}\right)N_2^2$ so that the short-circuit current is

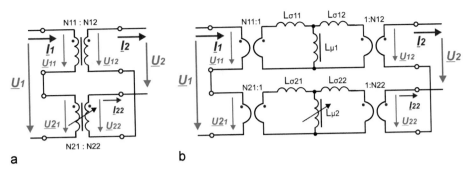

Fig. 10.183 Variable output voltage by variable counter-connection of a second transformer. (**a**) Principle structure, (**b**) electrical equivalent circuit diagram to explain the mode of operation

$$I_{K2} \approx \frac{U_{20}}{j\omega\left(L_{\sigma 2} + \frac{L_{\sigma}^*}{1 - \omega^2 L_{\sigma}^* N_3^2 C_1}\right)} = \frac{-jU_{20}\left(1 - \omega^2 L_{\sigma}^* N_3^2 C\right)}{\omega\left(L_{\sigma 2}\left(1 - \omega^2 L_{\sigma}^* N_3^2 C_1\right)\right) + L_{\sigma}^*}$$

will. With $C_1 = 0$ increasing capacitance, the short-circuit current decreases. At the same time, the voltage drop at C_1 or L^*_{σ} increases, depending on the current. The additional losses that then occur must be taken into account, as must any saturation phenomena/ ~losses at L^*_{σ}.

The combination of a transformer with a fixed magnetizing inductance with one with a variable magnetizing inductance is similar to an inductive voltage divider with isolated, transformed and summed output voltage (Fig. 10.7). In no-load operation, the transformed output voltages of the transformers add up. The following then applies

$$U_2 = U_{21} - U_{22} = \frac{L_{\mu 1} \cdot N_{11}^2}{L_{\mu 1} \cdot N_{11}^2 + L_{\mu 2} \cdot N_{21}^2}\cdot\left(\frac{N_{12}}{N_{11}}\right)\cdot U_1 - \frac{L_{\mu 2} \cdot N_{21}^2}{L_{\mu 1} \cdot N_{11}^2 + L_{\mu 2} \cdot N_{21}^2}\cdot\left(\frac{N_{22}}{N_{11}}\right)$$
$$\cdot U_1$$

If all numbers of turns are the same, you get, for example,

$$U_2 = U_{21}\frac{L_{\mu 1}}{L_{\mu 1} + L_{\mu 2}}\cdot U_1 - \frac{L_{\mu 2}}{L_{\mu 1} + L_{\mu 2}}\cdot U_1 \approx \left(1 - \frac{2L_{\mu 2}}{L_{\mu 1} + L_{\mu 2}}\right)\cdot U_1$$

In the case of a short circuit, the magnetizing reactance of the controlling transformer limits the current. As a result, a very high flux linkage builds up there, which would result in a large size. The application of the described operating principle will therefore be limited to small changes of the output voltage in the range of the voltage adjustment. Figure 10.184 shows the calculated load characteristics of the arrangement according to Fig. 10.183 for 3 different cases of the size of the magnetizing reactance. Linearity was assumed for the

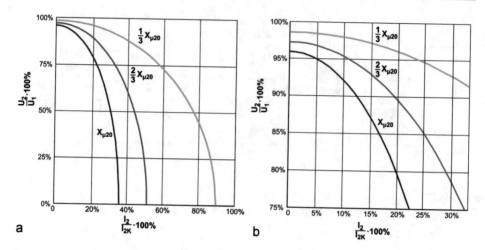

Fig. 10.184 Change in the load characteristic curves of the circuit according to Fig. 10.183 (**a**) changed magnetizing reactance $X_{\mu2}$ in relation to a value $X_{\mu20}$, (**b**) with extended scale for $X_{\mu20}/X_{\mu1} = 50$, $(X_{\sigma11} + X_{\sigma12})/X_{\mu1} = (X_{\sigma21} + X_{\sigma22})/X_{\mu20} = 0.01$, $N_{11}/N_{12} = N_{21}/N_{22} = 1$

controlled reactance as it can be achieved to a large extent by orthogonal premagnetization (see e.g. [43–45]. With the number of turns and cores used, it is possible to set the output characteristic curve.

Since energy is transferred from the primary to the secondary side via the leakage inductance, nonlinearities in the magnetization characteristic do not have a great effect on the output current and output voltage. For this reason, combinations of a linear transformer with a magnetic amplifier operated as a transformer can also be used [46]. The current-limiting leakage inductance does not change with this method. Various possibilities for implementing controllable transformers are described in [47–50].

The combination of switched and counter-switched winding components on controlled reluctances can also be used to build a "differential amplifier" (Fig. 10.185). By interconnecting the windings, an arrangement with 2 controlled reluctances $R_{m1}(I_0 + I_1)$ and $R_{m2}(I_0 - I_1)$ is obtained. The increase in the dependency $R_m(I)$ approaches zero when $I = 0$. If one assumes approximately that

$$R_{m1}(I) \approx R_{m0}\left(1 + (k \cdot I)^2\right) \text{ and } \Lambda_1(I) = \frac{1}{R_{m1}(I)}$$

with a symmetrical structure, one obtains the analytical relationships

$$R_{m1}(I) \approx R_{m0}\left(1 + (k \cdot I)^2\right) = R_{m0}\left(1 + (k \cdot (I_0 + I_1))^2\right)$$

and

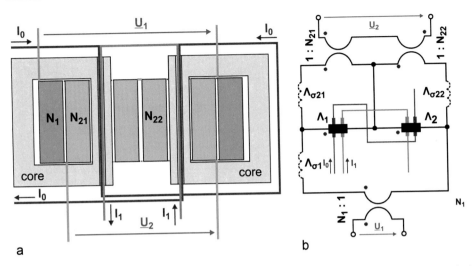

Fig. 10.185 Reluctance controlled "differential amplifier". (**a**) basic winding design, (**b**) electrical equivalent circuit diagram

$$R_{m1}(I) \approx R_{m0}\left(1 + (k \cdot I)^2\right) = R_{m0}\left(1 + (k \cdot (I_0 - I_1))^2\right)$$

Because the fluxes of the premagnetisation are added in one case and subtracted in the other case, the two reluctances R_{m1} and R_{m2} change in opposite directions. For the transmission of the transformer built up with these two reluctances, one obtains the following in small-signal operation at $N_{21} = N_{22}$

$$U_2 \approx -2kI_0I_1\frac{U_1}{N_1} \cdot N_{21}$$

This corresponds to a controllable transformer for the small-signal range. An AC voltage U_1 is multiplied by a proportionality factor and the product $I_0 \cdot I_1$. Assuming that I_0 is a direct current, this arrangement can in principle be used for multiplicative purposes in the small-signal range. This function is similar to a multiplier with a transistor differential amplifier [51].

10.16 Electronic "Direct Current Transformers"

Transformers are characterized by low-loss, bi-directional energy transmission. The voltage transmission ratio is largely adjustable by winding ratio and design. Apart from its insulation capability, a transformer is therefore similar to bi-directional DC–DC converters in its properties. The latter can additionally change the voltage ratio quickly via the control

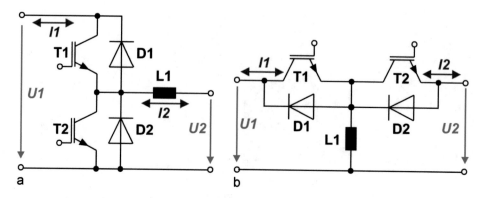

Fig. 10.186 Bi-directional buck converter as "DC transformer" (a) and bi-directional inverse converter as "DC transformer with sign inversion"

parameter 'D' (*duty cycle*) of the PWM signal (PWM: *pulse width modulation*). There are a large number of solutions in this field. Only the most important basic circuits that have achieved the greatest diffusion will be discussed here.

Figure 10.186 shows two converter types without galvanic isolation. The bi-directional high/low converter (a) is by far the most commonly used converter type. For example, it is included as a bridge branch in almost every single-phase or multi-phase drive inverter. T1 and T2 are controlled complementarily. The loads on the components are very moderate here and thus high efficiencies can be achieved. The output voltage has the same sign compared to the common GND. From left to right, the circuit has a buck converter, while from right to left it has a *boost converter*.

The converter shown in Fig. 10.186b has the inverted sign of the input voltage and is therefore called an inverse converter. Here, the load on the switches results from the sum of the input and output voltage. In terms of amount, the output voltage can be both lower and higher than the input voltage in both directions and is therefore called a *buck boost converter*. The main application is in the power supply technology of devices in order to generate a further negative operating voltage from one operating voltage. Both converters can be treated as first order systems during operation.

A DC coupling of input and output is mandatory for the two circuit variants mentioned above. This is not the case with the variants shown in Fig. 10.187. Here, the input and output are coupled via capacitors, which interrupt a DC path. Since a DC component is missing in the energy transmission, the converters can also be disconnected at the capacitors to insert an insulating transformer, which is also done for lower power ratings. However, the "hard" switching on and off of the semiconductor switches causes overvoltages at the switches due to the commutation inductors, which are difficult to control at higher currents. This limits the application to a few 100 W to a few kilowatts. In terms of control engineering, the requirements of the circuits in Fig. 10.187 are higher than those of the previous two, which only represent systems with one memory each (first

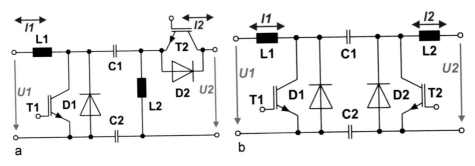

Fig. 10.187 Isolating DC/DC-converters: bi-directional high-low converter as consisting of sepic converter + zeta converter as "DC-transformer" (a) and bi-directional cuk converter as "low-high-setting DC-transformer with sign inversion"

order systems). In contrast, sepic/zeta converters and cuk converters represent third order systems, which react much more complex at load steps than first order systems.

If the capacitor C_2 is omitted, the function of the converters is basically unchanged. However, the input and output of the circuit are then connected together. The circuits both have step - steep down characteristics, like the inverse converter at similar voltage load of the switches.

Of interest are the input- and output-side general characteristics of the above-mentioned circuits. At weak loads, all DC–DC converters have a gaping current. This can change under load. In this case, the inverse converter has a gaping current on the input and output side, while the buck converter or boost converter, but also zeta/sepic converters, have a non-gaping current either on the output side or on the input side, since inductive current storage devices are in series. The cuk converter has an inductance on both the input and output side and therefore a non-stop current on both the input and output side under load. This is interesting for battery applications, since electrochemical energy storage devices (accumulators, batteries) age faster due to gaping current/high current ripple. And since accumulators are loaded and discharged much longer in many applications with high instantaneous power, value is placed on current smoothing during charging. Values between 1/3 and 3 are usually selected for the voltage ratio, since the semiconductor switches are loaded with both the highest voltage occurring and the highest current. With semiconductor switches, the switch losses increase disproportionately to the rated voltage. High voltage differences to be overcome therefore result in high switch losses.

If you do not want or cannot insert a transformer, the available voltage range can be increased by cascading, as shown as an example in Fig. 10.188. Input here is the left half of the circuit shown in Fig. 10.188a, while the right half of the circuit is connected 2x on the output side in series, which is easily possible using the capacitors $C_{1...3}$. T_1 is driven with the duty cycle D, while T_2 and T_3 are driven synchronously with the duty cycle (1-D). This measure doubles the range of the output voltage at "moderate" switch load, but at the price

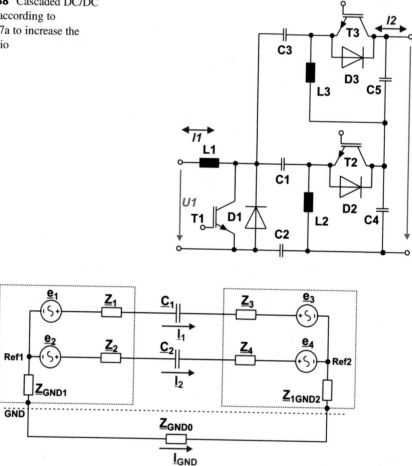

Fig. 10.188 Cascaded DC/DC converter according to Fig. 10.187a to increase the voltage ratio

Fig. 10.189 Equivalent circuit diagram describing the function of the two isolated DC/DC converters as insulating energy converters

of increased complexity and control engineering behavior. This procedure can be transferred to similar circuits.

The circuits shown in Fig. 10.187a and b are basically circuits, which isolate input and output galvanically, i.e. for direct current. Nevertheless, a considerable amount of alternating current flows through C2 during energy transport. This must be taken into account when assessing protection against contact or hazards in the event of a fault. For this purpose, the problem is to be examined more closely using the block diagram Fig. 10.198. The always present conductor of the earth potential to which persons in the vicinity are connected is shown separately here. This results in a 3-wire system. Energy can theoretically be transported in 3 wave channels (Fig. 10.189).

The two lines provided for the power line are interrupted by capacitors C_1 and C_2. The generation of the voltages in the DC/DC converter sections is symbolically represented by the 4 voltage sources $e_1(t)...e_4(t)$ with the associated impedances $\underline{Z}_1...{}_4$, which are measured relative to an internal reference point. These reference points are connected via the impedances $\underline{Z}_{GND1}$ and $\underline{Z}_{GND2}$. The distance between the two ground connection points itself has the impedance $\underline{Z}_{GND0}$. During normal operation, the currents $\underline{I}_1$, $\underline{I}_2$, and $\underline{I}_{GND}$ occur, which can be easily determined. Especially is

$$\underline{I}_{GND} = \frac{\left(\left(\frac{e_1 - e_3}{\underline{Z}_1 + \underline{Z}_3 + \underline{X}_{C1}}\right) + \left(\frac{e_2 - e_4}{\underline{Z}_2 + \underline{Z}_4 + \underline{X}_{C2}}\right)\right) \cdot (\underline{Z}_1 + \underline{Z}_3 + \underline{X}_{C1})//(\underline{Z}_2 + \underline{Z}_4 + \underline{X}_{C2})}{\underline{Z}_{GND0} + \underline{Z}_{GND1} + \underline{Z}_{GND2}}.$$

In the circuits shown in Fig. 10.191b,c, one can apply $e_2 = e_4 = 0$. $\underline{I}_{GND}$ is then

$$\underline{I}_{GND} = \frac{(\underline{e}_1 - \underline{e}_3) \cdot (\underline{Z}_1 + \underline{Z}_3 + \underline{X}_{C1})}{(\underline{Z}_{GND0} + \underline{Z}_{GND1} + \underline{Z}_{GND2})(\underline{Z}_1 + \underline{Z}_2 + \underline{Z}_3 + \underline{Z}_4 + \underline{X}_{C1} + \underline{X}_{C2})}$$

This means that $\underline{I}_{GND}$ is not equal to zero. When switching on, however, even $e_3 = 0$, so that a measurable inrush current in GND can be expected. *Thus, an alternating current flows via GND during operation, even though there is a DC blockade!* However, the current via GND becomes zero if $\underline{e}_1 = -\underline{e}_2$ and $\underline{e}_3 = -\underline{e}_4$, i.e. *if the voltage sources $\underline{e}_1...\underline{e}_4$ operate in pairs in opposite phase.*

→ This is the case when both "transmitter" and "receiver" circuits are bridge circuits. Since the mutual cancellation of the effect of the voltage sources can only be achieved to a residual error, the current via GND can be minimized by using a coupling element, which has a low impedance for the wave channel of lines 1 and 2, while the channels '1-GND' and '2-GND' each have a high impedance.

This task leads to so-called *common mode chokes* (CMC), as shown in Fig. 10.190. Due to the selected winding direction of the windings, the low leakage inductance between the windings is effective for differential voltages between (1) and (2). This hardly contributes to the increase in impedance ($Z_5 + Z_6$). With common mode voltages against GND, the magnetizing inductance acts in each line. Highly permeable core materials suitable for the frequency are therefore used for the construction of CMC. Under normal operating conditions, the current (interference current, touch current) via GND can therefore be kept very low.

In converters with purely transformer coupling (Fig. 10.191a), there is a limitation of the possible GND current via the size of the coupling capacitance of the windings. If necessary, the effect can be further increased by a common mode choke. In the series resonance bridge circuit (Fig. 10.191 b), the coupling capacitance naturally serves to transmit energy. Here the effect of a common mode choke is fully effective. However, when analyzing the possible errors, it should be noted that if a bridge output fails ($\underline{e}_2 = 0$), the common mode choke "sees" e.g. the voltage $\underline{e}_1$ as an unbalanced voltage against GND, so that with

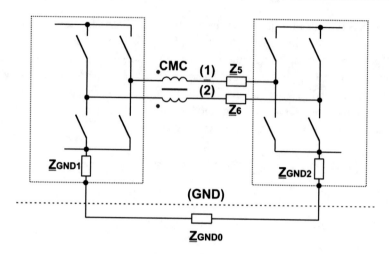

Fig. 10.190 Connection of 2 bi-directional DC/DC converters via a common mode choke (CMC) to minimize the HF residual current via GND

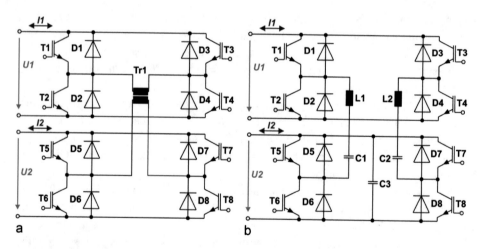

Fig. 10.191 Bidirectional DC/DC converter in bridge circuit. (**a**) Transformer insulation, (**b**) and capacitive insulation

or without saturation effects of the choke, the current I_{GND} can assume dangerous dimensions. This can be counteracted by monitoring and switching off in the event of a fault.

Since there is no "counter-capacitance" in the electric field with the appropriate coupling of network elements, voltage multiplication can only be achieved by cascading or parallel series connection of rectifier four poles. This circumstance is used by various rectifier circuits for voltage multiplication. This method is applied in Fig. 10.192. Two capacitively isolated active rectifiers are connected in series on the input side [52].

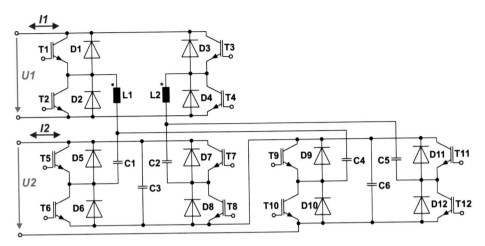

Fig. 10.192 Bidirectional DC/DC converter in bridge circuit with transformer as series resonance converter with voltage doubling

A highly efficient variant of a truly insulating DC/DC converter is shown in Fig. 10.193 with a bi-directional DC/DC converter, which is operated as a series resonant circuit inverter at D close to 0.5 at the resonant frequency or slightly below the resonant frequency of the series resonant circuit. Doing so practically eliminates the current limiting stray

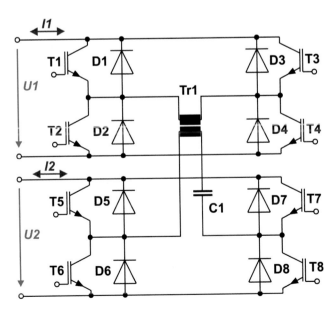

Fig. 10.193 Bidirectional resonant DC/DC converter in bridge circuit with transformer as bi-directional series resonant converter

reactance of the transformer. The energy can be transmitted in both directions with the transformer's voltage transformation ratio. Achievable efficiencies are 99%, since the reactive power requirement of the switch capacities can be covered, for example, by the magnetizing inductance, so that ZVS is possible. However, one loses the controllability of the converter due to these boundary conditions. If frequency or duty cycle is selected as the control parameter, no more ZVS is possible and the switching losses increase significantly.

10.17   Transformerless Grid Connected AC Current Feed in from DC Sources and Storages

Due to the modular design of the photovoltaic generators from cells with $U_{MPP} \sim 0.5$ V and typical nominal currents $I_{MPP} = 3...15$ A and the different installation conditions on buildings, there is a great variety of design options for systems even under *standard test conditions* (*STC*, means among others: 1000 W/m², 25 °C, terrestrial solar spectrum). But even with a fully dimensioned solar generator, high temperature coefficients in the range − 0.3...0.5%/K provide a wide output voltage range. At a usage temperature of −25 ° C... + 85 °C, this may result in an operating voltage range of +25% to −30% compared to STC. In addition, there are variations in the operating voltage range on the DC side, which result from the different shape of the load characteristics of the photovoltaic generator (Fig. 10.194). Since the maximum DC voltage (no-load) for the PV generator should not exceed 1000 V (now 1500 V in Germany and other countries) even at low temperatures, high and low voltage capable inverter topologies are usually required.

The earlier photovoltaic generators were operated via a isolating transformer, which resulted in additional losses in the transformer and thus lower efficiencies (−1... −1.5%). With a typical service life of at least 20 years and a typical annual yield of 800...900 kWh/kWp, this leads to a significant reduction in yield. For this reason, more and more inverters without insulating transformers were brought onto the market. Because the output current to the grid is controlled by pulse width modulation (PWM) in these inverters, effects on the environment are possible in addition to grid feedback. This is illustrated in

Fig. 10.194 Examples of standardized characteristic curves of photovoltaic generators made of poly-Si [53] and CIS [54]

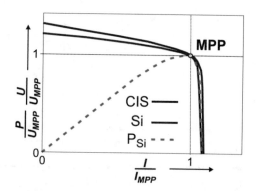

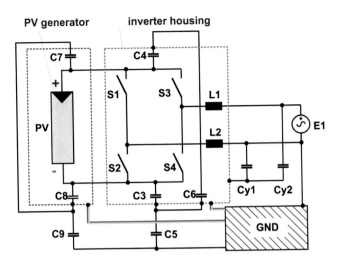

Fig. 10.195 Simplified diagram illustrating the effect of the earth capacitance of the PV generator [55]

Fig. 10.195. Schematically shown is a pulse inverter in a bridge circuit, whose DC link is connected to the PV generator. Depending on the selected pulse pattern for the switches, this can result in more or less strong potential jumps at the PV generator. The parasitic capacitances of the generator are reloaded, resulting in high-frequency leakage currents into the environment.

Depending on the generator design, elevation and surface wetting with water, the parasitic capacitances can be relatively large (2...100 nF/kWp, for thin-film modules up to 1 µF/kWp [55]), so that considerable RF leakage currents can flow into the design. When a transformer is inserted, the coupling capacitances of the windings are in series with the parasitic capacitances of the generator, thus reducing the leakage currents. However, it is also possible to minimize or eliminate the leakage currents by using suitable topologies and pulse patterns. In addition, filters are often used between the PV generator and inverter topology. These filters also have the task of protecting the power electronics from atmospherically caused overvoltages from the generator.

In some countries, the connection rules for the public grid require the PV generator to be earthed. In the USA, for example, there is also a requirement for permanent monitoring of this earthing [56, 57]. In addition, since the significant increase in the use of photovoltaic thin-film modules, the degradation properties of these modules, which in some cases obviously differ from crystalline silicon modules [58], have also become the focus of interest in the development of power electronic energy converters. It is obvious that PV modules of different construction show different degradation behaviour depending on the potential of the DC connections. In the case of energy converters with insulating transformers, this does not lead to any particular problems. In the case of highly efficient inverter arrangements without a transformer, a greater problem can arise if earthing of the

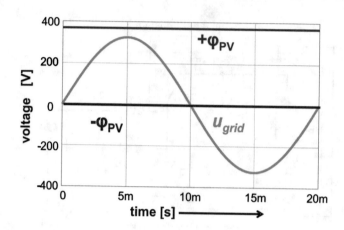

Fig. 10.196 Potential relations for the topologies according to [59, 60] in grid-connected operation

PV generator is required and this should also be freely selectable with regard to the polarity of the connections. Here a new driving force for the development of corresponding topologies has emerged.

Converter Structures to Meet the Generator Side Requirements
The problem solutions dealt with here relate to limiting or avoiding the HF leakage currents of the PV generator and the problem of earthing. Obviously, leakage currents over the frame of the PV generator can be practically avoided by defining the potentials of its DC connections. The solutions presented in [59, 60] follow this path. These are topologies with low or high/low transmission behavior. By means of a sufficiently large buffer capacitor, the voltage at the output of the PV generator is kept practically constant despite the fluctuating power output during single-phase connection, so that leakage currents via the parasitic ground capacitances of the generator cannot occur (Fig. 10.195). Figure 10.196 shows an example of the potential ratios when the negative pole is grounded.

Another possibility is to symmetrically fix the potentials of the DC connections of the PV generator. If a DC/DC converter is connected upstream of the DC/AC converter to compensate for the input voltage fluctuations, the symmetry of the connection potentials can be lost. However, according to current knowledge, this has no particular technical value. A symmetry only ensures a minimum voltage difference of the PV generator compared to earth potential. Figure 10.197 shows the corresponding ratios. The charging currents of the PV generator's earth capacitances are practically only determined by the residual "hum" voltage of the DC link capacitances.

The solution variant according to Fig. 10.195 has the disadvantage that the PV generator is not assigned to a preferred potential compared to earth. This can be remedied by the circuit topology shown in Fig. 10.198 with a bipolar DC/DC converter. The voltage level of the DC link is simultaneously decoupled from the PV generator voltage, thus extending the

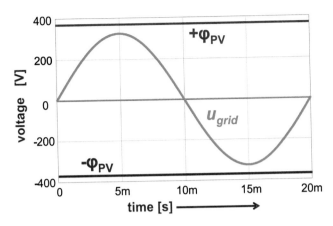

Fig. 10.197 Potential ratios for the topologies shown in Fig. 10.195 in line-coupled operation when $U_{PV} > 2 \cdot \hat{u}_{mains}$ and symmetrical ratios of the ground capacitances of the DC link are present

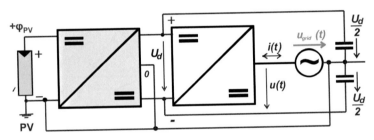

Fig. 10.198 One-sided grounding of the PV generator by inserting a DC/DC converter with bipolar output

input voltage range. Various solutions have been developed for the design of such bipolar converters [61]. Among them, there are transformerless variants, which enable grounding of both the positive and negative generator connection.

If we consider bridge circuits for single-phase grid feed-in of current, then with a completely symmetrical design (e.g. also 2 smoothing chokes in the two outputs) and symmetrical clocking, we obtain a state that avoids potential jumps at the PV generator (Fig. 10.199).

With such a "floating" voltage intermediate circuit, the (parasitic) capacitances of the PV generator C_{p1} and C_{p2} against earth are charged and discharged during the different switching states and store the charge absorbed, similar to a sample-and-hold circuit. The charging currents of C_{p1} and C_{p2} can easily be limited by current compensated chokes in the DC lines from C_1 to the PV generator. For the potentials of the DC connections can be determined (Fig. 10.200):

However, this only works with symmetrical control of the switch pairs $S_1\&S_4$ and $S_2\&S_3$. Especially with $\cos\varphi = 1$, however, an asymmetrical clocking would also be

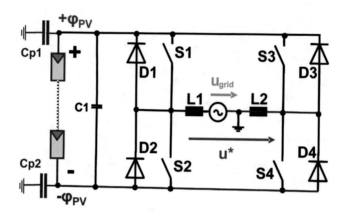

Fig. 10.199 Circuit diagram for a symmetrical bridge inverter (B2I) with "floating" intermediate voltage circuit

Fig. 10.200 Potential ratios on a symmetrical bridge inverter (B2I) with "floating" intermediate voltage circuit

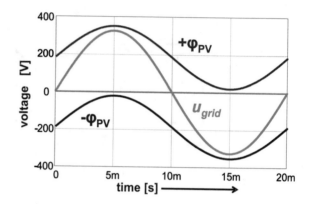

advantageous for the efficiency, since then the switching losses of only one switch become effective.

In order to achieve a behavior as shown in Fig. 10.201 in this case as well, the high-frequency clocking is carried out in such a way that the PV generator is disconnected from the grid symmetrically and synchronously by 2 switches during a grid half-period when it is switched off. The internal freewheeling and the polarity of the output voltage are ensured in the remaining inverter circuit. This divides the commutation voltage between 2 switches connected in series, so that the switching losses correspond to those of a single switch. As with the symmetrical clocking of the simple B2I, the ground capacitances are recharged similar to a sample-and-hold circuit when the two HF-clocked switches are closed, resulting in the potential curve already shown in Fig. 10.200. To this general principle, 2 circuit topologies have been published and used so far (Fig. 10.201). They led to the products with the highest efficiencies so far. In Fig. 10.201a, the RF clocking is performed in pairs via switches S_1&S_2 and S_3&S_4, while the internal freewheeling branch of the

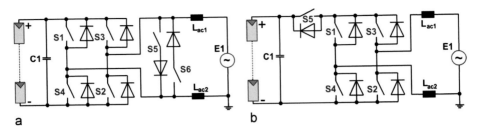

Fig. 10.201 Transformerless topologies for feeding energy from direct current sources into the grid. (**a**) Structure of the HERIC® topology [62], (**b**) structure of the H5® topology [63]

circuit is switched over with S_5 and S_6, depending on the output polarity. In Fig. 10.201b, the HF clocking is carried out by pairs S_2&S_5 or S_4&S_5, while the freewheeling branch is controlled by S_1 and S_3.

Stimulated by a high growth rate of the PV market, a variety of circuitry approaches have been developed that enable high efficiency in feeding active power into the interconnected grid [64]. Since 1995, the losses of commercial inverters have been reduced every year, even in the lower and medium power range, so that maximum efficiencies of 98...99% are standard for single-phase inverters. Inductive components are used in these circuits only to limit the current or current rise. Transformers are avoided as insulating elements with internal consumption. Falling costs for PV generators and decreasing feed-in tariffs will weaken the outstanding position of efficiency as a technical target parameter. The focus will be on manufacturing costs and *life cycle costs as* well as the ability to support the grid. This is also a major technical challenge with a high efficiency of 98–99%. Reactive power capability leads to other loads on the semiconductor elements. In order to guarantee low losses at low costs over the entire apparent power range of an inverter, new circuit technology solutions can be expected in the future.

10.18 Mains Pollutions of Power Converters (EMC) and Non-Linear Loads and Their Influence by Transformers

For the transmission and distribution of electrical energy, the standard 50/60 Hz sinusoidal alternating voltage is used. Only in the railway sector, one finds 50/3 Hz = 16 2/3 Hz. The advantage of a lower frequency is that the lines at the same voltage drop can be significantly longer than at 50 Hz with the same inductive reactance. This played a role in the beginning of the railway development. The disadvantage is that, in accordance with the laws of growth, the electromagnetic converters for energy conversion are much larger.

Sine and cosine function are orthogonal functions, i.e.

$$\int_{\varphi}^{\varphi+2\pi} \sin{(\omega t)} \cos{(\omega t)} d(\omega t) = 0$$

This also means: Each function sin(ωt + φ) or cos(ωt + φ) can be represented as a linear combination a*sin(ωt) + b*cos(ωt). Every periodic function can be represented by a Fourier series whose members form a complete orthogonal system, i.e. each summand is orthogonal to all others:

$$y(\omega t) = \sum_{\nu=1}^{\infty} (a_\nu \cdot \sin{(\nu \omega t)} + b_\nu \cdot \cos{(\nu \omega t)})$$

For a vector space, the orthogonality is defined by the scalar product of the vectors. This means in Cartesian coordinates

$$\vec{A} \cdot \vec{B} = |A| \cdot |B| \cdot \cos\gamma = \sum_{\nu=1}^{n} A_\nu \cdot B_\nu = 0$$

where γ is the angle enclosedby the two vectors $\vec{A}$ and $\vec{B}$. Under certain conditions, function spaces can be regarded as vector spaces. If one then extends the approach to continuous functions, one gets for a certain value interval

$$a \perp b \rightarrow \langle a(\omega t), b(\omega t) \rangle = \int_{\omega t_1}^{\omega t_2} a(\omega t) \cdot b(\omega t) \cdot d\omega t = 0$$

In this sense, frequency multiples of a sine- or cosine-shaped fundamental oscillation are orthogonal to each other in relation to multiples of a period of the fundamental oscillation. That means

$$\int_{\omega t}^{\omega t+2\pi} \sin{(\nu \omega t)} \cdot \cos{(\mu \omega t)} \cdot d(\omega t) = 0 \text{ for all } \nu \neq \mu.$$

Among other things, this is a basis for the representation of harmonic quantities as phasor quantities. Using Euler's formula, you can then write for a phasor $\underline{A}(t)$

$$\underline{A} = A \cdot \exp{(j(\omega t + \varphi))} = A \cdot (\cos{(\omega t + \varphi)} + j \sin{(\omega t + \varphi)})$$

Figure 10.202 illustrates this relationship for a phasor rotating at the angular frequency ω.

Fig. 10.202 Harmonic quantity as projection of a rotating phasor on the real and imaginary axis of the complex number plane

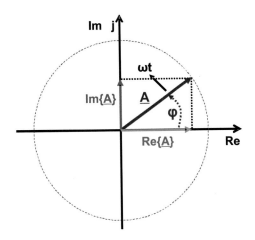

An arbitrarily shifted sine function results as a linear combination of sine and cosine function or as a projection on the real and imaginary axis. Every periodic function can be broken down into a Fourier series, i.e.:

$$f(t) = a_0 + \sum_{\nu=1}^{\infty} \left(a_\nu \cdot \sin\left(\nu \omega t\right) + b_\nu \cdot \cos\left(\nu \omega t\right) \right)$$

If you subtract the mean value a_0, you get a series that only contains sine and cosine functions. If you now form the scalar product of this series with an element of the orthogonal function system, you obtain

$$\int_{x}^{x+2\pi} \sin\left(\nu \omega t\right) f(t) dx = \int_{x}^{x+2\pi} \left[\sin\left(\nu \omega t\right) \cdot \sum_{\nu=1}^{\infty} \left(a_\nu \cdot \sin\left(\nu \omega t\right) + b_\nu \cdot \cos\left(\nu \omega t\right) \right) \right] d\omega t$$

$$= \int_{x}^{x+2\pi} a_\nu \cdot \sin^2\left(\nu \omega t\right) \cdot d\omega t = \pi \cdot a_\nu$$

The scalar products with summands of different frequencies deliver the result zero. For power converter circuits on the mains, sinusoidal voltage sources and non-sinusoidal mains currents can be assumed as a first approximation.

$$u(t) = \widehat{u} \cdot \sin(\omega t)$$

$$i(t) = \sum_{v=1}^{\infty} \widehat{i_v} \cdot e^{j(v\omega t + \varphi_v)} = \sum_{v=1}^{\infty} \widehat{i_v} \cdot (\cos(v\omega t + \varphi_v) + j\sin((v\omega t + \varphi_v)))$$

For the active power calculation, this means

\scale90%

$$P = \frac{1}{2\pi} \cdot \int_0^{2\pi} u(\omega t) \cdot i(\omega t) d\omega t = \frac{1}{2\pi} \cdot \int_0^{2\pi} \sum_{v=1}^{\infty} \widehat{u} \cdot \sin(\omega t) \cdot (a_v \cdot \sin(v\omega t) + b_v \cdot \cos(v\omega t)) \omega t$$

$$P = \frac{1}{2\pi} \cdot \int_0^{2\pi} \sum_{v=1}^{\infty} \widehat{u} \cdot a_v \cdot \sin^2(\omega t) \cdot d\omega t = \frac{\widehat{u}\widehat{i_1}}{2} \cos\varphi_1$$

With sinusoidal voltage sources, the active power is provided with the fundamental wave of the current. The measured apparent power results from the product of the effective values of current and voltage as the sum of the fundamental wave apparent power S_1 plus the distortion reactive power D.

$$S = U \cdot I = U \cdot \sqrt{\sum_{v=1}^{\infty} I_v^2} = \sqrt{S_1^2 + U^2 \cdot \sum_{v=2}^{\infty} I_v^2} = \sqrt{S_1^2 + D^2}$$

Under these conditions, the fundamental apparent power is composed of the active power P and the fundamental reactive power Q_1.

$$S_1^2 = P^2 + Q_1^2$$

This results in the general relationship for sinusoidal mains voltage

$$S^2 = S_1^2 + D^2 = P^2 + Q_1^2 + D^2 = (S_1 \cdot \cos\varphi_1)^2 + (S_1 \cdot \sin\varphi_1)^2 + D^2$$

In addition to the active power P and image power Q_1 the distortive power D also occurs (Fig. 10.203).

The effective factor $\cos\varphi_1$ only describes the ratio of active power and fundamental apparent power

$$\cos\varphi_1 = \frac{P}{S_1}$$

The ratio of active power to total apparent power is called the total power factor.

Fig. 10.203 Geometric illustration of the relationships between the terms active power P, fundamental reactive power Q_1, fundamental apparent power S_1, distortive (reactive power) D and total apparent power S

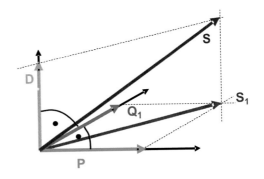

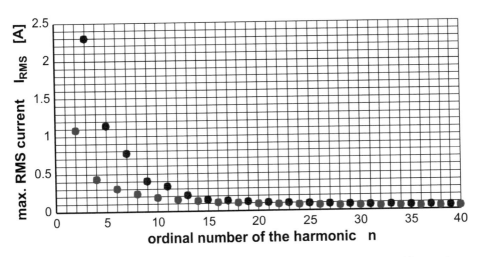

Fig. 10.204 Permissible single RMS values for harmonic currents of consumers with a maximum phase current of 16 A according to class A and D (n even: blue, n odd: red, DIN EN 61000–3-2 VDE 0838–2:2015–03)

$$\lambda = \frac{P}{S} = \frac{P}{S_1 + D}$$

The harmonic content of the currents in the network is limited by various characteristic values. The maximum effective harmonic currents are specified for individual device classes (see e.g. Fig. 10.204). The odd-numbered harmonics have different upper limit values because they have different consequences in the distribution system of electrical energy. Even-numbered components indicate asymmetries even with non-linear loads.

For a given Fourier series, one obtains for the RMS value by the orthogonality of the individual components the relationship

$$I_{RMS} = \sum_{v=1}^{n} \sqrt{I_v^2} = \sqrt{I_1^2 + I_2^2 + I_3^2 + \ldots + I_n^2}$$

If the RMS value of all permissible odd-numbered harmonics is calculated in Fig. 10.204, a harmonic current of 3.04 A would be obtained for a maximum effective current value of 16 A for a device. The RMS value of all harmonics in relation to the fundamental would then be

$$\frac{I_{RMS.OS}}{I_1} \cdot 100\% = \frac{\sum_{v=2}^{n} \sqrt{I_v^2}}{I_1} \cdot 100\% = \frac{\sqrt{I_{RMS}^2 - I_1^2}}{I_1} \cdot 100\% = 19.36\% = THDi.$$

This value is called the current harmonic distortion THDi (*total harmonic distortion*). Another measure to describe harmonics is the fundamental component g_i of the current

$$g_i = \frac{\sum_{v=2}^{n} \sqrt{I_v^2}}{\sum_{v=1}^{n} \sqrt{I_v^2}} \cdot 100\% = \frac{\sum_{v=2}^{n} \sqrt{I_v^2}}{I_{RMS}} \cdot 100\% = \frac{THDi}{1 + THDi} \cdot 100\%$$

If all permissible harmonics for a total effective value of 16 A are added to the phase shift, the current waveform shown in Fig. 10.205 is obtained. Although this has no general meaning, it illustrates the effect of possible mains current distortions.

For various equipment classes, power-dependent harmonic limits are applied (see IEC/EN 61000). The power quality requirements are described in the standard EN 50160 'Characteristics of voltage in public electricity supply networks'. The purpose of this standard is to define and describe the characteristics of the supply voltage with regard to frequency, level, waveform and symmetry of the three phase voltages. These characteristics change during normal operation of a network due to load fluctuations, interference from certain installations and the occurrence of faults, which are mainly caused by external events [65–67].

At a mains connection point, the currents absorbed cause additional harmonics of the voltage, which can be roughly calculated for a 1-phase connection using the mains replacement circuit according to VDE 0100 (see Fig. 10.206):

$$U_v = |Z_{mains}| \cdot I_v = \sqrt{R^2 + (n\omega L)^2} \cdot I_v = \sqrt{R^2 + (n\omega L)^2} \cdot I_v$$

$$\approx \sqrt{(0.4\ \Omega)^2 + (n \cdot 0.25\ \Omega)^2} \cdot I_v$$

With harmonics present at a grid connection point, a partial voltage is then obtained for a certain frequency n

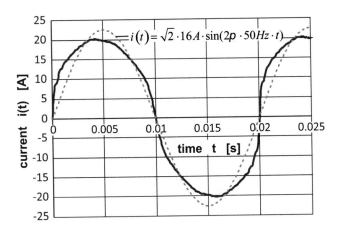

Fig. 10.205 Form of a current curve when summing up all limit values for harmonics with a phase shift of 0° between each other as an example of a possible (standard-compliant, extreme) current distortion

Fig. 10.206 Simplified equivalent circuit of a four-wire low-voltage network according to VDE 0100 with the typical sizes $R_1...R_3 = 0.24\ \Omega$, $R_N = 0.16\ \Omega$, $L_1...$ $L_3 = 477.5\ \mu H$ and $L_N = 318.3\ \mu H$

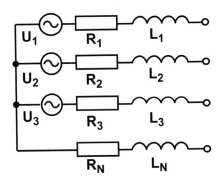

$$\underline{U}_{tot}(2\pi n f) = \underline{U}_{0n}(2\pi n f) + \underline{U}_n(2\pi n f)$$

Limit values for harmonics of the mains voltage without load are defined by the standard EN 50160. The grid operator is responsible for compliance with these limits. These are shown graphically in Fig. 10.207.

Again, the THD (*total harmonic distortions*) is a characteristic value that limits the injection of harmonics. It indicates the degree of distortion of the fundamental oscillation. It is defined via the RMS value of the signal and the RMS value of its fundamental oscillation as

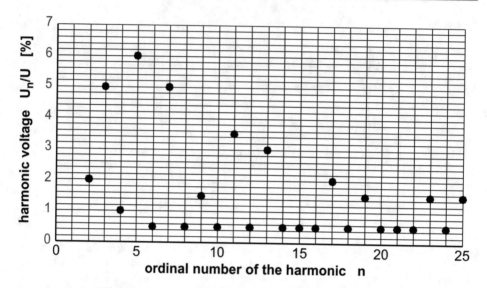

Fig. 10.207 Permissible effective proportion of harmonics in the effective value of the mains voltage to EN 50160 at the unloaded connection point

$$THD_U = \frac{\sqrt{U^2 - U_1^2}}{U_1} \cdot 100\%$$

The harmonic content THD of the supply voltage, formed from all harmonics up to atomic number 40, must not exceed a value of 8%.

Rectifiers with different loads are frequent non-linear loads that, together with magnetic components, cause harmonic oscillations in the network. Some important effects are discussed below. Figure 10.208 shows the diagram of a half-wave rectifier as an electrical equivalent circuit diagram of a transformer with currents converted to the secondary side and the associated characteristics. Without load ($i_{load} = 0$), the non-sinusoidal current, due to the non-linear magnetization characteristic, is symmetrical to zero. Half-wave rectification causes a secondary DC current component. This leads to a DC voltage component in the voltage drop of R_1. As a result, the mean value of the voltage at L_μ is no longer zero and the operating point on the magnetization curve shifts accordingly. As a result, the magnetizing current shows clear signs of saturation and the effective value of the current drawn increases considerably. A significantly increased heating is the result. Direct current components of absorbed load currents in the network are therefore limited to very small values. Although the transformer is connected to a sinusoidal voltage U_1, it shows saturation phenomena due to the non-linear load. Direct current premagnetisation takes place from the secondary side.

The single-phase connection of non-linear loads in a typical domestic installation can lead to such phenomena. Known representatives of these loads are e.g. hair dryers and drills

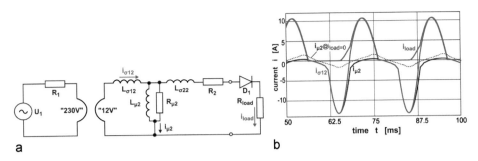

a   b

Fig. 10.208 Half-wave rectifier with ohmic load and upstream transformer. (**a**) Electrical equivalent circuit diagram, (**b**) Current curves: Load current i_{load} and secondary related magnetizing current $i_{\mu2}$ and primary current converted to the secondary side $i_{\sigma12}$ ($R_{1=}$ 18.4 Ω, $L_{\sigma12} = L_{\sigma22} = 1$ mH, $L_{\mu2}$ ($I_\mu = 0$) = 76 mH, $R_2 = 50$ mΩ, $R_{load} = 1.2$ Ω)

Fig. 10.209 Circuit diagram of a *low-cost* microwave oven with simple one-pulse rectification and capacitive smoothing

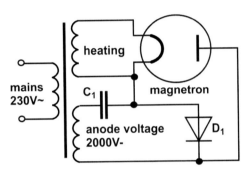

with simple diodes in the tap-changer. But also in some cheap versions of electronic devices, such as microwave ovens for heating food (see Fig. 10.209), such one-pulse rectifiers are occasionally found. Because these loads are only used for short periods of time, hardly ever simultaneously, the effects of such loads on upstream local power transformers are minimal. In principle, however, these loads lead to DC components on the mains supply lines, which also supply other loads. If inductive loads are connected there, DC bias is also caused in these loads. The result tends to be an increased mains current consumption with increased self-heating of transformers.

In the following, some basic circuits of rectifiers are described with their effects on the mains (Table 10.13).

The 2-pulse bridge rectifier B2 is the most frequently used rectifier for 1-phase applications. With a purely resistive load, a strongly pulsating DC voltage is generated at the load, but this produces an undistorted sinusoidal current on the AC side. The current form factor is 1.11.

Since the aim is usually a smooth direct current on the direct current side, various methods of current smoothing are used. A smoothing choke in series with the resistive load tends to always lengthen the current conduction of the rectifiers and finally leads to a

Table 10.13 Rectifier circuits with load current i_{load} and phase current $i_{1...3}$ at different load types $R_{load} = 16.9$ Ohm, $L = 0.5$ H, $C = 1$mF) on a 4-wire network according to Fig. 10.205

| Rectifier | Characteristic current courses | | |
	R load	RL load	RC load
B2U			
M3U			
B6U			

smooth DC current, which is proportional to the mean value of the output voltage of the rectifier with $I_d = U_d/R_{load}$. This is achieved with large chokes L with good approximation. The mains current is still an alternating current, but the current shape is almost rectangular. Corresponding harmonic oscillations are the result. The form factor of the current is close to 1.

A capacitively smoothed load current (R//C) always tends to shorten the current conduction phase. The sinusoidal mains current at $C = 0$ becomes shorter and shorter pulses with higher and higher amplitude as the capacitor increases, which together result in an alternating current on the mains side. The form factor of the current can reach very large values here and is typically 1.5...2 in the nominal case. Since the recharging of the smoothing capacitance always takes place near the maximum of the mains voltage and a large number of electronic devices are operated on the low-voltage network, synchronous rectification with capacitive smoothing has led to the shape of the mains voltage in low-voltage networks resembling more a trapezoidal than a sinusoidal oscillation. As a result, rectifiers with forced sinusoidal mains current consumption have been legally required since 2001, starting at a certain minimum power.

If 3 phases are connected together, each with a 1-pulse rectifier, a 3-pulse rectifier in mid-point connection (M3) is obtained. The load current at R-load is pulsed, but less than with the B2 circuit. The phase current is a pulsating direct current. As described above, such a rectifier generates a direct current component in the mains. This can have unpleasant consequences with transformers connected in parallel and other inductive loads based on a

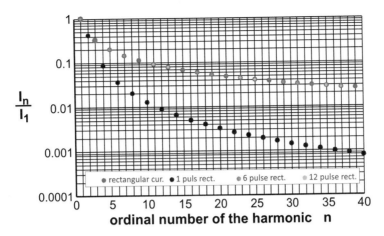

Fig. 10.210 Relative RMS currents of the harmonics in relation to the fundamental, forced by various rectifiers in the network

pure AC voltage due to the DC voltage drop at the "neutral" wire. Saturation phenomena in connected control transformers as a function of the load current of the M3 rectifier have been observed by the author. These led to a stronger heating of the transformers and finally to the triggering of the overtemperature sensors of the transformers. Although the M3 circuit is preferably used for low voltages and high currents, it should always be operated using its own converter transformers and never on the public power supply network. It is interesting to note that if the rectifiers have correspondingly short conducting phases (e.g. with RC load), a 150 Hz pulsation can be observed on the neutral conductor. With controlled rectifiers (thyristors), circuits can be built on this.

The most common rectifier circuit on the 3-phase network is the 6-pulse bridge circuit B6. As with all bridge circuits, alternating currents are generated on the mains side. These currents are load-dependent and are subject to harmonics. Since bridge circuits can be regarded as series circuits of 2 oppositely poled centre point circuits, the B6 produces twice the output voltage of an M3 with the same line voltage. The load current is accordingly twice as large with the same resistance. The ripple of the output voltage becomes smaller with the same memories 'L' or 'C'. The same applies to the effective value of the mains currents. This property can theoretically be used for switching at different mains voltages (e.g. 3–400 V to 3–200 V) at the same output voltage. However, it must be pointed out that M-circuits cause DC voltage components in the lines, which can then cause the effects described above of DC pre-magnetization, e.g. in transformers. Figure 10.210 shows the content of harmonic currents in the network as forced by different types of rectifiers.

A cascading of B6 circuits is also possible and is often used especially in the upper power range. One effect is the distribution of currents or voltages in or at the semiconductors. Usually, transformers with different switching groups and thus phase shifts are used. These lead to the fact that, for example, currents with different phase

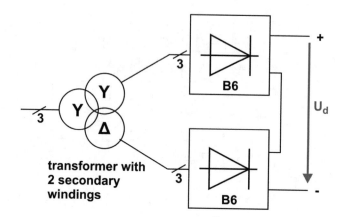

Fig. 10.211 Use of a 3-winding transformer with secondary windings in Y and D circuits to generate a 12-pulse DC voltage with 2 series-connected 6-pulse bridge rectifiers

positions are superimposed. The harmonics of the rectifiers have a fixed phase relationship to the fundamental oscillation. On the DC side, the voltages are added in the circuit Fig. 10.211. This results in a 12-pulse output voltage.

Superimposition with fundamental oscillations in opposite phase causes extinction. A combination of B6-switching with other transformer switching groups leads to 18- and 24-pulse DC voltages. In the spectrum of the harmonics, more and more components of higher frequencies are then "missing" and the THD decreases as the number of pulses in the DC voltage increases. Figure 10.212 shows the correspondingly altered spectra of the harmonics. Such rectifiers with a correspondingly increased effort are mainly used in the upper power range. Remaining harmonics can be eliminated from the spectrum with absorption circuit filters. Since some harmonics are already missing in the lower frequency

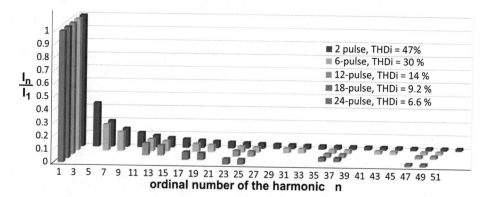

Fig. 10.212 Spectrum of harmonic currents in rectifiers with smooth DC current from B6 circuits combined with transformers of different switching groups to higher pulse count on the DC voltage side

range, the space required for these passive filters is relatively small. In this way, an electronic forcing of sinusoidal mains current consumption by pulse width modulation is avoided.

10.19   Pseudo Multilevel Converter

PWM-controlled power electronic converters are initially based on switched topologies, which are fed from a DC voltage source and whose inputs are alternately connected to the positive and negative pole. In this case, one speaks of a 2-level circuit. This can be realized very easily via a half-bridge branch. In the basic version, a half-bridge branch consists of a combination of a buck converter and a boost converter. This results in a bipolar, i.e. in both energy flow directions effective converter. The voltage switched with a switching frequency causes a switching frequency pulsating current at the load. The smoothing capacitor in parallel with the load must have a certain size in order to smooth the load voltage down to a residual ripple value. The size of this required capacitance is inversely proportional to the switching frequency. Therefore, the highest possible switching frequency is always aimed for. Inserting another voltage level in the middle between $+U_d$ and $-U_d$ has the same effect as frequency doubling: the current ripple is halved. Depending on the circuit topology, the switch loads and switching losses are also reduced. For this purpose, a large number of circuit concepts have been developed over the last decades, which lead to very complex circuit configurations that are also complex to control.

This subchapter shows how a similar circuit behaviour can be achieved by using inductive components. This means that the pulsation frequency of the output current is increased and the current ripple is reduced. The usual half-bridge branches and their control are sufficient [68]. The usual controller structures do not change. Phase shifted signals must be generated for the control. The starting point for this is the magnetically coupled chokes according to Sect. 10.8. It is therefore assumed that these chokes are available or can be set relatively freely in their required values. The basic design of a converter in a half-bridge circuit then has the appearance shown in Fig. 10.213. Two half bridges are connected to a coupled choke. The half-bridges are operated with the same duty cycle, but out of phase. As a result, the load current is divided by R_1 in half. The design of these chokes can be implemented as described in Sect 10.8. The switches T_1/T_2 and T_3/T_4 are switched alternatively. This produces the switching frequency voltages $\underline{U}_{11}$ and $\underline{U}_{12}$, whose moving average forms the output voltage U_1. Figure 10.213b illustrates these relationships. It should be noted that the displayed phasors $\underline{U}_{11}$ and $\underline{U}_{12}$ have a different frequency than the phasor $\underline{U}_1$. The position of the phasors relative to each other only applies to phasors of the same frequency. The representation in an overall picture is therefore arbitrary and chosen for reasons of clarity. In the example, a pulse pattern is impressed on the half bridges, forming a sinusoidal curve. The results were generated for illustration with a simulation using the following parameters.

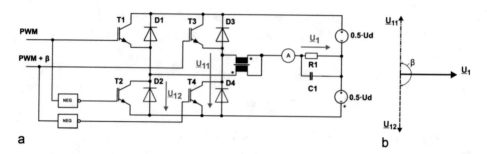

Fig. 10.213 Half-bridge circuit with phase-shifted control of two half-bridge branches. (**a**) Principle circuit forming a switch combination; (**b**) phasor diagram of voltage phasors $\underline{U}_{11}$ and $\underline{U}_{12}$ and $\underline{U}_1$ for visualization of the function

DC link voltage, U_d	600 V
Active load per phase, R_1	30 Ohm
Filter capacity, C_1	0.1 µF
Magnetising inductance, L_μ	50 mH
Coupling factor	0.95

Figure 10.214 shows the output voltage $u_1(t)$ at load resistor R_1 for a sinusoidal current with 50 Hz at different phase shifts β for out-of-phase operation. The case β = 0 corresponds to the operation of the circuit with parallel connected inputs of the coupled choke. Here too, the current is divided symmetrically between the two windings and thus between the storage inductors $L_{\sigma 1}$ and $L_{\sigma 1}$. The stray inductances connected in parallel then have an effective total inductance of 11.2 mH. At β = 180°, maximum compensation of the partial currents occurs, which results in a minimum current ripple. At β = 90°, a lower cancellation occurs, as can be seen in Fig. 10.2.

If you combine only this arrangement with a second one to form a two-phase system, you basically get a full bridge. Here you have again the possibility to switch the switches more or less out of phase. A maximum phase shift is obtained if one pair of half bridges is 180° out of phase and these are switched 90° out of phase [69]. In the graphic illustration, the phasor image Fig. 10.215a is then obtained. In relation to harmonics at the load $(R_1 + R_2)$, this has the same effect as a single-phase circuit with 4 half-bridges operating in phase offset mode. The maximum current ripple here is only half of the variant in Fig. 10.214c.

In a similar way, 3-phase systems can be set up, as shown in Fig. 10.216. The phase systems of load and switching electronics of the inverters are in principle independent. Nevertheless, it is useful to define an arbitrary but fixed relation. Otherwise, the changing phase relationships can cause subharmonics in the regulation. Figure 10.215a shows the common phasor image for both frequencies. It is characteristic that the half-bridge pairs are phase-shifted by 90°, and the pairs are phase-shifted by 180°. If one considers an idealized

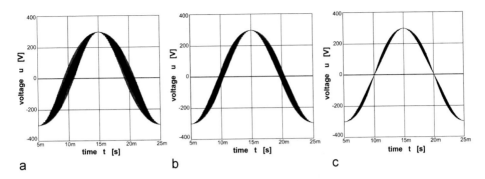

Fig. 10.214 Voltage curve at R1 for (**a**) β = 0°, (**b**) β = 90°, (**c**) β = 180° for f = 50 Hz

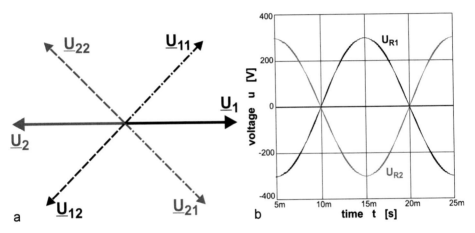

Fig. 10.215 Structure of a 2-phase system with 2 half-bridge pairs operating in phase offset mode. (**a**) Phasor pattern, (**b**) voltage curves at a resistive load

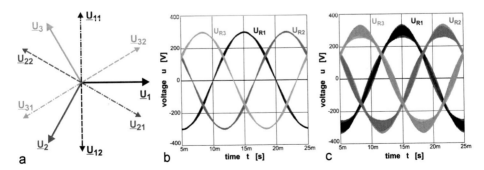

Fig. 10.216 Structure of a 3-phase voltage system consisting of 3 pairs of half-bridges switching in a phase-shifted manner. (**a**) Phasor image for the frequency of the 3-phase system, (**b**) result with half-bridge pairs switching in a 180° phase-shifted manner, (**c**) result when only one half-bridge switches per phase

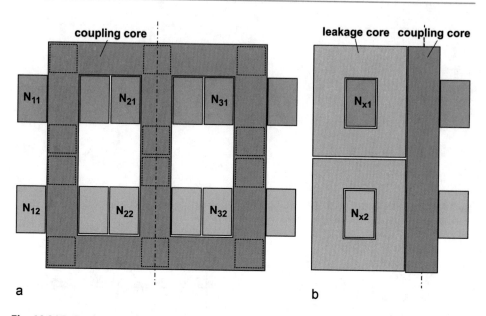

Fig. 10.217 Design proposal for a largely symmetrical 3-phase choke for phase-shifted operation of 2 half-bridge pairs for each phase. (**a**) Front view, (**b**) side view

stationary state here, the phase relationship between the half-bridge pairs is also not necessarily fixed. At 90°, however, minimal current ripple is achieved.

The ripple formation of the measures shown in Fig. 10.215 corresponds to 4 times the switching frequency of a switch and 5 virtual levels of the feeding voltages of the inverter. Physically, this is due to the effect of the chokes as autotransformers. It is possible to achieve the half-bridge branch shown in Fig. 10.213a or a version extended according to Fig. 10.215 as a configuration for a phase output of a 3-phase inverter. Via the 3-phase load, the phase-to-phase voltage is mapped to a phase voltage of a load connected in 'Y'. A big advantage is the modularity of the design and the similar control and regulation of all half bridges. Only a fraction of the total current has to be commutated and the commutation circuits are designed accordingly simple. The concept is particularly suitable for high currents and high power since the current is splitted and overvoltages at the single switches are lower. The additionally required filters become very small. By distributing the power over several chokes, their surface area is also relatively large in relation to the power. This facilitates cooling.

In principle, similar results can also be achieved with uncoupled chokes in phase-shifted operation. However, magnetic coupling, in conjunction with the design discussed in Sect. 10.8, allows extensive decoupling of alternating flux and pulsating unipolar flux. This opens up the possibility of using optimum core materials for the tasks of transformation and storage. The result are space-saving structures. Figure 10.217 shows a design proposal for a 3-phase choke with magnetic coupling and separation of DC and AC flux components. The proposed design avoids the use of air gaps and thus minimizes the field scattered into the

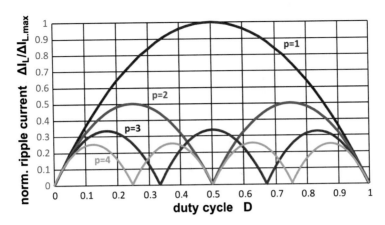

Fig. 10.218 Reduction of the ripple currents when changing the number of phases p for a phase-shifted operation with a buck converter

environment. Due to the limited saturation induction, the conductors of the magnetic field take up relatively much space. Complex structures are therefore difficult to minimize. Size and manufacturing effort therefore always have to be weighted against each other.

With an increasing number of phase-shifted half bridge branches, the current ripple in the load decreases inversely proportional. Theoretically, the number of half-bridges operated in parallel and phase shifted can increased at will. In practice, this will probably be limited to a maximum of 3–4 bridge branches. The absolute reduction of the current ripple also becomes smaller with each additional stage. However, the effort increases approximately linearly. Figure 10.218 shows the achievable current ripple for a single-phase application with different numbers of phases p in switching operation compared with the maximum ripple when using only one half-bridge branch. This means that in this circuit variant, the ripple at $D = 0.5$ forms the reference value. As Fig. 10.218 shows, cascading coupled chokes with $p = 2$ is an effective way of achieving compact arrangements even with higher phase counts ~2^k. 4-phase arrangements therefore can achieved in this way as well as by using a common coupling core.

Further investigations into the advantageous design of coupled chokes are contained in Sect. 10.8.

10.20 Energy Recovery from a Converter by Choke Transformer

Resonant circuits in the higher power range are usually based on 2 or more active switches. The circuit shown here concerns a simple design with only one switch. Design goal was a simple structured and robust controllable electronic current source for periodic capacitor charging for pulse applications. A secondary condition was the use of a thyristor as an extremely robust component. This circuit is equivalent to a switching relief when switching

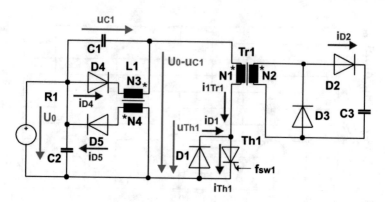

Fig. 10.219 Simplified equivalent circuit diagram of a resonant circuit inverter with only one thyristor switch [70, 71] for charging a capacitor

off a transistor. The power is drawn from the mains via a rectifier followed by a voltage intermediate circuit, which supplies the DC voltage U_0. The main part of the circuit is formed by a switched series resonant circuit, which is operated below its resonant frequency. The characteristic impedance of this resonant circuit limits the transmitted current and thus the short-circuit current of the circuit at constant input voltage. The resonant circuit is formed by a capacitor C_1 and the leakage inductance $L_{\sigma 1}$ of transformer Tr_1. The active switch is shown here as thyristor Th_1 with antiparallel diode. If a MOSFET is used, the intrinsic diode can be used. The MOSFET is then switched off without voltage in its conducting phase.

Figure 10.219 shows the simplified equivalent circuit diagram. At power-up, current flows through C_1, L_σ, the secondary rectifier and the capacitor to be charged. In the process, energy is transferred from the resonant circuit to the capacitor. Due to the "one-stroke operation", there is now the problem of restoring the initial conditions. In the circuit Fig. 10.219, this is solved by charging a storage choke L_1 in parallel during the oscillation process at C_1. As long as thyristor Th_1 and diode D_1 are conductive, choke L_1 takes up a voltage-time area $\Delta\Psi \approx U_0 \cdot T_0$ and is thus charged with current. T_0 is the duration of the natural oscillation of the oscillating circuit. Without L_1, the capacitor current would end after the conducting phase of D_1. However, in the circuit arrangement Fig. 10.219, C_1 can continue to release its energy and "charges" L_1 further. In doing so, it discharges to $U_{C1} = 0$, at which point there is a maximum energy content in L_1. L_1 can release this in the current direction specified by D_4 and in doing so recharges C_1. At the next power-up, the voltage to be switched is $U_{Th1} = U_0 + U_{C1}$. Consequently, the voltage driving the oscillation process when Th_1 is switched on becomes higher. If no energy is drawn from the system, e.g. if the load is short-circuited, the voltage at C_1 can increase indefinitely. A circuit constructed in this way would not be short-circuit proof and would not be structurally suitable as a capacitor charger. However, if the storage choke is designed slightly differently, stable operation can be achieved in a simple way. To do this, you apply another winding N_4 to the core of L_1. N_3 and N_4 then act like a transformer with a leakage inductance $L_{\sigma 2}$. If the capacitor voltage exceeds the value

$$u_{C1} = \frac{N_1}{N_2} \cdot U_0$$

the further energy stored in L_1 is fed back to the DC link. This limits the maximum voltage load of Th_1 and D_1 to

$$U_{Th1.\,max} = |U_{D1.\,max}| = \left(1 + \frac{N_1}{N_2}\right) \cdot U_0.$$

The circuit is controlled via the switching frequency and, in the case of a thyristor, with simple needle pulses. When using MOSFETs, the MOSFET must be switched on a little longer than $T_0/2$. Due to the intrinsic properties of the circuit, it is very robust and suitable up to several kJ/s.

At higher power levels, the unfavorable form factor of the current becomes noticeable. Push–pull circuits then have advantages in terms of efficiency and size.

Transformer Tr_1 is designed so that its leakage inductance $L_{\sigma 1}$ forms the resonant circuit inductance. Since a single-pulse circuit is used on the input side of the transformer, a single-pulse rectification with a freewheel is also implemented on the load side. Thus, energy is only transferred in the thyristor conduction phase. The blocking delay charge of D_1 leads to transient overvoltages at $Th_1//D_1$, which may have to be limited/damped by a snubber. With winding N_4, the storage choke L_1 still has the task of feeding back excess energy. The energy is stored in the magnetising inductance of L_1. For this purpose, an air gap in the core or an appropriate core material must be provided. The ideal target would be an effective leakage inductance $L_{\sigma 2} \to 0$. N_3 and N_4 must be coupled as closely as possible. Blocking delays of diodes D_4 and D_5 together with $L_{\sigma 2}$ cause transient overvoltages at the diodes, which disappear with $L_{\sigma 2} = 0$. The stationary voltage loads of D_4 and D_5 are

$$|U_{D4.\,max}| = \frac{N_1}{N_2} \cdot U_0$$

$$|U_{D5.\,max}| = \left(1 + \frac{N_2}{N_1}\right) \cdot U_0$$

The circuit is to be classified as a resonant single-ended flux transducer. The short-circuit current can be roughly calculated with the switching frequency f_{sw} to

$$I_{da} \approx 2U_0\left(1 + \frac{N_3}{N_4}\right) \cdot C_1 \cdot f_{sw} \cdot \frac{N_1}{N_2}$$

For a dimensioning example $L_1 = 1$ mH, characteristic dependencies are shown in Figs. 10.210 and 10.211.

The operating conditions of the circuit are:

$U_0 = 400$ V

$U_{C3} = 100$ V

$f_{sw} = 10$ kHz

$C_1 = 2.2$ μF

$C_3 = 1000$ μF

$L_{\sigma1} = 15$ μH

$N_1 = N_2, N_3 = N_4$

$$Z_w = \sqrt{\frac{L_{\sigma1}}{C_1}} = 2.611 \text{ Ohm}$$

Figure 10.210a shows the characteristic currents. The primary transformer current i_{1Tr1} is formed from thyristor and diode current. The capacitor is charged with the current i_{D5}. The current i_{D4} shows how the accumulator L_1 is first charged and then discharged again until the conduction condition of D_5 is reached. Then the current commutates from D_4 to D_5 and the remaining energy portion of L_1 is fed back to the DC link. Figure 10.210b shows the corresponding voltage curves at the thyristor and the voltage U_0-u_{C1}, which can be easily measured. When the thyristor begins to conduct, the recharging of C_1 is initiated. At the same time, the charging of L_1 begins. This is therefore the superimposed operation of 2 oscillating circuits with different natural frequencies. Besides the series resonant circuit $L_{\sigma1}C_1$, the parallel resonant circuit L_1C_1 is also effective. The figure formed from (u_{C1}; $Z_w \cdot C_1$) in the phase plane Figure 10.211a is therefore not composed of pure circular arcs, but is characteristic for each operating state. During the conducting phase of thyristor and diode, the voltage u_{Th1} is approximately zero. After the conducting phase of D_1, the voltage jumps to the current value of $U_0 - u_{C1}$ and then follows this voltage until the next switch-on of T_{h1}. One can clearly see the limiting effect of the combination ($N_3 + D_4$; $N_4 + D_5$) to 2400 V = 800 V.

The present example was given above all in order to show the targeted recovery of energy from a storage choke with coupled windings. This task occurs again and again in relief networks for power electronic switches (Fig. 10.220). In most of the PWM converters with transistors this energy recovery is intrinsic. and concerns the choke energy.

Figure 10.221 contains an example of a charging process of C_3 in addition to the representation of the phase plane on the basis of C_1. The charge current is initially relatively constant and quickly decays to zero near the open-circuit voltage, so that the voltage curve becomes correspondingly flat. In the example, the transformation ratio of the transformer is

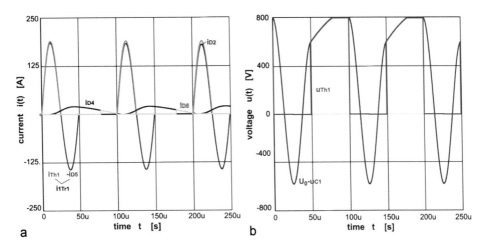

Fig. 10.220 Characteristic current and voltage curves under the operating conditions mentioned in the text: (**a**) current curves, (**b**) voltage curves

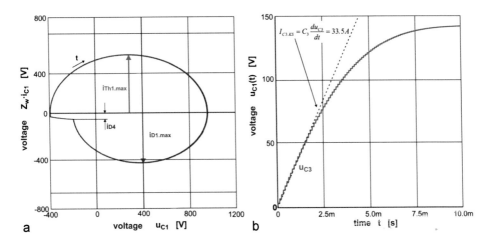

Fig. 10.221 Representation of capacitor current and capacitor voltage of C_1 in phase level (**a**) and example of a charging process of $C_3 = 1000\ \mu F$

1 in this example. With a different transformation ratio, one can adapt the target voltage. If a resonant circuit inductor is used instead of the transformer Tr_1, the useful power can also be decoupled in parallel to C_1, e.g. with a piezo oscillator. With the special design of L_1, fluctuations in the load can also be compensated and the circuit stabilized.

10.21 Inverter Based on a T-Filter

When designing chargers for capacitors or accumulators, it should be taken into account that the load ideally has an impedance of zero. Circuit concepts that are inherently short-circuit proof are therefore to be preferred. One such circuit is the Boucherot circuit (Fig. 10.222) [72–74]. If, with the aid of harmonic analysis, one considers only the fundamental of a voltage source over the RMS value of the fundamental, one obtains the fundamental of current I_1 the value

$$\underline{I}_1 = \frac{\underline{U}_1}{j\omega L_1 + \frac{R_1}{j\omega C_1 \cdot R_1 + 1}} \cdot \frac{1}{j\omega C_1 \cdot R_1 + 1} = \frac{\underline{U}_1}{R_1 \cdot (1 - \omega^2 L_1 C_1) + j\omega L_1}$$

The real part of the denominator of this expression becomes zero under the condition, i.e. at the resonant frequency of the oscillating circuit $L_1 C_1$. Relative to the load resistance R_1, the circuit then acts as an ideal AC source with the equivalent impedance $Z_{ers} \to \infty$ and the equivalent open circuit voltage $U_{0ers} \to \infty$. The current independent of R_1 is then

$$\underline{I}_1 = \frac{\underline{U}_1}{j\omega L_1}$$

This circuit is particularly short-circuit proof and therefore perfectly suited as charging circuit for memories. The short-circuit current can already be narrowly limited by the dimensioning. If the voltage source is formed by an inverter, a rectangular voltage is present. The circuit filters out the fundamental oscillation when operating at the resonance frequency. In the case of a short circuit, there is a triangular current waveform, so that the rectified averaged load current is somewhat higher than the current from the fundamental oscillation consideration.

For the resonance case, the output voltage is then dependent on resistor R_1 and input voltage U_1:

Fig. 10.222 Boucherot circuit as basic component of a T-filter

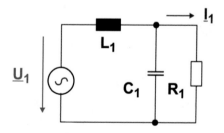

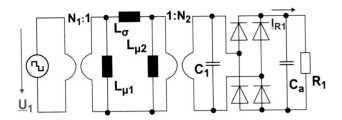

Fig. 10.223 Principle of a capacitor charger based on a Boucherot circuit

$$U_{R1} \cong \underline{I}_1 \cdot R_1 = \frac{U_1}{j\omega L_1} \cdot R_1$$

A PWM can be used to control the fundamental content of a square wave voltage. This results in the control concept of adjusting the output current of this resonant inverter when working at the resonant frequency by PWM. It is interesting to note that the output voltage can become much higher than the input voltage due to resonance boosting. This means that one is not initially bound to the transformation ratio of a transformer for voltage adjustment. This creates an additional design parameter for the output voltage. This is an important extension of the optimization space, especially for capacitor chargers with highly variable output and input voltage. A basic circuit for this is shown in Fig. 10.223. On a PWM inverter of constant frequency, the Boucherot circuit, formed from the leakage inductance of a transformer and C_2 (if necessary, the winding capacity of the secondary side), is formed. An uncontrolled rectifier with a capacitive load is connected downstream.

Several properties are achieved by the selected circuit principle

- Short-circuit safe (no open-circuit safety!)
- Output voltage as free parameter
- *Zero voltage switching* (ZVS) for the inverter switches under all load conditions
- Soft commutation of the diodes of the load rectifier

When generating high voltages, high transformation ratios of the transformer $ü = N_2/N_1$ are required. The winding capacity of the secondary side is transformed $\sim ü^2$ to the primary side. With an effective capacitance of e.g. 20 pF, this capacitance acts like a capacitance of 0.2 μF when viewed from the primary side at a transformation ratio $ü = 100$. This makes this normally disturbing capacitance interesting as a resonant circuit component. With oil-insulated transformers, for example, winding capacities with a very high quality can be achieved. The higher the quality, the higher the attainable output voltages. The output voltage must therefore be limited by control engineering. The load capacitance delays the possible voltage rise accordingly. A design with strong yoke scattering can be used as a transformer, as indicated in Fig. 10.223. $L_{\mu1}$ is thus practically parallel to the (limited) input voltage U_1. $L_{\mu2}$ on the other hand is connected in parallel to C_2 via the transformer. The voltage U_2 can theoretically become infinitely large. This means that $L_{\mu2}$ can go into

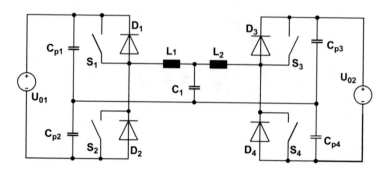

Fig. 10.224 Circuit diagram of a T-inverter

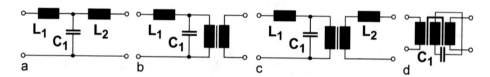

Fig. 10.225 T-filter in various designs. (**a**) Basic circuit, (**b**) use of a transformer with small dispersion, (**c**) use of a transformer with large dispersion, (**d**) transformer as integrated filter

saturation, while $L_{\mu 1}$ is practically stationary. This must be taken into account when dimensioning the transformer and predicting losses.

If you want to couple two batteries as storage for a bidirectional energy exchange, you reach the limits of the concept: the circuit topology is asymmetrical in such a way that a bidirectional energy flow is impossible. The inverter can be operated as a rectifier, but not the rectifier as an inverter. However, if the arrangement according to Fig. 10.222 is supplemented with an output-side choke L_2, a symmetrical circuit is obtained. If 2 bridge or half-bridge inverters are coupled, a bidirectional DC/DC converter is obtained (Fig. 10.224).

The coupling network can be designed with or without transformer (Fig. 10.225). If a transformer is inserted, it must be ensured that the arithmetic mean value of the magnetizing current of the transformer is zero. This can be achieved by a regulation or by a larger series capacitor ($>20 \times C_1$).

In principle, the filter can be integrated into a transformer. The transformation of the T-filter as a 3-port in a magnetic circuit leads to 3 disk windings in a transformer of the core type. The realization with a sheath type (ETD59, Fig. 10.226) is shown. On this core, 3 windings $N_1 = N_3 = 20$, $N_2 = 19$ were applied. Figure 10.226 a...f show the steps for morphological transformation from magnetic to electrical equivalent circuit. This is finally a double symmetrical T-filter or a symmetrical H-filter.

With the given data, one obtains $L_{\sigma 11} + L_{\sigma 12} = L_{\sigma 11} + L_{\sigma 12} = 21$ µH referred to N = 20. Figure 10.227 shows the measurement results for the impedance at winding N_1 at $C_1 = 10$ µF at no load of N_3 (parallel resonance of C_1 with the magnetizing inductance

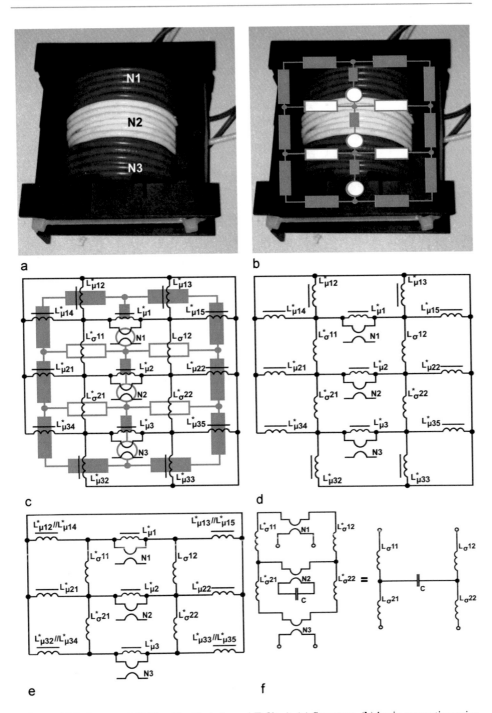

Fig. 10.226 Integrated H-filter (double-balanced T-filter). (**a**) Structure, (**b**) basic magnetic equivalent circuit diagram, (**c**) derivation of the electrical equivalent circuit diagram, (**d**) simplification of the electrical equivalent circuit diagram, (**e**) further simplification for $L_{\mu x} \ll L_{\sigma y}$ and effect diagram. (Source: Zacharias 2016)

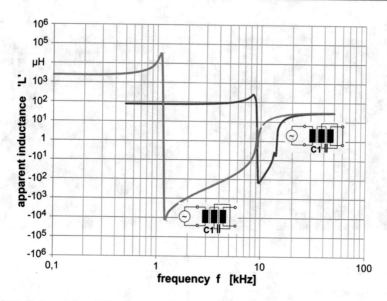

Fig. 10.227 Series and parallel resonance of the filter at 9.23 kHz at $C_1 = 10\ \mu F$ by measurement/ calculation of the apparent inductance

at approximately 1.1 kHz, series resonance of the filter at 9.1 kHz) and short circuit of N_3 (parallel resonance of the filter at 9.23 kHz), which show typical behavior of a symmetrical T-filter.

$$'L' = sign(\mathrm{Im}(\underline{Z})) \cdot \left(\frac{|\mathrm{Im}(\underline{Z})|}{\omega} \right)$$

The transmission element reproduces the characteristics of a balanced T-filter very well. The circuit works bidirectionally balanced. This means that the voltage can be lowered and raised in both directions. The connection of capacitor C_1 to the filter via a transformer opens up the possibility of adapting the required effective capacitance to the available capacitors according to the transformation ratio. A disadvantage of the circuit is that with a small duty cycle D, the required voltage is generated by resonance boosting. This leads to a reactive current load on the components and reduces the efficiency in the low load range. In the vicinity of the nominal range, the efficiencies are well over 90%. In the case of a 3-level inverter, one has the curve of the effective voltage shown in Fig. 10.228. With a duty cycle of D = 1, the conduction angle of a switch is then equal to π.

A simple control of the inverter thus initially assumes that the T-filter is controlled at its resonance frequency, whereby the fundamental oscillation component is changed via the control parameter 'D'. A realization for a simulation is shown in Fig. 10.229. The maximum voltage of U_1 is 100 V. The characteristic impedance is shown in the simulation

Fig. 10.228 Voltage curve and
duty cycle D for a 3-point
inverter

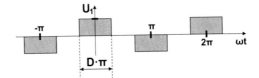

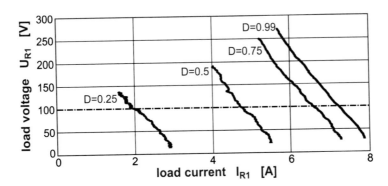

Fig. 10.229 Load characteristics of a T-converter

example $Z = \sqrt{\frac{L_1}{C_1}} = 10$ Ω. The load direct current was determined by charging a capacitor
and shown as a load characteristic in Fig. 10.229.

Figure 10.229 shows that the converter is short-circuit proof. Thus the transmission
behaviour is both step up and step down.. The load must have a low HF impedance. Areas
of application are therefore e.g. chargers, power supplies with DC voltage intermediate
circuit and inductive energy transmission. Operation is equally possible in both directions
due to the symmetry of the circuit.

10.22   Power-Modulated Electronic Power Source for Longitudinal Gas Flow CO_2 cw Lasers

CO_2 lasers are still used as a low-cost working tool in the range from 50 W to 10 kW
industrial application for material processing. The aim of many works was and is the
increase of the efficiency as well as the modulation of the output power and an improvement of the ignition behaviour of the laser at the beginning of the processing. This led to the
application of an alternating current excitation of the gas discharge in the medium and high
frequency range, which partly simplifies the laser design.

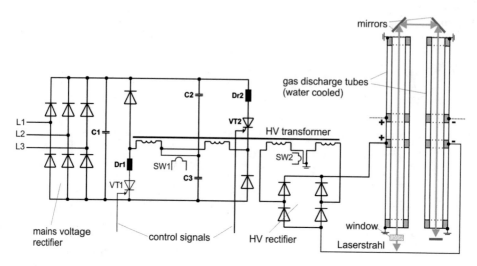

Fig. 10.230 Principle circuit of the power source with series resonant circuit inverter for CO_2 lasers for material processing

Functionality

The laser power source is implemented in the form of a series-compensated oscillating circuit converter (Fig. 10.230). First, a 6-pulse bridge is used to generate an intermediate circuit voltage U_d from the three-phase mains voltage and buffered with capacitor C_1. By alternately igniting the thyristors VT_1 and VT_2, the resonant circuit formed by the transformer Tr and the capacitors C_2 and C_3 is excited. The natural frequency of the resonant circuit in this case is about 5 kHz - depending on the type and load level. For an understanding of the circuit in Fig. 10.230, signal characteristics essential for the understanding of the circuit in Fig. 10.231 are included in Fig. 10.231. When thyristor VT_1 is ignited, the voltage across it breaks down and VT_1 takes over the current of diode VD_2. The choke Dr_1 serves to limit the rate of rise of the thyristor current.

After the current zero crossing in VT_1, the resonant circuit current changes from VT_1 to VD_1. A negative voltage is generated at VT_1, which leads to a reduction of the recovery time of the thyristor VT_1. In comparison with other circuits, this enables a reduction of the minimum permissible thyristor protection time t_s with this current source and thus a higher convertible active power. The transformer Tr in the resonant circuit is used for potential separation and for adapting the output voltage of the power source to the requirements of the low-pressure gas discharge of the CO_2 laser. To ensure the safe ignition of 2 separately connected discharge sections, the transformer is manufactured according to a special design specification as shown in Fig. 10.233. Due to the relatively high natural frequency of the resonant circuit, this high-voltage transformer can be designed with small dimensions and low mass. With an output power of the power source in continuous operation of 3.3 kW/10 kV, the mass of the transformer is only 4.5 kg.

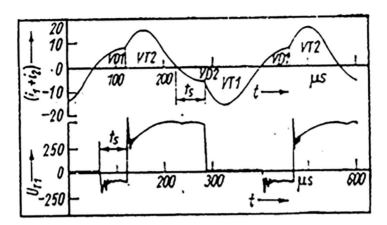

Fig. 10.231 Curves of transformer current ($i_1 + i_2$) and thyristor voltage U_{T1} at VT_1 and diode VD_1. (Source: Zacharias et al. 1986 [75])

In principle, oscillating circuit converters produce an approximately sinusoidal current through the transformer and the gas discharge paths. This results in relatively low radio interference. Output voltage and output current of the presented current source can be varied within wide limits by changing the thyristor protection time t_s. The current zero crossing of the thyristor is detected by evaluating the transformer current ($i_1 + i_2$) with the current transformer SW_1 and the time t_s is set as a function of the current setpoint and the actual current value determined with current transformer SW_2. In this way, sharply falling characteristics are produced within the natural output characteristic field. The natural output characteristics field, which is essential for the assessment, the control range and the control behavior of the current source, is shown in Fig. 10.232. The load characteristic curves of the current source were determined at resistive load in order to be able to represent the characteristic curve field as completely as possible. The real load caused by a gas discharge differs somewhat from this. In the present frequency range, it is comparable with the parallel connection of a capacitor with a non-linear loss resistance.

The output characteristic curves are strongly falling. This supports an ignition of the gas discharge. If there is no load, very high output voltages occur due to the low resonant circuit damping, which would eventually lead to the destruction of the transformer, the capacitors or the semiconductor valves. Therefore, the maximum output voltage is set to 25 kV by limiting the resonant circuit current.

Further limitations of the characteristic curve field result from the permissible minimum thyristor hold-off time as well as the minimum "lamp" voltage of the gas discharge. Operating points can be set within these limits.

For an output voltage of 10 kV, the control characteristic of the current source is shown in Fig. 10.233. The maximum output power of P = 2 × 2.5 kW at t_s = 25 μs is only achieved in pulsed or intermittent operation due to the thermal load of transformer and

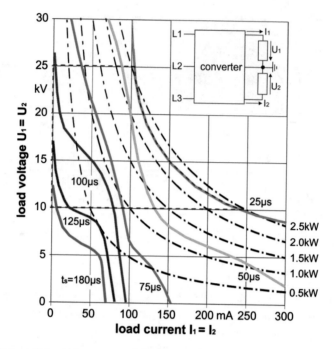

Fig. 10.232 "Natural" load characteristics of the power source with symmetrical ohmic load; glow voltage of the gas discharge: 10 kV, the ignition range for fast ignition is framed [75]

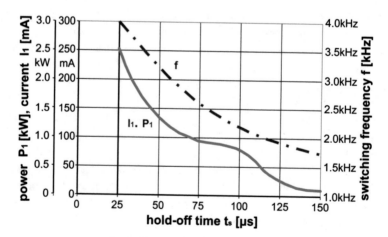

Fig. 10.233 Control characteristic of the current source with symmetrical load I_1, P_1, current or power in load resistance R_1; f inverter frequency [75]

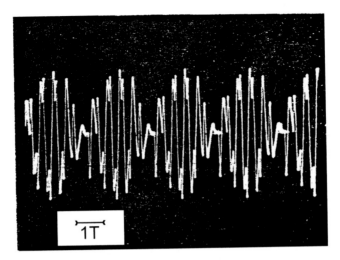

Fig. 10.234 Output current of the current source sinusoidally modulated with 400 Hz (scale: [x] = 1 ms/T, [y] = 100 mA/T). (Source: Zacharias et al. 1986 [75])

thyristors. In continuous operation, a maximum output power of 3.3 kW at 10 kV is delivered. If an efficiency of 4–8% is assumed for the laser, the power supply unit is suitable for CO$_2$ lasers with optical output powers of 130...260 W. The laser setup consisted of 6 gas discharge tubes with 2 discharge paths each, so that 6 synchronously operated power supply units were required. The gas discharge sections were connected "electrically in parallel" and "optically in series". Thus, a total optical output power of 1 kW could be achieved. In a number of applications of the laser beam for material processing, a fast changeability of the laser output power is required. When cutting complicated contours with sharp corners or brittle ceramic materials, the energy input is controlled by transition to pulsed laser radiation. This reduces the heat-affected zones, prevents burn-off and slag formation of the material and prevents heat-induced mechanical stresses in the base material. There are similar requirements for welding and surface treatment using laser radiation.

Figure 10.234 shows the load current at resistive load with a sinusoidal modulation at 400 Hz. With alternating current excitation, the laser changes from periodic pulse operation to quasi-continuous (cw) operation from frequencies of a few hundred Hertz. The optical output power of the laser pulses at twice the frequency of the current through the gas discharge. This pulsation of the output power decreases steadily with increasing frequency and amounts to only about 5–10% of the average power at f = 10 kHz. According to [76], a pulsation of 12–25% is to be expected with the present current source, at a pulsation frequency of 6–8 kHz. The time constants for the temperature transitions in the processed material depend on the type of material and the radius r of the incident laser beam. For mild steel and r = 0.25–1.5 mm, they are in the range of 0.8–25 ms.

Transformer Design

A special feature in the design is the transformer. Longitudinally flow gas discharge lasers of higher power are usually multi-folded lasers consisting of individual resonators each with two discharge paths. In order to be able to optimally control the parameters of the laser beam, in particular the laser output power, in accordance with the technological requirements, it is necessary to feed each resonator or gas discharge path separately, because only then is it possible to react to different shifts in the characteristic curves of the gas discharges of the individual resonators. The operation of the gas discharges with pure alternating voltage is not possible due to the given construction of the laser. The gas discharges are cooled with water through a double-walled tube. The water jacket represents a shunt for alternating voltage, so that the gas discharge only occurs in the vicinity of the electrodes. This requires prior rectification, so that a real ignition process is always necessary for the parallel start of the laser discharges. Due to the parasitic capacities of the water-cooled double-walled tube, the current is then partially smoothed. Within a resonator, it is important to ignite both discharge paths simultaneously if possible. With the current source described in EP0067464, it is in principle only possible to feed a single discharge path if its high efficiency is to be maintained, and not destroyed by series resistors that would allow several discharge paths to be fed in parallel. In addition, if both discharge paths of a resonator were fed in parallel, e.g. as mentioned above via series resistors, problems would arise with regard to their simultaneous ignition. These problems could be partially solved by increasing the open-circuit voltage. However, a deterioration of the overall efficiency would have to be accepted. The entire gas discharge laser would then be fed by a power source complex containing as many individual power sources as there are discharge paths. This disadvantage is avoided in the circuit arrangement in Fig. 10.230. The problem is solved by an electronically controllable power source consisting of a resonant circuit inverter, which contains a transformer consisting of a primary winding and two secondary windings as well as associated inverter control devices. It is characterized in that the primary winding of the transformer is arranged on the middle leg of a transformer core and the two secondary windings of the transformer are arranged on two symmetrical side legs of the transformer core and wound in the same direction. Each of the secondary windings supplies a gas discharge path via a rectifier. Figure 10.235 shows the basic structure of the transformer. Since the high-voltage windings are earthed at one end, the winding direction is always selected so that the start of the winding can be earthed. This makes the voltage distribution in the transformer easy to control.

The two primary windings N_{11}, N_{12} impress a magnetic flux into the middle leg, which is split symmetrically under otherwise symmetrical conditions. The secondary open-circuit voltages under these conditions are ($N_{11} = N_{12} = N_1$, $N_{21} = N_{22} = N_2$, $U_{11} = U_{12} = U_1$):

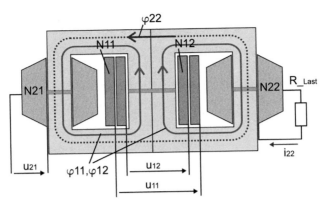

Fig. 10.235 Layout diagram of the high-voltage transformer for the circuit as shown in Fig. 10.230 with indicated unbalanced/unbalanced load [77]

$$U_{21} = U_{22} \approx 0.5 \frac{N_2}{N_1} U_1$$

If - as shown in Fig. 10.235 - a winding is loaded, it causes an opposite flux according to Lenz's rule, which increases the voltage induction in the other secondary winding. Thus, when one discharge is ignited, the ignition of the second discharge is also forced. The transformer is part of a resonant circuit inverter controlled by the switching frequency. Voltages and currents are therefore always extremely dependent on control parameters and load condition in relation to the resonant circuit parameters. In the present case, the circuit is controlled via the grace period of the thyristors as control parameters. The lower limit value can be conveniently monitored at the same time. In no-load operation, the resonant frequency is determined by the magnetizing inductance. In the short circuit, on the other hand, the resonant frequency of the oscillating circuit is determined in a first approximation from the parallel connection of magnetizing inductance and leakage inductance of the transformer. Experience has shown that with this (very simple and adaptable) control method, the two resonant frequencies should be separated by a factor of at most 3...5 in order to control abrupt load changes. Therefore, the magnetization inductance was artificially reduced by air gaps, as shown in Fig. 10.235. The advantage of this arrangement is that only one resonant circuit inverter is required to feed two discharge sections of a resonator without any loss of flexibility or efficiency. It is guaranteed that both discharge paths of a resonator ignite equally.

The transformer and high-voltage rectifier were placed in an oil vessel with appropriate bushings for the low-voltage and high-voltage connections and a membrane for the thermal expansion of the oil. Although the core 4xU80/46/18 used has a relatively large winding window, a trapezoidal winding cross section was used for the high-voltage windings to make better utilization of the winding space (Fig. 10.7). Due to different insulation protrusion, the field strengths at "sloping interfaces" can be controlled very well here. As

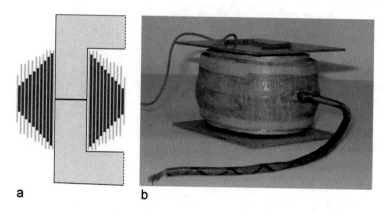

Fig. 10.236 High-voltage winding of the transformer. (**a**) Principle structure, (**b**) realization **b**. (Source Zacharias 2017)

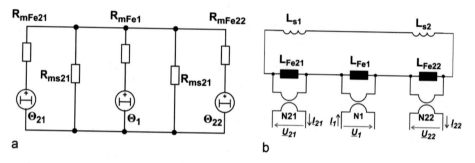

Fig. 10.237 Transformer equivalent circuit diagrams of the 3-winding transformer according to Fig. 10.235; $\underline{U}_1$: impressed voltage, $\underline{U}_{21}$, $\underline{U}_{22}$: load voltages. (**a**) Magnetic equivalent circuit diagram, (**b**) and electrical equivalent circuit diagram

a result, the voltages - especially in the short ignition range - were very well controllable. The insulation material was transformer paper/multilayer, which fills very well with oil when impregnated in a vacuum. The winding was finally coated with (1×) lacquer for stabilization. This condition is shown in the following Fig. 10.236.

Analytical Description of the Transformer

The following Fig. 10.237 shows the magnetic equivalent circuit diagram and the electrical equivalent circuit of the transformer of Fig. 10.235, whereby the primary winding, since closely coupled, is shown as a winding without its own leakage.

The electrical equivalent circuit diagram is shown below once again with ohmic load on the secondary side. The following observations on electrical behaviour refer to this diagram. In these considerations, the internal losses of the transformer and its winding capacities are not taken into account. This gives the possibility of a clear representation.

If the total leakage inductance normalized to N = 1 is applied to $L_S^* = L_{S1}^* + L_{S2}^*$, this quantity can be used as a reference value for further display. In addition, one can start from the relation

$$L_{Fe1}^* = 2L_{Fe21}^* = 2L_{Fe22}^*$$

This results from the symmetry of the transformer structure. A further specification to $L_{Fe1}^* = 48L_S^*$ corresponds to a ratio of 4.9 between series and parallel resonance frequency in the oscillating circuit converter of the presented example. These specifications enable clear, standardized representations to reflect the general conditions. From the electrical equivalent circuit, the Kirchhoff's mesh rule for the quantities transformed to the number of turns N = 1 results

$$- U_S^* + U_1^* - U_{21}^* - U_{22}^* = 0$$

For case R$_1$, R$_2 \to \infty$ for both windings, the secondary open circuit voltage is obtained to

$$U_{21}(I_{21}=0) = U_{22}(I_{22}=0) = \frac{N_{21}}{N_1} U_1 \cdot \frac{j\omega L_{Fe21}^*}{j\omega L_{Fe21}^* + j\omega L_{Fe22}^* + j\omega L_S^*} = \frac{24}{49} \frac{N_{21}}{N_1} U_1$$

$$\approx 0.5 \frac{N_{21}}{N_1} U_1.$$

If winding N$_{22}$ is short-circuited, you get from Fig. 10.238 for

$$U_{21}(U_{22}=0) = \frac{N_{21}}{N_1} U_1 \cdot \frac{j\omega L_{Fe21}^*}{j\omega L_{Fe21}^* + j\omega L_S^*} = \frac{24}{25} \frac{N_{21}}{N_1} U_1 = 0.96 \frac{N_{21}}{N_1} U_1$$

A relationship between output and input variables can be described for a fixed frequency ω simply by using the current divider rule. As an example, winding N$_{21}$ is considered here with loads on the secondary windings with R$_1$ and R$_2$ (Fig. 10.238):

$$\frac{I_{21}^*}{\frac{U_1^*}{\omega L_S^*}} = \frac{1}{\left(\frac{R_1^*}{\omega L_{21}^*} + j\right) \cdot \left(1 + \frac{L_{21}^*}{L_S^*} \frac{R_1^*}{R_1^* + j\omega L_{21}^*} + \frac{L_2^*}{L_S^*} \frac{R_2^*}{R_2^* + j\omega L_{22}^*}\right)}$$

With X$_S = \omega L_S$, the normalized dependencies of the current in winding N$_{21}$ under resistive load on both windings are obtained as shown in Fig. 10.239.

Figure 10.240 shows - also in logarithmic representation for the magnitude of the load resistances - the dependence of the output voltage at N$_{21}$. As in the investigation of the open-circuit voltage above, it can be seen that - as desired - the open-circuit voltage

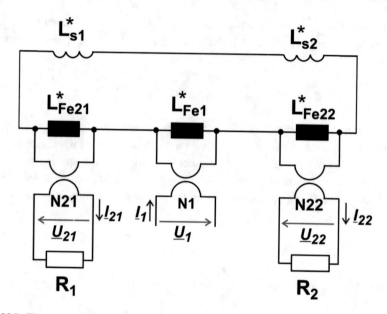

Fig. 10.238 Electrical equivalent circuit diagram of the transformer with secondary resistors as used for the measurements

$U_{21}(I_{21} = 0)$ increases when N_{22} is loaded. In the case of transformers of "normal" design, the loading of additional secondary windings would lead to a reduction of the open-circuit voltage.

For orientation, Fig. 10.240 shows the case of a symmetrical load as a line. The logarithmic representation of the load quantities R_1 and R_2 was chosen to cover a wide range of transformer states.

Usually, the load characteristics $U(I)$ are more common when representing the transmission behaviour of transformers. For this reason, Fig. 10.241 at the end of this chapter shows a set of load characteristics with the load resistance of the second secondary winding as a parameter. Here too, the case of symmetrical loading is shown for comparison. The diagrams were calculated for a fixed frequency and without consideration of parasitic capacitances. The operating frequency of the transformer is load-dependent due to the type of circuit described at the beginning and the selected control method. This is the main reason for the differences between the measured shape of the characteristic curves in Fig. 10.232 and the calculated characteristic curves in Fig. 10.241. The latter reflect the complex behaviour of the transformer.

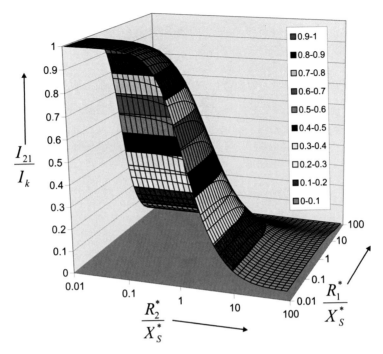

Fig. 10.239 Current in winding N_{21}, normalized to the short-circuit current (when loaded with resistive load on both windings)

10.23   Description of Resonant Circuits and Resonant Circuits in the Phase Plane

Resonant circuit inverters are systems with at least 2 inductive and capacitive storage elements. The solution of a differential equation of second order can be displayed as parameter curves $\left(u(t); \frac{du}{dt}(t)\right)$ in the so-called phase plane. A physical interpretation is the observation of the state of a memory. This way of representation allows the observation of a circuit in function as well as the construction of stationary solutions. An example is demonstrated below and shows how a modified resonant circuit inverter can also be described using this method.

Figure 10.242 shows an oscillating circuit, which is started from a charged capacitor (homogeneous solution of the differential equation). In the undamped case, in the normalized phase plane, one obtains

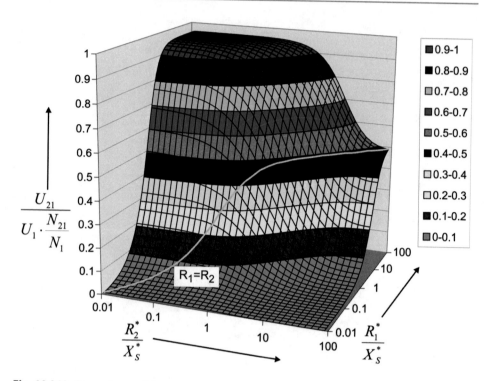

Fig. 10.240 Dependence of the output voltage at N_{21} on the quality of the load resistors

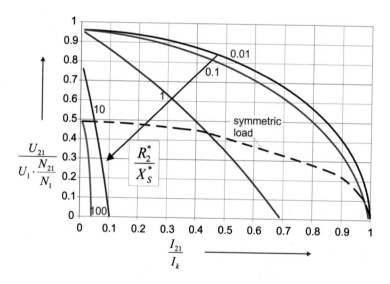

Fig. 10.241 Normalized load characteristics of a winding of the transformer with the normalized load resistance of the second secondary winding as parameter

Fig. 10.242 Series resonant circuit with marked counting directions

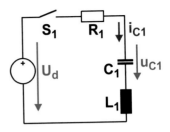

$$\left(u_C(t); k \cdot C \cdot \frac{du_C}{dt}(t) \right) = (u_C(t); Z_w \cdot i_C(t))$$

The trajectory in the phase plane is a circle around the coordinate origin at $U_d = 0$ and $R_1 = 0$ (Fig. 10.243a, b). In the damped case, a spiral shape results (Fig. 10.243c, d). If an oscillating circuit with discharged accumulators is connected to a voltage source, similar shapes are obtained (Fig. 10.243). The particulate solution U_d is the center of the trajectory. The representation in the phase plane is also very helpful in practical matters. Current and voltage at a capacitor can be set up very easily in the XY representation of oscilloscopes. A centered image, with all its deviations, can be observed much more easily during the transition to other load states than a two-line image with time deflection. Rare events are also easier to see.

From this, the construction of the trajectory in the $(u_C; Z_w \cdot i_C)$ plane can be derived in the undamped case. Figure 10.244 shows a resonant circuit inverter in a half-bridge circuit. Alternately $+0.5 \cdot U_d$ and $-0.5 \cdot U_d$ are applied to the resonant circuit and the load. In order to avoid damping and to be able to work with arcs, the load consists of a large capacitor with parallel load resistor and upstream bridge rectifier. The driving voltage for the current therefore consists of the difference between the source voltage and the load voltage U_{load}. If the thyristors Th_1 and Th_2 are excited at a frequency below half the resonant frequency of the LC resonant circuit, a partial oscillation can be carried out via the antiparallel diodes D_1 and D_2. The radius of the resulting circuits is a voltage when the current is normalized with the characteristic impedance Z_w of the oscillating circuit. In successive switching operations, the end point of a trajectory is again the beginning of a new one. It follows that the voltage at the load (U_3) cannot become greater than $0.5 \cdot U_d$.

In addition, a condition for the stationary state is thus given. Then one has cyclically the initial conditions constant. Figure 10.244 shows the current transient process when the load voltage (represented by a voltage source instead of a link $R_{load}//C_p$) is equal to $U_d/4$. For the secondary condition that the switching frequency is less than half the resonant frequency, the following is obtained for the peak currents of thyristor and diode

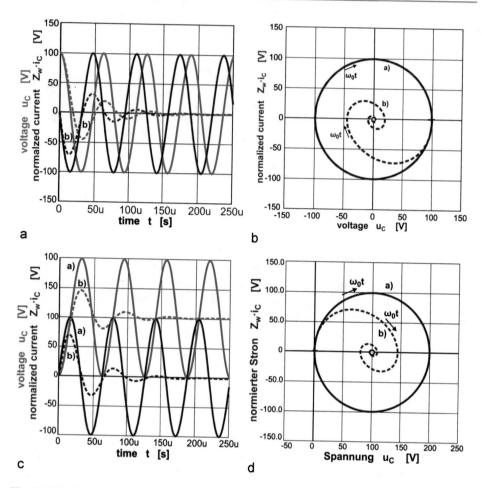

Fig. 10.243 Solutions of the oscillation differential equation according to Fig. 10.242 ($C_1 = 1\,\mu\text{F}$, $L_1 = 100\,\mu\text{H}$, $Z_w = 10\,\Omega$). (**a**) $U_{C1}(0) = 100$ V, time course $U_d = 0$, (**b**) $U_{C1}(0) = 100$ V, phase plane $U_d = 0$, (**c**) $U_{C1}(0) = 0$, time course $U_d = 100$ V, (**d**) phase plane $U_{C1}(0) = 0$, $U_d = 100$ V

$$\widehat{i}_{Th} = \frac{\frac{U_d}{2} + U_3}{Z_w} \quad \text{or} \quad \widehat{i}_{Th} = \frac{\frac{U_d}{2} - U_3}{Z_w}$$

The conducting time of thyristor and diode together always results in $t_{conducting} = T_0 = 2\pi\sqrt{LC}$. The period duration of a switching cycle is then

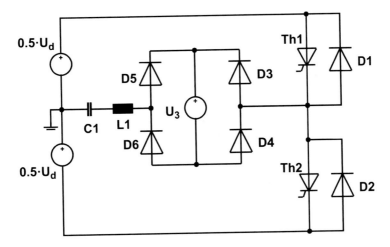

Fig. 10.244 Series resonant circuit inverter in half-bridge circuit

$$T_{sw} = t_{conducting} + 2 \cdot t_s = 2\pi\sqrt{L_1 C_1} + 2 \cdot t_s = T_0 + 2 \cdot t_s$$

If one wants to determine the average load current through the voltage source U_3, the averaging results in

$$\overline{I}_R = \frac{1}{T_{sw}} \int_0^{T_{sw}} i_R(t) \cdot dt = \frac{1}{T_{sw}}\left(\frac{\widehat{i_{Th}}}{\pi} \cdot \frac{T_0}{2} + \frac{\widehat{i_D}}{\pi} \cdot \frac{T_0}{2}\right) = \frac{T_0}{2 \cdot T_{sw}}\left(\widehat{i_{Th}} + \widehat{i_D}\right) = \frac{T_0}{2 \cdot (T_0 + 2 \cdot t_s)}$$

$$\times \left(\frac{U_d}{Z_w}\right).$$

This means that for a switching frequency lower than half the resonant frequency, the average load current from the load voltage is constant at $U_R/Z_w < 0.5 \cdot U_d/Z_w$, but can be adjusted via t_s. The circuit then operates like a constant current source and is well suited as a capacitor charger, for example. The thyristor and diode are operated with a relatively soft commutation. When the thyristor is switched on, the inductance L_1 limits the rate of current rise. The current is transferred to the antiparallel diode $u_{Th} \approx 0$. The diode is commutated with a limited current slope. The semiconductor losses are kept within narrow limits during this operation. At an excitation frequency $< f_0/2$, the trajectory in the phase plane is always composed of semicircles (Fig. 10.245).

If the switching frequency is increased beyond half the resonant frequency, the conditions change. The thyristors then each switch to a diode conducting phase. This also means that "hard" commutation occurs here. In the stationary case, the end of one half cycle is also equal to the beginning of the next. From this, the load characteristics can be derived in a similar way as described, but somewhat more complex. The conducting phase

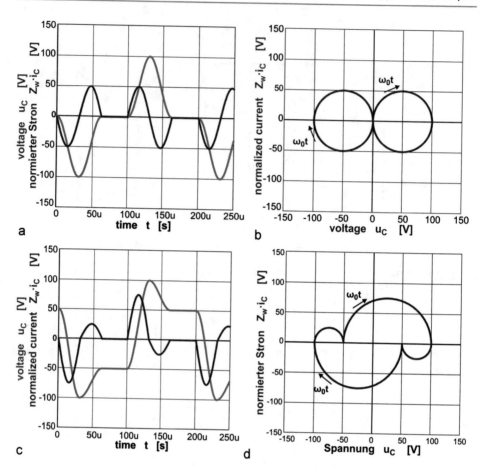

Fig. 10.245 Solutions of the oscillation differential equation of a oscillating circuit inverter according to Fig. 10.244. (**a**) $u_C(t)$-, $i_C(t)$ time histories, (**b**) u_C-i_C phase plane at load voltage $U_3 = 0$; (**c**) $u_C(t)$-, $i_C(t)$ time histories, (**d**) u_C-i_C phase plane at load voltage $U_3 = U_d/4$

of each diode has the length t_s. If the natural oscillation frequency is known, which in the undamped case is equal to the resonant frequency, the conduction angle can be calculated from this. The conduction duration of the thyristor t_1 is derived from the equilibrium condition for the steady-state case. This also means that operation of the circuit at a constant switching frequency f_{SW} is only equal to operation with a constant hold-off period t_s in the steady-state case. In the non-stationary case, there are fundamental differences, which cannot be compensated by switches that cannot be switched off, such as thyristors. It is therefore not advisable to operate resonant circuit inverters with frequency as controlling parameter. Even if the hold-off period has lost its original meaning with switches that can be switched off, experience has shown that its usefulness in the construction of stable control circuits is obvious (Fig. 10.246).

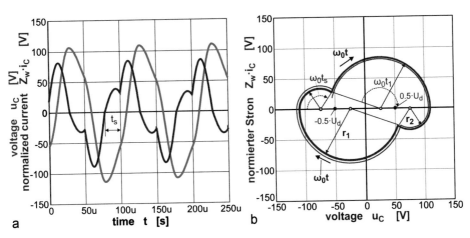

Fig. 10.246 Operation of a resonant circuit inverter according to Fig. 10.244 with a frequency $0.5 \cdot f_0 < f_{sw} < f_0$ and $U_3 = U_d/4$. (**a**) Curves of $i_{C1}(t)$ and $u_{C1}(t)$, (**b**) trajectory $\{i_{C1}(t); u_{C1}(t)\}$

From the conditions for the respective steady state, a number of trigonometric relationships can be derived, which allow the determination of the load characteristics for a counter-voltage load consisting of a bridge rectifier plus a large capacitor connected downstream. It is assumed that in the undamped case, the stationary trajectory is composed of circular arcs with centers $(U_d/2 - U_3; 0)$ and $(U_d/2 + U_3; 0)$ or $(-U_d/2 - U_3; 0)$ and $(-U_d/2 + U_3; 0)$.

Figure 10.247 shows the characteristic curves in standardized form for counter-voltage load and for resistive load. The case of resistance load is the result of simulation results. The hold-off period t_s after the zero crossing of the current, which is important for thyristors, is used as the control parameter. This is also useful because the oscillations are "free" oscillations. However, the oscillation period is only constant in the undamped

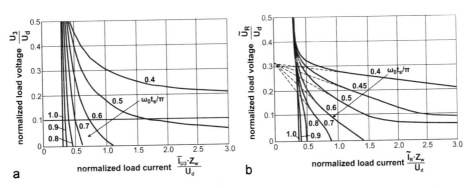

Fig. 10.247 Load characteristic curves of a resonant circuit inverter in a half-bridge circuit as a function of the hold-off period t_s. (**a**) With counter-voltage load U_3, (**b**) with resistive load

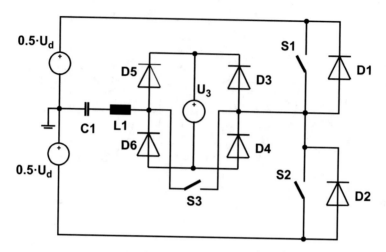

Fig. 10.248 Resonant circuit inverter in half-bridge circuit and short-circuitable counter-voltage load

case. In the more or less damped oscillation case, the duration of the natural oscillation depends on the damping.

If the hold-off period of the thyristors is kept constant, the frequency changes with the damping. When using transformers in oscillating circuit inverters, there is also the magnetizing inductance, which is then parallel to the load. Depending on the load, the natural oscillation frequency lies between the two frequencies $(L_s \cdot C)^{-0.5}$ and $(L_\mu \cdot C)^{-0.5}$. The lower the hold-off period, the closer the natural oscillation frequency can be approached. The originally strongly falling characteristic curve then becomes a practically horizontal line. This means that the "internal resistance" of the source is then zero when viewed from the load and the converter resembles a voltage source in its behaviour in relation to the load. When using a transformer, the output voltage is then only dependent on the transformation ratio. Very high efficiencies can be achieved at this operating point. These considerations are the starting point for a further modification of the resonant circuit inverter.

By means of a switch S_3 parallel to the input of the bridge rectifier, the current can be led past the load (Fig. 10.248). If the half-bridge inverter is simultaneously interrupted in its function, a capacitor charge can also be interrupted in this way without a half oscillation of the LC oscillating circuit being carried out completely via the load. Otherwise, the energy still stored in the inductor is always transferred to a capacitor switched on as load. After closing S_3, this is no longer the case.

In the phase plane, one can see that the oscillating circuit is additionally charged with energy by a jump from $U_{load} \rightarrow 0$. This means that this circuit can be put into a step-up function by additional periodic charging of the inductive component (Fig. 10.248). Here switches S_1 & S_2 are assumed, which are switched off when the current zero crossing is

Fig. 10.249 Signal characteristics in the circuit according to Fig. 10.247 using an example for the definition $D = 2 \cdot t_3/T_0$

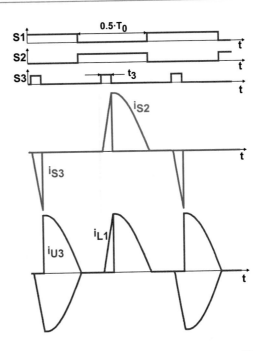

reached. The aim here is to switch off the switches at zero voltage on the switches if possible in order to keep the switch-off losses small. The circuit becomes practical with its two switches at resonance operation. For this to work technically, U_3 must be $\geq U_d/2$. The switches S_1 and S_2 switch to relatively low currents through the diodes. During their conducting phase, S_3 switches on and off again. When S_3 is switched on, a current is built up in inductance L_1, which is released again via load U_3 after S_3 is opened. If U_3 is designed as a parallel connection of R and C, energy is thus provided for consumption in R. Switch S_3 does not have to switch on a large current, but it must be possible to switch it off. How high the prevention of losses in the switches is depends on the characteristics of the switches used. Characteristic signal curves are shown in Fig. 10.249. It is assumed here that $D = 2 \cdot t_3/T_0$ can be a maximum of 0.5. Other concepts can also be implemented with the phase position of the control signals. In any case, switching on S_3 has the effect that inductance L_1 can be charged as with $U_3 = 0$. Since S_3 is closed when $i_{L1} = 0$, the driving voltage for the current i_{L1} results from a source voltage $0.5 \cdot U_d$ plus the current voltage at C_1. In the phase plane Fig. 10.250, the function can be followed accordingly. In the switch-on time of S_3, the driving voltage is $r_1 = | - 0.5 \cdot U_d + U_{C1}(i_{C1} = 0)|$. The center of the ringing circuit is at $(-0.5 \cdot U_d; 0)$. During the switch-off time, e.g. when $S_1 = ON$, the current is reduced by the voltage $-0.5 \cdot U_d + U_3$. This means that the centre of the the trajectory circuit is at $(-0.5 \cdot U_d + U_3; 0)$. The energy temporarily stored in the switch-on phase of S_3 is largely passed on to the load U_3 in the switch-off phase of S_3.

The values for Fig. 10.250 were determined for $Z_w = 10 \, \Omega$ and $D = 0.3$. For better orientation, the conducting phase of S_3 was marked in the current curve. The circuit shown

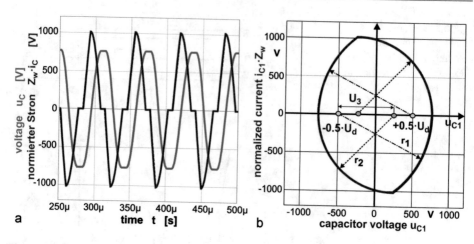

Fig. 10.250 Capacitor voltage u_{C1} and capacitor current i_{C1} when operating the circuit with $U_d = 1000$ V, $U_3 = 750$ V and $D = 0.3$ at $C_1 = 1$ μF and $L_1 = 100$ μH. (**a**) u_C-, i_C- time curves, (**b**) phase plane for the setup according to Fig. 10.248

in Fig. 10.247 thus works similar to a boost converter. When boosting, a constant switching frequency is set equal to the natural resonance or natural oscillation frequency of the oscillating circuit L_1C_1. Lowering is achieved by increasing the switching frequency above this frequency. In [78], several possibilities of execution of this principle are described. The step-up behavior can also be used for bidirectional operation, as shown in Fig. 10.251. In this symmetrical circuit, step-up and step-down behavior is present in both directions. The leakage inductance of the transformer serves as resonant circuit inductance. A bidirectionally acting switch combination on the respective secondary side of the transformer enables the corresponding load to be short-circuited. The circuit thus represents a bidirectional implementation of the circuit principle in Fig. 10.248.

10.24 Magnetic Stabilization of Heating Voltages

Thyratrons are still used in special applications such as radar, laser and accelerators. These gas-filled switching tubes are able to block very high voltages and, by avalanche multiplication of electrons from a heated cathode, to carry very high currents for short periods of time. This is achieved by filling the tube (Fig. 10.252) with a gas that is under low pressure. According to the Paschen curve of this gas, a high dielectric strength of the switching tube is thus ensured despite the small distance of the anode from the other electrodes. For high requirements on dielectric strength and pulse current, hydrogen, deuterium or, even more rarely, tritium is used as the gas. To create the required internal pressure of the tube, the gas is baked out of a metal-hydrogen compound (e.g. titanium hydride) (H_2 reservoir). A heated cathode provides the necessary electrons in the tube. A solid copper/molybdenum grid, negatively biased at -50 V... -200 V with respect to the cathode, prevents the electrons from drifting to the anode, which is also armored with molybdenum.

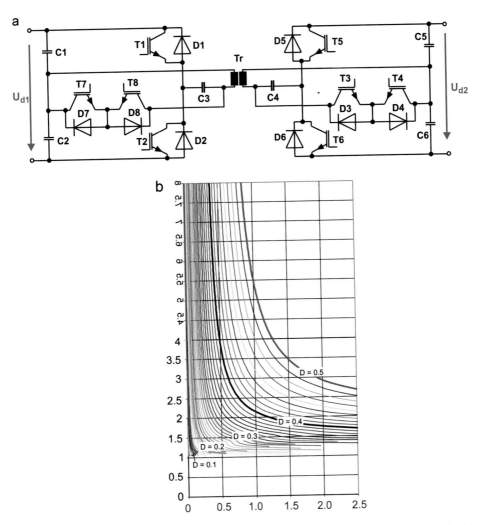

Fig. 10.251 Symmetrical design of a bidirectional DC/DC converter according to the principle shown in Fig. 10.248 based on [78]

A positive trigger pulse between grid and cathode initiates a current to the anode. The low gas pressure in the tube results in relatively large free path lengths of the electrons on their way to the anode. Shock processes with the gas molecules thus produce in a short time a multiple of the number of electrons that can be provided by glow emission. This determines the maximum peak current of the tube or the switchable charge quantity. Peak currents >10 kA are thus achieved, while the maximum continuous currents, which are limited by the glow emission, are only 0.3...0.5 A. At 300...500 kA/μs with maximum blocking voltages of 5 kV...130 kV and dv/dt values of −2... −5 kV/ns, the possible rate of current rise is significantly higher than values that can be achieved with semiconductor switches. A typical application example is the generation of short pulses in gas discharge lasers (Table 10.14).

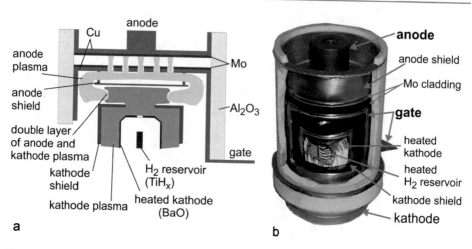

Fig. 10.252 H_2 thyratron as a high-power switching tube. (**a**) Schematic structure and plasma formation during the switch-on process, (**b**) view of a cut-open thyratron of type F-189 (glow cathode, cathode shield and H_2 reservoir are connected to the cathode terminal). (Source: Zacharias 2008)

The switching characteristics of the tube, such as maximum switching voltage, switching time and jitter, depend strongly on the two heating voltages (range typically 5...8 V) for reservoir and cathode heating. Therefore, they have to be stabilized against fluctuations of the mains voltage. With a tolerance of ±10% of the mains voltage against the respective nominal voltage, the heater voltage may only change less than ±1% [79]. Figure 10.253 shows an overview of the territorial distribution of the nominal grid voltages and grid frequencies in the countries of the world. Due to the different voltages and the corresponding tolerances, it is necessary to stabilize the voltages for cathode heating and hydrogen reservoir.

A heater isolated from the main circuit is not feasible due to the high inductively caused short-term voltage differences in the switching tube. The heating power supply is therefore galvanically connected to the pulse circuit of the thyratron and must be tolerant of voltage differences of several kilovolts for several 100 ns. This can only be reliably achieved with switching power supplies at relatively high effort and thus costs. A simple and also extremely robust possibility here is the magnetic voltage stabilization of the heating voltages. Figure 10.254 shows a proven principle.

The circuit works with a resonant circuit tuned to mains frequency, consisting of a foil capacitor and the magnetizing inductance of a transformer for the heating voltages. The exact tuning is done via an air gap in the core of the transformer. The heaters of reservoir and cathode provide the damping for the resonant circuit. The primary voltage at the transformer results from resonance above the input voltage and is limited by the saturation of the transformer core. The shape of the output voltage changes from a sinusoidal shape to

Table 10.14 Nominal values for the thyratron F-189 for sub-microsecond pulses

Maximum anode-cathode voltage	35 kV
High voltage segments	1
Peak current (<1 µs)	20 kA
Maximum average anode current	0.3 A
Maximum effective anode current	50 A
Grid control	+500 V/1.25 A
Nominal value heating	6.3 V/22 A
Nominal value H_2 reservoir	6.3 V/3 A
Reference types	F-189 (ITT) L-4189A (L3) GL-1689 (GL)

Fig. 10.253 Nominal voltages and mains frequencies in their worldwide use according to IEA (green: 220–240 V / 50 Hz, yellow: 220–240 V / 60 Hz, blue: 100–127 V / 50 Hz, red: 100–127 V / 60 Hz. (Source: IEA)

a rectangular shape when the input voltage is increased. The shape of the output voltage hardly changes when the input voltage is increased (Fig. 10.255).

This makes the effective output voltage of the circuit independent of the input voltage within wide limits. When the transformer is saturated, the capacitor reverses with a peak current that varies depending on the voltage reserve. This peak current is approximately the difference between the mains voltage and the capacitor voltage at the time of saturation and is also determined by the characteristic impedance of the resonant circuit. This is formed from the leakage inductance components $L_{\sigma 1}$ and $L_{\sigma 1}$, the saturated magnetizing inductance L_μ of the core and the capacitor C_1. For the characteristic impedance Z_W, one can write

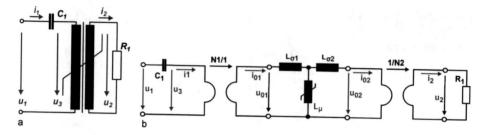

Fig. 10.254 Circuit for magnetic stabilization of the heating voltage of H_2 thyratrons. (**a**) Principle circuit diagram, (**b**) electrical equivalent circuit diagram

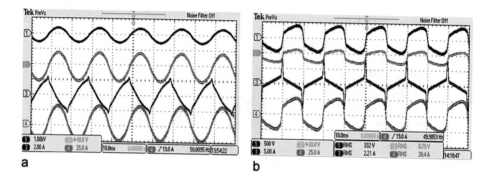

Fig. 10.255 Curves of input voltage of transformer (1), load voltage (2), input current (3) and load current (4) at (**a**) $U_1 = 160$ V and (**b**) $U_1 = 260$ V

$$Z_W = N_1 \sqrt{\frac{L_{\sigma 1} + \frac{L_{\sigma 2} \cdot L_{\mu SAT}}{L_{\sigma 2} + L_{\mu SAT}}}{C_1}}.$$

The form factor of the input current I_1, which increases with higher voltage, can be reduced by increasing the leakage inductance L_σ or by limiting the permissible input voltage range. The required type apparent power (rated power) of the transformer is given by

$$S_T = \frac{S_1 + S_2}{2} = \frac{1}{2}\left(U_{3.nom} \cdot I_1(U_1) + \frac{U_2^2}{R_1}\right)$$

The winding N_1 must therefore be oversized with regard to the wire cross section. The typical voltage tolerances of $\pm 10\%$ would result in an input voltage range of 207–253 V. However, since an adjustment range of the output voltage had to be provided and the input voltage also had to be adapted to the respective mains voltage via a transformer with winding taps, the increased input voltage tolerance of the constant results. The circuit is only suitable for low powers due to the mains feedback. In the given application, the total

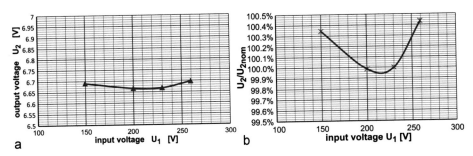

Fig. 10.256 Load voltage as a function of the mains voltage for a change of 150...260 V. (**a**) With voltage units, (**b**) in relative units

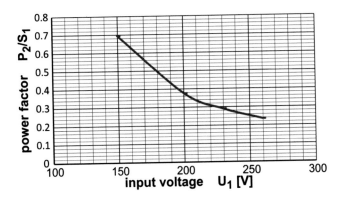

Fig. 10.257 Course of the total power factor at voltage increase for the circuit according to Fig. 10.254 at $C_1 = 1\ \mu F$, $L_{\mu 1} = 1.04\ H$

apparent power of the single-phase connected laser was a maximum of 3.2 kVA. By superimposing other current distortions and an upstream sinusoidal filter, all mains connection conditions could be met.

In the range shown for $U_1 = 150...260$ V, the voltage at the thyratron (here connections for heating and H_2 reservoir connected in parallel) changes only slightly, as shown in Fig. 10.256. The total change is about 0.5%.

The value 150 V also marks the beginning of the circuit's stabilizing effect. Due to the higher form factor of the current and the capacitive reactive power that increases with voltage increase, the power factor $\lambda = P/S$ for the circuit decreases with increasing mains voltage (Fig. 10.257). Because of these mains feedback effects, the application range of the circuit is limited to low power or a reduced control range.

If the transformer shows no signs of saturation, (at $L_\sigma \approx 0$) the load is connected via N_2 in parallel with the magnetizing inductance. If you look at the two poles from the load side, you get a source with an impedance

Fig. 10.258 Stationary load characteristic with parallel heating of cathode and H_2 reservoir

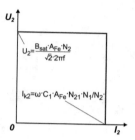

$$Z_{i2} = \underline{X}_{C1} // \underline{X}_{L\mu} \cdot \left(\frac{N_2^2}{N_1^2}\right) = \frac{j\omega L_\mu \cdot N_1^2}{1 - \omega^2 L_\mu \cdot N_1^2 \cdot C_1} \cdot \left(\frac{N_2^2}{N_1^2}\right) \Rightarrow \infty$$

and a short circuit current on the load side of

$$\underline{I}_{k2} = \frac{U_1}{\underline{X}_{C1}} \cdot \frac{N_1}{N_2} = j\omega C_1 \cdot \underline{U}_1 \cdot \frac{N_1}{N_2}.$$

This means that below transformer saturation, you are dealing with an "ideal" current source, while above saturation, you are dealing with an ideal voltage source, which has an effective value of

$$U_2 = \frac{B_{SAT} \cdot A_{Fe} \cdot N_2}{\sqrt{2} \cdot 2\pi f}$$

Figure 10.257 shows this connection schematically. The limitation of the short-circuit current is extremely advantageous for the application, as both heaters show pronounced PTC thermistor behaviour (Fig. 10.258). This means that the cold resistance at switch-on is considerably lower than the operating resistance, which only becomes apparent after a few minutes (approx. 8 min). The excess current at switch-on is thus automatically limited. The effective value of the output voltage remains almost constant over a wide range.

One can clearly see the "rectangular" shape of the characteristic curve in Fig. 10.259, which was estimated with the relationship shown in Fig. 10.258. At weaker loads, the oscillating circuit is increasingly less damped by the load resistance R_1 in parallel with the magnetizing inductance L_μ. This means that the load resistance R_1 becomes larger compared to the characteristic impedance Z_{W2} measured from the secondary side.

$$Z_{W2} = \sqrt{\frac{L_\mu}{N_1^2 \cdot C_1}} \cdot N_2^2$$

This has a particular effect on the current form factor on the primary side of the transformer. The voltage waveform changes from almost sinusoidal (Fig. 10.255a) to a more rectangular

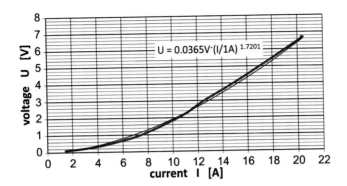

Fig. 10.259 Schematized load characteristic of the circuit

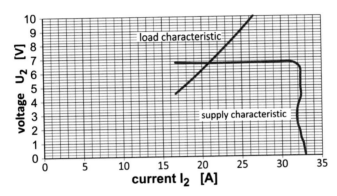

Fig. 10.260 Measured stationary supply characteristic and load cgaracteristic for an output voltage of 230 V~

voltage curve (Fig. 10.255b). The rapid "swinging" of the capacitor voltage u_{C1} occurs via the leakage inductance $L_{\sigma 1}$ plus the saturated magnetizing inductance L_μ. The smaller these values, the faster the oscillation. But the higher the current peaks in the primary current of the arrangement.

Figure 10.260 shows in a diagram the measured load characteristic of the constant $U_{20}(I_2)$ in comparison to the characteristic of the load. The basic shape of the characteristic curve shown in Fig. 10.259 is shown here.

If the circuit is connected to the same mains voltage at 60 Hz without changing C1, the resonant frequency is reached with a different number of primary windings:

$$f_0 = f_{mains} = \frac{1}{2\pi f \sqrt{L_1 \cdot C_1}} = \frac{1}{2\pi f \cdot N_1 \cdot \sqrt{L_\mu \cdot C_1}}$$

This means for the ratio of the two primary winding numbers

Fig. 10.261 Overall circuit of the "magnetic constant" with secondary taps for different load voltages and frequency switching

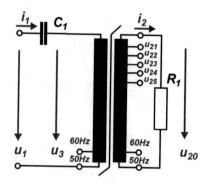

$$\frac{N_{1\_60\ Hz}}{N_{1\_50\ Hz}} = \frac{50\ Hz}{60\ Hz}.$$

Consequently, a tap for operation at 60 Hz must be led out of the transformer. To prevent the maximum adjustable output voltage U_{20} from changing as well, a connection for 60 Hz is also led out on the secondary side. The step of the output voltage changes slightly within an acceptable range (Fig. 10.261).

10.25 Oscillators with Non-Linear Magnetic Components

(a) *1-stroke "flyback Oscillator"*

The flyback converter is a feedback oscillation generator [80–83], which is used as an electrical oscillator circuit equipped with a non-linear magnetic transformer for the generation of pulses. In its simplest form, it consists only of a pulse transformer (not necessarily saturable) and an amplifying component such as a transistor. A distinction is made between flyback converters, which are operated in a self-oscillating manner (astable multivibrators) and those which are triggered once or periodically by an external trigger signal (monostable multivibrators). The simplest form is shown in Fig. 10.262.

A voltage is induced in winding N_2 by a change in the magnetic flux. This generates a base current of the transistor T_1 and brings the collector-emitter path into the conducting state. This continues until the base current is no longer large enough. The reason for this is either a fall below the necessary base-emitter voltage (case A; in the case of Si-transistors approx. 0.7 V) due to an excessive internal ohmic voltage drop in L_1 or a saturation of the core (case B).

By selecting the device parameters, the needle pulses, which are often short in time, can be selected over a very wide range of duty cycle and period duration. The magnetic response of a flyback converter to an induced voltage pulse is the emission of a magnetic field pulse. This can be used to drive mechanical systems with their own time constants,

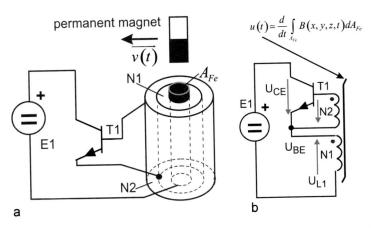

Fig. 10.262 Simplest form of a flyback oscillator externally triggered by a magnetic field. (**a**) Basic structure, (**b**) circuit diagram

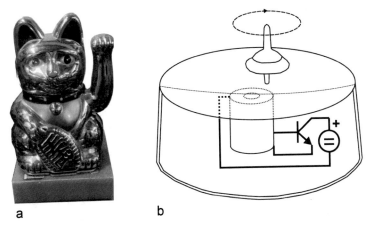

Fig. 10.263 Mechanical systems driven by external, synchronized magnetic pulses. (**a**) Pendulum motion of a Chinese "winky cat" (source: Zacharias 2018), (**b**) "perpetual gyroscope" [83–85]

such as a pendulum (Fig. 10.263a) or a mechanical gyroscope (Fig. 10.263b). Triggering and synchronization is achieved by the relative movement of a permanent magnet to the nearby coil, which creates a voltage induction in the pulse transformer.

In the case of the gyroscope, it is a magnet that is simply permanently magnetized perpendicular to the axis of rotation. During its rotational movement, it induces stresses in the magnetic field on a ferromagnetic rod core, which in turn induces the tilting oscillations described above. To ensure that the sensitivity of the arrangement is high enough, N_2 is greater than N_1 here. The base current is limited by the winding resistance of N_2. The induced voltages are shown in the following Fig. 10.264.

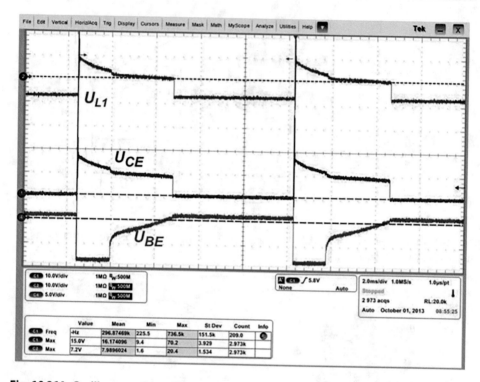

Fig. 10.264 Oscillograms of the collector-emitter voltage U_{CE}, the base-emitter voltage U_{BE} and the voltage U_{L1} induced in L_1 according to the counting arrow system shown in Fig. 10.262

In the switch-on phase of T_1, $U_{CE} = 0$ and $U_{BE} = 0.6$ V practically apply. If the base current is no longer sufficient, the collector current breaks down. This also leads to coupling. The negative current change reverses the voltage at the base-emitter path and the and the collector-emitter path is blocked. The stray component $L'_{\sigma 1}$ generates a considerable overvoltage (approx. 60 V), which is limited by the breakdown voltage U_{CB0}. Because of the low value of $L'_{\sigma 1}$ and the high voltage, the energy stored in this scattering component is quickly dissipated and the collector-emitter voltage drops rapidly. The remaining energy stored in the magnetic component is essentially dissipated via the base-emitter path. The base-emitter diode is operated in avelanche breakdown during this time. The corresponding measured voltage is -8.5 V for the transistor used. This is characterized by the temporarily constant voltage level $U_{BE} = -8.5$ V in the oscillogram shown above. At the end of the voltage plateau, only coil-internal and parasitic active components are responsible for converting the magnetically stored energy into heat. At L_1, the base-emitter voltage curve then appears reduced with the transmission ratio $N_1/N_2 = 1/3$. The model data were extracted from various test measurements without completely dismantling the object and plausibility was checked with generally available information. In the case considered here, case (A) is more likely to occur than case (B) (see Fig. 10.265).

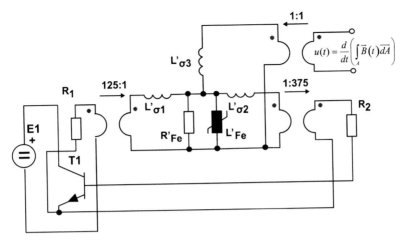

Fig. 10.265 Model for the calculation of a magnetic field-triggered flyback converter for driving a magnetic gyroscope in Fig. 10.264a ($R_1 = 120\ \Omega$, $R_2 = 1047\ \Omega$, $R'_{Fe} = 83.84\ m\Omega$, $N_2/N_1 = 3$, $L'_{Fe} = 8.6\ \mu H$, $L'_{Fe.sat} = 0.96\ \mu H$ at $\varphi_{sat} \sim 78\ \mu Vs$, $L'_{\sigma 1} = 0.256\ \mu H$, $L'_{\sigma 2} = 0.896\ \mu H$)

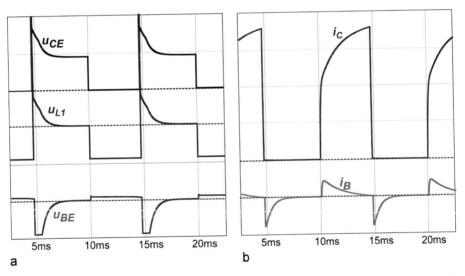

Fig. 10.266 Simulated curves of voltages ((**a**) 10 V/T) and currents ((**b**) 20 mA/T) of a monostable flyback converter triggered by an external field at 100 Hz

In Fig. 10.266, a simulation is shown where the monostable multivibrator according to Fig. 10.262 was simulated with the equivalent circuit diagram for the transformer in its function.

Figure 10.266a essentially reflects the ratios of the measured oscillograms in Fig. 10.264. Deviations are mainly due to the fact that the magnetization curve of a rod

Fig. 10.267 Circuit variant with reduced switch-off overvoltage at the transistor by du/dt negative feedback via capacitor [83]

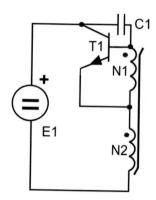

core looks different from that of a closed toroidal core, since the magnetic field lines also exit the core laterally, especially in the saturation region.

A high di/dt results in high induced voltages in the windings. For the simulation, it was assumed that the voltage at N_2 is limited by a base-emitter breakdown voltage of 8.5 V. This voltage is transformed (here 1:3) to the operating DC voltage of 9 V of the transistor. Because the leakage inductances of the transformer are also quickly rendered currentless, an additional voltage peak at U_{CE} is produced when the transformer is switched off, which can reach considerable magnitudes. The simulation also reflects these effects.

Many circuit variations have to do with limiting this peak. If the (parasitic) Miller capacity of the transistor is not sufficient, for example, an additional parallel capacitor between collector and base (Fig. 10.267) can be used to slow down the turn-off. Apart from the order of E_1 and the connected "working" winding, the circuit is identical to Fig. 10.262a.

However, the previously monostable flyback converter circuit can be modified so that a periodic self-start is initiated, thus creating a continous working multivibrator circuit (Fig. 10.268). One can see, for example, the periodic magnetic field pulses (Fig. 10.268a). The voltage divider $R_2/(R_{1+}R_2)$ is dimensioned in such a way that at the lowest voltage E_1, the transistor switches on safely.

This will then start the periodic tilting process. If the circuit is extended by an N_3 winding to extract energy, one obtains one of the simplest forms of insulating DC/DC converters, which still play a major role in low-cost devices. During the locking process of switch T_1, the energy is transferred to the secondary side. Due to the principle, this does not occur at a constant frequency. In the systematics of circuit topologies, the circuit corresponds to an (isolating) inverse converter with a voltage transformation ratio of

$$\frac{U_{N3}}{E_1} = \frac{N3}{N1} \cdot \frac{(\mp)\frac{t_{on}}{t_{on}+t_{off}}}{1 - \frac{t_{on}}{t_{on}+t_{off}}} = (\mp)\frac{N3}{N1} \cdot \frac{\frac{t_{on}}{t_{on}+t_{off}}}{1 - \frac{t_{on}}{t_{on}+t_{off}}} = (\mp)\frac{N3}{N1} \cdot \frac{t_{on}}{t_{off}}$$

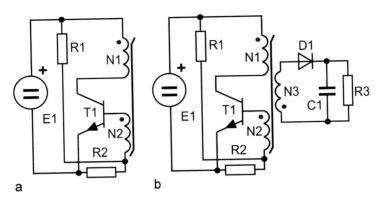

Fig. 10.268 Flyback converters as multivibrators/oscillators. (**a**) For providing periodic signals (frequency is voltage-dependent), (**b**) for energy transfer in the blocking phase of the transistor (flyback converter)

If one now sets the switch-on time $t_{on} = \frac{k_1}{E1}$, one obtains for the maximum switching frequency $f_{sw\_max} = \frac{E1}{k_1}$

$$f_{sw} = \cfrac{1}{\frac{1}{f_{sw\_max}} + t_{off}} = \cfrac{1}{\frac{1}{f_{sw\_max}} + \frac{N_3}{N_1}\frac{k_1}{U_{N3}}} = \cfrac{f_{sw\_max} \cdot U_{N3}}{U_{N3} + k_1 \cdot f_{sw\_max} \cdot \frac{N_3}{N_1}}$$

This means that at low output voltages, the switching frequency is low and increases with the output voltage until it finally reaches a maximum.

When using temperature-stable magnetic materials, these circuits are also extremely reliable and robust against interference. A disadvantage of the physical principles described above is the low flexibility with regard to voltage stabilization measures. However, very high voltages can easily be achieved, e.g. for Brownian tubes, electric shockers, electrostatic dust collectors, etc.

The "art" of designing the magnetic component is to set the energy stored in the switch-on phase at a maximum ratio compared to the energy stored in the leakage inductances. This means a close coupling of the windings, usually on a core with an air gap. The magnetizing inductance of the transformer is used as a storage element. This is not the case in the version of a saturation-controlled current flow transformer in mid-point connection described below.

(b) *2-stroke flyback converter*

As Fig. 10.269 shows, this circuit can be easily extended to push-pull circuits. Shown is such a DC/DC converter circuit with a center point circuit on both the input side (inverter) and the output side (rectifier). However, this circuit works as a current flow converter. This

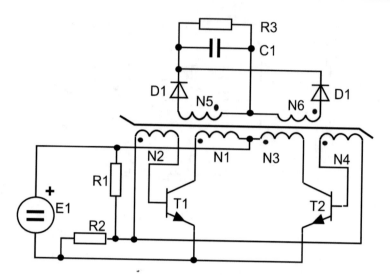

Fig. 10.269 Push-pull flyback converter with a saturable transformer core as control element

means that the energy is transferred during the turn-on phase of the transistors. Further variations are known and used [81]. Extensions to bridge circuits are possible without problems. Since the circuits are dominated by the magnetic core, reasonable/effective control methods are actually only conceivable via the magnetic circuit. However, these quickly cancel out the advantage of the simple design.

With the circuit as shown in Fig. 10.269, it is possible to operate without major problems in the range of 10... 1000 W of electrical power. Above 1 kW bridge circuits were and are used. Also (rectangular) alternating voltages are generated in this way for the smaller power requirement. The clear number of components makes the circuit repairable practically everywhere in the world.

10.26 Circuit Concepts for Generating Short High-Energy Pulses for Pulse Power Applications

Pulsed gas discharge lasers are excited by a pulsed high voltage discharge. A stationary, homogeneous discharge would be more advantageous. However, it is not possible to realize this because after a longer discharge period, arc or spark discharges occur, which lead to overheating of the gas and to destruction of the electrodes. Excitation must therefore take place in pulsed mode, whereby one discharge per pulse can be maintained until the transition to sparking begins. How long this takes depends on the gases used and their pressure. The gases are cooled during periodic pulse operation in order to keep the efficiency as high as possible. In order to achieve a uniform electrical breakdown of the gas in the entire volume, the gas volume is pre-ionised. UV radiation from spark discharges

Table 10.15 Overview of some pulsed lasers

Type	Active gas	Wavelength λ [nm]	Application examples
CO_2 TEA laser	CO_2	10,600	Micromachining, lettering
Titanium-sapphire laser	Xe flash lamp	670–1070	Scientific measurement technology
Nitrogen laser	N_2	337	Scientific measurements
Excimer laser	XeCl	308	Marking, medicine, microprocessing
-"-	KrF	248	Photolithography
-"-	ArF	193	Photolithography, ophthalmology
Fluor laser	F_2	157	Photolithography, ophthalmology

is used for this purpose, which are ignited synchronously. For special requirements, synchronously pulsed X-rays are also used for this purpose. Table 10.15 summarizes some information data on important gas discharge lasers.

The most commonly used discharge circuits are a simple C-C transfer circuit, a modified form of a spiker-sustainer circuit and a circuit named after its inventor (Blumlein) using saturable inductors or as a waveguide arrangement for N_2 lasers. The basic principle for all these circuits is a transfer of energy from a first to a second capacitor in a short time, with the discharge path being connected in parallel to the second capacitor. The transferred current is often used for pre-ionization by additional spark gaps.

Simple C-C Transfer Circuit

The simplest way of implementing a discharge circuit is the principle circuit shown in Fig. 10.270. First, the storage capacitor C_s is charged to the voltage U_{HV} via L_r and L_d by a charger with current source characteristics. When a switch is actuated, the storage capacitor C_s, charged to the required voltage, transfers its energy via L_s to a series-connected *peaking* capacitor C_p. The charging current of C_p flows through a larger number of parallel pins. A

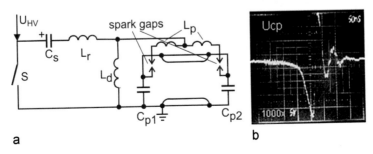

Fig. 10.270 Principle circuit of a pulsed laser with C-C transfer circuit. (**a**) Principle circuit, (**b**) voltage curve at the peaking capacitor C_p ($I_{peak.charge}$: 7kA, $I_{peak.discharge}$: 33kA, $U_{Cp.max} = 20kV$). (Source: Mallwitz 1994)

uniform current distribution is achieved by a small inductor L_p in series with each pin. The current through the pins with their spark gaps leads to UV radiation of the spark gaps. This ultraviolet radiation ionizes the space between the main electrodes evenly. As a result, when the breakdown voltage is reached sufficiently quickly, a uniform gas discharge occurs in the volume between the main electrodes. The discharge path of the laser is connected parallel to C_p.

All inductances shown are coreless. While L_d and L_p are coiled coils, L_s is the serial residual inductance of the construction. L_s is typically in a range of 80...120 nH due to its symmetrical construction and wide current leads.

In excimer lasers, the capacitors are recharged in a very short time (100...200 ns). At voltages in the kV range and recharging currents of several kA, extremely high current rise rates thus occur. When using the C-C transfer circuit variant, the switch must carry the full amount of current and voltage, so that the switch is subject to very high stresses, which can usually only be controlled by gas discharge switches with an acceptable service life. In the example in Fig. 10.270b, for example, the switch is loaded with an ideal switching capacity of 30 kV * 7 kA = 210 MVA at 500 kA/µs.

At the beginning of the transfer, C_p is discharged. The maximum current after closing the switch is

$$I_{max} = \frac{U_{HV}}{Z_w} = \frac{U_{HV}}{\sqrt{\frac{L_r}{\frac{C_s \cdot C_p}{C_s \cdot C_p}}}}$$

The reloading time of $C_s \rightarrow C_p$ is then

$$T_{transfer} = \pi \cdot \sqrt{L_r \frac{C_s \cdot C_p}{C_s + C_p}}$$

Theoretically, the maximum voltage of

$$U_{Cp.\,max} = \frac{2 \cdot U_{HV} \cdot C_s}{C_p + C_s}$$

is achieved at C_p. This means: With $C_s = C_p$, U_{cp} reaches the maximum charge voltage at C_s. For very small values of C_p, the maximum input voltage is doubled.

The maximum voltage slope at C_p is

$$\max\left(\frac{du_{Cp}}{dt}\right) = \frac{I_{\max}}{C_p} = U_{HV} \cdot \sqrt{\frac{C_s}{(C_p + C_s) \cdot C_p \cdot L_r}}$$

In the oscillogram shown in Fig. 10.270b, the gas discharge from C_p takes place via the extraordinarily low inductance of a wide discharge circuit with about 4...6 nH. Very short current pulses are thus achieved. Based on an estimate of the amplitudes of current and voltage in the gas discharge, the pumping capacity of the example shown is 660 MVA. Obviously only a small part of the power in the gas discharge is converted into active power. The discharge voltage is only weakly attenuated. For longer gas discharges, the circuit shown in Fig. 10.270a is modified so that $C_p \gg C_s$. C_p is then only used to ignite the gas discharge, while the discharge duration T_e is determined by L_r:

$$\frac{T_e}{2} \approx \sqrt{L_r \cdot C_s}$$

According to its function, this structure is called *spiker-sustainer circuit* and is used, for example, in CO_2-TEA lasers, long-pulse excimer lasers and solid-state lasers that are optically pumped through gas discharge tubes. Due to the low limit values of even a GTO thyristor compared to the current or voltage values of the discharge (see Table 10.16), direct application of this basic circuit with semiconductor switches at higher energies is therefore hardly possible. In addition to the high voltage load, the high current slopes are critical in semiconductor switches because of the low speeds of charge carriers in semiconductors. More recently, work on direct switching in these ultra-short times has been successfully carried out with unipolar switches (MOSFETs) [86].

Table 10.16 Comparison of gas discharge and semiconductor switches

Parameters/switch	Unit	H_2 thyratron	Thyristor	GTO thyristor
Type	–	LP189	D315CH32FD	H30K33YFH
Manufacturer	–	ITT (US)	Westcode (GB)	Meidensha (J)
U_{DRM}	kV	35	3.2	3.3
I_{FAV}	A	0.3	1065	300
I_{FRMS}	A	45	2100	500
di/dt_{crit}	kA/µs	$-(300...600)$	1	17 (21)
dv/dt_{crit}	kV/µs	–	0.2	1
t_{gd}	ns	5...10	–	3000
t_{rise}	ns	about 15...20	–	1500
t_q	µs	>50	100	40
P_{heat}	W	100...240	0	0

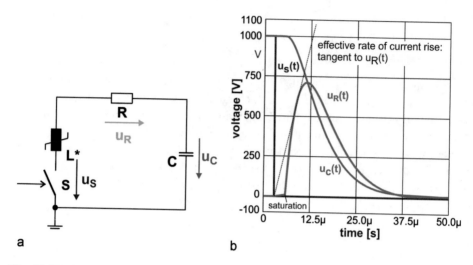

Fig. 10.271 Relief of a switch by saturable inductance in series ($U_c(0)$ = 1000 V, R = 1 Ω, L* = 300 μH/3 μH; Ψ_{SAT} = 3 mVs). (**a**) Circuit diagram, (**b**) voltage curves

Magnetic Pulse Compression

A similar circuit principle, which places less strain on the switch, works with saturable inductors. After applying a voltage, these reach saturation after a certain time. This means that they suddenly change from a high to a low inductance. If the ratio of inductance in the unsaturated state and in the saturated state is very high, this is the effect of a flux-controlled switch. In contrast to gas discharge and semiconductor switches, these switches are extremely insensitive to current and voltage gradients.

Figure 10.271 shows the simple way of relieving a switch by a saturable inductor connected in series. For the sake of clarity, a resistive load R is assumed as the load.

When the switch S closes, the voltage on it ideally collapses to 0 V. Because L* is in an unsaturated state, the current in the switch initially increases only with

$$\frac{di_s}{dt} = \frac{U_C(0)}{L(i_L = 0)}$$

If the flux linkage Ψ reaches the saturation value, the differential inductance of L* in the example decreases abruptly by a factor of 100. The steepness of the current rise thus increases by a factor of about 100.

The relief of the switch can be based on 2 mechanisms. The first variant concerns controlled bipolar components with a comparatively coarse structure. The transition from the blocked state to the conducting state is achieved by injection of charge carriers and emission of secondary charge carriers from an emitter zone in a solid state when this is reached by the "primary" charge carriers. Because of the low mobility of the charge carriers

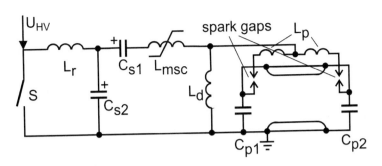

Fig. 10.272 C-C transfer - discharge circuit with dynamic voltage doubling according to Blumlein and with MSC as a supplement

in semiconductors (Si: 1400 cm²/(Vs) for electrons and 450 cm²/(Vs)) for holes, they reach only a low velocity. Therefore, it takes a certain time until the entire available semiconductor area is activated. If the current increases too quickly, it is concentrated on small areas. These are overheated and the component is usually destroyed. Thyristors show this property very distinctly and are therefore marked with a critical current slope $S_{i.crit}$ in the data sheet. However, in many pulsed applications (e.g. sterilization of food by repeating flash lamps emitting UV light), it is important that the current rise at the beginning of the pulse is very steep. This can be remedied by connecting the switch in series with a saturable choke. Until saturation, the current rise is kept low. An "effective" current slope is defined as a practicable method, which must be below the specified limit value. A tangent is applied to the current curve $i_R = u_R(t)/R$, which must pass through the switch-on time, as shown in Fig. 10.271b. This method is used in applications with a relatively low repetition rate (usually from "single shot operation" e.g. in *crow bars* up to several 10 Hz).

Bipolar components with more filigree control electrodes (GTO thyristors, IGCTs, IGBTs, etc.) usually limit the current themselves. Here, the current slope no longer leads to destruction. However, the "saturation voltage" is no longer a static value. It changes within a few µs from about ½...¼ of the switching voltage to its stationary value between 1...3 V. If you prevent a current increase during this time, you can drastically reduce the switch-on losses. The second possible relief effect thus relates to a reduction of the losses in the semiconductor switch. If the switch is decoupled from the main current by a saturable choke (*magnetic assist*) at the beginning of the switch-on phase, where it exhibits a high voltage drop, the heat converted in it is reduced dramatically.

The principle of the Blümlein circuit provides a further possibility of reducing the switch load (see Sect. 10.14). In the principle circuit shown in Fig. 10.272 (supplemented by a magnetic *switch circuit* MSC), energy transfer also takes place between capacitors. However, it has two storage capacitors (C_{S1} and C_{S2}), which are connected in parallel during charging, and a peaking capacitor. When the switch circuit is switched on, the capacitor C_{S2} is first recharged by oscillation in the oscillating circuit $L_r C_{S2}$. This means that there is a polarity change at C_{S2}. For the duration of this process, the MSC is in an unsaturated state, so that it is virtually current-impermeable. If C_{S2} is completely recharged, the MSC reaches saturation according to its design. Assuming that $L_{MSC}(0)$ is $\ll$ $L_{MSC}(I_{SAT})$, the energy transfer to the peaking capacitors C_p then begins. The storage capacitors C_{S1} and C_{S2} are now connected in series with the same polarity, so that twice the voltage is effective for charging C_p. This means a higher voltage and voltage rise rate over the discharge channel. For the same output parameters as in the previous variant, the load on the switch can thus be reduced (approximately halved), since only half of the storage charge has to be switched. A combination of this variant with a switching relief of S according to Fig. 10.271 is useful. With the increasing number of non-linear magnetic components, it is becoming more and more important to reproduce the initial conditions before a switching operation. When using modern amorphous and nanocrystalline ferromagnetic materials with Z-characteristics for the B(H) magnetization curve, however, this is possible with low control current or control power.

The realization of a discharge circuit with a semiconductor switch is usually not possible due to the high voltage levels in most pulse lasers. Both static and dynamic load characteristics do not allow a direct use in the concepts described above.

Charge Transformation and Subsequent Pulse Compression

Semiconductor switches are very attractive for many applications. The wear and tear caused by electrode burn-off in the thyratron represents a limiting factor for the service life in industrial use. Thyratrons require a heater for the glow emission of electrons from the cathode. In medical applications, for example, this means that the devices are not immediately ready for operation or permanent losses in the standby function. High repetition rates of the pulses also require components that can be switched off, which can only be realized by semiconductors. Figure 10.273 represents a solution approach for the use of semiconductor switches.

The circuit can be operated on the primary side with a simple C-C transfer as well as with a Blümlein circuit. To provide the required breakdown voltage of the gas discharge, a transformer is included in the transfer circuit, so that when S_1 is switched on, a transfer from C_1 to C_2 takes place via this transformer. The required voltage rise time is achieved on the secondary side with a circuit called "magnetic pulse compressor" with saturable

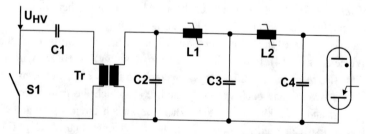

Fig. 10.273 Basic structure of a pulse compressor circuit with transformer

inductors. If the first capacitor is C_1, charged, L_1 saturates and the recharging to C_2 takes place. By suitable dimensioning of the saturable choke L_x, a reduction of the voltage rise time can be achieved for each such stage. The pre-ionization of the gas discharge path was only hinted at in Fig. 10.273, since it only plays a marginal role here.

One or more of these can be used depending on requirements. When saturation of an inductor with L_{SAT} is reached, the energy of the charged capacitor is transferred to the following capacitor. The transfer time is then C_1 and C_2 in the example

$$T_{transfer} = \pi \cdot \sqrt{L_{SAT1} \frac{C_1 \cdot C_2}{C_1 + C_2}}$$

The saturable choke L_2 of the following stage takes up a voltage-time area/flux linkage of

$$\Psi = \int u_L(t) \cdot dt = \frac{T_{transfer} \cdot U_{C2.\,max}}{2}$$

Since a certain capacitance is required for the capacitors, the reversal time and thus also the rise time depends mainly on the residual inductance of the magnetic switch. The requirements for an MSC in this circuit variant can be derived from these relationships:

- high inductance in unsaturated state
- low inductance in the saturated state,
- limited volume

A disadvantage of the transformer-pulse compressor circuit is the losses that occur during each transient process, especially at very high voltages and speeds of remagnetization. The implementation of the transformer is not unproblematic. For the very short pulse durations occurring here, transformation is only possible if the leakage inductance of the transformer is extremely low. A short transmission time keeps the number of switching chokes low. The same applies to the losses occurring in connection with them in the pulse compressor stages.

Marx Generator with Saturable Inductors

High voltages can also be generated - as usually used in surge voltage generators - even without a transformer. Figure 10.274 shows a circuit of a three-stage Marx generator (named after Erwin Otto Marx 1893–1980). The capacitors $C_{1...4}$, charged in parallel with a charging voltage U_L, are connected in series with the same polarity after simultaneous response of switches $S_{1...3}$. Their individual voltages add up, so that the charging voltage U_L is multiplied by the number of capacitors. With the Marx generator shown in Fig. 10.274, a tripling of the charging voltage is thus possible. The required voltage curve can be achieved at the connected gas discharge unit.

Fig. 10.274 Marx generator for
pulses with 3 switches and
4 capacitive storages

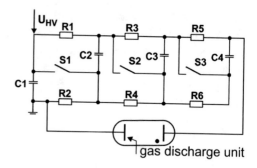

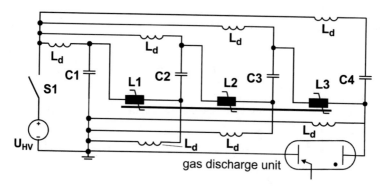

Fig. 10.275 Marx generator with saturable inductors as switches [87]

Usually a triggered spark gap is used as a switch for S_1, while the other switches are designed as ball spark gaps. The latter ignite when S_1 is triggered by exceeding their dielectric strength. Spark gaps are subject to electrode wear and require a certain grace period to re-solidify the insulating gap. This greatly limits the possible repetition rate.

Rothe/Lantis [88] proposed a variant with saturable inductors (MSC). Only one core is used for all inductors, which considerably reduces the effort required for circuit design. Furthermore, a largely synchronous saturation of all core areas can be assumed. This results in a design for this variant as shown in Fig. 10.275. If the MSCs are designed accordingly, the service life and switching behavior can be improved in contrast to spark gaps [86].

In the circuit shown in Fig. 10.275, the capacitors are charged with an appropriately adapted resonance charge. When the core sections of the common toroidal core reach saturation, $L_{1\ldots3}$ lose most of their inductance, so that the capacitors $C_{1\ldots4}$ are connected in series and supply the gas discharge with high voltage. This construction has several basic problems. First of all, the tuning and dielectric strength of the coreless coils for charging is problematic, because on the one hand synchronous charging of the capacitors is required and on the other hand the voltage load of the coils is very different, so that a similar structure is out of the question.

Fig. 10.276 Partial charge inversion for voltage doubling (The polarities shown are before charge inversion at C_2)

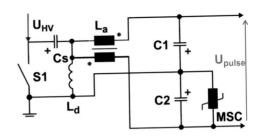

In addition, the design of the MSC can lead to problems. To accommodate the windings with high dielectric strength against each other, a suitably large core should be used, which has the necessary flux linkage. The problem for uniform switching is a symmetrical distribution of the windings around the core, taking into account the voltage distribution. When the core is "switched through", all 3 residual inductances of the windings $L_{1...3}$ are in series. This makes ultrashort pulses directly in the gas discharge practically impossible. At least one further stage for pulse compression would have to be connected downstream.

Use of a Partial Charge Inversion of Capacitors Connected in Series
In addition to the version of the Blumlein generator (Fig. 10.272) for voltage doubling, there is another basic variant that serves this purpose. It is the combination of a magnetic switch (MSC) for charge inversion with an asymmetrical but linear magnetic component. The principle is shown in Fig. 10.276.

C_S is first charged via a coreless coil L_d. When S is closed, energy is transferred from C_S to $C_1//C_2$. L_a consists of 2 closely coupled windings. Due to the selected winding sense, only the leakage inductance between the windings is effective for the transfer. If the MSC goes into saturation and thus starts the charge inversion, a "discharge inductance" L_a* acts between the capacitors, which is approximately 4 times the inductance of one winding. After the inversion, taking into account the losses in the MSC, one obtains the maximum output voltage $U_{HV2} \approx 2 \cdot U_{C1. max}$. A further energy transfer to a load can now take place.

Charge Transformation and with Subsequent Partial LC Inversion
Figure 10.277 shows the basic structure presented in [87]. The charged capacitor C_S discharges via the transformer when the switch S is closed and transfers its energy to the secondary-side capacitors $C_{1...4}$. The magnetically coupled chokes L_a and L_b act as current-compensated chokes during charge transfer, so that they counteract the charge with only a low inductance. C_1 and C_2 as well as C_3 and C_4 are parallel in pairs and have the polarity shown in Fig. 10.277. When the energy transfer is complete, the saturable chokes MSC_1 and MSC_2 cause the capacitors C_1 and C_3 to be recharged. The coupled chokes L_a and L_b act with almost 4 times the self-inductance of a single winding due to their winding direction and are therefore open switches for the time of charge inversion at C_1 and C_3. Thus, finally all capacitors with the same direction of the voltage drops are in series. The

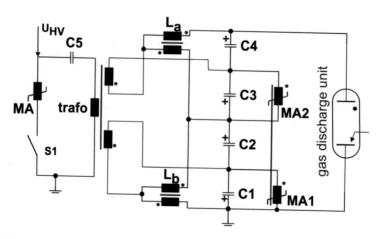

Fig. 10.277 Pulse generation by energy transformation and partial charge transformation according to [87] (polarity of the capacitors drawn before the charge inversion described in the text)

result is a multiplication of the voltage at the output in spite of a relatively low transformation ratio of the transformer.

The level of the output voltage is determined both by the number of secondary-side capacitors/windings and the transformer transformation ratio. The voltage pulse can be adapted to the application with a subsequent circuit. In series with the capacitors, there is no principle-conditional inductive component. Therefore, even very steep current pulses are possible at the output.

The saturable inductance MA serves as switch-on relief for the switch S to delay the current and thus reduce the switch-on losses. Very high voltages can be generated by an appropriate transformation ratio of the transformer or by parallel connection of several secondary windings with the same subsequent circuit design.

For the design according to Fig. 10.277 with a transformer transformation ratio of 1:1.4, 150 kV was achieved with an input voltage of 25 kV and a thyratron as switch [87]. A characteristic feature of this design is that it contains no active components apart from the switch. The distribution of the high voltages in the secondary circuit is relatively easy to control by distributing them over several inductive components. The design is therefore robust, but also has a large volume. Figure 10.278 shows a much more compact version using semiconductor diodes. Since these are flooded with charge carriers during the transfer of energy from C_5 to the secondary side, they may have to be assisted by a pre-magnetized choke in series for blocking.

The diodes D_1 and D_2 take over the functions of L_a and L_b in Fig. 10.277. In this circuit, the transformer itself takes over the function of the magnetic switching elements. The core of the transformer is tuned so that it goes into saturation after reaching the voltage maximum at the secondary capacitors. Via the residual inductance of the windings, the

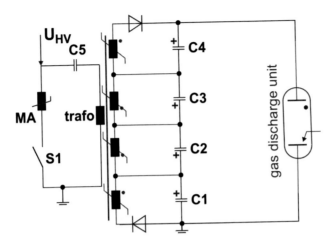

Fig. 10.278 Pulse generation by energy transformation and partial charge transformation with semiconductor elements on the secondary side

capacitors C_2 and C_3 invert their charge/sign. The diodes maintain the polarity at C_1 and C_4. As a result, all capacitors are connected in series with the same polarity. However, the diodes must be designed for relatively high pulse currents and at least half the maximum output voltage. The series connection of several capacitors always results in a relatively large inductance for energy transfer and may require subsequent pulse compression.

10.27 Pulse Laser with Semiconductor Switch

Pulsed laser applications are widespread: medicine, circuit manufacturing, mechanical engineering, packaging industry, etc. The activation of the optical amplifiers must be done in a short time. In general, the shorter the wavelength, the higher the necessary activation energy and the shorter the lifetime of the metastable energy levels. The wavelength range extends from the CO_2-TEA laser (10.6 μm) in the mid-infrared range to the F^{2+} laser (156 nm) in the short-wave UV range. Also, systems for the extreme UV range (EUV <50 nm) use the same excitation systems without representing laser systems themselves. Industrial systems require highly repeating solutions with long lifetime. Critical are the switches. Until the mid-1990s, only the use of gas discharge switches was possible. However, these switches wear out relatively quickly. Durable systems require semiconductor switches. Pulsed gas discharge is usually used to generate high-intensity UV radiation. Otherwise, the high power densities for excitation of UV emission can hardly be achieved. The gas discharge itself draws its energy from a capacitor (peaking capacitor), which must be charged sufficiently quickly to allow a uniform volume discharge to form in the pre-ionized gas between the discharge bars. The problem lies in the rapid charging of the peaking capacitors. If a switch is to be used in a CC transfer circuit, for example, the complete charging current flows through this switch. Assuming a storage capacitor C_s and a peaking capacitor C_p with a inductance L_r, this leads to a current pulse duration

$$t_p \approx \pi \sqrt{L_r \cdot \frac{C_s \cdot C_p}{C_s + C_p}}$$

with a voltage slope of

$$\max \left(\frac{du_{Cp}}{dt} \right) = \frac{I_{max}}{C_p} = U_{HV} \cdot \frac{U_{HV}}{\sqrt{L_r \frac{C_p}{C_s} \left(C_p + C_s \right)}}.$$

The maximum current during a discharge in resonance mode is

$$I_{max} = \frac{U_{HV}}{Z_w} = U_{HV} \sqrt{\frac{C_s \cdot C_p}{L_r \cdot (C_s + C_p)}} \text{with max} \left(\frac{di_{Lr}}{dt} \right) = \frac{U_{HV}}{L_r}$$

The magnitude of C_p is therefore decisive for the voltage rise. This is used in the *spiker-sustainer concept*. A semiconductor switch is to be used as a switch. This has the following advantages, especially for higher repetition rates [87]:

- Lifetimes of semiconductor switches are considerably higher than those of gas discharge switches when correctly dimensioned,
- Semiconductor switches that can be switched off enable high repetition rates up to the kHz range,
- No warm-up time is required, so that operational readiness can be achieved very quickly and
- No stand-by losses due to heating.

In the example described below [89], a GTO thyristor H30K33YFH from MEIDENSHA/Japan is used. Due to the fine gate structure, this switch has a critical current slope of 21 kA/μs at maximum blocking voltages of 3.3 kV. The concept shown in Fig. 10.279 was realized when setting up a test pattern for the investigation of a semiconductor-driven pulse forming network (PFN) [89]. It allows the switching of only a part of the total charge to be transferred and the individual optimization of each magnetic device. VC 6030F was used as the core material for the transformer, while materials with a Z-loop shape for the magnetization characteristic curve were used for the nonlinear magnetic components.

Figure 10.279a shows a dynamic voltage doubler according to Blumlein as input stage for pulse generation, which is fed from a source with a maximum of 3 kV. The magnetic switching choke MSC1 switches the output of the Blumlein generator to a transformer. This transforms voltage U_2 upwards by a factor of 6 and charges a capacitor C_{s3} as a buffer.

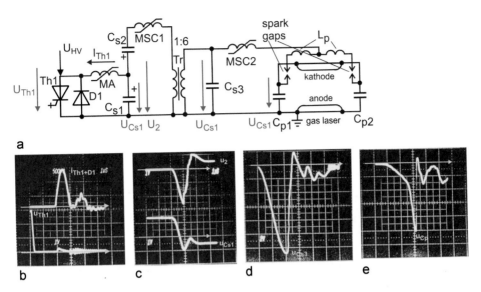

Fig. 10.279 Circuit of an implemented semiconductor switched pulser for a gas discharge (**a**), switch current (GTO + diode) 1.5 kA/T and (**b**) switch voltage 1 kV/T, (**c**)voltages at C_{s1} and $C_{s1} + C_{s2}$ (2 kV/T), (**d**) voltage at the buffer memory C_{s3} (2 kV/T, 200 ns/T) and (**e**) voltage at the peaking capacitor C_p (2 kV/T, 200 ns/T). (Source: Zacharias [87] and Mallwitz [89])

When MSC2 is switched through, the energy stored in C_{s3} is transferred to the peaking capacitor C_p, which consists of several capacitors connected in parallel to the discharge electrodes. During charging, the current flows through the spark gaps and leads to UV emission from them. This UV emission synchronously causes the pre-ionization of the electrode gap between the main electrodes. Figure 10.279b...e show the voltage curves for currents and voltages corresponding to this function.

Figure 10.279b shows current and voltage at the pair $Th_1 + D_1$. The conductive phase of the GTO thyristor Th_1 and the subsequent conductive phase of diode D_1 are clearly visible. The thyristor current starts with a delay after the switch-on time of the saturating choke MA (*magnetic assist*). As a result, almost the entire GTO area can be activated by the impressed gate current before the anode current rises. The voltage drop across the thyristor in forward direction then still reaches about 200 V. Without MA, approximately 800 V is measured. The switching relief MA thus achieves a significant reduction in losses. (This switching behaviour can also be observed with other bipolar components such as thyristors, IGBT's and bipolar transistors.) After switching on, the voltage at C_{s1} reverses and adds to the voltage drop U_{Cs2} (Fig. 10.279c). MSC1 is dimensioned so that it goes into saturation near the voltage maximum. This allows the charge of C_{s1} and C_{s2} to be transferred to the buffer C_{s3} (Fig. 10.279d). This is then connected via MSC2, which is present in the basic structure of the laser anyway. When the latter switches through, the charge transfer leads to a very rapid voltage increase at C_p (Fig. 10.279e). The structure from the switch to the buffer memory C_{s3} is critical with regard to the maximum acceptable inductances because of the

Table 10.17 Characteristic values of the transformer used Tr

Toroidal core	VC 6030F (141 × 83 × 26)
Ψ_{max}	312.5 µVs
Quantity	2
A_L	2.83 µH/Wdg.^{2}
N_1	5 (6× parallel)
N_2	30
HV cable	RG 58 C/U
$L_{\sigma1}$	38 nH

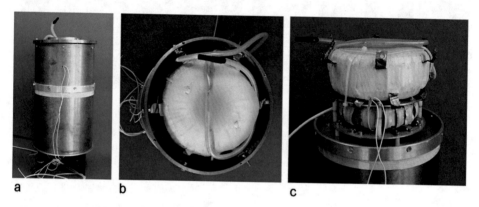

a   b   c

Fig. 10.280 Structure of the semiconductor pulser with saturable magnetic elements for pulse compression. (**a**) Overall structure, (**b**) top view of the transformer with HV connection, (**c**) partial view with removed coaxial outer conductor: transformer on top and MSC1 below for the low voltage side (Source: Zacharias [87] and Mallwitz [89])

comparatively low voltage level. Particularly in the case of the transformer, the problem is to ensure a very close coupling of the primary and secondary windings on the one hand and to guarantee high-voltage insulation on the other. In the present case, the transformer was constructed from two toroidal cores, with the windings being formed from a coaxial high-voltage cable. 30 turns of the (uninterrupted) inner conductor are used as the high-voltage winding. The shield conductor is cut after every 5 turns used as primary winding. All 6 partial windings are connected in parallel (Table 10.17). In this way, a low leakage inductance is achieved for the transformer construction. The overall construction up to the connection of C_{s3} is coaxial (Fig. 10.280).

The components of the pulser from Th$_1$ to transformer Tr were coaxially mounted in a brass tube (Fig. 10.280). The lines for supplying the premagnetising current are led out at the side. Transformer Tr is the uppermost assembly part in the tube, which is simultaneously the housing and return conductor. The tube is divided into 2 sections for assembly. The upper section contains the magnetic switch MSC2 and the transformer. Both consist of toroidal cores made of amorphous material. Due to the high voltages and the low number of windings, the requirements on the insulation of the tape layers of the core against each other

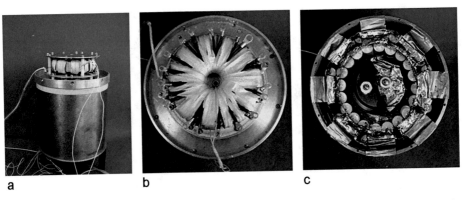

Fig. 10.281 Structure of the semiconductor pulser with saturable magnetic elements for pulse compression. (**a**) Lower part with MSC2, (**b**) top view MSC2, (**c**) C_{s1} (ring of film capacitors outside) and C_{s2} (MKV capacitor inside). (Source: Zacharias [87] and Mallwitz [89])

are relatively high. The induced voltage when circulating around the core is 600 V and must be safely insulated. In the present case, a simple oxide layer, which is otherwise used for this purpose, is not sufficient, so that a Hostaphan layer was inserted when winding the ferromagnetic amorphous tape. In order to ensure that all connections are adjacent to each other at the same potential, the 3-turn winding is composed of several partial windings with alternating clockwise and counterclockwise winding directions. This technique was also used for MSC1 and MSC2.

Figure 10.281 shows the further structure including the capacitors C_{s1} and C_{s2} (Fig. 10.281c). The capacitor C_{s1} is composed of several low-inductance MPP capacitors. C_{s2} consists of a single MKV capacitor. Figure 10.282a shows the structure of the lower range without phase conductor. The switching reactor MA is located directly below the capacitors. Here too, the lines for bias magnetization are led out at the side. The inner

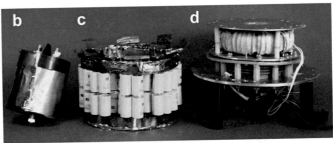

Fig. 10.282 Structure of the semiconductor pulser with saturable magnetic elements for pulse compression. (**a**) Lower structure without outer conductor, (**b**) C_{s2}, (**c**) C_{s1}, (**d**) semiconductor device + *magnetic assist* MA + driver under the mounting plate. (Source: Zacharias [87] and Mallwitz [89])

a   b   c

Fig. 10.283 Structure of the semiconductor pulser with saturable magnetic elements for pulse compression. (**a**), (**b**) Mounting platform for the semiconductor switches Th1 and D1 (formed by two diodes in series), (**c**) bottom side of the mounting platform with driver and pulse transformer for gate control. (Source: Zacharias [87] and Mallwitz [89])

structure is enclosed by a current transformer for measuring the current distributed over Th_1 and D_1. Thyristor Th_1 and D_1 are placed on the lowest mounting plate (Fig. 10.282d). The disk designs of the semiconductors require considerable pressing forces. Only these forces ensure sufficiently low contact resistances even inside the disc elements. Therefore, clamping bolts are arranged around the components. The triggering signal is supplied by a pulse transformer below the mounting plate. In contrast to normal thyristors, the gate impedance of GTO thyristors is very low. In the present case, a gate current of 240 A was required for a fast switch-on to ensure the critical current slope of 21 kA/µs in the main circuit.

In order to ensure a uniform activation of the large surface of the GTO thyristor, it has 2 gate connections. Due to short parallel gate connections, a low inductance is achieved and high values for the current slope di_g/dt. To generate the high gate current with a high slope, a driver unit has been developed in which the pulse is generated on a voltage level of 300 V (Fig. 10.283). The ignition pulse is then transformed by the ignition pulse transformer 15:1 to the level of the GTO thyristor.

The dynamic characteristics of the semiconductor switch resulted in a maximum height of 8 kA for the current pulse on the 3 kV level with a pulse duration of 2 µs. The gas discharge with synchronized pre-ionization required an approximately 10× higher (no-load) voltage and a pulse duration that was 10 times shorter. The adaptation was achieved by the transformer combined with the magnetic pulse compression.

Figure 10.284 shows another tested approach to voltage adjustment and pulse compression for a lower operating voltage (max. 15 kV). Here the transformer is part of the chain for pulse compression. A saturable transformer was used. All nonlinear magnetic components were realized from toroidal cores made of amorphous alloys. A further special feature in addition to the saturable transformer is a component, which will be referred to in the following as a "magnetic diode"(MD). This is a switching choke that is pre-saturated in

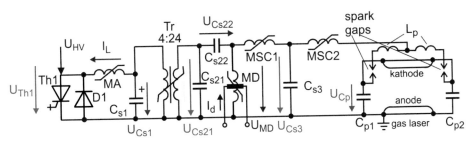

Fig. 10.284 Circuit arrangement of a semiconductor pulser with saturable transformer Tr as switch and a saturated choke MD as "magnetic pulse diode"

one direction by an external magnetizing current. When a current pulse in the saturation direction hits this component, only the residual inductance L_{sat} of this component is effective. In "reverse direction", the high inductance of the unsaturated state is effective, so the effect is like that of a diode. With a diode, however, high losses would have to be expected due to the carrier storage effect combined with high overvoltages.

When Th1 is ignited, energy is transferred from C_{s1} to $(C_{s21} + C_{s22})$. During this transfer, MD works in "forward direction". At the end of the transfer, the transformer core goes into saturation and the sign of the voltage at C_{s21} is reversed. This causes the voltage at MSC1 to rise until it saturates near the maximum. MD is polarized in the "blocking direction" as long as it is not saturated in the "blocking direction". When MSC1 saturates, an energy transfer $(C_{s21} + C_{s22}) \rightarrow C_{s3}$ starts. MSC2 is designed so that it saturates in the vicinity of the voltage maximum at C_{s3} and thus switches through. This starts the energy transfer to the peaking capacitors. Figure 10.285 shows significant current and voltage curves for the undervoltage and overvoltage side as examples. It becomes clear that the pulses become shorter and shorter with each transfer process. Some energy is lost during each transfer process - mainly due to the rapid remagnetization of the cores and the high current amplitudes in the windings of the inductive components.

Figure 10.286 shows the mechanical structure of the described pulser. A very compact design is possible. The only active element is the semiconductor module consisting of a thyristor with a low recovery time and an antiparallel diode. The energy transfer in several stages is carried out according to a flow chart, which is determined by the saturable chokes. Control is possible within narrow limits via the voltage U_{HV}. Although the initial conditions for all switch cores are established via external premagnetisation, this current is not used for control. Although this is theoretically possible, it has not proved practicable due to large tolerances. The current is used to "reset" the cores to the initial conditions after each pulse. The power loss depends on the repetition rate of the pulses and is dissipated from the semiconductor elements and the magnetic components by forced air cooling.

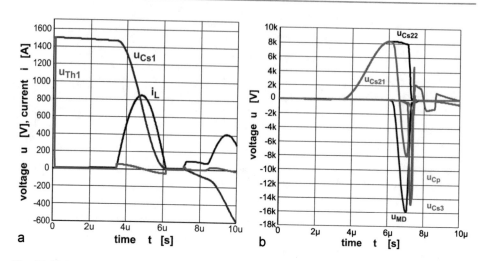

Fig. 10.285 Example of current and voltage curves on the pulser. (**a**) LV side, (**b**) magnetic pulse compression on the HV side at an ignition voltage of the laser of approximately 14 kV [87]

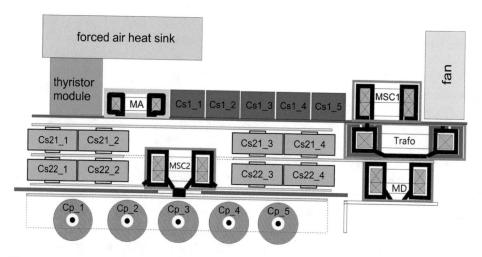

Fig. 10.286 Mechanical construction of the pulser according to Fig. 10.285 [87] (side view, gas discharge is located behind the peaking capacitors C_{px})

10.28 Magnetic Amplifier

In Sect. 5.6, the possibility of controlled modification of the characteristic curves of magnetic branches is already mentioned (see also [90]). The magnetic component based on this is called a magnetic amplifier or transductor. This subchapter provides an overview

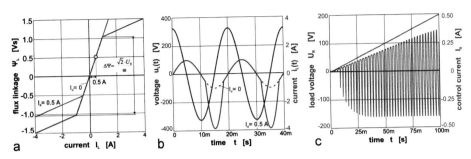

Fig. 10.287 Calculated current and voltage ratios of a loss-free non-linear inductor with and without DC bias. (**a**) From the viewpoint of the phase plane with the magnetization characteristic shown in red, (**b**) voltage and current at the inductor

of basic design approaches and applications. In the simplest case, the magnetic amplifier consists of a ferromagnetic core with a winding for alternating current, which is to be controlled, and for direct current, which serves as a control variable. In this case, the alternating flux is superimposed on the direct flux (Fig. 10.287). The magnetization characteristic B(H) remains the same for the sum of both fluxes (Fig. 10.287a). If only the alternating flux is considered, a shift occurs on the Θ axis (Fig. 10.287b). This leads to the core going into saturation at ever smaller AC magnetic forces, with the current in the AC winding rising rapidly. In the steady state, the average voltage drop across the AC winding is zero. If an ideal voltage source is loaded with an ideal non-linear, loss-free inductive component, the operating point on the magnetization curve would change due to the asymmetry of the magnetization curve until the following applies within one period of the voltage source

$$u_0(t) = L(i) \cdot \frac{di}{dt}$$

Figure 10.287b shows the simulation results (for the used model see Sect. 5.6) for an inductance connected to a sinusoidal voltage source 230 V/50 Hz with the magnetization curve shown in Fig. Figure 10.287a in red without DC bias and with 0.5 A. The shift of the operating point in the coordinate system $\Psi_L(i_L)$ is clearly visible. Since the change in the flux linkage remains the same, the result is a change in the curve shape of the current, as shown in Fig. 10.287b.

You can clearly see the partial "rectifying" effect, which changes the spectrum of the harmonics. This effect can be used for control purposes. One obtains a non-linear reactance that can be controlled by direct current. Since the cathode current of vacuum tubes is limited by glow emission, controlled mercury vapour rectifiers (thyratrons, mutators) and magnetic amplifiers were used in the past as control elements in the higher power range. Compared to the aforementioned, the latter are characterized by extreme robustness, applicability even in the higher frequency range and a comparatively high efficiency.

Applications have resulted, for example, in the modulation of long-wave transmitters, e.g. for oceans, for communication between continents and with ships, the control of submarines, torpedoes and rockets up to the more recent time range. Due to their mechanical-thermal robustness and resistance to many disturbing influences, magnetic amplifiers are still used in military technology and aerospace [91–95]. The main disadvantage is the weight and comparatively high space requirement. If one considers that inductive components become smaller and smaller with increasing frequency, new areas of application are again foreseeable. An advantage is that these components have no threshold voltage, as is the case with many transistors and diodes used in semiconductor technology. They can therefore also be used for low voltages and high currents. In recent decades, miniature and subminiature designs of magnetic amplifiers have also been developed.

In addition to higher operating frequencies, other aspects may also make magnetic amplifiers interesting for developers in the future. The service life and critical failure situations of magnetic amplifiers are determined by completely different factors than those of semiconductor components. For example, the sensitivity to overload is much lower than with semiconductor elements. In addition, unlike semiconductor components, magnetic amplifiers practically do not fail due to the ubiquitous cosmic radiation. This aspect becomes interesting when extremely high life expectancies are involved, such as in electrical supply engineering (20...50 years) for applications and/or these applications are used at several 1000 m above sea level. It is therefore worthwhile to summarize the main features of this technology, which has largely been forgotten, in this subchapter. Especially with Chapters 5–7 of this book, it is possible to simulate non-linear magnetic circuits in a simple way, which should make the application of this in detail complex technique easier.

The simplest structure for a magnetic amplifier, which is also the basis for the illustration in Fig. 10.287, is shown in Fig. 10.288. It is one winding of a transformer with 2 windings. One of them is used as controllable reactance in an AC load circuit. The other is used for control by means of direct current. Figure 10.288c shows the simulated effect of a

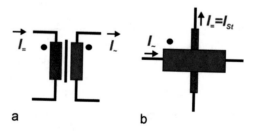

Fig. 10.288 Simple current-controlling magnetic amplifier as a pre-magnetized transformer. (**a**) Winding arrangement for direct current as control current and alternating current as current to be controlled on a common core, (**b**) schematic diagram, (**c**) control behaviour for toroidal core made of M400-50A with the same number of turns for a source 115 V/400 Hz and a load resistance 320 Ω at $N_\sim = N_=$

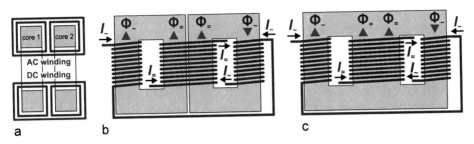

Fig. 10.289 Basic structure of a symmetrical current-controlling magnetic amplifier with common control winding. (**a**) From two toroidal cores, (**b**) from 2 separate 2xUU cores, (**c**) on a common core from two EE cores

partially saturated core with smooth DC current. Since the purpose of this is to compensate for magnetic force, the amount of DC current required is reduced by increasing the turns of the DC winding. To come close to the ideal "smooth DC current", a choke can be inserted in the control branch. The voltage induction in the arrangement acts in both directions. Supplying the DC winding from a DC voltage source with low impedance would correspond to the short circuit of a transformer, losing controllability. How large the impedance in the control circuit must be depends on the other parameters of the circuit and the application target. The required DC current generally becomes smaller, the greater the permeability of the core material in the unsaturated region and the shorter the path length of the core to be magnetized. As values of $\mu r = 20,000...100000$ are achieved with modern amorphous and nanocrystalline core materials, it is noticeable that the voltage or current form of such a simple magnetic amplifier is asymmetrical with respect to the t-axis. In the lower dynamic range, a "rectifying" effect can even be observed. This is unacceptable for many loads. Nevertheless, this principle can be found in suitable applications. It should be noted that a very similar effect is also observed with transformers with a single-pulse rectification at the output. The core is then pre-magnetized with direct current from the secondary side and tends to saturate earlier with a corresponding increase in self-heating.

The disadvantages of the simple current-controlling magnetic amplifier with regard to the unbalanced current characteristic can be avoided by designing the magnetic amplifier itself symmetrically (Fig. 10.289). The two AC windings can be connected either in parallel or in series. Figure 10.3c shows a design variant, which has all windings on one core, thus making it a very compact component. Air gaps to be overcome by the magnetic fluxes should be avoided as they worsen the transmission behaviour of the magnetic amplifier. In the case of laminated cores, for example, they must be alternately layered.

The winding connection and control behavior of such a current-controlling symmetrical magnetic amplifier is shown in Fig. 10.290. The same cores as for Fig. 10.288 were used as a basis. The resulting current curves at a resistive load are now symmetrical. As a special feature, it should be emphasized that the "amplitude modulation" causes only very slight distortions of the sinusoidal voltage of the source U_0.

Figure 10.291 shows an interesting application resulting from this. In line-frequency applications, a converter is often required, which draws a current with low harmonic

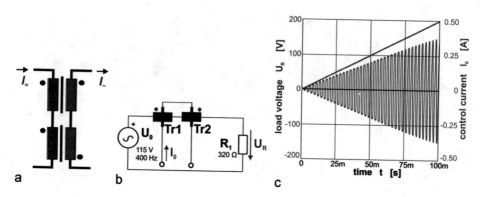

Fig. 10.290 Symmetrical current-controlling magnetic amplifier. (**a**) Winding circuit, (**b**) principle circuit, (**c**) control behaviour (simulated)

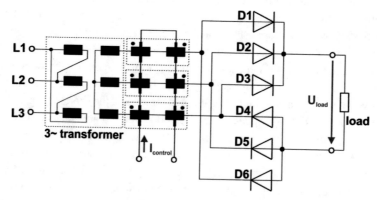

Fig. 10.291 Magnetic amplifier for current control for a downstream uncontrolled rectifier [96]

content from the mains as a voltage source, rectifies it and supplies a load with low impedance in a controllable manner. Thyristor-controlled power converters operate with phase-angle control and cause considerable harmonics in the grid, which then have to be filtered out, e.g. by suction circuits, in order to meet the conditions for a grid connection. The circuit arrangement shown in Fig. 10.291 avoids these problems to a large extent by connecting a balanced current-controlling magnetic amplifier upstream of a 3-phase bridge rectifier. The magnetic amplifiers connected in this way act in a similar way on the output side of similarly controllable alternating current sources when the current is sufficiently high. The current characteristic is largely sinusoidal. If the current is sufficiently high, the diodes always conduct 120°, so that a choke on the output side is largely unnecessary. A capacitor can be used to smooth the load current. The circuitry is based on a 6-pulse rectifier with an output current that can be controlled via the commutation inductors. A low-power DC/DC controller is required for control. The circuit is conditionally short-

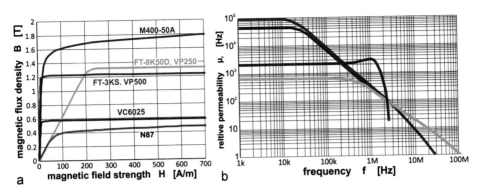

Fig. 10.292 Comparison of selected materials with regard to applications in magnetic amplifiers (grain-oriented sheet M400-50A, amorphous cobalt-containing material VC 6025 and nanocrystalline cobalt-free material. (**a**) Magnetization characteristics, (**b**) frequency dependence of the small-signal permeability [97–100])

circuit proof and extremely robust with a long expected useful lifetime. The weight might be a problem. But also at higher frequencies, interesting applications can be found here.

The previous presentations in this sub-chapter are based on the assumption of a grain-oriented sheet for the core material (M400-50A). Although this material has a high permeability in the unsaturated region of the B(H) curve, it tends to "creep" into saturation at high magnetic field strengths, as shown in Fig. 10.292. This is one of the reasons why it is not suitable for applications far above the mains frequency. Other materials have been developed for this purpose. As amorphous and nanocrystalline materials, which are produced as very thin metal strips with comparatively high specific resistance, materials for the higher frequency range are also available (Fig. 10.292). They are also characterized by a distinct "saturation kink", so that the polarization of the material above the magnetic field strength for saturation practically no longer increases. The material "switches" from the unsaturated to the saturated state. This is associated with a jump in the impedance of the corresponding magnetic component, which is particularly desirable with the voltage-controlling magnetic amplifiers described below. With the development of cobalt-free alloys with saturation induction around 1.2 T, cheaper materials than the cobalt-containing alloys are available. The magnetic properties are maintained up to a high frequency range. The drop in the measured permeability with frequency is mainly due to the effect of the induced eddy currents. The shape of the magnetization characteristic curve can be influenced by heat treatment under magnetic field. The Z-shape is particularly useful for applications in magnetic amplifiers. Although ferrites (here example N87) are suitable for HF, they do not have a pronounced saturation bend in the characteristic curve with only a low, temperature-dependent saturation induction (<0.5 T) and comparatively high coercivity. They are reserved for special circuits with controllable magnetic components. At high application frequencies, arrangements made of toroidal cores are preferred. At the transition from the saturated to the unsaturated state, this takes place first at the inner radius

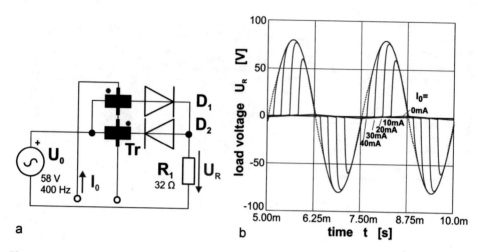

Fig. 10.293 AC power controller consisting of two voltage-controlling magnetic amplifiers. (**a**) Principle circuit, (**b**) control of the effective voltage via phase control

and then continues to the outer radius. The magnetic field strength is constant at a constant radius. This is not the case if the courses of the field lines with and without the core differ greatly. Switching losses are also lower with the shape of toroidal cores than with CC cores, for example.

If the two controlled simple magnetic amplifiers in Fig. 10.290 are operated in parallel and each is provided with a diode, a different category of magnetic amplifier is obtained (Fig. 10.293). These are called voltage-controlling magnetic amplifiers, which contain a rectifying element. This is an uncontrolled diode. By connecting the branches in parallel, positive and negative half-waves are influenced separately. This results in the connection of a single-phase AC power controller (Fig. 10.293a), which requires a symmetrical control variable. The control windings can therefore be connected in series.

Here too, the magnetization curve is shifted by the controlling direct current. In the unsaturated case, the core takes up a voltage-time surface that is changed by this, which is lost from the point of view of the load for the energy supply. This results in a control behavior similar to phase angle control with thyristors (Fig. 10.293b). Depending on the control current, a part of the sine wave can be cut off and thus the power converted in the load can be controlled. The same cores with the same magnetization length and the same saturation flux as in Fig. 10.290 were used for the calculations. The magnetization area was slightly increased with the same number of turns because of the material VP500 assumed here. It is noticeable that the control effect is achieved with a much lower current. The power amplification is much greater with voltage-controlling magnetic amplifiers than with current-controlling ones.

The high sensitivity to the control variable is once again clearly visible in Fig. 10.294. Here the output voltage and control current are used with the same number of turns as in

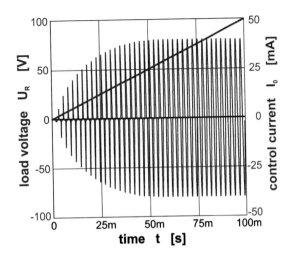

Fig. 10.294 Controlling the voltage at a resistor as shown in Fig. 10.293 with an AC power controller consisting of voltage-controlling magnetic amplifiers

Fig. 10.290. Since each core only acts in one current direction, the applied voltage U_0 was halved. The circuit works exactly like a thyristor-controlled AC power controller. In the mains frequency range, these transducers are no longer used - with some exceptions - because of the robust thyristors available. In the higher frequency range, however, applications are quite conceivable. With voltage-controlling magnetic amplifiers, good switching behaviour of the cores is important. Ideally, in the unsaturated case, the inductance should be towards infinity, whereas in the saturated case, it should be with the inductance without core. Reality deviates significantly from this. Nevertheless, this concept can be used to realize power converters with interesting properties.

In the previous considerations and calculations, the loss characteristics of the core were disregarded. For a given operating point, these losses can be represented by a (non-linear) resistor parallel to the non-linear inductance, which represents the proportion by polarization. Since the problem is of secondary importance and the resistance is an approximation, which can be quickly determined by experiments, it will not be discussed further here.

Figure 10.295 shows typical voltage setting configurations with voltage-controlling magnetic amplifiers. The starting point is Fig. 10.295a, which finds its modern equivalent in the thyristor in Fig. 10.295d. Although the control behavior is very similar, it is not the same. Although the transistor is controlled by a current, the switching time occurs differently depending on the effective applied voltage. This is due to the fact that current control causes a shift of the operating point on the $\Psi(I)$ characteristic. At higher voltages, the "blocking time" of the component is shortened. This must be taken into account during use and corrected by the control system. Figure 10.295b together with Fig. 10.295e provides a configuration that is used in some AC applications as both polarities of the current can be influenced. An example is the ring rectifier in welding applications. The same topology can also be used to represent a half-bridge branch as the basic topology of power electronic circuit technology.

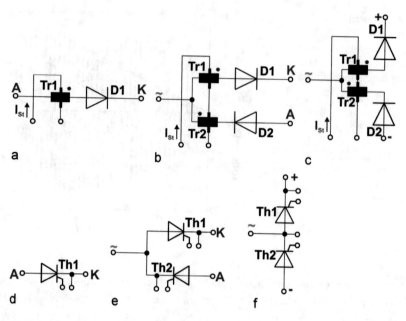

Fig. 10.295 Application of thyristors as successors of voltage-controlling magnetic amplifiers (transducers). (**a**), (**d**) Single-pole current control, (**b**), (**e**) combination for typical bipolar alternating current control, (**c**), (**f**) half-bridge circuit

An important aspect in the realization of the transducers, which is used to reduce mass and volume, shall only be pointed out here. If you magnetize the core of $\Theta_1 = 0$, you have only half of the maximum possible value available for the maximum change of the flux linkage $\Delta\Psi$. The volume can be halved by biasing the core with a bias current in the opposite direction. This can be added to the bias current or - electrically isolated - via a separate winding. Both can be found in the literature. The same effect would also be achieved by magnetic bias using permanent magnets. The problem here is that the transducer should not normally have an air gap in the main path of the magnetic flux. This option was not used in the past for this reason, and especially because of unavailable materials.

At the end of these chapters, the possibility of using voltage-controlling transducers in power supply technology is discussed. Figure 10.296 shows a 1-cycle current flow converter with demagnetization winding for the transformer. In a switched-mode power supply with only one output voltage U_1, this is controlled by the test ratio of transistor T_1. If several output voltages are to be controlled, this requires further secondary controller modules, which are designed as analog controllers or as switched DC/DC converters. In Fig. 10.296, a different approach is used, which can easily be extended to a number > 2 for the number of output voltages regulated. To regulate the voltage U_2, the induced voltage generated at N_{22} is used as the voltage system. If $N_{11} = N_{12}$, the duty cycle D of T_1 is limited to a maximum of 0.49, but changes with the reaction of controller 1. This would

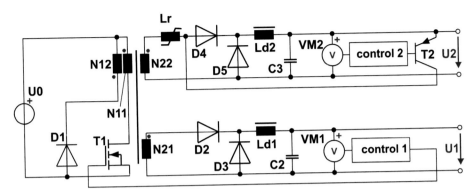

Fig. 10.296 Application of voltage-controlling magnetic amplifiers in switch mode power supplies SMPS with several separate output voltages

change U_2 without any further action. If, however, this voltage is measured and a current is generated from the control deflection, which is fed in L_r, it is possible to achieve a shift in the saturation point of this choke. Together with D_4, L_r forms a voltage-controlling transducer with which the output voltage can be controlled like a buck converter [101–103]. Compared to a simple rectifier, the basic circuit can only be supplemented by an inductor built on a small toroidal core. For core materials with μ_r up to 100,000, the control current, which is directly fed into the working winding here, is only a few milliamperes. The control is carried out downstream of controller 1 and compensates for the deviations caused by this controller from its own setpoint value. This structure can be extended by further potential-separated branches for power supply. Simple PI controllers are sufficient as controllers. The solution requires only a few components, is not susceptible to electromagnetic interference and is robust.

Transducers are not very common in modern applications of power electronic principles, although they are technically and economically interesting alternatives [104–107]. One reason is certainly their unusual function and the missing simulation tools. The present book simplifies these problems and opens new possibilities.

10.29 Rough Dimensioning of a Transformer

As was shown in Chap. 4, the dimensioning and even more the optimization of soft magnetic components with regard to losses is a multidimensional problem. Chap. 4 gives an orientation and overview of the nature and effect of the various influencing factors.

Since many geometry factors have an influence, one cannot avoid repeating the calculations. In order not to lose too much time when trying to find the initial approximation for an optimization, some steps can be shortened. A corresponding procedure for dimensioning a transformer is demonstrated in this subchapter. For a 2-channel *boost converter* that is to boost the voltage from 150 V to a maximum of 450 V, the

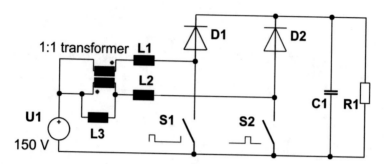

Fig. 10.297 Circuit for a 2-channel boost converter in interleaved operation with the storage chokes L_1 and L_2 and a transformer with close coupling for magnetic coupling of the two phases

corresponding transformer is to be dimensioned. More precisely, the question is: What maximum current is possible when using a T72 × 20 × 48 toroidal core made of N87, for example, if it is designed close to the optimum under typical boundary conditions. The associated simulation circuit is shown in Fig. 10.297, which shows the two storage chokes L_1 and L_2 and a transformer with close coupling for magnetic coupling of the two 180° offset driven phases. The inductance L_3 is chosen so that its value is large compared to the largest inductance that actually occurs in reality (in this case 100 H). L_3 only serves to determine the maximum flux linkage with $\psi = L_3 \cdot I_{L3}$.

The toroidal core T72 × 20 × 48 has the following mechanical properties according to the manufacturer's specifications.

Outer diameter (D_o)	72 mm
Inner diameter (D_i)	48 mm
Height (h)	20 mm
Core volume (V_{Fe})	43.42 cm³
Magnetisation cross section (A_{Fe})	237 mm²

The current density used generates a certain power loss density in the conductor material. The power loss must be dissipated so that the limit temperature of the insulation and the construction materials is not exceeded. The maximum current density is therefore mainly dependent on the cooling conditions. A conservative value for windings without special cooling measures is 2.5 A/mm². This value tends to be higher for smaller component volumes and correspondingly lower for larger ones. Assuming the current density of 2.5 A/mm², which is frequently used for transformers with free installation or higher rated power, the power loss density for the operating temperature of 100 °C is obtained:

$$p_V = \frac{P_V}{V} = \frac{I^2 \cdot \rho \cdot l}{l \cdot A_{Cu}^2} = \frac{I^2 \cdot \rho}{A_{Cu}^2} = J^2 \cdot \rho$$

$$p_V = 0.0173 \frac{\Omega \cdot mm^2}{m} \left(1 + 0.0041 K^{-1}(100\,^\circ C - 20\,^\circ C)\right) \left(2.5 \frac{A}{mm^2}\right)^2 = 143 \frac{mW}{cm^3}$$

Two bifilar windings with fill factor k_f and number of turns N_1 should fill the winding window to a maximum of 50%. The inner diameter of the wound core should therefore be

$$D_{iW} = \frac{D_i}{\sqrt{2}} = \frac{48\,mm}{\sqrt{2}} = 34\,mm$$

This results in the average coil length

$$l_m = \pi \frac{D_i - D_{iW}}{2} + (D_a - D_i) + 2 \ h = \left(\pi \frac{48 - 34}{2} + 72 - 48 + 2 \cdot 20\right) mm \approx 86 \ mm$$

The winding space can only be filled with a filling factor $k_f < 1$ with copper cross section. For compact windings, $k_f = 0.5$ has proven to be a useful starting value for calculations in the present exploratory status. For toroidal cores, the fill factor that can be achieved tends to be smaller due to the uneven distribution of the windings. The winding window can also only be used to about 50% of the total available area. This is especially true for closely coupled transformer windings. With a filling factor of the winding space of $k_f = 0.4$, the relationship to the number of turns is then obtained in the present case as

$$A_W \cdot k_f = 0.5 \cdot k_f \cdot \frac{\pi D_i^2}{4} = 2 \cdot N_1 \cdot A_{Cu}$$

The copper losses are then for both windings with $N_1 = N_2$

$$P_{VCu} = 143 \frac{mW}{cm^3} \cdot V_{Cu} = 143 \frac{mW}{cm^3} \cdot 2N_1 \cdot l_m \cdot 0.5 \cdot k_f \cdot \frac{\pi D_i^2}{2N_1 \cdot 4}$$

$$P_{VCu} = 143 \frac{mW}{cm^3} \cdot l_m \cdot 0.5 \cdot k_f \cdot \frac{\pi D_i^2}{4} = 4463 \ mW$$

Since the core losses are $\sim N^{-\alpha}$ and the winding losses are $\sim N^2$ (Fig. 4.7), it is a good approximation that at the minimum of the sum of both losses, the two losses P_{VCu} and P_{VFe} are more or less equal. The reason for this is the fact that α is close to 2. (In a detailed calculation, the relationship $P_{VCu}/P_{VFe} = \alpha/2$ applies to the location of the minimum losses. At 30 kHz for N87 $\alpha = 2.2$ can be derived from Fig. 10.298) This means that one can start with

Fig. 10.298 Specific losses of the material N87 with registered maximum power loss density of 103 mW/cm³. (Source: TDK/EPCOS)

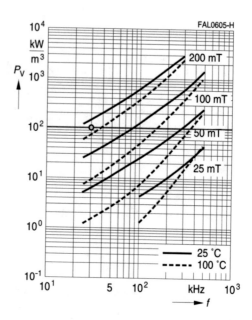

$$P_{VCu} = 4463\,\text{mW} = P_{VFe} = p_{VFe} \cdot V_{Fe}$$
$$p_{VFe} = \frac{4463\,\text{mW}}{43.42\,\text{cm}^3} = 103\,\frac{\text{mW}}{\text{cm}^3}$$

From the loss diagram Fig. 10.298 for material N87, the maximum flux density 0.224 T is obtained for the switching frequency 30 kHz and 100 °C by interpolation using the Steinmetz formula.

Thus, the maximum periodic flux with the magnetization cross section of 237 mm² is obtained to

$$\widehat{\Phi} = \widehat{B} \cdot A_{Fe} = 0.224\ \text{T} \cdot 237\,\text{mm}^2 = 0.000053\ \text{Vs} = 53\ \mu\text{Vs}$$

The current i_{L3} is an image of the flux linkage at the transformer. The current flux linkage is then $\psi = L_3\,i_{L3}$. Figure 10.299 shows a corresponding course of this flux linkage via the duty cycle when simulating the circuit Fig. 10.297. The maximum flux linkage that occurs can be obtained from this graphic as

$$\widehat{\Psi} = \frac{\Delta\Psi}{2} = \frac{1.775\ \text{mVs}}{2} = 0.8875\ \text{mVs}$$

From this, the required minimum number of turns can be calculated:

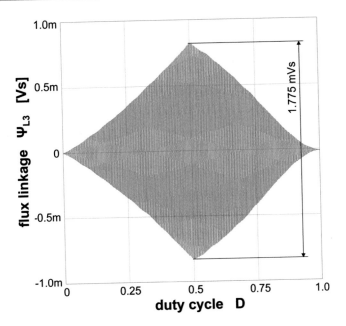

Fig. 10.299 Changing of the flux linkage with variable duty cycle D for the control of the circuit according to Fig. 10.297

$$N_1 = N_2 = \frac{\widehat{\Psi}}{\widehat{\Phi}} = \frac{887.5 \ \mu Vs}{53 \ \mu Vs} = 16.74 \approx 17$$

This gives you the possibility to calculate the conductor cross section roughly:

$$A_W = 0.5 \cdot \frac{\pi}{4} D_i^2 \cdot k_f = 2N_1 \cdot A_{Cu}$$

$$A_{Cu} = \frac{\pi}{16 \cdot N_1} D_i^2 \cdot k_f = \frac{\pi}{16 \cdot 17} 48^2 \ mm^2 \cdot 0.4 = 106 \ mm^2$$

The outer diameter of a copper conductor would then be

$$d_{Cu} = \sqrt{\frac{4}{\pi} A_{Cu}} = \sqrt{\frac{4}{\pi} 10.6 \ mm^2} = 3.674 \, mm$$

If the theoretically available copper cross section is used, a maximum current of 26.54 A per conductor results with the applied current density of 2.5 A/mm^2. The current drawn by the 2-channel boost converter could therefore be around 53 A with this component. The construction capacity of the transformer is (calculated for the rectangular voltage used)

$$S \approx U_1 I_1 = \widehat{\Psi} \cdot 4f \cdot I_1 = 0.8875 \text{ mVs} \cdot 4 \cdot 30 \text{ kHz} \cdot 26.5 \text{ A} = 2822 \text{ VA}$$

The step-up converter could transmit a power of

$$P_d = U_d \cdot 2 \cdot I_1 = 150 \text{ V} \cdot 53 \text{ A} = 7950 \text{ W}$$

At this power level, approximately $2 \cdot P_{VCu} = 8.92$ W is mathematically allotted to the transformer losses, which corresponds to 0.11% of the power transferred in the DC/DC converter. If D is always unequal to 0.5, i.e. if the voltage transmission ratio is different, the dimensioning can also be adapted.

When dimensioning DC chokes with AC component, the procedure is similar, but the saturation of the core material must be taken into account as a boundary condition. The input data for the calculation shown above is determined from the simulation of the circuit in the worst case when using strongly non-linear core materials (approximately). For toroidal cores without air gap, only the number of turns is a free parameter. For adjustable air gaps, these values are added as further parameters. The following boundary conditions from the application must be observed as conditions to be complied with:

maximum ripple current	$\Delta i_L < \Delta i_{L.\,max}$
maximum instantaneous choke current	$i_L(t)\big|_{max} \approx I_{d.\,max} + \frac{\Delta i_L}{2} < k_1 \cdot I_{SAT}$
balance of power losses	$P_{VCu} = P_{VFe}$

Especially for the determination of initial approximations for dimensioning, the algorithm shown above provides an effective aid.

10.30 Influence of Winding Material on Quality of HF Coils

It has been pointed out at various places in the book that, in addition to the core material, the winding material used also has a major influence on the properties of inductive components. In this section, the properties of the winding material and ways of taking them into account will be discussed using examples. For this purpose, the following procedure was followed. Six coil formers were made, each with $6 \times 6 = 36$ turns and various winding materials (Fig. 10.300).

This results in a solenoid coil in which a part of the generated magnetic field passes through the winding (Fig. 10.301a). The magnetic equivalent circuit shown in Fig. 10.301b can be derived from the field structure. The space filled by nonconducting and nonmagnetic material ($R_{m.air1}$ and $R_{m.air2}$) contributes to the storage of magnetic energy, but not to any losses that arise. Losses occur in the winding due to proximity effect ($R_{m.leak2}$) and skin effect ($R_{m.leak1}$). It is easy to imagine that eddy currents are induced in the windings ($R_{m.leak2}$), which ultimately account for the proximity effect. The portion of the skin effect is caused by non-linked magnetic flux within the conductors, which is not shown in

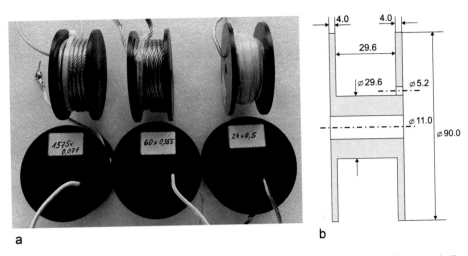

Fig. 10.300 Windings used for measurement with the same structure and different winding materials made of Cu strands with approximately the same cross-section: (**a**) photograph; (**b**) Dimensions of the coilformers

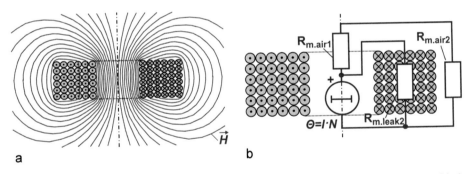

Fig. 10.301 Solenoid coil: (**a**) Cross-section of the principle structure of the solenoid coil with the resulting magnetic field; (**b**) Mapping of the magnetic field effects by magnetic resistors

Fig. 10.301b. In the electrical equivalent circuit (Fig. 10.302), the representative for this is found as L_{leak1} parallel to the loss resistance R_{skin}.

This results in the simplified electrical equivalent circuit shown in Fig. 10.302. The constant values are.

DC resistance of the Cu conductor	$R_{Cu.DC}$
Inductance of the air space	$L_{air} = L_{air1} \,/\!/\, L_{air2}$
Total development capacitance	C_w

The losses due to skin effect are transformer-coupled into the material within the conductors. The inductance $L_{leak1}(f)$ is essentially equal to the internal inductance of the

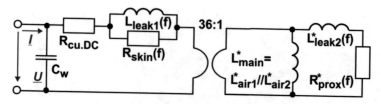

Fig. 10.302 Electrical equivalent circuit of a coil according to Fig. 10.301

conductors $L_i(f)$. The field caused by the main inductance transformerally causes the losses of the proximity effect $R^*_{prox}(f)$. Thereby the eddy currents counteract their cause, which is represented by the "leakage" inductance L^*_{leak2}. With the measuring device one can only determine current and voltage as well as their phase position. Thus there is inevitably a summary of the effects. Therefore, only the real part and imaginary part of the impedance as functions of the frequency f are available as measurement results. The structural considerations, which are summarized in the electrical equivalent circuit diagram Fig. 10.302, help in the further interpretation of the measurement results.

In Table 10.18 characteristic data of the investigated coils are summarized. Due to the differences in the structure of the conductor, small differences in the properties occur. However, these hardly influence the statements of the following considerations.

If the impedance locus curves are plotted, the picture shown in Fig. 10.303a is obtained for the frequency range up to 50 MHz. The measuring points seem to lie on a circle through the origin of the coordinates as in a parallel resonant circuit. The circle for the largest wire diameter of 0.5 mm has the largest diameter. The red circle for 0.071 mm wire diameter has the smallest diameter. Surprisingly, this means that in the range of the resonant frequencies (see Table 10.19), the conductor with the largest wire diameters has the lowest losses. If we look at the same measured values in the range up to 100 kHz, we get a completely different impression (see Fig. 10.303b). In this frequency range, the expected order can be found. As the frequency increases, the strands with the largest wire diameters have the greatest losses. These are expressed by the real component of the coil impedance.

The diagram Fig. 10.303b shows that the locus curve on the axis of the real part is shifted slightly by the DC resistor $R_{Cu.DC}$. The diagram Fig. 10.304 shows in normalized form that - starting from the DC resistor - the effective real component initially increases and continues to increase up to the resonant frequency. At the resonant frequency, inductive and capacitive reactive power compensate each other. The capacitive reactive power is provided by the winding capacitance C_w. From the comparison with the likewise normalized magnitude of the resistance increase due to the skin effect, it can be seen that the skin effect can only contribute to the measured resistance increase at higher frequencies.

The winding capacitance C_w is determined from the lowest resonant frequency assuming a constant inductance. Now that C_w and $R_{Cu.DC}$ are known, this can be used to further refine the analysis. If one subtracts the complex conductance $j\omega C_w$ from the measured values of the admittance locus curve $\underline{Y}(f) = \underline{Z}^{-1}(f)$, one obtains the admittance of the

Table 10.18 Overview of the determined properties of the 6 investigated coils (all values related to 36 turns)

litz wire		1575 × 0.071	630 × 0.1	200 × 0.2	100 × 0.3	60 × 0.355	24 × 0.5
turns	–	36	36	36	36	36	36
A_{Cu}	mm^2	6.24	4.95	6.28	7.07	5.94	4.71
l_{Cu}	m	6.78	6.75	7.12	7.12	6.65	7.15
R_{Cu}	mOhm	19.0	25.2	19.8	17.6	21.0	26.4
L_p(1 kHz)	µH	46.2	44.5	44.3	42.2	40.8	38.5
f_{res}	MHz	4.02	4.12	4.99	3.6	5.19	4.62
C_w	pF	33.6	34.2	22.7	45.7	23	31.2
L_i	µH	0.339	0.338	0.356	0.356	0.332	0.358

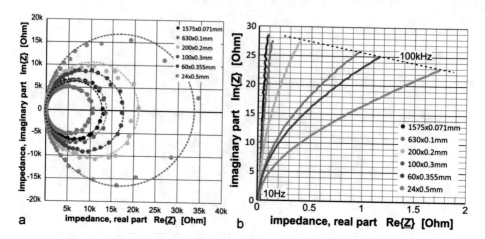

Fig. 10.303 Coil impedance Z(f) locus: (**a**) High frequency range up to 50 MHz; (**b**) How frequency range up to 100 kHz

Table 10.19 Winding material with optimum frequency range

Winding material	Frequency range
1575 × 0.071 mm	4 kHz … 1 MHz
630 × 0.1 mm	3 kHz … 500 kHz
200 × 0.2 mm	3.5 kHz … 150 kHz
100 × 0.3 mm	3 kHz … 50 kHz
60 × 0.355 mm	3.5 kHz … 40 kHz
24 × 0.5 mm	6 kHz … 15 kHz

assumed series connection of $R_s = R_{s.skin} + R_{s.prox}$ and L_s. From this it is easy to calculate the parameters for an assumed parallel set circuit:

$$\underline{Y}_1(f) = \underline{Z}^{-1}(f) - j\omega C_w$$
$$\underline{Z}_2(f) = \underline{Y}_1^{-1}(f) - R_{Cu.DC}$$

For the equivalent quantities of a parallel connection of L_p and R_p one obtains thus

$$R_p(f) = \frac{1}{\mathrm{Re}\left\{\frac{1}{\underline{Z}_2(f)}\right\}} \quad L_p(f) = \frac{1}{\mathrm{Im}\left\{\frac{1}{\underline{Z}_2(f)}\right\} \cdot j\omega}.$$

Thus, the components of the parallel circuit (see Fig. 10.305) can be calculated. The parallel loss resistance initially increases sharply in the lower frequency range with the proportionality $R_p \sim f^2$ for all winding materials investigated. In the higher frequency range, this increase decreases until the ranking of the magnitude of the loss resistances

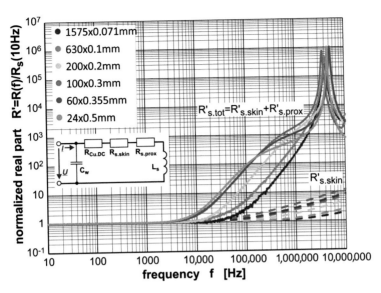

Fig. 10.304 Real part of the measured impedance locus curve Z(f) compared with the calculated course of the resistance increase due to the skin effect in normalized form

$R_p(f)$ then changes. It is assumed that the reason for this is the shielding effect of the conductor material for lower lying regions. This explains the behavior observed in Fig. 10.303.

It is interesting to note that the interpretation of the processed measurement results after taking into account C_w and $R_{Cu.DC}$ leads to a reduction in the size of L_p above a certain frequency (Fig. 10.306). It must be assumed that the losses in the proximity effect are caused by transformer effects. In the model (Fig. 10.302) this is considered by $R^*_{p.prox}//L^*_{main}$. To take into account the coupling of the affected area with the main inductance, the "leakage inductance" L^*_{leak2} is inserted in this model. At higher frequencies, when the magnitude of the corresponding reactance comes to the order of R^*_{prox}, the reactance of the series connection increases noticeably. Due to the parallel connection to L^*_{main}, this is observed in the measurement as an apparent reduction of the inductance. This can also be interpreted as a reduction of the real part of the permeability. The complex permeability thus has a smaller value. Correspondingly, one can write for the impedance of a real inductor with internal homogeneous field in complex notation

$$Z_{L0} = \frac{\omega \cdot l_{mag} \cdot N^2}{\underline{\mu}_r(\omega = 0) \cdot \mu_0 \cdot A_{mag}} \quad \text{with} \, \underline{\mu}_r(\omega = 0) = 1 \text{ and}$$

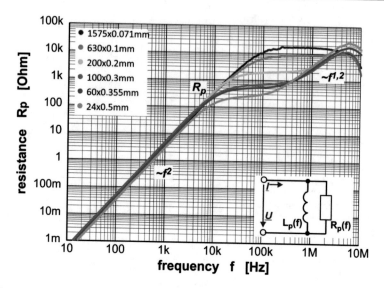

Fig. 10.305 Summarized loss resistance R_p after mathematical consideration of the effect of the winding capacitance C_w and the DC resistance $R_{Cu.DC}$

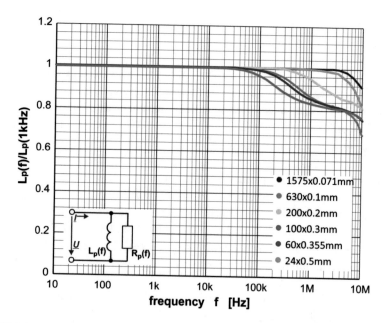

Fig. 10.306 Determined course of the resulting inductance L_p for different winding materials as a function of frequency in normalized form

$$\underline{Z}_L(\omega) = \frac{\omega \cdot l_{mag} \cdot N^2}{\underline{\mu}_r(\omega) \cdot \mu_0 \cdot A_{mag}}.$$

For the complex permeability, the relative value is then given by the expression

$$\underline{\mu}_r(\omega) = \frac{L(\omega)}{L_0}$$

In the test arrangement, it is obviously not a homogeneous field. Figure 10.306 shows the integral effect of all partial areas.

The quality factor Q is used to estimate the quality of an inductor. It results from the ratio of reactive power and active power for a component. In Fig. 10.307, the corresponding values for the coils investigated are plotted in one diagram. Here, at higher frequencies, clear advantages can be seen for stranded wires with thinner individual wires.

It should be noted that these specifications are not only related to the winding material. The quality factor Q is also dependent on the shape of the coil and thus of the magnetic field which passes through it. In transformers, i.e. inductive components with several coupled windings, the structure or the shape of the stray field area plays an important role.

If a quality factor Q = 50 is taken as a reference value, the following "optimum frequency ranges" would result for the coils investigated.

Above the resonance frequency, which is determined by (L_{main}; C_w), these analyses are hardly or not meaningful. The reason for this is the increasing "fuzziness" of the derived model (Fig. 10.302). Besides the resonance of Lmain with the winding capacitance Cw, there are others. They are caused by inductance parts and winding and layer capacitances.

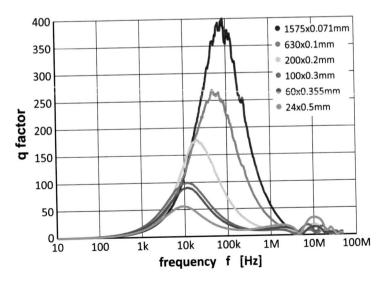

Fig. 10.307 Quality factor Q(f) of the tested coils as a function of frequency

10.31 Large-Signal Power Loss Measurements on Large Cores

The available material data on losses in ferromagnetic cores are measured under standardised conditions. These represent idealised conditions that are intended to make materials comparable. Transferability to real, large cores is only possible to a limited extent. For laminated cores, for example, the material data are determined with the Epstein frame on straight core sheets with an almost homogeneous field and then correction factors are introduced for the cores layered in a different way. The material data for ferrite cores are determined on small toroidal cores and then transferred to large cores by calculation. A few control values are given for a specific core type, which are then considered as assured values. For the concrete application of a certain core geometry, one has to rely on one's own calculations and control measurements. The present section is intended to show possibilities of how to proceed in this respect.

The following considerations are illustrated using the example of a core pair UR64/40/20/3C90 from Ferroxcube. According to the data sheet, this core has the following characteristics, among others: $A_{Fe} = 290 \ mm^2$, $l_{Fe} = 210 \ mm$, $V_{Fe} = 61{,}000 \ mm^3$.

One question is first: What does the small-signal measurement of the proportions of the complex permeability tell us? If one uses the given values for the complex permeability of the data sheet for 3C90, one can easily determine a small-signal equivalent circuit for a pair of cores for 1 turn. The measurements of $\mu_s{'}$ and $\mu_s{''}$ are those for a series equivalent circuit. The corresponding elements for the equivalent circuit are thus to be calculated as follows:

$$\underline{Z}_{1turn} = j\omega(\mu_s{'} - j\mu_s{''})\mu_0 \frac{A_{Fe}}{l_{Fe}} = j\omega\mu_s{'} \cdot \mu_0 \frac{A_{Fe}}{l_{Fe}} + \omega\mu_s{''}\mu_0 \frac{A_{Fe}}{l_{Fe}}$$

$$\underline{Z}_{1turn} = j\omega L_{s.1turn} + R_{s.1turn}$$

By comparing coefficients, one finds from this for the elements $L_{s.1turn}$ and $R_{s.1turn}$

$$L_{s.1turn} = \mu_s{'} \cdot \mu_0 \frac{A_{Fe}}{l_{Fe}}$$

$$R_{s.1turn} = \omega\mu_s{''}\mu_0 \frac{A_{Fe}}{l_{Fe}}$$

According to the rules of AC technology, these values (L_s; R_s) for a serial equivalent circuit can be converted into values (L_p; R_p) for an equivalent parallel equivalent circuit. Figure 10.308 contains both variants. Also according to the Steinmetz formula, the core losses are mainly dependent on voltage and frequency. In the parallel equivalent circuit diagram, the depiction of the losses in connection with the Steinmetz formula is therefore much clearer. The expected losses would then be

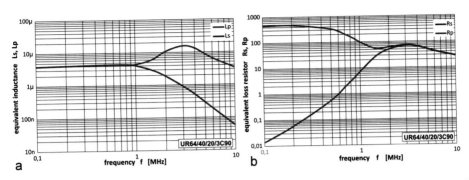

Fig. 10.308 Complex permeability with real part and imaginary part converted into serial and parallel equivalent circuit of an inductor ($L_s + R_s$ or L_p // R_p) with core UR64 with one turn based on small signal parameters of the ferrite 3C90: (**a**) Equivalent inductance (L_s; L_p); (**b**) Equivalent resistance (R_s; R_p)

$$P_{loss} = \frac{U^2_{1turn}}{R_{p.1turn}}$$

If you simply transfer this to large signals, you would be able to write at $B_{max} = 0.2$ T for one turn:

$$U = \frac{2\pi f}{\sqrt{2}} \cdot B_{max} \cdot A_{Fe} = \frac{2\pi \cdot 100 \text{ kHz}}{\sqrt{2}} \cdot 0.2 \text{ T} \cdot 290 \text{ mm}^2 = 25.77 \text{ V}$$

Further information on expected losses is available from the data sheet on core loss measurements on toroidal cores. In the present case from the data sheet of the UR64/40/20/3C90 core it is only announced that a core set can have a maximum of 7.3 W losses at 0.2 T at 25 kHz and 100 °C.

For one winding this means:

$$0.2 \text{ T} = \frac{\sqrt{2} \cdot U}{2\pi f \cdot A_{Fe}}$$

$$U = \frac{1}{\sqrt{2}} 0.2 \text{ T} \cdot 2\pi f \cdot A_{Fe} = \frac{1}{\sqrt{2}} 0.2 \text{ T} \cdot 2 \pi \cdot 25 \text{ kHz} \cdot 290 \text{ mm}^2 = 8.14 \text{ V}$$

This then results in an equivalent loss resistance of

$$7.3 \text{ W} = \frac{U^2}{R_p} = 7.3 \text{ W} = \frac{(8.14 \text{ V})^2}{R_p}$$

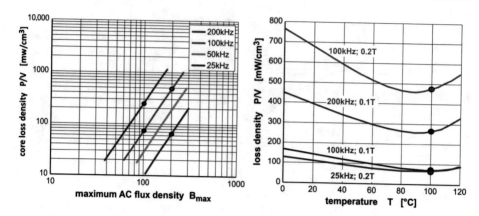

Fig. 10.309 Power dissipation density and its temperature dependence for ferrite 3C90 [according to data sheet, Ferroxcube 2004]: (**a**) Power dissipation density as a function of magnetic flux density and frequency at 100 °C; (**b**) Temperature dependence of power dissipation density for selected points in (a)

$$R_p > \frac{(8.14 \ \text{V})^2}{7.3 \ \text{W}} = 9.077 \ \text{Ohm}$$

This value is obviously determined for large signal stress and is considerably lower than the value of approx. 420 Ohm determined from the values for small signal stress. The material used in the core is 3C90. In the data sheet of Ferroxcube you can find more information about the magnetisation losses of the material. The following picture shows the loss diagram for 3C90 on toroidal cores:

For the operating point (0.2 T, 25 kHz), the value 65 mW/cm^3 can be taken from the diagram in Fig. 10.309a. For a core pair UR64 ($V_{\text{Fe}} = 61 \ \text{cm}^3$) this means a power loss of 3.965 W at 100 °C. This corresponds to a parallel loss resistance for the conditions (0.2 T; 25 kHz, 100 °C) of

$$R_p = \frac{(8.14 \ \text{V})^2}{3.965 \ \text{W}} = 167 \ \text{Ohm} > 9.077 \ \text{Ohm}$$

This value is greater than the value calculated from the data sheet of the UR64 core. According to this calculation, the expected losses are only 54% of the maximum value given in the data sheet for core UR64. As can also be seen in the Fig. 10.309b, the specific losses change strongly with temperature. With a determination of the Steinmetz formula oriented to the centre of the characteristic curve field, one finds approximately

$$\frac{P}{V} = k_e \cdot \left(\frac{B_{\max}}{1\ T}\right)^{\alpha} \cdot \left(\frac{f}{1\ kHz}\right)^{\beta} \approx 22.84\,\frac{mW}{cm^3} \cdot \left(\frac{B_{\max}}{1\ T}\right)^{2.726} \cdot \left(\frac{f}{1\ kHz}\right)^{1.614}$$

Here the exponents are in the range $\alpha = 2.676...\underline{2.726}...2.78$ and $\beta = 1.437...\underline{1.614}.... 1849$.

In the vicinity of an operating point, this dependency can be used for extrapolation. In the present example, a dependency averaged over a range of values was used. Of course, it is also possible to determine the approximation functions around an operating point. In principle, extrapolations are always problematic. However, they can be used with the appropriate caution for qualitative statements. The determined approximations themselves only represent a plausible analytical approximation of a mathematical expression to measured data. In this sense, a plausible approximation of the behaviour of an assumed loss resistance based on the data sheets would be the following expression

$$R_p \sim R_{p.ref} \cdot \left(\frac{f}{f_{ref}}\right)^{1.112} \cdot \left(\frac{U}{U_{ref}}\right)^{-0.726}$$

For the operating point (0.2 T; 25 kHz), the expected dependence is as follows

$$R_p \sim 16.7\ \text{Ohm} \cdot \left(\frac{f}{25\ kHz}\right)^{1.112} \cdot \left(\frac{U}{8.14\ V}\right)^{-0.726}$$

At constant frequency, the "loss resistance" R_p of the core consequently decreases with increasing voltage. The $I_{Rp}(U_{1turn})$ characteristic could be used for an approximate simulation of the losses that occur.

For a fixed frequency of 100 kHz, one obtains for the loss resistance $R_p(100\ °C)$

$$R_p \sim 16.7\ \text{Ohm} \cdot \left(\frac{100\ kHz}{25\ kHz}\right)^{1.112} \cdot \left(\frac{U}{8.14\ V}\right)^{-0.726} = 78.02\ \text{Ohm} \cdot \left(\frac{U}{8.14\ V}\right)^{-0.726}$$

The graph shows that the equivalent resistance R_p changes strongly with the modulation of the core material. With small voltages per turn or low flux densities, the value of R_p in the present case reaches values between 400 Ohm and 500 Ohm. In the present case, this value corresponds to the range of values shown in Fig. 10.308 for small signal measurement. However, it also shows that the value found there is not suitable for calculations in the large signal range. Because of the model errors (see e.g. the variations in the exponents of the Steinmetz formula), it is not possible to derive usable values for the large signal range from the measurements of R_p in the small signal range (Fig. 10.310).

For a rough determination of the $I_{Rp}(U_{1turn})$ characteristic, one can use as a further approximation at 100 kHz

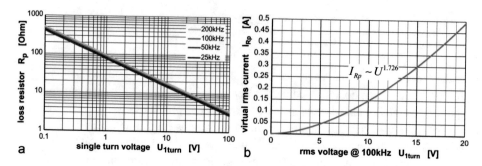

a

b

Fig. 10.310 Calculated voltage dependence of the equivalent loss resistance: (**a**) depending on the single turn voltage and frequency, (**b**) virtual I(U) characteristic of R_p at 100 kHz

$$I_p = \frac{U_{1turn}}{R_p} = \frac{U_{1turn}}{78.02 \ \text{Ohm} \cdot \left(\frac{U_{1turn}}{8.14 \ \text{V}}\right)^{-0.726}} = \frac{U_{1turn} \cdot \left(\frac{U_{1turn}}{8.14 \ \text{V}}\right)^{0.726}}{78.02 \ \text{Ohm}}$$

This is a formula that can be used for simulations. However, it only describes the losses resulting from a sinusoidal voltage waveform. But this is also the case with the power loss data from the data sheets. The order of magnitude and the qualitative dependence of these characteristic curves calculated from measurement data from the data sheet are well represented in the present case (Fig. 10.311).

The Steinmetz formula is too imprecise an approximation for large value ranges and has a high parameter sensitivity. This means that measurements in the large signal range must be used for reliable statements regarding the losses. The loss values from the data values of the material are determined on small toroidal cores. To what extent a transfer for the loss

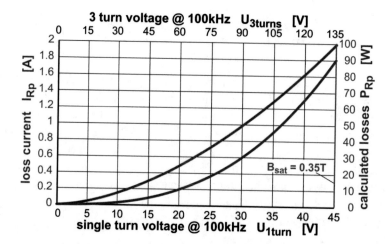

Fig. 10.311 Equivalent characteristic curves for the loss resistance R_p and calculated losses P_{Rp} as a function of a sinusoidal voltage with a frequency of 100 kHz at 100 °C

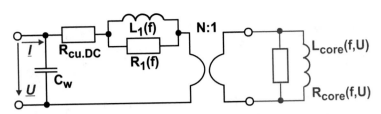

Fig. 10.312 Electrical equivalent circuit diagram to illustrate the measurement situation when determining losses with large signals

calculation of cores with a deviating shape is possible is unclear and requires further investigation. However, an application in finite element calculations seems to be possible.

Large-signal measurements that are as simple as possible are very useful in the application area of interest, for example, for comparing core shapes or materials. The present chapter shows one possibility for sinusoidal voltages by means of several application examples based on the proposal in Sect. 7.4 of this book. The method, which is easy to implement in the laboratory, was originally designed by the author to compare cores from different manufacturers under conditions relevant to the application. Figure 10.312 shows the measurement problem to be solved schematically. The model representation was chosen in such a way that the losses related to the winding are shown on the left side and those related to the core on the right side of the ideal transformer. Physically, the model represents several effects that can be represented by connecting L and C in parallel. This concerns in particular the skin effect and proximity effect. The skin effect takes place in the conductors of a winding. The proximity effect concerns adjacent conductors, but also interactions with the stray field of the core. In addition, a series connection with an ideal inductance results if the inductance is considered as an effect of the coreless space together with the core. Let everything be summed up and transformed to the primary side of the ideal transformer in Fig. 10.312. If one makes the losses on the left side as small as possible, the measurement essentially provides the core losses. If the winding capacitance C_w is sufficiently low with respect to the measurement frequency, the inductance of the structure valid for the operating point or the permeance of the core is also obtained if the number of turns is taken into account.

Only voltage U and current I can be observed at the terminals of the arrangement. The "magnetic world" is connected to the "electrical world" via an ideal transformer with N windings. In order to determine the core-related data as accurately as possible, the DC and AC resistance of the winding must be kept as low as possible. In Fig. 10.312, the DC resistance $R_{Cu.DC}$ is represented by a fixed resistance. In order to represent the physical relationships (skin and proximity effect) structurally correctly, the AC resistance is represented by a parallel connection of a frequency-dependent inductance L_1 with a frequency-dependent resistor R_1. These are transformer-generated losses in the winding due to the inner and outer magnetic field of the conductors. Also, the chosen measuring frequency must be well below the natural resonance frequency of the assembly. The aim

Fig. 10.313 Winding arrangement for the determination of core losses for a core UR64: (**a**) Total assembly; (**b**) Parts of the assembly

must be to keep these quantities of the winding as low as possible and / or to know them exactly. $R_{Cu.DC}$ can be kept very small by choosing the largest possible conductor cross-section. To minimise the skin effect, choose an HF stranded wire with the thinnest possible individual conductors. Figure 10.313 shows a version of the measuring windings for the UR64. On each leg of the core, four times 3 windings á 1575 × 0.071 mm high-frequency stranded wire are connected in parallel. This means a DC resistance of 1.035 mOhm for one partial winding including the connection length. In order to achieve the most even possible magnetic flux through the entire core, all 8 partial windings are connected in parallel. This results in a calculated total DC resistance of 0.129 mOhm for the entire winding. The frequency at which the skin thickness is equal to half the diameter for 0.071 mm thick Cu wires is equal to

$$s = \frac{D}{2} = \frac{1}{\sqrt{\pi f_{crit} \sigma \mu_0}}$$

$$f_{crit} = \frac{4}{\pi D^2 \sigma \mu_0} = 3.517 \ \text{MHz}$$

Actually, the coil should not show any increase in resistance below this frequency because of the skin effect. However, as the measurement results described below show, this is not the case.

At higher frequencies, it must be taken into account that additional losses and voltage drops occur in the windings, which lead to errors in the determination of the power loss. Therefore, next the parameters of the model in Fig. 10.312 are determined approximately by measurement. Figure 10.313 shows the corresponding results for a single winding with 3 turns and a winding with 4 windings connected in parallel.

What is surprising about the measurements documented in Fig. 10.314 is that despite the selected small wire diameter of 0.071 mm for a single conductor, an increase in the

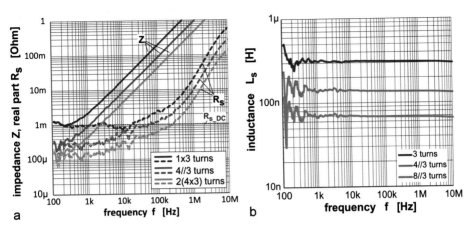

Fig. 10.314 Analysis of coreless windings (3 turns of 1575 × 0.071 mm stranded copper wire) for power dissipation measurement: (**a**) Parameters for a serial equivalent circuit for a single winding (1) of 3 turns, the parallel connection of (4) windings of 3 turns each and both of those in parallel; (**b**) Inductance for windings consisting of 1, 4 and 8 sub-windings connected in parallel

measured real component of the impedance of the winding without core can already be observed at approx. 10 kHz.

With high-frequency stranded wire, however, the individual wires lie close together and cause eddy currents in the neighbouring wires. These are probably responsible for the observed changes in the loss resistance. Physically, skin effect and proximity effect are manifestations of the effect of eddy currents. A mapping via equivalent circuit diagrams is carried out in both cases via (frequency-dependent) inductances and resistances connected in parallel. The winding causes a largely inductive voltage drop (here with 4 parallel partial windings, for example, $\phi = 89.2°$). This voltage drop reduces the core load due to the magnetic field. This error can be roughly determined to be

$$|\Delta \underline{U}_L| = |\underline{U}_C - \underline{U}_L| < |\underline{I} \cdot \underline{Z}_w|$$

This allows the relative measurement error for the measurement of the voltage at the "core inductance" to be estimated at

$$F_{rel} < \frac{|\underline{I} \cdot \underline{Z}_w|}{|\underline{U}_C|} \cdot 100\%$$

This procedure of taking the "coreless field" into account corresponds methodically approximately to that implemented in the Epstein frame.

Several parallel-connected partial windings with 3 turns each on one leg are magnetically coupled. Without a core, the coupling factor is significantly smaller than 1. Thus, the measured inductance of 4 parallel-connected windings is ~136nH compared to 306nH.

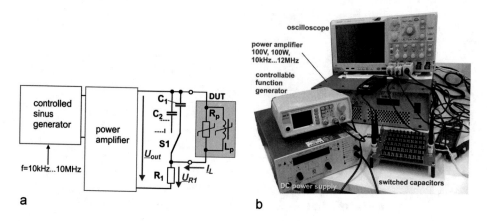

Fig. 10.315 Presented method for the determination of core losses: (**a**) principle circuit; (**b**) experimental setup with DC power supply for premagnetization

With uncoupled windings this value would be 76.5nH. If the two windings are each connected in parallel with 4 parallel windings of 3 turns each, one can assume uncoupled windings. The same applies to the real part of the impedances, which are partly due to the effect of magnetic couplings.

Figure 10.315 shows the basic measurement set-up of the power loss measurement arrangement. The frequency-variable sinusoidal voltage of a generator is amplified to the required voltage level with a broadband power amplifier and fed to the network for measurement. The resistor R_1 serves as a reference element and for current measurement. The capacitance C_1 and the device under test (DUT) form a parallel resonant circuit. When the parallel resonant circuit is in resonance, the phase shift between U_{out}, U_{R1} and U_{C1} is zero. The following relationship then applies

$$f_{res} = \frac{1}{2\pi\sqrt{L^* \cdot C_1}}$$

From this, the large-signal inductance can be determined as

$$L^* = \frac{1}{\left(2\pi f_{res}\right)^2 \cdot C_1}$$

Assuming sinusoidal quantities of voltages and currents, which implies a sufficiently linear calculation, the following relationship applies to the power loss in R_p

$$P_{loss} = I_{R1} \cdot U_{C1} = \frac{U_{R1}}{R_1} \cdot (U_{out} - U_{R1})$$

The power loss is thus traced back to the measurement of 2 RMS voltages. It is useful to observe the current I_{DUT}. The current also represents a load variable for the capacitors and switches used. With the current - as shown above - an estimation of the error magnitude in the loss measurement can be carried out. With strong saturation phenomena, the current shape deviates more and more from the sinusoidal shape and the error of the measurement results increases. In most of the measuring range of interest, however, the sinusoidal voltage is supported by the capacitance C_1, so that the voltage shape at the DUT only deviates noticeably from the sinusoidal shape in the case of strong current distortions. Since the non-linearities of L_{DUT} cause the effective inductance to change when the voltage U_{C1} is increased, the capacitance must always be readjusted for measurements at a fixed frequency. Otherwise, new resonance frequencies will always result. The setup is very simple and can be realised quickly in a laboratory. Since R_1 and C_1 are included in the measurement results, narrow-tolerance, broadband components must be used. Since the measured power loss corresponds to the total power loss of the parallel resonant circuit, capacitors with low losses should be used for C_1.

Why such electrical measurements? Thermal measurements prevent electric measurement errors of phase and magnitude too and are theoretically more accurate. However, since they have specific sources of error, this "apparent" advantage only really comes into play with non-sinusoidal loading of inductive components. An electrical measurement requires the lowest amplitudes and phase errors in the measurement over a very high frequency bandwidth. The same generally applies to inductive components of high quality. The phase shift between voltage and current is approximately 90°. The smallest measurement errors, which inevitably occur with small measured quantities, for example, have a large effect on the total measurement error. Calorimetric measurements have similar problems, as the effect of small losses are correspondingly difficult to observe. The proposed method adopts a transformation of the measurement problem that avoids the usual problems associated with measurement errors. First, the voltage at the unknown inductive component is supported by the parallel capacitor. On the one hand, this means that the deviation from the sinusoidal shape only becomes apparent in the case of strong saturation phenomena. On the other hand, this only makes it possible to use the method with sinusoidal quantities. The adjustment of the bridge to the condition $\phi(U_{out}) = \phi(U_{R1})$ is significantly less subject to errors than amplitude and phase measurements and requires only a few seconds. In contrast, calorimetric measurements require long times because of the thermal time constants of the components. Assuming that the thermal time constant in a simple model is $\tau_{th} = R_{th} \cdot C_{th}$ and that one must wait at least 5 times the time constant to achieve a deviation of $<1\%$ from the final value, the quantity required for the device can be estimated approximately. Figure 10.316 shows a dependence of the thermal resistance on the volume of the core of an inductive device. Ferroxcube [108] gives $c_s = 700...800$ Ws/

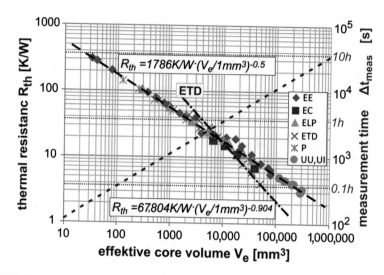

Fig. 10.316 Relationship between core volume, thermal resistance and required measurement time for measuring power dissipation according to 5τth rule (cf. Fig. 8.17)

(kg·K) for MnZn ferrites and 750 Ws/(kg·K) for NiZn ferrites. With a density $\rho = 4.8...5$ g/cm^3, the thermal time constant is roughly calculated as follows

$$\tau_{th} = R_{th}(V_e) \cdot C_{th}(V_e) = R_{th}(V_e) \cdot \rho \cdot V_e \cdot c_s$$

$$\Delta t_{meas} = 5 \cdot \tau_{th} \approx 32.82s \cdot \left(\frac{V_e}{1 \text{ mm}^3}\right)^{0,5}$$

Over the entire measurement period, the conditions of the measurement must be kept constant. This is a distinct advantage of using electrical measurement methods where this is not the case.

An evaluation of measurements provides the results presented in the graph Fig. 10.318. The exponential approach of the Steinmetz formula describes the measured values very well. The properties extrapolated from the material properties for 3C90 are not met. On the one hand, the measured values would have to be higher than the values of the data sheet for 3C90, because the measurements were made at room temperature and those of the data sheet refer to 100 °C. On the other hand, the exponents in the exponent formula derived from the values of the data sheet differ obviously. On the other hand, the exponents in the Steinmetz formula differ significantly in both cases. The latter changes, however, if one takes the values from Fig. 10.309 for 25 °C and 100 kHz as the basis for the calculation. Instead of 2.726, one then obtains the value 2.21 for the exponent. The value determined experimentally at 25 °C is and is thus much closer to the measured value 2.185. How far the values of the data sheet measured for small toroidal cores can be transferred to large cores is

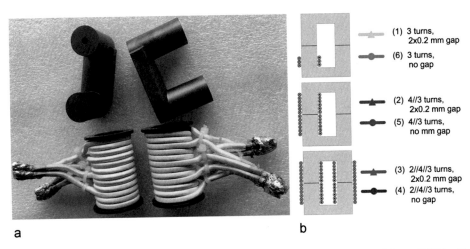

Fig. 10.317 Ferrite core U64 with windungs prepared for lor loss measurement: (**a**) core UR64 with coils, consisting of 4 turns in parallel (1575 × 0.071 mm Cu litz wire); (**b**) 6 winding arrangements used for loss measurement

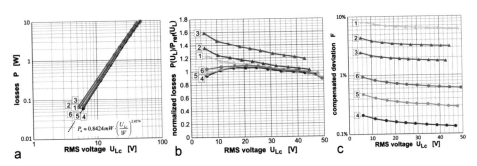

Fig. 10.318 Measured losses at a UR64 core with 3 turns at 25 °C and (sinusoidal) 100 kHz: (**a**) losses depending on voltage for different windung designs (see Fig. 10.317); (**b**) measurement results related to approximation function in 'a'; (**c**) compensated deviation using the according parts of the impedance of the coreless coils

still under discussion in the literature. Therefore, such measurements as described here are useful for own developments and dimensioning.

The described considerations are applicable for a reduction of the influence of the winding losses. If one starts from Fig. 10.312, one must know the series resistance of the winding $Rs(\omega)$ effective at the measuring frequency and use it for the calculation of the core losses according to the following formula.

$$P_{loss.core} = P_{loss.total} - I_{meas}^2 \cdot R_s(\omega) \approx P_{loss.total} - (\omega C_1 \cdot U_{C1})^2 \cdot R_s(\omega)$$

For larger loss fractions or low Q-values this approximation can be improved by

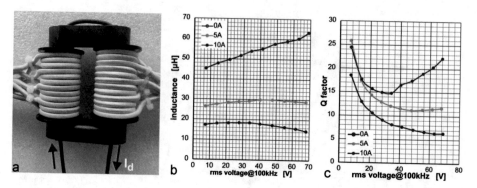

Fig. 10.319 Measurement of core losses at orthogonal premagnetization: (**a**) - setup with UR64 and pre-magnetization current I_d, (**b**) - change of inductance at different pre-magnetization current I_d, (**c**) - variation of Q factor at different pre-magnetization current I_d

$$P_{loss.core} \approx P_{loss.total} - \left[\left(\frac{U_{R1}}{R_1} \right)^2 + (\omega C_1 \cdot U_{C1})^2 \right] \cdot R_s(\omega).$$

Because of the non-linearity of the inductive and loss components, it always remains an approximation. But it allows the best possible approximation to the sought values.

A frequently asked problem is the question of the change in losses with a DC premagnetisation. Figure 10.319 shows as an example the core UR64 with an orthogonal premagnetisation by a DC current flowing through the red line through the core. The symmetrical magnetisation characteristic of the inductance is thus preserved. The inductance itself changes as a function of the modulation (Fig. 10.319b). The same applies to the quality factor Q (Fig. 10.319c).

If we now look at the change in losses (Fig. 10.320), we find that the greater the core's exitationn in terms of flux density, the smaller the changes due to pre-magnetisation.

In this chapter an easy-to-implement electrical method for measuring arbitrary core losses was presented. The reactive power during measurement is fully compensated. The amplifier as the power source only has to cover the losses in the inductance and in the measuring resistor R_1. The method is using sinusoidal exitation voltage. An automation of the adjustment at a given capacitor is possible, but more difficult. Additionally it changes the measuring frequency. Automation is easier if you change the frequency. This can be done with a simple PLL. With suitable winding material and winding design, the winding influence is low. The correction calculations to take winding losses into account still need to be improved.

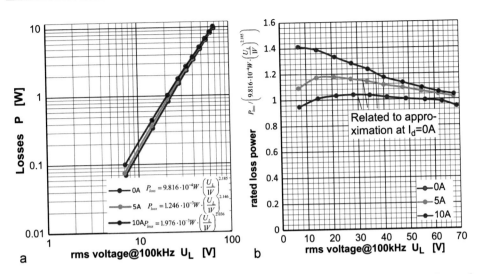

Fig. 10.320 Core losses of a core UR64/3C90 at 100 kHz without and with orthogonal premagnetization (I_d = 0A, 5A, and 10A, no gap): (**a**) measured losses in linear scale; (**b**) measured losses in double logarithmic scale

10.32 Modelling the Dependence of the Inductance of Air Gap Reactors

Data from Hitachi Metals [109, 110] (AMCC series) is used as an example of the methodology employed. These cores are offered in a series of CC-cut tape cores. The inductance can be adjusted by inserting air gaps between the C core halfs. Usually 2 windings connected in series are placed on the two legs of the core (Fig. 10.321). This "forces" the magnetic field lines into the core material as far as possible.

The core material is an amorphous material bonded in thin layers to the halves of the core. Figure 10.322 shows the magnetisation curve of alloy 2605SA1 as an example. The maximum permeability at H = 0 here is 104. An advantageous property for many applications is that the transition to saturation is "gentle". Reactions to the applications are then generally not so strong in the vicinity of the saturation induction. Due to the high permeability of the material, even the smallest air gaps have the effect of reducing the inductance of the arrangement due to surface roughness in the air gap. In this section, the effect of the air gap and a way to mathematically approximate these effects for the AMCC core series is described.

With the AMCC core series, a wide range can be realised for the area product '$A_{Fe} \cdot A_{Cu}$', as Fig. 10.323 shows. The ratio between AMCC1000 and AMCC4, for example, is 400. This makes it possible to realise inductive components in a wide power range.

Fig. 10.321 Example of a
choke realised with an
AMCC320 core

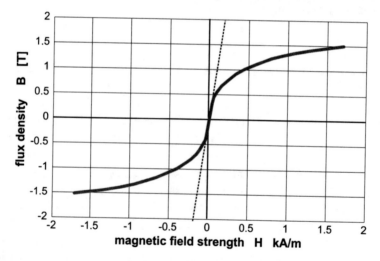

Fig. 10.322 DC B(H) curve measured at a AMCC6,3 with zero gap (POWERLITE® C-Cores
Metglas® amorphous Alloy 2605SA1) [111]

The geometrical parameters of the cores offered by Hitachi are not constant in their
ratios, as can be seen in Fig. 10.324. The magnetic form factor l_{Fe}/A_{Fe}, for example, varies
between 1.9 cm^{-1} and 8.2 cm^{-1}. The aspect ratio of the magnetisation cross-section varies
between 1.5 and 3.5. This information is important insofar as an attempt is to be made to
analytically describe the dependence of the A_L value or the permeance of the cores with air
gap for the entire core series.

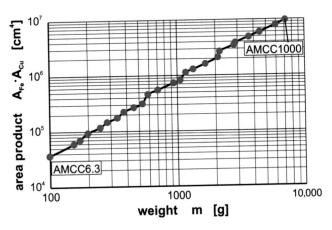

Fig. 10.323 DC B(H) curve measured at a AMCC6.3 with zero gap (POWERLITE® C-Cores Metglas®)

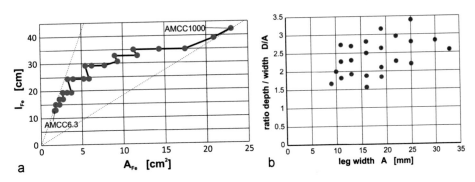

Fig. 10.324 Geometrical properties of the core family: (a) Relationship between l_{Fe} and A_{Fe} for cores without air gap; (b) Values for the aspect ratio of the edges of the iron cross-section

If one plots all the data on permeances $\Lambda(I = 0) = A_L$, one obtains the representation Fig. 10.325. All dependencies show the same tendency. Theoretically, with a given air gap δ and concentrated flux, the permeance should result in

$$\Lambda = A_L = \frac{1}{\frac{l_{Fe}}{\mu_r \mu_0 A_{Fe}} + \frac{\delta}{\mu_0 A_{Fe}}}.$$

If one uses this formula, one finds strong deviations especially for large air gaps and small core sizes. On the basis of these "official" values of the manufacturer, an analysis was therefore carried out in order to find a generally applicable relationship for this core family.

For further approximation to reality, the fact is used that as the air gap increases, the magnetic field in the vicinity of the air gap expands further and further into space. This

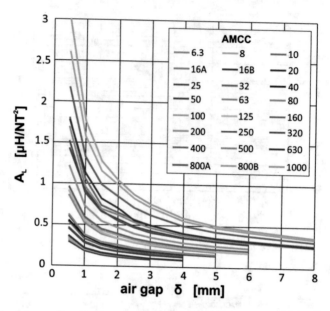

Fig. 10.325 A_L values as a function of air gap for different core sizes

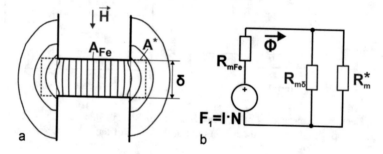

Fig. 10.326 Visualisation of the modelling to describe the influence of the air gap: a sketch of the image for the magnetic field in and around the air gap; b simplified magnetic equivalent circuit diagram

corresponds to an additional magnetic resistance R^*_m, which is connected in parallel with the reluctance $R_{m\delta}$ of the air gap. In order to obtain a simple model of the problem, which should actually be described using a 3D field model, it is now assumed that $R_{m\delta}$ and R^*_m are directly connected in parallel. Figure 10.326 serves as an illustration.

We assume that the additional parallel reluctance has the same length as the air gap. Then a free parameter whose size changes with the air gap would be the "effective" cross-sectional area A^* of this additional reluctance. One can then write

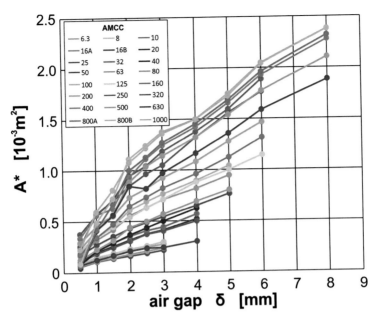

Fig. 10.327 Fictitious additional air gap cross-section for different cores as a function of the air gap length δ

$$A_L = \Lambda = \frac{1}{\frac{l_{Fe}}{\mu_r \mu_0 A_{Fe}} + \frac{2\delta}{\mu_0 (A_{Fe} + A^*)}} = \frac{\mu_0}{\frac{l_{Fe}}{\mu_r A_{Fe}} + \frac{2\delta}{(A_{Fe} + A^*)}} \cdot$$

If one equips the magnetic circuit with an air gap, one inserts a section with considerably lower magnetic conductivity (permeability). This makes the effective reluctance larger and the effective inductance smaller. A directly related process is that the magnetic field in the vicinity of the air gap expands into the surrounding space. If we now assume that the magnetisation cross-section AFe in the ferromagnetic part and in the air gap is the same, while fictitiously the additional resistance Rm* has the length δ, we obtain an "additional air gap area" per air gap with 2 air gaps

$$A^* = \frac{2\delta}{\mu_0 \frac{N^2}{L} - \frac{l_{Fe}}{\mu_r A_{Fe}}} - A_{Fe}$$

As can be seen from Fig. 10.327, the course A*(δ) apparently follows a power function A (x) ~ δ^x. Deviations occur with smaller air gaps. For sufficiently large air gaps, x = 0.7 proves to be a good solution.

Dividing the values A* by (δ/1 mm)$^{0.7}$ gives the plots in Fig. 10.328. They make it clear that an exponent x = 0.7 is a good approximation for a quite large range of values, since most of the points for a core each lie on a horizontal line.

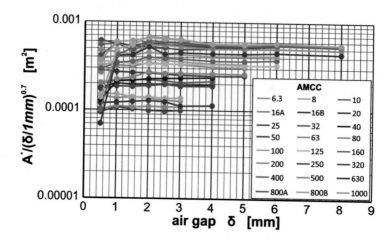

Fig. 10.328 Plot of the results of a division of A*(δ, A_Fe) by δ^0.7 as a function of the air gap δ (**a**) and the magnetisation cross-section A_Fe (**b**)

The area A* is provided with differently proportioned edges depending on the size. The perimeter of a magnetising surface is $(2 \cdot A + 2 \cdot D)$ [110]. Dividing A* by $(2 \cdot A + 2 \cdot D)$ gives results that are summarised in Fig. 10.329. The calculation results give an impression of how the magnetic field around the core expands into space. These are "reasoned estimates", not "exact ranges".

In order to also take into account the variance of the residual data in Fig. 10.329, the following function is inserted multiplicatively.

$$f(A_{Fe}) = \left(1 - \exp\left(-\frac{A_{Fe}}{9 \ \text{cm}^2}\right)\right)$$

The approximation function that can be used to describe the size of A*(δ) is then

$$A^* = 6.5 \ \text{cm}^2 \left(1 - \exp\left(-\frac{A_{Fe}}{9 \ \text{cm}^2}\right)\right) \cdot \left(\frac{\delta}{1 \ \text{mm}}\right)^{0.7}.$$

If one compares the values for A* determined from measurements with those of the approximating formula, one obtains the diagrams in Fig. 10.330. The general tendency is also well illustrated here. The partly divergent order of the curves is probably due to the different geometrical parameters of the cores (compare Fig. 10.325).

Finally, an approximation function for the description of the A_L values of the AMCC cores of sizes 6.3...1000 can be found in this way:

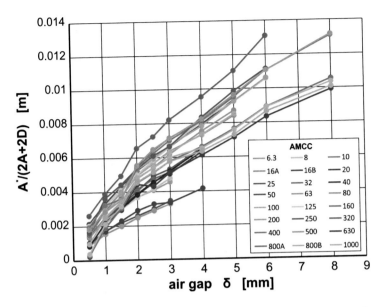

Fig. 10.329 Results of the division of the virtual additional area A* by the corresponding perimeter of the magnetisation area A_{Fe}

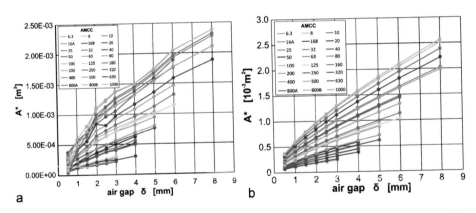

Fig. 10.330 Comparison of the experimental results for the virtual additional surface (**a**) and the computational results of the data model (**b**)

$$\frac{L(\delta=0)}{N^2} = A_L(I=0) = \cfrac{\mu_0}{\cfrac{l_{Fe}}{\mu_r \cdot A_{Fe}} + \cfrac{2 \cdot \delta}{A_{Fe} + 6.5 \ \text{cm}^2 \cdot \left(1 - \exp\left(\frac{-A_{Fe}}{9 \ \text{cm}^2}\right)\right) \cdot \left(\frac{\delta}{1 \ \text{mm}}\right)^{0.7}}}$$

A comparison of the approaches is shown in Fig. 10.331 for a selection of cores. The dashed lines show the results of the simplified calculation formulas. The solid lines show

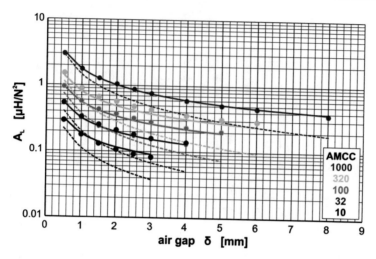

Fig. 10.331 Comparison of measured values (dots), results of the simplified calculation (dashed) and the extended calculation with the assumption of a calculation formula dependent on the magnetisation cross-section and the air gap (solid line)

the calculation results when considering the virtual area $A^*(\delta)$. Obviously, the approximation to the measured data when considering the correction function $A^*(\delta)$ is a significant improvement. Further improvements can probably be achieved if further parameter variations in Fig. 10.325 are taken into account.

The procedure described in this section shows that the rules of network analysis can be used to obtain an overview of the structure of relationships. This can be used beneficially for a mathematical approximation. Such approximations are typically less parameter-sensitive than general function systems without structure fitting. These approximations do not represent generally valid formulas. They must be parameterised anew for each core form.

10.33   Dimensioning of a Choke for a Buck Converter with AMCC Core

A very common application of inductive components are chokes for smoothing a current. Here, the differential or small-signal inductance acts to suppress strong current changes. At the same time, the maximum "working current" is usually much higher than the current amplitude of the high-frequency alternating component. Such a component must be dimensioned in such a way that not only the heat development is controlled, but also the saturation of the core material is prevented. This chapter will deal with an example from this subject area.

Figure 10.332 shows the circuit of a buck converter. The current ripple, which can be expressed via the current change, changes as a function of the duty cycle according to

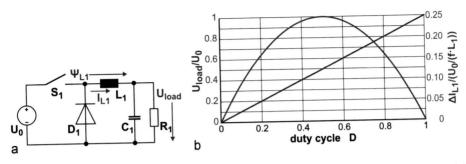

Fig. 10.332 Application situation of a storage and smoothing choke in a buck converter: (**a**) Principle circuit; (**b**) Variation of the output voltage and the current ripple with the duty cycle

$$\Delta I = \frac{U_d \cdot T}{L} D(D-1)$$

In the selected example, a smoothing reactor is to be dimensioned for a step-down converter. The following values are specified.

permanent DC current	$I_{DC} = 45$ A
Maximum DC current	$I_{shortDC.max} = 63.45$ A
Peak current	$I_{peak} = I_{shortDC} + \Delta I_{max}/2 = 63.45$ A $+ \Delta I_{max}/2$
Spec. resistivity at 20 °C	$\rho_{Cu20°C} = 0.0178$ Ohm·mm^2/m
Spec. resistivity at 100 °C	$\rho_{Cu100°C} = 0.0234$ Ohm·mm^2/m
Ambient temperature	$T_{amb.max} = 40$ °C
Maximum device temperature	$T_{max} = 100$ °C
Switching frequency	$f_{sw} = 16$ kHz
DC voltage	$U_{DC} = 1050$ V
Nominal inductance	$L = 750$ µH
Effective core permeability	$\mu_{Fe} = 8000$

With the buck converter, the maximum of ΔI occurs at $D = 0.5$ and then amounts to

$$\Delta I_{max} = \frac{U_d \cdot T}{4 \cdot L} = \frac{U_d}{4 \cdot L \cdot f} = \frac{1050 \text{ V}}{4 \cdot 750 \text{ µs} \cdot 16 \text{ kHz}} = 21.88 \text{ A}$$

With a triangular current, this corresponds to a maximum effective value of $I_{16kHz} = 6.32$A. The peak current at which the inductance may not yet reach saturation results in $I_{peak} = I_{shortterm} + \Delta I_{max}/2 = 63.45$A $+ 10.94$A $= 74.39$A.

The electric current and magnetic flux must be conducted through electric and magnetic conductors. Excessive electric current densities cause excessive heating. Excessive magnetic flux densities lead to saturation of core sections, which then lose their original function. Both must be avoided. The AMCC core family has a wide range for the cross-

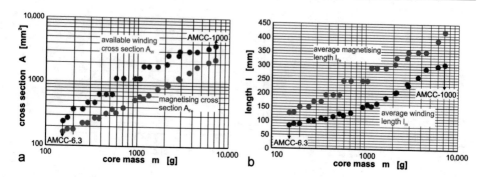

Fig. 10.333 Geometrical parameters of the AMCC core family: (**a**) Available cross-sections for core and winding; (**b**) Average lengths for one winding and for the magnetic field lines in the core

sections provided, as can be seen in Fig. 10.333. Of course, large areas always mean large weights for the components. Therefore, the smallest core that just meets the requirements must be selected from the cores offered. A possible way to do this is offered.

An inductor is to be dimensioned in such a way that it has both low losses and low volume for the most unfavourable operating case (i.e.: choke current $I_L = I_{max}$, flux linkage $\Psi \leq 1.3$ T). Because of the switching frequency of 16 kHz, AMCC is to be used as the core material. The value of $\mu_{Fe} = 8000$ shall be assumed as the value for the effective permeability of the material of the ribbon cores. The value of the inductance is set via parallel air gaps in both legs of the core. The inductance can then be approximately calculated as follows

$$L \approx \frac{\mu_0 \cdot N^2 \cdot A_{Fe}}{\frac{l_{Fe}}{\mu_{Fe}} + 2\delta}$$

As Sect. 10.32 shows, a better numerical representation of the relationship between the inductance and the geometric quantities is given by

$$\frac{L(A_{Fe}; l_{Fe}; \delta)}{N^2} = \frac{\mu_0}{\frac{l_{Fe}}{\mu_r \cdot A_{Fe}} + \frac{2 \cdot \delta}{A_{Fe} + 6.5 \text{ cm}^2 \cdot \left(1 - \exp\left(\frac{-A_{Fe}}{9 \text{ cm}^2}\right)\right) \cdot \left(\frac{\delta}{1 \text{ mm}}\right)^{0.7}}}$$

The specified maximum induction $B_{max} = 1.3$ T represents a limit value that should not be exceeded. Ultimately, this material value determines how much energy can be stored magnetically in a component. If one can determine the corresponding flux for a component, one has a condition that ensures that the component does not go into saturation. The following figure Fig. 10.334 shows for the core AMCC-100 the behaviour of the core at large DC exitation. In the double logarithmic scale, the "kink points" of all curves can be

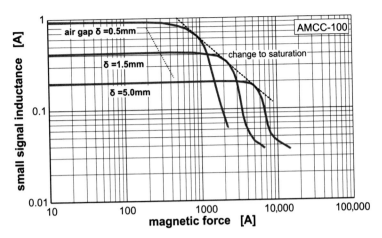

Fig. 10.334 Dependence of the small-signal inductance on the DC magnetisation for the AMCC-100 core with different air gaps (after [112])

connected by a straight line. They mark the transition of the core material into saturation. However, this is only an approximate solution.

To determine the point of flux linkage for saturation, one can write approximately

$$\Psi_{sat} = N \cdot B_{sat} \cdot A_{Fe} \approx L(A_{Fe}; l_{Fe}; \delta) \cdot I_{sat}$$

$$I_{peak} = I_{sat} \approx \frac{N \cdot B_{sat} \cdot A_{Fe}}{L(A_{Fe}; l_{Fe}; \delta)}.$$

L should be equal to 750 µH in the present case.

Then

$$I_{peak} = I_{sat} \approx \frac{N \cdot B_{sat} \cdot A_{Fe}}{750 \ \mu H}.$$

This defines the necessary number of turns to prevent saturation

$$N \approx \frac{750 \ \mu H \cdot I_{peak}}{B_{sat} \cdot A_{Fe}}$$

At the same time, the winding losses are

$$P_{V.Cu} = \rho_{Cu} \cdot \frac{l_m}{k_f \cdot A_W} \cdot (N \cdot I_{rms})^2$$

For the specific conduction losses, the value valid for the effective frequency at the maximum temperature of 100 °C is to be used. For the intended winding material HF stranded wire with a single wire thickness of 0.2 mm, the dependence shown in Fig. 10.335 was determined for the form of an air core coil of this core family. According to this, the

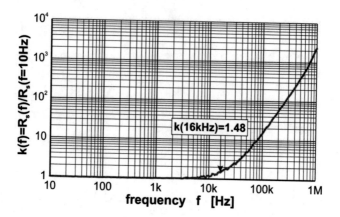

Fig. 10.335 Measured change in series resistance of a winding without core with the proportions of a winding of the AMCC core family when using HF stranded wire with 0.2 mm single wire diameter gauge

resistance coefficient of this stranded wire at 16 kHz increases by a factor of 1.48 compared to the amount at DC.

$$\rho_{Cu.16 \ kHz.100 \ °C} = 1.48 \cdot 0.0234 \frac{Ohm \cdot mm^2}{m}$$

Thus, to take into account the DC component and the 16 kHz component of the winding losses, one must apply

$$P_{V.Cu}(A_W \cdot A_{Fe}) = \rho_{Cu} \cdot \frac{l_m}{k_f \cdot A_W} N^2 \cdot \left(\rho_{Cu100 \ °C} \cdot I^2_{DC \ max} + \rho_{Cu100 \ °C} \cdot 1.48 \cdot I^2_{16 \ kHz}\right).$$

The utilisation factor for the winding room is usually made up of at least 2 factors. Factor $k_{f0} = A_{W.bobbin}/A_{W.core}$ results from the use of winding bodies. The other factor k_{fw} results from the filling factor of the winding material. Both together result in the total filling factor $k_{ftot} = k_{f0} \cdot k_{fw}$. Figure 10.336 shows this relationship for a value of $k_{fw} = 0.5$.

When estimating the core losses, one can use the Steinmetz formula with [113].

$$P_{V.Fe} = l_{Fe} \cdot A_{Fe} \cdot k_e \cdot \left(\frac{\widehat{B}}{1 \ T}\right)^\alpha \left(\frac{f}{1 \ kHz}\right)^\beta = l_{Fe} \cdot A_{Fe} \cdot 46,605 \frac{mW}{cm^3} \cdot \left(\frac{\widehat{B}}{1T}\right)^{1.74} \left(\frac{f}{1 \ kHz}\right)^{1.51}$$

The maximum peak value of the flux linkage occurs at D = 0.5 and amounts to $\max\left(\widehat{\Psi}\right) = \frac{U_d}{8f} = \frac{1050 \ V}{8 \cdot 16 \ kHz} = 0.0082$ Vs.

This gives the formula for different core dimensions

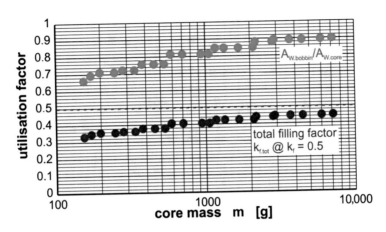

Fig. 10.336 Utilisation factor for the winding window for AMCC cores as a function of core weight

$$P_{V.Fe} = l_{Fe} \cdot A_{Fe} \cdot 46{,}605 \frac{mW}{cm^3} \cdot \left(\frac{0.0082 \, Vs}{N \cdot A_{Fe} \cdot 1 \, T} \right)^{1.74} \left(\frac{16 \ kHz}{1 \ kHz} \right)^{1.51}$$

Replacing the number of turns with the value found above gives.

$$P_{V.Fe} = l_{Fe} \cdot A_{Fe} \cdot 46.605 \frac{mW}{cm^3} \cdot \left(\frac{0.082 \ Vs \cdot B_{sat} \cdot A_{Fe}}{750 \ \mu H \cdot I_{peak} \cdot A_{Fe} \cdot 1T} \right)^{1.74} \left(\frac{16 \ kHz}{1 \ kHz} \right)^{1.51}$$

If one calculates the occurring power losses for the different AMCC cores, one obtains the figure Fig. 10.337 under the secondary condition of $B_{max} = 1.3$ T. There, in addition to these two types of losses and their sum, the calculated number of turns for the respective

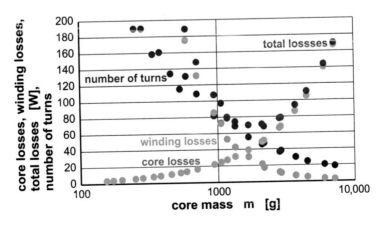

Fig. 10.337 Winding numbers and winding and core losses for different sizes of AMCC cores

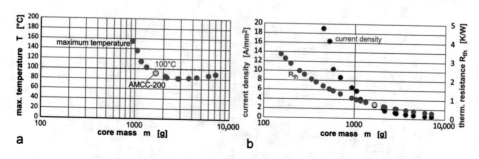

Fig. 10.338 Parameters of the solution of the dimensioning problem: (**a**) Thermal resistance and associated current density for the assumptions made; (**b**) Calculated maximum temperature in the component with the core AMCC-200 as solution for the task with the boundary conditions

core size is also entered. If the cores are too small, the number of windings increases drastically in order to avoid saturation. But the calculated winding losses increase even faster. The apparent solutions thus become irrelevant.

If a certain temperature is not to be exceeded, a relation between core size and thermal resistance R_{th} must be known and applied. The internal temperature of the component is then calculated with the formula

$$T = T_{amb} + R_{th} \cdot P_{loss.tot} = 40 \ \degree C + R_{th} \cdot P_{loss.tot}.$$

Usually the R_{th} of a core family is not known. One can help oneself by determining it experimentally or by extracting it from a similar core family. In the present case, the following was determined from [114] for similar core materials with similar sizes.

$$R_{th} \approx 1912 \frac{K}{W} \left(\frac{V_{core}}{1 \ mm^3} \right)^{-0.6368}$$

In Fig. 10.338b the resulting thermal resistances for the cores AMCC-6.3...AMCC-1000 are depicted. This allows the maximum temperature of the choke to be calculated or estimated as in Fig. 10.338a. With the core AMCC-200 (1.87 kg, 70 W) the specifications can be met. The calculated temperature is 92 °C. Since the thermal resistance R_{th} can be reduced with forced cooling, smaller cores with higher number of turns can be used with different cooling. The increase in the number of turns must be (partially) compensated for with an increased air gap. But this measure is limited in reality, as Sect. 10.32 shows. At the same time, the resulting current density is shown in Fig. 10.338b. For the AMCC-200 core, 2.82A/mm² is calculated for the given boundary conditions. This is within the usual range for such components. It should be noted that the requirement for minimum losses with this calculation method would lead to a core with a higher weight (AMCC320, 2.17 kg, 68 W).

Für ausgewählten Kern AMCC-200 berechnet man als erforderliche Windungszahl $N = 55$. Daraus lässt sich der erforderliche A_L-Wert berechnen zu

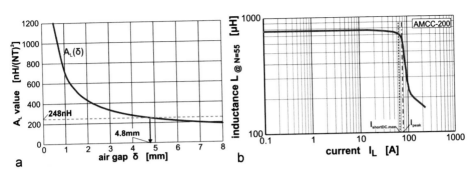

Fig. 10.339 Setting the A_L value sought via the air gap: (**a**) Dependence of the A_L value on the air gap; (**b**) Course of the small-signal inductance of an AMCC-200 choke as a function of the current through the inductance

$$A_{L.AMCC320} = \frac{L}{N^2} = \frac{750 \ \mu H}{55^2} = 248 \ nH$$

This value must be set via the air gap of the selected core. For this purpose, the approximated dependence (see Sect. 10.32) of the A_L value on the air gap δ is shown in Fig. 10.339a for the core AMCC-200. The target 248nH leads here to an air gap δ = 4.8 mm, which must be inserted into each leg of the core.$A_{L.AMCC200}(\delta) = \dfrac{\mu_0}{\frac{29.2 \ cm}{8000 \cdot 7.79 \ cm^2} + \frac{2 \cdot \delta}{7.79 \ cm^2 + 6.5 \ cm^2 \cdot \left(1 - \exp\left(\frac{-7.79 \ cm^2}{9 \ cm^2}\right)\right) \cdot \left(\frac{\delta}{1 \ mm}\right)^{0.7}}}.$

With these data it is possible to determine the dependence of the inductance on the current from [115] by interpolation and scaling (Fig. 10.339b). The values of the maximum currents are entered for orientation. The presented methodology is a powerful approximation method. More accurate results can be obtained by iteration or by solving the non-linear system of equations.

10.34 Current Ripple Cancellation in DC/DC and DC/AC Converters

Around 90% of all power electronic applications are based on the principle of pulse width modulation (PWM). Since the following explanations refer to this, this principle is explained again in Fig. 10.340. A voltage (in the example $-u_D(t)$) is applied to an inductance L through a switch S. The current i increases. This increases the current $i_L(t)$ in the choke L inserted to limit the current increase. When the converter changes the switched voltage, the current is driven further by the choke 'L'. Due to the associated energy output of the current storage 'L', the current decreases. This results in a triangular-shaped current (Fig. 10.340b) that oscillates around its mean value. A constant/stationary

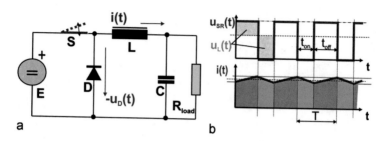

Fig. 10.340 Principle of pulse width modulation using the example of a buck converter: (**a**) Circuit diagram; (**b**) Current and voltage curves

mean value of the current requires that the voltage mean value at the choke L is zero. Changing the duty cycle changes the output voltage and current ripple (see Sect. 10.33).

If a second converter is constructed (Fig. 10.341) so that it produces exactly current rises with the same Δi and inverse rise at the according cycle and the mean value of the current is eliminated, the alternating current components of converter SR1 and SR2 cancel each other out. The separation of the mean value of the current in the output of converter 2 can be done with a capacitor of adapted size, since capacitors cannot carry direct current. This idea is described, for example, in [116].

The implementation of this concept is shown in Fig. 10.342. Here, a buck converter (SR1) in a half-bridge circuit has been supplemented with an inversely controlled half-bridge (SR2). The output voltage at the load Z_{load} is set according to the duty cycle of the control signal. The load is supplemented by a parallel capacitor, as corresponds to most applications of the buck converter. This represents a short circuit for higher frequency (ripple) currents. This means that the HF impedance of the load becomes almost zero. This means that SR1 and SR2 are largely decoupled in their operation. This means that if a current is built up in L_1* in phase opposition to the current in L_1, it can partially or completely compensate for the current i_{L1}. The capacitors C_1* and C_1** galvanically isolate the power converter SR2 from the power converter SR1. The supply voltage $U_{d1}*$ can thus theoretically be selected independently of U_{d1}.

The additional blue inserted power converter SR2 (T_3;D_3;T_4;D_4) supplies alternating current through the inserted capacitors C_1* and C_1**. If one assumes a sufficient size, the AC voltage at these capacitors is low and the current courses can be calculated in a simplified way.

Change in current when T_2 is switched on:

$$\Delta i_{L1}|_{T2=ein} = -\frac{u_{Zload}}{L_1}t_{1on} = -\frac{u_{Zload} \cdot T}{L_1}D_{T2}$$

When switching off T_2 and switching on T_1, the following then applies accordingly

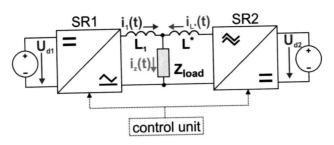

Fig. 10.341 Principle of current ripple lowering and suppression in PWM converters

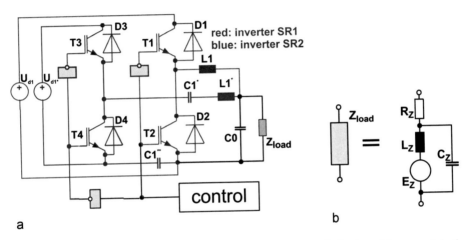

a   b

Fig. 10.342 Example of a bidirectional buck converter with active elimination of the current ripple:
(**a**) circuit with galvanically isolated DC source to supply the compensation converter; (**b**) approach
of a general load

$$\Delta i_{L1}|_{T2=off} = \frac{U_{d1} - u_{Zload}}{L_1} t_{T2off} = \frac{(U_{d1} - u_{Zload}) \cdot T}{L_1}(1 - D_{T2}) = \frac{(U_{d1} - u_{Zload}) \cdot T}{L_1} D_{T1}.$$

From this, it is easy to calculate the well-known formula for the current ripple of a buck
converter

$$\Delta i_{L1} = \frac{(U_{d1} - (1 - D_{T2})U_{d1}) \cdot T}{L_1}(1 - D_{T2}) = D_{T2}(1 - D_{T2})\frac{U_{d1} \cdot T}{L_1}$$

$$= (1 - D_{T1})D_{T1}\frac{U_{d1} \cdot T}{L_1}$$

If T_1 is switched off and T_4 is switched on, the following applies accordingly to the branch
with $L_1{}^*$, taking into account the fact that $C_1{}^*$ and $C_1{}^{**}$ must carry the mean value of the
PWM-converted voltage between $U_{d1}{}^*$ and u_1

Fig. 10.343 Example of
elimination of harmonics from
the output current of a
PWM-controlled SR1 converter
with an inverse-controlled SR2
converter (compare Fig. 10.345)

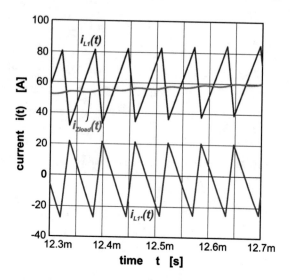

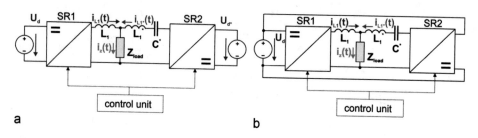

Fig. 10.344 General design examples of the principle presented: (**a**) Design with two galvanically isolated sources; (**b**) Design with supply of the two converters from the same source

$$\Delta i_{L1*}|_{T4 = on} = -\frac{u_{AC} + u_{C1*}}{L_1^*} T \cdot D_{T4}$$

If you add the currents through L_1 and L_1^* in the interval in which applies (T_1 = off and T_4 = on), you get

$$\Delta i_{L1}|_{T1 = aus} + \Delta i_{L1*}|_{T4 = ein} = \left(\frac{U_{d1}}{L1} - \frac{U_{d1*}}{L1^*}\right) \cdot D_1(1 - D_1) \cdot T$$

This means that the addition of the currents is exactly zero when $U_{d1}/L_1 = U_{d1*}/L_1^*$. Figure 10.343 shows an example of how the currents I_{L1} and i_{L1*} form and largely compensate for the switching frequency components in the resulting mains current i_{Zload}.

A general representation of the solution is shown in Fig. 10.344. Power converter SR1 is the PWM-controlled unit with the current-storing choke 'L_1' and the general load Z_{load}. The cancelling current is fulfilled by another PWM-controlled converter SR2 in connection with the current storage choke L_1^* and a capacitor C^*. The inductance L_1^* and C^* have the task of generating a current curve i_{L1}^* from the switched output voltage of SR2 in such a way that for a sufficient averaging period the following applies

$$i_{L1*}(t) = \overline{i_{L1}(t)} - i_{L1}(t).$$

If these properties are fulfilled, the following then applies for the load current because of Kirchhoff's node rule

$$i_Z(t) = i_{L1}(t) + i_{L1*}(t) = \overline{i_{L1}(t)}$$

This corresponds to a cancellation of the harmonics of $i_{L1}(t)$ in the current $i_{Zload}(t)$, as Fig. 10.343 shows. The two voltage sources U_{d1} and U_{d1}^* supply the two converters SR1 and SR2 galvanically separated in Fig. 10.344a. DC isolation is provided by the two capacitors C_1^* and C_1^{**}. Thus, the voltage supply is possible from separate as well as from a common voltage source U_d, as fig. 10.344b shows.

By adjusting the ratios of U_d, $L_1 = L_1^*$ and C^*, a current $i_{L1}^*(t)$ can be derived from the PWM pattern of SR1 that fulfils the above condition. A special application of the principle according to Fig. 10.334 for sinusoidal energy supply and energy extraction from the mains is shown in Fig. 10.335. Here the power converter SR1 is designed as a half-bridge circuit with a divided voltage source. By differentiating the output voltage with a part of the DC input voltage, this circuit can supply voltages with positive and negative polarity. It is the basis of most single-phase and multi-phase mains-coupled converters. Such circuits are used for active frontend rectifiers and grid-connected suppliers of AC power from wind turbines; battery storage or photovoltaic systems. It is clear from Fig. 10.345b that the compensation device almost completely succeeds in eliminating the high-frequency alternating component of the load current.

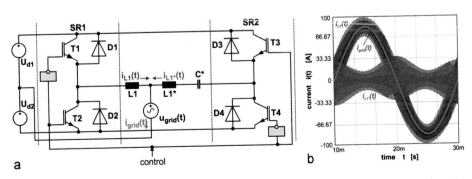

Fig. 10.345 Circuit for sinusoidal current injection into a single-phase grid with the compensation principle according to the idea: (**a**) Circuit diagram with a half-bridge and divided supply voltage; (**b**) Examples for the current characteristics $i_{L1}(t)$, $i_{L1}^*(t)$ and $i_{grid}(t)$

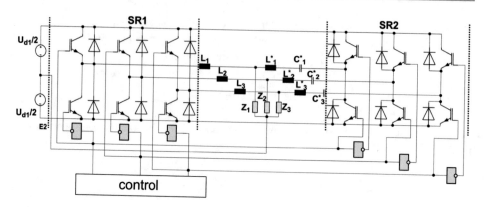

Fig. 10.346 3-phase bridge inverter with the current ripple compensation principle by a second (DC/AC) converter

The half-bridge circuit according to Fig. 10.345 is the basic component for an overwhelming part of single-phase and multi-phase circuits for bidirectional DC/AC conversion. Figure 10.346 shows an example of the compensation device for a 3-phase circuit. Here the circuit arrangement Fig. 10.345 is simply extended by 2 further half-bridge branches. In this way, a 3-phase load can be supplied. No neutral wire is required in the delta connection. A star connection of the load can also be operated without a neutral wire. With unbalanced loads, a neutral wire can be created by connecting the star point of the load to the connection point of the divided voltage U_{d1}. However, the interconnection of the two sources $U_{d1}/2$ must be capable of this.

In [117, 118], further approaches to reduce current ripple in 1-phase and multiphase systems are presented. In contrast, the method presented in the present chapter is much more effective. In [119] an overview of the topic "active filtering" in networks is given, so that one can get an impression of possible applications of this principle as well.

The inverter circuit in Fig. 10.345 provides 2 voltage levels from which the load voltage is generated by PWM. Especially to reduce the voltage stress on the semiconductors and to reduce the physical size of L_1, the circuit in Fig. 10.345 can be extended to a higher number of levels. This is made possible by the possible isolated operation of SR2. A possible version is shown in Fig. 10.347. The converter SR2 can be operated from half the DC voltage $U_{d1}/2$. However, this would lead to an asymmetrical load of the two voltages $U_d/2$ forming the DC source. According to the calculations carried out at the beginning, however, a supply from the full DC voltage U_{d1} is also possible by selecting the inductance $L_1^* = 2 \cdot L_1$. The provision of a separate (regulated) DC voltage is also conceivable. The additional effort is limited because only the losses in SR2 have to be covered. Many multilevel topologies for inverters can be traced back to the interconnection of several half-bridge circuits. Often, the principle circuit described can also be applied here. The possibility of separating the operating voltage supply for the main converter SR1 and the compensation converter SR2 proves to be particularly advantageous.

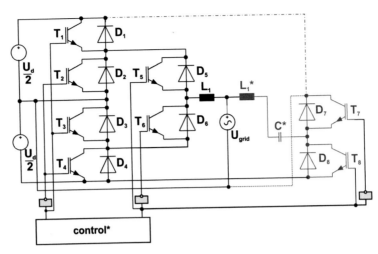

Fig. 10.347 Application of the described current ripple suppression method to a 3 level inverter (active neutral point clamped (ANPC) inverter)

For a brief investigation of the properties of this method, Fig. 10.342 is used as a starting point. The losses of the choke L_1 and the saturation induction essentially determine the size of the choke. For further considerations, the following quantities are to be assumed.

DC supply voltage	U_d
Switchung frequency	fsw
Maximum DC current	I_{DC}
Maximum current change	ΔI
Winding window	A_W
Filling factor	k_f
average winding length	l_m
Magnetisation cross-section	A_{Fe}
Magnetisation length	l_{Fe}

The maximum current change is related to other variables and is maximum at $D = 0.5$

$$\Delta I = \frac{U_d}{4L \cdot f_{sw}}$$

From this, the amplitude of the AC flux linkage can be calculated with

$$\widehat{\Psi} = N_{L1} \cdot A_{Fe} \cdot \widehat{B}_{HF} = L \cdot \frac{\Delta I}{2} = \frac{U_d}{8 \cdot f_{sw}}$$

For the choke L1, due to the assumed high direct current, the following condition results

$$\Psi_{max} = L\left(I_{DC} + \frac{\Delta I}{2}\right) \le N_{L1} \cdot B_{max} \cdot A_{Fe}$$

The winding losses result for L_1 from

$$P_{loss.Cu1} = \frac{l_{m1}}{A_{W1}} \cdot \frac{N_{L1}^2}{k_f}\left(\rho_{Cu} \cdot I_{DC}^2 + \rho_{CuHF} \cdot \frac{1}{12}(\Delta I)^2\right)$$

The core losses can be estimated with the Steinmetz formula to be

$$P_{loss.Fe1} \approx A_{Fe1} \cdot l_{Fe1} \cdot k_e \cdot \left(\frac{\widehat{B}_{HF}}{1T}\right)^\alpha \cdot \left(\frac{f_{sw}}{1kHz}\right)^\beta = A_{Fe1} \cdot l_{Fe1} \cdot k_e \cdot \left(\frac{U_d}{N_{L1} \cdot A_{Fe1} \cdot 1T}\right)^\alpha$$
$$\cdot \left(\frac{f_{sw}}{1kHz}\right)^\beta$$

Since the choke $L_1{}^*$ only has to carry the alternating part of the current, its geometric size will be much smaller than that of L_1. The calculations so far lead to a complete cancellation of the current ripple if one assumes idealised components. However, this is not the case due to device tolerances, temperature dependencies and non-linearities. The reduction of the current ripple $\Delta I/I_{DC}$ is only possible down to a residual ε. On the other hand, this also means that the proposed combination of L_1 and $L_1{}^*$ emulates an apparent choke of

$$L_{virt} = \frac{U_d}{4 \cdot \varepsilon \cdot \Delta I \cdot f_{sw}} = \frac{L_1}{\varepsilon}$$

This would require for L1 a number of turns of at least

$$N_{L10} \ge \frac{L_{virt}\left(I_{DC} + \frac{\Delta I}{2}\right)}{B_{max} \cdot A_{Fe}} = \frac{L\left(I_{DC} + \frac{\Delta I}{2}\right)}{\varepsilon \cdot B_{max} \cdot A_{Fe}}$$

The number of turns is included in the winding losses by the power of two. Then the following would apply

$$P_{loss.Cu1} \approx \frac{l_{m1}}{A_{W1}} \cdot \frac{N_{L10}^2}{\varepsilon^2 k_f}\left(\rho_{Cu} \cdot I_{DC}^2\right)$$

This requires a much larger core with a larger winding window. Consequently, the main effect of this proposal for current ripple compensation is the emulation of a significantly larger choke. The success is bigger the more similar the core materials are in terms of linearity and temperature performance. Because the control of the SR2 current converter consists practically only of negated control signals of the SR1 current converter, it can be realised with simple logic circuits. The control of the converter for a specific purpose remains unaffected by this. This means that it is even possible to subsequently improve a power converter with regard to its current ripple by adding an appropriately dimensioned SR2 power converter. Because of the additional effort, it is worthwhile to use it especially for higher power ratings or in applications where a low current ripple is combined with the requirement for a very low weight. The construction power of the additional current converter is significantly lower than that of the main current converter. If one assumes L1* for the rms current of the choke

$$I_{L1*} = \frac{1}{\sqrt{3}} \cdot \frac{\Delta I}{2},$$

one obtains an approximate switching capacity of

$$S_{nom.SR2} = \max{(I_{L1*})} \cdot \frac{U_d}{2} \approx \frac{1}{\sqrt{3}} \cdot \frac{\Delta I}{2} \cdot \frac{U_d}{2} = 0.144 \cdot \Delta I \cdot U_d.$$

This is much less than the switching capacity of the main power converter, which can be estimated as

$$S_{nom.SR1} \approx \max{(I_{DC})} \cdot U_d$$

Das ungefähre Verhältnis der Bauleistungen beträgt folglich nur

$$\frac{S_{nom.SR2}}{S_{nom.SR1}} \approx \frac{0.144 \cdot \Delta I}{\max{(I_{DC})}}$$

Consequently, one can achieve a great effect with a relatively small additional effort.

10.35   Controllable Inductors in Parallel Operation

In recent decades, amorphous and nanocrystalline materials have been developed that have properties that also make them interesting in modern power electronics and in connection with power electronics circuit technology. In this chapter, controllable inductive components with nanocrystalline and amorphous materials are presented and compared with the use of grain-oriented electrical steel. The "main windings" are connected in parallel. The materials made of amorphous and nanocrystalline materials are characterised by very low coercivity and a permeability that continuously decreases with increasing magnetic field strength. Both are very advantageous in the realisation of controllable inductive components.

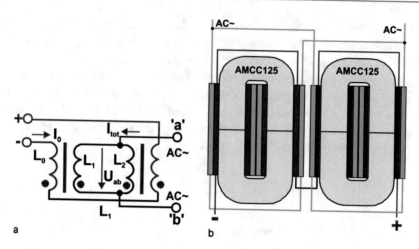

Fig. 10.348 Winding arrangement for generating a DC-controllable AC choke with amorphous AMCC cores: a winding diagram, b geometric layout

Design of a Controllable Inductor with Wide Control Range and High Linearity

Figure 10.348 shows the design of a magnetically controllable inductor for alternating current applications. Two chokes of the same design are connected in parallel, their cores being premagnetised in opposite directions. For this purpose, each core carries 2 windings: one winding for alternating current (AC) and one winding for direct current (DC). The winding for alternating current must be designed for the maximum operating current. The direct current winding must be able to carry the current that is continuously required with minimum inductance.

For experimental purposes, in the present example, half of the winding space was used for the AC winding and half for the DC winding in order to determine the characteristics at maximum modulation. The DC windings are connected in series. Since the AC windings are connected in parallel, alternating voltages are induced in the DC windings, which are largely compensated for by the windings being connected in opposite directions. As a result, the total coupling factor for the control winding and the working winding is practically zero. Ideally, no AC voltages are induced from the AC winding in the control winding. Due to the DC bias, the magnetisation of the two cores are shifted against each other. The resulting inductance L_{tot} results from the parallel connection of the two individual inductances.

In order to better understand the effect of this connection, one can use Fig. 10.349. This picture, which can be used for the simulation of the circuit according to the explanations in Sect. 5.6, shows 2 cores premagnetised with the magnetic force $I_0 \cdot N = I_0 \cdot N_1 = I_0 \cdot N_2$ with AC windings connected in parallel (see Fig. 10.348). It must be pointed out that the shown combination of inductances and current sources is not accepted in some simulation programs and can lead to numerical problems. Inserting the loss resistor parallel to the

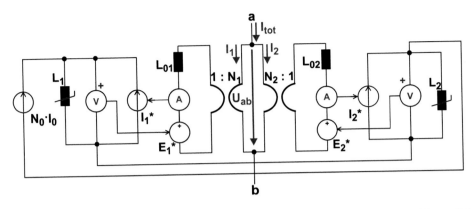

Fig. 10.349 Simulation circuit for the behaviour of 2 premagnetised inductors connected in parallel

non-linear inductances then usually solves the problem. The inductances L_{01} and L_{02} represent the inductance of the winding without core.

The small signal behaviour is the easiest way to explain the function of the premagnetised chokes connected in parallel: Let each core be premagnetised in a different direction. The differential inductance then results from the parallel connection of the individual small-signal inductances L_{diff}.

$$L_{diff.total} = \left(\frac{1}{L_{diff1}(I_1 + I_0)} + \frac{1}{L_{diff1}(I_1 - I_0)} \right)^{-1}$$

The integration over the parallel-connected inductances then yields the flux linkages we are looking for:

$$\Psi = \int \left(L_{diff1} // L_{diff1} \right) dI$$

For illustration purposes, both relationships are shown in Fig. 10.350. In addition to a premagnetisation current of zero, two further premagnetisation currents are taken into account there.

Fig. 10.350 shows how the inductance changes with this control method for the use of a core AMCC125. The permeance λ is equal to the respective AL value at a corresponding current. It can be seen that a large change in inductance can be achieved with a relatively low current. In order to be able to make statements independent of the number of windings, the dependence on the magnetic force was shown. Considering that the winding window of the AMCC125 has 1075 mm^2, a maximum current density of 3.5 A/mm^2 and a fill factor of 30% results in a maximum maximum flux of 2180 A per winding. While Fig. 10.350a shows the differential permeance of the arrangement, Fig. 10.350b shows the corresponding magnetic flux of the arrangement. The magnetic saturation flux of the

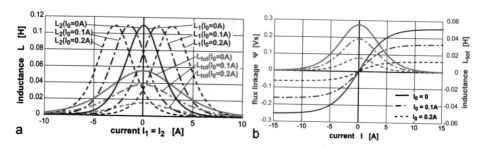

Fig. 10.350 Effect of the parallel connection of 2 premagnetised windings: (**a**) Change in the small-signal inductance, (**b**) Change in the magnetisation characteristic of the parallel connection (AMCC-100, $\mu_r = 8000$, $N_1 = N_2 = 70$, $N_0 = 1000$, $2 \cdot \delta = 0.1$ mm)

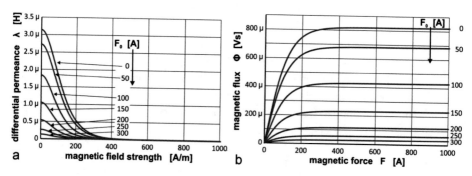

Fig. 10.351 Control characteristics of a magnetic component with an AMCC125 core: (**a**) Differential permeance (diff. A_L value) and (**b**) magnetic flux as a function of the magnetic force

arrangement is influenced by the control performance. The similarity to the characteristic curves of MOSFETs is striking, in which the saturation current of the drain I_{Dsat} at large current values and the R_{DSon} at small currents can be controlled with the gate voltage. Essential differences result on the one hand from the fact that the characteristic curves shown in Fig. 10.351b are symmetrical in the first and third quadrants (compare Fig. 10.350b) In addition, the vertical axis of the diagram represents the current in the time domain in the case of the MOSFET, whereas in the case of the transducer under consideration it represents the voltage integrated over time. This has consequences for the different applications of the components.

The saturation flux in Fig. 10.351b can be approximated in a wide range when using AMCC or similar materials with

$$\Phi_{max}(F_0) = 818\mu Vs \cdot \cosh^{-1}\left(\frac{F_0}{255A}\right)$$

This function obviously tends towards zero at high magnetic forces. In reality, however, the relative permeability does not become smaller than 1. This results in a limit value $\Phi_{max} > 0$.

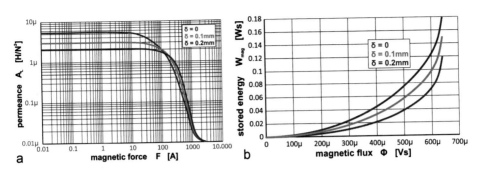

Fig. 10.352 Changes in permeance (**a**) and energy storage capacity (**b**) of a controlled AMCC125 core when inserting small air gaps

The higher the permeability of a material, the less control current is needed to change the inductance, but the lower the storable energy density in this core material. It is therefore obvious to insert an air gap to achieve a greater storage capacity. This reduces the permeance (the A_L value) of the core and increases the linear utilisation range (Fig. 10.352a) and the maximum storable energy increases (Fig. 10.352b).

With the insertion of an air gap, the control current to be used increases, as shown in Fig. 10.353a. There, the maximum permeance of the core is shown as a function of the controlling magnetic force. This occurs with this material at low control currents at $I_{AC} \sim 0$. In the present practical examples, the control of the saturation flux can be approximated very well by a cosine hyperbolic function. This approximation works particularly well in the lower range of the controlling magnetic force. The relative sensitivity of the change in permeance to the controlling current is shown in Fig. 10.353b.

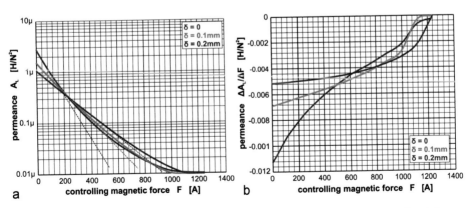

Fig. 10.353 Veränderungen Änderung der maximalen Permeanz durch die steuernde Durchflutung: (**a**) als absolute Größe und (**b**) als relative Größe

Fig. 10.354 Symbolic representation of a controllable inductance: (**a**) basic symbol, (**b**) equivalent circuit with parasitic elements

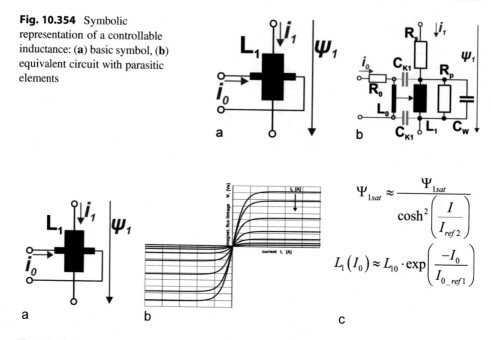

Fig. 10.355 Control characteristics of a controllable inductance according to Fig. 10.354: (**a**) principle circuit, (**b**) characteristic curve, (**c**) analytical description possibility

In Fig. 10.354 the previous statements were summarised in a symbolic way. The controllable inductance was shown as an isolated controllable two-port. Figure 10.354a shows the basic schematic structure of an inductor L_1 controlled by an isolated control branch. Figure 10.354b also contains the most important parasitic elements for a dynamic description.

With this method of representation, a characteristic field of an active quadripole can be described similar to a MOS transistor, as Fig. 10.355 shows. Formally and in terms of shape, much is reminiscent of MOS transistors. As many things in magnetic components are dual to corresponding electrical components, such characteristics can also be found here. At control current $I_0 = 0$, the saturation flux is maximum. When the control current is increased, the saturation flux decreases. The controlled inductance is basically an alternating current component. The polarity of the control current is irrelevant. The design described above achieves a minimal feedback effect from the working circuit into the control circuit.

$$\Psi_{1sat} \approx \frac{\Psi_{1sat}}{\cosh^2\left(\frac{I}{I_{ref2}}\right)}$$

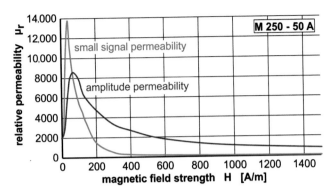

Fig. 10.356 Permeability values for the electrical steel sheet M250-50A. (according to [120])

$$L_1(I_0) \approx L_{10} \cdot \exp\left(\frac{-I_0}{I_{0\_ref1}}\right)$$

What the characteristic curves look like exactly depends very much on the core material used. Many of the commutation curves of soft magnetic materials have the shape of a double S-curve. The consequence is that the maximum permeability is not reached at current $I_1 = 0$, but at $|I_1| > 0$. When two inductors with mutually shifted characteristics are connected in parallel, this leads to further non-linearities in the curves. Figure 10.356 shows such an example for an electrical steel sheet M250 - 50A. Starting from an approximate value $\mu r = 2000$ at $I_{L1} = 0$, the differential permeability increases to more than 13,000 and then decreases to small values at correspondingly high polarisations. With parallel connection and premagnetisation, these high small-signal permeabilities are weakened in their effect by shearing. The same effect is observed when an air gap is inserted.

This can create the problem that satisfying higher linearity requirements leads to higher required control currents. The selection of suitable materials is therefore of particular importance.

For transformers, a controllable inductance can be used as magnetising inductance (see Sect. 10.15). Figure 10.357 shows the basic structure of an electrically controllable transformer and the resulting characteristic curves.

In this transformer, which consists of 3 cores, the input voltage is divided according to the reactance of the primary windings. With unloaded secondary voltage U_s, the two secondary winding voltages U_{21} and U_{22} are obtained from the transformed parts of the total primary voltage. The voltage division takes place via the magnetising inductances of the upper and lower transformer. While the primary voltages add, the secondary voltages subtract. This leads to a variable secondary total voltage. The internal voltage drop of the transformer results from the leakage reactances of the windings and is increased by the magnetising inductance. At maximum output voltage U_s, the output impedance becomes

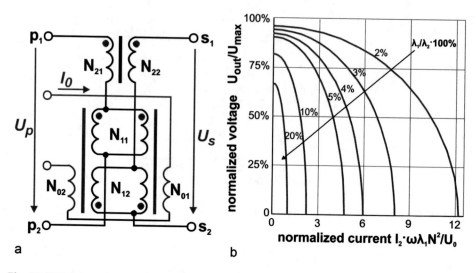

Fig. 10.357 Magnetically controllable transformer as a possible application magnetically controllable inductances: (**a**) Circuit arrangement, (**b**) Principle curve of the load characteristic at different permeances Λ_1 und $N_{11} = N_{12} = N_{21} = N_{22} = N = 22$, $\lambda_2 = 5.44\ \mu H/N^2$

minimum and the short-circuit current is therefore maximum. At minimum output voltage, the output impedance is maximum and the short-circuit current is therefore minimum.

In principle, this circuit allows a phase inversion of the secondary voltage with suitable dimensioning of the number of windings. This means that for one current range, a variable AC output voltage is obtained in phase with the AC input voltage, and for another control current range, a voltage with opposite phase position (see laboratory setup at Fig. 10.358).

Another obvious application is the construction of an electromagnetically controllable filter. The design induces two voltages in series in the control windings, both in phase and rotated by 180°. The resulting induced voltage is thus practically zero. This means that the actuator "variable filter" is largely reaction-free. Figure 10.359 shows the circuit of such an arrangement with measured impedances dependent on the frequency.

Such an arrangement can be used, for example, as a harmonic absorber filter to filter out harmonic currents in the mains. For this purpose, series resonant circuits are used between the lines, which are more or less set to the harmonic frequency as the resonant frequency. Since the harmonics in the network practically only contain reactive line components, this filter circuit covers the corresponding reactive power demand of the harmonics for a load or in a certain network area. The control range required for this is considerably smaller than that shown in Fig. 10.359. To limit the setting range, simply connect the minimum required inductance L_0 in series (Fig. 10.360a). With the same current range as in Fig. 10.359 only the desired range for setting a certain impedance is set. In this way a certain harmonic mains current (here the fifth harmonic for a 50 Hz mains) can be set adaptively to the prevailing conditions in the mains. The impedance of the oscillating circuit at 250 Hz can be continuously adjusted according to the requirements.

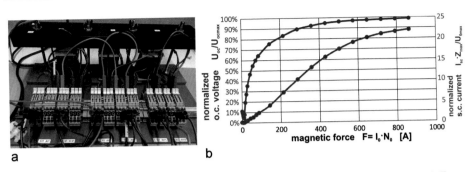

a   b

Fig. 10.358 Laboratory set-up of a controllable transformer according to Fig. 10.357: (**a**) Transformer consisting of 3 core pairs AMCC125; (**b**) Measured characteristics for open-circuit voltage and short-circuit current

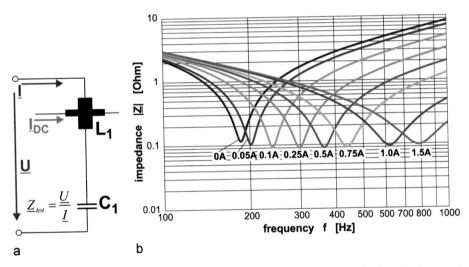

a   b

Fig. 10.359 Magnetically controllable filter with controllable inductance: (**a**) principle circuits; (**b**) measured frequency-dependent impedance curves

In the present case, cut strip-wound cores SU102a (Waasner company) with an air gap of 0.2 mm were used for the controllable choke. The material C5 has similar properties as described above. However, the resulting non-linearities are largely suppressed by the shearing of the core with the air gap and could no longer be found in the effect of the circuit. Figure 10.361 shows how with a small change of the DC current in a total development resistance of 4.8 ohms, a change of the AC current by a factor of 9.5 is achieved.

The quality with which the desired effects are achieved depends on the core materials used. Therefore, some other relevant material problems are also discussed here. First of all, it should be noted once again that rolled ferromagnetic sheets in particular have a magnetisation characteristic that has a strongly varying differential permeability. The

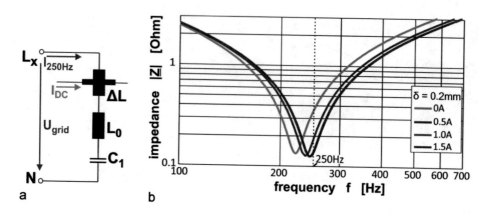

a   b

Fig. 10.360 Harmonic absorber filter for the fifth harmonic in a 50 Hz network: (**a**) Principle circuit; (**b**) Impedance curve of the filter with different control current in the vicinity of the 250 Hz

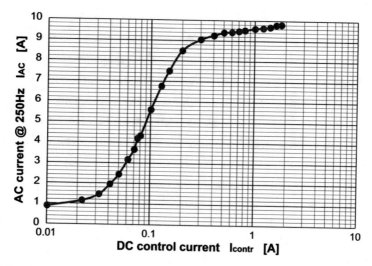

Fig. 10.361 Change of AC current at a frequency of 250 Hz with a suction circuit filter of adjustable frequency

differential permeability is responsible for the small-signal inductance or small-signal permeance.

However, there are also dynamic parameters that are of interest for the operation of the components. The control loop is inductive and can be characterised by a non-linear inductance. The reaction to the control signal depends on how fast and how the material of the core to be controlled can change its properties. There are serious differences between materials in this ability. The ferromagnetic properties of a material are the result of the superposition of different material properties. These have different dynamics. For a controllable element, a differential permeability that decreases monotonically with the control

Fig. 10.362 Change of the small signal impedance after switching off a control current with and without air gap

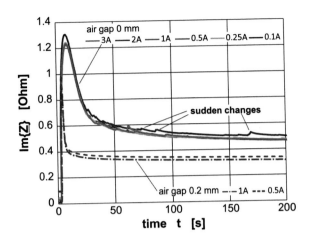

signal is advantageous. This property is found, for example, in many ferrites as well as in amorphous and nanocrystalline metallic materials. However, the rate of change of the material properties also seems to depend on the amplitude of the control signal. This is illustrated in Fig. 10.362. Here it was investigated for an electrical steel sheet how fast the small-signal inductance behaves after a control current is switched off. The behaviour with and without an air gap was compared. It can be determined that with this material without an air gap, the original value in the relaxed state is only reached after many minutes. Even after inserting a small air gap, this process accelerates considerably. The causes are probably accommodation processes in the core material. This is also indicated by sudden changes in the small-signal impedance (measured at 250 Hz). These occur repeatedly after other times and are probably due to spontaneous rearrangement processes in the crystal structure of the material. In particular, the overshooting of the curve Im{Z(t)} after switching off the control current is surprising. It apparently disappears to a large extent when the large signal is applied. The measurement results in Fig. 10.361 were obtained without noticeable delay.

To explore the properties of different materials for controlled devices, different materials and the same conditions were compared.

In Fig. 10.363, the nanocrystalline material VP500 was loaded or unloaded with a DC magnetic force. The small signal inductance was measured as indicator. The partly very slow changes of the measured inductance are surprising. The VP500 material reacts particularly slowly to small modulations when the DC current is switched on. The recovery phase, on the other hand, seems to have approximately the same time constants for all magnitudes of the direct current. In this case, the DC bias voltage was switched off with a free-wheeling diode for the current (L/R = τ < 0.1 s).

Each material behaves differently in this context, which is probably related to the different types of magnetic properties that dominate in each case. The behaviour with small and large signals is also different. Figure 10.364 shows a comparison of typical

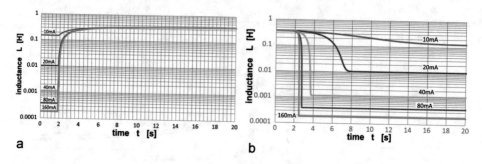

Fig. 10.363 Change in the measured inductance of an arrangement according to Fig. 10.348a when the control current changes abruptly: (**a**) Switching on the DC current; (**b**) Switching off the control current

Fig. 10.364 Basic relaxation performance of various ferromagnetic materials after application of a DC magnetizing current to the materials in comparison. Material C5 [121] with and without air gap

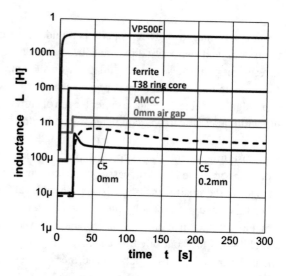

characteristics of the measured inductance after switching off a magnetic DC field for different core materials. Ferrite and the amorphous metal AMCC seem to be particularly fast in this test. The present work can only be a prelude to further investigations into the problem of control properties. The grain-oriented material C5 from Waasner Co., for example, exhibits pronounced and quite long-lasting accommodation processes. However, similar processes can also be observed, for example, in ferrites or AMCC, even if they are not as pronounced and have a much shorter effect. The relaxation processes could be significantly shortened by introducing air gaps.

The clarification of the dynamics of the magnetic field in the core materials used is thus becoming more and more important for the realisation of controllable magnetic devices, depending on the application. A number of observations lead to the assumption that the materials behave similarly to a non-Newtonian fluid. This means, for example, that their viscosity depends on the alternating load. However, these are currently unconfirmed conjectures that still need to be investigated.

Controllable inductors are possible complementary components of power electronics that can be controlled in their inductive properties by an isolated circuit with direct current. The component itself is a pure AC component. This makes very durable and reliable actuators available in power engineering, which practically do without semiconductor electronics. They expand the possibilities for controlling the energy transmission.

References

1. Rogowski, W.; Steinhaus, W.: Die Messung der magnetischen Spannung. In: Archiv für Elektrotechnik. Band 1, Nr. 4, April 1912, S. 141–150
2. Hain, St.; Bakran, M. M.: New Rogowski coil design with a high DV/DT immunity and high bandwidth. Power Electronics and Applications (EPE), 2013 15th European Conference on Power Electronics
3. Rohde, K.: Auslegung und Design von magnetischen Kreisen mit Luftspalt. Abschlussarbeit Bachelor. Universität Kassel 2017
4. Zhang, Y.: Entwicklung und Aufbau einer neuen DC-Netznachbildung. Diplomarbeit II, Universität Kassel, 2011
5. Wendt, M.: Ermittlung der Verlustleistungen in einem Synchron-Tiefsetzsteller mit Niedervolt-GaN-HFETs. Dissertation. Universität Kassel 2015
6. Kleeb, Th.: Hybrid and coreless magnetics for future power electronics. ECPE JointResearch Programme 2016. Report. Universität Kassel 2017
7. Rint, C.: Handbuch für den Hochfrequenz- und Elektrotechniker, Bd. 5, S. 363 ff., Hüthig-Verlag, Heidelberg 1981, Tietze, U.; Schenk, Ch.: Halbleiterschaltungstechnik. Springer-Verlag, S. 383 ff., 10. Auflage 1993)
8. Tietze, U.; Schenk, Ch.: Halbleiter-Schaltungstechnik. 10. Auflage 1993
9. Calvert, J. B.: Electronics. University of Denver 2008
10. Zacharias, P.; Kleeb, Th.; Fenske, F.; Wende, J.; Pfeiffer, J: Controlled magnetic devices in power electronic applications. EPE'17 ECCE Europe
11. Küster, P.: Verluste von ferritischen Reluktanzabschnitten. Masterarbeit Universität Kassel 2018
12. Samuel Vasconcelos Araújo, Peter Zacharias: Analysis on the potential of Silicon Carbide MOSFETs and other innovative semiconductor technologies in the photovoltaic branch. EPE 2009
13. DE 10 2011 056 667 A1
14. DE000019934767A1
15. Chapman, D.: Power Quality Application Guide. Copper Development Association. 2001
16. TÜV Süddeutschland: Oberschwingungen in Starkstromnetzen - Ursache, Analyse, Abhilfe. München 2002
17. DIN EN 50160
18. VDEW-Richtlinie Grundsätze für die Beurteilung von Netzrückwirkungen, 3. Überarbeitete Ausgabe 1992
19. Fenske, F.; Faßhauer, M., Zacharias, P.: Bericht zum Projekt Reactiv Power Control (NR2-RPC, FKZ: 0324106A, BMWi). Universität Kassel 2018

20. Liu, Jae: Investigation of Multiphase Power Converter using Integrated Coupled Inductor Regarding Electric Vehicle Applicaticon. Dissertation, Universität Kassel 2013

21. Моин, В. С.: Стабилизированные Транзисторные Преобразователы. Москва: Атомэнергоиздат, 1986, 376 с. (Moin, V. S.: Stabilisierte Transistorumrichter. Moskau: Atomenergoizdat 1986, 376 S. (russ.))

22. Lee, J., Cha, H.; Shin, D.; Lee, K.; Yoo, D.; Yoo, J.: Analysis and Design of Coupled Inductors for Two-Phase Interleaved DC-DC Converters. Journal of Power Electronics, Vol. 13, No. 3, May 2013

23. Chang Sung Corporation (CSC): Magnetic Powder Cores. CSC Catalogue 2017

24. DE 10 2009 052 461 A1

25. Wintrich, A.; Nicolai, U.; Tursky, W.; Reimann, T: Applikationshandbuch Leistungshalbleiter. SEMIKRON 2015

26. Wittig, B.,: Verbesserung des Schalt- und Betriebsverhaltens von Leistungs-MOSFETs mit niedriger Spannungsfestigkeit und hoher Stromtragfähigkeit durch Optimierung der Treiberschaltung. Dissertation. Universität Kiel 2012

27. Wendt, M.: Ermittlung der Verlustleistung in einem Synchron-Tiefsetzsteller mit Niedervolt-GaN-HFETs. Dissertation, Universität Kassel

28. Mecke, H.: Betriebsverhalten und Berechnung von Transformatoren für das Lichtbogenschweißen. Habilitationsschrift, TH Magdeburg 1979

29. Rogowski, W., Über das Streufeld und den Streuinduktionskoeffizienten eines Transformators mit Scheibenwicklung und geteilten Endspulen, (Dissertation), VdI, Mitteilung über Forschungsarbeiten auf dem Gebiet des Ingenieurwesens, 1909

30. Doebbelin, R & Lindemann, Andreas. (2010). Leakage Inductance Determination for Transformers with Interleaving of Windings. Piers Online. 6. 527-531. https://doi.org/10. 2529/PIERS091220093021.

31. Gyimesi, M.; Ostergaard, D.: Inductance Computation by Incremental Finite Element Analysis, IEEE Transaction on Magnetics, Vol. 35, S. 1119-1122, Mai 1999

32. Ouyang, Z.; Zhang, J.; Hurley, W. G.: Calculation of Leakage Inductance for High-Frequency Transformers. IEEE Transaction on Power Electronics, VOL. 30, No. 10, October 2015

33. Petrov, G. N.: Transformatoren, Band 1, Theoretische Grundlagen. Staatlicher Energetischer Verlag, Moskau 1934 (Петров, Г. Н. Трансформаторы, том 1, основы теории, Государственное Энергетическое Издательство, Москва 1934)

34. Kapp, G.: Transformatoren für Wechselstrom und Drehstrom. Springer-Verlag, Berlin Heidelberg, 1907

35. Philippow, E.: Taschenbuch Elektrotechnik, Bd. 1

36. Küchler, R.: Die Transformatoren. Grundlagen für ihre Berechnung und Konstruktion. Springer, 2. Auflage 1966

37. Klaus K. Streng: Formeln, die man sucht. Der Junge Funker, Bd. 24, Militärverlag der DDR 1979

38. Albach, M.: Induktivitäten in der Leistungselektronik. Springer Vieweg 2017

39. Davis, W. A.; Agarwal, K.: Radio Frequency Circuit Design. John Wiley & Sons, 2001, chapter 6: Transmission Line Transformers

40. Guanella, G. "New method of impedance matching in radio-frequency circuits." Brown Boveri Review, September 1944: 329–32

41. Beckers, F. J. C. M.: Pulsed power driven industrial plasma processing. PhD thesis, Technische Universiteit Eindhoven 2015

42. Pemen, A. J. M.; van Heesch, E. J. M.; Yan, K.; Huijbrechts, P. A. H. J.; Zacharias, P.: A repetitive high-voltage pulse source for pulsed corona treatment of gases. Proc. 2000 24th International Power Modulator Symposium, United States

43. US7061356

44. Divan, D.; Sastry, J.: Controllable Network Transformers. 2008 IEEE Power Electronics Specialists Conference, 15-19 June 2008

45. Zacharias, P.; Kleeb, Th.; Fenske, F.; Wende, J.; Pfeiffer, J.: Controlled Magnetic Devices in Power Electronic Applications. EPE 2017, Warszow

46. US4907246

47. US4841428

48. US5363035

49. US6137391

50. Бамдас, А. М.; Шапиро, С. В.; Трансформаторы, регулируемые подмагничиванием. Библиотека по автоматике, быпуск 147, изд. Энергия, Москва 1965 (russ.: Bamdas, A. M.; Schapiro, S. V.: Durch Vormagnetisierung gesteuerte Transformatoren, Bibliothek der Automatisierungstechnik, Ausg. 147, Verlag Energija, Moskau 1965)

51. Tietze, U.; Schenk, Ch.: Halbleiterschaltungstechnik. 10. Auflage, S. 350 ff., Springer-Verlag 1978

52. US 8023288 B2

53. Kyocera, Datenblatt KC60

54. Wächter, R.; Powalla, M.: Pilotproduktion von CIS-Dünnschichtsolarmodulen: Status und TCO-Aspekte. FVS-Workshop 2002, Proceedings S. 126

55. Bendel, C. et al.: Sicherheitsaspekte bei dezentralen netzgekoppelten Energieerzeugungsanlagen -SIDENA – Abschlussbericht BMU-Projekt FKZ 0329900C, Kassel 2006

56. Real Decreto 1663/2000, de 29 de septiembre, sobre conexión de instalaciones fotovoltaicas a la red de baja tensión;

57. M. Earley et al., National Electrical Code Handbook, 10th Edition, National Fire Protection Association, USA, 2005, pp. 1022. - UL 1741 §31

58. H. Schmidt, B. Burger, K. Kiefer: "Welcher Wechselrichter für welche Modultechnologie?", 21. Symposium Photovoltaische Solarenergie, Bad Staffelstein, März 2006.

59. DE 10 2005 046 379 A1, DE 10 2007 030 577.1, WO 2005/122371 A2, DE 197 32 218 C1, EP 1 950 876 A2

60. DE 196 42 522 C1, WO 03/0412248 A2

61. DE 10 2007 028077.9, US 2007 0047277A1, EP Az. 070 15324.2

62. DE 10 22 192A1

63. DE 10 2004 030 912B3

64. VDN: Richtlinie für Anschluss und Parallelbetrieb von Erzeugungsanlagen am Mittelspannungsnetz 12. 9. 2007

65. EN 55011...14

66. CISPR 11-16, 20, 25, 32, 35

67. IEC 61000-1-x...61000-4-x

68. Моин, В. С.: Стабилизированные Транзисторные Преобразователи. Москва, Энергоатомиздат 1986 (Moin, V. S.: Stabilisierte Transistorumrichter. Moskau, Energoatomizdat 1986)

69. Liu, J.: Investigation of Multiphase Power Converter using Integrated Coupled Inductor Regarding Electric Vehicle Applicaticon. Dissertation, University Kassel Germany 2013

70. DD 291 295

71. DD 291 296

72. Rindt, C.: Handbuch für den Hochfrequenz- und Elektrotechniker. Band 2, 5. Hüthig & Pflaum 1979

73. Hiersig, H. M. (Hrsg.): VDI-Lexikon Energietechnik. Springer-Verlag Berlin-Heidelberg GmbH, Berlin 1994

74. Hauffe, G.: Zur Theorie der Boucherot-Schaltung. Archiv für Elektrotechnik, June 1938, Volume 32, Issue 6, pp 398–400

75. Zacharias, P.; Schiedung, H.; Reißmüller, R.: Leistungsmodulierbare elektronische Stromquelle für CO_2-Laser. ELEKTRIE. Berlin 40(1986) 447-448

76. Reißmüller, R.: Laserversuchsanlage. Diplomarbeit, TH-Magdeburg 1985

77. WP H01S/ DD 261 250

78. EP2144359

79. Data sheets: EG&G, E2V, GL, L3, ITT, Maxwell

80. National Arnold Magnetics. User Hand Book 1995

81. В. С. Моин: Стабилизированные Транзисторные Преобразователи. Москва, Энергоатомиздат 1986 (Moin, V. S.: Stabilisierte Transistorumrichter. Moskau, Atomenergoizdat 1986)

82. Speiser A.P.: Sperrschwinger. In: Impulsschaltungen. Springer, Berlin, Heidelberg (1963)

83. US3783550

84. Paul P. Lin and Chunliang Zhang: Educational Resonator Gyro System Driven by Electro-Magnetic Force and Systems Engineering Process for Design. Applied Mechanics and Materials (Volumes 105 - 107), Pages 1916-1919, 10.4028/www.scientific.net/AMM.105-107.1916

85. Nan-Chyuan Tsai*, Jiun-Sheng Liou, Chih-Che Lin, Tuan Li: Design of Electromagnetic Drive Module for Micro-gyroscope. World Academy of Science, Engineering and Technology 46 2010, pp. 868-873

86. Rothe, Dietmar E.; Lantis, Robert: Magnetically switch voltage multipliers for high-prf, megavolt pulsed power, 20. Power modulation Symposium Myrtle Beach, SC (June 1992)

87. Zacharias, P.: unpublished own works, Lambda Physik GmbH/Coherent AG 1992 - 2000

88. EG&E, Electronic Components: Krytrons, Sprytrons, MiniTriggert Spark Gaps, Transformers and Detonators, product overview 4/92

89. Mallwitz, R.: GTO-Thyristoren in Hochspannungsanwendungen. Diplomarbeit, TU Magdeburg 1994

90. Zacharias, P.: Controllable inductance as actuator in power electronics. EPE 2021 ECCE Europe

91. Alexanderson, E. F. W., "Transoceanic Radio Communication," General Electric Review, October 1920, pp. 794-797.

92. Mali, P.: magnetic amplifiers - principles and applications. John F. Rider Publisher, New York 1960

93. Chute, George M., "Magnetic Amplifiers," Electronics in Industry, 1970, New York: McGraw-Hill, Inc., pp. 344-351.

94. Trinkaus, George, "The Magnetic Amplifier: A Lost Technology of the 1950s," Nuts & Volts, February 2006, pp. 68-71.

95. Austrin, L.: On Magnetic Amplifiers in Aircraft Applications. PhD thesis. Royal Institute of Technology Sweden. Stockholm 2007

96. Transduktor - "der netzrückwirkungsfreie Thyristorsteller". B + S Transformatoren GmbH, Hauptkatalog, 1997

97. Weichmagnetische Werkstoffe und Halbzeuge. Firmenschrift VACUUMSCHMELZE GMBH & CO. KG2002

98. Nanocrystalline soft material FINEMET®. Firmenschrift HL-FM9-E Hitachi Inc. 2010

99. Power Electronics Components Catalog: Metglas® AMCC cut Cores & FINEMET® F3CC Cores, Hitachi 2010

100. Toroidal strip-wound cores of VITROVAC® 6025Z. PV-007, VAC Vacuumschmelze 10/89

101. Schaltnetzteile mit Transduktorregelung. VAC Vaccumschmelze 1989

102. Melkonyan, A.: High Efficiency Power Supply using new SiC Devices. PhD thesis. Universität Kassel, 2006

103. Mammano, B.: Magnetic Amplifier Control for Simple, Low Cost Secondary Regulation. Unitrode Corp., Texas Instruments 2001

104. Sun, N.; Chen, D. Y.; Lee, F. C.; Gradzki; P. M.; Knights, M. A.: Forward Converter Regulator Using Controlled Transformer. IEEE Trans. on Power Electronics, Vol. 11, No. 2, March 1996

105. Smith, K. M.; Smedley, K. M.: Intelligent Magnetic-Amplifier-Controlled Soft-Switching Method for Amplifiers and Inverters. IEEE Trans. on Power Electronics, Vol. 13, No. 1, Jan. 1998

106. Grätzer, D.; Loges, W.: Transduktorregler in Schaltnetzteilen. Technische Informationsschrift, VAC Vacuumschmelze, TB-410-1, 1990

107. VITROVAC 6025Z - Ringbandkerne für Transduktordrosseln. VAC Vacuumschmelze, PK-002, 1990

108. Ferroxcube : Soft ferrites. Catalogue 2008

109. Hitachi Metals America Ltd.: POWERLITE® Inductor Cores - Magnetization Curves. www.hitachimetals.com

110. High-Grade Metals Company Soft Magnetic Materials and Components Business Unit: Metglas AMCC® Series Cut Core - FINEMET® F3CC Series Cut Core. Catalog No. HJ-B11. www.hitachi-metals.co.jp

111. Metglas® Inc.: Powerlite® Inductor Cores. Technical Bulletin, ref:PLC05092011

112. Metglas® Inc.: Magnetic Alloy 2605SA1 (iron-based). ref:2605SA106192009

113. Hitachi: POWERLITE® Inductor Cores. Technical Bulletin ref: PLC05092011

114. Sekels: Magnetwerkstoffe und Systeme. Firmenschrift,(Magnetic Materials and Systems. Company Magazine, 36p., p. 27)

115. POWERLITE Inductor cores. Magnetization curves - Technical Bulletin.

116. Carsten, Bruce.: Ripple cancellation with fast load response for switch mode voltage regulators with synchronous rectification. US000005929692A 27.07.1999

117. P. Zacharias and A. Aganza-Torres: Comparison and optimization of magnetically coupled and non-coupled magnetic devices in interleaved operation. EPE 2020, Lyon

118. P. Zacharias, A. Aganza-Torres and M. Münch Direct Harmonic Compensation for Grid-Connected DC/AC Converter. NEIS Conference Hamburg 2020

119. H. Akagi, Active Harmonic Filters. in Proceedings of the IEEE, vol. 93, no. 12, pp. 2128-2141, Dec. 2005, doi: https://doi.org/10.1109/JPROC.2005.859603.

120. Waasner: Werkstoff- Kennlinien, 2011;

121. SEKELS: Amorphous C-C Cores. SEKELS GmbH, Issue 11 2013

Related Literature

Chapter 1

[1.7]   Adolf J. Schwab: Begriffswelt der Feldtheorie, Springer Verlag, ISBN 3-540-42018-5

[1.8]   Philippow, E.: Taschenbuch der Elektrotechnik, bd.1 Carl Hanser Verlag München Wien 1981

[1.9]   Wolfgang Nolting: Grundkurs Theoretische Physik 3: Elektrodynamik. Springer, Berlin 2007, ISBN 978-3-540-71251-0

[1.10] Meschede, D.: Gerthsen Physik. Springer-Verag, 25. Auflage 2015

[1.11] Ørsted, H. Chr.: Der Geist in der Natur, Leipziger Verlagsbuchhandlung Carl B. Lorck, 1854

[1.12] Lamont: Handbuch des Erdmagnetismus. 1923, Reprint 2009, Books on Demand

[1.13] Küpfmüller, Kohn, K.: Theoretische Elektrotechnik und Elektronik. 16. Auflage. SprK.;inger Verlag, 2005

[1.14] Vömel, M.; Zastrow, D.: Aufgabensammlung Elektrotechnik: .2 Magnetisches Feld und Wechselstrom. Springer Berlin, 2016

Chapter 2

[2.7]   Pichler, F.: Historische elektrische Apparate und Maschinen. Nachbauten von Franz Mock, Mechanicus aus Krems, Telegraphie, Radio, Fernsehen, Physikalische Geräte, Elektrische Maschinen, in: Schriftenreihe Geschichte der Naturwissenschaften und der Technik 17, Linz 2009

[2.8]   Belevitch, V.: Classical Network Theory. Holden-Day, San Francisco, 1968

[2.9]   Georg, O.: Elektromagnetische Felder und Netzwerke. Springer Berlin Heidelberg 1999

[2.10] Rebhan, E.: Theoretische Physik: Elektrodynamik Springer Spektrum. 2007

[2.11] Meschede, D.: Gerthsen Physik. Springer-Verlag, 25. Auflage 2015

[2.12] Schätzing, W.: FEM für Praktiker 4. Elektrotechnik. Basiswissen und Arbeitsbeispiele zu FEM-Anwendungen in der Elektrotechnik. Lösungen mit dem Programm ANSYS® Rev. 12, 3. Auflage 2013.

© Springer Fachmedien Wiesbaden GmbH, part of Springer Nature 2022
P. Zacharias, *Magnetic Components*,
https://doi.org/10.1007/978-3-658-37206-4

[2.13] Feuerbacher, B.: Tutorium Elektrodynamik: Elektro- und Magnetostatik - endlich ausführlich erklärt, Springer Spektrum 2016

[2.14] Pavel Kabos, P.; Stalmachov, V. S.: Magnetostatic Waves and Their Application, 1993, Springer Netherlands; Chapman & Hall

[2.15] Touzani, R.; Jacques Rappaz, J.: Mathematical Models for Eddy Currents and Magnetostatics, Springer Netherlands 2013

[2.16] Vagner, I. D.; Lembrikov, B. I.; Wyder, P. R.: Electrodynamics of Magnetoactive Media. Springer Science & Business 2003

Chapter 3

[3.15] VACUUMSCHMELZE: Ringbandkerne 1976

[3.16] SUMIDA Components GmbH: Catalogue 2017

[3.17] TRIDELTA-Weichferrite: Produktkatalog 2013

[3.18] Micrometals/Arnold Powder Cores: Product Information 2017

[3.19] Kleeb, Th.; Araújo, S.; Zacharias, P.: Characterization of magnetic materials for power electronic devices. Report, ECPE Joint Research Program 2011

[3.20] Kuklinski, P. u.a.:Werkstoffe der Elektrotechnik und Elektronik. Deutscher Verlag für Grundstoffindustrie Leipzig 1980

[3.21] Döring, E.: Werkstoffkunde der Elektrotechnik. Vieweg Verlag Braun-schweig/ Wiesbaden

[3.22] Rindt, C.: Handbuch für Hochfrequenz- und Elektrotechniker, Band 1 Hüthig-Verlag Heidelberg

[3.23] Lindner, Brauer, Lehmann: Taschenbuch der Elektrotechnik und Elektronik. Fachbuchverlag Leipzig-Köln

[3.24] F. Fiorillo, Measurement and characterization of magnetic materials, Elsevier Academic Press, 2004

[3.25] Bertotti, G.: Hysteresis in Magnetism: For Physicists, Materials Scientists, and Engineers, San Diego: Academic, 1998.

Chapter 4

[4.11] Steinmetz, Charles P. (1892). "On the law of hysteresis". Trans. AIEE. 9 (2): 3–62. doi: https://doi.org/10.1109/PROC.1984.12842.

[4.12] Wallmeier, P.: Improved Analytical Modelling of Conductive Losses in Gapped High-Frequency Inductors. IEEE Transactions on Industry Applications, Vol. 37, No. 4, 2001

[4.13] Ferreira, J. A.: Improved Analytical Modelling of Conductive Losses in Magnetic Components. IEEE Transactions on Power Electronics, Vol. 9, No. 1, 1994

[4.14] Baguley, C. A.; Carsten, B.; Madawala, U. K.: An Investigation into the Impact of DC Bias Conditions on Ferrite Core Losses. 33rd Annual Conference of the IEEE Industrial Electronics Society (IECON), Taipei 2007

[4.15] Li, J.; T. Abdallah, T.; Sullivan, C. R.: Improved Calculation of Core Loss With Nonsinusoidal Waveforms. IEEE Industry Applications Society Annual Meeting, Oct. 2001, pp. 2203–2210.

[4.16] Sudhoff, S. D.: Power Magnetic Devices - A Multi-Objective Design Approach. Wiley, 2014

[4.17] Mühlethaler et al.: "Core Losses Under the DC Bias Condition Based on Steinmetz
        Parameters". IEEE Transactions on Power Electronics. 27 (2): 953, 2012

[4.18] Nan, X.; Sulivan, C. R.: Simplified High Accuracy Calculation of Eddy-Current Losses
        in Round Wire Windings. IEEE Power Electronics Specialist Conference, June 2004,
        pp. 873 – 879

[4.19] Reinert, J.; Brockmeyer, A.; De Doncker, R. W.: "Calculation of losses in ferro- and
        ferrimagnetic materials based on the modified Steinmetzequation", in Proceedings of
        34th Annual Meeting of the IEEE Industry Applications Society, 1999, pp. 2087–92
        vol. 3.

[4.20] Carpenter, K. H.: "Simple models for dynamic hysteresis which addfrequency-
        dependent losses to static models", IEEE Transactions on Magnetics, vol. 34, no.
        3, pp. 619–22, 1998

[4.21] Mühlethaler, J, Biela J.; Kolar, J. W.; Ecklebe, A: Improved Core-Loss Calculation for
        Magnetic Components Employed in Power Electronic Systems. IEEE Transactions on
        Power Electronics, Vol. 27, No. 2, February 2012.

Chapter 5

[5.9]   Jiles, D.: Introduction to Magnetism and Magnetic Materials, 2nd ed. London: Chap-
        man & Hall, 1998

[5.10] Carpenter, C. J.: Magnetic equivalent circuits. Proceedings of the Institution of Electri-
        cal Engineers, Volume: 115, Issue: 10, 1968

[5.11] Deskur, J.: Models of magnetic circuits and their equivalent electrical diagrams.
        COMPEL - The international journal for computation and mathematics in electrical
        and electronic engineering, Vol. 18 No. 4, pp. 600-610, 1999

[5.12] Hammond, E.: Applied Elektromagnetism. Pergamon Press. 1971

[5.13] Cristaldi, L.; Leva, S.; P. Morando, A. P.: Electric Circuit Representation of a Magnetic
        Circuit with Hysteresis. In: Wiak S., Krawczyk A., Trlep M. (eds) Computer Engineer-
        ing in Applied Electromagnetism. pp 261-266, Springer, Dordrecht 2005,

Chapter 6

[6.10] von Ardenne, M.: Verstärkermesstechnik. Julius-Springer-Verlag Berlin 1929

[6.11] Мураховская, М. А.: Обобщенный метод расчета индуктивности рассеяния
        обмоток трансформаторов : Диссертация кандидата технических наук / М. А.
        Мураховская, Моск. энерг. ин-т (МЭИ) 1963. (Murachovskaya, M. A.:
        Verallgemeinerte   Methode   zur   Berechnung   der   Streuinduktivität   von
        Transformatorwicklungen. Dissertation. Moskauer Energetisches Institut (MEI). 1963)

[6.12] Петров, Г. Н.: Трансформаторы. Государственное энергетическое издательство.
        Москва. Ленинград. 1934 (Petrov, G. N.: Transformatoren. Staatlicher Energieverlag
        (Gosenergoizdat). Moskau, Leningrad 1934)

[6.13] Doebbelin, R.; Teichert, C.; Benecke, M., Lindemann, A.: Computerized Calculation of
        Leakage Inductance Values of Transformers. PIERS Online Vol. 5 No. 8 2009 pp:
        721-726

[6.14] Петров, Г. Н.: К теории расчета индуктивности рассеяния трансформаторов.
        журнал "Электричество", 3, 1948, с. 30 - 35, Госэнергоиздат. Petrov, G. N.,

"Berechnung der Streuung von Transformatoren (russ.), Journal Elektrichestvo, No. 3, 1948, S. 30-35, Gosenergoizdat.

[6.15]   J. Koch und K. Ruschmeyer: Permanentmagnete I; Grundlagen. Boysen und Maasch, Valvo Hamburg 1983

[6.16]   Friebe, J.: Permanentmagnetische Vormagnetisierung von Speicherdrosseln in Stromrichtern. Dissertation, Universität Kassel 2014;

[6.17]   VACUUMSCMELZE: Firmenschrift: Nanocrystalline VITROPERM, EMC Products, VACChokesandCoresDatasheet.pdf

Chapter 7

[7.7]   Boll, R.: Weichmagnetische Werkstoffe. VACUUMSCMELZE (VAC), 1990

[7.8]   Cedighian, S.: Die weichmagnetischen Werkstoffe. VDI-Verlag 1973

[7.9]   Michalowsky, L.; Schneider, J.: Magnettechnik – Grundlagen, Werkstoffe, Anwendungen. Vulkan-Verlag Essen, 2006.

[7.10]   Van den Bossche, A.: Inductors and Transformers for Power Electronics, St Lucide Pr, 2005

[7.11]   Mühlethaler, J., Biela, J. and Kolar, J. W.: Improved Core Loss Calculation for Magnetic Components Employed in Power Electroic Systems, Applied Power Electronics Conference and Exhibition, pp. 1729-1736, March 2011

[7.12]   Sullivan, C. R., Harris, J. H. and Herbert, E.: Core Loss Predictions for General PWM Waveforms from a Simplified Set of Measured Data, Applied Power Electronics Conference and Exhibition, pp. 1048-1055, February 2010

[7.13]   Mühlethaler, J., Biela, J., Kolar, J. W. and Ecklebe, A.: Core Losses under DC bias Conditions based on Steinmetz Parameters, The 2010 International Power Electronics Conference, pp. 2430-2437, June 2010

[7.14]   Prabhakaran, S.: Impedance-analyzer measurements of high-frequency power passives: techniques for high power and low impedance, Industry Applications Conference, Vol. 2, pp. 1360-1367, Oct. 2002

[7.15]   Ferreira, J. A.: Analytical computation of ac resistance of round and rectangular litz wire windings, IEE Proceedings B, Electric Power Applications, Vol. 139, pp. 21-25, Jan. 1992

[7.16]   Albach, M.: Two-dimensional calculation of winding losses in transformers, Power Electronics Specialists Conference, Vol. 3, pp. 1639-1644, 2000

[7.17]   Albach, M.: Grundlagen und Dimensionierung von Induktivitäten, ECPE Cluster Seminar: Induktivitäten in der Leistungselektronik, October 2011

[7.18]   Hitachi Metals, Power Electronics Components Catalog – Metglass AMCC Series Cut Cores – Finemet F3CC Series Cut Cores, 2010

[7.19]   Rossmanith H., Doebroenti, M., Albach, M. and Exner, D.: Measurement and Characterization of High Frequency Losses in Nonideal Litz Wires, IEEE Transactions on Power Electronics, Vol. 26 no 11, 2011

[7.20]   Pflier, P. M.: Elektrische Messgeräte und Messverfahren. Springer Berlin Heidelberg 1951, S. 55-56

Chapter 8

[8.14] Schwab, A. J.: Elektroenergiesysteme – Erzeugung, Transport, Übertragung und Verteilung elektrischer Energie. Springer, 2006

[8.15] Pierce, L. W.: Transformer Design and Application Considerations for Nonsinusoidal Load Currents, IEEE Trans. On. Industry Applications. Vol. 32 (3). pp. 633-645, 1996

[8.16] Harlow, J. H.: Electric Power Transformer Engineering. 2nd ed., CRC Press 2007

[8.17] Hurley, W. G.: Transformers and Inductors for Power Electronics: Theory, Design and Applications. Wiley 2013

[8.18] Fyvie J.: Design Aspects of Power Transformers and Reactors. abramis, 2016

[8.19] Vosen, H.: Kühlung und Belastbarkeit von Transformatoren. VDE, Berlin 1997

Chapter 9

[9.29] Pawlowski, J.: Die Ähnlichkeitstheorie in der physikalisch-technischen Forschung. Grundlagen und Anwendung. Springer-Verlag Berlin Heidelberg 1971

[9.30] Kaul, Th.: Multiple lineare Regression & High Performance Computing. Books on Demand 2015

[9.31] Hedderich, J.; Prof. Dr. Lothar Sachs, L.: Angewandte Statistik. Springer Berlin Heidelberg 2018

[9.32] Blobel, V.: Statistische und numerische Methoden der Datenanalyse. Teubner 1998

Chapter 10.1

[10.1.1] Mühl, Th.: Elektrische Messtechnik. 5. Aufl. 2017. Springer-Vieweg 2017

[10.1.2] Philippow, E. Taschenbuch Elektrotechnik. Band 1. Verlag Technik Berlin 1968

[10.1.3] Schwab, A. J.: Elektroenergiesysteme. 2. Auflage. Springer, 2009

[10.1.4] Firmenschriften: Isabellenhütte, Vectron, T&M RESEARCH PRODUCTS, Powertek, Ohm-Labs,

[10.1.5] Pfeifer, T; Profos, P.: Handbuch der industriellen Messtechnik. Oldenbourg Wissenschaftsverlag 1994

[10.1.6] Bode, P. A.: Current measurement applications handbook. Zetex Semiconductors AN 39

[10.1.7] Costa, F.; Poulichet, P.; Mazaleyrat, F.; Labouré, E.: The Current Sensors in Power Electronics, a Review. EPE Journal. 11 (1): 7–18, 2001

[10.1.8] Spaziani, L.: Using Copper PCB Etch for Low Value Resistance". Texas Instruments. DN-71, 1997

[10.1.9] DIN EN 60751

Chapter 10.2

[10.2.1] Mühl, Th.: Elektrische Messtechnik. 5. Aufl. 2017. Springer-Vieweg 2017

[10.2.2] Philippow, E. Taschenbuch Elektrotechnik. Band 1. Verlag Technik Berlin 1968

[10.2.3] Schwab, A. J.: Elektroenergiesysteme. 2. Auflage. Springer, 2009

[10.2.4] Pfeifer, T,; Profos, P.: Handbuch der industriellen Messtechnik. Oldenbourg Wissenschaftsverlag 1994

[10.2.5] Firmenschriften von: Honeywell, VAC, Murata Power Solutions, Siemens, EPCOS, LEM International SA, VAC, ZES, Tektronix; YOKOGAWA, Pearson

[10.2.6] Hartmann, M.; Biela, J.; Ertl, H.; Member, IEEE, Kolar, J. W.: Wideband Current Transducer for Measuring AC Signals With Limited DC Offset. IEEE Trans. on Power Electronics, Vol. 24, No. 7, July 2009

[10.2.7] IEC 61869

Chapter 10.3

[10.3.3] Mühl, Th.: Elektrische Messtechnik. 5. Aufl. 2017. Springer-Vieweg 2017

[10.3.4] Philippow, E. Taschenbuch Elektrotechnik. Band 1. Verlag Technik Berlin 1968

[10.3.5] Schwab, A. J.: Elektroenergiesysteme. 2. Auflage. Springer, 2009

Chapter 10.5

[10.5.4] Sedra, A. S.; Smith, K. C.:Microelectronic Circuits, 2nd ed. (New York: Holt, Rinehart and Winston, 1987), pp. 40, 79, 111-113

[10.5.5] Horowitz, P.; Hill, W.: The Art of Electronics, 2nd ed. (Cambridge: Cambridge Univ. Press, 1989, pp. 266, 281.

[10.5.6] Fenske, F., Dissertation, unveröff. Skript, Universität Kassel)

Chapter 10.7

[10.7.6] VDEW-Richtlinie Anschluss von USV-Anlagen in Drehstromtechnik im Leistungsbereich von 10 kVA bis 1 MVA an das öffentliche Netz, 1. Ausgabe 1995

[10.7.7] VDEW-Richtlinie Anschluss von primär getakteten Schaltnetzteilen mit B6-Drehstromeingang, 1. Ausgabe 1992

[10.7.8] VDEW-Richtlinie Anschluss und Parallelbetrieb von Eigenerzeugungsanlagen am Mittelspannungsnetz, 2. Ausgabe 1998

Chapter 10.8

[10.8.5] Kleeb, Th.; Araújo, S., Zacharias, P: Characterization of magnetic materials for power electronic devices. Report, European Center of Power Electronics Joint Research Program 2011/2012

[10.8.6] Araujo, S.; P. Zacharias: Study on the Perspectives of Wide-Band Gap Power Devices in Electronic-Based Power Conversion for Renewable Systems. Report, European Center of Power Electronics Joint Research Program 2011/2012

[10.8.7]   Kleeb, Th.: Investigation on Performance Advantage of Functionally Integrated Magnetic Components in Decentralised Power Electronic Applications. Dissertation, Universität Kassel 2016

[10.8.8]   Mu, M.; Lee, F. C.; Jiao, Y.; Lu, S.: Analysis and design of coupled inductor for interleaved multiphase three-level DC-DC converters. 2015 IEEE Applied Power Electronics Conference and Exposition (APEC)

[10.8.9]   Kolar, J. W.; Krismer, F.; Leibl, M.; Neumayr, D.; Schrittwieser, L.; Bortis, D.: Impact of Magnetics on Power Electronics Converter Performance. PSMA Workshop 2017

Chapter 10.10

[10.10.4]   Alickovic, J.: Maximierung der Leistungsdichte von selbstgeführten hochfrequenten Energiewandlern auf Basis ultraschneller Wide-Band-Gag Bauelemente. Dissertation, Universität Kassel 2018

[10.10.5]   Pengfei Tu, Peng Wang, Xiaolei Hu, Chen Qi, Shan Yin; Zagrodnik, M. A.: Evaluation of IGBT Turn-on Loss with Double Pulse Testing. 2016 IEEE 11th Conference on Industrial Electronics and Applications (ICIEA)

[10.10.6]   CREE/Wolfspeed: CPWR-AN09, REV, SiC Mosfet double Pulse Fixture, "SiC Isolated Gate Driver" Application Note CPWR-AN10

Chapter 10.13

[10.13.1]   ÖVE/ÖNORM E 8001-1: Errichtung von elektrischen Anlagen mit Nennspannungen bis 1000 V AC und 1500 V DC. Teil 1: Begriffe und Schutz gegen elektrischen Schlag (Schutzmaßnahmen)

[10.13.2]   Michael Johnston: Installations and Inspections of Corner-grounded Systems, IEAI Magazine January 16, 2002

[10.13.3]   Tōru Takenaka: Siemens in Japan: von der Landesöffnung bis zum Ersten Weltkrieg. Franz Steiner Verlag, 1996

[10.13.4]   IEC60364-1 / VDE0100-100

[10.13.5]   Data Bulletin 2700DB0202R03/12, 09/2012: Corner-Grounded Delta (Grounded B Phase) Systems

[10.13.6]   Grounding methods in mission critical facilities. White Paper WP027004EN Effective December 2013

[10.13.7]   IEEET STANDARD 142-2007—IEEE Recommended Practice for Grounding of Industrial and Commercial Power Systems

[10.13.8]   IEEE STANDARD 1100-2005—IEEE Recommended Practice for Powering and Grounding Electronic Equipment

[10.13.9]   EN61558 / VDE0570

[10.13.10]   IEC 61557

[10.13.11]   DIN VDE 0100-410

[10.13.12]   NFPA 70: National Electrical Code (NEC)

[10.13.13]   Rene Flosdorff, Günther Hilgarth: Elektrische Energieverteilung. Teubner, 2003

Chapter 10.15

[10.15.10]   Knights, M.A.; Erickson; K.L.: Controlled Transformer. APEC 1991

[10.15.11]   Witulski, A. F.: Modeling and Design of Transformers and Coupled Inductors. APEC 1993

[10.15.12]   Ramay, R. A.: On the Mechanics of Magnetic Amplifier Operation. AIEE Transactions 1951, pp. 1214 - 1223

[10.15.13]   Zhu, G.; Wang, K.: Modeling and Design Considerations of Coupled Inductor Converters. APEC 2010, pp.

[10.15.14]   Perdigão, M. S.; Marcos Alonso, J. M.; Vaquero, D. G.; Saraiva,E. S.: Magnetically Controlled Electronic Ballasts With Isolated Output: The Variable Transformer Solution. IEEE TRANSACTIONS ON INDUSTRIAL ELECTRONICS, VOL. 58, No. 9, 2011

[10.15.15]   Mandache L.; Al-Haddad, K: High Precision Modeling of Saturable Transformers used as Voltage Regulators. IEEE ISIE 2006, Montreal, pp. 2695 – 2699

[10.15.16]   Бамдас, А. М., Шапиро, С. В. Стабилизаторы с подмагничиваниемыми трансформаторами. Библиотека по автоматике, быпуск 153, изд. Энергия, Москва 1965 (russ.: Bamdas, A. M.; Schapiro, S. V.: Stabilisatoren mit vormagnetisierten Transformatoren, Bibliothek der Automatisierungstechnik, Ausg. 153, Verlag Energija, Moskau 1965)

Chapter 10.17

[10.17.13]   Zacharias, P.: Wechselrichter für die Solartechnik. Symposium PV-Solarenergie. Staffelstein 2008

[10.17.14]   DE 42 43 206 A1, WO 2006/00652 A1, WO 2006/032694 A1, US 7 319 313 B2

[10.17.15]   Burger, B.: Photovoltaic Inverters for Grid Connection. Power Electronics for Photovoltaics. Intersolar Conference Munich 2008

[10.17.16]   US 4 670 828

[10.17.17]   Calais, M.; Agelidis, V. G. and Meinhardt, M.; "Multilevel Converters Single-Phase Grid Converter for Photovoltaic System", J, Solar energy, Vol. 66, No. 5, pp. 325-335, 1999

[10.17.18]   Myrzik, J.: Topologische Untersuchungen zur Anwendung von tief-hochsetzenden Stellern für Wechselrichter . Dissertation. Kassel 2000

[10.17.19]   DE 10 2007 038 960.6

[10.17.20]   Zacharias, P. (Herausgeber): Use of Electronic-Based Power Conversion for Distributed and Renewable Energy Sources. ISET 2008, Kassel

[10.17.21]   Li, Quan, Wolfs, P.: A Review of the Single Phase PV Module Integrated Converter Topologies with Three Different DC link Configurations. IEEE Trans. on Power Electronics, V.23,3(2008), pp.1320

[10.17.22]   Zacharias, P.: Topologische Ansätze für Wechselrichter in netzgekoppelten Photovoltaikanlagen / Topological Approaches for Grid-Connected Photovoltaic Inverters. ETG Konferenz, München 2008

[10.17.23]   ISET: 99% Wandlerwirkungsgrad für Photovoltaikwechselrichter erreicht. Pressemitteilung ISET. März 2008

Chapter 10.22

[10.22.4] Eichler, J.; Eichler, H. -J.: Laser - Bauformen, Strahlführung, Anwendungen. Springer 2010
[10.22.5] Materialbearbeitung mit CO_2-Hochleistungslasern. VDI-Berichte 535. Düsseldorf: VDI-Verlag 1984

Chapter 10.23

[10.23.2] Unbehauen, H.; Ley, F.: Regelungs- und Steuerungstechnik. Springer 2007
[10.23.3] Boyce, W. E; Diprima, R. C.: Elementary Differential Equations and Boundary Value Problems (4th ed.). John Wiley & Sons 1986

Chapter 10.24

[10.24.2] Pirrie, C.A.; Menown, H.: The Evolution of the Hydrogen Thyratron. Marconi Applied Technologies Ltd, Chelmsford, U.K., PMS 2000
[10.24.3] Pirrie, C. A.; Culling, P. D.; Menown, H.; Nicholls, N. S.: Performance of a compact four gap thyratron in a high voltage, high repetition rate test circuit. IEEE Conference Record of the 1988 Eighteenth Power Modulator Symposium
[10.24.4] E.A. Farrell: A high voltage DC buck regulator using a crossatron switch tube. Nineteenth IEEE Symposium on Power Modulators 1990
[10.24.5] Zhilichev; Y.: Models of Ferroresonant Transformers. IEEE Transactions on Power Delivery, Volume: 29 Issue: 6, 2014, pp. 2631 - 2639

Chapter 10.26

[10.26.3] Schumacher, R.W.: Crossatron switch applications in power conditioning and pulsed-power-modulator systems, 23rd Intersociety Energy Convession Engeneering Conference Vol.3, Book-N-2: 10272C, p.699-707
[10.26.4] Pemen, A. J. M.; van Heesch, E. J. M.; Yan, K. P.; Hujprechts, P. A. H. J; Zacharias, P.: Cracking of heavy tar components in biogas by means of pulsed corona discharges. Proc. XIII International Conference on Gas Discharges and their Applications. 2, 3-8 September 2000, Glasgow, U.K., 2000, pp. 724-727
[10.26.5] Beckers, F.J.C.M.: Pulsed power driven industrial plasma processing. PhD thesis, Technische Universiteit Eindhoven 2015
[10.26.6] Discharge-Pumped Excimer Laser Research in Japan. Institute of Laser Technology, Japan 1992
[10.26.7] Blom, P. P. M: High Power Pulsed Corona. PhD thesis, TU Eindhoven 1997

Index

© Springer Fachmedien Wiesbaden GmbH, part of Springer Nature 2022
P. Zacharias, *Magnetic Components*,
https://doi.org/10.1007/978-3-658-37206-4

Printed in the United States
by Baker & Taylor Publisher Services